weighted matrix must be diagonal if use recursive

# Lessons in Estimation Theory for Signal Processing, Communications, and Control

## PRENTICE HALL SIGNAL PROCESSING SERIES

*Alan V. Oppenheim, Series Editor*

ANDREWS & HUNT *Digital Image Restoration*
BRACEWELL *Two Dimensional Imaging*
BRIGHAM *The Fast Fourier Transform and Its Applications*
BURDIC *Underwater Acoustic System Analysis, 2/E*
CASTLEMAN *Digital Image Processing*
CROCHIERE & RABINER *Multirate Digital Signal Processing*
DUDGEON & MERSEREAU *Multidimensional Digital Signal Processing*
HAYKIN *Advances in Spectrum Analysis and Array Processing. Vols. I, II, & III*
HAYKIN, ED. *Array Signal Processing*
JOHNSON & DUDGEON *Array Signal Processing*
KAY *Fundamentals of Statistical Signal Processing*
KAY *Modern Spectral Estimation*
KINO *Acoustic Waves: Devices, Imaging, and Analog Signal Processing*
LIM *Two-Dimensional Signal and Image Processing*
LIM, ED. *Speech Enhancement*
LIM & OPPENHEIM, EDS. *Advanced Topics in Signal Processing*
MARPLE *Digital Spectral Analysis with Applications*
MCCLELLAN & RADER *Number Theory in Digital Signal Processing*
MENDEL *Lessons in Estimation Theory for Signal Processing, Communications, and Control*
NIKIAS & PETROPULU *Higher-Order Spectra Analysis*
OPPENHEIM & NAWAB *Symbolic and Knowledge-Based Signal Processing*
OPPENHEIM, WILLSKY, WITH YOUNG *Signals and Systems*
OPPENHEIM & SCHAFER *Digital Signal Processing*
OPPENHEIM & SCHAFER *Discrete-Time Signal Processing*
PHILLIPS & NAGLE *Digital Control Systems Analysis and Design, 3/E*
PICINBONO *Random Signals and Systems*
RABINER & GOLD *Theory and Applications of Digital Signal Processing*
RABINER & JUANG *Fundamentals of Speech Recognition*
RABINER & SCHAFER *Digital Processing of Speech Signals*
ROBINSON & TREITEL *Geophysical Signal Analysis*
STEARNS & DAVID *Signal Processing Algorithms in Fortran and C*
THERRIEN *Discrete Random Signals and Statistical Signal Processing*
TRIBOLET *Seismic Applications of Homomorphic Signal Processing*
VAIDYANATHAN *Multirate Systems and Filter Banks*
WIDROW & STEARNS *Adaptive Signal Processing*

# Lessons in Estimation Theory for Signal Processing, Communications, and Control

Jerry M. Mendel

*Department of Electrical Engineering*
*University of Southern California*
*Los Angeles, California*

PRENTICE HALL PTR
Englewood Cliffs, New Jersey 07632

**Library of Congress Cataloging-in-Publication Data**

Mendel, Jerry M.
[Lessons in digital estimation theory]
Lessons in estimation theory for signal processing, communications, and control / Jerry M. Mendel.
p. cm.—(Prentice Hall signal processing series)
Previous edition published under the title: Lessons in digital estimation theory.
Includes bibliographical references and index.
ISBN 0-13-120981-7
1. Estimation theory. I. Title. II. Series.
QA276.8.M46 1995
519.5'44—dc20 94-15781
CIP

Editorial/production supervision: *Jane Bonnell*
Cover design: *Douglas DeLuca*
Acquisitions editor: *Karen Gettman*
Manufacturing manager: *Alexis R. Heydt*


Prentice-Hall, Inc.
A Simon & Schuster Company
Englewood Cliffs, New Jersey 07632

Previous edition published under the title of *Lessons in Digital Estimation Theory*.

The publisher offers discounts on this book when ordered in bulk quantities. For more information, contact:

Corporate Sales Department
Prentice Hall PTR
113 Sylvan Avenue
Englewood Cliffs, NJ 07632

Phone: 800-382-3419 or 201-592-2498
Fax: 201-592-2249
email: dan_rush@prenhall.com



Printed in the United States of America

10 9 8 7 6 5 4 3 2 1

ISBN 0-13-120981-7

Prentice-Hall International (UK) Limited, *London*
Prentice-Hall of Australia Pty. Limited, *Sydney*
Prentice-Hall Canada Inc., *Toronto*
Prentice-Hall Hispanoamericana, S.A., *Mexico*
Prentice-Hall of India Private Limited, *New Delhi*
Prentice-Hall of Japan, Inc., *Tokyo*
Simon & Schuster Asia Pte. Ltd., *Singapore*
Editora Prentice-Hall do Brasil, Ltda., *Rio de Janeiro*

*To my wife Letty*

# Contents

# Preface

Estimation theory is widely used in many branches of science and engineering. No doubt, one could trace its origin back to ancient times, but Karl Friederich Gauss is generally acknowledged to be the progenitor of what we now refer to as estimation theory. R. A. Fisher, Norbert Wiener, Rudolph E. Kalman, and scores of others have expanded upon Gauss's legacy and have given us a rich collection of estimation methods and algorithms from which to choose. This book describes many of the important estimation methods and shows how they are interrelated.

Estimation theory is a product of need and technology. Gauss, for example, needed to predict the motions of planets and comets from telescopic measurements. This "need" led to the method of least squares. Digital computer technology has revolutionized our lives. It created the need for recursive estimation algorithms, one of the most important ones being the Kalman filter. Because of the importance of digital technology, this book presents estimation from a discrete-time viewpoint. In fact, it is this author's viewpoint that *estimation theory is a natural adjunct to classical digital signal processing.* It produces time-varying digital filter designs that operate on random data in an optimal manner.

Although this book is entitled "Estimation *Theory* . . . ," computation is essential in order to be able to use its many estimation algorithms. Consequently, computation is an integral part of this book. It is this author's viewpoint that, whenever possible, *computation should be left to the experts.* Consequently, I have linked computation into MATLAB® (MATLAB is a registered trademark of The MathWorks, Inc.) and its associated toolboxes. A small number of important estimation M-files, which do not presently appear in any MathWorks toolbox, have been included in this book; they can be found in Appendix B.

This book has been written as a collection of lessons. It is meant to be an *introduction* to the general field of estimation theory and, as such, is not encyclopedic in content or in references. The supplementary material, which has been included at the end of many lessons, provides additional breadth or depth to those lessons. This book can be used for self-study or in a one-semester course.

Each lesson begins with a *summary* that describes the main points of the lesson and also lets the reader know exactly what he or she will be able to do as a result of completing the lesson. Each lesson also includes a small collection of multiple-choice *summary questions*, which are meant to test the reader on whether or not he or she has grasped the lesson's key points. Many of the lessons include a section entitled "Computation." When I decided to include material about computation, it was not clear to me whether such material should be collected together in one place, say at the rear of the book in an appendix, or whether it should appear at the end of each lesson, on demand so to speak. I sent letters to more than 50 colleagues and former students asking them what their preference would be. The overwhelming majority recommended having discussions about computation at the end of each lesson. I would like to thank the following for helping me to make this decision: Chong-Yung Chi, Keith Chugg, Georgios B. Giannakis, John Goutsias, Ioannis Katsavounidis, Bart Kosko, Li-Chien Lin, David Long, George Papavassilopoulos, Michael Safonov, Mostafa Shiva, Robert Scholtz, Ananthram Swami, Charles Weber, and Lloyd Welch.

Approximately one-half of the book is devoted to parameter estimation and the other half to state estimation. For many years there has been a tendency to treat state estimation, especially Kalman filtering, as a stand-alone subject and even to treat parameter estimation as a special case of state estimation. Historically, this is incorrect. In the musical *Fiddler on the Roof*, Tevye argues on behalf of *tradition* . . . "Tradition!" Estimation theory also has its tradition, and it begins with Gauss and parameter estimation. In Lesson 2 we show that state estimation is a special case of parameter estimation; i.e., it is the problem of estimating random parameters when these parameters change from one time instant to the next. Consequently, the subject of state estimation flows quite naturally from the subject of parameter estimation.

There are four supplemental lessons. Lesson A is on sufficient statistics and statistical estimation of parameters and has been written by Professor Rama Chellappa. Lessons B and C are on higher-order statistics. These three lessons are on parameter estimation topics. Lesson D is a review of state-variable models. It has been included because I have found that some people who take a course on estimation theory are not as well versed as they need to be about state-variable models in order to understand state estimation.

This book is an outgrowth of a one-semester course on estimation theory taught at the University of Southern California since 1978, where we cover all its contents at the rate of two lessons a week. We have been doing this since 1978. I wish to thank Mostafa Shiva, Alan Laub, George Papavassilopoulos, and Rama Chellappa for encouraging me to convert the course lecture notes into a book. The result was the first version of this book, which was published in 1987 as *Lessons in Digital Estimation Theory*. Since that time the course has been taught many times and additional materials have been included. Very little has been deleted. The result is this new edition.

Most of the book's important results are summarized in theorems and corollaries. In order to guide the reader to these results, they have been summarized for easy reference in Appendix A.

Problems are included for all the lessons (except Lesson 1, which is the Introduction), because this is a textbook. The problems fall into three groups. The first group contains problems that ask the reader to fill in details, which have been "left to the reader as an exercise." The second group contains problems that are related to the material in the lesson. They range from theoretical to easy computational problems, easy in the sense that the computations can be carried out by hand. The third group contains computational problems that can only be carried out using a computer. Many of the problems were developed by students in my Fall 1991 and Spring 1992 classes at USC on Estimation Theory. For these problems, the name(s) of the problem developer(s) appears in parentheses at the beginning of each problem. The author wishes to thank all the problem developers. Solutions to the problems can be obtained by contacting the Prentice Hall PTR Editorial Department at 201-816-4116.

While writing the first edition of the book, the author had the benefit of comments and suggestions from many of his colleagues and students. I especially want to acknowledge the help of Georgios B. Giannakis, Guan-Zhong Dai, Chong-Yung Chi, Phil Burns, Youngby Kim, Chung-Chin Lu, and Tom Hebert. While writing the second edition of the book, the author had the benefit of comments and suggestions from Georgios B. Giannakis, Mithat C. Dogan, Don Specht, Tom Hebert, Ted Harris, and Egemen Gonen. Special thanks to Mitsuru Nakamura for writing the estimation algorithm M-files that appear in Appendix B; to Ananthram Swami for generating Figures B-4, B-5, and B-7; and to Gent Paparisto for helping with the editing of the galley proofs.

Additionally, the author wishes to thank Marcel Dekker, Inc., for permitting him to include material from J. M. Mendel, *Discrete Techniques of Parameter Estimation: The Equation Error Formulation*, 1973, in Lessons 1–3, 5–9, 11, 18, and 23; Academic Press, Inc., for permitting him to include material from J. M. Mendel, *Optimal Seismic Deconvolution: An Estimation-based Approach*, copyright © 1983 by Academic Press, Inc., in Lessons 11–17, 19–21, and 25; and the Institute of Electrical and Electronic Engineers (IEEE) for permitting him to include material from J. M. Mendel, *Kalman Filtering and Other Digital Estimation Techniques: Study Guide*, © 1987 IEEE, in Lessons 1–3, 5–26, and D. I hope that the readers do not find it too distracting when I reference myself for an item such as a proof (e.g., the proof of Theorem 17-1). This is done only when I have taken material from one of my former publications (e.g., any one of the preceding three), to comply with copyright law, and is in no way meant to imply that a particular result is necessarily my own.

I am very grateful to my editor Karen Gettman and to Jane Bonnell and other staff members at Prentice Hall for their help in the production of this book.

Finally, I want to thank my wife, Letty, to whom this book is dedicated, for providing me, for more than 30 years, with a wonderful environment that has made this book possible.

JERRY M. MENDEL
*Los Angeles, California*

# LESSON 1

# Introduction, Coverage, Philosophy, and Computation

## SUMMARY

The objectives of this first lesson are to (1) relate estimation to the larger issue of modeling, (2) describe the coverage of the book, and (3) explain why the discrete-time (digital) viewpoint is emphasized.

Estimation is one of four modeling problems. The other three are representation, measurement, and validation. We shall cover a wide range of estimation techniques, which are summarized in Table 1-1. The techniques are for parameter or state estimation or a combination of the two, as applied to either linear or nonlinear models.

The discrete-time viewpoint is emphasized in this book because (1) many real data are collected in a digitized manner, so they are in a form ready to be processed by discrete-time estimation algorithms, and (2) the mathematics associated with discrete-time estimation theory is simpler than for continuous-time estimation theory. For estimation purposes, our modeling philosophy is to discretize the model at the front end of the problem.

We view (discrete-time) estimation theory as the extension of classical signal processing to the design of discrete-time (digital) filters that process *uncertain* data in an *optimal* manner. Estimation theory can, therefore, be viewed as a natural adjunct to digital signal processing theory.

Estimation algorithms process data and, as such, must be implemented on a digital computer. Our computation philosophy is, whenever possible, leave it to the experts. We explain how this book can be used with MATLAB® and pertinent toolboxes (MATLAB is a registered trademark of The MathWorks, Inc.).

When you complete this lesson you will (1) have a big picture of the contents of the book and (2) be able to explain the modeling and estimation theory philosophies emphasized in the book.

## INTRODUCTION

This book is all about estimation theory. It is useful, therefore, for us to understand the role of estimation in relation to the more global problem of modeling. Figure 1-1 decomposes modeling into four problems: representation, measurement, estimation, and validation. The following discussion is taken from Mendel (1973). The *representation problem* deals with how something should be modeled. We shall be interested only in mathematical models. Within this class of models we need to know whether the model should be static or dynamic, linear or nonlinear, deterministic or random, continuous or discretized, fixed or varying, lumped or distributed, in the time domain or in the frequency domain, etc. Algebraic equations, ordinary differential equations, partial differential equations, finite-difference equations, transfer functions, impulse responses, and state-variable models are some examples of mathematical models.

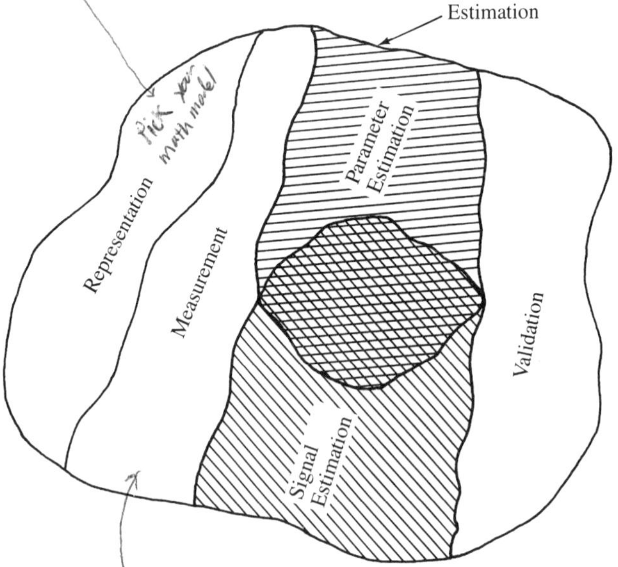


**Figure 1-1** Modeling problem (reprinted from Mendel, 1973, p. 4, by courtesy of Marcel Dekker, Inc.).

In order to verify a model, physical quantities must be measured. We distinguish between two types of physical quantities, *signals* and *parameters*. Parameters express a relation between signals. For example, in Newton's law $F(t) = MA(t)$, force $F(t)$ and acceleration $A(t)$ are viewed as signals, whereas mass $M$ is viewed as a parameter. The coefficients that appear in ordinary differential or difference equations, or in transfer functions, are further examples of parameters.

Not all signals and parameters are measurable. The *measurement problem* deals with which physical quantities should be measured and how they should be measured. Physical quantities are measured by sensors. High-quality sensors measure quantities very accurately. Most sensors, however, introduce errors into the measurement process, and these errors may also have to be modeled. The most common type of sensor error is additive measurement noise, which is usually modeled as a random process.

The *estimation problem* deals with the determination of those physical quantities that cannot be measured from those that can be measured. We shall distinguish between the estimation of signals (i.e., states) and the estimation of parameters. Because a subjective decision must sometimes be made to classify a physical quantity as a signal or a parameter, there is some overlap between signal estimation and parameter estimation. Returning to Newton's law, for example, when the mass $M$ is a constant, it is always treated as a parameter. When it varies in time, as in the case of a rocket whose mass decreases as its fuel is burned, then $M = M(t)$, and the mass may be treated as a signal or as a function of a collection of parameters, e.g.,

$$M(t) = \sum_{i=1}^{L} C_i \phi_i(t)$$

known basis functions

In this algebraic model for mass, the "signal" mass has been expressed in terms of $L$ parameters, $C_1, C_2, \ldots, C_L$.

After a model has been completely specified, through choice of an appropriate mathematical representation, measurement of measurable signals, estimation of nonmeasurable signals, and estimation of its parameters, the model must be checked out. The *validation problem* deals with demonstrating confidence in the model. Often, statistical tests involving confidence limits are used to validate a model.

In this book we shall be interested in parameter estimation, state estimation, and combined state and parameter estimation. In Lesson 2 we provide seven examples each of which can be categorized either as a parameter or state estimation problem. Here we just mention that the problem of identifying the sampled values of a linear and time-invariant system's impulse response from input/output data is one of parameter estimation, whereas the problem of reconstructing a state vector associated with a dynamical system, from noisy measurements, is one of state estimation (state estimates might be needed to implement a linear-quadratic-Gaussian optimal control law or to perform postflight data analysis or signal processing such as deconvolution).

## COVERAGE

This book focuses on a wide range of estimation techniques that can be applied either to linear or nonlinear models. Both parameter and state estimation techniques are treated. Some parameter estimation techniques are for random parameters, others are for deterministic parameters, and some are for either random or deterministic parameters; however, state estimation techniques are for random states. Table 1-1 summarizes the book's coverage in terms of a large body of estimation techniques. Supplementary lessons, which are collected at the rear of the book, provide additional points of view or some important background material.

Four lessons (Lessons 3, 4, 5, and 8) are devoted to least-squares estimation because it is a very basic and important technique and, under certain often-occurring conditions, other techniques reduce to it. Consequently, once we understand the nuances of least squares and have established that a different technique has reduced to least squares, we do not have to restudy the nuances of that technique.

**TABLE 1-1** ESTIMATION TECHNIQUES

I. LINEAR MODELS
   A. Parameter Estimation
      1. Deterministic Parameters
         a. Weighted least squares (batch and recursive processing)
         b. Best linear unbiased estimation (BLUE)
         c. Maximum likelihood
      2. Random Parameters
         a. Mean squared
         b. Maximum a posteriori
         c. BLUE
         d. Weighted least squares
   B. State Estimation
      1. Mean-squared prediction
      2. Mean-squared filtering (Kalman filter/Kalman–Bucy filter)
      3. Mean-squared smoothing

II. NONLINEAR MODELS
   A. Parameter Estimation
      Iterated least squares for deterministic parameters
   B. State Estimation
      Extended Kalman filter
   C. Combined State and Parameter Estimation
      1. Extended Kalman filter
      2. Maximum likelihood

In order to fully study least-squares estimators, we must establish their small and large sample properties. What we mean by such properties is the subject of Lessons 6 and 7.

Having spent four lessons on least-squares estimation, we cover best linear unbiased estimation (BLUE) in one lesson, Lesson 9. We are able to do this because BLUE is a special case of least squares.

In order to set the stage for maximum-likelihood estimation, which is covered in Lesson 11, we describe the concept of likelihood and its relationship to probability in Lesson 10.

Lesson A, which is a supplemental one, is on the subject of sufficient statistics and statistical estimation of parameters. Lessons B and C cover higher-order statistics. Each of these lessons, which can be covered after Lesson 11, will provide you with new and different perspectives on estimation theory.

Lesson 12 provides a transition from our study of estimation of deterministic parameters to our study of estimation of random parameters. It provides much useful information about elements of Gaussian random variables. To some readers, this lesson may be a review of material already known to them.

Lesson 13 focuses on mean-squared estimation of random parameters, whereas Lesson 14 focuses on maximum a posteriori estimation of such parameters. In Lesson 13, no probability model is assumed given for the random parameters, whereas in Lesson 14 a probability model is assumed given for these parameters. Best linear unbiased and weighted least-squares estimation are also revisited in these lessons, and we learn conditions under which mean-squared, maximum a posteriori, best-linear unbiased, and weighted least-squares estimates of random parameters are identical. Although a lot of emphasis is given to the important special case of the linear and Gaussian model, for which the mean-squared and maximum a posteriori estimators are linear transformations of the data, the more general case is also treated. Detection and its relation to estimation are covered in Lesson 14.

Lesson 15 provides a transition from our study of parameter estimation to our study of state estimation. It provides much useful information about elements of discrete-time Gauss–Markov random processes and also establishes the *basic state-variable model* and its statistical properties, for which we derive a wide variety of state estimators. To some readers, this lesson may be a review of material already known to them.

Supplementary Lesson D provides a lot of background material on state-variable models, and their relationships to input-output models. If you need a refresher on state-variable models, it is advisable to read Lesson D before you read Lesson 15.

Lessons 16 through 21 cover state estimation for the Lesson 15 basic state-variable model. Prediction is treated in Lesson 16. The important innovations process is also covered in this lesson. Filtering is the subject of Lessons 17, 18, and 19. The mean-squared state filter, commonly known as the Kalman filter, is developed in Lesson 17. Seven examples that illustrate some interesting numerical and theoretical aspects of Kalman filtering are presented in Lesson 18. Lesson 19 establishes a bridge between mean-squared estimation and mean-squared digital signal processing. It shows how the steady-state Kalman filter is related to a digital Wiener filter. The latter is widely used in digital signal processing. Smoothing is the subject of Lessons 20 and 21. Fixed-interval, fixed-point, and fixed-lag smoothers are developed in Lessons 20 and 21. Lesson 21 also presents some applications that illustrate interesting numerical and theoretical aspects of fixed-interval smoothing. These applications are taken from the field of digital signal processing and include minimum-variance deconvolution and maximum-likelihood deconvolution.

Lesson 22 shows how to modify results given in Lessons 16, 17, 19, 20, and 21 from the basic state-variable model to a state-variable model that includes the following effects:

1. either nonzero-mean noise processes or known bias functions or both in the state or measurement equations,
2. correlated noise processes,
3. colored noise processes, and
4. perfect measurements.

Lesson 23 provides a transition from our study of estimation for linear models to estimation for nonlinear models. Because many real-world systems are continuous time in nature and nonlinear, this lesson explains how to linearize and discretize a nonlinear differential equation model.

Lesson 24 is devoted primarily to the extended Kalman filter (EKF), which is a form of the Kalman filter that has been "extended" to nonlinear dynamical systems of the type described in Lesson 23. The EKF is related to the method of iterated least squares (ILS), the major difference between the two being that the EKF is for dynamical systems, whereas ILS is not. This lesson also shows how to apply the EKF to parameter estimation, in which case states and parameters can be estimated simultaneously and in real time.

The problem of obtaining maximum-likelihood estimates of a collection of parameters that appears in the basic state-variable model is treated in Lesson 25. The solution involves state and parameter estimation, but calculations can only be performed off-line after data from an experiment have been collected.

The Kalman–Bucy filter, which is the continuous-time counterpart to the Kalman filter, is derived from two different viewpoints in Lesson 26. This is the only lesson that deviates from our modeling philosophy, in that it focuses on a continuous-time model and develops a continuous-time state estimator. We include it because of the tremendous importance of the Kalman–Bucy filter in stochastic optimal control theory.

## PHILOSOPHY

The discrete-time (digital) viewpoint is emphasized throughout this book. Our estimation algorithms are digital in nature; many are recursive. The reasons for the discrete-time viewpoint are:

1. many real data are collected in a digitized manner, so they are in a form ready to be processed by digital estimation algorithms, and
2. the mathematics associated with discrete-time estimation theory is simpler than that associated with continuous-time estimation theory.

Regarding reason 2, we mention that very little knowledge about random processes is needed to derive discrete-time estimation algorithms, because discrete-time random processes can be treated as vectors of random variables. Much more knowledge about random processes is needed to design continuous-time estimation algorithms.

Suppose our underlying model is continuous time in nature. We are faced with two choices: develop a continuous-time estimation theory and then implement the resulting estimators on a digital computer (i.e., discretize the continuous-time estimation algorithm), or discretize the model and develop a discrete-time estimation theory that leads to estimation algorithms readily implemented on a digital computer. If both approaches lead to algorithms that are implemented digitally, then we advocate the *principle of simplicity* for their development, and this leads us to adopt the second choice. *For estimation, our* ***modeling philosophy*** *is, therefore, discretize the model at the front end of the problem.*

Estimation theory has a long and glorious history, from Gauss to Fisher to Wiener to Kalman (e.g., see Sorenson, 1970); however, it has been greatly influenced by technology, especially the computer. Although much of estimation theory was developed in the mathematics, statistical, and control theory literatures, we shall adopt the following viewpoint toward that theory: *estimation theory is the extension of classical signal processing to the design of digital filters that process uncertain data in an optimal manner.* In fact, estimation algorithms are just filters that transform input streams of numbers into output streams of numbers.

Most of classical digital filter design (e.g., Oppenheim and Schafer, 1989; Hamming, 1983; Peled and Liu, 1976) is concerned with designs associated with deterministic signals, e.g., low-pass and bandpass filters, and, over the years specific techniques have been developed for such designs. The resulting filters are usually "fixed" in the sense that their coefficients do not change as a function of time. Estimation theory, on the other hand, frequently leads to filter structures that are time varying. These filters are designed (i.e., derived) using time-domain performance specifications (e.g., smallest error variance), and, as mentioned previously, process random data in an optimal manner. *Our* ***philosophy about estimation theory*** *is that it can be viewed as a natural adjunct to digital signal processing theory.*

**EXAMPLE 1-1**

At one time or another we have all used the sample mean to compute an "average." Suppose we are given a collection of $k$ measured values of quantity $X$, that is, $x(1), x(2), \ldots, x(k)$. The sample mean of these measurements, $\bar{x}(k)$, is

$$\bar{x}(k) = \frac{1}{k}\sum_{j=1}^{k} x(j) \tag{1-1}$$

This equation is one in which all $k$ measurements are processed at one time; i.e., it is a *batch equation* for computing the sample mean. A recursive formula for the sample mean is obtained from (1-1), as follows:

$$\begin{aligned}\bar{x}(i+1) &= \frac{1}{i+1}\sum_{j=1}^{i+1} x(j) = \frac{1}{i+1}\left[\sum_{j=1}^{i} x(j) + x(i+1)\right] \\ \bar{x}(i+1) &= \frac{i}{i+1}\bar{x}(i) + \frac{1}{i+1}x(i+1)\end{aligned} \tag{1-2}$$

This recursive version of the sample mean is used for $i = 0, 1, \ldots, k-1$, by setting $\bar{x}(0) = 0$.

Observe that *the sample mean, as expressed in (1-2), is a time-varying recursive digital filter whose input is measurement* $x(i+1)$. In later lessons we show that the sample mean is also an optimal estimation algorithm; thus, although the reader many not have been aware of it, the sample mean, which he or she has been using since early schooldays, is an estimation algorithm. □

## COMPUTATION

Estimation theory leads to estimation algorithms, as well as to an understanding of their performance. Estimation algorithms process data and, as such, must be implemented on a digital computer. Our ***computation philosophy*** is *whenever* *possible, leave it to the experts*; hence, we describe how to implement estimation algorithms, and even modeling and associated analysis operations, using MATLAB®, pertinent toolboxes, and SIMULINK® (SIMULINK is a registered trademark of The MathWorks, Inc.).

MATLAB is a large collection of more than 500 subroutines, which are called *M-files*. They are easily linked together by the end user. Toolboxes are collections of special-purpose area-specific M-files that have been prepared by experts. The toolboxes that are needed in order to implement much of the material in this book are *Control Systems, Optimization*, and *Hi-Spec*™ (Hi-Spec is a trademark of United Signals and Systems, Inc.). Additionally, the *System Identification* toolbox will be useful to the end user whose specific objective is to estimate parameters or signals in very specific models (e.g., autoregressive models).

The intent of the computation section is to direct the reader to very specific existing M-files. Occasionally, when no M-file exists for a very important estimation algorithm, we provide one for it.

## SUMMARY QUESTIONS*

1. The modeling problem decomposes into:
   (a) two problems
   (b) three problems
   (c) four problems
2. Our modeling philosophy is to:
   (a) discretize at the front end of the problem
   (b) work with the continuous-time model for as long as possible
   (c) discretize the filter and just ignore the model
3. Our philosophy about estimation theory is that:
   (a) it is control theory
   (b) it is a branch of mathematics
   (c) it is a natural adjunct to digital signal processing
4. Our philospohy about computation is to:
   (a) write your own programs
   (b) leave it to the experts
   (c) use your friend's programs

*Answers to all *Summary Questions* are given in Appendix C.

# LESSON 2

# The Linear Model

## SUMMARY

The main purpose of this lesson is to introduce a generic model that is linear in the unknown parameters, $\mathbf{Z}(k) = \mathbf{H}(k)\boldsymbol{\theta} + \mathbf{V}(k)$. In this model, which we refer to as a *generic linear model*, vector $\boldsymbol{\theta}$ contains the unknown deterministic or random parameters that will be estimated using one or more of this book's techniques. Examples are given which demonstrate how to obtain a generic linear model for real problems.

Some estimation notation is also introduced: $\hat{x}$ denotes an estimate of $x$ and $\tilde{x}$ denotes the error in estimation, i.e., $\tilde{x} = x - \hat{x}$.

The generic linear model is the starting point for the derivation of many classical parameter estimation techniques.

When you complete this lesson you will be able to (1) show how a wide range of applications can be put into the form of a generic linear model, and, after seeing how to do this for the text's examples, you will be able to do it for your own applications; and (2) recognize and understand the commonly used notation in estimation theory.

## INTRODUCTION

In order to estimate unknown quantities (i.e., parameters or signals) from measurements and other given information, we must begin with model representations and express them in such a way that attention is focused on the explicit relationship between the unknown quantities and the measurements. Many familiar models are linear in the unknown quantities (denoted $\boldsymbol{\theta}$), and can be expressed as *make your system be modelled by*

$$\mathbf{Z}(k) = \mathbf{H}(k)\theta + \mathbf{V}(k) \qquad (2\text{-}1)$$

In this model, $\mathbf{Z}(k)$, which is $N \times 1$, is called the *measurement vector*; $\theta$, which is $n \times 1$, is called the *parameter vector*; $\mathbf{H}(k)$, which is $N \times n$ is called the *observation matrix*; and, $\mathbf{V}(k)$, which is $N \times 1$, is called the *measurement noise* vector. Usually, $\mathbf{V}(k)$ is random. By convention, the argument "$k$" of $\mathbf{Z}(k)$, $\mathbf{H}(k)$, and $\mathbf{V}(k)$ denotes the fact that the last measurement used to construct (2-1) is the $k$th. All other measurements occur "before" the $k$th.

Strictly speaking, (2-1) represents an "affine" transformation of parameter vector $\theta$ rather than a linear transformation. We shall, however, adhere to traditional estimation-theory literature by calling (2-1) a "linear model."

*H is given or measured ahead of time (e.g. known)*

## EXAMPLES

Some examples that illustrate the formation of (2-1) are given in this section. What distinguishes these examples from one another are the nature of and interrelationships between $\theta$, $\mathbf{H}(k)$ and $\mathbf{V}(k)$. The following situations can occur.

**A.** $\theta$ is deterministic
   1. $\mathbf{H}(k)$ is deterministic.
   2. $\mathbf{H}(k)$ is random.
      a. $\mathbf{H}(k)$ and $\mathbf{V}(k)$ are statistically independent.
      b. $\mathbf{H}(k)$ and $\mathbf{V}(k)$ are statistically dependent.

**B.** $\theta$ is random
   1. $\mathbf{H}(k)$ is deterministic.
   2. $\mathbf{H}(k)$ is random.
      a. $\mathbf{H}(k)$ and $\mathbf{V}(k)$ are statistically independent.
      b. $\mathbf{H}(k)$ and $\mathbf{V}(k)$ are statistically dependent.

### EXAMPLE 2-1   Impulse Response Identification

It is well known that the output of a single-input single-output, linear, time-invariant, discrete-time system is given by the following convolution-sum relationship:

$$y(k) = \sum_{i=-\infty}^{\infty} h(i)u(k - i) \qquad (2\text{-}2)$$

where $k = 1, 2, \ldots, N$, $h(i)$ is the system's impulse response (IR), $u(k)$ is its input, and $y(k)$ its output. If $u(k) = 0$ for $k < 0$, the system is causal, so that $h(i) = 0$ for $i \leq 0$, and $h(i) \simeq 0$ for $i > n$, then

$$y(k) = \sum_{i=1}^{n} h(i)u(k - i) \qquad (2\text{-}3)$$

*e.g. no poles*

Equation (2-3) is known as a *finite impulse response (FIR) model*. Besides being the fundamental input/output relationship for all linear, causal, time-invariant systems, (2-3) is also a time-series model known in the time-series literature as a *moving average* (MA). Observe from (2-3) that $y(k)$ is a linear combination of exactly $n$ values of signal $u(\cdot)$, that is,

$u(k-1), u(k-2), \ldots, u(k-n)$. The $h(i)$'s, which are known as the *MA coefficients*, weight these $n$ signals to form their weighted average. As time index $k$ ranges over its admissible set of values (e.g., $k = 1, 2, \ldots$) the window of $n$ values of $u(\cdot)$ moves from left to right; hence the name "moving average." Note that MA models do not have to be causal. An example of a noncausal MA model is

$$y(k) = \sum_{i=-m}^{n} h(i)u(k-i)$$

Signal $y(k)$ is measured by a sensor that is corrupted by additive measurement noise, $\nu(k)$, i.e., we only have access to measurement $z(k)$, where

$$z(k) = y(k) + \nu(k) \tag{2-4}$$

and $k = 1, 2, \ldots, N$. Equations (2-3) and (2-4) combine to give the "signal-plus-noise" model, depicted in Figure 2-1, which is the starting point for much analysis and design in signal processing and communications.

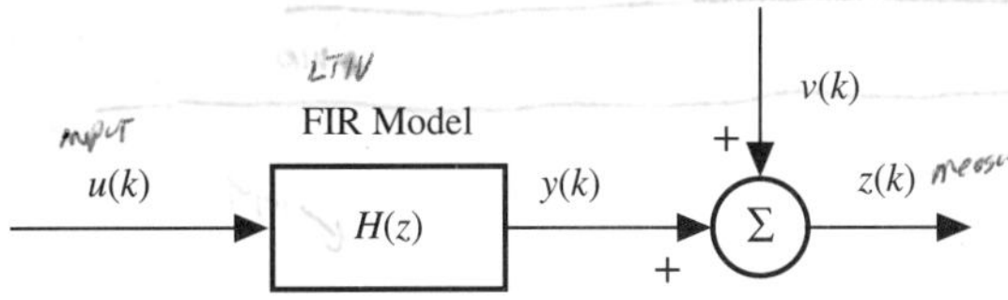


**Figure 2-1** Widely used signal-plus-noise model.

We now collect (concatenate) the $N$ measurements for the FIR model as follows:

$$\underbrace{\begin{pmatrix} z(N) \\ z(N-1) \\ \vdots \\ z(n) \\ \vdots \\ z(2) \\ z(1) \end{pmatrix}}_{\mathbf{Z}(N)} = \underbrace{\begin{pmatrix} u(N-1) & u(N-2) & u(N-3) & \ldots & u(N-n) \\ u(N-2) & u(N-3) & u(N-4) & \ldots & u(N-n-1) \\ \vdots & \vdots & \vdots & \ddots & \vdots \\ u(n-1) & u(n-2) & u(n-3) & \ldots & u(0) \\ \vdots & \vdots & \vdots & \ddots & \vdots \\ u(1) & u(0) & 0 & \ldots & 0 \\ u(0) & 0 & 0 & \ldots & 0 \end{pmatrix}}_{\mathbf{H}(N-1)} \times \underbrace{\begin{pmatrix} h(1) \\ h(2) \\ \vdots \\ h(n) \end{pmatrix}}_{\boldsymbol{\theta}} + \underbrace{\begin{pmatrix} \nu(N) \\ \nu(N-1) \\ \vdots \\ \nu(n) \\ \vdots \\ \nu(2) \\ \nu(1) \end{pmatrix}}_{\mathbf{V}(N)} \tag{2-5}$$

Clearly, (2-5) is in the form of (2-1).

In this application the $n$ sampled values of the IR, $h(i)$, play the role of unknown parameters, i.e., $\theta_1 = h(1), \theta_2 = h(2), \ldots, \theta_n = h(n)$, and these parameters are deterministic.

Our objective will be to estimate the unknown $\theta_i$'s. If input $u(k)$ is deterministic and is known ahead of time (or can be measured) without error, then $\mathbf{H}(N-1)$ is deterministic so that we are in case A.1. Often, however, $u(k)$ is random so that $\mathbf{H}(N-1)$ is random; but $u(k)$ is in no way related to measurement noise $\nu(k)$, so we are in case A.2.a.

Observe, from (2-5), that it is our convention to put the last available measurement at the very top of the concatenated measurement vector $\mathbf{Z}(N)$. Other authors sometimes prefer to put the last available measurement at the very bottom of this vector. Doing this, of course, changes the specific structures of $\mathbf{H}(N-1)$ and $\mathbf{V}(N)$; however, the two results are related by the invertible antidiagonal identity matrix. □

**EXAMPLE 2-2 Identification of the Coefficients of a Finite-difference Equation**

Suppose a linear, time-invariant, discrete-time system is described by the following $n$th-order finite-difference equation:

$$y(k) + \alpha_1 y(k-1) + \cdots + \alpha_n y(k-n) = u(k-1) \tag{2-6}$$

This model is often referred to as an all-pole or autoregressive (AR) model. It occurs in many branches of engineering and science, including speech modeling and geophysical modeling. Suppose, also, that $N$ perfect measurements of signal $y(k)$ are available. Parameters $\alpha_1, \alpha_2, \ldots, \alpha_n$ are unknown and are to be estimated from the data. To do this, we can rewrite (2-6) as

$$y(k) = -\alpha_1 y(k-1) - \cdots - \alpha_n y(k-n) + u(k-1) \tag{2-7}$$

and collect $y(1), y(2), \ldots, y(N)$ as we did in Example 2-1. Doing this, we obtain

$$\underbrace{\begin{pmatrix} y(N) \\ y(N-1) \\ \vdots \\ y(n) \\ \vdots \\ y(2) \\ y(1) \end{pmatrix}}_{\mathbf{Z}(N)} = \underbrace{\begin{pmatrix} y(N-1) & y(N-2) & y(N-3) & \ldots & y(N-n) \\ y(N-2) & y(N-3) & y(N-4) & \ldots & y(N-n-1) \\ \vdots & \vdots & \vdots & \ddots & \vdots \\ y(n-1) & y(n-2) & y(n-3) & \ldots & y(0) \\ \vdots & \vdots & \vdots & \ddots & \vdots \\ y(1) & y(0) & 0 & \ldots & 0 \\ y(0) & 0 & 0 & \ldots & 0 \end{pmatrix}}_{\mathbf{H}(N-1)}$$

$$\times \underbrace{\begin{pmatrix} -\alpha_1 \\ -\alpha_2 \\ \vdots \\ -\alpha_n \end{pmatrix}}_{\boldsymbol{\theta}} + \underbrace{\begin{pmatrix} u(N-1) \\ u(N-2) \\ \vdots \\ u(n) \\ \vdots \\ u(1) \\ u(0) \end{pmatrix}}_{\mathbf{V}(N-1)} \tag{2-8}$$

which, again, is in the form of (2-1).

Observe, in (2-8), that system input $u(\cdot)$ plays the role of the generic measurement noise. This is why this noise is called "generic"; it does not have to correspond to additive measurement noise, as the model in (2-1) would have us believe. Observe, also, that by iterating (2-7) for $k = 1, 2, \ldots$, we can determine that $y(k)$ depends on the following collection of input values: $u(0), u(1), \ldots, u(k-1)$. In short, the output of this system, $y(k)$, depends on all the past inputs up to and including the one at time $t_{k-1}$.

In this example $\boldsymbol{\theta} = \text{col}(-\alpha_1, -\alpha_2, \ldots, -\alpha_n)$, and these parameters are deterministic. If input $u(k-1)$ is deterministic, then the system's output $y(k)$ will also be deterministic, so that both $\mathbf{H}(N-1)$ and $\mathbf{V}(N-1)$ are deterministic. This is a very special case of case A.1, because usually $\mathbf{V}$ is random. If, however, $u(k-1)$ is random, then $y(k)$ will also be random; but the elements of $\mathbf{H}(N-1)$ will now depend on those in $\mathbf{V}(N-1)$, because $y(k)$ depends upon $u(0), u(1), \ldots, u(k-1)$. In this situation we are in case A.2.b.

Before leaving this example, we want to provide the model in (2-6) with another useful interpretation. Suppose we want to predict a future value of a stationary discrete-time random process using a finite set of past samples of the process, i.e., we want to predict the value of $y(k)$ using $y(k-1), y(k-2), \ldots, y(k-n)$. Letting $\hat{y}(k)$ denote the predicted value of $y(k)$, then the structure of a *linear predictor* is

$$\hat{y}(k) = \sum_{i=1}^{n} a_i y(k-i) \tag{2-9}$$

Alternatively, we can write this equation as

$$y(k) = \sum_{i=1}^{n} a_i y(k-i) + \tilde{y}(k) \tag{2-10}$$

where $\tilde{y}(k)$ is the error in prediction, i.e., $\tilde{y}(k) = y(k) - \hat{y}(k)$. Comparing (2-10) and (2-7), we see that they are similar. The prediction error in (2-10) plays the role of the system input in (2-7), and the coefficients of the linear prediction filter play the role of the (negative) AR coefficients. Consequently, some of the parameter estimation techniques that we will discuss in this book (e.g., least squares) are also useful for design of linear prediction filters.

Finally, Problems 2-6 and 2-7 demonstrate that the second-order statistics of $y(k)$ and even higher-order statistics of $y(k)$ also satisfy AR equations; hence, the AR coefficients can not only be estimated using the raw data, but they can also be estimated using statistics (i.e., nonlinear transformations of the data) associated with the data. □

### EXAMPLE 2-3 Function Approximation

In a function approximation problem, we wish to fit a given set of data, $(\mathbf{x}_1, f(\mathbf{x}_1)), (\mathbf{x}_2, f(\mathbf{x}_2)), \ldots, (\mathbf{x}_N, f(\mathbf{x}_N))$, by the approximating function

$$f(\mathbf{x}) \approx \hat{f}(\mathbf{x}) = \sum_{j=1}^{n} \theta_j \phi_j(\mathbf{x}) \tag{2-11}$$

where $\phi_j(\mathbf{x})$ are a set of $n$ prespecified *basis functions*. These could be multidimensional orthogonal polynomials, trigonometric functions, Walsh functions, spline functions, radial basis functions (Powell, 1987) or even fuzzy basis functions (Wang and Mendel, 1992). Each of the different basis functions provides a different nonlinear structure for $\phi(\cdot)$. If only imperfect (i.e., noisy) values of $f(\mathbf{x}_i)$ are known, that is, $f_m(\mathbf{x}_i)$, where

$$f_m(\mathbf{x}_i) = f(\mathbf{x}_i) + e(\mathbf{x}_i), \tag{2-12}$$

then collecting these $N$ measurements, as before, we obtain

$$\begin{bmatrix} f_m(\mathbf{x}_N) \\ f_m(\mathbf{x}_{N-1}) \\ \cdots \\ f_m(\mathbf{x}_1) \end{bmatrix} = \begin{bmatrix} \phi_1(\mathbf{x}_N) & \phi_2(\mathbf{x}_N) & \ldots & \phi_n(\mathbf{x}_N) \\ \phi_1(\mathbf{x}_{N-1}) & \phi_2(\mathbf{x}_{N-1}) & \ldots & \phi_n(\mathbf{x}_{N-1}) \\ & \cdots & & \\ \phi_1(\mathbf{x}_1) & \phi_2(\mathbf{x}_1) & \ldots & \phi_n(\mathbf{x}_1) \end{bmatrix} \begin{bmatrix} \theta_1 \\ \theta_2 \\ \cdots \\ \theta_n \end{bmatrix} + \begin{bmatrix} e(\mathbf{x}_N) \\ e(\mathbf{x}_{N-1}) \\ \cdots \\ e(\mathbf{x}_1) \end{bmatrix} \tag{2-13}$$

Once again, we have been led to (2-1), and we are in case A.1. □

**EXAMPLE 2-4 State Estimation**

State-variable models are widely used in control and communication theory and in signal processing. Often, we need the entire state vector of a dynamical system in order to implement an optimal control law for it or to implement a digital signal processor. Usually, we cannot measure the entire state vector, and our measurements are corrupted by noise. In state estimation, our objective is to estimate the entire state vector from a limited collection of noisy measurements.

Here we consider the problem of estimating $n \times 1$ state vector $\mathbf{x}(k)$, at $k = 1, 2, \ldots, N$ from a scalar measurement $z(k)$, where $k = 1, 2, \ldots, N$. The model for this example is

$$\mathbf{x}(k+1) = \mathbf{\Phi}\mathbf{x}(k) + \gamma u(k) \tag{2-14}$$

$$z(k) = \mathbf{h}'\mathbf{x}(k) + \nu(k) \tag{2-15}$$

We are keeping this example simple by assuming that the system is time invariant and has only one input and one output; however, the results obtained in this example are easily generalized to time-varying and multichannel systems or multichannel systems alone. For a review of state-variable models and their relationship to transfer function and difference equation models, see Lesson D.

If we try to collect our $N$ measurements as before, we obtain

$$\left.\begin{aligned} z(N) &= \mathbf{h}'\mathbf{x}(N) + \nu(N) \\ z(N-1) &= \mathbf{h}'\mathbf{x}(N-1) + \nu(N-1) \\ &\cdots \\ z(1) &= \mathbf{h}'\mathbf{x}(1) + \nu(1) \end{aligned}\right\} \tag{2-16}$$

Observe that a different (unknown) state vector appears in each of the $N$ measurement equations; thus, there does not appear to be a common "$\boldsymbol{\theta}$" for the collected measurements. Appearances can sometimes be deceiving.

So far, we have not made use of the state equation. Its solution can be expressed as (see Lesson D for a derivation of this equation)

$$\mathbf{x}(k) = \mathbf{\Phi}^{k-j}\mathbf{x}(j) + \sum_{i=j+1}^{k} \mathbf{\Phi}^{k-i}\gamma u(i-1) \tag{2-17}$$

where $k \geq j+1$. *We now focus our attention on the value of* $\mathbf{x}(j)$ *at* $j = k_1$, *where* $1 \leq k_1 \leq N$. Our goal is to express $z(N), z(N-1), \ldots, z(1)$, as given in (2-16), in terms of $\mathbf{x}(k_1)$. This has to be done carefully, because (2-17) can only be used, as is, for $k = k_1+1, k_1+2, \ldots, N$. Consequently, using (2-17), we can express $\mathbf{x}(k_1+1), \ldots, \mathbf{x}(N-1), \mathbf{x}(N)$ as an explicit function of $\mathbf{x}(k_1)$ as

$$\mathbf{x}(k) = \mathbf{\Phi}^{k-k_1}\mathbf{x}(k_1) + \sum_{i=k_1+1}^{k} \mathbf{\Phi}^{k-i}\gamma u(i-1) \tag{2-18}$$

where $k = k_1 + 1, k_1 + 2, \ldots, N$. In order to do the same for $\mathbf{x}(1), \mathbf{x}(2), \ldots, \mathbf{x}(k_1 - 1)$, we solve (2-17) for $\mathbf{x}(j)$ and set $k = k_1$:

$$\mathbf{x}(j) = \mathbf{\Phi}^{j-k_1}\mathbf{x}(k_1) - \sum_{i=j+1}^{k_1} \mathbf{\Phi}^{j-i}\boldsymbol{\gamma}u(i-1) \tag{2-19}$$

where $j = k_1 - 1, k_1 - 2, \ldots, 2, 1$. Using (2-18) and (2-19), we can reexpress (2-16) as

$$\left.\begin{aligned}
z(k) &= \mathbf{h}'\mathbf{\Phi}^{k-k_1}\mathbf{x}(k_1) + \mathbf{h}' \sum_{i=k_1+1}^{k} \mathbf{\Phi}^{k-i}\boldsymbol{\gamma}u(i-1) + \nu(k) \\
k &= N, N-1, \ldots, k_1 + 1 \\
z(k_1) &= \mathbf{h}'\mathbf{x}(k_1) + \nu(k_1) \\
z(l) &= \mathbf{h}'\mathbf{\Phi}^{l-k_1}\mathbf{x}(k_1) - \mathbf{h}' \sum_{i=l+1}^{k_1} \mathbf{\Phi}^{l-i}\boldsymbol{\gamma}u(i-1) + \nu(l) \\
l &= k_1 - 1, k_1 - 2, \ldots, 1
\end{aligned}\right\} \tag{2-20}$$

These $N$ equations can now be collected together to give

$$\underbrace{\begin{pmatrix} z(N) \\ z(N-1) \\ \vdots \\ z(1) \end{pmatrix}}_{\mathbf{Z}(N)} = \underbrace{\begin{pmatrix} \mathbf{h}'\mathbf{\Phi}^{N-k_1} \\ \mathbf{h}'\mathbf{\Phi}^{N-1-k_1} \\ \vdots \\ \mathbf{h}'\mathbf{\Phi}^{1-k_1} \end{pmatrix}}_{\mathbf{H}(N,k_1)} \underbrace{\mathbf{x}(k_1)}_{\boldsymbol{\theta}} + \underbrace{\mathbf{M}(N,k_1) \begin{pmatrix} u(N-1) \\ u(N-2) \\ \vdots \\ u(0) \end{pmatrix} + \begin{pmatrix} \nu(N) \\ \nu(N-1) \\ \vdots \\ \nu(1) \end{pmatrix}}_{\mathbf{V}(N,k_1)} \tag{2-21}$$

where the exact structure of matrix $\mathbf{M}(N, k_1)$ is not important to us at this point. Observe that the state at $k = k_1$ plays the role of parameter vector $\boldsymbol{\theta}$ and that both $\mathbf{H}$ and $\mathbf{V}$ are different for different values of $k_1$. Observe, also, that the "generic" measurement noise $\mathbf{V}(N, k_1)$ is a linear transformation of the system's input history as well as the history of the additive measurement noise.

If $\mathbf{x}(0)$ and the system input $u(k)$ are deterministic, then $\mathbf{x}(k)$ is deterministic for all $k$. In this case $\boldsymbol{\theta}$ is deterministic, but $\mathbf{V}(N, k_1)$ is a superposition of deterministic and random components. On the other hand, if either $\mathbf{x}(0)$ or $u(k)$ is random, then $\boldsymbol{\theta}$ is a vector of *random parameters*. This latter situation is the more usual one in state estimation. It corresponds to case B.1.

While it is true that we have indeed shown that state estimation can be cast into the form of our generic linear model, it is also true that every time we focus our attention on a different value of time $k_1$ in (2-21) we must recompute $\mathbf{H}(N, k_1)$ and $\mathbf{M}(N, k_1)$. This is terribly inefficient, especially if our ultimate goal is to estimate the state vector at a multitude of values of $k_1$. In Lessons 16–22 we approach state estimation from a computationally more efficient point of view. □

**EXAMPLE 2-5 A Nonlinear Model**

Many of the estimation techniques that are described in this book in the context of linear model (2-1) can also be applied to the estimation of unknown signals or parameters in nonlinear models, when such models are suitably linearized. Suppose, for example, that

$$z(k) = f(\theta, k) + \nu(k) \tag{2-22}$$

where $k = 1, 2, \ldots, N$, and the structure of nonlinear function $f(\theta, k)$ is known explicitly. To see the forest from the trees in this example, we assume $\theta$ is a scalar parameter.

Let $\theta^*$ denote a *nominal value* of $\theta$, $\delta\theta = \theta - \theta^*$, and $\delta z = z - z^*$, where

$$z^*(k) = f(\theta^*, k) \tag{2-23}$$

Observe that the *nominal measurements*, $z^*(k)$, can be computed once $\theta^*$ is specified, because $f(\cdot, k)$ is assumed to be known.

Using a first-order Taylor series expansion of $f(\theta, k)$ about $\theta = \theta^*$, it is easy to show that

$$\delta z(k) = \left.\frac{\partial f(\theta, k)}{\partial \theta}\right|_{\theta=\theta^*} \delta\theta + \nu(k) \tag{2-24}$$

where $k = 1, 2, \ldots, N$. It is easy to see how to collect these $N$ equations to give

$$\underbrace{\begin{pmatrix} \delta z(N) \\ \delta z(N-1) \\ \vdots \\ \delta z(1) \end{pmatrix}}_{\mathbf{Z}(N)} = \underbrace{\begin{pmatrix} \partial f(\theta^*, N)/\partial\theta^* \\ \partial f(\theta^*, N-1)/\partial\theta^* \\ \vdots \\ \partial f(\theta^*, 1)/\partial\theta^* \end{pmatrix}}_{\mathbf{H}(N, \theta^*)} \delta\theta + \underbrace{\begin{pmatrix} \nu(N) \\ \nu(N-1) \\ \vdots \\ \nu(1) \end{pmatrix}}_{\mathbf{V}(N)} \tag{2-25}$$

in which $\partial f(\theta^*, k)/\partial\theta^*$ is short for "$\partial f(\theta, k)/\partial\theta$ evaluated at $\theta = \theta^*$."

Observe that (2-25) is linear in $\delta\theta$ and that $\mathbf{H}$ depends on $\theta^*$. We will discuss different ways for specifying $\theta^*$ in Lesson 24. □

**EXAMPLE 2-6 Deconvolution** (Mendel, 1983)

In Example 2-1 we showed how a convolutional model could be expressed as the linear model $\mathbf{Z} = \mathbf{H}\boldsymbol{\theta} + \mathbf{V}$. In that example we assumed that both input and output measurements were available, and we wanted to estimate the sampled values of the system's impulse response. Here we begin with the same convolutional model, written as

$$z(k) = \sum_{i=1}^{k} \mu(i)h(k-i) + \nu(k) \tag{2-26}$$

where $k = 1, 2, \ldots, N$. Noisy measurements $z(1), z(2), \ldots, z(N)$ are available to us, and we assume that we know the system's impulse response $h(j)$, $\forall j$. What is not known is the input to the system $\mu(1), \mu(2), \ldots, \mu(N)$. *Deconvolution is the signal processing procedure for removing the effects of* h(j) *and* v(j) *from the measurements so that one is left with an estimate of* $\mu$(j) (see Figure 2-2).

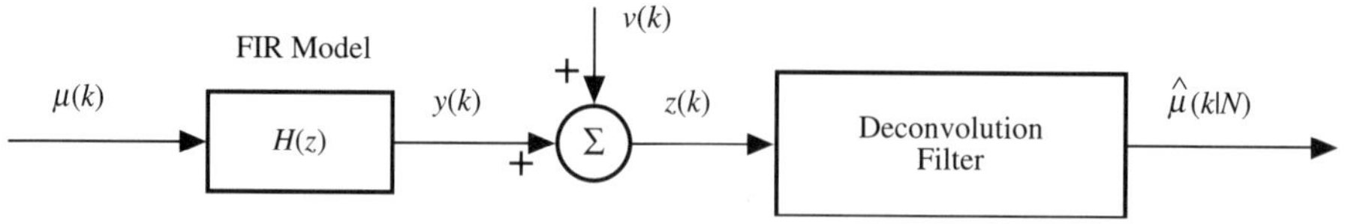


**Figure 2-2** Deconvolution filter provides an estimate of input to the system.

In deconvolution we often assume that input $\mu(j)$, is white noise, but not necessarily Gaussian. This type of deconvolution problem occurs in reflection seismology and digital communications.

We now collect the $N$ measurements, but in such a way that $\mu(1), \mu(2), \ldots, \mu(N)$ are treated as the unknown parameters. Doing this, we obtain the following linear deconvolution model:

$$\underbrace{\begin{pmatrix} z(N) \\ z(N-1) \\ \vdots \\ z(2) \\ z(1) \end{pmatrix}}_{\mathbf{Z}(N)} = \underbrace{\begin{pmatrix} h(N-1) & h(N-2) & \ldots & h(1) & h(0) \\ h(N-2) & h(N-3) & \ldots & h(0) & 0 \\ \vdots & \vdots & \ddots & \vdots & \vdots \\ h(1) & h(0) & \ldots & 0 & 0 \\ h(0) & 0 & \ldots & 0 & 0 \end{pmatrix}}_{\mathbf{H}(N-1)}$$

$$\times \underbrace{\begin{pmatrix} \mu(1) \\ \mu(2) \\ \vdots \\ \mu(N-1) \\ \mu(N) \end{pmatrix}}_{\boldsymbol{\theta}} + \underbrace{\begin{pmatrix} \nu(N) \\ \nu(N-1) \\ \vdots \\ \nu(2) \\ \nu(1) \end{pmatrix}}_{\mathbf{V}(N)} \tag{2-27}$$

We shall often refer to $\boldsymbol{\theta}$ as $\boldsymbol{\mu}$.

We see that deconvolution corresponds to case B.1. Put another way, we have shown that *the design of a deconvolution signal-processing filter is isomorphic to the problem of estimating random parameters in a linear model.* Note, however, that the dimension of $\boldsymbol{\theta}$, which is $N \times 1$, increases as the number of measurements increases. In all other examples $\boldsymbol{\theta}$ was $n \times 1$ where $n$ is a fixed integer. We return to this point in Lesson 13 where we discuss convergence of estimates of $\boldsymbol{\mu}$ to their true values. □

**EXAMPLE 2-7 Volterra Series Representation of a Nonlinear System** (Adapted from Hsia, 1977, pp. 141–142)

The input/output relationship for a single-input, single-output nonlinear system can be expressed explicitly as the following Volterra series (Volterra, 1959; Eykhoff, 1974; Schetzen, 1974):

$$\begin{aligned} y(t) = \int_{-\infty}^{t} g_1(\tau)u(t-\tau)d\tau + \int_{-\infty}^{t}\int_{-\infty}^{t} g_2(\tau_1,\tau_2)u(t-\tau_1)u(t-\tau_2)d\tau_1 d\tau_2 + \cdots \\ + \int_{-\infty}^{t}\cdots\int_{-\infty}^{t} g_n(\tau_1,\tau_2,\cdots,\tau_n)\prod_{i=1}^{n} u(t-\tau_i)d\tau_i + \cdots \end{aligned} \tag{2-28}$$

where $g_n(\tau_1, \tau_2, \ldots, \tau_n)$ is the $n$th-order Volterra kernel. Clearly, linear systems are a special case of a Volterra series. The challenge before us is to identify the Volterra kernels from input/output measurements of both $u(t)$ and $y(t)$, where both measurements might be corrupted by additive measurement noise. To do this, we can approximate the Volterra series by its discretized form:

$$y(k) = \sum_{i=0}^{p} h(i)u(k-i) + \sum_{i=0}^{p}\sum_{j=0}^{p} h(i,j)u(k-i)u(k-j) + \cdots \tag{2-29}$$

for $k \geq p$. To arrive at (2-29) from (2-28), we have set $t_k = kT$, where $T$ is sampling time, let $h(i) = g_1(\tau = iT)T$, $h(i,j) = g_2(\tau_1 = iT, \tau_2 = jT)T^2$, etc., and have assumed that both

$h(i)$ and $h(i, j)$ can be truncated when $t > pT$. A more general model would use different truncation limits for $h(i)$ and $h(i, j)$. Sampling time $T$ has been dropped for notational simplicity.

Let us assume, for purposes of this example, that the Volterra series is truncated at two terms. Our objective is to estimate $h(i)$ and $h(i, j)$ from measurements of $u(k)$ and $y(k)$, $k = p + 1, \ldots, N$. Observe, from (2-29), that $y(k)$ can be written as

$$y(k) = \mathbf{h}'(k)\boldsymbol{\theta}, \qquad k = p + 1, \ldots, N \tag{2-30}$$

where

$$\boldsymbol{\theta} = \text{col}[h(0), h(1), \ldots, h(p), h(0, 0), h(0, 1), \ldots, h(p, p)] \tag{2-31}$$

and

$$\mathbf{h}(k) = \text{col}[u(k), u(k-1), \ldots, u(k-p), u^2(k), u(k)u(k-1), \ldots, u^2(k-p)] \tag{2-32}$$

It is now a simple matter to collect the $N - p$ measurements, as

$$\mathbf{Y}(N) = \mathbf{H}(N)\boldsymbol{\theta} \tag{2-33}$$

where the exact structures of $\mathbf{Y}(N)$ and $\mathbf{H}(N)$ are easy to deduce from (2-30) and (2-32). One of the interesting features of this model is that it treats a two-dimensional array of unknowns [i.e., the $h(i, j)$'s] in an ordered way (known as a lexicographical ordering) as a one-dimensional vector of unknowns. Note, also, that this model contains a lot of unknown parameters, that is, $p^2 + 3p + 2, \ldots$ or does it? A more careful examination of $h(k)$, in (2-32), reveals that it always contains the pair of terms $u(k - i)u(k - j)$ and $u(k - j)u(k - i)(i, j = 0, 1, \ldots, p)$ [e.g., $u(k)u(k - 1)$ and $u(k - 1)u(k)$]; hence, we are unable to uniquely distinguish between $h(i, j)$ and $h(j, i)$, which means that, in practice, $\boldsymbol{\theta}$ must be reformulated to account for this. In general, we will only be able to estimate $[h(i, j) + h(j, i)]/2$; however, if $h(i, j)$ is symmetric, so that $h(i, j) = h(j, i)$, then we can estimate $h(i, j)$.

When perfect measurements are made of the system's input, we are in case A.1; otherwise, as in Example 2-1, we are in case A.2.a. □

## NOTATIONAL PRELIMINARIES

Equation (2-1) can be interpreted as a *data-generating model*; it is a mathematical representation that is associated with the data. Parameter vector $\boldsymbol{\theta}$ is assumed to be unknown and is to be estimated using $\mathbf{Z}(k)$, $\mathbf{H}(k)$, and possibly other a priori information. We use $\hat{\boldsymbol{\theta}}(k)$ to denote the estimate of constant parameter vector $\boldsymbol{\theta}$. Argument $k$ in $\hat{\boldsymbol{\theta}}(k)$ denotes the fact that the estimate is based on measurements up to and including the $k$th. In our preceding examples, we would use the following notation for $\hat{\boldsymbol{\theta}}(k)$:

Example 2-1 [see (2-5)]: $\hat{\boldsymbol{\theta}}(N)$ with components $\hat{h}(i|N)$

Example 2-2 [see (2-8)]: $\hat{\boldsymbol{\theta}}(N)$ with components $-\hat{\alpha}_i(N)$

Example 2-3 [see (2-13)]: $\hat{\boldsymbol{\theta}}(N)$

Example 2-4 [see (2-21)]: $\hat{\mathbf{x}}(k_1|N)$

Example 2-5 [see (2-25)]: $\widehat{\delta\theta}(N)$

Example 2-6 [see (2-27)]: $\hat{\theta}(N)$ with components $\hat{\mu}(i|N)$

Example 2-7 [see (2-33)]: $\hat{\theta}(N)$ with components $\hat{h}(i|N)$ and $\hat{h}(i, j|N)$

The notation used in Examples 1, 4, 6, and 7 is a bit more complicated than that used in the other examples, because we must indicate the time point at which we are estimating the quantity of interest (e.g., $k_1$ or $i$) as well as the last data point used to obtain this estimate (e.g., $N$). We often read $\hat{\mathbf{x}}(k_1|N)$ as "the estimate of $\mathbf{x}(k_1)$ conditioned on $N$" or as "**x** hat at $k_1$ conditioned on $N$."

In state estimation (or deconvolution), three situations are possible depending upon the relative relationship of $N$ to $k_1$. For example, when $N < k_1$, we are estimating a future value of $\mathbf{x}(k_1)$, and we refer to this as a *predicted estimate.* When $N = k_1$, we are using all past measurements and the most recent measurement to estimate $\mathbf{x}(k_1)$. The result is referred to as a *filtered estimate.* Finally, when $N > k_1$, we are estimating an earlier value of $\mathbf{x}(k_1)$ using past, present, and future measurements. Such an estimate is referred to as a *smoothed* or *interpolated* estimate. Figure 2-3 further clarifies the distinctions between the notions of *past, present,* and *future* measurements. It is the use of different collections of these measurements that distinguishes filtering from prediction and smoothing from filtering. Prediction and filtering can be done in real time, whereas smoothing can never be done in real time. We will see that the impulse responses of predictors and filters are causal, whereas the impulse response of a smoother is noncausal.

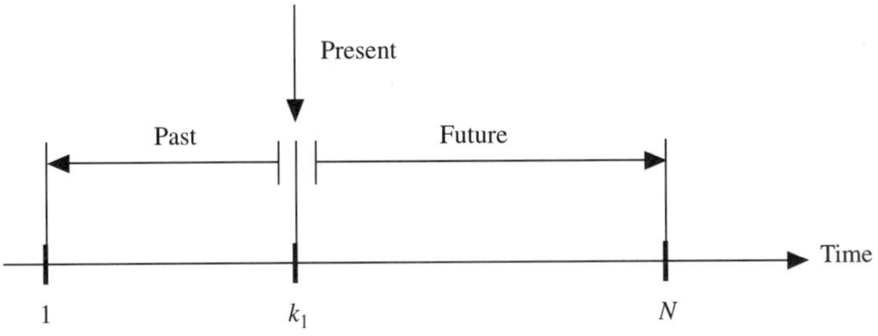


**Figure 2-3** Past, present, and future measurements in relation to time point $k_1$.

We use $\tilde{\theta}(k)$ to denote *estimation error,* i.e.,

$$\tilde{\theta}(k) = \theta - \hat{\theta}(k) \tag{2-34}$$

In state estimation, $\tilde{\mathbf{x}}(k_1|N)$ denotes state estimation error, and $\tilde{\mathbf{x}}(k_1|N) = \mathbf{x}(k_1) - \hat{\mathbf{x}}(k_1|N)$. In deconvolution, $\tilde{\mu}(i|N)$ is defined in a similar manner.

Very often we use the following *estimation model* for $\mathbf{Z}(k)$:

$$\hat{\mathbf{Z}}(k) = \mathbf{H}(k)\hat{\theta}(k) \tag{2-35}$$

To obtain (2-35) from (2-1), we assume that $\mathbf{V}(k)$ is zero-mean random noise that cannot be measured. In some applications (e.g., Example 2-2) $\hat{\mathbf{Z}}(k)$ represents a predicted value of $\mathbf{Z}(k)$. Associated with $\hat{\mathbf{Z}}(k)$ is the error $\tilde{\mathbf{Z}}(k)$, where

$$\tilde{\mathbf{Z}}(k) = \mathbf{Z}(k) - \hat{\mathbf{Z}}(k) \tag{2-36}$$

satisfies the equation [obtained by combining (2-36), (2-1), (2-35), and (2-34)]

$$\tilde{\mathbf{Z}}(k) = \mathbf{H}(k)\tilde{\theta}(k) + \mathbf{V}(k) \tag{2-37}$$

In those applications where $\hat{\mathbf{Z}}(k)$ is a predicted value of $\mathbf{Z}(k)$, $\tilde{\mathbf{Z}}(k)$ is known as a *prediction error*. Other names for $\tilde{\mathbf{Z}}(k)$ are *equation error* and *measurement residual*.

In the rest of this book we develop specific structures for $\hat{\theta}(k)$. These structures are referred to as *estimators*. *Estimates* are obtained whenever data are processed by an estimator. Estimator structures are associated with specific estimation techniques, and these techniques can be classified according to the natures of $\theta$ and $\mathbf{H}(k)$ and what a priori information is assumed known about noise vector $\mathbf{V}(k)$ or $\theta$ (if $\theta$ is random). See Lesson 1 for an overview of all the different estimation techniques that are covered in this book.

## COMPUTATION

If you want to generate data for different types of models, this can be accomplished using M-files from the *Contol System*, *System Identification*, and *Hi-Spec* toolboxes. For example, various types of single-input, single-output *finite-difference equation models* are available in one M-file in *Hi-Spec*. These include the all-pole AR model, the all-zero MA model, and the pole-zero ARMA model.

**armasyn**: generates ARMA synthetics for the model

$$\begin{aligned} z(k) &= \frac{B(z)}{A(z)}u(k) + g(k) \\ &= \frac{B(z)}{A(z)}u(k) + \frac{B_n(z)}{A_n(z)}w(k) \end{aligned}$$

The input, $u(k)$, to this model can be deterministic or it can be random. In the latter case, the M-file **rpiid** generates i.i.d. random sequences, with any one of the following density functions: single-sided exponential, double-sided exponential (Laplacian), Gaussian, Bernoulli–Gaussian, and uniform. The noise, $g(k)$, can be white or colored. Different types of coloring can be acheived by appropriate choices for $B_n(z)$ and $A_n(z)$. $w(k)$ can be chosen to be Gaussian or non-Gaussian and can be generated using **rpiid**.

A convolutional model can also be implemented using **armasyn**, because the coefficients of an MA($q$) model equal the sampled values of the impulse response of that model.

Generating time responses for state-variable models is discussed in Lesson 15.

## Supplementary Material

# CONVOLUTIONAL MODEL IN REFLECTION SEISMOLOGY

The convolutional model in (2-3) is, as mentioned in the text, the fundamental input-output relationship for all linear, causal, time-invariant systems. Here we demonstrate how it can be obtained in a nonobvious situation, *reflection seismology.* The following discussion will also illustrate the interplay between the representation, measurement, and estimation subproblems, which are depicted in Figure 1-1, for a specific application.

Reflection seismology (Mendel, 1983), by far the most widely used geophysical technique for petroleum exploration, generates pictures of Earth's subsurface from surface measurements. A source of seismic energy, such as an explosive, is located at Earth's surface near an array of sensors. The source produces waves that go down into the earth where they are reflected off and transmitted through interfaces owing to the impedance mismatches between different geological layers. The reflected waves are then transmitted back to the surface, where they are recorded by sensors (geophones on land or hydrophones in water). By repeating this procedure at many source and sensor locations, we can produce an image of the subsurface of Earth.

To generate this reflector image, many levels of signal processing are performed that attempt to remove various undesired aspects of the raw data. One such undesired aspect is the duration of the signal (called the *source wavelet*) produced by the seismic source. Ideally, we would like the source wavelet to be an impulse function; but physical devices cannot achieve the ideal. Due to the finite duration of the source wavelet, reflected signals from two contiguous reflectors may overlap and interfere. The effects of the finite duration of the source wavelet must therefore be unraveled from the data.

If we regard the transmission of seismic waves as a relatively linear process, we can regard the measured signal $z(t)$ (at any of the surface sensors) as a convolution of the source wavelet $h(t)$ with a reflection signal (the impulse response of the earth) $\mu(t)$, or

$$z(t) = \int_0^t \mu(\tau)h(t-\tau)d\tau + \nu(t) \tag{2-38}$$

where $\nu(t)$ is an error or noise term. The object of *deconvolution* is to remove the effects of the source wavelet and the noise term from the measured signal so that one is left with the desired reflection signal, or at least an estimate thereof.

Equation (2-38) can be discretized to give

$$z(k) = y_I(k) + \nu(k) = \sum_{j=1}^{k} \mu(j)h(k-j) + \nu(k) \tag{2-39}$$

where $k = 1, 2, \dots, N$. In this model $y_I(k)$ is the ideal seismic trace (i.e., the noise-free signal); $\nu(k)$ is "measurement" noise, which accounts for physical effects not explained by $y_I(k)$, as well as sensor noise; $h(i)$, $i = 0, 1, \dots, n$, is an FIR sequence associated with the seismic source wavelet that is usually much shorter than $N$ and is band limited in frequency; and $\mu(j) j = 1, 2, \ldots,$ is the reflectivity sequence (i.e., the desired signal; the earth's "message"), which is frequently modeled as a random sequence.

The convolution summation model in (2-39) can be derived from physical principles and some simplifying assumptions, such as normal incidence, each layer is homogeneous and isotropic, small strains, and *pressure and velocity satisfy a one-dimensional wave equation.* Signal $y_I(k)$, which is recorded as $z(k)$ by a seismic sensor, is a superposition of source wavelets reflected from the interfaces of subsurface layers. The $\mu(j)$ are related to interface reflection coefficients.

When comparing (2-39) with (2-3) and (2-4), we are led to the following important system interpretation for the seismic trace model (Mendel, 1977a; Mendel and Kormylo, 1978): signal $y_I(k)$ can be thought of as the output of a linear time-invariant system whose impulse response is $h(i)$ and whose input sequence is the reflectivity sequence $\mu(i)$.

From a physical point of view, the interpretation of $\mu(i)$ as the system input and of $h(i)$ as the system impulse response is *counterintuitive.* In a physically *intuitive* model, $h(i)$ would be associated with the input to the earth system that is characterized by the sequence $\mu(i)$; for is it not true that during the seismic experiment energy [characterized by $h(i)$] is imparted into the earth [characterized by $\mu(i)$], and reflected energy [$y_I(k)$] is recorded? Of course (2-39) can also be written as

$$z(k) = y_I(k) + \nu(k) = \sum_{j=1}^{k} h(j)\mu(k-j) + \nu(k) \tag{2-40}$$

where we have made the additional (reasonable) modeling assumptions that $h(0) = 0$ and $\mu(0) = 0$. Now when we compare (2-40) and (2-3) and (2-4), we can think of $y_I(k)$ as the output of a linear time-invariant system whose impulse response is $\mu(i)$ and whose input sequence is $h(i)$; this is a physically intuitive model.

From a mathematical point of view, the counterintuitive approach is more useful than the intuitive approach. This is because we wish to associate the superposition model in (2-39) with a linear dynamical system, and it is mathematically more convenient for such a system to be characterized by a finite-dimensional system, as is the case when $h(i)$ is assumed to be its impulse response, than as an infinite-dimensional system, as would be the case if $\mu(i)$—a sequence of time-delayed spikes—were assumed to be its impulse response. Additionally, owing to the frequently assumed random nature of the reflectivity sequence, it is much more convenient to treat it as a system input.

Although a layered earth is a distributed-parameter system (i.e., it is described by the wave equation that is a partial differential equation), its output equation (2-39) can be associated with a lumped-parameter system. The distributed-parameter and lumped-parameter systems are equivalent from an input/output point of view. Although this is very useful for signal-processing purposes, we must bear in mind that we have lost the ability to reconstruct the internal states of the layered system. The lumped-parameter model does not retain any of the physics associated with the wave motion that produces $y_I(k)$.

## SUMMARY QUESTIONS

1. The generic linear model $\mathbf{Z}(k) = \mathbf{H}(k)\boldsymbol{\theta} + \mathbf{V}(k)$ is:
   **(a)** a linear transformation of $\boldsymbol{\theta}$
   **(b)** an affine transformation of $\boldsymbol{\theta}$
   **(c)** a nonlinear transformation of $\boldsymbol{\theta}$
2. Which of the following applications is one in which $\boldsymbol{\theta}$ is deterministic?
   **(a)** impulse response identification
   **(b)** identification of the coefficients in a finite-difference equation
   **(c)** function approximation
   **(d)** state estimation
   **(e)** nonlinear model
   **(f)** deconvolution
   **(g)** Volterra series representation of a nonlinear system
3. In the application of identifying the coefficients of a finite-difference equation, when the system input is random, $\mathbf{H}$ and $\mathbf{V}$ are:
   **(a)** independent
   **(b)** dependent
   **(c)** uncorrelated
4. An AR($n$) model is characterized by:
   **(a)** a number of parameters that depends on the specific state-variable model chosen to represent it
   **(b)** $n$ parameters
   **(c)** $n$ numerator parameters and $n + 1$ denominator parameters
5. In the application of deconvolution, $\boldsymbol{\theta}$ is:
   **(a)** of varying dimension
   **(b)** of fixed dimension
   **(c)** deterministic
6. The linear prediction problem is analogous to:
   **(a)** estimating impulse response coefficients
   **(b)** deconvolution
   **(c)** estimating AR coefficients
7. The mathematical model that underlies the reflection seismology convolutional model is the:
   **(a)** lossy wave equation
   **(b)** heat equation
   **(c)** lossless wave equation
8. "Past" measurements always:
   **(a)** lie to the right of present time
   **(b)** lie to the left of present time
   **(c)** include the measurement made at the present time
9. An *estimator*, as distinct from an *estimate*, is:
   **(a)** a filter structure
   **(b)** a model
   **(c)** a numerical value of an estimate
10. Filtering of data occurs when there are:
    **(a)** the same number of measurements as unknown parameters

(b) more measurements than unknown parameters
(c) fewer measurements than unknown parameters

11. The hat notation (i.e., ˆ) denotes:
(a) estimation error
(b) estimate
(c) nominal value

12. The tilde notation (i.e., ˜) denotes:
(a) estimation error
(b) estimate
(c) nominal value

13. Signal $\tilde{\mathbf{Z}}(k)$ is known as:
(a) estimation error
(b) measurement error
(c) measurement residual

## PROBLEMS

**2-1.** Suppose $z(k) = \theta_1 + \theta_2 k + \nu(k)$, where $z(1) = 0.2$, $z(2) = 1.4$, $z(3) = 3.6$, $z(4) = 7.5$, and $z(5) = 10.2$. What are the explicit structures of $\mathbf{Z}(5)$ and $\mathbf{H}(5)$?

**2-2.** According to thermodynamic principles, pressure $P$ and volume $V$ of a given mass of gas are related by $PV^{\gamma} = C$, where $\gamma$ and $C$ are constants. Assume that $N$ measurements of $P$ and $V$ are available. Explain how to obtain a linear model for estimation of parameters $\gamma$ and $\ln C$.

**2-3.** (Mendel, 1973, Exercise 1-16(a), p. 46). Suppose we know that a relationship exists between $y$ and $x_1, x_2, \ldots, x_n$ of the form

$$y = \exp(a_1 x_1 + a_2 x_2 + \cdots + a_n x_n)$$

We desire to estimate $a_1, a_2, \ldots, a_n$ from measurements of $y$ and $\mathbf{x} = \text{col}(x_1, x_2, \ldots, x_n)$. Explain how to do this.

**2-4.** (Mendel, 1973, Exercise 1-17, pp. 46-47). The efficiency of a jet engine may be viewed as a linear combination of functions of inlet pressure $p(t)$ and operating temperature $T(t)$; i.e.,

$$E(T) = C_1 + C_2 f_1[p(t)] + C_3 f_2[T(t)] + C_4 f_3[p(t), T(t)] + \nu(t)$$

where the structures of $f_1$, $f_2$, and $f_3$ are known a priori and $\nu(t)$ represents modeling error of known mean and variance. From tests on the engine a table of values of $E(t)$, $p(t)$, and $T(t)$ is given at discrete values of $t$. Explain how $C_1$, $C_2$, $C_3$, and $C_4$ are estimated from these data.

**2-5.** Consider the problem of identifying the $n \times 1$ initial condition vector $\mathbf{x}(0)$ of the linear, time-invariant discrete-time system $\mathbf{x}(k+1) = \mathbf{\Phi}\mathbf{x}(k)$ from the $N$ measurements $z(k) = \mathbf{h}'\mathbf{x}(k) + \nu(k)$, $k = 1, 2, \ldots, N$. Show how this can be cast into the format of (2-1).

**2-6.** Beginning with the AR model in (2-6), and letting autocorrelation function $r_y(k-m) = \mathbf{E}\{y(n-k)y(n-m)\}$, show that the autocorrelation function satisfies the following homogeneous AR equation:

$$\sum_{i=0}^{n} \alpha_i r_y(i - m) = 0, \qquad \text{if } m > 0$$

In this equation, $\alpha_0 = 1$. Then explain how this correlation-based model can be used to estimate the unknown AR coefficients.

**2-7.** Lessons B and C introduce the reader to the world of higher-order statistics. If signals are non-Gaussian, then more than just second-order statistics are needed to describe it. For example, the third-order "cumulant" of signal $y(k)$, denoted $C_{3,y}(m_1, m_2)$, is the following triple correlation function: $C_{3,y}(m_1, m_2) = \mathbf{E}\{y(k)y(k+m_1)y(k+m_2)\}$. As in Problem 2-6, let us begin with the AR model in (2-6). In Example C-4 of Lesson C, it is shown that $C_{3,y}(m_1, m_2)$ also satisfies a homogeneous AR equation:

$$\sum_{i=0}^{n} \alpha_i C_{3,y}(\tau - i, k_0) = 0, \qquad \text{for } \tau > 0$$

where $k_0$ is a parameter whose choice is discussed in Example C-4. Explain how this cumulant-based AR model can be used to estimate the unknown AR coefficients.

When would you use the correlation-based model described in Problem 2-6 versus the cumulant-based model described in this problem?

**2-8.** Preprocessing of data can lead to the generic linear model in (2-1). Suppose, for example, you are given the equation $\mathbf{Y}(k) = \mathbf{H}(k)\boldsymbol{\theta}$, but the elements in both $\mathbf{Y}(k)$ and $\mathbf{H}(k)$ have to be estimated from the given data. Problems 2-6 and 2-7 give two situations where this can occur. Let $\hat{\mathbf{Y}}(k)$ and $\hat{\mathbf{H}}(k)$ denote these estimated quantities, where, for example, $\hat{\mathbf{H}}(k) = \mathbf{H}(k) + \mathbf{E}_{\mathbf{H}}(k)$. Obtain the generic linear model that exists between $\hat{\mathbf{Y}}(k)$ and $\hat{\mathbf{H}}(k)$ and discuss its "new twist."

**2-9.** (Li-Chien Lin, Fall 1991) Consider the differential equation

$$\frac{d^2x(t)}{dt^2} - 3\frac{dx(t)}{dt} + 2x(t) = 0$$

Given measured values of $x(t)$, explain how to obtain a linear model for estimation of initial values $x(0)$ and $dx(0)/dt$.

**2-10.** (Keith M. Chugg, Fall 1991) The error rate of a digital binary communication system can be reduced by introducing redundancy to the data through the use of a well-designed *error-correcting code*. A simple yet effective error-correction technique is known as linear block coding. With this technique a $k_c$-bit data word is used to determine a linear combination of basis code words. The $n_c$-bit codeword, $\mathbf{c}$, is determined by

$$\mathbf{c} = \sum_{i=1}^{k_c} d_i \mathbf{b}_i$$

where $d_i$ is the $i$th bit of the data word and $\mathbf{b}_i$ is the $i$th basis codeword for $i = 1, 2, \ldots, k_c$. All the variables are binary, so all sums are done modulo 2.

The received $n_c$-tuple is $\mathbf{c}$ plus a binary noise vector. The redundancy is then used to estimate the data $\mathbf{d} = \text{col}(d_1, d_2, \ldots, d_{k_c})$. Express the linear block coding problem in the form of the generic linear model. Identify all parameters and dimensions.

**2-11.** (Sam Heidari, Fall 1991) The captain of a spacecraft wishes to estimate the constant acceleration and initial velocity of his spacecraft based on $N$ measurements of the distance traveled, $x(t)$, at times $t = 1, 2, \ldots, N$ sec. The measured values of distance are known to be corrupted by additive noise. Obtain a linear model for estimation of the acceleration and initial velocity. Note that at time zero the distance traveled by the spacecraft is zero.

# LESSON 3

# Least-squares Estimation: Batch Processing

## SUMMARY

The main purpose of this lesson is the derivation of the classical *batch* formula of (weighted) least squares. The term *batch* means that all measurements are collected together and processed simultaneously. A second purpose of this lesson is to demonstrate that least-squares estimates may change in numerical value under changes of scale. One way around this difficulty is to use normalized data.

Least-squares estimates require no assumptions about the nature of the generic linear model. Consequently, the formula for the least-squares estimator (LSE) is easy to derive. We will learn in Lesson 8, that the price paid for ease in derivation is difficulty in performance evaluation.

The supplementary material at the end of this lesson contrasts least squares, total least squares, and constrained total least squares. The latter techniques are frequently more powerful than least squares, especially in signal-processing applications.

When you complete this lesson you will be able to (1) derive and use the classical batch formulas of (weighted) least squares; these are theoretical formulas that should not be programmed as is for digital computation; numerically well behaved linear algebra programs are commercially available for computational purposes; (2) explain the sensitivity of (weighted) least squares to scale change; and (3) explain the difference between least squares, total least squares, and constrained total least squares.

## INTRODUCTION

The method of least squares dates back to Karl Gauss around 1795 and is the cornerstone for most estimation theory, both classical and modern. It was invented by Gauss at a time when he was interested in predicting the motion of planets and comets using telescopic measurements. The motions of these bodies can be completely characterized by six parameters. The estimation problem that Gauss considered was one of inferring the values of these parameters from the measurement data.

We shall study least-squares estimation from two points of view: the classical batch-processing approach, in which all the measurements are processed together at one time, and the more modern recursive processing approach, in which measurements are processed only a few (or even one) at a time. The recursive approach has been motivated by today's high-speed digital computers; however, as we shall see, the recursive algorithms are outgrowths of the batch algorithms.

The starting point for the method of least squares is the linear model

$$\mathbf{Z}(k) = \mathbf{H}(k)\boldsymbol{\theta} + \mathbf{V}(k) \tag{3-1}$$

where $\mathbf{Z}(k) = \text{col}\,(z(k), z(k-1), \ldots, z(k-N+1))$, $z(k) = \mathbf{h}'(k)\boldsymbol{\theta} + \nu(k)$, and the estimation model for $\mathbf{Z}(k)$ is

$$\hat{\mathbf{Z}}(k) = \mathbf{H}(k)\hat{\boldsymbol{\theta}}(k) \tag{3-2}$$

We denote the (weighted) least-squares estimator of $\boldsymbol{\theta}$ as $[\hat{\boldsymbol{\theta}}_{\text{WLS}}(k)]\hat{\boldsymbol{\theta}}_{\text{LS}}(k)$. In this lesson and the next two we shall determine explicit structures for this estimator.

## Gauss: A Short Biography

Carl Friederich Gauss was born on April 30, 1777, in Braunschweig, Germany. Although his father wanted him to go into a trade, his mother had the wisdom to recognize Gauss's genius, which manifested itself at a very early age, and saw to it that he was properly schooled.

E.T. Bell (1937), in his famous essay on Gauss, refers to him as the "Prince of Mathematicians." Courant (1969) states that "Gauss was one of the three most famous mathematicians in history, in company with Archimedes and Newton."

Gauss invented the method of least squares at the age of 18. This work was the beginning of a lifelong interest in the theory of observation. He attended the University of Göttingen from October 1795 to September 1798. Some say that these three years were his most prolific. He received his doctor's degree in absentia from the University of Helmstedt in 1799. His doctoral dissertation was the first proof of the fundamental theorem of algebra.

In 1801 he published his first masterpiece, *Arithmetical Researches*, which revolutionized all of arithmetic and established number theory as an organic branch of mathematics. In 1809 he published his second masterpiece, *Theory of the Motion of Heavenly Bodies Revolving Round the Sun in Conic Sections*, in which he predicted the orbit of Ceres. E.T. Bell felt that Gauss's excursions into astronomical works was a waste of 20 years, during which time he could have been doing more pure mathematics. This is an interesting point of view (not shared by Courant); for what we take as very important to us in estimation theory, least squares and its applications, has been viewed by some mathematicians as a diversion.

In 1812 Gauss published another great work on the hypergeometric series from which developed many applications to differential equations in the nineteenth century. Gauss invented the electric telegraph (working with Wilhelm Weber) in 1833. He made major contributions in geodesy, the theories of surfaces, conformal mapping, mathematical physics (particularly electromagnetism, terrestrial magnetism, and the theory of attraction according to Newtonian law), analysis situs, and the geometry associated with functions of a complex variable.

Gauss was basically a loner. He published his results only when they were absolutely polished, which made his publications extremely difficult to understand, since so many of the details had been stripped away. He kept a diary throughout his lifetime in which he briefly recorded all of his "gems." This diary did not become known until many years after his death. It established his precedence for results associated with the names of many other famous mathematicians (e.g., he is now credited with being one of the founders of non-Euclidean geometry). For discussions of an interesting feud between Gauss and Legendre over priority to the method of least squares, see Sorenson (1970).

Gauss died at 78 on February 23, 1855. As Bell says, "He lives everywhere in mathematics."

*Let $err = \tilde{z}(k) = \tilde{z}(k) - \hat{z}(k)$ = err estimaty $z$*

*$\tilde{\theta}(k) = \theta - \hat{\theta}(k)$ = err in estimate.*

## NUMBER OF MEASUREMENTS

*prediction err, or measurement residual* $\rightarrow$ *$\tilde{z}(k) = z - \hat{z} = H(k)\theta + V(k) - H(k)\hat{\theta}(k)$ $= H(k)\tilde{\theta}(k) + V(k)$   See eq. 2-37*

Suppose that $\theta$ contains $n$ parameters and $\mathbf{Z}(k)$ contains $N$ measurements. If $N < n$, we have fewer measurements than unknowns and (3-1) is an underdetermined system of equations that does not lead to unique values for $\theta_1, \theta_2, \ldots, \theta_n$. If $N = n$, we have exactly as many measurements as unknowns, and as long as the $n$ measurements are linearly independent, so that $\mathbf{H}^{-1}(k)$ exists, we can solve (3-1) for $\theta$, as

$$\theta = \mathbf{H}^{-1}(k)\mathbf{Z}(k) - \mathbf{H}^{-1}(k)\mathbf{V}(k) \tag{3-3}$$

Because we cannot measure $\mathbf{V}(k)$, it is usually neglected in the calculation of (3-3). For small amounts of noise this may not be a bad thing to do, but for even moderate amounts of noise this will be quite bad. Finally, if $N > n$, we have more measurements than unknowns, so (3-1) is an overdetermined system of equations. The extra measurements can be used to offset the effects of the noise; i.e., they let us "filter" the data. Only this last case is of real interest to us. Some discussions on the underdetermined case are given in Lesson 4.

*goal: get algorithm for $\hat{\theta}$*

*$\hat{\theta} \leftarrow \leftarrow z$.*

## OBJECTIVE FUNCTION AND PROBLEM STATEMENT

A *direct approach* for obtaining $\hat{\theta}(k)$ is to choose it so as to minimize the sum of the squared errors between its components and the respective components of $\theta$, i.e., to minimize

*algorithm #1 is to minimize this* $\longrightarrow$

$$\sum_{i=1}^{n}[\theta_i - \hat{\theta}_i(k)]^2 = |\theta - \hat{\theta}(k)|^2$$

*But This can't be solved :(next page)*

The solution to this minimization problem is $\hat{\theta}(k) = \theta$, which, of course is a useless result, because, if we knew $\theta$ ahead of time, we would not need to estimate it.

*algorithm #2*    A less direct approach for obtaining $\hat{\theta}(k)$ is based on minimizing the objective function   *$\tilde{z} = z - H(k)\hat{\theta}(k) \leftarrow$ minimize this (instead of $\tilde{\theta}$) (here $w(k)=1$)*

*more general form introduce weights:*
*weighted LSE estimate*

$$J[\hat{\theta}(k)] = w(1)\tilde{z}^2(1) + w(2)\tilde{z}^2(2) + \cdots + w(k - N + 1)\tilde{z}^2(k - N + 1)$$

$$= \tilde{\mathbf{Z}}'(k)\mathbf{W}(k)\tilde{\mathbf{Z}}(k) \quad \underline{w = N \times N} \tag{3-4}$$

$$= [\mathbf{Z}(k) - \hat{\mathbf{Z}}(k)]'\mathbf{W}(k)[\mathbf{Z}(k) - \hat{\mathbf{Z}}(k)]$$

where

$$\tilde{\mathbf{Z}}(k) = \text{col}\,[\tilde{z}(k), \tilde{z}(k - 1), \ldots, \tilde{z}(k - N + 1)] \tag{3-5}$$

and *weighting matrix* $\mathbf{W}(k)$ must be symmetric and positive definite, for reasons explained later.

No general rules exist for how to choose $\mathbf{W}(k)$. The most common choice is a diagonal matrix such as

$$\mathbf{W}(k) = \text{diag}\,[\mu^{-(N-1)}, \mu^{-(N-2)}, \ldots, \mu^{-1}, 1]\mu^k$$

$$= \text{diag}\,[\mu^{k-N+1}, \mu^{k-N+2}, \ldots, \mu^k]$$

When $|\mu| < 1$ so that $1/\mu > 1$, recent errors (and associated measurements) [e.g., $\tilde{z}(k), \tilde{z}(k - 1)$] are weighted more heavily than past ones [e.g., $\tilde{z}(k - N + 2), \tilde{z}(k - N + 1)$]. Such a choice for $\mathbf{W}(k)$ provides the weighted least-squares estimator with an "aging" or "forgetting" factor. When $|\mu| > 1$, recent errors are weighted less

*"forgetting"*

heavily than past ones. Finally, if $\mu = 1$, so that $\mathbf{W}(k) = \mathbf{I}$, then all errors are weighted by the same amount. When $\mathbf{W}(k) = \mathbf{I}$, $\hat{\theta}(k) = \hat{\theta}_{LS}(k)$, whereas for all other $\mathbf{W}(k)$, $\hat{\theta}(k) = \hat{\theta}_{WLS}(k)$. Note also that if $\mathbf{W}(k) = c\mathbf{I}$, where $c$ is a constant, then $\hat{\theta}(k) = \hat{\theta}_{LS}(k)$ (see Problem 3-2).

Our objective is to determine the $\hat{\theta}_{WLS}(k)$ that minimizes $J[\hat{\theta}(k)]$.

*Solution!*

## DERIVATION OF ESTIMATOR

To begin, we express (3-4) as an explicit function of $\hat{\theta}(k)$, using (3-2):

$$J[\hat{\theta}(k)] = \tilde{\mathbf{Z}}'(k)\mathbf{W}(k)\tilde{\mathbf{Z}}(k) = [\mathbf{Z}(k) - \hat{\mathbf{Z}}(k)]'\mathbf{W}(k)[\mathbf{Z}(k) - \hat{\mathbf{Z}}(k)]$$

$$= [\mathbf{Z}(k) - \mathbf{H}(k)\hat{\theta}(k)]'\mathbf{W}(k)[\mathbf{Z}(k) - \mathbf{H}(k)\hat{\theta}(k)]$$

$$= \mathbf{Z}'(k)\mathbf{W}(k)\mathbf{Z}(k) - 2\mathbf{Z}'(k)\mathbf{W}(k)\mathbf{H}(k)\hat{\theta}(k) \tag{3-6}$$

$$+ \hat{\theta}'(k)\mathbf{H}'(k)\mathbf{W}(k)\mathbf{H}(k)\hat{\theta}(k)$$

Next, we take the vector derivative of $J[\hat{\theta}(k)]$ with respect to $\hat{\theta}(k)$, but before doing this recall from vector calculus, that:

If $\mathbf{m}$ and $\mathbf{b}$ are two $n \times 1$ nonzero vectors, and $\mathbf{A}$ is an $n \times n$ symmetric matrix, then

$$\frac{d}{d\mathbf{m}}(\mathbf{b}'\mathbf{m}) = \mathbf{b} \tag{3-7}$$

and

$$\frac{d}{d\mathbf{m}}(\mathbf{m}'\mathbf{A}\mathbf{m}) = 2\mathbf{A}\mathbf{m} \tag{3-8}$$

Using these formulas, we find that

$$\frac{dJ[\hat{\theta}(k)]}{d\hat{\theta}(k)} = -2[\mathbf{Z}'(k)\mathbf{W}(k)\mathbf{H}(k)]' + 2\mathbf{H}'(k)\mathbf{W}(k)\mathbf{H}(k)\hat{\theta}(k) \tag{3-9}$$

Setting $dJ[\hat{\theta}(k)]/d\hat{\theta}(k) = \mathbf{0}$, we obtain the following formula for $\hat{\theta}_{\text{WLS}}(k)$:

$$\hat{\theta}_{\text{WLS}}(k) = [\mathbf{H}'(k)\mathbf{W}(k)\mathbf{H}(k)]^{-1}\mathbf{H}'(k)\mathbf{W}(k)\mathbf{Z}(k) \tag{3-10}$$

Note, also, that

$$\hat{\theta}_{\text{LS}}(k) = [\mathbf{H}'(k)\mathbf{H}(k)]^{-1}\mathbf{H}'(k)\mathbf{Z}(k) \tag{3-11}$$

By substituting (3-10) into (3-6), we obtain the minimum value of $J[\hat{\theta}(k)]$:

$$\begin{aligned} J[\hat{\theta}_{\text{WLS}}(k)] &= \mathbf{Z}'(k)\mathbf{W}(k)[\mathbf{Z}(k) - \mathbf{H}(k)\hat{\theta}_{\text{WLS}}(k)] \\ &= \mathbf{Z}'(k)\mathbf{W}(k)\mathbf{Z}(k) - \hat{\theta}'_{\text{WLS}}(k)[\mathbf{H}'(k)\mathbf{W}(k)\mathbf{H}(k)]\hat{\theta}_{\text{WLS}}(k) \end{aligned} \tag{3-12}$$

### Comments

1. Matrix $\mathbf{H}'(k)\mathbf{W}(k)\mathbf{H}(k)$ must be nonsingular for its inverse to exist. Matrix $\mathbf{H}'(k)\mathbf{W}(k)\mathbf{H}(k)$ is said to be nonsingular if it has an inverse satisfying

$$\begin{aligned} [\mathbf{H}'(k)\mathbf{W}(k)\mathbf{H}(k)]^{-1}[\mathbf{H}'(k)\mathbf{W}(k)\mathbf{H}(k)] &= [\mathbf{H}'(k)\mathbf{W}(k)\mathbf{H}(k)][\mathbf{H}'(k)\mathbf{W}(k)\mathbf{H}(k)]^{-1} \\ &= \mathbf{I} \end{aligned}$$

If $\mathbf{W}(k)$ is positive definite, then it can be written uniquely as $\mathbf{W}(k) = \mathbf{L}'(k)\mathbf{L}(k)$, where $\mathbf{L}(k)$ is a lower triangular matrix with positive diagonal elements. Consequently, we can express $\mathbf{H}'(k)\mathbf{W}(k)\mathbf{H}(k)$ as

$$\mathbf{H}'(k)\mathbf{W}(k)\mathbf{H}(k) = \mathbf{A}'(k)\mathbf{A}(k), \qquad \text{where } \mathbf{A}(k) = \mathbf{L}(k)\mathbf{H}(k)$$

If $\mathbf{A}(k)$ has linearly independent columns, i.e., is of maximum rank, then $\mathbf{A}'(k)\mathbf{A}(k)$ is nonsingular. Finally, rank $[\mathbf{L}(k)\mathbf{H}(k)]$ = rank $[\mathbf{H}(k)]$, because $\mathbf{L}(k)$ is nonsingular. Consequently, if $\mathbf{H}(k)$ is of maximum rank, then $\mathbf{A}'(k)\mathbf{A}(k)$ is nonsingular, so $[\mathbf{A}'(k)\mathbf{A}(k)]^{-1} = [\mathbf{H}'(k)\mathbf{W}(k)\mathbf{H}(k)]^{-1}$ exists. The two conditions that have fallen out of this analysis are that $\mathbf{W}(k)$ must be positive definite and $\mathbf{H}(k)$ must be of maximum rank.

**2.** How do we know that $\hat{\theta}_{WLS}(k)$ minimizes $J[\hat{\theta}(k)]$? We compute $d^2 J[\hat{\theta}(k)]/d\hat{\theta}^2(k)$ and see if it is positive definite [which is the vector calculus analog of the scalar calculus requirement that $\hat{\theta}$ minimizes $J(\hat{\theta})$ if $dJ(\hat{\theta})/d\hat{\theta} = 0$ and $d^2 J(\hat{\theta})/d\hat{\theta}^2$ is positive]. Doing this, we see that

$$\frac{d^2 J[\hat{\theta}(k)]}{d\hat{\theta}^2(k)} = 2\mathbf{H}'(k)\mathbf{W}(k)\mathbf{H}(k) > 0$$ (3-13)

here this means LHS is positive definite

$\neq H(k)$

because $\mathbf{H}'(k)\mathbf{W}(k)\mathbf{H}(k)$ is invertible.

**3.** Estimator $\hat{\theta}_{WLS}(k)$ processes the measurements $\mathbf{Z}(k)$ linearly; thus, it is referred to as a *linear estimator*. It processes the data contained in $\mathbf{H}(k)$ in a very complicated and nonlinear manner.

**4.** When (3-9) is set equal to zero, we obtain the following system of *normal equations*:

$$[\mathbf{H}'(k)\mathbf{W}(k)\mathbf{H}(k)]\hat{\theta}_{WLS}(k) = \mathbf{H}'(k)\mathbf{W}(k)\mathbf{Z}(k)$$ (3-13)

$n \times n$   $n \times 1$

This is a system of $n$ linear equations in the $n$ components of $\hat{\theta}_{WLS}(k)$.

In practice, we do not compute $\hat{\theta}_{WLS}(k)$ using (3-10), because computing the inverse of $\mathbf{H}'(k)\mathbf{W}(k)\mathbf{H}(k)$ is fraught with numerical difficulties. Instead, the normal equations are solved using stable algorithms from numerical linear algebra. Golub and Van Loan (1989) have an excellent chapter, entitled "Orthogonalization and Least Squares Methods," devoted to numerically sound ways for computing $\hat{\theta}_{WLS}(k)$ from (3-13) (see, also, Stewart, 1973; Bierman, 1977; and Dongarra, et al., 1979). They state that "One tactic for solution [of (3-13)] is to convert the original least squares problem into an equivalent, easy-to-solve problem using orthogonal transformations. Algorithms of this type based on Householder and Givens transformations ... compute the factorization $\mathbf{H}'(k)\mathbf{W}(k)\mathbf{H}(k) = \mathbf{Q}(k)\mathbf{R}(k)$, where $\mathbf{Q}(k)$ is orthogonal and $\mathbf{R}(k)$ is upper triangular."

In Lesson 4 we describe how (3-10) can be computed using the very powerful singular-value decomposition (SVD) method. SVD can be used for both the overdetermined and underdetermined situations.

Based on this discussion, we must view (3-10) as a useful *"theoretical"* formula and not as a useful computational formula.

**5.** Using the fact that $\tilde{\mathbf{Z}}(k) = \mathbf{Z}(k) - \mathbf{H}(k)\hat{\theta}_{WLS}(k)$, equation (3-13) can also be reexpressed as

$$\tilde{\mathbf{Z}}'(k)\mathbf{W}(k)\mathbf{H}(k) = \mathbf{0}'$$ (3-14)

which can be viewed as an *orthogonality condition* between $\tilde{\mathbf{Z}}(k)$ and $\mathbf{W}(k)\mathbf{H}(k)$. Orthogonality conditions play an important role in estimation theory. We shall see many more examples of such conditions throughout this book. For a very lucid discussion on the least-squares orthogonality principle, see Therrien (1992, pp. 525–528). See, also, Problems 3-12 and 3-13.

**6.** Estimates obtained from (3-10) will be random! This is because $\mathbf{Z}(k)$ is random, and in some applications even $\mathbf{H}(k)$ is random. It is therefore instructive to view (3-10) as a complicated transformation of vectors or matrices of random variables into the vector of random variables $\hat{\boldsymbol{\theta}}_{\text{WLS}}(k)$. In later lessons, when we examine the properties of $\hat{\boldsymbol{\theta}}_{\text{WLS}}(k)$, these will be statistical properties because of the random nature of $\hat{\boldsymbol{\theta}}_{\text{WLS}}(k)$.

**7.** The assumption that $\boldsymbol{\theta}$ is deterministic was never made during our derivation of $\hat{\boldsymbol{\theta}}_{\text{WLS}}(k)$; hence, *(3-10) and (3-11) also apply to the estimation of random parameters*. We return to this important point in Lesson 13. If $\boldsymbol{\theta}$ is random, then a performance analysis of $\hat{\boldsymbol{\theta}}_{\text{WLS}}(k)$ is much more difficult than when $\boldsymbol{\theta}$ is deterministic. See Lesson 8 for some performance analyses of $\hat{\boldsymbol{\theta}}_{\text{WLS}}(k)$.

**EXAMPLE 3-1** (Mendel, 1973, pp. 86–87)

Suppose we wish to calibrate an instrument by making a series of uncorrelated measurements on a constant quantity. Denoting the constant quantity as $\theta$, our measurement equation becomes

$$z(k) = \theta + \nu(k) \tag{3-15}$$

where $k = 1, 2, \ldots, N$. Collecting these $N$ measurements, we have

$$\begin{pmatrix} z(N) \\ z(N-1) \\ \vdots \\ z(1) \end{pmatrix} = \begin{pmatrix} 1 \\ 1 \\ \vdots \\ 1 \end{pmatrix} \theta + \begin{pmatrix} \nu(N) \\ \nu(N-1) \\ \vdots \\ \nu(1) \end{pmatrix} \tag{3-16}$$

Clearly, $\mathbf{H} = \text{col}\,(1, 1, \ldots, 1)$; hence

$$\hat{\theta}_{\text{LS}}(N) = \frac{1}{N}\sum_{i=1}^{N} z(i) \tag{3-17}$$

which is the sample mean of the $N$ measurements. We see, therefore, that *the sample mean is a least-squares estimator.* □

**EXAMPLE 3-2** (Mendel, 1973)

Figure 3-1 depicts simplified third-order pitch-plane dynamics for a typical, high-performance, aerodynamically controlled aerospace vehicle. Cross-coupling and body-bending effects are neglected. Normal acceleration control is considered with feedback on normal acceleration and angle-of-attack rate. Stefani (1967) shows that if the system gains are chosen as

$$K_{N_i} = \frac{C_2}{100 M_\delta Z_\alpha} \tag{3-18}$$

$$K_{\dot{\alpha}} = \frac{C_1 - 100\left(\dfrac{Z_\alpha 1845}{\mu}\right) + M_\alpha}{100 M_\delta} \tag{3-19}$$

and

$$K_{N_a} = \frac{C_2 + 100 M_\alpha}{100 M_\delta Z_\alpha} \tag{3-20}$$

then

$$\frac{N_a}{N_i}(s) = \frac{C_2}{s^3 + 100 s^2 + C_1 s + C_2} \tag{3-21}$$

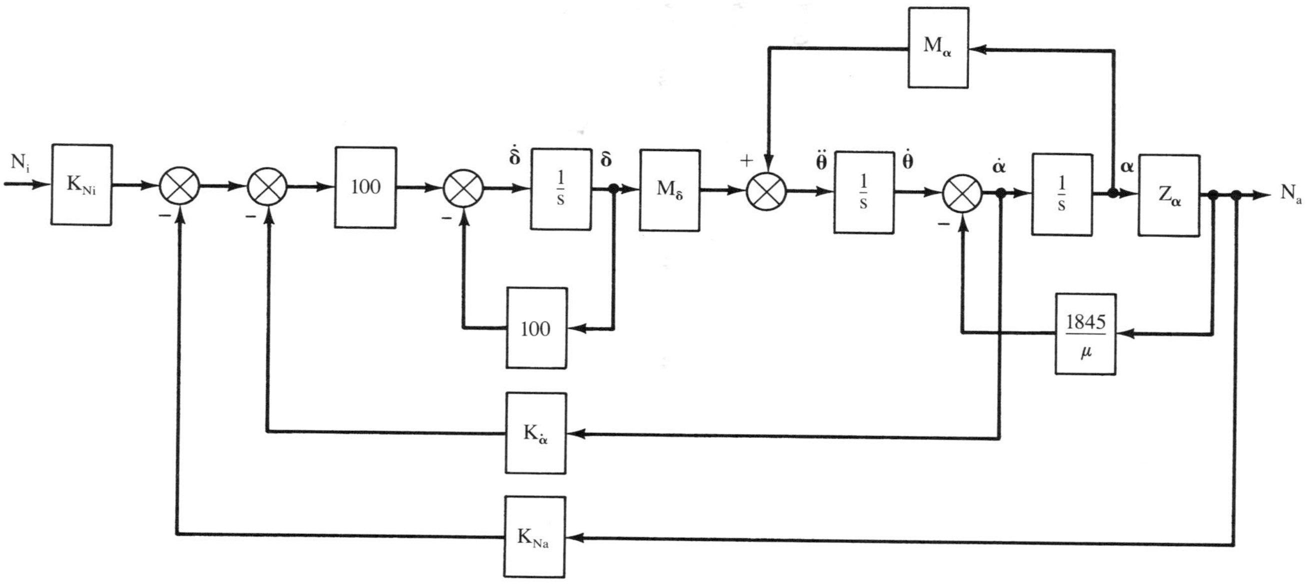


**Figure 3-1** Pitch-plane dynamics and nomenclature: $N_i$, input normal acceleration along the negative $Z$ axis; $K_{N_i}$, gain on $N_i$; $\delta$, control-surface deflection; $M_\delta$, control-surface effectiveness; $\ddot{\theta}$, rigid-body acceleration; $\alpha$, angle of attack; $M_\alpha$, aerodynamic moment effectiveness; $K_{\dot{\alpha}}$, control gain on $\dot{\alpha}$; $Z_\alpha$, normal acceleration force coefficient; $\mu$, axial velocity; $N_a$, system-achieved normal acceleration along the negative $Z$ axis; $K_{N_a}$, control gain on $N_a$ (reprinted from Mendel, 1973, p. 33, by courtesy of Marcel Dekker, Inc.).

Stefani assumes $Z_\alpha 1845/\mu$ is relatively small, and chooses $C_1 = 1400$ and $C_2 = 14,000$. The closed-loop response resembles that of a second-order system with a bandwidth of 2 Hz and a damping ratio of 0.6 that responds to a step command of input acceleration with zero steady-state error.

In general, $M_\alpha$, $M_\delta$, and $Z_\alpha$ are dynamic parameters and all vary through a large range of values. Also, $M_\alpha$ may be positive (unstable vehicle) or negative (stable vehicle). System response must remain the same for all values of $M_\alpha$, $M_\delta$, and $Z_\alpha$; thus, it is necessary to estimate these parameters so that $K_{N_i}$, $K_{\dot\alpha}$, and $K_{N_a}$ can be adapted to keep $C_1$ and $C_2$ invariant at their designed values. For present purposes we shall assume that $M_\alpha$, $M_\delta$, and $Z_\alpha$ are frozen at specific values.

From Figure 3-1,

$$\ddot\theta(t) = M_\alpha\alpha(t) + M_\delta\delta(t) \tag{3-22}$$

and

$$N_a(t) = Z_\alpha\alpha(t) \tag{3-23}$$

Our attention is directed at the estimation of $M_\alpha$ and $M_\delta$ in (3-22). We leave it as an exercise for the reader to explore the estimation of $Z_\alpha$ in (3-23).

Our approach will be to estimate $M_\alpha$ and $M_\delta$ from the equation

$$\ddot\theta_M(k) = M_\alpha\alpha(k) + M_\delta\delta(k) + \nu_{\ddot\theta}(k) \tag{3-24}$$

where $\ddot\theta_m(k)$ denotes the measured value of $\ddot\theta(k)$ that is corrupted by measurement noise $\nu_{\ddot\theta}(k)$. We shall assume (somewhat unrealistically) that $\alpha(k)$ and $\delta(k)$ can both be measured perfectly. The concatenated measurement equation for $N$ measurements is

$$\begin{pmatrix} \ddot\theta_m(k) \\ \ddot\theta_m(k-1) \\ \vdots \\ \ddot\theta_m(k-N+1) \end{pmatrix} = \begin{pmatrix} \alpha(k) & \delta(k) \\ \alpha(k-1) & \delta(k-1) \\ \vdots & \vdots \\ \alpha(k-N+1) & \delta(k-N+1) \end{pmatrix} \times \begin{pmatrix} M_\alpha \\ M_\delta \end{pmatrix} + \begin{pmatrix} \nu_{\ddot\theta}(k) \\ \nu_{\ddot\theta}(k-1) \\ \vdots \\ \nu_{\ddot\theta}(k-N+1) \end{pmatrix} \tag{3-25}$$

Hence, the *least-squares* estimates of $M_\alpha$ and $M_\delta$ are

$$\begin{pmatrix} \hat M_\alpha(k) \\ \hat M_\delta(k) \end{pmatrix} = \begin{pmatrix} \sum_{j=0}^{N-1} \alpha^2(k-j) & \sum_{j=0}^{N-1} \alpha(k-j)\delta(k-j) \\ \sum_{j=0}^{N-1} \alpha(k-j)\delta(k-j) & \sum_{j=0}^{N-1} \delta^2(k-j) \end{pmatrix}^{-1} \times \begin{pmatrix} \sum_{j=0}^{N-1} \alpha(k-j)\ddot\theta_m(k-j) \\ \sum_{j=0}^{N-1} \delta(k-j)\ddot\theta_m(k-j) \end{pmatrix} \quad \square \tag{3-26}$$

## FIXED AND EXPANDING MEMORY ESTIMATORS

Estimator $\hat{\theta}_{WLS}(k)$ uses the measurements $z(k - N + 1), z(k - N + 2), \ldots, z(k)$. When $N$ is fixed ahead of time, $\hat{\theta}_{WLS}(k)$ uses a fixed window of measurements, a window of length $N$, and $\hat{\theta}_{WLS}(k)$ is then referred to as a *fixed-memory estimator*. The batch weighted least-squares estimator, obtained in this lesson, has a fixed memory.

A second approach for choosing $N$ is to set it equal to $k$; then $\hat{\theta}_{WLS}(k)$ uses the measurements $z(1), z(2), \ldots, z(k)$. In this case, $\hat{\theta}_{WLS}(k)$ uses an expanding window of measurements, a window of length $k$, and $\hat{\theta}_{WLS}(k)$ is then referred to as an *expanding-memory estimator*. The recursive weighted least-squares estimator, obtained in Lesson 5, has an expanding memory.

## SCALE CHANGES AND NORMALIZATION OF DATA

Least-squares (LS) estimates may not be invariant under changes of scale. One way to circumvent this difficulty is to use normalized data.

As long as the elements of $\mathbf{Z}(k)$, the $z(k - j)$'s ($j = 0, 1, \ldots, N - 1$), are scalars, there is no problem with changes in scale. For example, if measurements of velocity are in miles per hour or are scaled to feet per second, we obtain the same weighted least-squares or unweighted least-squares estimates. Reasons for this are explored in Problems 3-7 and 3-8. If, on the other hand, the elements of $\mathbf{Z}(k)$, the $z(k - j)$'s ($j = 0, 1, \ldots, N - 1$), are vectors (see the section in Lesson 5 entitled "Generalization to Vector Measurements"), then scaling of measurements can be a serious problem, as we now demonstrate.

*Now at each k sample you take more than 1 measurements.*

Assume that observers $A$ and $B$ are observing a process; but observer $A$ reads the measurements in one set of units and $B$ in another. Let $\mathbf{M}$ be a diagonal matrix of scale factors relating $A$ to $B$; $\mathbf{Z}_A(k)$ and $\mathbf{Z}_B(k)$ denote the total measurement vectors of $A$ and $B$, respectively. Then $Z_B = M Z_A$

$$\mathbf{Z}_B(k) = \mathbf{H}_B(k)\theta + \mathbf{V}_B(k) = \mathbf{M}\mathbf{Z}_A(k) = \mathbf{M}\mathbf{H}_A(k)\theta + \mathbf{M}\mathbf{V}_A(k) \quad (3\text{-}27)$$

which means that

$$\mathbf{H}_B(k) = \mathbf{M}\mathbf{H}_A(k) \quad (3\text{-}28)$$

Let $\hat{\theta}_{A,WLS}(k)$ and $\hat{\theta}_{B,WLS}(k)$ denote the WLSEs associated with observers $A$ and $B$, respectively, where $\hat{\theta}_{A,WLS}(k) = [\mathbf{H}'_A(k)\mathbf{W}_A(k)\mathbf{H}_A(k)]^{-1}\mathbf{H}'_A(k)\mathbf{W}_A(k)\mathbf{Z}_A(k)$ and $\hat{\theta}_{B,WLS}(k) = [\mathbf{H}'_B(k)\mathbf{W}_B(k)\mathbf{H}_B(k)]^{-1}\mathbf{H}'_B(k)\mathbf{W}_B(k)\mathbf{Z}_B(k)$; then $\hat{\theta}_{B,WLS}(k) = \hat{\theta}_{A,WLS}(k)$ if $\mathbf{M}\mathbf{W}_B(k)\mathbf{M} = \mathbf{W}_A(k)$, or

$$\mathbf{W}_B(k) = \mathbf{M}^{-1}\mathbf{W}_A(k)\mathbf{M}^{-1} \quad (3\text{-}29)$$

It seems a bit peculiar though to have different weighting matrices for the two WLSEs. In fact, if we begin with $\hat{\theta}_{A,LS}(k)$, then it is impossible to obtain $\hat{\theta}_{B,LS}(k)$

such that $\hat{\theta}_{B,\text{LS}}(k) = \hat{\theta}_{A,\text{LS}}(k)$. The reason for this is simple. To obtain $\hat{\theta}_{A,\text{LS}}(k)$, we set $\mathbf{W}_A(k) = \mathbf{I}$, in which case (3-29) reduces to $\mathbf{W}_B(k) = (\mathbf{M}^{-1})^2 \neq \mathbf{I}$.

Next, let $\mathbf{N}_A$ and $\mathbf{N}_B$ denote diagonal normalization matrices for $\mathbf{Z}_A(k)$ and $\mathbf{Z}_B(k)$, respectively. We shall assume that our data is always normalized to the same set of numbers, i.e., that

$$\boxed{\mathbf{N}_A\mathbf{Z}_A(k) = \mathbf{N}_B\mathbf{Z}_B(k)} \tag{3-30}$$

- do this w/ vector measurements

Observe that

$$\mathbf{N}_A\mathbf{Z}_A(k) = \mathbf{N}_A\mathbf{H}_A(k)\theta + \mathbf{N}_A\mathbf{V}_A(k) \tag{3-31}$$

and

$$\mathbf{N}_B\mathbf{Z}_B(k) = \mathbf{N}_B\mathbf{M}\mathbf{Z}_A(k) = \mathbf{N}_B\mathbf{M}\mathbf{H}_A(k)\theta + \mathbf{N}_B\mathbf{M}\mathbf{V}_A(k) \tag{3-32}$$

From (3-30), (3-31), and (3-32), we see that

$$\mathbf{N}_A = \mathbf{N}_B\mathbf{M} \tag{3-33}$$

We now find that

$$\hat{\theta}_{A,\text{WLS}}(k) = (\mathbf{H}'_A\mathbf{N}_A\mathbf{W}_A\mathbf{N}_A\mathbf{H}_A)^{-1}\mathbf{H}'_A\mathbf{N}_A\mathbf{W}_A\mathbf{N}_A\mathbf{Z}_A(k) \tag{3-34}$$

and

$$\hat{\theta}_{B,\text{WLS}}(k) = (\mathbf{H}'_A\mathbf{M}\mathbf{N}_B\mathbf{W}_B\mathbf{N}_B\mathbf{M}\mathbf{H}_A)^{-1}\mathbf{H}'_A\mathbf{M}\mathbf{N}_B\mathbf{W}_B\mathbf{N}_B\mathbf{M}\mathbf{Z}_A(k) \tag{3-35}$$

Substituting (3-33) into (3-35), we then find

$$\hat{\theta}_{B,\text{WLS}}(k) = (\mathbf{H}'_A\mathbf{N}_A\mathbf{W}_B\mathbf{N}_A\mathbf{H}_A)^{-1}\mathbf{H}'_A\mathbf{N}_A\mathbf{W}_B\mathbf{N}_A\mathbf{Z}_A(k) \tag{3-36}$$

Comparing (3-36) and (3-34), we conclude that $\hat{\theta}_{B,\text{WLS}}(k) = \hat{\theta}_{A,\text{WLS}}(k)$ if $\mathbf{W}_B(k) = \mathbf{W}_A(k)$. This is precisely the result we were looking for. It means that, under proper normalization, $\hat{\theta}_{B,\text{WLS}}(k) = \hat{\theta}_{A,\text{WLS}}(k)$ and, as a special case, $\hat{\theta}_{B,\text{LS}}(k) = \hat{\theta}_{A,\text{LS}}(k)$.

One way to normalize data is to divide all values by the maximum value of the data. Another way is to work with percentage values, since a percentage value is a ratio of numerator to denominator quantities that have the same units; hence, percentage value is unitless.

## COMPUTATION

See Lesson 4 for how to compute $\hat{\theta}_{\text{LS}}(k)$ or $\hat{\theta}_{\text{WLS}}(k)$ using the singular-value decomposition and the pseudoinverse.

Total least squares estimates, which are described in the Supplementary Material at the end of this lesson, can be computed using the following M-file in *Hi-Spec*:

**tls:** Total least squares solution to an overdetermined system of linear equations.

## Supplementary Material

### LEAST SQUARES, TOTAL LEAST SQUARES, AND CONSTRAINED TOTAL LEAST SQUARES

Consider the overdetermined linear system of equations

$$\mathbf{Ax} = \mathbf{b} \tag{3-37}$$

which must be solved for $\mathbf{x}$. Numerical linear algebra specialists (e.g., Stewart, 1973, and Golub and Van Loan, 1989) provide the following very interesting interpretation for the least-squares solution to (3-37). Suppose there are "errors" associated with the numbers that are entered into the vector $\mathbf{b}$, i.e., $\mathbf{b} \to \mathbf{b} + \boldsymbol{\Delta}\mathbf{b}$, in which case (3-37) actually is

$$\mathbf{Ax} = \mathbf{b} + \boldsymbol{\Delta}\mathbf{b} \tag{3-38}$$

This equation can now be expressed as our generic linear model, as

$$\mathbf{b} = \mathbf{Ax} - \boldsymbol{\Delta}\mathbf{b} \tag{3-39}$$

Here $\mathbf{b}$, $\mathbf{A}$, $\mathbf{x}$, and $-\boldsymbol{\Delta}\mathbf{b}$ play the roles of $\mathbf{Z}(k)$, $\mathbf{H}(k)$, $\boldsymbol{\theta}$, and $\mathbf{V}(k)$, respectively. In our derivation of the least-squares estimator, we minimized $\tilde{\mathbf{Z}}'(k)\tilde{\mathbf{Z}}(k) = [\mathbf{Z}(k) - \mathbf{H}(k)\hat{\boldsymbol{\theta}}(k)]'[\mathbf{Z}(k) - \mathbf{H}(k)\hat{\boldsymbol{\theta}}(k)]$, which, in the notation of (3-39) means that we minimized $[\mathbf{b} - \mathbf{Ax}]'[\mathbf{b} - \mathbf{Ax}] = \boldsymbol{\Delta}\mathbf{b}'\boldsymbol{\Delta}\mathbf{b}$; i.e., the least-squares solution of (3-38) finds the $\mathbf{x}$ that minimizes the errors in the vector $\mathbf{b}$.

Many signal-processing problems lead to a linear system of equations like (3-37). In these problems the elements of $\mathbf{b}$ must first be estimated directly from the data; e.g., they could be autocorrelations, cross-correlations, higher-order statistics, etc. As such, the estimated statistics are themselves in error; hence, in these problems there usually is an error associated with $\mathbf{b}$. Unfortunately, in these same problems there usually are also errors associated with the numbers entered into matrix $\mathbf{A}$ (see Problem 2-8). These elements may also be autocorrelations or higher-order statistics; hence, in many signal-processing problems (3-37) actually becomes

$$(\mathbf{A} + \boldsymbol{\Delta}\mathbf{A})\mathbf{x} = \mathbf{b} + \boldsymbol{\Delta}\mathbf{b} \tag{3-40}$$

Solving for $\mathbf{x}$ using least squares ignores the errors in $\mathbf{A}$.

Equation (3-40) can be reexpressed as

$$(\mathbf{A} + \boldsymbol{\Delta}\mathbf{A}|\mathbf{b} + \boldsymbol{\Delta}\mathbf{b})\begin{bmatrix} \mathbf{x} \\ \hline -1 \end{bmatrix} = \mathbf{0} \tag{3-41}$$

Golub and Van Loan (1980, 1989) determined a solution $\mathbf{x}$ of (3-41) that minimizes the *Frobenius norm* of $(\boldsymbol{\Delta}\mathbf{A}|\boldsymbol{\Delta}\mathbf{b})$. [Note that the Frobenius norm of the real $L \times P$ matrix $\mathbf{M}$ is defined by

$$\|\mathbf{M}\|_F \triangleq \left(\sum_{i=1}^{L}\sum_{j=1}^{P}|m_{ij}|^2\right)^{1/2} = (\text{tr}\,\mathbf{MM}')^{1/2}]$$

They called this the "total least squares (TLS)" solution, $\mathbf{x}_{TLS}$. Their solution involves the singular-value decomposition of $(\mathbf{\Delta A}|\mathbf{\Delta b})$. Van Huffel and Vandewalle (1991) provide a comprehensive treatise on TLS.

One major assumption made in TLS is that the errors in **A** are independent, as are the errors in **b**; i.e., the elements of $\mathbf{\Delta A}$, $\Delta A_{ij}$ are totally independent, as are the elements of $\mathbf{\Delta b}$, $\Delta b_i$. Unfortunately, this is not the situation in most signal-processing problems, where matrix **A** has a specific structure (e.g., Toeplitz, block Toeplitz, Hankel, block Hankel, and circulant). Additionally, **b** may also contain elements that also appear in **A**. Consequently, the errors $\mathbf{\Delta A}$ and $\mathbf{\Delta b}$ are not usually independent in signal-processing applications. Yet, TLS is widely used in such applications and often (if not always) gives much better results than does least squares. Apparently, doing something to account for errors in both **A** and **b** is better than only accounting for errors in **b**.

Abatzoglou, Mendel, and Harada (1991) developed a variation of TLS that is appropriate for the just described situation in which elements of $\mathbf{\Delta A}$ and $\mathbf{\Delta b}$ are dependent. The dependency can be represented as linear constraints among the elements of $\mathbf{\Delta A}$ and $\mathbf{\Delta b}$. They determine a solution **x** to (3-41) that in essence minimizes a norm of $(\mathbf{\Delta A}|\mathbf{\Delta b})$ subject to linear constraints between the elements of $\mathbf{\Delta A}$ and $\mathbf{\Delta b}$. They called this the "constrained total least squares (CTLS)" solution $\mathbf{x}_{CTLS}$. Their solution requires mathematical programming; hence, it is computationally more intensive than the TLS solution; but examples demonstrate that it outperforms TLS.

When the signal-to-noise ratio is high, it is frequently true that elements of $\mathbf{\Delta A}$ and $\mathbf{\Delta b}$ are very small, in which case $\mathbf{x}_{LS}$, $\mathbf{x}_{TLS}$, and $\mathbf{x}_{CTLS}$ give essentially the same results. When the signal-to-noise ratio is low so that elements of $\mathbf{\Delta A}$ and $\mathbf{\Delta b}$ are large, there will be a significant payoff when using TLS or CTLS.

## SUMMARY QUESTIONS

**1.** The method of least squares is credited to:
  **(a)** Lagrange
  **(b)** Rayleigh
  **(c)** Gauss

**2.** A weighted least-squares estimator reduces to a least-squares estimator when:
  **(a)** past measurements are weighted more heavily than present measurements
  **(b)** past measurements are weighted the same as present measurements
  **(c)** past measurements are weighted less heavily than present measurements

**3.** The normal equations:
  **(a)** should be programmed for solution using a matrix inversion routine
  **(b)** should be solved using Gaussian elimination
  **(c)** should be solved using stable algorithms from numerical linear algebra that use orthogonal transformations

**4.** When $N$ is set equal to $k$, then $\hat{\theta}_{WLS}(N)$ is known as a:
  **(a)** fixed-memory estimator
  **(b)** expanding-memory estimator
  **(c)** recursive estimator

5. Least-squares estimates may not be invariant under scale change. One way to circumvent this difficulty is to use:
(a) normalized data
(b) squared data
(c) redundant data

6. Let $\mathbf{N}_A$ and $\mathbf{N}_B$ denote symmetric normalization matrices for $\mathbf{Z}_A(k)$ and $\mathbf{Z}_B(k)$, respectively. The condition that our data are always normalized to the same set of numbers is:
(a) $\mathbf{N}_A\mathbf{Z}_A(k) = \mathbf{N}_B\mathbf{Z}_B(k)$
(b) $\mathbf{Z}_A(k)\mathbf{N}_B = \mathbf{N}_A\mathbf{Z}_B(k)$
(c) $\mathbf{Z}_A(k)\mathbf{N}_A = \mathbf{N}_B\mathbf{Z}_B(k)$

7. Weighted least-squares estimates are:
(a) deterministic
(b) random
(c) a mixture of deterministic and random

8. Gauss was worried that his discovery of the method of least squares would be credited to:
(a) Laplace
(b) Legendre
(c) Lagrange

9. Consider the equation $\mathbf{Ax} = \mathbf{b}$. Which of the following statements are true?
(a) least-squares is associated with accounting for the errors in $\mathbf{A}$, that is, $\boldsymbol{\Delta}\mathbf{A}$
(b) least-squares is associated with accounting for the errors in $\mathbf{b}$, that is, $\boldsymbol{\Delta}\mathbf{b}$
(c) TLS accounts for errors in both $\mathbf{A}$ and $\mathbf{b}$, but it assumes all elements of $\boldsymbol{\Delta}\mathbf{A}$ are independent, as are all elements of $\boldsymbol{\Delta}\mathbf{b}$
(d) TLS accounts for errors in both $\mathbf{A}$ and $\mathbf{b}$, but it assumes some elements of $\boldsymbol{\Delta}\mathbf{A}$ are dependent, as are some elements of $\boldsymbol{\Delta}\mathbf{b}$
(e) CTLS accounts for errors in both $\mathbf{A}$ and $\mathbf{b}$, and it assumes elements of $\boldsymbol{\Delta}\mathbf{A}$ are nonlinearly related, as are some elements of $\boldsymbol{\Delta}\mathbf{b}$
(f) CTLS accounts for errors in both $\mathbf{A}$ and $\mathbf{b}$, and it assumes elements of $\boldsymbol{\Delta}\mathbf{A}$ are linearly related, as are some elements of $\boldsymbol{\Delta}\mathbf{b}$
(g) TLS solution $\mathbf{x}_{\text{TLS}}$ can be obtained using singular-value decomposition

## PROBLEMS

**3-1.** Derive the formula for $\hat{\theta}_{\text{WLS}}(k)$ by completing the square on the right-hand side of the expression for $J[\hat{\theta}(k)]$ in (3-6). In the calculus approach to deriving $\hat{\theta}_{\text{WLS}}(k)$, described in the text, invertibility of $\mathbf{H}'(k)\mathbf{W}(k)\mathbf{H}(k)$ is needed in order to solve (3-9) for $\hat{\theta}_{\text{WLS}}(k)$. Is this condition still needed in this problem's algebraic derivation? If so, how does it occur in the algebraic derivation of $\hat{\theta}_{\text{WLS}}(k)$? What is the main advantage of the algebraic derivation over the calculus derivation?

**3-2.** In the text we stated that $\hat{\theta}_{\text{LS}}(k)$ is obtained when the weighting matrix is chosen as $\mathbf{W}(k) = \mathbf{I}$. More generally, $\hat{\theta}_{\text{LS}}(k)$ is obtained when the weighting matrix is chosen as $\mathbf{W}(k) = c\mathbf{I}$, where $c$ is a constant. Verify the truth of this statement. What happens to $J[\hat{\theta}_{\text{WLS}}(k)]$ in (3-12), and is it important?

**3-3.** (Prof. G. B. Giannakis) Show that weighted least squares can be interpreted as least squares applied to filtered versions of $\mathbf{Z}(k)$ and $\mathbf{H}(k)$. [*Hint:* Decompose $\mathbf{W}(k)$ as $\mathbf{W}(k) = \mathbf{L}'(k)\mathbf{D}(k)\mathbf{L}(k)$, where $\mathbf{L}(k)$ is lower triangular and $\mathbf{D}(k)$ is diagonal.] What is the "filter"? What happens if $\mathbf{W}(k)$ is itself diagonal?

**3-4.** Here we explore the estimation of $Z_\alpha$ in (3-23). Assume that $N$ noisy measurements of $N_a(k)$ are available, i.e., $N_{a_m}(k) = Z_\alpha \alpha(k) + \nu_{N_a}(k)$. What is the formula for the least-squares estimator of $Z_\alpha$?

**3-5.** Here we explore the simultaneous estimation of $M_\alpha$, $M_\delta$, and $Z_\alpha$ in (3-22) and (3-23). Assume that $N$ noisy measurements of $\ddot{\theta}(k)$ and $N_a(k)$ are available, i.e., $\ddot{\theta}_m(k) = M_\alpha \alpha(k) + M_\delta \delta(k) + \nu_{\ddot{\theta}}(k)$ and $N_{a_m}(k) = Z_\alpha \alpha(k) + \nu_{N_a}(k)$. Determine the least-squares estimator of $M_\alpha$, $M_\delta$, and $Z_\alpha$. Is this estimator different from $\hat{M}_{\alpha_{LS}}$ and $\hat{M}_{\delta_{LS}}$ obtained just from $\ddot{\theta}_m(k)$ measurements and $\hat{Z}_{\alpha_{LS}}$ obtained just from $N_{a_m}(k)$ measurements?

**3-6.** In a *curve-fitting problem* we wish to fit a given set of data $z(1), z(2), \ldots, z(N)$ by the approximating function (see, also, Example 2-3)

$$z(k) \simeq \hat{z}(k) = \sum_{j=1}^{n} \hat{\theta}_j \phi_j(k)$$

where $\phi_j(k)$ $(j = 1, 2, \ldots, n)$ are a set of prespecified basis functions.

**(a)** Obtain a formula for $\hat{\boldsymbol{\theta}}_{LS}(N)$ that is valid for any set of basis functions.

**(b)** The simplest approximating function to a set of data is the straight line. In this case, $\hat{z}(k) = \hat{\theta}_1 + \hat{\theta}_2 k$, which is known as the least-squares or regression line. Obtain closed-form formulas for $\hat{\theta}_{1,LS}(N)$ and $\hat{\theta}_{2,LS}(N)$.

**3-7.** Suppose $z(k) = \theta_1 + \theta_2 k$, where $z(1) = 3$ miles per hour and $z(2) = 7$ miles per hour. Determine $\hat{\theta}_{1,LS}$ and $\hat{\theta}_{2,LS}$ based on these two measurements. Next, redo these calculations by scaling $z(1)$ and $z(2)$ to the units of feet per second. Are the least-squares estimates obtained from these two calculations the same? Use the results developed in the section entitled "Scale Changes and Normalization of Data" to explain what has happened here.

**3-8.** **(a)** Under what conditions on scaling matrix $\mathbf{M}$ is scale invariance preserved for a least-squares estimator?

**(b)** If our original model is nonlinear in the measurements [e.g., $z(k) = \theta z^2(k-1) + \nu(k)$], can anything be done to obtain invariant WLSEs under scaling?

**3-9.** (Tony Hung-yao Wu, Spring 1992) Sometimes we need to fit data that are exponential in nature. This requires the approximating function to be of the form $y = be^{ax}$. An easy way to fit this model to data is to work with the logarithm of the model, i.e., $\ln y = \ln b + ax$. Obtain least-squares estimates for $a$ and $\ln b$ for the following data:

| $i$ | $x(i)$ | $y(i)$ |
|---|---|---|
| 1 | 1.00 | 5.10 |
| 2 | 1.25 | 5.79 |
| 3 | 1.50 | 6.53 |
| 4 | 1.75 | 7.45 |
| 5 | 2.00 | 8.46 |

**3-10.** (Ryuji Maeda, Spring 1992) In an experiment to determine Planck's constant $h$ using the photoelectric effect, the following linear relationship is obtained between retarding potential $V$ and frequency of the incident light $\nu$:

$$V = \frac{h}{e}\nu - \frac{h\nu_0}{e} + n_\nu$$

where $e$ is the charge of an electron (assume its value is known), $n_\nu$ is the measurement noise associated with measuring $V$, and $h\nu_0$ is the work function.

**(a)** The following measurements are made: $V_1, \ldots, V_N$, and $\nu_1, \ldots, \nu_N$. Derive the formula for $\hat{h}_{\text{LS}}$.

**(b)** Suppose the following data have been obtained from an experiment in which sodium is used as the "target": $V_0 = 4.4 \times 10^{14}$ Hz and $e = 1.6 \times 10^{-19}$.

| $\nu \times 10^{-14}$ (Hz) | $V$ (V) |
|---|---|
| 6.9 | 1.1 |
| 9.0 | 1.8 |
| 12.1 | 3.1 |

Find the numerical value of $\hat{h}_{\text{LS}}$ for these data.

**3-11.** (Liang-Jin Lin, Spring 1992) Supose that $Y$ is modeled as a quadratic function of $X$, i.e.,

$$Y(X) = aX^2 + bX + c$$

and the measurements of $Y$, denoted $Z$, are subject to measurement errors.

**(a)** Given a set of measurements $\{X(i), Z(i)\}$, $i = 1, 2, \ldots, N$, determine a formula for col $(\hat{a}_{\text{LS}}, \hat{b}_{\text{LS}}, \hat{c}_{\text{LS}})$.

**(b)** Calculate col $(\hat{a}_{\text{LS}}, \hat{b}_{\text{LS}}, \hat{c}_{\text{LS}})$ for the following data. (*Note:* The actual values of parameters $a$, $b$, and $c$ used to generate these data are 0.5, 1.5, and 3, respectively.)

$$\{X(i), Z(i)\} = \{(0, 2.66), (0.2, 3.65), (0.5, 3.84), (0.8, 4.83), (1, 5.1), (1.4, 6.2), (1.8, 6.91), (2.5, 10.16)\}$$

**3-12.** Let $\mathbf{H}(k)$ be expressed as $\mathbf{H}(k) = (\mathbf{h}_1(k)|\mathbf{h}_2(k)|\ldots|\mathbf{h}_n(k))$.

**(a)** Show that, for least-squares estimates,

$$\hat{\mathbf{Z}}(k) = \sum_{i=1}^{n} \hat{\theta}_{i,\text{LS}}(k)\mathbf{h}_i(k)$$

**(b)** Explain why $\hat{\mathbf{Z}}(k)$ lies in the subspace defined by the $n$ vectors $\mathbf{h}_1(k), \mathbf{h}_2(k), \ldots, \mathbf{h}_n(k)$.

**(c)** Does $\mathbf{Z}(k)$ lie in $\hat{\mathbf{Z}}(k)$'s subspace?

**(d)** Provide a three-dimensional diagram to clarify parts (b) and (c) for $n = 2$.

**(e)** Explain why $\hat{\mathbf{Z}}(k)$ is the orthogonal projection of $\mathbf{Z}(k)$ onto the subspace spanned by the $\mathbf{h}_i(k)$'s. Use your diagram from part (d) to do this.

**(f)** How do the results from parts (a)–(e) assist your understanding of least-squares estimation?

**3-13.** The derivation of the least-squares estimator makes extensive use of the approximation of $\mathbf{Z}(k)$ as $\hat{\mathbf{Z}}(k) = \mathbf{H}(k)\hat{\boldsymbol{\theta}}_{LS}(k)$.

**(a)** Show that $\hat{\mathbf{Z}}(k)$ can be expressed as $\hat{\mathbf{Z}}(k) = \mathbf{P}_{\mathbf{H}(k)}(k)\mathbf{Z}(k)$, where the matrix $\mathbf{P}_{\mathbf{H}(k)}(k)$ is called a *projection matrix*. Why is $\mathbf{P}_{\mathbf{H}(k)}(k)$ called a projection matrix? (*Hint:* See Problem 3-12.)

**(b)** Prove that $\mathbf{P}_{\mathbf{H}(k)}(k)$ is an idempotent matrix.

**(c)** Show that $\tilde{\mathbf{Z}}(k) = (\mathbf{I} - \mathbf{P}_{\mathbf{H}(k)}(k))\mathbf{Z}(k) \triangleq \mathbf{P}^{\perp}_{\mathbf{H}(k)}(k)\mathbf{Z}(k)$. Matrix $\mathbf{P}^{\perp}_{\mathbf{H}(k)}(k)$ is associated with the complementary (orthogonal) subspace to $\mathbf{H}(k)$ (see Problem 3-12). Prove that $\mathbf{P}^{\perp}_{\mathbf{H}(k)}(k)$ is also idempotent.

**(d)** Prove that $\mathbf{P}_{\mathbf{H}(k)}(k)\mathbf{P}^{\perp}_{\mathbf{H}(k)}(k) = \mathbf{P}^{\perp}_{\mathbf{H}(k)}(k)\mathbf{P}_{\mathbf{H}(k)}(k) = \mathbf{0}$.

**(e)** How do the results from parts (a)–(d) further assist your understanding of least-squares estimation?

**3-14.** Consider the generic model in (2-1) subject to the restrictions (constraints) $\mathbf{C}\boldsymbol{\theta} = \mathbf{r}$, where $\mathbf{C}$ is a $J \times n$ matrix of known constants and is of rank $J$, and $\mathbf{r}$ is a vector of known constants. This is known as the *restricted least-squares* problem (Fomby et al., 1984). Let $\hat{\boldsymbol{\theta}}^{*}_{LS}(k)$ denote the restricted least-squares estimator of $\boldsymbol{\theta}$. Show that

$$\hat{\boldsymbol{\theta}}^{*}_{LS}(k) = \hat{\boldsymbol{\theta}}_{LS}(k) + [\mathbf{H}'(k)\mathbf{H}(k)]^{-1}\mathbf{C}'\{\mathbf{C}[\mathbf{H}'(k)\mathbf{H}(k)]^{-1}\mathbf{C}'\}^{-1}[\mathbf{r} - \mathbf{C}\hat{\boldsymbol{\theta}}_{LS}(k)]$$

(*Hint:* Use the method of Lagrange multipliers.)

# LESSON 4

# Least-squares Estimation: Singular-value Decomposition

## SUMMARY

The purpose of this lesson is to show how least-squares estimates can be computed using the singular-value decomposition (SVD) of matrix $\mathbf{H}(k)$. This computation is valid for both the overdetermined and underdetermined situations and for the situations when $\mathbf{H}(k)$ may or may not be of full rank. The SVD of $\mathbf{H}(k)$ also provides a practical way to test the rank of $\mathbf{H}(k)$.

The SVD of $\mathbf{H}(k)$ is also related to the pseudoinverse of $\mathbf{H}(k)$. When $\mathbf{H}(k)$ is of maximum rank, we show that the pseudoinverse of $\mathbf{H}(k)$ reduces to the formula that we derived for $\hat{\theta}_{LS}(k)$ in Lesson 3.

When you complete this lesson you will be able to (1) compute $\hat{\theta}_{LS}(k)$ using one of the most powerful algorithms from numerical linear algebra, the SVD, and (2) understand the relationship between the pseudoinverse of $\mathbf{H}(k)$ and the SVD of $\mathbf{H}(k)$.

## INTRODUCTION

The singular-value decomposition (SVD) of a matrix is a very powerful tool in numerical linear algebra. Among its important uses are the determination of the rank of a matrix and numerical solutions of linear least-squares (or weighted least-squares) problems. It can be applied to square or rectangular matrices, whose elements are either real or complex.

According to Klema and Laub (1980), "The SVD was established for real square matrices in the 1870's by Beltrami and Jordan [see, e.g., MacDuffee (1933, p. 78)], for complex square matrices by Autonne (1902), and for general rectangular matrices by Eckart and Young (1939) (the Autonne–Eckart–Young theorem)." The SVD is presently one of the major computational tools in linear systems theory and signal processing.

As in Lesson 3, our starting point for (weighted) least squares is the generic linear model:

$$\mathbf{Z}(k) = \mathbf{H}(k)\theta + \mathbf{V}(k) \tag{4-1}$$

Recall that by choosing $\hat{\theta}(k)$ such that $J[\hat{\theta}(k)] = \tilde{\mathbf{Z}}'(k)\tilde{\mathbf{Z}}(k) = [\mathbf{Z}(k) - \mathbf{H}(k)\hat{\theta}(k)]'$ $[\mathbf{Z}(k) - \mathbf{H}(k)\hat{\theta}(k)]$ is minimized, we obtained $\hat{\theta}_{LS}(k)$, as

$$\hat{\theta}_{LS}(k) = [\mathbf{H}'(k)\mathbf{H}(k)]^{-1}\mathbf{H}'(k)\mathbf{Z}(k) \tag{4-2}$$

where $J = (z - H\hat{\theta})(z - H\hat{\theta})$
$= z'z - z'H(H'H^{-1})H'z$
$= z'z - z'H\hat{\theta} \longrightarrow$

$$J[\hat{\theta}_{LS}(k)] = \mathbf{Z}'(k)[\mathbf{Z}(k) - \mathbf{H}(k)\hat{\theta}_{LS}(k)] \tag{4-3}$$

Note that $J[\hat{\theta}_{LS}(k)]$ in (4-3) was derived using (4-2); hence, it requires $\mathbf{H}(k)$ to be of maximum rank. In this lesson we show how to compute $\hat{\theta}_{LS}(k)$ even if $\mathbf{H}(k)$ is not of maximum rank.

In Lesson 3 we advocated obtaining $\hat{\theta}_{LS}(k)$ [or $\hat{\theta}_{WLS}(k)$] by solving the normal equations $\mathbf{H}'(k)\mathbf{H}(k)\hat{\theta}_{LS}(k) = \mathbf{H}'(k)\mathbf{Z}(k)$. SVD does not obtain $\hat{\theta}_{LS}(k)$ by solving the normal equations. When $\mathbf{H}(k)$ is of maximum rank, it obtains $\hat{\theta}_{LS}(k)$ by computing $[\mathbf{H}'(k)\mathbf{H}(k)]^{-1}\mathbf{H}'(k)$, a matrix that is known as the *pseudoinverse* [of $\mathbf{H}(k)$]. Even when $\mathbf{H}(k)$ is not of maximum rank, we can compute $\hat{\theta}_{LS}(k)$ using the SVD of $\mathbf{H}(k)$.

## SOME FACTS FROM LINEAR ALGEBRA

Here we state some facts from linear algebra that will be used at later points in this lesson [e.g., Strang (1988) and Golub and Van Loan (1989)].

**[F1]** A *unitary matrix* $\mathbf{U}$ is one for which $\mathbf{U}'\mathbf{U} = \mathbf{I}$. Consequently, $\mathbf{U}^{-1} = \mathbf{U}'$.

**[F2]** Let $\mathbf{A}$ be a $K \times M$ matrix and $r$ denote its *rank*; then $r \leq \min(K, M)$.

**[F3]** Matrix $\mathbf{A}'\mathbf{A}$ has the same rank as matrix $\mathbf{A}$.

**[F4]** The *eigenvalues of a real and symmetric matrix* are nonnegative if that matrix is positive semidefinite. Suppose the matrix is $M \times M$ and its rank is $r$, where $r < M$; then the first $r$ eigenvalues of this matrix are positive, whereas the last $M - r$ eigenvalues are zero. If the matrix is of maximum rank, then $r = M$.

**[F5]** If two matrices $\mathbf{V}_1$ and $\mathbf{V}_2$ are *orthogonal*, then $\mathbf{V}_1'\mathbf{V}_2 = \mathbf{V}_2'\mathbf{V}_1 = \mathbf{0}$.

## SINGULAR-VALUE DECOMPOSITION

**Theorem 4-1** [Stewart, 1973; Golub and Van Loan, 1989; Haykin, 1991 (Chapter 11)]. *Let* $\mathbf{A}$ *be a* $K \times M$ *matrix, and* $\mathbf{U}$ *and* $\mathbf{V}$ *be two unitary matrices. Then*

$u + v = $ unitary

$$U'AV = \left[\begin{array}{c|c} \Sigma & \mathbf{0} \\ \hline \mathbf{0} & \mathbf{0} \end{array}\right] \quad \text{This is called USVD of A''} \qquad (4\text{-}4)$$

*where*

$$\Sigma = \text{diag}(\sigma_1, \sigma_2, \ldots, \sigma_r) \qquad (4\text{-}5)$$

$A = k \times m$

$A'A = m \times m$

*and*

$$\sigma_1 \geq \sigma_2 \geq \cdots \geq \sigma_r > 0 \qquad (4\text{-}6)$$

*The $\sigma_i$'s are the  singular values of $\mathbf{A}$ and r is the rank of $\mathbf{A}$.*

Before providing a proof of this very important (SVD) theorem, we note that (1) Theorem 4-1 is true for both the overdetermined ($K > M$) and underdetermined ($K < M$) cases, (2) because $\mathbf{U}$ and $\mathbf{V}$ are unitary, another way of expressing the SVD in (4-4) is

$u'Av = [\;] $ of $uu'Avv' = A = u[\;]v'$
$$A = U\left[\begin{array}{c|c} \Sigma & \mathbf{0} \\ \hline \mathbf{0} & \mathbf{0} \end{array}\right]V' \qquad (4\text{-}7)$$

and, (3) the zero matrices that appear in (4-4) or (4-7) allow for matrix $\mathbf{A}$ to be less than full rank (i.e., *rank deficient*); if matrix $\mathbf{A}$ is of maximum rank, then $\mathbf{U'AV} = \Sigma$, or $\mathbf{A} = \mathbf{U\Sigma V'}$.

*Proof of Theorem 4-1 (overdetermined case).* In the overdetermined case $K > M$, we begin by forming the $M \times M$ symmetric positive semidefinite matrix $\mathbf{A'A}$, whose eigenvalues are greater than or equal to zero (see [**F4**]). Let these eigenvalues be denoted as $\sigma_1^2, \sigma_2^2, \ldots, \sigma_M^2$, where (see [**F4**])

so $A'A$ & find $\lambda$'s & order
Let $\sigma_j = \sqrt{\lambda_j}$

$$\sigma_1 \geq \sigma_2 \geq \cdots \geq \sigma_r > 0 \quad \text{and} \quad \sigma_{r+1} = \sigma_{r+2} = \cdots = \sigma_M = 0 \qquad (4\text{-}8)$$

Let $\mathbf{v}_1, \mathbf{v}_2, \ldots, \mathbf{v}_M$ be the set of $M \times 1$ orthonormal eigenvectors of matrix $\mathbf{A'A}$ that are associated with eigenvalues $\sigma_1^2, \sigma_2^2, \ldots, \sigma_M^2$, respectively; i.e.,

$$\mathbf{A'Av}_i = \sigma_i^2 \mathbf{v}_i, \qquad i = 1, 2, \ldots, M \qquad (4\text{-}9)$$

Collect the $M$ eigenvectors into the $M \times M$ matrix $\mathbf{V}$; i.e.,

$$\mathbf{V} = (\mathbf{v}_1 | \mathbf{v}_2 | \cdots | \mathbf{v}_M) \qquad (4\text{-}10)$$

Matrix $\mathbf{V}$ is unitary.

From (4-9), (4-8), and (4-10), we can reexpress the complete system of $M$ equations in (4-9) as

$$\mathbf{A'AV} = (\mathbf{A'Av}_1 | \mathbf{A'Av}_2 | \cdots | \mathbf{A'Av}_M) = (\sigma_1^2 \mathbf{v}_1 | \cdots | \sigma_r^2 \mathbf{v}_r | \mathbf{0} | \cdots | \mathbf{0}) \qquad (4\text{-}11)$$

Note that the last $M - r$ columns of $\mathbf{A'AV}$ are zero vectors, and, if $\mathbf{A'A}$ is of maximum rank so that $r = M$, then $\mathbf{A'AV} = (\sigma_1^2 \mathbf{v}_1 | \cdots | \sigma_r^2 \mathbf{v}_M)$.

Next, form the matrix $\mathbf{V'A'AV}$. In order to understand the structure of this matrix, we interrupt our proof with an example.

**EXAMPLE 4-1**

Consider the case when $\mathbf{A'A}$ is $3 \times 3$ and $r = 2$, in which case (4-11) becomes $\mathbf{A'AV} = (\sigma_1^2 \mathbf{v}_1 | \sigma_2^2 \mathbf{v}_2 | \mathbf{0})$, so that

$$\mathbf{V'A'AV} = (\mathbf{v}_1 | \mathbf{v}_2 | \mathbf{v}_3)'(\sigma_1^2 \mathbf{v}_1 | \sigma_2^2 \mathbf{v}_2 | \mathbf{0}) = \begin{bmatrix} v_{11} & v_{21} & v_{31} \\ v_{12} & v_{22} & v_{32} \\ v_{13} & v_{23} & v_{33} \end{bmatrix}' (\sigma_1^2 \mathbf{v}_1 | \sigma_2^2 \mathbf{v}_2 | \mathbf{0})$$

$$= \begin{bmatrix} v_{11} & v_{12} & v_{13} \\ v_{21} & v_{22} & v_{23} \\ v_{31} & v_{32} & v_{33} \end{bmatrix} (\sigma_1^2 \mathbf{v}_1 | \sigma_2^2 \mathbf{v}_2 | \mathbf{0}) = \begin{bmatrix} \mathbf{v}_1' \\ \mathbf{v}_2' \\ \mathbf{v}_3' \end{bmatrix} (\sigma_1^2 \mathbf{v}_1 | \sigma_2^2 \mathbf{v}_2 | \mathbf{0})$$

$$= \begin{bmatrix} \sigma_1^2 \mathbf{v}_1' \mathbf{v}_1 & \sigma_2^2 \mathbf{v}_1' \mathbf{v}_2 & 0 \\ \sigma_1^2 \mathbf{v}_2' \mathbf{v}_1 & \sigma_2^2 \mathbf{v}_2' \mathbf{v}_2 & 0 \\ \sigma_1^2 \mathbf{v}_3' \mathbf{v}_1 & \sigma_2^2 \mathbf{v}_3' \mathbf{v}_2 & 0 \end{bmatrix} = \begin{bmatrix} \sigma_1^2 & 0 & 0 \\ 0 & \sigma_2^2 & 0 \\ 0 & 0 & 0 \end{bmatrix} \triangleq \begin{bmatrix} \mathbf{\Sigma}^2 & \mathbf{0} \\ \mathbf{0'} & \mathbf{0} \end{bmatrix}$$

where we have made use of the orthonormality of the three eigenvectors. ☐

Returning to the proof of Theorem 4-1, we can now express $\mathbf{V'A'AV}$ as

$$\mathbf{V'A'AV} = \begin{bmatrix} \mathbf{\Sigma}^2 & \mathbf{0}_{12} \\ \mathbf{0}_{21} & \mathbf{0}_{22} \end{bmatrix} = \begin{bmatrix} r \times r & r \times (M-r) \\ (M-r) \times r & (M-r) \times (M-r) \end{bmatrix} \tag{4-12}$$

where

$$\mathbf{\Sigma}^2 = \text{diag}(\sigma_1^2, \sigma_2^2, \ldots, \sigma_r^2), \tag{4-13}$$

$\mathbf{\Sigma}^2$ is $r \times r$, and zero matrices $\mathbf{0}_{12}$, $\mathbf{0}_{21}$ and $\mathbf{0}_{22}$ are $r \times (M - r)$, $(M - r) \times r$, and $(M - r) \times (M - r)$, respectively.

Because the right-hand side of (4-12) is in partitioned form, we partition matrix $\mathbf{V}$, as

$$\mathbf{V} = (\mathbf{V}_1 | \mathbf{V}_2) \tag{4-14}$$

where $\mathbf{V}_1$ is $M \times r$ and $\mathbf{V}_2$ is $M \times (M - r)$. Because the eigenvectors $\mathbf{v}_i$ are orthonormal, we see that

$$\mathbf{V}_1' \mathbf{V}_2 = \mathbf{0} \tag{4-15}$$

We now equate the left-hand side of (4-12), expressed in partitioned form, to its right-hand side to see that

$$\mathbf{V}_1' \mathbf{A'AV}_1 = \mathbf{\Sigma}^2 \tag{4-16a}$$

and

$$\mathbf{V}_2' \mathbf{A'AV}_2 = \mathbf{0}_{22} \tag{4-16b}$$

Equation (4-16b) can only be true if

$$\mathbf{AV}_2 = \mathbf{0} \tag{4-17}$$

Defining the $K \times r$ matrix $\mathbf{U}_1$ as

$$\mathbf{U}_1 = \mathbf{AV}_1 \mathbf{\Sigma}^{-1}, \tag{4-18}$$

(4-16a) can be reexpressed as

$$\mathbf{U}_1'\mathbf{U}_1 = \mathbf{I} \tag{4-19}$$

which means that matrix $\mathbf{U}_1$ is unitary.

Next, we create $K \times M$ matrix $\mathbf{U}$, where

$$\mathbf{U} \triangleq (\mathbf{U}_1|\mathbf{U}_2) \tag{4-20}$$

in which

$$\mathbf{U}_1'\mathbf{U}_2 \triangleq \mathbf{0} \tag{4-21a}$$

and

$$\mathbf{U}_2'\mathbf{U}_2 \triangleq \mathbf{I} \tag{4-21b}$$

From (4-19), (4-21a), (4-21b), and (4-20), we observe that the matrix $\mathbf{U}$ (just like matrix $\mathbf{V}$) is unitary.

Now we are ready to examine the matrix $\mathbf{U}'\mathbf{AV}$, which appears on the left-hand side of (4-4):

$$\begin{aligned}\mathbf{U}'\mathbf{AV} &= \left[\begin{array}{c}\mathbf{U}_1'\\ \hline \mathbf{U}_2'\end{array}\right]\mathbf{A}(\mathbf{V}_1|\mathbf{V}_2) = \left[\begin{array}{c|c}\mathbf{U}_1'\mathbf{AV}_1 & \mathbf{U}_1'\mathbf{AV}_2\\ \hline \mathbf{U}_2'\mathbf{AV}_1 & \mathbf{U}_2'\mathbf{AV}_2\end{array}\right]\\ &= \left[\begin{array}{c|c}(\mathbf{\Sigma}^{-1}\mathbf{V}_1'\mathbf{A}')\mathbf{AV}_1 & \mathbf{U}_1'(\mathbf{0})\\ \hline \mathbf{U}_2'(\mathbf{U}_1\mathbf{\Sigma}) & \mathbf{U}_2'(\mathbf{0})\end{array}\right] = \left[\begin{array}{c|c}\mathbf{\Sigma}^{-1}(\mathbf{V}_1'\mathbf{A}'\mathbf{AV}_1) & \mathbf{0}\\ \hline \mathbf{0} & \mathbf{0}\end{array}\right]\\ &= \left[\begin{array}{c|c}\mathbf{\Sigma} & \mathbf{0}\\ \hline \mathbf{0} & \mathbf{0}\end{array}\right]\end{aligned} \tag{4-22}$$

To obtain the first partitioned matrix on the second line of (4-22), we used (4-18) and its transpose and (4-17). To obtain the second partitioned matrix on the second line of (4-22), we used (4-21a). Finally, to obtain the third partitioned matrix on the second line of (4-22), we used (4-16a). □

We leave the proof of this theorem in the underdetermined case as Problem 4-7.

### Comments

1. The $\sigma_i$'s $(i = 1, 2, \ldots, r)$ are called the *singular values* of matrix $\mathbf{A}$.
2. From (4-11), **[F3]** and **[F4]**, we see that the rank of matrix $\mathbf{A}$ equals the number of its positive singular values; hence, SVD provides a practical way to determine the rank of a matrix.
3. If $\mathbf{A}$ is a symmetric matrix, then its singular values equal the absolute values of the eigenvalues of $\mathbf{A}$ (Problem 4-4).
4. Vectors $\mathbf{v}_1, \mathbf{v}_2, \ldots, \mathbf{v}_M$, which were called the set of orthonormal eigenvectors of matrix $\mathbf{A}'\mathbf{A}$, are also known as the *right singular vectors* of $\mathbf{A}$.
5. Let $\mathbf{U} = (\mathbf{u}_1|\mathbf{u}_2|\cdots|\mathbf{u}_K)$. Problem 4-7 provides the vectors $\mathbf{u}_1, \mathbf{u}_2, \ldots, \mathbf{u}_K$ with a meaning comparable to that just given to the vectors $\mathbf{v}_1, \mathbf{v}_2, \ldots, \mathbf{v}_M$.

## USING SVD TO CALCULATE $\hat{\theta}_{LS}(k)$

Here we show that $\hat{\theta}_{LS}(k)$ can be computed in terms of the SVD of matrix $\mathbf{H}(k)$. To begin, we rederive $\hat{\theta}_{LS}(k)$ in a way that is much more general than the Lesson 3 derivation, but one that is only possible now after our derivation of the SVD of a matrix.

**Theorem 4-2.** *Let the SVD of* $\mathbf{H}(k)$ *be given as in (4-4). Even if* $\mathbf{H}(k)$ *is not of maximum rank, then*

$$\hat{\theta}_{LS}(k) = \mathbf{V}\left[\begin{array}{c|c} \mathbf{\Sigma}^{-1} & \mathbf{0} \\ \hline \mathbf{0} & \mathbf{0} \end{array}\right]\mathbf{U}'\mathbf{Z}(k) \tag{4-23}$$

*where*

$$\mathbf{\Sigma}^{-1} = \text{diag}\,(\sigma_1^{-1}, \sigma_2^{-1}, \ldots, \sigma_r^{-1}) \tag{4-24}$$

*and r equals the rank of* $\mathbf{H}(k)$.

*Proof.* We begin with $J[\hat{\theta}(k)]$ from the second line of (3-6) (Stewart, 1973):

$$J[\hat{\theta}(k)] = [\mathbf{Z}(k) - \mathbf{H}(k)\hat{\theta}(k)]'[\mathbf{Z}(k) - \mathbf{H}(k)\hat{\theta}(k)] = \|\mathbf{Z}(k) - \mathbf{H}(k)\hat{\theta}(k)\|^2 \tag{4-25}$$

Let the SVD of $\mathbf{H}(k)$ be expressed as in (4-4). Using the unitary natures of matrices $\mathbf{U}(k)$ and $\mathbf{V}(k)$, we see that (for notational simplicity, we now suppress the dependency of all quantities on "$k$")

$$J[\hat{\theta}(k)] = \|\mathbf{Z} - \mathbf{H}\hat{\theta}\|^2 = \|\mathbf{U}'(\mathbf{Z} - \mathbf{H}\mathbf{V}\mathbf{V}'\hat{\theta})\|^2 = \|\mathbf{U}'\mathbf{Z} - \mathbf{U}'\mathbf{H}\mathbf{V}\mathbf{V}'\hat{\theta}\|^2 \tag{4-26}$$

Let

$$\mathbf{V}'\hat{\theta} \triangleq \mathbf{m} = \left[\begin{array}{c} \mathbf{m}_1 \\ \hline \mathbf{m}_2 \end{array}\right] \tag{4-27}$$

and

$$\mathbf{U}'\mathbf{Z} \triangleq \mathbf{c} = \left[\begin{array}{c} \mathbf{c}_1 \\ \hline \mathbf{c}_2 \end{array}\right] \tag{4-28}$$

where the partitioning of $\mathbf{m}$ and $\mathbf{c}$ conform with the SVD of $\mathbf{H}$ given in (4-4). Substituting (4-4), (4-27), and (4-28) into (4-26), we find that

$$J(\hat{\theta}) = \left\| \left[\begin{array}{c} \mathbf{c}_1 \\ \hline \mathbf{c}_2 \end{array}\right] - \left[\begin{array}{c|c} \mathbf{\Sigma} & \mathbf{0} \\ \hline \mathbf{0} & \mathbf{0} \end{array}\right]\left[\begin{array}{c} \mathbf{m}_1 \\ \hline \mathbf{m}_2 \end{array}\right] \right\|^2 = \left\| \left[\begin{array}{c} \mathbf{c}_1 - \mathbf{\Sigma}\mathbf{m}_1 \\ \hline \mathbf{c}_2 \end{array}\right] \right\|^2 \tag{4-29}$$

Clearly, for $J(\hat{\theta})$ to be a minimum, in which case $\hat{\theta} = \hat{\theta}_{LS}(k)$, $\mathbf{c}_1 - \mathbf{\Sigma}\mathbf{m}_1 = \mathbf{0}$, from which it follows that

$$\mathbf{m}_1 = \mathbf{\Sigma}^{-1}\mathbf{c}_1 \tag{4-30}$$

Notice in (4-29) that $J(\hat{\theta})$ does not depend upon $\mathbf{m}_2$; hence, the actual value of $\mathbf{m}_2$ is arbitrary in relation to our objective of obtaining the smallest value of $J(\hat{\theta})$. Consequently, we can choose $\mathbf{m}_2 = \mathbf{0}$. Using $\mathbf{m}_1$ in (4-30) and this zero value of $\mathbf{m}_2$ in (4-27), along with the unitary nature of $\mathbf{V}$, we find that

$$v\hat{\theta} = m = \begin{pmatrix} \Sigma^{-1}c_1 \\ 0 \end{pmatrix}$$
$$vv\hat{\theta} = \hat{\theta} = v\begin{pmatrix} \Sigma^{0}_{00} \end{pmatrix}\begin{pmatrix} c_1 \\ c_L \end{pmatrix}$$
$$\begin{vmatrix} c_1 \\ c_L \end{vmatrix} = u'z$$

$$\hat{\theta}_{LS} = \mathbf{Vm} = \mathbf{V}\left[\frac{\Sigma^{-1}\mathbf{c}_1}{\mathbf{0}}\right] = \mathbf{V}\left[\begin{array}{c|c}\Sigma^{-1} & \mathbf{0} \\ \hline \mathbf{0} & \mathbf{0}\end{array}\right]\left[\begin{array}{c}\mathbf{c}_1 \\ \mathbf{c}_2\end{array}\right] = \mathbf{V}\left[\begin{array}{c|c}\Sigma^{-1} & \mathbf{0} \\ \hline \mathbf{0} & \mathbf{0}\end{array}\right]\mathbf{U}'\mathbf{Z} \tag{4-31}$$

where we have also used (4-28). $\square$

Equation (4-23) can also be written as

$$\hat{\theta}_{LS}(k) = \mathbf{H}^+(k)\mathbf{Z}(k) = (H'H)^{-1}H'z \tag{4-32}$$

where $\mathbf{H}^+(k)$ is the *pseudoinverse* of $\mathbf{H}(k)$:

$$\mathbf{H}^+(k) = \mathbf{V}\left[\begin{array}{c|c}\Sigma^{-1} & \mathbf{0} \\ \hline \mathbf{0} & \mathbf{0}\end{array}\right]\mathbf{U}' \tag{4-33}$$

For discussions about the pseudoinverse and its relationship to $\hat{\theta}_{LS}(k)$, see the Supplementary Material at the end of this lesson.

**Theorem 4-3.** *In the overdetermined case, when $N > n$*

here $(H'H)^{-1} = \sum_{1}^{r} \frac{v_i}{\sigma_i^2} v_i'$

In overdetermined case does need $u_i'$ at all!

$$\hat{\theta}_{LS}(k) = \sum_{i=1}^{r}\frac{\mathbf{v}_i(k)}{\sigma_i^2(k)}\mathbf{v}_i'(k)\mathbf{H}'(k)\mathbf{Z}(k) \tag{4-34}$$

*Proof.* Using the first term on the right-hand side of (4-31), we see that

$$\hat{\theta}_{LS} = \mathbf{V}_1\Sigma^{-1}\mathbf{U}_1'\mathbf{Z} \tag{4-35}$$

But, from (4-18), we see that $\mathbf{U}_1' = (\mathbf{H}\mathbf{V}_1\Sigma^{-1})'$; hence,

$$\hat{\theta}_{LS} = \mathbf{V}_1\Sigma^{-2}\mathbf{V}_1'\mathbf{H}'\mathbf{Z} \tag{4-36}$$

which can also be expressed as in (4-34). $\square$

method

A computational procedure for obtaining $\hat{\theta}_{LS}(k)$ is now clear: (1) compute the SVD of $\mathbf{H}(k)$, and (2) compute $\hat{\theta}_{LS}(k)$ using (4-34). The generalization of this procedure to $\hat{\theta}_{WLS}(k)$ is explored in Problem 4-10. Of course, when we know that we are in the overdetermined situation so that we can use (4-34), we only have to determine matrices $\mathbf{V}$ and $\Sigma$; matrix $\mathbf{U}$ does not have to be computed to determine $\hat{\theta}_{LS}(k)$.

For numerical methods to accomplish step 1, see Golub and Van Loan (1989) or Haykin (1991). See also the section in this lesson entitled "Computation."

**EXAMPLE 4-2** (Todd B. Leach, Spring 1992)

Here we compute $\hat{\theta}_{LS}(k)$ using SVD for the overdetermined situation when

$$\mathbf{H} = \begin{bmatrix} 1 & 0 & 1 \\ 0 & 1 & 1 \\ 1 & 0 & 1 \\ 0 & 1 & 1 \end{bmatrix} \tag{4-37}$$

We begin by determining the eigenvalues of $\mathbf{H}'\mathbf{H}$, where

$$\mathbf{H}'\mathbf{H} = \begin{bmatrix} 2 & 0 & 2 \\ 0 & 2 & 2 \\ 2 & 2 & 4 \end{bmatrix} \tag{4-38}$$

Observe that

$$\det(\sigma^2\mathbf{I} - \mathbf{H}'\mathbf{H}) = 0 \Rightarrow \begin{vmatrix} \sigma^2 - 2 & 0 & -2 \\ 0 & \sigma^2 - 2 & -2 \\ -2 & -2 & \sigma^2 - 4 \end{vmatrix} = \sigma^2(\sigma^2 - 2)(\sigma^2 - 6) = 0 \tag{4-39}$$

so that the eigenvalues of $\mathbf{H}'\mathbf{H}$ are at $\sigma^2 = 6, 2, 0$, which means that the rank of $\mathbf{H}$ is 2. Next, we determine only those right singular vectors of $\mathbf{H}'\mathbf{H}$ that comprise $\mathbf{V}_1$ by solving for $\mathbf{v}_i$ $(i = 1, 2)$ from (4-9). For example, for $\sigma^2 = 6$ we obtain

$$(\mathbf{H}'\mathbf{H} - 6\mathbf{I})\mathbf{v}_1 = \mathbf{0} \Rightarrow \begin{bmatrix} -4 & 0 & 2 \\ 0 & -4 & 2 \\ 2 & 2 & -2 \end{bmatrix} \mathbf{v}_1 = \mathbf{0} \Rightarrow \mathbf{v}_1 = \text{col}\,(6^{-1/2}, 6^{-1/2}, 2 \times 6^{-1/2}) \tag{4-40}$$

Additionally,

$$\mathbf{v}_2 = \text{col}\,(2^{-1/2}, -2^{-1/2}, 0) \tag{4-41}$$

We are now able to use (4-34) to compute $\hat{\theta}_{LS}(k)$ as

$$\hat{\theta}_{LS}(k) = \left( \frac{\mathbf{v}_1\mathbf{v}_1'}{6} + \frac{\mathbf{v}_2\mathbf{v}_2'}{2} \right) \mathbf{H}'\mathbf{Z}(k) = \begin{bmatrix} 1/3 & -1/6 & 1/3 & -1/6 \\ -1/6 & 1/3 & -1/6 & 1/3 \\ 1/6 & 1/6 & 1/6 & 1/6 \end{bmatrix} \mathbf{Z}(k) \quad \square \tag{4-42}$$

## COMPUTATION

The following M-files are found directly in MATLAB and will let the end user perform singular-value decomposition, compute the pseudoinverse of a matrix, or obtain the batch WLS estimator.

**svd**: Computes the matrix singular-value decomposition.

**pinv**: Moore–Penrose pseudoinverse of a matrix **A**. Computation is based on **svd(A)**.

**lscov**: Least-squares solution in the presence of known covariance. Solves $\mathbf{Ax} = \mathbf{B}$ using weighted least squares for an *overdetermined* system of equations, $\mathbf{x} = \mathbf{lscov}(\mathbf{A}, \mathbf{B}, \mathbf{V})$, where $\mathbf{V} \equiv \mathbf{W}$ and $\mathbf{W}$ is our WLSE weighting matrix.

### Supplementary Material

## PSEUDOINVERSE

The inverse of a square matrix is familiar to anyone who has taken a first course in linear algebra. Of course, for the inverse to exist, the square matrix must be of maximum rank. SVD can be used to test for this. The pseudoinverse is a generalization of the concept of matrix inverse to rectangular matrices. We use $\mathbf{H}^+$ to denote the $n \times N$ pseudoinverse of $N \times n$ matrix $\mathbf{H}$. $\mathbf{H}^+$ is defined as the matrix that satisfies the following four properties:

$$\begin{aligned} &\text{(a)}\quad \mathbf{H}\mathbf{H}^+\mathbf{H} = \mathbf{H} \\ &\text{(b)}\quad \mathbf{H}^+\mathbf{H}\mathbf{H}^+ = \mathbf{H}^+ \\ &\text{(c)}\quad (\mathbf{H}^+\mathbf{H})' = \mathbf{H}^+\mathbf{H} \\ &\text{(d)}\quad (\mathbf{H}\mathbf{H}^+)' = \mathbf{H}\mathbf{H}^+ \end{aligned} \tag{4-43}$$

Observe that if $\mathbf{H}$ is square and invertible, then $\mathbf{H}^+ = \mathbf{H}^{-1}$. Matrix $\mathbf{H}^+$ is unique (Greville, 1959) and is also known as the "generalized inverse" of $\mathbf{H}$. According to Greville (1960), the notion of the pseudoinverse of a rectangular or singular matrix (was) introduced by Moore (1920, 1935) and later rediscovered independently by Bjerhammar (1951a, 1951b) and Penrose (1955).

**Theorem 4-4.** *When matrix* $\mathbf{H}$ *is of full rank and* $N > n$ *(overdetermined case), so that* $r = min\,(N, n) = n$, *then*

$$\mathbf{H}^+ = (\mathbf{H}'\mathbf{H})^{-1}\mathbf{H}' \tag{4-44a}$$

*When* $N < n$ *(underdetermined case), so that* $r = min\,(N, n) = N$, *then*

$$\mathbf{H}^+ = \mathbf{H}'(\mathbf{H}\mathbf{H}')^{-1} \tag{4-44b}$$

*Proof.* Note that matrix $\mathbf{H}$ must be of full rank for $(\mathbf{H}'\mathbf{H})^{-1}$ or $(\mathbf{H}\mathbf{H}')^{-1}$ to exist. To prove this theorem, show that $\mathbf{H}^+$ in (4-44) satisfies the four defining properties of $\mathbf{H}^+$ given in (4-43). □

We recognize the pseudoinverse of $\mathbf{H}$ in (4-44a) as the matrix that premultiplies $\mathbf{Z}(k)$ in the formula that we derived for $\hat{\theta}_{LS}(k)$, Eq. (3-11). Note that, when we express $\hat{\theta}_{LS}(k)$ as $\hat{\theta}_{LS}(k) = \mathbf{H}^+(k)\mathbf{Z}(k)$, (4-44b) lets us compute the least-squares estimate in the underdetermined case.

Next we demonstrate how $\mathbf{H}^+$ can be computed using the SVD of matrix $\mathbf{H}$.

**Theorem 4-5.** *Given an* $N \times n$ *matrix* $\mathbf{H}$ *whose SVD is given in (4-7). Then*

$$\mathbf{H}^+ = \mathbf{V}\left[\begin{array}{c|c} \boldsymbol{\Sigma}^{-1} & \mathbf{0} \\ \hline \mathbf{0} & \mathbf{0} \end{array}\right]\mathbf{U}' \tag{4-45}$$

*where*

$$\boldsymbol{\Sigma}^{-1} = \text{diag}\,(\sigma_1^{-1}, \sigma_2^{-1}, \ldots, \sigma_r^{-1}) \tag{4-46}$$

*and* $r$ *equals the rank of* $\mathbf{H}$.

*Proof.* Here we derive (4-45) for (4-44a); the derivation for (4-44b) is left as a problem. Our goals are to express $(\mathbf{H}'\mathbf{H})^{-1}$ and $\mathbf{H}'$ in terms of the quantities that appear in the SVD of $\mathbf{H}$. Equation (4-45) will then follow directly from (4-44a).

Observe, from (4-16a) and the unitary nature of $\mathbf{V}_1$, that matrix $\mathbf{H}'\mathbf{H}$ can be expressed as

$$\mathbf{H}'\mathbf{H} = \mathbf{V}_1\mathbf{\Sigma}^2\mathbf{V}_1' \tag{4-47}$$

Hence,

$$(\mathbf{H}'\mathbf{H})^{-1} = (\mathbf{V}_1')^{-1}\mathbf{\Sigma}^{-2}\mathbf{V}_1^{-1} = \mathbf{V}_1\mathbf{\Sigma}^{-2}\mathbf{V}_1' \tag{4-48}$$

where we have used [**F1**]. Next, solve (4-18) for matrix $\mathbf{H}$, again using the unitary nature of $\mathbf{V}_1$, and then show that

$$\mathbf{H}' = \mathbf{V}_1\mathbf{\Sigma}\mathbf{U}_1' \tag{4-49}$$

Finally, substitute (4-48) and (4-49) into (4-44a) to see that

$$(\mathbf{H}'\mathbf{H})^{-1}\mathbf{H}' = (\mathbf{V}_1\mathbf{\Sigma}^{-2}\mathbf{V}_1')(\mathbf{V}_1\mathbf{\Sigma}\mathbf{U}_1') = \mathbf{V}_1\mathbf{\Sigma}^{-1}\mathbf{U}_1' = \mathbf{V}\left[\begin{array}{c|c} \mathbf{\Sigma}^{-1} & \mathbf{0} \\ \hline \mathbf{0} & \mathbf{0} \end{array}\right]\mathbf{U}' = \mathbf{H}^{+} \ \square \tag{4-50}$$

## SUMMARY QUESTIONS

1. The SVD of a full rank matrix $\mathbf{A}$ decomposes that matrix into the product:
   **(a)** of three matrices, $\mathbf{U}'\mathbf{A}\mathbf{V}$
   **(b)** of three matrices $\mathbf{U}\mathbf{\Sigma}\mathbf{V}'$
   **(c)** $\mathbf{L}\mathbf{L}'$
2. In the SVD of matrix $\mathbf{A}$, which of the following are true?
   **(a)** $\mathbf{U}$ and $\mathbf{V}$ are unitary
   **(b)** $\mathbf{A}$ must be of full rank
   **(c)** $\mathbf{\Sigma}$ is a diagonal matrix whose elements may be negative
   **(d)** $\mathbf{\Sigma}$ is a diagonal matrix whose elements must all be positive
   **(e)** The elements of $\mathbf{\Sigma}$ are the eigenvalues of $\mathbf{A}$
   **(f)** The elements of $\mathbf{\Sigma}$ are the singular values of $\mathbf{A}$
3. The rank of matrix $\mathbf{A}$ equals:
   **(a)** trace ($\mathbf{A}$)
   **(b)** the number of positive singular values of $\mathbf{A}'\mathbf{A}$
   **(c)** the number of positive singular values of $\mathbf{A}$
4. When $\mathbf{A}$ is symmetric, then:
   **(a)** its singular values equal the absolute values of its eigenvalues
   **(b)** $\mathbf{A}^{-1} = \mathbf{A}$
   **(c)** $\mathbf{A}\mathbf{B} = \mathbf{B}\mathbf{A}$
5. $\hat{\theta}_{LS}(k)$ can be computed by computing the:
   **(a)** inverse of $\mathbf{H}(k)$
   **(b)** SVD of $\mathbf{H}(k)$ and then using (4-34)
   **(c)** the eigenvalues of $\mathbf{H}(k)$ and then using (4-34)
6. For the pseudoinverse of $\mathbf{H}$, $\mathbf{H}^{+}$, which of the following are true? $\mathbf{H}^{+}$:
   **(a)** is the same as $\mathbf{H}^{-1}$, if $\mathbf{H}$ is invertible
   **(b)** satisfies five properties

**(c)** can be computed using the SVD of $\mathbf{H}$
**(d)** satisfies four properties

**7.** $\hat{\theta}_{LS}(k)$ can be expressed in terms of $\mathbf{H}^+$ as:
**(a)** $(\mathbf{H}\mathbf{H}^+\mathbf{H})\mathbf{Z}(k)$
**(b)** $(\mathbf{H}^+\mathbf{H})^{-1}\mathbf{H}^+\mathbf{Z}(k)$
**(c)** $(\mathbf{H}^+\mathbf{H}\mathbf{H}^+)\mathbf{Z}(k)$

## PROBLEMS

**4-1.** (David Adams, Spring 1992) Find the SVD of the matrix

$$\mathbf{A} = \begin{bmatrix} 5 & 0 & 1 & 0 \\ 1 & 0 & 5 & 0 \\ 0 & 4 & 0 & -4 \end{bmatrix}$$

**4-2.** Compute the SVD of the following observation matrix, $\mathbf{H}(k)$, and show how it can be used to calculate $\hat{\theta}_{LS}(k)$:

$$\begin{bmatrix} 1 & 2 & 1 \\ 0 & 1 & 1 \\ 1 & 2 & 1 \\ 2 & 1 & 1 \end{bmatrix}$$

**4-3.** (Brad Verona, Spring 1992) Given the linear measurement equation $\mathbf{Z}(k) = \mathbf{H}(k)\theta + \mathbf{V}(k)$, where $\mathbf{Z}(k)$ is a $3 \times 1$ measurement vector, $\theta$ is a $2 \times 1$ parameter vector, $\mathbf{V}(k)$ is a $3 \times 1$ noise vector, and $\mathbf{H}(k)$ is the $3 \times 2$ observation matrix

$$\mathbf{H} = \begin{bmatrix} 1 & -1 \\ -1 & 1 \\ 2 & -2 \end{bmatrix}$$

Find the least-squares estimator of $\theta$ using SVD and the pseudoinverse.

**4-4.** Prove that if $\mathbf{A}$ is a symmetric matrix then its singular values equal the absolute values of the eigenvalues of $\mathbf{A}$.

**4-5.** (Chanil Chung, Spring 1992) Let $\mathbf{A}$ be an $m \times m$ Hermitian matrix with orthonormal columns that is partitioned as follows:

$$\mathbf{A} = \begin{bmatrix} \mathbf{A}_1 \\ \hline \mathbf{A}_2 \end{bmatrix}$$

Show that if $\mathbf{A}_1$ has a singular vector $\mathbf{u}$ with singular value $\gamma$, then $\mathbf{u}$ is also a singular vector of $\mathbf{A}_2$ with singular value $\sigma$, where $\gamma^2 + \sigma^2 = 1$.

**4-6.** (Patrick Lippert, Spring 1992) Singular matrices have long been a source of trouble in the field of robotic manipulators. Consider the planar robot in Figure P4-6, which consists of a revolute joint and two translational joints (commonly known as an RTT arm).

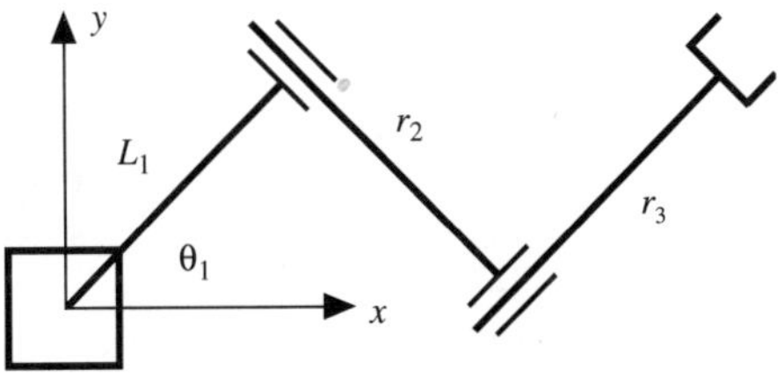

**Figure P4-6**

The following matrix $\mathbf{P}_T$ describes the position vector of the tip of the arm with respect to the base of the robot:

$$\mathbf{P}_T = \begin{bmatrix} (L_1 + r_3)\cos\theta_1 + r_2 \sin\theta_1 \\ (L_1 + r_3)\sin\theta_1 - r_2\cos\theta_1 \\ 0 \end{bmatrix}$$

Here we have three degrees of freedom in a system that only has two degrees of freedom (i.e., the arm can only move in the $x$ and $y$ directions). The problem arises when we try to compute the necessary joint velocities to match a desired Cartesian velocity in the following equation:

$$\frac{d\boldsymbol{\theta}(t)}{dt} = \mathbf{J}^{-1}(\boldsymbol{\theta})\mathbf{V}(t)$$

where $\mathbf{J}(\boldsymbol{\theta})$ is the Jacobian matrix of the position vector, $d\boldsymbol{\theta}(t)/dt = \text{col}\,[d\theta_1(t)/dt, dr_2(t)/dt, dr_3(t)/dt]$ and $\mathbf{V}(t) = \text{col}\,[dx(t)/dt, dy(t)/dt, dz(t)/dt]$.

**(a)** Find the Jacobian matrix $\mathbf{J}(\boldsymbol{\theta})$.

**(b)** Show that the pseudoinverse of $\mathbf{J}(\boldsymbol{\theta})$ is singular.

**4-7.** Prove SVD Theorem 4-1 for the *underdetermined case*. Let $\mathbf{U} = (\mathbf{u}_1|\mathbf{u}_2|\cdots|\mathbf{u}_K)$; explain why the $\mathbf{u}_j$'s are the *left singular vectors* of $\mathbf{A}$ and are also the eigenvectors of $\mathbf{AA}'$.

**4-8.** State and prove the SVD theorem for complex data.

**4-9.** Let $\mathbf{A}$ denote a rectangular $K \times M$ matrix. For large values of $K$ and $M$, storage of this matrix can be a serious problem. Explain how SVD can be used to approximate $\mathbf{A}$ and how the approximation can be stored more efficiently than storing the entire $\mathbf{A}$ matrix. This procedure is known as *data compression.*

**4-10.** Demonstrate how SVD can be used to compute the *weighted least-squares estimate* of $\boldsymbol{\theta}$ in the generic linear model (4-1).

**4-11.** (Charles Pickman, Spring 1992) In many estimation problems that use the linear model $\mathbf{Z}(k) = \mathbf{H}(k)\boldsymbol{\theta} + \mathbf{V}(k)$, $\mathbf{H}(k)$ is assumed constant and full rank. In practice, the elements of $\mathbf{H}(k)$ are measured (or estimated) and include perturbations due to measurement errors. The effects of these perturbations on the estimates need to be determined.

**(a)** Determine $\hat{\boldsymbol{\theta}}_{\text{LS}}(k)$ given the following SVD of $4 \times 3$ matrix $\mathbf{H}(k)_{4\times3}$ and a laboratory measurement of $\mathbf{Z}(k) = \text{col}\,(1, 1, 1, 1)$: $\mathbf{H}(k)_{4\times3} = \mathbf{U}(k)_{4\times4}\boldsymbol{\Sigma}(k)_{4\times3}\mathbf{V}(k)_{3\times3}$, where $\mathbf{U}(k)_{4\times4}$, $\boldsymbol{\Sigma}(k)_{4\times3}$, and $\mathbf{V}(k)_{3\times3}$ are given, respectively as:

$$\begin{bmatrix} 1/\sqrt{2} & -1/2 & 0 & -1/2 \\ 1/2 & 1/\sqrt{2} & -1/2 & 0 \\ 0 & 1/2 & 1/\sqrt{2} & -1/2 \\ 1/2 & 0 & 1/2 & 1/\sqrt{2} \end{bmatrix}, \quad \begin{bmatrix} 4/\sqrt{2} & 0 & 0 \\ 0 & 2/\sqrt{2} & 0 \\ 0 & 0 & \sqrt{2} \\ 0 & 0 & 0 \end{bmatrix}, \quad \begin{bmatrix} 0.6 & 0 & 0.8 \\ 0 & 1 & 0 \\ -0.8 & 0 & 0.6 \end{bmatrix}$$

(*Hint:* Carry the $1\sqrt{2}$ and $1/2$ terms throughout all the calculations until the final calculation.)

**(b)** If the perturbations are on the order of the $\boldsymbol{\Sigma}$ matrix diagonal values, $\sigma_{ii}$, and directly subtract from all diagonal values, then the **H** matrix may be rank deficient (a near zero diagonal term exists). What is a reasonable rank to use, given a measurement error on the order of 0.1, 1, 1.5, or 3?

**(c)** There is a certain amount of error associated with $\hat{\theta}_{\text{LS}}(k)$:

$$e_{\text{LS}} = (\mathbf{Z}(k) - \mathbf{H}(k)\hat{\theta}_{\text{LS}}(k))'(\mathbf{Z}(k) - \mathbf{H}(k)\hat{\theta}_{\text{LS}}(k))$$

Decompose this error into a summation of inner products. Show the derivation and resulting equation.

**(d)** The error $e_{\text{LS}}$ is a minimum given the available $\mathbf{H}(k)$ and $\mathbf{Z}(k)$. Calculate its value when the rank of $\mathbf{H}(k)$, $\rho(\mathbf{H}) = 3$, using $\boldsymbol{\Sigma}_{4\times 3}$ given in part (a). Then assume the $\mathbf{H}(k)$ matrix $\sigma_{ii}$ values are perturbed due to measurement errors. Recalculate $e_{\text{LS}}$ when $\rho(\mathbf{H}) = 2$ and $\rho(\mathbf{H}) = 1$. Show results for all three cases. Compare the three cases.

**4-12.** Develop a formula for computing $J(\hat{\theta}_{\text{LS}})$ using SVD.

**4-13.** Prove the truth of the following pseudoinverse facts and related facts:

**(a)** $\mathbf{H}'\mathbf{H}\mathbf{H}^+ = \mathbf{H}'$

**(b)** $\mathbf{H}^+\mathbf{H}\mathbf{H}' = \mathbf{H}'$

**(c)** $\mathbf{H}'\mathbf{H}'^+\mathbf{H}^+ = \mathbf{H}^+$

**(d)** $\mathbf{H}^+\mathbf{H}'^+\mathbf{H}' = \mathbf{H}^+$

**(e)** $\mathbf{H}^{++} = \mathbf{H}$

**4-14.** (Michiel van Nieustadt, Spring 1992) A standard problem in radar theory is the estimation of the number of independent sources in a signal. This is usually done by estimating the rank of a matrix.

Consider $p$ sources and $m$ sensors, where $m > p$ and $p$ is unknown. The $j$th source sends out a signal with unknown frequency $f_j$, i.e., $s_j(k) = \sin(f_j k)$. The $i$th sensor receives the sum of the $p$ signals each combined with an unknown scaling factor $a_{ij}$. These signal sums are also corrupted by zero-mean Gaussian noise:

$$r_i(k) = \sum_{j=1}^{p} a_{ij} s_j(k) + \sigma n_i(k), \qquad i = 1, 2, \ldots, m$$

where $k$ denotes discrete time, and $n_i(k)$ are unity variance white noises. This equation can also be expressed in matrix form as

$$\mathbf{r}(k) = \mathbf{A}\mathbf{s}(k) + \sigma\mathbf{n}(k)$$

The number of signals can now be estimated as the rank of the matrix $\mathbf{R} = \mathbf{E}\{\mathbf{r}(k)\mathbf{r}'(k)\}$, which is approximated by the sample covariance

$$\hat{\mathbf{R}} = \frac{1}{N}\sum_{k=1}^{N} \mathbf{r}(k)\mathbf{r}'(k)$$

If **S** denotes the covariance matrix of the signals **s**, then

$$\mathbf{R} = \mathbf{A}\mathbf{S}\mathbf{A}' + \sigma^2\mathbf{I}$$

Assume that **A** has full rank $p$.

**(a)** Show that $\mathbf{R}$ has $m - p$ eigenvalues equal to $\sigma^2$ and $p$ eigenvalues greater than $\sigma^2$. Consequently, the rank of $\mathbf{S}$ equals $p$ and can be estimated by looking at the smallest $k$ equal eigenvalues of $\hat{\mathbf{R}}$ and estimating $p = m - k$.

**(b)** Let $p = 2$, $m = 4$, $N = 20$, $f = (1, 2)$, and

$$\mathbf{A} = \begin{bmatrix} 1 & 12 \\ 14 & 25 \\ 37 & 12 \\ 11 & 13 \end{bmatrix}$$

Use SVD to find the eigenvalues of $\mathbf{RR}'$ and then of $\mathbf{R}$ when $\sigma = 0.1$.

**(c)** Repeat part (b) when $\sigma = 1.0$. Comment on the effect of noise.

# LESSON 5

# Least-squares Estimation: Recursive Processing

Uses Batch solution, B-r process in real time.

This process's data as it becomes available.

## SUMMARY

Modern digital computers have caused us to take a second look at the classical batch formulas of (weighted) least squares. These formulas can be made recursive in time by using simple vector and matrix partitioning techniques. The purpose of this lesson is to derive two recursive (weighted) least-squares estimators, referred to as the *information* and *covariance* forms.

The information form is most readily derived directly from the batch formulas in Lesson 3, whereas the covariance form is derived from the information form using a very important identity from matrix algebra, known as the matrix inversion lemma.

The information form is often more useful than the covariance form in theoretical analyses; however, the covariance form is more useful than the information form in most on-line applications, where speed of computation is often the most important consideration.

When you complete this lesson you will be able to derive and use two recursive (weighted) least-squares digital filtering algorithms. These algorithms are easily programmed for digital computations. After understanding these real-time least-squares algorithms, you will be able to implement them for your own real-time applications.

## INTRODUCTION

In Lesson 3 we assumed that $\mathbf{Z}(k)$ contained $N$ elements, where $N > \dim \theta = n$. Suppose we decide to add more measurements, increasing the total number of them from $N$ to $N'$. Formula (3-10) would not make use of the previously calculated value of $\hat{\theta}$ that is based on $N$ measurements during the calculation of $\hat{\theta}$ that is based on $N'$ measurements. This seems quite wasteful. We intuitively feel that it should be possible to compute the estimate based on $N'$ measurements from the estimate based on $N$ measurements, and a modification of this earlier estimate to account for the $N' - N$ new measurements. In this lesson we shall justify our intuition.

## RECURSIVE LEAST SQUARES: INFORMATION FORM

To begin, we consider the case when one additional measurement $z(k+1)$ made at $t_{k+1}$ becomes available:

$$z(k+1) = \mathbf{h}'(k+1)\theta + \nu(k+1) \tag{5-1}$$

When this equation is combined with our earlier linear model, we obtain a new linear model,

$$\mathbf{Z}(k+1) = \mathbf{H}(k+1)\theta + \mathbf{V}(k+1) \tag{5-2}$$

where

$$\mathbf{Z}(k+1) = \text{col}\,(z(k+1)|\mathbf{Z}(k)) \tag{5-3}$$

$$\mathbf{H}(k+1) = \begin{pmatrix} \mathbf{h}'(k+1) \\ \hline \mathbf{H}(k) \end{pmatrix} \tag{5-4}$$

and

$$\mathbf{V}(k+1) = \text{col}\,(\nu(k+1)|\mathbf{V}(k)) \tag{5-5}$$

Using (3-10) and (5-2), it is clear that

$$\hat{\theta}_{\text{WLS}}(k+1) = [\mathbf{H}'(k+1)\mathbf{W}(k+1)\mathbf{H}(k+1)]^{-1}\mathbf{H}'(k+1)\mathbf{W}(k+1)\mathbf{Z}(k+1) \tag{5-6}$$

To proceed further, we must assume that $\mathbf{W}$ is diagonal, i.e.,

$$\mathbf{W}(k+1) = \text{diag}\,(w(k+1)|\mathbf{W}(k)) \tag{5-7}$$

- We shall now show that it is possible to determine $\hat{\theta}_{\text{WLS}}(k+1)$ from $\hat{\theta}_{\text{WLS}}(k)$ and $z(k+1)$.

**Theorem 5-1** (Information Form of Recursive LSE). *A recursive structure for $\hat{\theta}_{\text{WLS}}(k)$ is*

$$\hat{\theta}_{\text{WLS}}(k+1) = \hat{\theta}_{\text{WLS}}(k) + \mathbf{K}_\mathbf{W}(k+1)[z(k+1) - \mathbf{h}'(k+1)\hat{\theta}_{\text{WLS}}(k)] \tag{5-8}$$

*where*

$$\mathbf{K}_\mathbf{W}(k+1) = \mathbf{P}(k+1)\mathbf{h}(k+1)w(k+1) \tag{5-9}$$

*and*

$$\mathbf{P}^{-1}(k+1) = \mathbf{P}^{-1}(k) + \mathbf{h}(k+1)w(k+1)\mathbf{h}'(k+1) \tag{5-10}$$

*These equations are initialized by* $\hat{\theta}_{WLS}(n)$ *and* $\mathbf{P}^{-1}(n)$, *where* $\mathbf{P}(k)$ *is defined later in (5-13), and are used for* $k = n, n+1, \ldots, N-1$.

*Proof.* Substitute (5-3), (5-4), and (5-7) into (5-6) (sometimes dropping the dependence upon $k$ and $k+1$, for notational simplicity) to see that

$$\hat{\theta}_{WLS}(k+1) = [\mathbf{H}'(k+1)\mathbf{W}(k+1)\mathbf{H}(k+1)]^{-1}[\mathbf{h}wz + \mathbf{H}'\mathbf{W}\mathbf{Z}] \tag{5-11}$$

Express $\hat{\theta}_{WLS}(k)$ as

$$\hat{\theta}_{WLS}(k) = \mathbf{P}(k)\mathbf{H}'(k)\mathbf{W}(k)\mathbf{Z}(k) \tag{5-12}$$

where

$$\mathbf{P}(k) = [\mathbf{H}'(k)\mathbf{W}(k)\mathbf{H}(k)]^{-1} \tag{5-13}$$

From (5-12) and (5-13) it follows that

$$\mathbf{H}'(k)\mathbf{W}(k)\mathbf{Z}(k) = \mathbf{P}^{-1}(k)\hat{\theta}_{WLS}(k) \tag{5-14}$$

Set $k = k+1$ in (5-13) to show, after a little algebra, that

$$\mathbf{P}^{-1}(k+1) = \mathbf{P}^{-1}(k) + \mathbf{h}(k+1)w(k+1)\mathbf{h}'(k+1) \tag{5-15}$$

It now follows that

$$\begin{aligned}\hat{\theta}_{WLS}(k+1) &= \mathbf{P}(k+1)[\mathbf{h}wz + \mathbf{P}^{-1}(k)\hat{\theta}_{WLS}(k)] \\ &= \mathbf{P}(k+1)\{\mathbf{h}wz + [\mathbf{P}^{-1}(k+1) - \mathbf{h}w\mathbf{h}']\hat{\theta}_{WLS}(k)\} \\ &= \hat{\theta}_{WLS}(k) + \mathbf{P}(k+1)\mathbf{h}w[z - \mathbf{h}'\hat{\theta}_{WLS}(k)] \\ &= \hat{\theta}_{WLS}(k) + \mathbf{K}_W(k+1)[z - \mathbf{h}'\hat{\theta}_{WLS}(k)]\end{aligned} \tag{5-16}$$

which is (5-8) when gain matrix $\mathbf{K}_W$ is defined as in (5-9).

Based on preceding discussions about $\dim \theta = n$ and $\dim \mathbf{Z}(k) = N$, we know that the first value of $N$ for which (3-10) can be used is $N = n$; thus, (5-8) must be initialized by $\hat{\theta}_{WLS}(n)$, which is computed using (3-10). Equation (5-10) is also a recursive equation for $\mathbf{P}^{-1}(k+1)$, which is initialized by $\mathbf{P}^{-1}(n) = \mathbf{H}'(n)\mathbf{W}(n)\mathbf{H}(n)$. □

## Comments

**1.** Equation (5-8) can also be expressed as

$$\hat{\theta}_{WLS}(k+1) = [\mathbf{I} - \mathbf{K}_W(k+1)\mathbf{h}'(k+1)]\hat{\theta}_{WLS}(k) + \mathbf{K}_W(k+1)z(k+1) \tag{5-17}$$

which demonstrates that *the recursive least-squares estimator (LSE) is a time-varying digital filter that is excited by random inputs* (i.e., the measurements), one whose plant matrix $[\mathbf{I} - \mathbf{K}_W(k+1)\mathbf{h}'(k+1)]$ may itself be random, because $\mathbf{K}_W$ and $\mathbf{h}(k+1)$ may be random. The random natures of $\mathbf{K}_W$ and $(\mathbf{I} - \mathbf{K}_W\mathbf{h}')$ make the analysis of this filter exceedingly difficult. If $\mathbf{K}_W$ and

**h** are deterministic, then stability of this filter can be studied using Lyapunov stability theory.

2. In (5-8), the term $\mathbf{h}'(k+1)\hat{\boldsymbol{\theta}}_{\text{WLS}}(k)$ is a prediction of the actual measurement $z(k+1)$. Because $\hat{\boldsymbol{\theta}}_{\text{WLS}}(k)$ is based on $\mathbf{Z}(k)$, we express this predicted value as $\hat{z}(k+1|k)$, i.e.,

$$\hat{z}(k+1|k) = \mathbf{h}'(k+1)\hat{\boldsymbol{\theta}}_{\text{WLS}}(k) \tag{5-18}$$

*$\hat{z}_{k+1} \equiv$* $\qquad$ *$\hat{z}_{k+1|k} = h_{k+1}' \hat{\theta}_k$*

so that

$$\hat{\boldsymbol{\theta}}_{\text{WLS}}(k+1) = \hat{\boldsymbol{\theta}}_{\text{WLS}}(k) + \mathbf{K}_{\text{W}}(k+1)[z(k+1) - \hat{z}(k+1|k)] \tag{5-19}$$

*$= \hat{\theta}_{wls}(k) + K_w(k+1)\,\tilde{z}(k+1|k)$*

Consequently, $\hat{\boldsymbol{\theta}}_{\text{WLS}}(k+1)$ combines the just-computed $\hat{\boldsymbol{\theta}}_{\text{WLS}}(k)$ with a linear transformation of the prediction error $z(k+1) - \hat{z}(k+1|k)$. This structure is prevalent throughout all recursive estimation theory.

3. Two recursions are present in our recursive LSE. The first is the *vector recursion* for $\hat{\boldsymbol{\theta}}_{\text{WLS}}$ given by (5-8). Clearly, $\hat{\boldsymbol{\theta}}_{\text{WLS}}(k+1)$ cannot be computed from this expression until measurement $z(k+1)$ is available. The second is the *matrix recursion* for $\mathbf{P}^{-1}$ given by (5-10). Observe that values for $\mathbf{P}^{-1}$ (and subsequently $\mathbf{K}_{\text{W}}$) can be precomputed before measurements are made.

4. A digital computer implementation of (5-8)–(5-10) proceeds as follows:

*1st* $\qquad$ *2nd* $\qquad$ *3rd* $\qquad$ *4th*

*compute* $\qquad$ $\mathbf{P}^{-1}(k+1) \rightarrow \mathbf{P}(k+1) \rightarrow \mathbf{K}_{\text{W}}(k+1) \rightarrow \hat{\boldsymbol{\theta}}_{\text{WLS}}(k+1)$

5. Equations (5-8)–(5-10) can also be used for $k = 0, 1, \ldots, N-1$ using the following values for $\mathbf{P}^{-1}(0)$ and $\hat{\boldsymbol{\theta}}_{\text{WLS}}(0)$:

$$\mathbf{P}^{-1}(0) = \frac{1}{a^2}\mathbf{I}_n + \mathbf{h}(0)w(0)\mathbf{h}'(0) \tag{5-20}$$

*n×n* $\quad$ *n×n* $\quad$ *n×1 1×1 1×n*

and

$$\hat{\boldsymbol{\theta}}_{\text{WLS}}(0) = \mathbf{P}(0)\left[\frac{1}{a}\boldsymbol{\epsilon} + \mathbf{h}(0)w(0)z(0)\right] \tag{5-21}$$

*n×1* $\qquad$ *n×n* $\quad$ *$\frac{1}{a}$ n×1 n×1 1×1 1×1*

In these equations (which are derived at the end of this lesson in the Supplementary Material) $a$ is a very large number, $\boldsymbol{\epsilon}$ is a very small number, $\boldsymbol{\epsilon}$ is $n \times 1$, and $\boldsymbol{\epsilon} = \text{col}(\epsilon, \epsilon, \ldots, \epsilon)$. When these initial values are used in (5-8)–(5-10) for $k = 0, 1, \ldots, n-1$, the resulting values obtained for $\hat{\boldsymbol{\theta}}_{\text{WLS}}(n)$ and $\mathbf{P}^{-1}(n)$ are the very same ones that are obtained from the batch formulas for $\hat{\boldsymbol{\theta}}_{\text{WLS}}(n)$ and $\mathbf{P}^{-1}(n)$.

*— note this is special*

• Often $z(0) = 0$, or there is no measurement made at $k = 0$, so that we can set $z(0) = 0$. In this case we can set $w(0) = 0$ so that $\mathbf{P}^{-1}(0) = \mathbf{I}_n/a^2$ and $\hat{\boldsymbol{\theta}}_{\text{WLS}}(0) = a\boldsymbol{\epsilon}$. By choosing $\boldsymbol{\epsilon}$ on the order of $1/a^2$, we see that (5-8)–(5-10) can be initialized by setting $\hat{\boldsymbol{\theta}}_{\text{WLS}}(0) = \mathbf{0}$ and $\mathbf{P}(0)$ equal to a diagonal matrix of very large numbers. This is very commonly done in practice.

6. The reason why the results in Theorem 5-1 are referred to as the information form of the recursive LSE is deferred until Lesson 11, where connections

are made between least-squares and maximum-likelihood estimators (see the section entitled The Linear Model [$\mathbf{H}(k)$ Deterministic], pp. 152–154).

## MATRIX INVERSION LEMMA

Equations (5-10) and (5-9) require the inversion of $n \times n$ matrix $\mathbf{P}$. If $n$ is large, then this will be a costly computation. Fortunately, an alternative is available, one that is based on the following *matrix inversion lemma.*

**Lemma 5-1.** *If the matrices* $\mathbf{A}$, $\mathbf{B}$, $\mathbf{C}$, *and* $\mathbf{D}$ *satisfy the equation*

$$\mathbf{B}^{-1} = \mathbf{A}^{-1} + \mathbf{C}'\mathbf{D}^{-1}\mathbf{C} \tag{5-22}$$

*where all matrix inverses are assumed to exist, then*

$$\mathbf{B} = \mathbf{A} - \mathbf{AC}'(\mathbf{CAC}' + \mathbf{D})^{-1}\mathbf{CA} \tag{5-23}$$

*Proof.* This is a constructive proof of the matrix inversion lemma. Premultiply (5-22) by $\mathbf{B}$ to obtain

$$\mathbf{I} = \mathbf{BA}^{-1} + \mathbf{BC}'\mathbf{D}^{-1}\mathbf{C} \tag{5-24}$$

Postmultiply this equation by $\mathbf{A}$ and then postmultiply the new result by $\mathbf{C}'$ to obtain

$$\mathbf{A} = \mathbf{B} + \mathbf{BC}'\mathbf{D}^{-1}\mathbf{CA} \tag{5-25}$$

and

$$\mathbf{AC}' = \mathbf{BC}' + \mathbf{BC}'\mathbf{D}^{-1}\mathbf{CAC}' = \mathbf{BC}'\mathbf{D}^{-1}(\mathbf{D} + \mathbf{CAC}') \tag{5-26}$$

Now postmultiply this equation by $(\mathbf{D} + \mathbf{CAC}')^{-1}$ and the resulting equation by $\mathbf{CA}$ to obtain

$$\mathbf{AC}'(\mathbf{D} + \mathbf{CAC}')^{-1}\mathbf{CA} = \mathbf{BC}'\mathbf{D}^{-1}\mathbf{CA} \tag{5-27}$$

Subtract (5-27) from $\mathbf{A}$ to obtain

$$\mathbf{A} - \mathbf{AC}'(\mathbf{D} + \mathbf{CAC}')^{-1}\mathbf{CA} = \mathbf{A} - \mathbf{BC}'\mathbf{D}^{-1}\mathbf{CA} \tag{5-28}$$

and substitute (5-25) for the first term on the right-hand side of this equation to obtain

$$\mathbf{A} - \mathbf{AC}'(\mathbf{D} + \mathbf{CAC}')^{-1}\mathbf{CA} = \mathbf{B} \tag{5-29}$$

which is the desired result stated in (5-22). □

Observe that if $\mathbf{A}$ and $\mathbf{B}$ are $n \times n$ matrices, $\mathbf{C}$ is $m \times n$, and $\mathbf{D}$ is $m \times m$, then to compute $\mathbf{B}$ from (5-23) requires the inversion of one $m \times m$ matrix. On the other hand, to compute $\mathbf{B}$ from (5-22) requires the inversion of one $m \times m$ matrix and two $n \times n$ matrices [$\mathbf{A}^{-1}$ and $(\mathbf{B}^{-1})^{-1}$]. When $m < n$, it is definitely advantageous to compute $\mathbf{B}$ using (5-23) instead of (5-22). Observe, also, that in the special case when $m = 1$ matrix inversion in (5-23) is replaced by division.

# RECURSIVE LEAST SQUARES: COVARIANCE FORM

*If #measurements < #parameters use this else use other recursive (see Section 1064)*

**Theorem 5-2** (Covariance Form of Recursive LSE).  *Another recursive structure for* $\hat{\theta}_{WLS}(k)$ *is*

$$\hat{\theta}_{WLS}(k+1) = \hat{\theta}_{WLS}(k) + \mathbf{K}_W(k+1)[z(k+1) - \mathbf{h}'(k+1)\hat{\theta}_{WLS}(k)] \qquad (5\text{-}30)$$

*where*

$$\mathbf{K}_W(k+1) = \mathbf{P}(k)\mathbf{h}(k+1)\left[\mathbf{h}'(k+1)\mathbf{P}(k)\mathbf{h}(k+1) + \frac{1}{w(k+1)}\right]^{-1} \qquad (5\text{-}31)$$

*and*

$$\mathbf{P}(k+1) = [\mathbf{I} - \mathbf{K}_W(k+1)\mathbf{h}'(k+1)]\mathbf{P}(k) \qquad (5\text{-}32)$$

*These equations are initialized by* $\hat{\theta}_{WLS}(n)$ *and* $\mathbf{P}(n)$ *and are used for* $k = n, n + 1, \ldots, N - 1.$

*Proof.* We obtain the results in (5-31) and (5-32) by applying the matrix inversion lemma to (5-10), after which our new formula for $\mathbf{P}(k + 1)$ is substituted into (5-9). To accomplish the first part of this, let $\mathbf{A} = \mathbf{P}(k)$, $\mathbf{B} = \mathbf{P}(k + 1)$, $\mathbf{C} = \mathbf{h}'(k + 1)$, and $\mathbf{D} = 1/w(k + 1)$. Then (5-10) looks like (5-22), so, using (5-23), we see that

$$\mathbf{P}(k + 1) = \mathbf{P}(k) - \mathbf{P}(k)\mathbf{h}(k+1)[\mathbf{h}'(k+1)\mathbf{P}(k)\mathbf{h}(k+1)$$
$$+ w^{-1}(k+1)]^{-1}\mathbf{h}'(k+1)\mathbf{P}(k) \qquad (5\text{-}33)$$

Consequently,

$$\begin{aligned}
\mathbf{K}_W(k + 1) &= \mathbf{P}(k + 1)\mathbf{h}(k + 1)w(k + 1) \\
&= [\mathbf{P} - \mathbf{Ph}(\mathbf{h}'\mathbf{Ph} + w^{-1})^{-1}\mathbf{h}'\mathbf{P}]\mathbf{h}w \\
&= \mathbf{Ph}[\mathbf{I} - (\mathbf{h}'\mathbf{Ph} + w^{-1})^{-1}\mathbf{h}'\mathbf{Ph}]w \\
&= \mathbf{Ph}(\mathbf{h}'\mathbf{Ph} + w^{-1})^{-1}(\mathbf{h}'\mathbf{Ph} + w^{-1} - \mathbf{h}'\mathbf{Ph})w \\
&= \mathbf{Ph}(\mathbf{h}'\mathbf{Ph} + w^{-1})^{-1}
\end{aligned}$$

which is (5-31). To obtain (5-32), express (5-33) as

$$\begin{aligned}
\mathbf{P}(k + 1) &= \mathbf{P}(k) - \mathbf{K}_W(k+1)\mathbf{h}'(k+1)\mathbf{P}(k) \\
&= [\mathbf{I} - \mathbf{K}_W(k+1)\mathbf{h}'(k+1)]\mathbf{P}(k) \quad \square
\end{aligned}$$

## Comments

**1.** The recursive formula for $\hat{\theta}_{WLS}$, (5-30), is unchanged from (5-8). Only the matrix recursion for $\mathbf{P}$, leading to gain matrix $\mathbf{K}_W$ has changed. A digital computer implementation of (5-30)–(5-32) proceeds as follows: $\mathbf{P}(k) \rightarrow \mathbf{K}_W(k + 1) \rightarrow \hat{\theta}_{WLS}(k + 1) \rightarrow \mathbf{P}(k + 1)$. This order of computations differs from the preceding one.

2. When $z(k)$ is a scalar, the covariance form of the recursive LSE requires no matrix inversions and only one division.
3. Equations (5-30)–(5-32) can also be used for $k = 0, 1, \ldots, N-1$ using the values for $\mathbf{P}(0)$ and $\hat{\boldsymbol{\theta}}_{\text{WLS}}(0)$ given in (5-20) and (5-21).
4. The reason why the results in Theorem 5-2 are referred to as the covariance form of the recursive LSE is deferred to Lesson 9, where connections are made between least-squares and best linear unbiased minimum-variance estimators (see p. 130).

**EXAMPLE 5-1**

To illustrate some of this lesson's results, we shall obtain a recursive algorithm for the least-squares estimator of the scalar $\theta$ in the instrument calibration example (Lesson 3). Gain $\mathbf{K}_{\mathbf{W}}(k+1)$ is computed using (5-9) and $\mathbf{P}(k+1)$ is computed using (5-13). Generally, we do not compute $\mathbf{P}(k+1)$ using (5-13); but the simplicity of our example allows us to use this formula to obtain a closed-form expression for $\mathbf{P}(k+1)$ in the most direct way. Recall that $\mathbf{H} = \text{col}\,(1, 1, \ldots, 1)$, which is a $k \times 1$ vector, and $h(k+1) = 1$; thus, setting $\mathbf{W}(k) = \mathbf{I}$ and $w(k+1) = 1$ (to obtain the recursive LSE of $\theta$) in the preceding formulas, we find that

$$\mathbf{P}(k+1) = [\mathbf{H}'(k+1)\mathbf{H}(k+1)]^{-1} = \frac{1}{k+1} \tag{5-34}$$

and

$$\mathbf{K}_{\mathbf{W}}(k+1) = \mathbf{P}(k+1) = \frac{1}{k+1} \tag{5-35}$$

Substituting these results into (5-8), we then find

$$\hat{\theta}_{\text{LS}}(k+1) = \hat{\theta}_{\text{LS}}(k) + \frac{1}{k+1}[z(k+1) - \hat{\theta}_{\text{LS}}(k)]$$

or

$$\hat{\theta}_{\text{LS}}(k+1) = \frac{k}{k+1}\hat{\theta}_{\text{LS}}(k) + \frac{1}{k+1}z(k+1) \tag{5-36}$$

Formula (5-36), which can be used for $k = 0, 1, \ldots, N-1$ by setting $\hat{\theta}_{\text{LS}}(0) = 0$, lets us reinterpret the well-known sample mean estimator as a time-varying digital filter [see, also, equation (1-2)]. We leave it to the reader to study the stability properties of this first-order filter.

Usually, it is in only the simplest of cases that we can obtain closed-form expressions for $\mathbf{K}_{\mathbf{W}}$ and $\mathbf{P}$ and subsequently $\hat{\boldsymbol{\theta}}_{\text{LS}}$ [or $\hat{\boldsymbol{\theta}}_{\text{WLS}}$]; however, we can always obtain values for $\mathbf{K}_{\mathbf{W}}(k+1)$ and $\mathbf{P}(k+1)$ at successive time points using the results in Theorems 5-1 or 5-2. □

## WHICH FORM TO USE

We have derived two formulations for a recursive least-squares estimator, the information and covariance forms. In on-line applications, where speed of computation is often the most important consideration, the covariance form is preferable to the information form. This is because a smaller matrix usually needs to be inverted in the covariance form. As mentioned in Comment 2, when $z(k)$ is a scalar, a division replaces matrix inversion. If a vector of measurements is made at every time point, so that $z(k)$ becomes an $m \times 1$ vector $\mathbf{z}(k)$ (see the next section), then the covariance form requires inversion of an $m \times m$ matrix. When $m < n$, as is frequently the case, the covariance form is preferable to the information form. Be advised though that there are some applications for which sensors are abundant, so that $m > n$. In those applications the information form is computationally more efficient than the covariance form.

Fast fixed-order recursive least-squares algorithms, which require on the order of $n$ flops per iteration (rather than the $n^2$ flops per iteration required by the covariance form), have been developed for the linear prediction problem, which was described in Example 2-2. In linear prediction, (2-9) is used to "predict" $y(k)$ from the earlier samples $y(k-1), y(k-2), \ldots$. Because $y(k)$ is a function of earlier samples of $y(\cdot)$, (2-9) is referred to as a forward prediction model, and $\tilde{y}(k)$ is the forward prediction error.

In the linear prediction problem $\hat{\theta}_{LS}(k)$ contains much more structure than for other problems, because the elements of $\mathbf{Z}(k)$ also appear in $\mathbf{H}(k)$, and successive rows of $\mathbf{H}(k)$ are shifted versions of their immediate predecessors [see $\mathbf{Z}$ and $\mathbf{H}$ in (2-8)]. For extensive discussions on the prediction problem, use of linear predictors in different disciplines, and fast fixed-order recursive least-squares algorithms, see Haykin (1991).

More recently, fast fixed-order recursive least-squares algorithms that are valid for *all* least-squares problems have been developed. They are based on the Givens rotation (Golub and Van Loan, 1989) and can be implemented using systolic arrays. For details see Haykin (1991) and the references therein.

The information form is often more useful than the covariance form in analytical studies. For example, it is used to derive the initial conditions for $\mathbf{P}^{-1}(0)$ and $\hat{\theta}_{WLS}(0)$, which are given in (5-20) and (5-21) (see Supplementary Material). The information form is also to be preferred over the covariance form during the start-up of recursive least squares. We demonstrate why this is so next.

We consider the case when

$$\mathbf{P}(0) = a^2\mathbf{I}_n \tag{5-37}$$

where $a^2$ is a very, very large number. Using the information form, we find that, for $k = 0$, $P^{-1}(1) = \mathbf{h}(1)w(1)\mathbf{h}'(1) + 1/a^2\mathbf{I}_n$, and, therefore, $\mathbf{K_W}(1) = [\mathbf{h}(1)w(1)\mathbf{h}'(1) + 1/a^2\mathbf{I}_n]^{-1}\mathbf{h}(1)w(1)$. No difficulties are encountered when we compute $\mathbf{K_W}(1)$ using the information form.

Using the covariance form, we find, first, that $\mathbf{K_W}(1) \simeq a^2\mathbf{h}(1)[\mathbf{h}'(1)a^2\mathbf{h}(1)]^{-1} = \mathbf{h}(1)[\mathbf{h}'(1)\mathbf{h}(1)]^{-1}$, and then that

$$\mathbf{P}(1) = \{\mathbf{I} - \mathbf{h}(1)[\mathbf{h}'(1)\mathbf{h}(1)]^{-1}\mathbf{h}'(1)\}a^2 \tag{5-38}$$

However, this matrix is singular. To see this, postmultiply both sides of (5-38) by $\mathbf{h}(1)$ to obtain

$$\mathbf{P}(1)\mathbf{h}(1) = \{\mathbf{h}(1) - \mathbf{h}(1)[\mathbf{h}'(1)\mathbf{h}(1)]^{-1}\mathbf{h}'(1)\mathbf{h}(1)\}a^2 = \mathbf{0} \tag{5-39}$$

Neither $\mathbf{P}(1)$ nor $\mathbf{h}(1)$ equals zero; hence, $\mathbf{P}(1)$ must be a singular matrix for $\mathbf{P}(1)\mathbf{h}(1)$ to equal zero. In fact, once $\mathbf{P}(1)$ becomes singular, all other $\mathbf{P}(j)$, $j \geq 2$, will be singular.

In Lesson 9 we shall show that when $\mathbf{W}^{-1}(k) = \mathbf{E}\{\mathbf{V}(k)\mathbf{V}'(k)\} = \mathbf{R}(k)$ then $\mathbf{P}(k)$ is the covariance matrix of the estimation error, $\tilde{\theta}(k)$. This matrix must be positive definite, and it is not possible to maintain this property if $\mathbf{P}(k)$ is singular; hence, it is advisable to initialize the recursive least-squares estimator using the information form. However, it is also advisable to switch to the covariance formulation as soon after initialization as possible in order to reduce computing time.

## GENERALIZATION TO VECTOR MEASUREMENTS

A vector of measurements can occur in any application where it is possible to use more than one sensor; however, it is also possible to obtain a vector of measurements from certain types of individual sensors. In spacecraft applications, it is not unusual to be able to measure attitude, rate, and acceleration. In electrical systems applications, it is not uncommon to be able to measure voltages, currents, and power. Radar measurements often provide information about range, azimuth, and elevation. Radar is an example of a single sensor that provides a vector of measurements.

In the vector measurement case, (5-1) is changed from $z(k+1) = \mathbf{h}'(k+1)\theta + \nu(k+1)$ to

$$\mathbf{z}(k+1) = \overline{\mathbf{H}}(k+1)\theta + \mathbf{v}(k+1) \tag{5-40}$$

where $\mathbf{z}$ is now an $m \times 1$ vector, $\overline{\mathbf{H}}$ is $m \times n$ and $\mathbf{v}$ is $m \times 1$.

We leave it to the reader to show that all the results in Lessons 3 and 4 are unchanged in the vector measurement case; but some notation must be altered (see Table 5-1).

**TABLE 5-1** TRANSFORMATIONS FROM SCALAR TO VECTOR MEASUREMENT SITUATIONS, AND VICE VERSA

| Scalar measurement | Vector of measurements |
|---|---|
| $z(k+1)$ | $\mathbf{z}(k+1)$, an $m \times 1$ vector |
| $\nu(k+1)$ | $\mathbf{v}(k+1)$, an $m \times 1$ vector |
| $w(k+1)$ | $\mathbf{w}(k+1)$, an $m \times m$ matrix |
| $\mathbf{h}'(k+1)$, a $1 \times n$ matrix | $\overline{\mathbf{H}}(k+1)$, an $m \times n$ matrix |
| $\mathbf{Z}(k)$, an $N \times 1$ vector | $\mathbf{Z}(k)$, an $Nm \times 1$ vector |
| $\mathbf{V}(k)$, an $N \times 1$ vector | $\mathbf{V}(k)$, an $Nm \times 1$ vector |
| $\mathbf{W}(k)$, an $N \times N$ matrix | $\mathbf{W}(k)$, an $Nm \times Nm$ matrix |
| $\mathbf{H}(k)$, an $N \times n$ matrix | $\mathbf{H}(k)$, an $Nm \times n$ matrix |

Source: Reprinted from Mendel, 1973, p. 110. Courtesy of Marcel Dekker, Inc., NY.

## COMPUTATION

No M-files exist in any toolbox to perform recursive weighted least-squares estimation for our generic linear model; hence, we have written our own recursive weighted least-squares estimation M-file, **rwlse**. A complete description and listing of this M-file can be found in Appendix B.

### Supplementary Material

## DERIVATION OF START-UP CONDITIONS FOR RECURSIVE ALGORITHMS

Here we provide a derivation for the start-up values for $\mathbf{P}^{-1}(0)$ and $\hat{\boldsymbol{\theta}}_{\text{WLS}}(0)$, which are given in (5-20) and (5-21), respectively. Our derivation is adopted from Mendel (1973, pp. 101–106) and is presented here because the derivation illustrates interesting applications of both the batch and recursive weighted least-squares algorithms.

We begin by introducing $n$ artificial measurements (this is a "thought experiment"; the measurements do not actually exist) $z^a(-1) = \epsilon, z^a(-2) = \epsilon, \ldots, z^a(-n) = \epsilon$, in which $\epsilon$ is a very small number. Observe that these artificial measurements all occur to the left of time point zero. We also assume that the model for $z^a(j)$ is

$$z^a(j) = \frac{\theta_j}{a}, \qquad j = 1, 2, \ldots, n \tag{5-41}$$

where $a$ is a very large number. Observe that for $a \gg$ it doesn't matter what values $\theta_j$ have; for, in this case, $\theta_j/a \ll$, and this very small number is $\epsilon$.

From this description of the artificial measurements, we see that $\mathbf{Z}^a(-1) = \mathbf{H}^a(-1)\boldsymbol{\theta}$, where

$$\mathbf{Z}^a(-1) = \text{col}\,[Z^a(-1), Z^a(-2), \ldots, Z^a(-n)] = \text{col}\,(\epsilon, \epsilon, \ldots, \epsilon) \tag{5-42}$$

and

$$\mathbf{H}^a(-1) = \frac{\mathbf{I}_n}{a} \tag{5-43}$$

where $\mathbf{I}_n$ is the $n \times n$ identity matrix.

We begin by batch processing the $n$ artificial measurements, using a least-squares algorithm, because there is no reason to weight any of these measurements more or less heavily than others, since they are all numerically equal to $\epsilon$. Values of $\hat{\boldsymbol{\theta}}^a_{\text{WLS}}(-1)$ and $\mathbf{P}^a(-1)$ are found from (3-11) and (5-13) to be

$$\hat{\boldsymbol{\theta}}^a_{\text{LS}}(-1) = [\mathbf{H}^{a\prime}(-1)\mathbf{H}^a(-1)]^{-1}\mathbf{H}^{a\prime}(-1)\mathbf{Z}^a(-1) = a\boldsymbol{\epsilon} \tag{5-44}$$

and

$$\mathbf{P}^a(-1) = [\mathbf{H}^{a\prime}(-1)\mathbf{H}^a(-1)]^{-1} = a^2\mathbf{I}_n \tag{5-45}$$

When $z(0)$ is added to the $n$ artificial measurements, our concatenated measurement equation becomes

$$\mathbf{Z}^a(0) = \mathbf{H}^a(0)\boldsymbol{\theta} + \mathbf{V}^a(0) \tag{5-46}$$

where

$$\mathbf{Z}^a(0) = \left[\frac{z(0)}{\mathbf{Z}^a(-1)}\right], \qquad \mathbf{V}^a(0) = \left[\frac{\nu(0)}{\mathbf{0}}\right], \quad \text{and} \quad \mathbf{H}^a(0) = \left[\frac{\mathbf{h}'(0)}{\mathbf{H}^a(-1)}\right] \tag{5-47}$$

Using the first line of (5-16) and the inverse of (5-15), we find that

$$\hat{\theta}^a_{\text{WLS}}(0) = \mathbf{P}^a(0)\{\mathbf{h}(0)w(0)z(0) + [\mathbf{P}^a(-1)]^{-1}\hat{\theta}^a_{\text{LS}}(-1)\} \tag{5-48}$$

and

$$\mathbf{P}^a(0) = \{[\mathbf{P}^a(-1)]^{-1} + \mathbf{h}(0)w(0)\mathbf{h}'(0)\}^{-1} \tag{5-49}$$

Substituting $\hat{\theta}^a_{\text{LS}}(-1)$ and $\mathbf{P}^a(-1)$ into these expressions, we find that

$$\hat{\theta}^a_{\text{WLS}}(0) = \mathbf{P}^a(0)\left[\frac{\epsilon}{a} + \mathbf{h}(0)w(0)z(0)\right] \tag{5-50}$$

and

$$\mathbf{P}^a(0) = \left[\frac{\mathbf{I}_n}{a^2} + \mathbf{h}(0)w(0)\mathbf{h}'(0)\right]^{-1} \tag{5-51}$$

Next, when $z(1)$ is added to the $n$ artificial measurements and $z(0)$, our concatenated measurement equation becomes

$$\mathbf{Z}^a(1) = \mathbf{H}^a(1)\theta + \mathbf{V}^a(1) \tag{5-52}$$

where

$$\mathbf{Z}^a(1) = \left[\frac{z(1)}{\mathbf{Z}^a(0)}\right], \qquad \mathbf{V}^a(1) = \left[\frac{\nu(1)}{\mathbf{V}^a(0)}\right], \quad \text{and} \quad \mathbf{H}^a(1) = \left[\frac{\mathbf{h}'(1)}{\mathbf{H}^a(0)}\right] \tag{5-53}$$

Values of $\hat{\theta}^a_{\text{WLS}}(1)$ and $\mathbf{P}^a(1)$ are computed in a manner similar to the way in which $\hat{\theta}^a_{\text{WLS}}(0)$ and $\mathbf{P}^a(0)$ were computed, i.e.,

$$\hat{\theta}^a_{\text{WLS}}(1) = \mathbf{P}^a(1)\{\mathbf{h}(1)w(1)z(1) + [\mathbf{P}^a(0)]^{-1}\hat{\theta}^a_{\text{WLS}}(0)\} \tag{5-54}$$

and

$$\mathbf{P}^a(1) = \{[\mathbf{P}^a(0)]^{-1} + \mathbf{h}(1)w(1)\mathbf{h}'(1)\}^{-1} \tag{5-55}$$

Substituting $\hat{\theta}^a_{\text{WLS}}(0)$ and $\mathbf{P}^a(0)$ into these expressions, we find that

$$\hat{\theta}^a_{\text{WLS}}(1) = \mathbf{P}^a(1)\left[\frac{\epsilon}{a} + \sum_{j=0}^{1}\mathbf{h}(j)w(j)z(j)\right] \tag{5-56}$$

and

$$\mathbf{P}^a(1) = \left[\frac{\mathbf{I}_n}{a^2} + \sum_{j=0}^{1}\mathbf{h}(j)w(j)\mathbf{h}'(j)\right]^{-1} \tag{5-57}$$

Comparing the equations for $\hat{\theta}^a_{\text{WLS}}(0)$ and $\hat{\theta}^a_{\text{WLS}}(1)$ and $\mathbf{P}^a(0)$ and $\mathbf{P}^a(1)$, we claim (by analytical extension, or proof by induction) that these results generalize to

$$\hat{\theta}^a_{WLS}(\ell+1) = \mathbf{P}^a(\ell+1)\left[\frac{\boldsymbol{\epsilon}}{a} + \sum_{j=0}^{\ell+1} \mathbf{h}(j)w(j)z(j)\right] \quad (5\text{-}58)$$

and

$$\mathbf{P}^a(\ell+1)\left[\frac{\mathbf{I}_n}{a^2} + \sum_{j=0}^{\ell+1} \mathbf{h}(j)w(j)\mathbf{h}'(j)\right]^{-1} \quad (5\text{-}59)$$

Next, when just the measurements $z(0), z(1), \ldots, z(\ell+1)$ are used (i.e., the artificial measurements are not used),

$$\begin{aligned}\hat{\theta}_{WLS}(\ell+1) &= \mathbf{P}(\ell+1)[\mathbf{H}'(\ell+1)\mathbf{W}(\ell+1)\mathbf{Z}(\ell+1)] \\ &= \mathbf{P}(\ell+1)\sum_{j=0}^{\ell+1} \mathbf{h}(j)w(j)z(j) \quad (5\text{-}60)\end{aligned}$$

and

$$\mathbf{P}(\ell+1) = [\mathbf{H}'(\ell+1)\mathbf{W}(\ell+1)\mathbf{H}(\ell+1)]^{-1} = \left[\sum_{j=0}^{\ell+1} \mathbf{h}(j)w(j)\mathbf{h}'(j)\right]^{-1} \quad (5\text{-}61)$$

Comparing $\hat{\theta}^a_{WLS}(\ell+1)$ with $\hat{\theta}_{WLS}(\ell+1)$ and $\mathbf{P}^a(\ell+1)$ with $\mathbf{P}(\ell+1)$, under the original assumption that $a$ is very large and $\epsilon$ is very small, we conclude that, for $a \gg$ and $\epsilon \ll$, $\mathbf{P}^a(\ell+1) \to \mathbf{P}(\ell+1)$ and $\hat{\theta}^a_{WLS}(\ell+1) \to \hat{\theta}_{WLS}(\ell+1)$. In this way we have shown that our recursive weighted least-squares algorithm can be initialized at $k = 0$ with $\mathbf{P}^{-1}(0)$ and $\hat{\theta}_{WLS}(0)$ given in (5-20) and (5-21), respectively.

## SUMMARY QUESTIONS

1. Recursive least-squares estimators are recursive in:
   **(a)** number of measurements
   **(b)** time
   **(c)** number of unknown parameters
2. How many recursions are there in the recursive LSE?
   **(a)** three
   **(b)** two
   **(c)** one
3. The difference between the *information* and *covariance* forms of the recursive LSE is in:
   **(a)** both the $\mathbf{K}_W$ and $\mathbf{P}$ formulas
   **(b)** just the $\mathbf{K}_W$ formula
   **(c)** just the $\mathbf{P}$ formula
4. The matrix inversion lemma lets us invert an $n \times n$ matrix by inverting $m \times m$ matrices, where:
   **(a)** $m > n$

**(b)** $m < n$
**(c)** $m = n$

**5.** In on-line applications, where speed of computation is often the most important consideration, which form of the LSE should be used?
**(a)** information
**(b)** batch
**(c)** covariance

**6.** Which recursive form of the LSE is often more useful in theoretical studies?
**(a)** information
**(b)** adaptive
**(c)** covariance

**7.** The plant matrix of the recursive LSE may be random because:
**(a)** $z(k+1)$ is random
**(b)** $\mathbf{P}$ may be random
**(c)** $\mathbf{h}$ may be random

**8.** The computer implementation $\mathbf{P}^{-1}(k+1) \rightarrow \mathbf{P}(k+1) \rightarrow \mathbf{K}_\mathbf{W}(k+1) \rightarrow \hat{\theta}_{\text{WLS}}(k)$ is associated with the:
**(a)** covariance form of the recursive LSE
**(b)** batch LSE
**(c)** information form of the LSE

**9.** Recursive least-squares estimators:
**(a)** must always be initialized by $\hat{\theta}_{\text{WLS}}(n)$ and $\mathbf{P}(n)$ [or $\mathbf{P}^{-1}(n)$]
**(b)** must always be initialized by $\hat{\theta}_{\text{WLS}}(0)$ and $\mathbf{P}(0)$ [or $\mathbf{P}^{-1}(0)$]
**(c)** may be initialized by $\hat{\theta}_{\text{WLS}}(0)$ and $\mathbf{P}(0)$ [or $\mathbf{P}^{-1}(0)$]

## PROBLEMS

**5-1.** Prove that, once $\mathbf{P}(1)$ becomes singular, all other $\mathbf{P}(j)$, $j \geq 2$, will be singular.

**5-2.** (Mendel, 1973, Exercise 2-12, p. 138) The following weighting matrix weights past measurements less heavily than the most recent measurements:

$$\mathbf{W}(k+1) = \text{diag}\,(w(k+1), w(k), \ldots, w(1)) \triangleq \left(\begin{array}{c|c} \overline{w}(k+1) & \mathbf{0}' \\ \hline \mathbf{0} & \beta^{t_{k+1}-t_k}\mathbf{W}(k) \end{array}\right)$$

where

$$0 < \beta < 1$$

**(a)** Show that $w(j) = \overline{w}(j)\beta^{t_{k+1}-t_j}$ for $j = 1, \ldots, k+1$.
**(b)** How must the equations for the recursive weighted least-squares estimator be modified for this weighting matrix? The estimator thus obtained is known as a *fading memory estimator* (Morrison, 1969).

**5-3.** We showed in Lesson 3 that it doesn't matter how we choose the weights, $w(j)$, in the method of least squares, because the weights cancel out in the batch formula for $\hat{\theta}_{\text{LS}}(k)$. On the other hand, $w(k+1)$ appears explicitly in the recursive WLSE, but only in the formula for $\mathbf{K}_\mathbf{W}(k+1)$, i.e.,

$$\mathbf{K}_\mathbf{W}(k+1) = \mathbf{P}(k)\mathbf{h}(k+1)[\mathbf{h}'(k+1)\mathbf{P}(k)\mathbf{h}(k+1) + w^{-1}(k+1)]^{-1}$$

It would appear that if we set $w(k+1) = w_1$, for all $k$, we would obtain a $\hat{\theta}_{LS}(k+1)$ value that would be different from that obtained by setting $w(k+1) = w_2$, for all $k$. Of course, this cannot be true. Show, *using the formulas for the recursive* WLSE, how they can be made independent of $w(k+1)$, when $w(k+1) = w$, for all $k$.

**5-4.** For the data in the accompanying table, do the following:

**(a)** Obtain the least-squares line $\hat{y}(t) = a + bt$ by means of the batch processing least-squares algorithm;

**(b)** Obtain the least-squares line by means of the recursive least-squares algorithm, using the recursive start-up technique (let $a = 10^8$ and $\epsilon = 10^{-16}$).

| $t$ | $y(t)$ |
|---|---|
| 0 | 1 |
| 1 | 5 |
| 2 | 9 |
| 3 | 11 |

**5-5.** (G. Caso, Fall 1991) In communications, the channel output is often modeled as

$$u(k) = \sum_{j=0}^{L-1} h(j)d(k-j) + \nu(k)$$

where $\{d(k)\}$ is the data symbol sequence, $\{h(j)\}$ is the channel impulse response, and $\{\nu(k)\}$ is additive noise. To investigate the effects of intersymbol interference caused by the channel filtering and to reduce the variance due to the additive noise, it is desirable to process the received signal in an FIR filter (an *equalizer*) before the data symbols are detected; i.e., we form

$$y(k) = \sum_{j=0}^{m-1} w(j)u(k-j)$$

where $\{w(j)\}$ is a set of unknown tap weights that must be determined. A common method used to determine the optimal tap weights is for the transmitter to send a sequence of $N$ agreed upon symbols $\{d(1), d(2), \ldots, d(N)\}$ so that the demodulator can determine the optimal tap weights.

**(a)** A reasonable criterion for the determination of the weights $\{w(j)\}$ is to minimize the squared error between $d(k)$ and $y(k)$ for $k = 1, 2, \ldots, N$. Show that this leads to the same cost function with the standard linear model, in which $\mathbf{Z}(N) \to \mathbf{D}(N)$, $\hat{\theta} \to \hat{\mathbf{W}}(N)$, and $\mathbf{H}(N) \to \mathbf{U}(N)$, where

$$\begin{aligned}
\mathbf{D}(N) &= \text{col}\,(d(N), \ldots, d(1)) \\
\hat{\mathbf{W}}(N) &= \text{col}\,(\hat{w}(0|N), \ldots, \hat{w}(m-1|N)); \\
\mathbf{u}(k) &= \text{col}\,(u(k), \ldots, u(k-m+1)); \\
\mathbf{U}(N) &= \begin{bmatrix} \mathbf{u}'(N) \\ \mathbf{u}'(N-1) \\ \vdots \\ \mathbf{u}'(1) \end{bmatrix}
\end{aligned}$$

and that the optimal solution for the tap weights is, therefore, given by

$$\hat{\mathbf{W}}(N) = [\mathbf{U}'(N)\mathbf{U}(N)]^{-1}\mathbf{U}'(N)\mathbf{D}(N)$$

**(b)** To evaluate the solution in real time, a recursive algorithm must be used. By analogy with the developments in this lesson, determine the covariance form of the RLSE for the evaluation of the tap weights.

**(c)** What are the *conceptual differences* between this filter design problem and a parameter estimation problem?

**5-6.** (T. A. Parker and M. A. Ranta, Fall 1991) (Computer Project) This problem investigates the sensitivity of linear least-squares estimation to noise. Use the recursive linear least-squares method to estimate the slope and intercept of the line $y = 5x + n$, where $n$ is white Gaussian noise with zero mean and variance $\sigma^2$. First, implement the recursive linear least-squares algorithm on the computer with no noise present to verify that the program works correctly. Then examine how the estimates of the slope and intercept are affected as the variance of the noise increases. Perform 30 iterations of the algorithm for noise variances of 0.1, 0.5, 1.0, and 4.0. Draw conclusions.

**5-7.** Suppose that at each sampling time $t_{k+1}$ there are $q$ sensors or groups of sensors that provide our vector measurement data. These sensors are corrupted by noise that is uncorrelated from one sensor group to another. The $m$-dimensional vector $\mathbf{z}(k+1)$ can be represented as

$$\mathbf{z}(k+1) = \text{col}\,(\mathbf{z}_1(k+1), \mathbf{z}_2(k+1), \ldots, \mathbf{z}_q(k+1))$$

where

$$\mathbf{z}_i(k+1) = \mathbf{H}_i(k+1)\boldsymbol{\theta} + \mathbf{v}_i(k+1)$$

$\dim \mathbf{z}_i(k+1) = m_i \times 1$,

$$\sum_{i=1}^{q} m_i = m$$

$\mathbf{E}\{\mathbf{v}_i(k+1)\} = \mathbf{0}$, and

$$\mathbf{E}\{\mathbf{v}_i(k+1)\mathbf{v}'_j(k+1)\} = \mathbf{R}_i(k+1)\delta_{ij}$$

An alternative to processing all $m$ measurements in one batch (i.e., simultaneously) is available, and is one in which we freeze time at $t_{k+1}$ and recursively process the $q$ batches of measurements one batch at a time. Data $\mathbf{z}_1(k+1)$ are used to obtain an estimate (for notational simplicity, we omit the subscript WLS or LS on $\hat{\boldsymbol{\theta}}$) $\hat{\boldsymbol{\theta}}_1(k+1)$ with $\hat{\boldsymbol{\theta}}_1(k) \triangleq \hat{\boldsymbol{\theta}}(k)$ and $\mathbf{z}(k+1) \triangleq \mathbf{z}_1(k+1)$. When these calculations are completed $\mathbf{z}_2(k+1)$ is processed to obtain the estimate $\hat{\boldsymbol{\theta}}_2(k+1)$. Estimate $\hat{\boldsymbol{\theta}}_1(k+1)$ is used to initialize $\hat{\boldsymbol{\theta}}_2(k+1)$. Each set of data is processed in this manner until the final set $\mathbf{z}_q(k+1)$ has been included. Then time is advanced to $t_{k+2}$ and the cycle is repeated. This type of processing is known as *cross-sectional processing*. It is summarized in Figure P5-7.

A very large computational advantage exists for cross-sectional processing if $m_i = 1$. In this case the matrix inverse $[\mathbf{H}(k+1)\mathbf{P}(k)\mathbf{H}'(k+1) + \mathbf{w}^{-1}(k+1)]^{-1}$ needed in (5-25) is replaced by the division $[\mathbf{h}'_i(k+1)\mathbf{P}_i(k)\mathbf{h}_i(k+1) + 1/w_i(k+1)]^{-1}$.

**(a)** Prove that, using cross-sectional processing $\hat{\boldsymbol{\theta}}_q(k+1) = \hat{\boldsymbol{\theta}}(k+1)$. [*Hint:* Express $\mathbf{I} - \mathbf{KH}$ as $\mathbf{P}(k+1)\mathbf{P}^{-1}(k)$ and develop the equations for $\hat{\boldsymbol{\theta}}_i(k+1)$ in terms of these quantities; then back-substitute.]

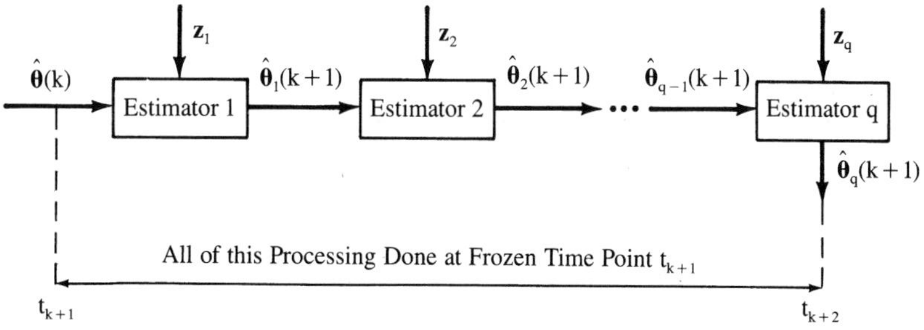


**Figure P5-7** Cross-sectional processing of $m$ measurements.

**(b)** Give the equations for $\hat{\theta}_i(k+1)$, $i = 1, 2, \ldots, q$.

**5-8.** In Lesson 3 we assumed that $\hat{\theta}$ is determined for a fixed value of $n$. In many system modeling problems, we are interested in a preliminary model in which dimension $n$ is a variable. One approach is to recompute $\hat{\theta}$ by means of Equation (3-10) for different values of $n$. This may be very costly, especially when $n$ is large, because the number of flops to compute $\hat{\theta}$ is on the order of $n^3$. A second approach, the one we explore in this problem, is to obtain $\hat{\theta}$ for $n = n_1$ and to use the estimate in a computationally effective manner to obtain $\hat{\theta}$ for $n = n_2$, where $n_2 > n_1$. These estimators are recursive in the dimension of $\theta$.

Suppose we are given a linear model $\mathbf{Z}(k) = \mathbf{H}_1(k)\theta_1 + \mathbf{V}(k)$ with $n$ unknown parameters $\theta_1$, data $\{\mathbf{Z}(k), \mathbf{H}_1(k)\}$, and the LSE of $\theta_1$, $\hat{\theta}^*_{1,\text{LS}}(k)$, where

$$\hat{\theta}^*_{1,\text{LS}}(k) = [\mathbf{H}_1'(k)\mathbf{H}_1(k)]^{-1}\mathbf{H}_1'(k)\mathbf{Z}(k)$$

We extend this model to include $\ell$ additional parameters, $\theta_2$, so that our model is given by

$$\mathbf{Z}(k) = \mathbf{H}_1(k)\theta_1 + \mathbf{H}_2(k)\theta_2 + \mathbf{V}(k)$$

For this model, data $\{\mathbf{Z}(k), \mathbf{H}_1(k), \mathbf{H}_2(k)\}$ are available. We wish to compute the least-squares estimate of $\theta_1$ and $\theta_2$ for the $n+\ell$ parameter model using the previously computed $\hat{\theta}^*_{1,\text{LS}}(k)$.

**(a)** Derive the following "multistage" algorithm (Åström, 1968; Mendel, 1975):

$$\begin{aligned}
\hat{\theta}_{1,\text{LS}}(k) &= \hat{\theta}^*_{1,\text{LS}}(k) - \mathbf{G}(k)\mathbf{C}(k)\mathbf{H}_2'(k)[\mathbf{Z}(k) - \mathbf{H}_1(k)\hat{\theta}^*_{1,\text{LS}}(k)] \\
\hat{\theta}_{2,\text{LS}}(k) &= \mathbf{C}(k)\mathbf{H}_2'(k)[\mathbf{Z}(k) - \mathbf{H}_1(k)\hat{\theta}^*_{1,\text{LS}}(k)] \\
\mathbf{G}(k) &= [\mathbf{H}_1'(k)\mathbf{H}_1(k)]^{-1}\mathbf{H}_1'(k)\mathbf{H}_2(k) \\
\mathbf{C}(k) &= [\mathbf{H}_2'(k)\mathbf{H}_2(k) - \mathbf{H}_2'(k)\mathbf{H}_1(k)\mathbf{G}(k)]^{-1}
\end{aligned}$$

[*Hints:* (1) Use the block decomposition method for inverting $\mathbf{H}'\mathbf{H}$, where $\mathbf{H} = (\mathbf{H}_1|\mathbf{H}_2)$; (2) Use the matrix inversion lemma 5-1.]

**(b)** Show what happens to the multistage algorithm when the regressors (which are contained in $\mathbf{H}_1$ and $\mathbf{H}_2$) are orthogonal.

**(c)** How do we obtain orthogonal regressors?

# LESSON 6

# Small-sample Properties of Estimators*

## SUMMARY

This lesson begins with a question: "How do we know whether or not the results obtained from the LSE, or for that matter any estimator, are good?" We learn that, to answer this question, we must make use of the fact that all estimators represent transformations of random data; hence, $\hat{\theta}(k)$ is itself random, so that its properties must be studied from a statistical viewpoint. This fact and its consequences, which seem so obvious to us today, are due to the eminent statistician R. A. Fisher.

The purpose of this lesson is to introduce two widely used small-sample properties of estimators, *unbiasedness* and *efficiency*. The phrase *small sample* means *any number of measurements* (e.g., 1, 2, 100, $10^4$, or even an infinite number).

An estimator is unbiased if its mean value is tracking the unknown parameter at *every* value of time. Dispersion about the mean is measured by error variance. Efficiency is related to how small the error variance will be. Associated with efficiency is the very famous Cramer–Rao inequality, which places a lower bound on the error variance, a bound that does not depend on a particular estimator.

When you complete this lesson you will be able to (1) Define and explain the two small-sample properties of estimators known as *unbiasedness* and *efficiency*, and (2) test different estimators for unbiasedness and efficiency.

---

*This lesson was written using some suggestions and inputs from Dr. Georgios B. Giannakis, Department of Electrical Engineering, University of Virginia, Charlottesville, VA.

## INTRODUCTION

How do we know whether or not the results obtained from the LSE, or for that matter any estimator, are good? To answer this question, we make use of the fact that all estimators represent transformations of random data. For example, our LSE, $[\mathbf{H}'(k)\mathbf{W}(k)\mathbf{H}(k)]^{-1}\mathbf{H}'(k)\mathbf{W}(k)\mathbf{Z}(k)$, represents a linear transformation on $\mathbf{Z}(k)$. Other estimators may represent nonlinear transformations of $\mathbf{Z}(k)$. *The consequence of this is that $\hat{\theta}(k)$ is itself random.* Its properties must therefore be studied from a statistical viewpoint.

In the estimation literature, it is common to distinguish between small-sample and large-sample properties of estimators. The term *sample* refers to the number of measurements used to obtain $\hat{\theta}$, i.e., the dimension of $\mathbf{Z}$. The phrase *small sample* means *any number of measurements* (e.g., 1, 2, 100, $10^4$, or even an infinite number), whereas the phrase *large sample* means *an infinite number of measurements.* Large-sample properties are also referred to as asymptotic properties. It should be obvious that *if an estimator possesses a small-sample property, it also possesses the associated large-sample property; but the converse is not always true.*

Why bother studying large-sample properties of estimators if these properties are included in their small-sample properties? Put another way, why not just study small-sample properties of estimators? For many estimators it is relatively easy to study their large-sample properties and virtually impossible to learn about their small-sample properties. For example, except for special cases it is impossible to determine probability density functions of correlated non-Gaussian data; however, since most estimators are formed by some sort of averaging, the central limit theorem guarantees, under fairly general conditions, that asymptotic distributions are Gaussian. An analogous situation occurs in stability theory, where most effort is directed at infinite-time stability behavior rather than at finite-time behavior.

Although large sample means an infinite number of measurements, estimators begin to enjoy their large-sample properties for much fewer than an infinite number of measurements. How few usually depends on the dimension of $\theta$, $n$, the memory of the estimators, and in general on the underlying, albeit unknown, probability density function.

A thorough study into $\hat{\theta}$ would mean determining its probability density function $p(\hat{\theta})$. Usually, it is too difficult to obtain $p(\hat{\theta})$ for most estimators (unless $\hat{\theta}$ is multivariate Gaussian); thus, it is customary to emphasize the first- and second-order statistics of $\hat{\theta}$ (or its associated error $\tilde{\theta} = \theta - \hat{\theta}$), the mean and covariance.

We shall examine the following small- and large-sample properties of estimators: unbiasedness and efficiency (small sample) and asymptotic unbiasedness, consistency, asymptotic normality, and asymptotic efficiency (large sample). Small sample properties are the subject of this lesson, whereas large-sample properties are studied in Lesson 7.

**Definition 6-1.** *Estimator $\hat{\theta}(k)$ is an unbiased estimator of deterministic $\theta$ if*

$$\mathbf{E}\{\hat{\theta}(k)\} = \theta, \qquad \text{for all } k \tag{6-1}$$

*or of random $\theta$ if*

$$\mathbf{E}\{\hat{\theta}(k)\} = \mathbf{E}\{\theta\}, \qquad \text{for all } k \ \square \tag{6-2}$$

In terms of estimation error, $\tilde{\theta}(k)$, unbiasedness means, that

$$\mathbf{E}\{\tilde{\theta}(k)\} = \mathbf{0}, \qquad \text{for all } k \tag{6-3}$$

**EXAMPLE 6-1**

In the instrument calibration example of Lesson 3, we determined the following LSE of $\theta$:

$$\hat{\theta}_{\text{LS}}(N) = \frac{1}{N}\sum_{i=1}^{N} z(i) \tag{6-4}$$

where

$$z(i) = \theta + \nu(i) \tag{6-5}$$

Suppose $\mathbf{E}\{\nu(i)\} = 0$ for $i = 1, 2, \ldots, N$; then

$$\mathbf{E}\{\hat{\theta}_{\text{LS}}(N)\} = \frac{1}{N}\sum_{i=1}^{N}\mathbf{E}\{z(i)\} = \frac{1}{N}\sum_{i=1}^{N}\theta = \theta$$

which means that $\hat{\theta}_{\text{LS}}(N)$ is an unbiased estimator of $\theta$. $\square$

Many estimators are linear transformations of the measurements; i.e.,

$$\hat{\theta}(k) = \mathbf{F}(k)\mathbf{Z}(k) \tag{6-6}$$

In least squares, we obtained this linear structure for $\hat{\theta}(k)$ by solving an optimization problem. Sometimes, we begin by assuming that (6-6) is the desired structure for $\hat{\theta}(k)$. We now address the question "when is $\mathbf{F}(k)\mathbf{Z}(k)$ an unbiased estimator of deterministic $\theta$?"

**Theorem 6-1.** *When* $\mathbf{Z}(k) = \mathbf{H}(k)\theta + \mathbf{V}(k)$, $\mathbf{E}\{\mathbf{V}(k)\} = \mathbf{0}$, *and* $\mathbf{H}(k)$ *is deterministic, then* $\hat{\theta}(k) = \mathbf{F}(k)\mathbf{Z}(k)$ [*where* $\mathbf{F}(k)$ *is deterministic*] *is an unbiased estimator of* $\theta$ *if*

$$\mathbf{F}(k)\mathbf{H}(k) = \mathbf{I}, \qquad \text{for all } k \tag{6-7}$$

Note that this is the first place where we have had to assume any a priori knowledge about the nature of noise $\mathbf{V}(k)$.

*Proof.* From the model for $\mathbf{Z}(k)$ and the assumed structure for $\hat{\theta}(k)$, we see that

$$\hat{\theta}(k) = \mathbf{F}(k)\mathbf{H}(k)\theta + \mathbf{F}(k)\mathbf{V}(k) \tag{6-8}$$

Because $\mathbf{F}(k)$ and $\mathbf{H}(k)$ are deterministic and $\mathbf{V}(k)$ is zero mean, it then follows that

$$\mathbf{E}\{\hat{\theta}(k)\} = \mathbf{F}(k)\mathbf{H}(k)\theta \tag{6-9}$$

Assuming the truth of (6-7), it must be that

$$\mathbf{E}\{\hat{\theta}(k)\} = \theta \tag{6-10}$$

which, of course, means that $\hat{\theta}(k)$ is an unbiased estimator of $\theta$. □

**EXAMPLE 6-2**

Matrix $\mathbf{F}(k)$ for the WLSE of $\theta$ is $[\mathbf{H}'(k)\mathbf{W}(k)\mathbf{H}(k)]^{-1}\mathbf{H}'(k)\mathbf{W}(k)$. Observe that this $\mathbf{F}(k)$ matrix satisfies (6-7); thus, when $\mathbf{H}(k)$ is *deterministic the WLSE of* $\theta$ *is unbiased.* Unfortunately, in many interesting applications $\mathbf{H}(k)$ is random, and we cannot apply Theorem 6-1 to study the unbiasedness of the WLSE. We return to this issue in Lesson 8. □

Suppose that we begin by assuming a linear recursive structure for $\hat{\theta}$ :

$$\hat{\theta}(k+1) = \mathbf{A}(k+1)\hat{\theta}(k) + \mathbf{b}(k+1)z(k+1) \tag{6-11}$$

We then have the following counterpart to Theorem 6-1.

**Theorem 6-2.** *When* $z(k+1) = \mathbf{h}'(k+1)\theta + v(k+1)$, $\mathbf{E}\{v(k+1)\} = 0$, $\mathbf{h}(k+1)$ *is deterministic and* $\mathbf{E}\{\hat{\theta}(k+1)\} = \mathbf{E}\{\hat{\theta}(k)\}$ *for any value of* k, *say* $k^*$, *then* $\hat{\theta}(k+1)$ *given by (6-11) in an unbiased estimator of* $\hat{\theta}$ *if*

$$\mathbf{A}(k+1) = \mathbf{I} - \mathbf{b}(k+1)\mathbf{h}'(k+1) \tag{6-12}$$

*where* $\mathbf{A}(k+1)$ *and* $\mathbf{b}(k+1)$ *are deterministic.* □

We leave the proof of this result to the reader. Unbiasedness means that our recursive estimator does not have two independent design matrices (degrees of freedom), $\mathbf{A}(k+1)$ and $\mathbf{b}(k+1)$. Unbiasedness constrains $\mathbf{A}(k+1)$ to be a function of $\mathbf{b}(k+1)$. When (6-12) is substituted into (6-11), we obtain the following important structure for an unbiased linear recursive estimator of $\theta$:

$$\hat{\theta}(k+1) = \hat{\theta}(k) + \mathbf{b}(k+1)[z(k+1) - \mathbf{h}'(k+1)\hat{\theta}(k)] \tag{6-13}$$

Our recursive WLSE of $\theta$ has this structure; thus, as long as $\mathbf{h}(k+1)$ is deterministic, it produces unbiased estimates of $\theta$. Many other estimators that we shall study will also have this structure. h(k+1) = random maybe biased & may NOT be,

## EFFICIENCY

Did you hear the story about the conventioners who all drowned in a lake that was on the average 6 ft deep? The point of this rhetorical question is that unbiasedness by itself is not terribly meaningful. We must also study the dispersion about the mean, the variance. If the conventioners had known that the variance about the 6 ft average depth was 120 ft, they might not have drowned!

Ideally, we would like our estimator to be unbiased and to have the smallest possible error variance. We consider the case of a scalar parameter first.

**Definition 6-2.** *An unbiased estimator,* $\hat{\theta}(\mathrm{k})$ *of* $\theta$ *is said to be more efficient than any other unbiased estimator,* $\hat{\hat{\theta}}(\mathrm{k})$, *of* $\theta$, *if*

$$\mathrm{Var}(\hat{\theta}(k)) \le \mathrm{Var}(\hat{\hat{\theta}}(k)), \qquad \text{for all } k \ \square \tag{6-14}$$

Very often, it is of interest to know if $\hat{\theta}(k)$ satisfies (6-14) for *all* other unbiased estimators, $\hat{\hat{\theta}}(k)$. This can be verified by comparing the variance of $\hat{\theta}(k)$ with the smallest error variance that can ever be attained by any unbiased estimator. Theorem 6-3 provides a lower bound for $\mathbf{E}\{\tilde{\theta}^2(k)\}$ when $\theta$ is a scalar deterministic parameter. Theorem 6-4 generalizes these results to the case of a vector of deterministic parameters.

**Theorem 6-3** (Cramer–Rao inequality). *Let* $\mathbf{Z}$ *denote a set of data* [i.e., $\mathbf{Z} = \mathrm{col}\,(z_1, z_2, \ldots, z_k)$] *which is characterized by the probability density function* $\mathrm{p}(\mathbf{Z};\theta) \triangleq \mathrm{p}(\mathbf{Z})$. *If* $\hat{\theta}(\mathrm{k})$ *is an unbiased estimator of deterministic* $\theta$, *then the variance of the unbiased estimator* $\mathbf{E}\{\tilde{\theta}^2(\mathrm{k})\}$ *is bounded from below as*

$$\mathbf{E}\{\tilde{\theta}^2(k)\} \ge \frac{1}{\mathbf{E}\left\{\left[\dfrac{\partial}{\partial\theta}\ln p(\mathbf{Z})\right]^2\right\}}, \qquad \text{for all } k \tag{6-15}$$

*Two other ways for expressing (6-15) are*

$$\mathbf{E}\{\tilde{\theta}^2(k)\} \ge \frac{1}{\displaystyle\int_{-\infty}^{\infty}\left[\frac{\partial p(\mathbf{Z})}{\partial\theta}\right]^2 \frac{1}{p(\mathbf{Z})}d\mathbf{Z}}, \qquad \text{for all } k \tag{6-16}$$

*and*

$$\mathbf{E}\{\tilde{\theta}^2(k)\} \ge \frac{1}{-\mathbf{E}\left\{\dfrac{\partial^2 \ln p(\mathbf{Z})}{\partial\theta^2}\right\}}, \qquad \text{for all } k \tag{6-17}$$

*where* $\mathrm{d}\mathbf{Z}$ *is short for* $\mathrm{dz_1 dz_2 \ldots dz_k}$. $\square$

Inequalities (6-15), (6-16), and (6-17) are named after Cramer and Rao, who discovered them. They are functions of $k$ because $\mathbf{Z}$ is. The denominator of (6-15) is called *Fisher's information*, $J(\theta)$ (a short biography of R. A. Fisher appears at the end of this section). This name was originally given because the denominator of (6-15) contains the relative rate (derivative) at which the probability density function changes with respect to the data. Note that the greater the expectation of a change is at a given value, say $\hat{\theta}$, the easier it is to distinguish $\hat{\theta}$ from neighboring values $\theta$, and hence the more precisely $\theta$ can be estimated at $\theta = \hat{\theta}$.

Of course, for the results in Theorem 6-3 to be applicable, the derivatives in (6-15)–(6-17) must exist and be absolutely integrable. The reason for the latter requirement is explained in the proof of Theorem 6-3, just below (6-34).

The generalization of (6-15) to a biased estimator is given in Problem 6-4.

Before proving this theorem, it is instructive to illustrate its use by means of an example.

**EXAMPLE 6-3**

We are given $M$ statistically independent observations of a random variable $z$ that is known to have a Cauchy distribution, i.e.,

$$p(z_i) = \frac{1}{\pi[1 + (z_i - \theta)^2]} \tag{6-18}$$

Parameter $\theta$ is unknown and will be estimated using $z_1, z_2, \ldots, z_M$. We shall determine the lower bound for the error variance of *any* unbiased estimator of $\theta$ using (6-15). Observe that we are able to do this without having to specify an estimator structure for $\hat{\theta}$. Without further explanation, we calculate

$$p(\mathbf{Z}) = \prod_{i=1}^{M} p(z_i) = \frac{1}{\pi^M \prod_{i=1}^{M} [1 + (z_i - \theta)^2]} \tag{6-19}$$

$$\ln p(\mathbf{Z}) = -M \ln \pi - \sum_{i=1}^{M} \ln[1 + (z_i - \theta)^2] \tag{6-20}$$

and

$$\frac{\partial \ln p(\mathbf{Z})}{\partial \theta} = \sum_{i=1}^{M} \frac{2(z_i - \theta)}{1 + (z_i - \theta)^2} \tag{6-21}$$

so that

$$\mathbf{E}\left\{\left[\frac{\partial}{\partial \theta} \ln p(\mathbf{Z})\right]^2\right\} = \mathbf{E}\left\{\left[\sum_{i=1}^{M} \frac{2(z_i - \theta)}{1 + (z_i - \theta)^2}\right]\left[\sum_{j=1}^{M} \frac{2(z_j - \theta)}{1 + (z_j - \theta)^2}\right]\right\} \tag{6-22}$$

Next, we must evaluate the right-hand side of (6-22). This is tedious to do, but can be accomplished, as follows. Note that

$$\sum_i \sum_j = \sum_i \sum_{\substack{j \\ i \neq j}} + \sum_i \sum_{\substack{j \\ i = j}} = \mathbf{TA} + \mathbf{TB} \tag{6-23}$$

Consider **TA** first; i.e.,

$$\mathbf{TA} = \mathbf{E}\left\{\sum_{i=1}^{M} \frac{2(z_i - \theta)}{1 + (z_i - \theta)^2}\right\} \mathbf{E}\left\{\sum_{j=1}^{M} \frac{2(z_j - \theta)}{1 + (z_j - \theta)^2}\right\} \tag{6-24}$$

where we have made use of statistical independence of the measurements. Observe that

$$\mathbf{E}\left\{\frac{z_i - \theta}{1 + (z_i - \theta)^2}\right\} = \int_{-\infty}^{\infty} \frac{y}{1 + y^2} \frac{1}{\pi} \frac{1}{1 + y^2} dy = 0 \tag{6-25}$$

where $y = z_i - \theta$. The integral is zero because the integrand is an odd function of $y$. Consequently,

$$\mathbf{TA} = 0 \tag{6-26}$$

Next, consider **TB**, i.e.,

$$\mathbf{TB} = \mathbf{E}\left\{\sum_{i=1}^{M} \frac{4(z_i - \theta)^2}{[1 + (z_i - \theta)^2]^2}\right\} \tag{6-27}$$

which can also be written as

$$\mathbf{TB} = 4\sum_{i=1}^{M} \mathbf{TC} \tag{6-28}$$

where

$$\mathbf{TC} = \mathbf{E}\left\{\frac{(z_i - \theta)^2}{[1 + (z_i - \theta)^2]^2}\right\} \tag{6-29}$$

or

$$\mathbf{TC} = \int_{-\infty}^{\infty} \frac{y^2}{(1 + y^2)^2} \frac{1}{\pi} \frac{1}{(1 + y^2)} dy \tag{6-30}$$

Integrating (6-30) by parts twice, we find that

$$\mathbf{TC} = \frac{1}{8\pi} \int_{-\infty}^{\infty} \frac{dy}{1 + y^2} = \frac{1}{8} \tag{6-31}$$

because the integral in (6-31) is the area under the Cauchy probability density function, which equals $\pi$. Substituting (6-31) into (6-28), we determine that

$$\mathbf{TB} = \frac{M}{2} \tag{6-32}$$

Thus, when (6-23) is substituted into (6-22) and that result is substituted into (6-15), we find that

$$\mathbf{E}\{\tilde{\theta}^2(M)\} \geq \frac{2}{M}, \qquad \text{for all } M \tag{6-33}$$

Observe that the Cramer–Rao bound depends on the number of measurements used to estimate $\theta$. For large numbers of measurements, this bound equals zero. □

*Proof of Theorem 6-3.* Because $\hat{\theta}(k)$ is an unbiased estimator of $\theta$,

$$\mathbf{E}\{\tilde{\theta}(k)\} = \int_{-\infty}^{\infty} [\hat{\theta}(k) - \theta] p(\mathbf{Z}) d\mathbf{Z} = 0 \tag{6-34}$$

Differentiating (6-34) with respect to $\theta$, we find that [this is where we need absolute integrability, because we interchange the order of differentiation and integration (see, also, Problem 6-2)]

$$\int_{-\infty}^{\infty} [\hat{\theta}(k) - \theta] \frac{\partial p(\mathbf{Z})}{\partial \theta} d\mathbf{Z} - \int_{-\infty}^{\infty} p(\mathbf{Z}) d\mathbf{Z} = 0$$

which can be rewritten as

$$1 = \int_{-\infty}^{\infty} [\hat{\theta}(k) - \theta] \frac{\partial p(\mathbf{Z})}{\partial \theta} d\mathbf{Z} \tag{6-35}$$

As an aside, we note that

$$\frac{\partial}{\partial \theta} \ln p(\mathbf{Z}) = \frac{\partial p(\mathbf{Z})}{\partial \theta} \frac{1}{p(\mathbf{Z})} \tag{6-36}$$

so that

$$\frac{\partial p(\mathbf{Z})}{\partial \theta} = p(\mathbf{Z}) \frac{\partial}{\partial \theta} \ln p(\mathbf{Z}) \tag{6-37}$$

Substitute (6-37) into (6-35) to obtain

$$1 = \int_{-\infty}^{\infty} \left[ (\hat{\theta}(k) - \theta) \sqrt{p(\mathbf{Z})} \right] \left[ \sqrt{p(\mathbf{Z})} \frac{\partial}{\partial \theta} \ln p(\mathbf{Z}) \right] d\mathbf{Z} \tag{6-38}$$

Recall the Schwarz inequality

$$\left[ \int_{-\infty}^{\infty} a(\mathbf{Z}) b(\mathbf{Z}) d\mathbf{Z} \right]^2 \leq \left[ \int_{-\infty}^{\infty} a^2(\mathbf{Z}) d\mathbf{Z} \right] \left[ \int_{-\infty}^{\infty} b^2(\mathbf{Z}) d\mathbf{Z} \right] \tag{6-39}$$

where equality is achieved when $b(\mathbf{Z}) = ca(\mathbf{Z})$ in which $c$ is a constant that is not dependent upon $\mathbf{Z}$. Next, square both sides of (6-38) and apply (6-39) to the new right-hand side to see that

$$1 \leq \left[ \int_{-\infty}^{\infty} [\hat{\theta}(k) - \theta]^2 p(\mathbf{Z}) d\mathbf{Z} \right] \left[ \int_{-\infty}^{\infty} \left[ \frac{\partial}{\partial \theta} \ln p(\mathbf{Z}) \right]^2 p(\mathbf{Z}) d\mathbf{Z} \right]$$

or

$$1 \leq \mathbf{E}\{\tilde{\theta}^2(k)\} \mathbf{E} \left\{ \left[ \frac{\partial}{\partial \theta} \ln p(\mathbf{Z}) \right]^2 \right\} \tag{6-40}$$

Finally, to obtain (6-15), solve (6-40) for $\mathbf{E}\{\tilde{\theta}^2(k)\}$.

To obtain (6-16) from (6-15), observe that

$$\begin{aligned} \mathbf{E} \left\{ \left[ \frac{\partial}{\partial \theta} \ln p(\mathbf{Z}) \right]^2 \right\} &= \int_{-\infty}^{\infty} \left[ \frac{\partial p(\mathbf{Z})}{\partial \theta} \frac{1}{p(\mathbf{Z})} \right]^2 p(\mathbf{Z}) d\mathbf{Z} \\ &= \int_{-\infty}^{\infty} \left[ \frac{\partial p(\mathbf{Z})}{\partial \theta} \right]^2 \frac{1}{p(\mathbf{Z})} d\mathbf{Z} \end{aligned} \tag{6-41}$$

To obtain (6-41), we have also used (6-36).

To obtain (6-17), we begin with the identity

$$\int_{-\infty}^{\infty} p(\mathbf{Z}) d\mathbf{Z} = 1$$

and differentiate it twice with respect to $\theta$, using (6-37) after each differentiation, to show that

$$\int_{-\infty}^{\infty} p(\mathbf{Z}) \frac{\partial^2 \ln p(\mathbf{Z})}{\partial \theta^2} d\mathbf{Z} = -\int_{-\infty}^{\infty} \left[ \frac{\partial \ln p(\mathbf{Z})}{\partial \theta} \right]^2 p(\mathbf{Z}) d\mathbf{Z}$$

which can also be expressed as

$$\mathbf{E}\left\{ \left[ \frac{\partial}{\partial \theta} \ln p(\mathbf{Z}) \right]^2 \right\} = -\mathbf{E}\left\{ \frac{\partial^2 \ln p(\mathbf{Z})}{\partial \theta^2} \right\} \quad (6\text{-}42)$$

Substitute (6-42) into (6-15) to obtain (6-17). □

It is sometimes easier to compute the Cramer–Rao bound using one form [i.e., (6-15) or (6-16) or (6-17)] than another. The logarithmic forms are usually used when $p(\mathbf{Z})$ is exponential (e.g., Gaussian).

**Corollary 6-1.** *If the lower bound is achieved in Theorem 6-3, then*

$$\tilde{\theta}(k) = \frac{1}{c(\theta)} \frac{\partial \ln p(\mathbf{Z})}{\partial \theta} \quad (6\text{-}43)$$

*where* $c(\theta)$ *can depend on* $\theta$ *but not on* $\mathbf{Z}$.

*Proof.* In deriving the Cramer–Rao bound, we used the Schwarz inequality (6-39) for which equality is achieved when $b(\mathbf{Z}) = c(\theta)a(\mathbf{Z})$. In our case, $a(\mathbf{Z}) = [\hat{\theta}(k) - \theta]\sqrt{p(\mathbf{Z})}$ and $b(\mathbf{Z}) = \sqrt{p(\mathbf{Z})}\partial \ln p(\mathbf{Z})/\partial\theta$. Setting $b(\mathbf{Z}) = c(\theta)a(\mathbf{Z})$, we obtain (6-43). □

For another interpretation of Corollary 6-1, see Problem 6-3. We turn next to the general case of a vector of parameters.

**Definition 6-3.** *An unbiased estimator,* $\hat{\boldsymbol{\theta}}(k)$, *of vector* $\boldsymbol{\theta}$ *is said to be more efficient than any other unbiased estimator,* $\hat{\hat{\boldsymbol{\theta}}}(k)$, *of* $\boldsymbol{\theta}$, *if*

$$\mathbf{E}\{[\boldsymbol{\theta} - \hat{\boldsymbol{\theta}}(k)][\boldsymbol{\theta} - \hat{\boldsymbol{\theta}}(k)]'\} \leq \mathbf{E}\{[\boldsymbol{\theta} - \hat{\hat{\boldsymbol{\theta}}}(k)][\boldsymbol{\theta} - \hat{\hat{\boldsymbol{\theta}}}(k)]'\} \quad (6\text{-}44)$$

For a vector of parameters, we see that a more efficient estimator has the smallest error covariance among all unbiased estimators of $\boldsymbol{\theta}$, "smallest" in the sense that $\mathbf{E}\{[\boldsymbol{\theta} - \hat{\boldsymbol{\theta}}(k)][\boldsymbol{\theta} - \hat{\boldsymbol{\theta}}(k)]'\} - \mathbf{E}\{[\boldsymbol{\theta} - \hat{\hat{\boldsymbol{\theta}}}(k)][\boldsymbol{\theta} - \hat{\hat{\boldsymbol{\theta}}}(k)]'\}$ is negative semidefinite.

The generalization of the Cramer–Rao inequality to a vector of parameters is given next.

**Theorem 6-4** (Cramer–Rao inequality for a vector of parameters). *Let* $\mathbf{Z}$ *denote a set of data as in Theorem 6-3, which is characterized by the probability density function* $p(\mathbf{Z}; \boldsymbol{\theta}) \stackrel{\Delta}{=} p(\mathbf{Z})$. *If* $\hat{\boldsymbol{\theta}}(k)$ *is an unbiased estimator of deterministic* $\boldsymbol{\theta}$, *then the covariance matrix of the unbiased estimator* $\mathbf{E}\{\tilde{\boldsymbol{\theta}}(k)\tilde{\boldsymbol{\theta}}'(k)\}$ *is bounded from below as*

$$\mathbf{E}\{\tilde{\boldsymbol{\theta}}(k)\tilde{\boldsymbol{\theta}}'(k)\} \geq \mathbf{J}^{-1}, \qquad \text{for all } k \quad (6\text{-}45)$$

*where* $\mathbf{J}$ *is the* Fisher information matrix,

$$\mathbf{J} = \mathbf{E}\left\{\left[\frac{\partial}{\partial \boldsymbol{\theta}} \ln p(\mathbf{Z})\right]\left[\frac{\partial}{\partial \boldsymbol{\theta}} \ln p(\mathbf{Z})\right]'\right\} \tag{6-46}$$

*which can also be expressed as*

$$\mathbf{J} = -\mathbf{E}\left\{\frac{\partial^2}{\partial \boldsymbol{\theta}^2} \ln p(\mathbf{Z})\right\} \tag{6-47}$$

*Equality holds in (6-45) if and only if*

$$\frac{\partial}{\partial \boldsymbol{\theta}} \ln p(\mathbf{Z}) = \mathbf{C}(\boldsymbol{\theta})\tilde{\boldsymbol{\theta}}(k) \tag{6-48}$$

*where* $\mathbf{C}(\boldsymbol{\theta})$ *is a matrix that doesn't depend upon* $\mathbf{Z}$. □

Of course, for this theorem to be applicable, the vector derivatives in (6-46) and (6-47) must exist and the norm of $\partial p(\mathbf{Z})/\partial \boldsymbol{\theta}$ must be absolutely integrable. The reason for the latter requirement is explained in the proof of Theorem 6-4, just before (6-60).

A complete proof of this result is given in the Supplementary Material at the end of this lesson. Although the proof is similar to our proof of Theorem 6-3, it is a bit more intricate because of the vector nature of $\boldsymbol{\theta}$.

Inequality (6-45) demonstrates that any unbiased estimator can have a covariance no smaller than $\mathbf{J}^{-1}$. Unfortunately, $\mathbf{J}^{-1}$ is not a greatest lower bound for the error covariance. Other bounds exist that are tighter than (6-45) [e.g., the Bhattacharyya bound (Van Trees, 1968)], but they are even more difficult to compute than $\mathbf{J}^{-1}$.

**Corollary 6-2.** *Let* $\mathbf{Z}$ *denote a set of data as in Theorem 6-3 and* $\hat{\theta}_i(k)$ *be any unbiased estimator of deterministic* $\theta_i$ *based on* $\mathbf{Z}$. *Then*

$$\mathbf{E}\{\tilde{\theta}_i^2(k)\} \geq (\mathbf{J}^{-1})_{ii}, \qquad i = 1, 2, \ldots, n \text{ and all } k \tag{6-49}$$

*where* $(\mathbf{J}^{-1})_{ii}$ *is the iith element in matrix* $\mathbf{J}^{-1}$.

*Proof.* Inequality (6-45) means that $\mathbf{E}\{\tilde{\boldsymbol{\theta}}(k)\tilde{\boldsymbol{\theta}}'(k)\} - \mathbf{J}^{-1}$ is a positive semidefinite matrix, i.e.,

$$\mathbf{a}'[\mathbf{E}\{\tilde{\boldsymbol{\theta}}(k)\tilde{\boldsymbol{\theta}}'(k)\} - \mathbf{J}^{-1}]\mathbf{a} \geq 0 \tag{6-50}$$

where $\mathbf{a}$ is an arbitrary nonzero vector. Choosing $\mathbf{a} = \mathbf{e}_i$ (the $i$th unit vector), we obtain (6-49). □

Results similar to those in Theorem 6-4 and Corollary 6-2 are also available for a vector of random parameters (e.g., Sorenson, 1980, pp. 99–100). Let $p(\mathbf{Z}, \boldsymbol{\theta})$ denote the joint probability density function between $\mathbf{Z}$ and $\boldsymbol{\theta}$. The Cramer–Rao inequality for random parameters is obtained from Theorems 6-3 and 6-4 by replacing $p(\mathbf{Z})$ by $p(\mathbf{Z}, \boldsymbol{\theta})$. Of course, the expectation is now with respect to $\mathbf{Z}$ and $\boldsymbol{\theta}$.

**EXAMPLE 6-4**

Let us consider our generic linear model

$$\mathbf{Z}(k) = \mathbf{H}(k)\boldsymbol{\theta} + \mathbf{V}(k) \tag{6-51}$$

where $\mathbf{H}(k)$ and $\boldsymbol{\theta}$ are deterministic, and we now assume that $\mathbf{V}(k) \sim N(\mathbf{V}(k); \mathbf{m}_\mathbf{V}, \mathbf{R})$. If both $\mathbf{m}_\mathbf{V}$ and $\mathbf{R}$ are unknown, we can define a new parameter vector $\boldsymbol{\phi}$, where

$$\boldsymbol{\phi} = \text{col}\,(\boldsymbol{\theta}, \mathbf{m}_\mathbf{V}, \text{elements of } \mathbf{R}) \tag{6-52}$$

In this case (e.g., see Lesson 12),

$$\mathbf{Z}(k) \sim N(\mathbf{Z}(k); \mathbf{m}_\mathbf{Z}(\boldsymbol{\phi}), \mathbf{P}_\mathbf{Z}(\boldsymbol{\phi})) \tag{6-53}$$

and the computation of the Fisher information matrix, $\mathbf{J}(\boldsymbol{\phi})$, is quite complicated. Kay (1993, pp. 73–76) gives a complete derivation of the following result for $\mathbf{J}(\boldsymbol{\phi})$:

$$\begin{aligned}[\mathbf{J}(\boldsymbol{\phi})]_{ij} &= \left[\frac{\partial \mathbf{m}_\mathbf{Z}(\boldsymbol{\phi})}{\partial \phi_i}\right]' \mathbf{P}_\mathbf{Z}^{-1}(\boldsymbol{\phi}) \left[\frac{\partial \mathbf{m}_\mathbf{Z}(\boldsymbol{\phi})}{\partial \phi_j}\right] \\ &\quad + \frac{1}{2} \times \text{tr}\left[\mathbf{P}_\mathbf{Z}^{-1}(\boldsymbol{\phi}) \frac{\partial \mathbf{P}_\mathbf{Z}(\boldsymbol{\phi})}{\partial \phi_i} \mathbf{P}_\mathbf{Z}^{-1}(\boldsymbol{\phi}) \frac{\partial \mathbf{P}_\mathbf{Z}(\boldsymbol{\phi})}{\partial \phi_j}\right]\end{aligned} \tag{6-54}$$

Here we shall be less ambitious and shall assume that the statistics of $\mathbf{V}(k)$ are known, i.e., $\mathbf{V}(k) \sim N(\mathbf{V}(k); \mathbf{0}, \mathbf{R})$; hence,

$$\mathbf{Z}(k) \sim N(\mathbf{Z}(k); \mathbf{H}(k)\boldsymbol{\theta}, \mathbf{R}) \tag{6-55}$$

In this case $\boldsymbol{\phi} = \boldsymbol{\theta}$. To compute the Fisher information matrix, $\mathbf{J}(\boldsymbol{\theta})$, we could use (6-54), because its second term equals zero, and it is easy to compute $\partial \mathbf{m}_\mathbf{Z}(\boldsymbol{\phi})/\partial \phi_i$. The result is

$$\mathbf{J}(\boldsymbol{\theta}) = \mathbf{H}'(k)\mathbf{R}^{-1}\mathbf{H}(k) \tag{6-56}$$

Alternatively, we can return to (6-47) to compute $\mathbf{J}(\boldsymbol{\theta})$. To begin, we need $p(\mathbf{Z})$; but this is given in (6-55) (see Lesson 12 for much discussion of multivariate Gaussian random variables). Consequently,

$$p(\mathbf{Z}) = \frac{1}{\sqrt{(2\pi)^N |\mathbf{R}|}} \times \exp\{-\frac{1}{2} \times [\mathbf{Z}(k) - \mathbf{H}(k)\boldsymbol{\theta}]'\mathbf{R}^{-1}[\mathbf{Z}(k) - \mathbf{H}(k)\boldsymbol{\theta}]\} \tag{6-57}$$

so that

$$\frac{\partial \ln p(\mathbf{Z})}{\partial \boldsymbol{\theta}} = \mathbf{H}'(k)\mathbf{R}^{-1}[\mathbf{Z}(k) - \mathbf{H}(k)\boldsymbol{\theta}] \tag{6-58}$$

and

$$\frac{\partial^2 \ln p(\mathbf{Z})}{\partial \boldsymbol{\theta}^2} = -\mathbf{H}'(k)\mathbf{R}^{-1}\mathbf{H}(k) \tag{6-59}$$

Substituting (6-59) into (6-47), we obtain $\mathbf{J}(\boldsymbol{\theta})$ as stated in (6-56). □

### Comments

**1.** Clearly, to compute the Cramer–Rao lower bound, we need to know the probability density function $p(\mathbf{Z})$. Many times we do not know this information; hence, in those cases we cannot evaluate this bound. If the data are multivariate

Gaussian or if they are independent and identically distributed with known distribution, then we can evaluate the Cramer–Rao lower bound. Chapter 3 of Kay (1993) has many interesting signal–processing examples of the computation of the Cramer–Rao lower bound.

**2.** Of course, whether or not the Cramer–Rao lower bound is actually achieved depends on the specific estimator. The best linear unbiased estimator, described in Lesson 9, achieves the Cramer–Rao lower bound *by design* (see Theorem 9-4). Under certain conditions, a least-squares estimator is also efficient (see Theorem 8-3).

## Fisher: A Short Biography

Sir Ronald Aylmer Fisher was born in East Finchley, London, in 1890. He studied at Cambridge University in England. Throughout his life his scientific interest was professionally channeled into statistics and genetics.

When Fisher began his career, it was generally assumed that small samples are not important and that the method of moments was appropriate to the fitting of any frequency distribution, even though this method was chosen without any theoretical justification. His daughter, Joan Fisher Box, states (1978)

> Fisher suggested an alternative. In the Bayesian formula, the posterior distribution is given by multiplying together the prior probability of the parameter and the probability, for any given value of the parameter, of obtaining the actual data observed. The latter function Fisher called *likelihood.* In his first mathematical paper, published in 1912, in which he provided a mathematical proof of the formula for the Student's distribution, he was able to show that progress could be made by studying the likelihood itself without reference to the prior probability and, in particular, that the estimates obtained by maximizing the likelihood have desirable properties. The method of maximum likelihood provides a direct solution to the problem of fitting frequency curves of known form, without introducing any arbitrary assumptions.

In 1914, Fisher gave the exact distribution of the correlation function. In later years he derived other distributions, such as those for the regression coefficient, partial correlation coefficient, and multiple correlation coefficient. His 1922 paper "On the Mathematical Foundations of Theoretical Statistics" (*Phil. Trans.*, **A**, **222**: 309–368) stands as a monument in the development of mathematical statistics. In it he introduced and provided technical definitions for the terms consistency, sufficiency, efficiency, and likelihood, and, according to his daughter,

> He disentangled ... two concepts, defining on the one hand the *hypothetical infinite population* whose *parameters* we should like to know and on the other hand the *sample* whose *statistics* are our estimates of the parameter values.... It is the first time *information* enters into the technical vocabulary.... He began with the notion that there was a precise *amount of information* in a sample and the proposition that it was the job of the statistician to ensure a minimal *loss of information* in the reduction of the data.... In 1925, Fisher elaborated ("Theory of Statistical Estimation," *Proc. Camb. Phil. Soc.*, **22**: 700–725) and expanded the results obtained in 1922. He showed that, among

all efficient statistics, that derived by maximum likelihood loses less information asymptotically than any other.

Fisher was also the first, according to his daughter, "to take the apparantly risky step of assuming general error normality, and, as it turned out, was able to make tremendous advances which had very general practical applicability, because most of the results obtained on the assumption of normality were rather insensitive to the assumption itself." He also did monumental work on tests of significance and design of experiments. His works on genetics we leave to the reader to explore.

His daughter states it best, when she says

> It is difficult now to imagine the field of mathematical statistics as it existed in the first decade of the twentieth century.... The whole field was like an unexplored archeological site, its structure hardly perceptible above the accretions of rubble, its treasures scattered throughout the literature.... [Fisher] recognized the fundamental issues and set himself the task of finding solutions. In the process he defined the aims and scope of modern statistics and introduced many of the important concepts, together with much of the vocabulary we use to frame them.

In short, R. A. Fisher played a pivotal role in developing theories and methods of science during the first half of the twentieth century that affect everyone doing research today.

## Supplementary Material

### PROOF OF THEOREM 6-4

Here we provide the details to the proof of (6-45) in Theorem 6-4, which is the Cramer–Rao inequality for a vector of parameters. Our proof follows the one given in Sorenson (1980, pp. 94–95).

Following exactly the same steps that were taken to reach (6-35), we obtain [this is where we need absolute integrability of the norm of $\partial p(\mathbf{Z})/\partial \boldsymbol{\theta}$, because of an interchange of the order of differentiation and integration]

$$\mathbf{I} = \int_{-\infty}^{\infty} [\hat{\boldsymbol{\theta}}(k) - \boldsymbol{\theta}] \left[ \frac{\partial \ln p(\mathbf{Z})}{\partial \boldsymbol{\theta}} \right]' p(\mathbf{Z}) d\mathbf{Z} \tag{6-60}$$

At this point, the proof of Theorem 6-4 departs rather dramatically from the proof of Theorem 6-3. Multiply (6-60) by $\mathbf{a}'\mathbf{J}^{-1}\mathbf{a}$, where $\mathbf{a}$ is an arbitrary vector and $\mathbf{J}$ is the Fisher information matrix, which is defined in (6-47), to see that

$$\mathbf{a}'\mathbf{J}^{-1}\mathbf{a} = \int_{-\infty}^{\infty} \mathbf{a}'[\hat{\boldsymbol{\theta}}(k) - \boldsymbol{\theta}] \left[ \frac{\partial \ln p(\mathbf{Z})}{\partial \boldsymbol{\theta}} \right]' \mathbf{J}^{-1}\mathbf{a}p(\mathbf{Z}) d\mathbf{Z} \tag{6-61}$$

Squaring both sides of this equation, using the fact that the square of a scalar can be treated as the product of the scalar and its transpose, or vice versa, and applying the Schwarz inequality (6-39) to the result, we obtain

$$
\begin{aligned}
(\mathbf{a}'\mathbf{J}^{-1}\mathbf{a})^2 &\le \int_{-\infty}^{\infty} \{\mathbf{a}'[\hat{\boldsymbol{\theta}}(k) - \boldsymbol{\theta}]\}^2 p(\mathbf{Z})d\mathbf{Z} \int_{-\infty}^{\infty} \{\left[\frac{\partial \ln p(\mathbf{Z})}{\partial \boldsymbol{\theta}}\right]' \mathbf{J}^{-1}\mathbf{a}\}^2 p(\mathbf{Z})d\mathbf{Z} \\
&\le \mathbf{a}'\mathbf{E}\{\tilde{\boldsymbol{\theta}}(k)\tilde{\boldsymbol{\theta}}'(k)\}\mathbf{a}\mathbf{a}'\mathbf{J}^{-1} \int_{-\infty}^{\infty} \{\left[\frac{\partial \ln p(\mathbf{Z})}{\partial \boldsymbol{\theta}}\right]\left[\frac{\partial \ln p(\mathbf{Z})}{\partial \boldsymbol{\theta}}\right]' p(\mathbf{Z})d\mathbf{Z}\mathbf{J}^{-1}\mathbf{a} \\
&\le \mathbf{a}'\mathbf{E}\{\tilde{\boldsymbol{\theta}}(k)\tilde{\boldsymbol{\theta}}'(k)\}\mathbf{a}\mathbf{a}'\mathbf{J}^{-1}\mathbf{J}\mathbf{J}^{-1}\mathbf{a} \\
&\le \mathbf{a}'\mathbf{E}\{\tilde{\boldsymbol{\theta}}(k)\tilde{\boldsymbol{\theta}}'(k)\}\mathbf{a}(\mathbf{a}'\mathbf{J}^{-1}\mathbf{a}) \qquad (6\text{-}62)
\end{aligned}
$$

Hence,

$$
\mathbf{E}\{\tilde{\boldsymbol{\theta}}(k)\tilde{\boldsymbol{\theta}}'(k)\} \ge \mathbf{J}^{-1}, \qquad \text{for all } k \qquad (6\text{-}63)
$$

Note that the alternative expression for $\mathbf{J}$, which is given in (6-47), is derived along the same lines as the derivation of (6-42). Finally, note that (6-48) follows from the condition of equality in the Schwarz inequality.

## SUMMARY QUESTIONS

**1.** If an estimator possesses a large-sample property:
- **(a)** it never possesses the corresponding small-sample property
- **(b)** it sometimes possesses the corresponding small-sample property
- **(c)** it always possesses the corresponding small-sample property

**2.** A "thorough" study into $\hat{\boldsymbol{\theta}}$ means determining its:
- **(a)** probability density function
- **(b)** mean vector
- **(c)** covariance matrix

**3.** If $\mathbf{E}\{\hat{\boldsymbol{\theta}}(k)\} = \boldsymbol{\theta}$ for all $k$, then $\hat{\boldsymbol{\theta}}(k)$ is said to be:
- **(a)** tracking $\boldsymbol{\theta}$
- **(b)** efficient
- **(c)** unbiased

**4.** *Efficiency* is a small-sample property of an estimator that involves:
- **(a)** first-order statistics
- **(b)** second-order statistics
- **(c)** third-order statistics

**5.** To evaluate the *Cramer–Rao bound*, we need to know the:
- **(a)** first- and second-order statistics of the measurements
- **(b)** probability density function of a measurement
- **(c)** joint probability density function of the measurements

**6.** The Cramer–Rao inequality bounds the estimation variance:
- **(a)** from below
- **(b)** from above
- **(c)** in two directions

**7.** If $\hat{\boldsymbol{\theta}}(k)$ is a linear transformation of the measurements, then a sufficient condition for $\hat{\boldsymbol{\theta}}(k)$ to be an unbiased estimator of $\boldsymbol{\theta}$ is:
- **(a)** $\mathbf{H}(k)\mathbf{F}(k) = \mathbf{I}$

**(b)** $\mathbf{F}(k)\mathbf{H}(k) = \mathbf{I}$
**(c)** $\mathbf{F}'(k)\mathbf{H}(k) = \mathbf{I}$

**8.** The proof of the Cramer–Rao inequality uses which inequality?
**(a)** triangle
**(b)** Schwartz
**(c)** Cauchy

**9.** The Cramer–Rao inequality:
**(a)** depends on an estimator structure
**(b)** provides a bound that can never be computed
**(c)** does not depend on an estimator structure

## PROBLEMS

**6-1.** Prove Theorem 6-2, which provides an unbiasedness constraint for the two design matrices that appear in a linear *recursive* estimator. During your proof you will reach the following statement: $\mathbf{h}'(k+1)[\boldsymbol{\theta} - \mathbf{E}\{\hat{\boldsymbol{\theta}}(k^*)\}] = 0$ for all $k$. Be sure to explain why this can only be zero if $\boldsymbol{\theta} - \mathbf{E}\{\hat{\boldsymbol{\theta}}(k^*)\} = \mathbf{0}$. (*Hint:* Return to the derivation of the WLSE in Lesson 3.)

**6-2.** The condition $\mathbf{E}\{\partial \ln p(\mathbf{Z})/\partial\theta\} = 0$ is known as a *regularity condition* (Kay, 1993). Prove that the regularity condition is satisfied if the order of differentiation with respect to $\theta$ and integration with respect to $\mathbf{Z}$ of $p(\mathbf{Z})$ can be interchanged.

**6-3.** Show that (6-43) in Corollary 6-1 can be expressed as $\partial \ln p(\mathbf{Z})/\partial\theta = J(\theta)\tilde{\theta}(k)$, where $J(\theta)$ is *Fisher's information*, which is given by the denominator of (6-15).

**6-4.** Suppose $\hat{\theta}(k)$ is a biased estimator of deterministic $\theta$, with bias $B(\theta)$.
**(a)** Show that

$$\mathbf{E}\{\tilde{\theta}^2(k)\} \geq \frac{\left[1 + \dfrac{\partial B(\theta)}{\partial\theta}\right]^2}{\mathbf{E}\left\{\left[\dfrac{\partial}{\partial\theta}\ln p(\mathbf{Z})\right]^2\right\}}, \qquad \text{for all } k$$

**(b)** (Sam Heidari, Fall 1991) Show that if $-2 < \partial B(\theta)/\partial\theta < 0$ then $\hat{\theta}(k)$ may be a more efficient estimator than an unbiased estimator.

**6-5.** Here we consider the Cramer–Rao inequality and the Fisher information matrix when it is desired to estimate a function of an unknown deterministic parameter or parameters.
**(a)** Let $\theta$ be the basic parameter, and $\phi = f(\theta)$ be a scalar differentiable function of $\theta$. Show that (Kay, 1993) if $\mathbf{E}\{\hat{\phi}\} = \phi = f(\theta)$ then

$$\mathbf{E}\{\tilde{\phi}^2\} \geq -\frac{(\partial f(\theta)/\partial\theta)^2}{\mathbf{E}\{\partial^2 \ln p(\mathbf{Z})/\partial\theta^2\}}$$

**(b)** Let $\boldsymbol{\theta}$ be the basic parameter vector, and $\boldsymbol{\phi} = \mathbf{f}(\boldsymbol{\theta})$ be a vector differentiable function of $\boldsymbol{\theta}$. Show that (Kay, 1993) if $\mathbf{E}\{\hat{\boldsymbol{\phi}}\} = \boldsymbol{\phi} = \mathbf{f}(\boldsymbol{\theta})$ then

$$\mathbf{E}\{\tilde{\boldsymbol{\phi}}\tilde{\boldsymbol{\phi}}'\} \geq \frac{\partial\mathbf{f}(\boldsymbol{\theta})}{\partial\boldsymbol{\theta}}\mathbf{J}^{-1}(\boldsymbol{\theta})\left[\frac{\partial\mathbf{f}(\boldsymbol{\theta})}{\partial\boldsymbol{\theta}}\right]'$$

where $\partial\mathbf{f}(\boldsymbol{\theta})/\partial\boldsymbol{\theta}$ is the Jacobian matrix of $\mathbf{f}(\boldsymbol{\theta})$ with respect to $\boldsymbol{\theta}$.

**(c)** Suppose $\phi = \sin\theta$, $z(k) = h(k)\theta + \nu(k)(k = 1, 2, \ldots, N)$, $h(k)$ is deterministic, and white $\nu(k) \sim N(\nu(k); 0, r)$. Determine the Cramer–Rao lower bound on $\phi$.

**6-6.** (G. Caso, Fall 1991)

**(a)** Show that, if the measurement vector $\mathbf{Z}$ is a discrete-valued vector of random variables [i.e., it is described by a probability mass function $P(\mathbf{Z})$], then the Cramer–Rao inequality becomes

$$\mathbf{E}\{\tilde{\theta}^2(k)\} \geq \frac{1}{\mathbf{E}\left[\frac{\partial \ln P(\mathbf{Z})}{\partial \theta}\right]^2}$$

*Hint:* If $f(\mathbf{x})$ and $g(\mathbf{x})$ are scalar-valued functions, then

$$\left(\sum_{\mathbf{x}\in\mathbf{X}} f(\mathbf{x})g(\mathbf{x})\right)^2 \leq \left(\sum_{\mathbf{x}\in\mathbf{X}} f^2(\mathbf{x})\right)\left(\sum_{\mathbf{x}\in\mathbf{X}} g^2(\mathbf{x})\right)$$

**(b)** Using the result obtained in part (a), determine the Cramer–Rao bound for the unbiased estimate of $\lambda$ from a set of independent measurements, $z$, where

$$P(z = k) = \frac{\lambda^k e^{-\lambda}}{k!}, \qquad k = 0, 1, \ldots$$

**6-7.** We are given a random sample $\{x_1, x_2, \ldots, x_N\}$. Consider the following estimator for $\mu$,

$$\hat{\mu}(N) = \frac{1}{N + a}\sum_{i=1}^{N} x_i$$

where $a \geq 0$. For what value(s) of $a$ is $\hat{\mu}(N)$ an unbiased estimator of $\mu$?

**6-8.** Suppose $z_1, z_2, \ldots, z_N$ are random samples from an arbitrary distribution with unknown mean, $\mu$, and variance, $\sigma^2$. Reasonable estimators of $\mu$ and $\sigma^2$ are the sample mean and sample variance,

$$\bar{z} = \frac{1}{N}\sum_{i=1}^{N} z_i$$

and

$$s^2 = \frac{1}{N}\sum_{i=1}^{N}(z_i - \bar{z})^2$$

Is $s^2$ an unbiased estimator of $\sigma^2$? [*Hint:* Show that $\mathbf{E}\{s^2\} = (N - 1)\sigma^2/N$.]

**6-9.** (Mendel, 1973, first part of Exercise 2-9, p. 137). Show that if $\hat{\theta}$ is an unbiased estimate of $\theta$, $a\hat{\theta} + b$ is an unbiased estimate of $a\theta + b$.

**6-10.** Suppose that $N$ independent observations $(x_1, x_2, \ldots, x_N)$ are made of a random variable $X$ that is Gaussian, i.e.,

$$p(x_i|\mu, \sigma^2) = \frac{1}{\sqrt{2\pi}\,\sigma}\exp\left[\frac{-(x_i - \mu)^2}{2\sigma^2}\right]$$

In this problem only $\mu$ is unknown. Derive the Cramer–Rao lower bound of $\mathbf{E}\{\tilde{\mu}^2(N)\}$ for an unbiased estimator of $\mu$.

**6-11.** Repeat Problem 6-10, but in this case assume that only $\sigma^2$ is unknown; i.e., derive the Cramer–Rao lower bound of $\mathbf{E}\{[\widetilde{\sigma^2}(N)]^2\}$ for an unbiased estimator of $\sigma^2$.

**6-12.** Repeat Problem 6-10, but in this case assume both $\mu$ and $\sigma^2$ are unknown; i.e., compute $\mathbf{J}^{-1}$ when $\boldsymbol{\theta} = \text{col}\,(\mu, \sigma^2)$.

**6-13.** (Andrew D. Norte, Fall 1991) An exponentially distributed random variable $x$ is defined over the sample space of a particular chance experiment. We perform $N$ independent trials of the chance experiment and observe $N$ independent values of the random variable $x$. The probability density function of $x$ is

$$p(x) = \alpha e^{-\alpha x}, \qquad x > 0 \quad \text{and} \quad p(x) = 0, \qquad x \leq 0$$

Use the Cramer–Rao bound to determine the lower bound for $\mathbf{E}\{\tilde{\alpha}^2\}$ for any unbiased estimator of the parameter $\alpha$.

**6-14.** (Geogiang Yue and Sungook Kim, Fall 1991) We are given a random sample sequence $X_1, X_2, \ldots, X_n$.

**(a)** Find the constraint on the $a_i$ such that the $\hat{\mu}(n)$ given next is an unbiased estimator of the population mean $\mu$:

$$\hat{\mu}(n) = \sum_{i=1}^{n} a_i X_i$$

**(b)** Determine the $a_i$ such that $\hat{\mu}(n)$ is the most efficient (minimum variance) linear estimator of the population mean $\mu$.

**6-15.** (James Leight, Fall 1991) In radar detection, input measurements are typically Rayleigh distributed when a target is present. The Rayleigh density function is

$$p(x) = \frac{x}{\alpha^2} \exp\left(\frac{-x^2}{2\alpha^2}\right), \qquad x \geq 0$$

Assume that we are given $N$ statistically independent measurements, $X_1, X_2, \ldots, X_N$, and we are given an unbiased estimator for $\alpha$. What is the Cramer–Rao bound for this estimator?

**6-16.** In this problem you will be asked to show that *a biased estimator may be more "efficient" than an unbiased estimator* (Stoica and Moses, 1990). Let $\{y(t)\}_{t=1}^{N}$ be a sequence of i.i.d. zero-mean Gaussian random variables with finite variance $\sigma^2$. Consider the following estimator of $\sigma^2$:

$$\hat{\sigma}^2 = \frac{\alpha}{N} \sum_{t=1}^{N} y^2(t), \qquad \text{where } \alpha > 0$$

Unless $\alpha = 1$, this estimator of $\sigma^2$ is biased.

**(a)** Prove that $\text{MSE}(\hat{\sigma}^2) = \sigma^4[\alpha^2(1 + 2/N) + (1 - 2\alpha)]$ (*Hint:* $\mathbf{E}\{y^4(t)\} = 3\sigma^4$.)

**(b)** Prove that $\min_\alpha \text{MSE}(\hat{\sigma}^2) = 2\sigma^4/(N + 2)$.

**(c)** Derive the Cramer–Rao lower bound of any unbiased estimator of $\sigma^2$.

**(d)** Explain why the preceding italicized statement is true.

**(e)** What is the significance of this result to you?

**6-17.** A very popular model that occurs in signal-processing and communication applications is $z(k) = s(k; \theta) + n(k)$, $k = 1, 2, \ldots, N$, where $s(k; \theta)$ is a signal that depends on deterministic parameter $\theta$ and $n(k)$ is Gaussian white noise with variance $\sigma_n^2$. Show (Kay, 1993) that the Cramer–Rao lower bound is

$$\frac{\sigma_n^2}{\displaystyle\sum_{k=0}^{N} \left[\frac{\partial s(k; \theta)}{\partial \theta}\right]^2}$$

# LESSON 7

# Large-sample Properties of Estimators*

## SUMMARY

Large-sample properties of an estimator mitigate the effect of the criterion function that is used to derive the estimator. The purpose of this lesson is to introduce four widely used large-sample properties of estimators, *asymptotic unbiasedness, consistency, asymptotic normality*, and *asymptotic efficiency*. The phrase *large sample* means an infinite number of measurements, although, in practice, estimators enjoy large-sample properties for much fewer than an infinite number of measurements.

Recall, from Lesson 6, that if an estimator enjoys a small-sample property then it automatically enjoys the associated large-sample property; however, the converse is not always true. Recall, also, that for many estimators it is relatively easy to study their large-sample properties and virtually impossible to learn about their small-sample properties because it is virtually impossible to write down their probability density functions.

Asymptotic unbiasedness and efficiency are limiting forms of their small-sample counterpart unbiasedness and efficiency. The importance of an estimator being asymptotically normal (Gaussian) is that its entire probabilistic description is then known, and it can be entirely characterized just by its asymptotic first- and second-order statistics. Consistency is a form of convergence of $\hat{\theta}(k)$ to $\theta$; it is synonymous with convergence in probability. One reason for the importance of consistency in estimation theory is that any continuous function of a consistent estimator is itself a consistent estimator, i.e., "consistency carries over."

* This lesson was written using many suggestions and inputs from Dr. Georgios B. Giannakis, Department of Electrical Engineering, University of Virginia, Charlottesville, VA.

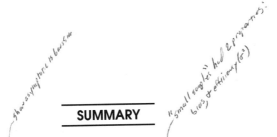

When you complete this lesson you will be able to (1) define and explain the four large sample properties of estimators known as *asymptotic unbiasedness, consistency* (an important mode of stochastic convergence), *asymptotic normality*, and *asymptotic efficiency*, and (2) test different estimators for these large-sample properties.

## INTRODUCTION

To begin, we reiterate the fact that "if an estimator possesses a small-sample property, it also possesses the associated large-sample property; but the converse is not always true." In this lesson we shall examine the following large-sample properties of estimators: asymptotic unbiasedness, consistency, asymptotic normality, and asymptotic efficiency. The first and fourth properties are natural extensions of the small-sample properties of unbiasedness and efficiency in the limiting situation of an infinite number of measurements. The second property is about convergence of $\hat{\theta}(k)$ to $\theta$. The third property is about the asymptotic distribution of $\hat{\theta}(k)$.

Note that asymptotic properties mitigate the effect of the criterion function that is used to derive an estimator, e.g., if a least-squares estimator, a maximum-likelihood estimator, and a maximum a posteriori estimator are all asymptotically normal, then we can compare them, even though they have been designed based on different criteria. As another example, suppose that we have designed estimators based on minimizing the mean-squared error, the least-squares error, and the fourth-order moment of the estimation error. It is questionable as to whether we should compare their small-sample properties (apples versus oranges), but, if all these estimators are asymptotically normal, we can certainly compare them in a meaningful way in terms of their asymptotic relative efficiencies.

Of course, we use large-sample properties on the premise that they will approximate the estimator's behavior for sufficiently large but finite sample sizes.

The downside of conclusions based on large samples is that in most cases it is difficult to estimate well the accuracy of the approximation, because it is difficult to determine how many measurements are actually needed to achieve a large-sample property. This is where Monte Carlo simulations with increasing sample sizes come in very handy.

## STOCHASTIC CONVERGENCE

Large-sample properties of estimators deal with a sequence of random variables, $\hat{\theta}(1), \hat{\theta}(2), \ldots, \hat{\theta}(k), \hat{\theta}(k+1), \ldots$, which necessitates studies of stochastic convergence; hence, we need to understand in what sense sequences of random variables converge. Because the study of stochastic convergence is covered in a course on random processes, it should be familiar to most readers of this book; hence, we only summarize and contrast the different types of stochastic convergence in this section.

The concepts of convergence of a stochastic sequence may be concerned with the convergence of individual sample paths (i.e., the results of a probability experiment), or the convergence of the probabilities of some sequence of events determined by the entire ensemble, or, most frequently, both. Four popular forms of stochastic convergence are defined next.

**Convergence in Distribution** (Rohatgi, 1976)

Let $\{F_k\}$ be a sequence of probability distribution functions. If there exists a distribution function $F$ such that

$$\lim_{k \to \infty} F_k(X) = F(X) \tag{7-1}$$

at every point $X$ at which $F$ is continuous, we say that $F_k$ converges in distribution to $F$. If $\{X_k\}$ is a sequence of random variables and $\{F_k\}$ is the corresponding sequence of distribution functions, we say that $X_k$ converges in distribution to $X$ if there exists a random variable $X$ with distribution function $F$ such that $F_k$ converges in distribution to $F$. We write $X_k \to^d X$.

**Convergence in Probability** (Rohatgi, 1976)

Let $\{X_k\}$ be a scalar sequence of random variables defined on some probability space. Sequence $\{X_k\}$ converges in probability to the random variable $X$ if, for every $\epsilon > 0$,

$$\lim_{k \to \infty} P\{|X_k - X| > \epsilon\} = 0 \tag{7-2}$$

We write $X_k \to^P X$ or $\operatorname{plim} X_k = X$. Vector sequences of random variables converge in probability to a constant vector if and only if the corresponding component-wise convergences occur (Serfling, 1980).

**Convergence with Probability 1** (Rohatgi, 1976)

Let $\{X_k\}$ be a scalar sequence of random variables defined on some probability space $(\mathbf{\Omega}, \mathcal{F}, \mathbf{P})$. We say that $X_k$ converges with probability 1 (or *strongly*, *almost surely*, *almost everywhere*) to a random variable $X$ if and only if

$$P\{\omega \in \mathbf{\Omega} : \lim_{k \to \infty} X_k(\omega) = X(\omega)\} = 1 \tag{7-3}$$

and we write $X_k \to^{\text{WP1}} X$ or p1-lim $X_k = X$.

Note that $X_k \to^{\text{WP1}} X$ if and only if $\lim_{k \to \infty} P\{\sup_{m \geq k} |X_m - X| > \epsilon\} = 0$ for all $\epsilon > 0$.

**Convergence in rth Mean** (Rohatgi, 1976)

Let $\{X_k\}$ be a scalar sequence of random variables defined on some probability space such that $\mathbf{E}\{|X_k|^r\} < \infty$, for some $r > 0$. We say that $X_k$ converges in the $r$th mean to a random variable $X$ if

$$\lim_{k \to \infty} \mathbf{E}\{|X_k - X|^r\} = 0 \tag{7-4}$$

and we write $X_k \to^r X$.

### Relationships among the Different Types of Convergence

Many interesting relationships exist among the different types of stochastic convergence. Serfling (1980), Rohatgi (1976), and Grimmett and Stirzaker (1982) are excellent references for such results. Here we briefly state some of the more well-known relationships. Many other interesting relationships can be found in these references.

1. If $X_k \rightarrow^{\text{WP1}} X$, then $X_k \rightarrow^{P} X$. Convergence in probability is sometimes called *weak consistency*, whereas convergence with probability 1 is called *strong consistency*. Obviously, if we can prove strong consistency, then we have also proved weak consistency.
2. If $X_k \rightarrow^{r} X$, then $X_k \rightarrow^{P} X$. The most important case of this for us is when $r = 2$. We then say that mean-squared convergence implies convergence in probability. This is discussed more fully in this lesson's section entitled "Consistency." Convergence in a mean-squared sense is sometimes referred to as *consistency in m.s.s.*
3. Convergence in probability implies convergence in distribution. Relationships 1 and 2 therefore imply that convergence with probability 1, or in $r$th mean, also implies convergence in distribution. Consequently, we can show (e.g., Arnold, 1990, p. 240) that (a) if $X_k \rightarrow^{d} X$ and $g(x)$ is continuous for all $x$ in the range of $X$, then $g(X_k) \rightarrow^{d} g(X)$, and (b) if $X_k \rightarrow^{d} X$, $Y_k \rightarrow^{P} b$, and $g(x, y)$ is continuous at $(x, b)$ for all $x$ in the range of $X$, then $g(X_k, Y_k) \rightarrow^{d} g(X, b)$. Multivariate versions of these two results are also true (e.g., Arnold, 1990, p. 251).
4. Convergence in probability "sufficiently fast" implies convergence with probability 1; i.e., if

$$\sum_{k=1}^{\infty} P(|X_k - X| > \epsilon) < \infty, \qquad \text{for every } \epsilon > 0 \tag{7-5}$$

then $X_k \rightarrow^{\text{WP1}} X$.

## ASYMPTOTIC UNBIASEDNESS

**Definition 7-1.** *Estimator* $\hat{\theta}(k)$ *is an asymptotically unbiased estimator of deterministic* $\theta$ *if*

$$\lim_{k\to\infty} \mathbf{E}\{\hat{\theta}(k)\} = \theta \tag{7-6}$$

*or of random* $\theta$ *if*

$$\lim_{k\to\infty} \mathbf{E}\{\hat{\theta}(k)\} = \mathbf{E}\{\theta\} \;\square \tag{7-7}$$

Figure 7-1 depicts an example in which the asymptotic mean of $\hat{\theta}$ has converged to $\theta$ (note that $p_{200}$ is centered about $m_{200} = \theta$). Clearly, $\hat{\theta}$ is biased but is asymptotically unbiased.

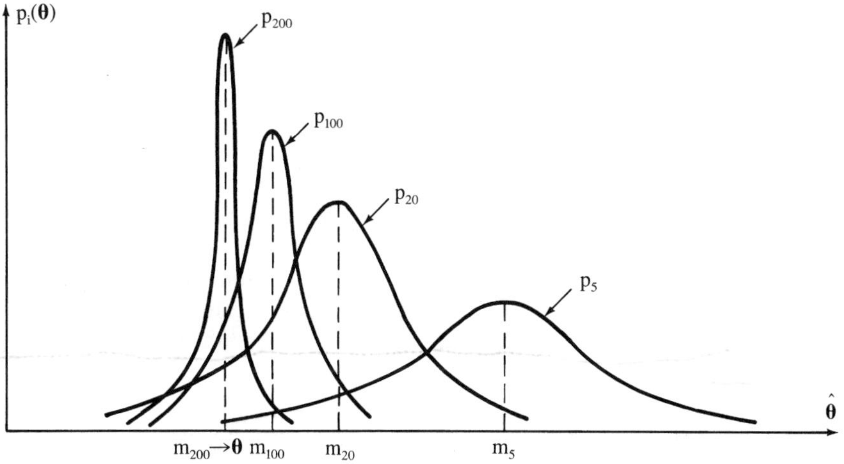


**Figure 7-1** Probability density function for estimate of scalar $\theta$ as a function of number of measurements; e.g., $p_{20}$ is the p.d.f. for 20 measurements.

**EXAMPLE 7-1**

Recall our linear model $\mathbf{Z}(k) = \mathbf{H}(k)\boldsymbol{\theta} + \mathbf{V}(k)$ in which $\mathbf{E}\{\mathbf{V}(k)\} = \mathbf{0}$. Let us assume that each component of $\mathbf{V}(k)$ is uncorrelated and has the same variance $\sigma_v^2$. In Lesson 8, we determine an unbiased estimator for $\sigma_v^2$. Here, on the other hand, we just assume that

$$\hat{\hat{\sigma}}_v^2(k) \triangleq \frac{\tilde{\mathbf{Z}}'(k)\tilde{\mathbf{Z}}(k)}{k} \tag{7-8}$$

where $\tilde{\mathbf{Z}}(k) = \mathbf{Z}(k) - \mathbf{H}(k)\hat{\boldsymbol{\theta}}_{LS}(k)$. We leave it to the reader to show that (see Lesson 8)

$$\mathbf{E}\{\hat{\hat{\sigma}}_v^2(k)\} = \frac{k-n}{k}\sigma_v^2 \tag{7-9}$$

Observe that $\hat{\hat{\sigma}}_v^2(k)$ is not an unbiased estimator of $\sigma_v^2$; but it is an asymptotically unbiased estimator, because $\lim_{k\to\infty} [(k-n)/k]\sigma_v^2 = \sigma_v^2$. □

## CONSISTENCY

We now direct our attention to the issue of stochastic convergence. In principle, we would like to establish the strongest possible convergence results for an estimator, i.e., convergence with probability 1 or mean-squared convergence. In practice, this is often extremely difficult or impossible to do; hence, we frequently focus on convergence in probability because, as we shall explain in more detail later, it is relatively easy to do.

**Definition 7-2.** *The probability limit of* $\hat{\theta}$(k) *is the point* $\theta^*$ *on which the distribution of our estimator collapses (for discussions on asymptotic distributions, see the section in this lesson entitled "Asymptotic Distributions"). We abbreviate "probability limit of* $\hat{\theta}$(k)*" by* plim $\hat{\theta}$(k). *Mathematically speaking,*

$$\text{plim}\,\hat{\theta}(k) = \theta^* \leftrightarrow \lim_{k\to\infty} P\{|\hat{\theta}(k) - \theta^*| > \epsilon\} = 0 \qquad (7\text{-}10)$$

*for every* $\epsilon > 0$. □

**Definition 7-3.** $\hat{\theta}$(k) *is a consistent estimator of* $\theta$ *if*

$$\text{plim}\,\hat{\theta}(k) = \theta \quad \square \qquad (7\text{-}11)$$

Note that "consistency" means the same thing as "convergence in probability." For an estimator to be consistent, its probability limit $\theta^*$ must equal its true value $\theta$. Note, also, that a consistent estimator need not be unbiased or asymptotically unbiased.

A reason for focusing attention on consistency over other modes of stochastic convergence is that plim $(\cdot)$ can be treated as an operator. For example, suppose $X_k$ and $Y_k$ are two random sequences, for which plim $X_k = X$ and plim $Y_k = Y$; then (see Tucker, 1962, and Rohatgi, 1976, for simple proofs of these and other facts)

$$\text{plim}\, X_k Y_k = (\text{plim}\, X_k)(\text{plim}\, Y_k) = XY \qquad (7\text{-}12)$$

and

$$\text{plim}\left(\frac{X_k}{Y_k}\right) = \frac{\text{plim}\, X_k}{\text{plim}\, Y_k} = \frac{X}{Y} \qquad (7\text{-}13)$$

Additionally, suppose $\mathbf{A}_k$ and $\mathbf{B}_k$ are two commensurate matrix sequences, for which plim $\mathbf{A}_k = \mathbf{A}$ and plim $\mathbf{B}_k = \mathbf{B}$ (note that plim $\mathbf{A}_k$, for example, denotes a matrix of probability limits, each element of which is a probability limit); then (Goldberger, 1964)

$$\text{plim}\, \mathbf{A}_k\mathbf{B}_k = (\text{plim}\, \mathbf{A}_k)(\text{plim}\, \mathbf{B}_k) = \mathbf{AB} \qquad (7\text{-}14)$$

$$\text{plim}\, \mathbf{A}_k^{-1} = (\text{plim}\, \mathbf{A}_k)^{-1} = \mathbf{A}^{-1} \qquad (7\text{-}15)$$

as long as the requisite inverses exist, and

$$\text{plim}\, \mathbf{A}_k^{-1}\mathbf{B}_k = \mathbf{A}^{-1}\mathbf{B} \qquad (7\text{-}16)$$

The treatment of plim $(\cdot)$ as an operator often makes the study of consistency quite easy. We shall demonstrate the truth of this in Lesson 8 when we examine the consistency of the least-squares estimator.

A second reason for the importance of consistency is the property that "consistency carries over"; i.e., *any continuous function of a consistent estimator is itself a consistent estimator* [see Tucker, 1967, for a proof of this property, which relies heavily on the preceding treatment of plim $(\cdot)$ as an operator].

**EXAMPLE 7-2**

Suppose $\hat{\theta}$ is a consistent estimator of $\theta$. Then $1/\hat{\theta}$ is a consistent estimator of $1/\theta$, $(\hat{\theta})^2$ is a consistent estimator of $\theta^2$, and $\ln\hat{\theta}$ is a consistent estimator of $\ln\theta$. These facts are all due to the consistency carry-over property. □

The reader may be scratching his or her head at this point and wondering about the emphasis placed on these illustrative examples. Isn't, for example, using $\hat{\theta}$ to estimate $\theta^2$ by $(\hat{\theta})^2$ the "natural" thing to do? The answer is "Yes, but only if we know ahead of time that $\hat{\theta}$ is a consistent estimator of $\theta$." If you do not know this to be true, then there is no guarantee that $\widehat{\theta^2} = (\hat{\theta})^2$. In Lesson 11 we show that maximum-likelihood estimators are consistent; thus $\widehat{(\theta^2)}_{ML} = (\hat{\theta}_{ML})^2$. We mention this property about maximum-likelihood estimators here, because we must know whether or not an estimator is consistent before applying the consistency carry-over property. Not all estimators are consistent!

Finally, this carry-over property for consistency does not necessarily apply to other properties. For example, if $\hat{\boldsymbol{\theta}}(k)$ is an unbiased estimator of $\boldsymbol{\theta}$, then $\mathbf{A}\hat{\boldsymbol{\theta}}(k) + \mathbf{b}$ will be an unbiased estimator of $\mathbf{A}\boldsymbol{\theta} + \mathbf{b}$; but $\hat{\theta}^2(k)$ will not be an unbiased estimator of $\theta^2$.

How do you determine whether an estimator is consistent? Often, the direct approach, which makes heavy use of plim $(\cdot)$ operator algebra, is possible. Sometimes, an indirect approach is used, one that examines whether both the bias in $\hat{\theta}(k)$ and variance of $\hat{\theta}(k)$ approach zero as $k \to \infty$. To understand the validity of this indirect approach, we digress to provide more details about mean-squared convergence and its relationship to convergence in probability.

**Definition 7-4.** *According to (7-4),* $\hat{\theta}(k)$ *converges to* $\theta$ *in a mean-squared sense, if*

$$\lim_{k\to\infty} \mathbf{E}\{[\hat{\theta}(k) - \theta]^2\} = 0 \quad \square \tag{7-17}$$

**Theorem 7-1.** *If* $\hat{\theta}(k)$ *converges to* $\theta$ *in mean square, then it converges to* $\theta$ *in probability.*

*Proof* (Papoulis, 1991, p. 114). Recall the Tchebycheff inequality

$$P[|x - a| \geq \epsilon] \leq \frac{\mathbf{E}\{|x - a|^2\}}{\epsilon^2} \tag{7-18}$$

Let $a = 0$ and $x = \hat{\theta}(k) - \theta$ in (7-18), and take the limit as $k \to \infty$ on both sides of (7-18) to see that

$$\lim_{k\to\infty} P[|\hat{\theta}(k) - \theta| \geq \epsilon] \leq \lim_{k\to\infty} \frac{\mathbf{E}\{[\hat{\theta}(k) - \theta]^2\}}{\epsilon^2} \tag{7-19}$$

Using the fact that $\hat{\theta}(k)$ converges to $\theta$ in mean square, we see that

$$\lim_{k\to\infty} P[|\hat{\theta}(k) - \theta| \geq \epsilon] = 0 \tag{7-20}$$

Thus, $\hat{\theta}(k)$ converges to $\theta$ in probability. □

Recall, from probability theory, that although mean-squared convergence implies convergence in probability the converse is not true.

**EXAMPLE 7-3**   (Kmenta, 1971, p. 166)

Let $\hat{\theta}(k)$ be an estimator of $\theta$, and let the probability density function of $\hat{\theta}(k)$ be

| $\hat{\theta}(k)$ | $p(\hat{\theta}(k))$ |
|---|---|
| $\theta$ | $1 - \frac{1}{k}$ |
| $k$ | $\frac{1}{k}$ |

$= \sum \hat{\theta}_{(k)} p(\hat{\theta})$

$E(\hat{\theta}) = \theta(1 - \frac{1}{k}) + k(\frac{1}{k}) = \theta + 1 - \frac{\theta}{k}$

In this example $\hat{\theta}(k)$ can only assume two different values, $\theta$ and $k$. Obviously, $\hat{\theta}(k)$ is consistent, because as $k \to \infty$ the probability that $\theta(k)$ equals $\theta$ approaches unity; i.e., $\text{plim}\,\hat{\theta}(k) = \theta$. Observe, also, that $\mathbf{E}\{\hat{\theta}(k)\} = 1 + \theta(1 - 1/k)$, which means that $\hat{\theta}(k)$ is biased and even asymptotically biased.

Now let us investigate the mean-squared error between $\hat{\theta}(k)$ and $\theta$; i.e.,

$$\lim_{k \to \infty} \mathbf{E}\{[\hat{\theta}(k) - \theta]^2\} = \lim_{k \to \infty}\left[ (\theta - \theta)^2\left(1 - \frac{1}{k}\right) + (k - \theta)^2 \frac{1}{k} \right]$$

$$= \lim_{k \to \infty} k \to \infty$$

In this pathological example, the mean-squared error is diverging to infinity; but $\hat{\theta}(k)$ converges to $\theta$ in probability. $\square$ *while* $p(\theta)_{k=\infty} = 1$

*Sufficient condition but not necessary.*

**Theorem 7-2.**   *Let $\hat{\theta}(k)$ denote an estimator of $\theta$. If bias $\hat{\theta}(k)$ and variance $\hat{\theta}(k)$ both approach zero as $k \to \infty$, then the mean-squared error between $\hat{\theta}(k)$ and $\theta$ approaches zero, and therefore $\hat{\theta}(k)$ is a consistent estimator of $\theta$.*

*Proof.*  From elementary probability theory, we know that

$$\mathbf{E}\{[\hat{\theta}(k) - \theta]^2\} = [\text{bias}\,\hat{\theta}(k)]^2 + \text{variance } \hat{\theta}(k) \qquad (7\text{-}21)$$

If, as assumed, bias $\hat{\theta}(k)$ and variance $\hat{\theta}(k)$ both approach zero as $k \to \infty$, then

$$\lim_{k \to \infty} \mathbf{E}\{[\hat{\theta}(k) - \theta]^2\} = 0 \qquad (7\text{-}22)$$

which means that $\hat{\theta}(k)$ converges to $\theta$ in mean square. Thus, by Theorem 7-1, $\hat{\theta}(k)$ also converges to $\theta$ in probability. $\square$

The importance of Theorem 7-2 is that it provides a constructive way to test for consistency.

**EXAMPLE 7-4**

Here we show that the sample mean is a consistent estimator of the population mean. Suppose $z_1, z_2, \ldots, z_N$ are a random sample from a population with mean $\mu$ and variance $\sigma^2$. The sample mean, denoted $\bar{z}$, is

$$\bar{z} = \frac{1}{N} \sum_{i=1}^{N} z_i \qquad (7\text{-}23)$$

*If bias $\to 0$ to $\sigma^2 \to 0$ then wh MSE $\to 0$ of $\hat{\theta}(k)$ is consistent*

Clearly, $\mathbf{E}\{\bar{z}\} = \mu$, which means that $\bar{z}$ has zero bias, and

$$\text{var}\,(\bar{z}) = \text{var}\,(\frac{1}{N}\sum_{i=1}^{N} z_i) = \frac{1}{N^2}\text{var}\left(\sum_{i=1}^{N} z_i\right) = \frac{1}{N^2}\sum_{i=1}^{N}\text{var}(z_i) = \frac{1}{N^2}(N\sigma^2) = \frac{\sigma^2}{N} \quad (7\text{-}24)$$

which means that var $(\bar{z})$ approaches zero as $N \to \infty$. Applying Theorem 7-2 to the sample mean, we see that $\bar{z}$ converges to $\mu$ in mean square; hence, the sample mean is a consistent estimator of the population mean. □

• We could ask, "If we have to prove mean-squared convergence in order to establish convergence in probability, then why not focus at the very beginning on mean-squared convergence instead of convergence in probability?" The answer is that a *general* carry-over property does not exist for mean-squared convergence (or for convergence with probability 1), whereas it does exist for convergence in probability; hence, if we can prove (either directly or indirectly) that $\hat{\theta}$ converges to $\theta$ in probability, then we know, by the consistency carry-over property, that $f(\hat{\theta})$ converges in probability to $f(\theta)$ for very general functions $f(\cdot)$. It may also be true that $f(\hat{\theta})$ converges in mean square or with probability 1 to $f(\theta)$; but *this must be verified for each specific* $f(\cdot)$. See Ljung (1987) for many interesting and important convergences with probability 1 results.

## ASYMPTOTIC DISTRIBUTIONS

According to Kmenta (1971, pp. 162–163),

> if the distribution of an estimator tends to become more and more similar in form to some specific distribution as the sample size increases, then such a specific distribution is called the *asymptotic distribution* of the estimator in question.... What is meant by the asymptotic distribution is not the ultimate form of the distribution, which may be degenerate, but the form that the distribution tends to put on in the last part of its journey to the final collapse (if this occurs).

Consider the situation depicted in Figure 7-1, where $p_i(\hat{\theta})$ denotes the probability density function associated with estimator $\hat{\theta}$ of the scalar parameter $\theta$, based on $i$ measurements. As the number of measurements increases, $p_i(\hat{\theta})$ changes its shape (although, in this example, each density function is Gaussian). The density function eventually centers itself about the true parameter value $\theta$, and the variance associated with $p_i(\hat{\theta})$ tends to get smaller as $i$ increases. Ultimately, the variance will become so small that in all probability $\hat{\theta} = \theta$. The asymptotic distribution refers to $p_i(\hat{\theta})$ as it evolves from $i = 1, 2, \ldots$, etc., especially for large values of $i$.

The preceding example illustrates one of the three possible cases that can occur for an asymptotic distribution, the case when an estimator has a distribution of the same form regardless of the sample size, and this form is known (e.g., Gaussian). Some estimators have a distribution that, although not necessarily always of the same form, is also known for every sample size. For example, $p_5(\hat{\theta})$ may be uniform, $p_{20}(\hat{\theta})$ may be Rayleigh, and $p_{200}(\hat{\theta})$ may be Gaussian. Finally, for some estimators the distribution is not necessarily known for every sample size, but is known only for $k \to \infty$.

Asymptotic distributions, like other distributions, are characterized by their moments. We are especially interested in their first two moments, the asymptotic mean and variance.

**Definition 7-5.** *The asymptotic mean is equal to the asymptotic expectation,* $\lim_{k\to\infty} \mathbf{E}\{\hat{\theta}(k)\}$. □

As noted in Goldberger (1964, p. 116), if $\mathbf{E}\{\hat{\theta}(k)\} = m$ for all $k$, then $\lim_{k\to\infty} \mathbf{E}\{\hat{\theta}(k)\} = \lim_{k\to\infty} m = m$. On the other hand, suppose that

$$\mathbf{E}\{\hat{\theta}(k)\} = m + k^{-1}c_1 + k^{-2}c_2 + \cdots \tag{7-25}$$

where the $c$'s are finite constants; then

$$\lim_{k\to\infty} \mathbf{E}\{\hat{\theta}(k)\} = \lim_{k\to\infty} \{m + k^{-1}c_1 + k^{-2}c_2 + \cdots\} = m \tag{7-26}$$

Thus if $\mathbf{E}\{\hat{\theta}(k)\}$ is expressible as a power series in $k^0, k^{-1}, k^{-2}, \ldots$, the asymptotic mean of $\hat{\theta}(k)$ is the leading term of this power series; as $k \to \infty$, the terms of "higher order of smallness" in $k$ vanish.

**Definition 7-6.** *The asymptotic variance, which is short for "variance of the asymptotic distribution" is not equal to* $\lim_{k\to\infty} \text{var}\,[\hat{\theta}(k)]$ *(which is sometimes called the limiting variance). It is defined as*

$$\text{asymptotic var}\,[\hat{\theta}(k)] = \frac{1}{k} \lim_{k\to\infty} \mathbf{E}\left\{k[\hat{\theta}(k) - \lim_{k\to\infty} \mathbf{E}\{\hat{\theta}(k)\}]^2\right\} \quad \square \tag{7-27}$$

Kmenta (1971, p. 164) states

> The asymptotic variance... is *not* equal to $\lim_{k\to\infty} \text{var}(\hat{\theta})$. The reason is that in the case of estimators whose variance decreases with an increase in $k$, the variance will approach zero as $k \to \infty$. This will happen when the distribution collapses on a point. But, as we explained, the asymptotic distribution is not the same as the collapsed (degenerate) distribution, and its variance is *not* zero.

Goldberger (1964, p. 116) notes that if $\mathbf{E}\{[\hat{\theta}(k) - \lim_{k\to\infty} \mathbf{E}\{\hat{\theta}(k)\}]^2\} = v/k$ for *all* values of $k$, then asymptotic var $[\hat{\theta}(k)] = 1/k \lim_{k\to\infty} v = v/k$. On the other hand, suppose that

$$\mathbf{E}\{[\hat{\theta}(k) - \lim_{k\to\infty} \mathbf{E}\{\hat{\theta}(k)\}]^2\} = k^{-1}v + k^{-2}c_2 + k^{-3}c_3 + \cdots \tag{7-28}$$

where the $c$'s are finite constants; then

$$\text{asymptotic var}\,[\hat{\theta}(k)] = \frac{1}{k} \lim_{k \to \infty} (v + k^{-1}c_2 + k^{-2}c_3 + \cdots)$$

$$= \frac{v}{k} \quad \left(\text{This} = f(k) \neq 0\right) \atop \text{eg. 'on the way' renormalizable.}$$

(7-29)

Thus, if the variance of each $\hat{\theta}(k)$ is expressible as a power series in $k^{-1}, k^{-2}, \ldots$, the asymptotic variance of $\hat{\theta}(k)$ is the leading term of this power series; as $k$ goes to infinity the terms of "higher order of smallness" in $k$ vanish. Observe that if (7-29) is true then the asymptotic variance of $\hat{\theta}(k)$ decreases as $k \to \infty \ldots$, which corresponds to the situation depicted in Figure 7-1. In both of the cases just discussed, the "limiting variance" equals zero.

Extensions of our definitions of asymptotic mean and variance to sequences of random *vectors* [e.g., $\hat{\theta}(k), k = 1, 2, \ldots$] are straightforward and can be found in Goldberger (1964, p. 117), where he states, "The asymptotic expectation of a (random) vector is simply the vector of asymptotic expectations of the random variables that are the elements of the vector," and "the asymptotic covariance matrix of a random vector is simply the matrix of asymptotic variances and covariances of the random variables that are the elements of the vector."

## ASYMPTOTIC NORMALITY

Central limit theorems play an important role in determining if $\hat{\theta}(k)$ is asymptotically normal. Knowing that $\hat{\theta}(k)$ is asymptotically normal means that we can completely specify its probability density function by its asymptotic mean vector and covariance matrix, quantities that are usually easy to determine (or estimate from the data). If $\hat{\theta}(k)$ is to be used in decision-making procedures such as classification, then the asymptotic normality of $\hat{\theta}(k)$ lets us develop optimal procedures. Maximum-likelihood estimates are, as we shall see in Lesson 11, asymptotically normal. This is one reason for the popularity of such estimates; i.e., once you know that an estimate is maximum likelihood, you do not have to reestablish its asymptotic distribution. The same cannot be said for least-squares estimates.

### Central Limit Theorems

Larson and Shubert (1979, p. 357) state that "Under a wide range of circumstances the distribution of sums of random variables (appropriately normalized) converges to a Gaussian distribution as the number of terms in the sum increases without limit." This is known as the:

**Central Limit Theorem.** *Let* $X_1, \ldots, X_k$ *be i.i.d. random variables with mean* $\mu$ *and finite variance* $\sigma^2$. *Let* $Z_k = k^{1/2}(\overline{X}_k - \mu)/\sigma$, *where* $\overline{X}_k = 1/k \times \sum_{i=1}^{k} X_i$. *Then* $Z_k \to^d Z \sim N(Z; 0, 1)$. $\square$

A multivariate version of the Central Limit Theorem is:

**Multivariate Central Limit Theorem.** *Let* $\mathbf{X}_1, \ldots, \mathbf{X}_k$ *be a sequence of i.i.d.* p*-dimensional random vectors with mean vector* $\boldsymbol{\mu}$ *and finite covariance matrix* $\boldsymbol{\Sigma}$*. Let* $\mathbf{Y}_k = k^{1/2}(\overline{\mathbf{X}}_k - \boldsymbol{\mu})$*, where* $\overline{\mathbf{X}}_k = 1/k \times \sum_{i=1}^{k} \mathbf{X}_i$*. Then* $\mathbf{Y}_k \rightarrow^d \mathbf{Y} \sim N(\mathbf{Y}; \mathbf{0}, \boldsymbol{\Sigma})$. □

The central limit theorem still is true under a variety of weaker conditions than those just stated. Here are some other cases for which it is true when the random variables $X_1, X_2, \ldots, X_k, \ldots$ are independent (Larson and Shubert, 1979, pp. 359–360):

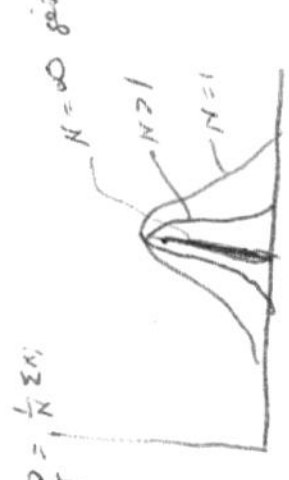

> The random variables $X_1, X_2, \ldots, X_k, \ldots$ have the same mean and variance but are not identically distributed.
>
> The random variables $X_1, X_2, \ldots, X_k, \ldots$ have different means and the same variance and are not identically distributed.
>
> The random variables $X_1, X_2, \ldots, X_k, \ldots$ have variances $\sigma_1^2, \sigma_2^2, \ldots, \sigma_k^2, \ldots$. If there exist positive constants $\epsilon$, $M$ such that $\epsilon < \sigma_i^2 < M$ for all $i$, then the distribution of the standardized sum converges to the standard Gaussian; this says in particular that the variances must exist and be neither too large nor too small.

If $Z$ is a Gaussian random variable with $\mu = 0$ and $\sigma^2 = 1$, then $Z$ is called the *standard Gaussian random variable*. Its density function is $p(z) = 1/\sqrt{2\pi} \times \exp(-z^2/2)$.

Other central limit theorems also allow the possibility that the $X_i$ are not even independent.

## Asymptotic Normality

It is generally agreed, as stated in Serfling (1980, p. 20), that "The most important special case of convergence in distribution consists of convergence to a normal distribution. A sequence of random variables converges in distribution to $N(\mu, \sigma^2), \sigma > 0$, if, equivalently, the sequence $\{(X_k - \mu)/\sigma\}$ converges in distribution to $N(0, 1)$." The latter can be verified by using a theorem due to Slutsky stated on page 19 of Serfling.

Serfling states further that

> More generally, a sequence of random variables $\{X_k\}$ is *asymptotically normal* with "mean" $\mu_k$ and "variance" $\sigma_k^2$ if $\sigma_k > 0$ for all $k$ sufficiently large and
>
> $$(X_k - \mu_k)/\sigma_k \rightarrow^d N(0, 1)$$
>
> We write "$X_k$ is $AN(\mu_k, \sigma_k^2)$." Here $\{\mu_k\}$ and $\{\sigma_k\}$ are sequences of constants. It is not necessary that $\mu_k$ and $\sigma_k^2$ be the mean and variance of $X_k$, nor even that $X_k$ possess such moments. Note that if $X_k$ is $AN(\mu_k, \sigma_k^2)$, it does not necessarily follow

that $\{X_k\}$ converges in distribution to anything. It is $(X_k - \mu_k)/\sigma_k$ that is converging to something.

For random vectors, Serfling (1980, p. 21) states

A sequence of random *vectors* $\{\mathbf{X}_k\}$ is asymptotically (multivariate) normal with "mean vector" $\boldsymbol{\mu}_k$ and "covariance matrix" $\boldsymbol{\Sigma}_k$ if $\boldsymbol{\Sigma}_k$ has nonzero diagonal elements for all $k$ sufficiently large, and for every vector $\boldsymbol{\lambda}$ such that $\boldsymbol{\lambda}'\boldsymbol{\Sigma}_k\boldsymbol{\lambda} > 0$ for all $k$ sufficiently large, the sequence $\boldsymbol{\lambda}'\mathbf{X}_k$ is $AN(\boldsymbol{\lambda}'\boldsymbol{\mu}_k, \boldsymbol{\lambda}'\boldsymbol{\Sigma}_k\boldsymbol{\lambda})$. We write "$\mathbf{X}_k$ is $AN(\boldsymbol{\mu}_k, \boldsymbol{\Sigma}_k)$." Here $\{\boldsymbol{\mu}_k\}$ is a sequence of vector constants and $\{\boldsymbol{\Sigma}_k\}$ a sequence of covariance matrix constants. . . . [Note that] $\mathbf{X}_k$ is $AN(\boldsymbol{\mu}_k, c_k^2\boldsymbol{\Sigma}_k)$ if and only if

$$(\mathbf{X}_k - \boldsymbol{\mu}_k)/c_k \to^d N(\mathbf{0}, \boldsymbol{\Sigma})$$

Here $\{c_k\}$ is a sequence of real constants and $\boldsymbol{\Sigma}$ a covariance matrix.

**Definition 7-7** (Fomby et al., 1984, p. 54). *An estimator* $\hat{\theta}(k)$ *is defined to be* consistent uniformly asymptotically normal *when* (i) *it is consistent,* (ii) $k^{1/2}[\hat{\theta}(k) - \theta]$ *converges in distribution to* $N(\mathbf{0}, \boldsymbol{\Sigma})$, *and* (iii) *if the convergence is uniform over any compact subspace of the parameter set.* □

## ASYMPTOTIC EFFICIENCY

**Definition 7-8** (Fomby et al., 1984, p. 54). *Let* $\hat{\theta}(k)$ *and* $\hat{\hat{\theta}}(k)$ *be consistent uniformly asymptotically normal estimators with asymptotic covariance matrices* $\boldsymbol{\Sigma}/k$ *and* $\boldsymbol{\Omega}/k$, *respectively. Then* $\hat{\theta}(k)$ *is said to be* asymptotically efficient relative *to* $\hat{\hat{\theta}}(k)$ *if the matrix* $\boldsymbol{\Omega} - \boldsymbol{\Sigma}$ *is positive semidefinite.* □

**Definition 7-9** (Fomby et al., 1984, p. 54). *A consistent uniformly asymptotically normal estimator is said to be* asymptotically efficient *if it is asymptotically efficient relative to any other consistent uniformly asymptotically normal estimator.* □

Once an asymptotically efficient estimator is obtained, many estimators can be obtained that are asymptotically efficient; hence, asymptotically efficient estimators are not unique, and they need not even be asymptotically unbiased.

**Definition 7-10** (Fomby et al., 1984, p. 55). *A sufficient condition for a consistent uniformly asymptotically normal estimator to be asymptotically efficient is that its asymptotic covariance matrix equals the lower bound*

$$\frac{1}{k} \times \lim_{k\to\infty} k\mathbf{J}^{-1}(\theta)$$

*where* $\mathbf{J}(\theta)$ *is the Fisher information matrix given in (6-46) or (6-48).* □

The lower bound in Definition 7-10 is frequently refrerred to as the *asymptotic Cramer–Rao lower bound.*

Under mild conditions, maximum-likelihood estimators can be shown to be asymptotically efficient (see Lesson 11), which is yet another reason for their popularity. Consequently (Fomby et al., 1984, p. 59), there are two ways to show that a consistent uniformly asymptotically normal estimator is asymptotically efficient: (1) prove that the estimator in question has the same asymptotic distribution as the maximum-likelihood estimator, or (2) show that the variance of the estimator attains the asymptotic Cramer–Rao lower bound.

## SUMMARY QUESTIONS

**1.** If $\hat{\theta}(k)$ converges to $\theta$ in a mean-squared sense, then:
**(a)** it always converges to $\theta$ in probability
**(b)** it sometimes converges to $\theta$ in probability
**(c)** it never converges to $\theta$ in probability

**2.** If $\hat{\theta}(k)$ is a consistent estimator of $\theta$, then the probability limit of $\hat{\theta}(k)$:
**(a)** sometimes equals $\theta$
**(b)** is proportional to $\theta$
**(c)** always equals $\theta$

**3.** If $\text{plim}\, X_k = X$ and $\text{plim}\, Y_k = Y$, then $\text{plim}\, X_k^2 Y_k^{-3}$ equals:
**(a)** $(XY)^{-1}$
**(b)** $X^2 Y^{-3}$
**(c)** $(XY)^{2/3}$

**4.** Any continuous function of a consistent estimator is:
**(a)** itself a consistent estimator
**(b)** unbiased
**(c)** efficient

**5.** If the bias and variance of $\hat{\theta}(k)$ both approach zero as $k \to \infty$, then $\hat{\theta}(k)$:
**(a)** is efficient
**(b)** converges to $\theta$ in probability
**(c)** converges to $\theta$ in distribution

**6.** An asymptotic distribution is the:
**(a)** ultimate form of the distribution
**(b)** initial distribution
**(c)** form that the distribution tends to put on in the last part of its journey to the final collapse

**7.** The importance of knowing that $\hat{\theta}(k)$ is asymptotically Gaussian is, we can:
**(a)** then assert that $\hat{\theta}(k)$ is asymptotically efficient
**(b)** completely specify its probability density function by its asymptotic mean and covariance matrix, quantities that are usually easy to determine
**(c)** then assert that $\hat{\theta}(k)$ is unbiased

**8.** The condition $\text{plim}\, [\mathbf{H}'(k)\mathbf{V}(k)/k] = \mathbf{0}$ means that:
**(a)** $\mathbf{H}'(k)\mathbf{V}(k)/k$ converges to zero in probability
**(b)** $\mathbf{H}'(k)$ and $\mathbf{V}(k)$ are asymptotically orthogonal
**(c)** $\mathbf{H}(k)$ and $\mathbf{V}(k)$ are asymptotically uncorrelated

# PROBLEMS

**7-1.** We are given a random sample $\{X_1, X_2, \ldots, X_N\}$. Consider the following estimator for $\mu$,

$$\hat{\mu}(N) = \frac{1}{N+a}\sum_{i=1}^{N} X_i$$

where $a \geq 0$.

**(a)** For what value(s) of $a$ is $\hat{\mu}(N)$ an asymptotically unbiased estimator of $\mu$?

**(b)** Prove that $\hat{\mu}(N)$ is a consistent estimator of $\mu$ for all $a \geq 0$.

**(c)** Compare the results obtained in part **(a)** with those obtained in Problem 6-7.

**7-2.** Suppose $Z_1, Z_2, \ldots, Z_N$ are random samples from an arbitrary distribution with unknown mean, $\mu$, and variance, $\sigma^2$. Reasonable estimators of $\mu$ and $\sigma^2$ are the sample mean and sample variance

$$\bar{z} = \frac{1}{N}\sum_{i=1}^{N} Z_i$$

and

$$s^2 = \frac{1}{N}\sum_{i=1}^{N}(Z_i - \bar{z})^2$$

**(a)** Is $s^2$ an asymptotically unbiased estimator of $\sigma^2$? [*Hint:* Show that $\mathbf{E}\{s^2\} = (N-1)\sigma^2/N$].

**(b)** Compare the result in part **(a)** with that from Problem 6-8.

**(c)** We can show that the variance of $s^2$ is

$$\text{var}\,(s^2) = \frac{\mu_4 - \sigma^4}{N} - \frac{2(\mu_4 - 2\sigma^4)}{N^2} + \frac{\mu_4 - 3\sigma^4}{N^3}$$

Explain whether or not $s^2$ is a consistent estimator of $\sigma^2$.

**7-3.** (Todd B. Leach, Spring 1992) You are given $n$ independently identically distributed random variables $X_i$. Each random variable is sampled and the samples are arranged in ascending order such that $X_1 \leq X_2 \leq \cdots \leq X_{n-1} \leq X_n$. Suppose further that each random variable is uniformly distributed between 0 and $a$. Two estimators are defined for parameter $a$:

$$\hat{a}_1 = \max(X_i) \quad \text{and} \quad \hat{a}_2 = \min(X_i)$$

Are $\hat{a}_1$ and $\hat{a}_2$ unbiased and/or consistent estimators of $a$? [*Hint:* The random variables $Y_k$ corresponding to the $k$th ordered values in the sequence $X_1 \leq X_2 \leq \cdots \leq X_{n-1} \leq X_n$ are called the *order statistics* of the random variables $X_k$. See Papoulis (1991, pp. 185–186).]

**7-4.** (Patrick Lippert, Spring 1992) Let $Z$ be a random variable with mean $\alpha$. Given $N$ independent observations $\{Z(1), Z(2), \ldots, Z(N)\}$, we use the following estimator of the mean:

$$\hat{\alpha}(N) = \frac{\ln a}{N}\sum_{i=1}^{N} Z(i), \qquad \text{where } a \text{ is a known constant}$$

**(a)** Find the asymptotic mean of $\hat{\alpha}(N)$.

**(b)** Find the asymptotic variance of $\hat{\alpha}(N)$.

**(c)** For what value of $a$ is $\hat{\alpha}(N)$ a consistent estimator of $\alpha$?

[*Note:* $\hat{\alpha}(N)$ is a maximum-likelihood estimator of $\alpha$ when $p[z(i)] = (1/\alpha)a^{-z(i)/\alpha}$. See Lesson 11 for maximum-likelihood estimation.]

**7-5.** (Brad Verona, Spring 1992) You are given a random sample of measurements, $Z(1), Z(2), \ldots, Z(N)$, of random variable $Z$, which has mean $\mu$ and variance $\sigma^2$. It is suggested that a "good" estimator for $\mu$ is given by

$$\hat{\mu}(N) = \frac{a}{2N+1}\sum_{i=1}^{N} Z(i) + \frac{b}{N^2}\sum_{i=1}^{N} Z(i)$$

where $a > 0$ and $b > 0$ are scalar design parameters.

**(a)** For what value(s) of $a$ and $b$, if any, is this estimator asymptotically unbiased?

**(b)** Given the parameter values $a = b = 2$, is this estimator consistent?

**7-6.** Random variable $X \sim N(x; \mu, \sigma^2)$. Consider the following estimator of the population mean obtained from a random sample of $N$ observations of $X$,

$$\hat{\mu}(N) = \bar{x} + \frac{a}{N}$$

where $a$ is a finite constant, and $\bar{x}$ is the sample mean.

**(a)** What are the asymptotic mean and variance of $\hat{\mu}(N)$?

**(b)** Is $\hat{\mu}(N)$ a consistent estimator of $\mu$?

**(c)** Is $\hat{\mu}(N)$ asymptotically efficient?

**7-7.** Let $X$ be a Gaussian variable with mean $\mu$ and variance $\sigma^2$. Consider the problem of estimating $\mu$ from a random sample of observations $X_1, X_2, \ldots, X_N$. Three estimators are proposed:

$$\hat{\mu}_1(N) = \frac{1}{N}\sum_{i=1}^{N} X_i$$

$$\hat{\mu}_2(N) = \frac{1}{N+1}\sum_{i=1}^{N} X_i$$

$$\hat{\mu}_3(N) = \frac{1}{2}x_1 + \frac{1}{2N}\sum_{i=2}^{N} X_i$$

You are to study unbiasedness, efficiency, asymptotic unbiasedness, consistency, and asymptotic efficiency for these estimators. Show the analysis that allows you to complete the following table.

| | Estimator | | |
|---|---|---|---|
| Properties | $\hat{\mu}_1$ | $\hat{\mu}_2$ | $\hat{\mu}_3$ |
| Small sample | | | |
| Unbiasedness | | | |
| Efficiency | | | |
| Large sample | | | |
| Unbiasedness | | | |
| Consistency | | | |
| Efficiency | | | |

Entries to this table are *yes* or *no*.

**7-8.** (Chanil Chung, Spring 1992) $X$ is a Gaussian random variable, i.e., $X \sim N(x; \mu, \sigma^2)$. Two estimators of $\mu$ are

$$\hat{\mu}_1(N) = \frac{1}{N}\sum_{i=1}^{N} X_i \qquad \text{and} \qquad \hat{\mu}_2(N) = \frac{\alpha}{N-\alpha}\left(\sum_{i=2}^{N} X_i\right) - \frac{\alpha}{N}X_1$$

Estimator $\hat{\mu}_1(N)$ has been analyzed in Example 7-4 and has been shown to be a consistent estimator of $\mu$.

**(a)** For what value of $\alpha$ is $\hat{\mu}_2(N)$ asymptotically unbiased?

**(b)** Assuming $\hat{\mu}_2(N)$ is asymptotically unbiased, is $\hat{\mu}_2(N)$ consistent?

**(c)** Is $\hat{\mu}_2(N)$ asymptotically efficient?

**7-9.** (Michiel van Nieustadt, Spring 1992) Consider the exponentially distributed random variables $X_i$, with probability density function

$$p(X) = ae^{-a(x-b)}, \qquad \text{where } X \in (b, \infty)$$

$X$ can be interpreted as the lifetime of an object that is being used from time $b$ on; $b$ can be viewed as the time of birth. We have available samples $\{X_i, i = 1, 2, \ldots, N\}$ and are concerned with estimation of the parameter $b$.

**(a)** Show that $\min(X_i)$ is an asymptotically unbiased estimator of $b$.

**(b)** Show that the estimator $\min(X_i)$ is consistent.

**(c)** Show that the estimator $\min(X_i)$ is *not* asymptotically efficient.

**7-10.** If $\operatorname{plim} X_k = X$ and $\operatorname{plim} Y_k = Y$, where $X$ and $Y$ are constants, then $\operatorname{plim}(X_k + Y_k) = X + Y$ and $\operatorname{plim} cX_k = cX$, where $c$ is a constant. Prove that $\operatorname{plim} X_k Y_k = XY$ {*Hint:* $X_k Y_k = \frac{1}{4}[(X_k + Y_k)^2 - (X_k - Y_k)^2]$}.

LESSON 8

# Properties of Least-squares Estimators

## SUMMARY

The purpose of this lesson is to apply results from Lessons 6 and 7 to least-squares estimators. First, small-sample properties are studied; then consistency is studied.

Generally, it is very difficult to establish small- or large-sample properties for least-squares estimators, except in the very special case when $\mathbf{H}(k)$ and $\mathbf{V}(k)$ are statistically independent. While this condition is satisfied in the application of identifying an impulse response, it is violated in the important application of identifying the coefficients in a finite-difference equation.

We also point out that many large-sample properties of LSEs are determined by establishing that the LSE is equivalent to another estimator for which it is known that the large-sample property holds true.

In Lesson 3 we noted that "Least-squares estimators require no assumptions about the nature of the generic model. Consequently, the formula for the LSE is easy to derive." The price paid for not making assumptions about the nature of the generic linear model is great difficulty in establishing small- or large-sample properties of the resulting estimator.

When you complete this lesson you will be able to (1) explain the limitations of (weighted) least-squares estimators, so you will know when they can and cannot be reliably used, and (2) describe why you must be very careful when using (weighted) least-squares estimators.

## INTRODUCTION

In this lesson we study some small- and large-sample properties of least-squares estimators. Recall that in least squares we estimate the $n \times 1$ parameter vector $\boldsymbol{\theta}$ of the linear model $\mathbf{Z}(k) = \mathbf{H}(k)\boldsymbol{\theta} + \mathbf{V}(k)$. We will see that most of the results in this lesson require $\mathbf{H}(k)$ to be deterministic or $\mathbf{H}(k)$ and $\mathbf{V}(k)$ to be statistically independent. In some applications one or the other of these requirements is met; however, there are many important applications where neither is met.

## SMALL-SAMPLE PROPERTIES OF LEAST-SQUARES ESTIMATORS

In this section (parts of which are taken from Mendel, 1973, pp. 75–86) we examine the bias and variance of weighted least squares and least-squares estimators.

To begin, we recall Example 6-2, in which we showed that, when $\mathbf{H}(k)$ *is deterministic, the WLSE of* $\boldsymbol{\theta}$ *is unbiased.* We also showed [after the statement of Theorem 6-2 and Equation (6-13)] that our recursive WLSE of $\boldsymbol{\theta}$ has the requisite structure of an unbiased estimator, but that unbiasedness of the recursive WLSE of $\boldsymbol{\theta}$ also requires $\mathbf{h}(k+1)$ to be deterministic.

When $\mathbf{H}(k)$ is random, we have the following important result:

**Theorem 8-1.** *The WLSE of* $\boldsymbol{\theta}$,

$$\hat{\boldsymbol{\theta}}_{WLS}(k) = [\mathbf{H}'(k)\mathbf{W}(k)\mathbf{H}(k)]^{-1}\mathbf{H}'(k)\mathbf{W}(k)\mathbf{Z}(k) \qquad (8\text{-}1)$$

*is unbiased if* $\mathbf{V}(k)$ *is zero mean and if* $\mathbf{V}(k)$ *and* $\mathbf{H}(k)$ *are statistically independent.*

Note that this is the first place (except for Example 6-2) where, in connection with least squares, we have had to assume any a priori knowledge about noise $\mathbf{V}(k)$.

*Proof.* From (8-1) and $\mathbf{Z}(k) = \mathbf{H}(k)\boldsymbol{\theta} + \mathbf{V}(k)$, we find that

$$\begin{aligned} \hat{\boldsymbol{\theta}}_{WLS}(k) &= (\mathbf{H}'\mathbf{W}\mathbf{H})^{-1}\mathbf{H}'\mathbf{W}(\mathbf{H}\boldsymbol{\theta} + \mathbf{V}) \\ &= \boldsymbol{\theta} + (\mathbf{H}'\mathbf{W}\mathbf{H})^{-1}\mathbf{H}'\mathbf{W}\mathbf{V}, \qquad \text{for all } k \end{aligned} \qquad (8\text{-}2)$$

where, for notational simplification, we have omitted the functional dependences of $\mathbf{H}$, $\mathbf{W}$, and $\mathbf{V}$ on $k$. Taking the expectation on both sides of (8-2), it follows that

$$\mathbf{E}\{\hat{\boldsymbol{\theta}}_{WLS}(k)\} = \boldsymbol{\theta} + \mathbf{E}\{(\mathbf{H}'\mathbf{W}\mathbf{H})^{-1}\mathbf{H}'\}\mathbf{W}\mathbf{E}\{\mathbf{V}\}, \qquad \text{for all } k \qquad (8\text{-}3)$$

In deriving (8-3) we have used the fact that $\mathbf{H}(k)$ and $\mathbf{V}(k)$ are statistically independent [recall that if two random variables, $a$ and $b$, are statistically independent $p(a,b) = p(a)p(b)$; thus, $\mathbf{E}\{ab\} = \mathbf{E}\{a\}\mathbf{E}\{b\}$ and $\mathbf{E}\{g(a)h(b)\} = \mathbf{E}\{g(a)\}\mathbf{E}\{h(b)\}$]. The second term in (8-3) is zero, because $\mathbf{E}\{\mathbf{V}\} = \mathbf{0}$, and therefore

$$\mathbf{E}\{\hat{\boldsymbol{\theta}}_{WLS}(k)\} = \boldsymbol{\theta}, \qquad \text{for all } k \ \square \qquad (8\text{-}4)$$

This theorem only states *sufficient conditions* for unbiasedness of $\hat{\boldsymbol{\theta}}_{WLS}(k)$, which means that, if we do not satisfy these conditions, we *cannot* conclude anything about whether $\hat{\theta}_{WLS}(k)$ is unbiased or biased. To obtain *necessary conditions* for unbiasedness, assume that $\mathbf{E}\{\hat{\theta}_{WLS}(k)\} = \theta$ and take the expectation on both sides of (8-2). Doing this, we see that

$$\mathbf{E}\{(\mathbf{H}'\mathbf{W}\mathbf{H})^{-1}\mathbf{H}'\mathbf{W}\mathbf{V}\} = \mathbf{0} \tag{8-5}$$

Letting $\mathbf{M} = (\mathbf{H}'\mathbf{W}\mathbf{H})^{-1}\mathbf{H}'\mathbf{W}$ and $\mathbf{m}_i'$ denote the $i$th row of matrix $\mathbf{M}$, (8-5) can be expressed as the following collection of *orthogonality conditions*:

$$\mathbf{E}\{\mathbf{m}_i'\mathbf{V}\} = 0, \qquad \text{for } i = 1, 2, \ldots, N \tag{8-6}$$

Orthogonality [recall that two random variables $a$ and $b$ are orthogonal if $\mathbf{E}\{ab\} = 0$] is a weaker condition than statistical independence, but is often more difficult to verify ahead of time than independence, especially since $\mathbf{m}_i'$ is a very nonlinear transformation of the random elements of $\mathbf{H}$.

**EXAMPLE 8-1**

Recall the impulse response identification Example 2-1, in which $\theta = \text{col}[h(1), h(2), \ldots, h(n)]$, where $h(i)$ is the value of the sampled impulse response at time $t_i$. System input $u(k)$ may be deterministic or random.

If $\{u(k), k = 0, 1, \ldots, N-1\}$ is deterministic, then $\mathbf{H}(N-1)$ [see Equation (2-5)] is deterministic, so that $\hat{\theta}_{WLS}(N)$ is an unbiased estimator of $\theta$. Often, we use a random input sequence for $\{u(k), k = 0, 1, \ldots, N-1\}$, such as from a random number generator. This random sequence is in no way related to the measurement noise process, which means that $\mathbf{H}(N-1)$ and $\mathbf{V}(N)$ are statistically independent, and again $\hat{\theta}_{WLS}(N)$ will be an unbiased estimate of the impulse response coefficients. We conclude, therefore, that *WLSEs of impulse response coefficients are unbiased.* □

**EXAMPLE 8-2**

As a further illustration of an application of Theorem 8-1, let us take a look at the weighted least-squares estimates of the $n$ $\alpha$-coefficients in the Example 2-2 AR model. We shall now demonstrate that $\mathbf{H}(N-1)$ and $\mathbf{V}(N-1)$, which are defined in Equation (2-8), are dependent, which means of course that we cannot apply Theorem 8-1 to study the unbiasedness of the WLSEs of the $\alpha$-coefficients.

We represent the explicit dependences of $\mathbf{H}$ and $\mathbf{V}$ on their elements in the following manner:

$$\mathbf{H} = \mathbf{H}[y(N-1), y(N-2), \ldots, y(0)] \tag{8-7}$$

and

$$\mathbf{V} = \mathbf{V}[u(N-1), u(N-2), \ldots, u(0)] \tag{8-8}$$

Direct iteration of difference equation (2-7) for $k = 1, 2, \ldots, N-1$, reveals that $y(1)$ depends on $u(0)$, $y(2)$ depends on $u(1)$ and $u(0)$, and finally that $y(N-1)$ depends on $u(N-2), \ldots, u(0)$; thus,

$$\mathbf{H}[y(N-1), y(N-2), \ldots, y(0)] = \mathbf{H}[u(N-2), u(N-3), \ldots, u(0), y(0)] \tag{8-9}$$

Comparing (8-8) and (8-9), we see that $\mathbf{H}$ and $\mathbf{V}$ depend on similar values of random input $u$; hence, they are statistically dependent. □

**EXAMPLE 8-3** "AR(1) channel"

We are interested in estimating the parameter $a$ in the following first-order system:

$$y(k+1) = -ay(k) + u(k) \tag{8-10}$$

where $u(k)$ is a zero-mean white noise sequence. One approach to doing this is to collect $y(k+1), y(k), \ldots, y(1)$ as follows:

$$\underbrace{\begin{bmatrix} y(k+1) \\ y(k) \\ y(k-1) \\ \cdots \\ y(1) \end{bmatrix}}_{\mathbf{Z}(k+1)} = \underbrace{\begin{bmatrix} -y(k) \\ -y(k-1) \\ -y(k-2) \\ \cdots \\ -y(0) \end{bmatrix}}_{\mathbf{H}(k)} a + \underbrace{\begin{bmatrix} u(k) \\ u(k-1) \\ u(k-2) \\ \cdots \\ u(0) \end{bmatrix}}_{\mathbf{V}(k)} \tag{8-11}$$

and, to obtain $\hat{a}_{\text{LS}}$. To study the bias of $\hat{a}_{\text{LS}}$, we use (8-2) in which $\mathbf{W}(k)$ is set equal to $\mathbf{I}$, and $\mathbf{H}(k)$ and $\mathbf{V}(k)$ are defined in (8-11). We also set $\hat{\theta}_{\text{LS}}(k) = \hat{a}_{\text{LS}}(k+1)$. The argument of $\hat{a}_{\text{LS}}$ is $k+1$ instead of $k$, because the argument of $\mathbf{Z}$ in (8-11) is $k+1$. Doing this, we find that

$$\hat{a}_{\text{LS}}(k+1) = a - \frac{\sum_{i=0}^{k} u(i)y(i)}{\sum_{j=0}^{k} y^2(j)} \tag{8-12}$$

Thus,

$$\mathbf{E}\{\hat{a}_{\text{LS}}(k+1)\} = a - \sum_{i=0}^{k} \mathbf{E}\left\{ \frac{u(i)y(i)}{\sum_{j=0}^{k} y^2(j)} \right\} \tag{8-13}$$

Note, from (8-10), that $y(j)$ depends at most on $u(j-1)$; therefore,

the $k^{th}$ term in the sum (5-> 

$$\mathbf{E}\left\{ \frac{u(k)y(k)}{\sum_{j=0}^{k} y^2(j)} \right\} = \mathbf{E}\{u(k)\}\mathbf{E}\left\{ \frac{y(k)}{\sum_{j=0}^{k} y^2(j)} \right\} = 0 \tag{8-14}$$

because $\mathbf{E}\{u(k)\} = 0$. Unfortunately, all the remaining terms in (8-13), i.e., for $i = 0, 1, \ldots, k-1$, will not be equal to zero; consequently,

$$\mathbf{E}\{\hat{a}_{\text{LS}}(k+1)\} = a + 0 + k \quad \text{nonzero terms} \tag{8-15}$$

Unless we are very lucky so that the $k$ nonzero terms sum identically to zero, $\mathbf{E}\{\hat{a}_{\text{LS}}(k+1)\} \neq a$, which means, of course, that $\hat{a}_{\text{LS}}$ is biased.

The results in this example generalize to higher-order difference equations, so we can conclude that *least-squares estimates of coefficients in an AR model are biased.* □

Next we proceed to compute the covariance matrix of $\tilde{\theta}_{\text{WLS}}(k)$, where

$$\tilde{\theta}_{\text{WLS}}(k) = \theta - \hat{\theta}_{WLS}(k) \tag{8-16}$$

**Theorem 8-2.** *If* $\mathbf{E}\{\mathbf{V}(k)\} = \mathbf{0}$, $\mathbf{V}(k)$ *and* $\mathbf{H}(k)$ *are statistically independent, and*

$$\mathbf{E}\{\mathbf{V}(k)\mathbf{V}'(k)\} = \mathbf{R}(k) \tag{8-17}$$

*then*

$$\text{cov}\,[\tilde{\theta}_{\text{WLS}}] = \mathbf{E_H}\{(\mathbf{H'WH})^{-1}\mathbf{H'WRWH}(\mathbf{H'WH})^{-1}\} \tag{8-18}$$

*Proof.* Because $\mathbf{E}\{\mathbf{V}(k)\} = \mathbf{0}$ and $\mathbf{V}(k)$ and $\mathbf{H}(k)$ are statistically independent, $\mathbf{E}\{\tilde{\theta}_{\text{WLS}}(k)\} = \mathbf{0}$, so

$$\text{cov}\,[\tilde{\theta}_{\text{WLS}}(k)] = \mathbf{E}\{\tilde{\theta}_{\text{WLS}}(k)\tilde{\theta}'_{\text{WLS}}(k)\} \tag{8-19}$$

Using (8-2) in (8-16), we see that

$$\tilde{\theta}_{\text{WLS}}(k) = -(\mathbf{H}'\mathbf{W}\mathbf{H})^{-1}\mathbf{H}'\mathbf{W}\mathbf{V} \tag{8-20}$$

Hence

$$\text{cov}\,[\tilde{\theta}_{\text{WLS}}(k)] = \mathbf{E}_{\mathbf{H},\mathbf{V}}\{(\mathbf{H}'\mathbf{W}\mathbf{H})^{-1}\mathbf{H}'\mathbf{W}\mathbf{V}\mathbf{V}'\mathbf{W}\mathbf{H}(\mathbf{H}'\mathbf{W}\mathbf{H})^{-1}\} \tag{8-21}$$

where we have made use of the fact that $\mathbf{W}$ is a symmetric matrix and the transpose and inverse symbols may be permuted. From probability theory (e.g., Papoulis, 1991), recall that

$$\mathbf{E}_{\mathbf{H},\mathbf{V}}\{\cdot\} = \mathbf{E}_{\mathbf{H}}\{\mathbf{E}_{\mathbf{V}|\mathbf{H}}\{\cdot|\mathbf{H}\}\} \tag{8-22}$$

Applying (8-22) to (8-21), we obtain (8-18). □

Because (8-22) (or a variation of it) is used many times in this book, we provide a proof of it in the supplementary material at the end of this lesson.

As it stands, Theorem 8-2 is not too useful because it is virtually impossible to compute the expectation in (8-18), due to the highly nonlinear dependence of $(\mathbf{H}'\mathbf{W}\mathbf{H})^{-1}\mathbf{H}'\mathbf{W}\mathbf{R}\mathbf{W}\mathbf{H}(\mathbf{H}'\mathbf{W}\mathbf{H})^{-1}$ on $\mathbf{H}$. The following special case of Theorem 8-2 is important in practical applications in which $\mathbf{H}(k)$ is deterministic and $\mathbf{R}(k) = \sigma_\nu^2\mathbf{I}$. Note that in this case our generic model in (2-1) is often referred to as the *classical linear regression model* (Fomby et al., 1984), provided that

$$\lim_{k\to\infty}\left[\frac{\mathbf{H}'(k)\mathbf{H}(k)}{k}\right] = \mathbf{Q} \tag{8-23}$$

where $\mathbf{Q}$ is a finite and nonsingular matrix.

**Corollary 8-1.** *Given the conditions in Theorem 8-2, and that $\mathbf{H}(k)$ is deterministic, and the components of $\mathbf{V}(k)$ are independent and identically distributed with zero-mean and constant variance $\sigma_\nu^2$, then*

$$\text{cov}\,[\tilde{\theta}_{\text{LS}}(k)] = \sigma_\nu^2[\mathbf{H}'(k)\mathbf{H}(k)]^{-1} \tag{8-24}$$

*Proof.* When $\mathbf{H}$ is deterministic, $\text{cov}\,[\tilde{\theta}_{\text{WLS}}(k)]$ is obtained from (8-18) by deleting the expectation on its right-hand side. To obtain $\text{cov}\,[\tilde{\theta}_{\text{LS}}(k)]$ when $\text{cov}\,[\mathbf{V}(k)] = \sigma_\nu^2\mathbf{I}$, set $\mathbf{W} = \mathbf{I}$ and $\mathbf{R}(k) = \sigma_\nu^2\mathbf{I}$ in (8-18). The result is (8-24). □

**Theorem 8-3.** *If* $\mathbf{E}\{\mathbf{V}(\text{k})\} = \mathbf{0}$, $\mathbf{H}(\text{k})$ *is deterministic, and the components of* $\mathbf{V}(\text{k})$ *are independent and identically distributed with constant variance* $\sigma_\nu^2$, *then* $\hat{\theta}_{\text{LS}}(\text{k})$ *is an efficient estimator within the class of linear estimators.*

*Proof.* Our proof of efficiency does not compute the Fisher information matrix, because we have made no assumption about the probability density function of $\mathbf{V}(k)$.

Consider the following alternative linear estimator of $\boldsymbol{\theta}$:

$$\boldsymbol{\theta}^*(k) = \hat{\boldsymbol{\theta}}_{\mathrm{LS}}(k) + \mathbf{C}(k)\mathbf{Z}(k) \tag{8-25}$$

where $\mathbf{C}(k)$ is a matrix of constants. Substituting (3-11) and (2-1) into (8-25), it is easy to show that

$$\boldsymbol{\theta}^*(k) = (\mathbf{I} + \mathbf{CH})\boldsymbol{\theta} + [(\mathbf{H}'\mathbf{H})^{-1}\mathbf{H}' + \mathbf{C}]\mathbf{V} \tag{8-26}$$

Because $\mathbf{H}$ is deterministic and $\mathbf{E}\{\mathbf{V}(k)\} = \mathbf{0}$,

$$\mathbf{E}\{\boldsymbol{\theta}^*(k)\} = (\mathbf{I} + \mathbf{CH})\boldsymbol{\theta} \tag{8-27}$$

Hence, $\boldsymbol{\theta}^*(k)$ is an unbiased estimator of $\boldsymbol{\theta}$ if and only if $\mathbf{CH} = \mathbf{0}$. We leave it to the reader to show that

$$\mathbf{E}\{[\boldsymbol{\theta}^*(k)-\boldsymbol{\theta}][\boldsymbol{\theta}^*(k)-\boldsymbol{\theta}]'\} = \sigma_v^2(\mathbf{H}'\mathbf{H})^{-1} + \sigma_v^2\mathbf{CC}' = \mathrm{cov}\,[\tilde{\boldsymbol{\theta}}_{\mathrm{LS}}(k)] + \sigma_v^2\mathbf{CC}' \tag{8-28}$$

Because $\mathbf{CC}'$ is a positive semidefinite matrix, $\mathrm{cov}\,[\tilde{\boldsymbol{\theta}}_{\mathrm{LS}}(k)] \le \mathbf{E}\{[\boldsymbol{\theta}^*(k) - \boldsymbol{\theta}][\boldsymbol{\theta}^*(k) - \boldsymbol{\theta}]'\}$; hence, we have now shown that $\hat{\boldsymbol{\theta}}_{\mathrm{LS}}(k)$ satisfies Definition 6-3, which means that $\hat{\boldsymbol{\theta}}_{\mathrm{LS}}(k)$ is an efficient estimator. □

Usually, when we use a least-squares estimation algorithm, we do not know the numerical value of $\sigma_v^2$. If $\sigma_v^2$ is known ahead of time, it can be used directly in the estimate of $\boldsymbol{\theta}$. We show how to do this in Lesson 9. Where do we obtain $\sigma_v^2$ in order to compute (8-24)? We can estimate it!

**Theorem 8-4.** *If* $\mathbf{E}\{\mathbf{V}(\mathrm{k})\} = \mathbf{0}$, $\mathbf{V}(\mathrm{k})$ *and* $\mathbf{H}(\mathrm{k})$ *are statistically independent [or* $\mathbf{H}(\mathrm{k})$ *is deterministic], and the components of* $\mathbf{V}(\mathrm{k})$ *are independent and identically distributed with zero mean and constant variance* $\sigma_v^2$*, then an unbiased estimator of* $\sigma_v^2$ *is*

$$\hat{\sigma}_v^2(k) = \frac{\tilde{\mathbf{Z}}'(k)\tilde{\mathbf{Z}}(k)}{k-n} \tag{8-29}$$

*where*

$$\tilde{\mathbf{Z}}(k) = \mathbf{Z}(k) - \mathbf{H}(k)\hat{\boldsymbol{\theta}}_{\mathrm{LS}}(k) \tag{8-30}$$

So compute $\hat{\theta}_{LS}(k)$ then estimate $\hat{\sigma}_v^2$ then estimate $\mathrm{cov}(\tilde{\theta}_{LS}(k))$.

*Proof.* We shall proceed by computing $\mathbf{E}\{\tilde{\mathbf{Z}}'(k)\tilde{\mathbf{Z}}(k)\}$ and then approximating it as $\tilde{\mathbf{Z}}'(k)\tilde{\mathbf{Z}}(k)$, because the latter quantity can be computed from $\mathbf{Z}(k)$ and $\hat{\boldsymbol{\theta}}_{\mathrm{LS}}(k)$, as in (8-30).

First, we compute an expression for $\tilde{\mathbf{Z}}(k)$. Substituting both the linear model for $\mathbf{Z}(k)$ and the least-squares formula for $\hat{\boldsymbol{\theta}}_{\mathrm{LS}}(k)$ into (8-30), we find that

$$\begin{aligned}\tilde{\mathbf{Z}} &= \mathbf{H}\boldsymbol{\theta} + \mathbf{V} - \mathbf{H}(\mathbf{H}'\mathbf{H})^{-1}\mathbf{H}'(\mathbf{H}\boldsymbol{\theta} + \mathbf{V}) \\ &= \mathbf{V} - \mathbf{H}(\mathbf{H}'\mathbf{H})^{-1}\mathbf{H}'\mathbf{V} \\ &= [\mathbf{I}_k - \mathbf{H}(\mathbf{H}'\mathbf{H})^{-1}\mathbf{H}']\mathbf{V}\end{aligned} \tag{8-31}$$

where $\mathbf{I}_k$ is the $k \times k$ identity matrix. Let

$$\mathbf{M} = \mathbf{I}_k - \mathbf{H}(\mathbf{H}'\mathbf{H})^{-1}\mathbf{H}' \tag{8-32}$$

Matrix $\mathbf{M}$ is idempotent, i.e., $\mathbf{M}' = \mathbf{M}$ and $\mathbf{M}^2 = \mathbf{M}$; therefore,

$$\begin{aligned} \mathbf{E}\{\tilde{\mathbf{Z}}'\tilde{\mathbf{Z}}\} &= \mathbf{E}\{\mathbf{V}'\mathbf{M}'\mathbf{M}\mathbf{V}\} = \mathbf{E}\{\mathbf{V}'\mathbf{M}\mathbf{V}\} \\ &= \mathbf{E}\{\text{tr}\,\mathbf{M}\mathbf{V}\mathbf{V}'\} \end{aligned} \tag{8-33}$$

Recall the following well-known facts about the trace of a matrix:

1. $\mathbf{E}\{\text{tr}\,\mathbf{A}\} = \text{tr}\,\mathbf{E}\{\mathbf{A}\}$
2. $\text{tr}\,c\mathbf{A} = c\,\text{tr}\,\mathbf{A}$, where $c$ is a scalar
3. $\text{tr}\,(\mathbf{A} + \mathbf{B}) = \text{tr}\,\mathbf{A} + \text{tr}\,\mathbf{B}$
4. $\text{tr}\,\mathbf{I}_N = N$
5. $\text{tr}\,\mathbf{AB} = \text{tr}\,\mathbf{BA}$

Using these facts, we now continue the development of (8-33) for the case when $\mathbf{H}(k)$ is deterministic [we leave the proof for the case when $\mathbf{H}(k)$ is random as an exercise], as follows:

$$\begin{aligned} \mathbf{E}\{\tilde{\mathbf{Z}}'\tilde{\mathbf{Z}}\} &= \text{tr}[\mathbf{M}\mathbf{E}\{\mathbf{V}\mathbf{V}'\}] = \text{tr}\,\mathbf{M}\mathbf{R} = \text{tr}\,\mathbf{M}\sigma_v^2 \\ &= \sigma_v^2\,\text{tr}\,\mathbf{M} = \sigma_v^2\,\text{tr}\,[\mathbf{I}_k - \mathbf{H}(\mathbf{H}'\mathbf{H})^{-1}\mathbf{H}'] \\ &= \sigma_v^2 k - \sigma_v^2\,\text{tr}\,\mathbf{H}(\mathbf{H}'\mathbf{H})^{-1}\mathbf{H}' \\ &= \sigma_v^2 k - \sigma_v^2\,\text{tr}\,(\mathbf{H}'\mathbf{H})(\mathbf{H}'\mathbf{H})^{-1} \\ &= \sigma_v^2 k - \sigma_v^2\,\text{tr}\,\mathbf{I}_n = \sigma_v^2(k - n) \end{aligned} \tag{8-34}$$

Solving this equation for $\sigma_v^2$, we find that

$$\sigma_v^2 = \frac{\mathbf{E}\{\tilde{\mathbf{Z}}'\tilde{\mathbf{Z}}\}}{k - n} \tag{8-35}$$

Although this is an exact result for $\sigma_v^2$, it is not one that can be evaluated, because we cannot compute $\mathbf{E}\{\tilde{\mathbf{Z}}'\tilde{\mathbf{Z}}\}$

Using the structure of $\sigma_v^2$ as a starting point, we estimate $\sigma_v^2$ by the simple formula

$$\hat{\sigma}_v^2(k) = \frac{\tilde{\mathbf{Z}}'(k)\tilde{\mathbf{Z}}(k)}{k - n} \tag{8-36}$$

To show that $\hat{\sigma}_v^2(k)$ is an unbiased estimator of $\sigma_v^2$, we observe that

$$\begin{aligned} \mathbf{E}\{\hat{\sigma}_v^2(k)\} &= \frac{\mathbf{E}\{\tilde{\mathbf{Z}}'\tilde{\mathbf{Z}}\}}{k - n} \\ &= \sigma_v^2 \end{aligned} \tag{8-37}$$

where we have used (8-34) for $\mathbf{E}\{\tilde{\mathbf{Z}}'\tilde{\mathbf{Z}}\}$. □

# LARGE-SAMPLE PROPERTIES OF LEAST-SQUARES ESTIMATORS

Many large sample properties of LSEs are determined by establishing that the LSE is equivalent to another estimator for which it is known that the large-sample property holds true. In Lesson 11, for example, we will provide conditions under which the LSE of $\boldsymbol{\theta}$, $\hat{\boldsymbol{\theta}}_{LS}(k)$, is the same as the maximum-likelihood estimator of $\boldsymbol{\theta}$, $\hat{\boldsymbol{\theta}}_{ML}(k)$. Because $\hat{\boldsymbol{\theta}}_{ML}(k)$ is consistent, asymptotically efficient, and asymptotically Gaussian, $\hat{\boldsymbol{\theta}}_{LS}(k)$ inherits all these properties.

**Theorem 8-5.** *If*

$$\text{plim}\left[\frac{\mathbf{H}'(k)\mathbf{H}(k)}{k}\right] = \Sigma_{\mathbf{H}} \tag{8-38}$$

$\Sigma_{\mathbf{H}}^{-1}$ *exists, and*

$$\text{plim}\left[\frac{\mathbf{H}'(k)\mathbf{V}(k)}{k}\right] = \mathbf{0} \tag{8-39}$$

*then*

$$\text{plim}\,\hat{\boldsymbol{\theta}}_{LS}(k) = \boldsymbol{\theta} \tag{8-40}$$

Assumption (8-38) postulates the existence of a probability limit for the second-order moments of the variables in $\mathbf{H}(k)$, as given by $\boldsymbol{\Sigma}_{\mathbf{H}}$. Note that $\mathbf{H}'(k)\mathbf{H}(k)/k$ is the sample covariance matrix of the population covariance matrix $\boldsymbol{\Sigma}_{\mathbf{H}}$. Assumption (8-39) postulates a zero probability limit for the correlation between $\mathbf{H}(k)$ and $\mathbf{V}(k)$. $\mathbf{H}'(k)\mathbf{V}(k)$ can be thought of as a "filtered" version of noise vector $\mathbf{V}(k)$. For (8-39) to be true, "filter $\mathbf{H}'(k)$" must be stable. If, for example, $\mathbf{H}(k)$ is deterministic and $\sigma^2_{\mathbf{V}_i(k)} < \infty$, then (8-39) will be true. See Problem 8-10 for more details about (8-39).

*Proof.* Beginning with (8-2), but for $\hat{\boldsymbol{\theta}}_{LS}(k)$ instead of $\hat{\boldsymbol{\theta}}_{WLS}(k)$, we see that

$$\hat{\boldsymbol{\theta}}_{LS}(k) = \boldsymbol{\theta} + (\mathbf{H}'\mathbf{H})^{-1}\mathbf{H}'\mathbf{V} \tag{8-41}$$

Operating on both sides of this equation with plim and using properties (7-15), (7-16), and (7-17), we find that

$$\begin{aligned}
\text{plim}\,\hat{\boldsymbol{\theta}}_{LS}(k) &= \boldsymbol{\theta} + \text{plim}\left(\frac{\mathbf{H}'\mathbf{H}}{k}\right)^{-1}\frac{\mathbf{H}'\mathbf{V}}{k} \\
&= \boldsymbol{\theta} + \text{plim}\left(\frac{\mathbf{H}'\mathbf{H}}{k}\right)^{-1}\text{plim}\frac{\mathbf{H}'\mathbf{V}}{k} \\
&= \boldsymbol{\theta} + \Sigma_{\mathbf{H}}^{-1}\cdot\mathbf{0} \\
&= \boldsymbol{\theta}
\end{aligned}$$

which demonstrates that, under the given conditions, $\hat{\boldsymbol{\theta}}_{LS}(k)$ is a consistent estimator of $\boldsymbol{\theta}$. □

In some important applications Eq. (8-39) does not apply, e.g., Example 8-2. Theorem 8-5 then does not apply, and the study of consistency is often quite complicated in these cases. The *method of instrumental variables* (Durbin, 1954; Kendall and Stuart, 1961; Young, 1984; Fomby et al., 1984) is a way out of this difficulty. For least squares the normal equations (3-13) are

$$(\mathbf{H}'(k)\mathbf{H}(k))\hat{\boldsymbol{\theta}}_{LS}(k) = \mathbf{H}'(k)\mathbf{Z}(k) \tag{8-42}$$

In the method of instrumental variables, $\mathbf{H}'(k)$ is replaced by the instrumental variable matrix $\mathbf{H}'_{\mathbf{IV}}(k)$ so that (8-42) becomes

$$(\mathbf{H}'_{\mathbf{IV}}(k)\mathbf{H}(k))\hat{\boldsymbol{\theta}}_{\mathbf{IV},LS}(k) = \mathbf{H}'_{\mathbf{IV}}(k)\mathbf{Z}(k) \tag{8-43}$$

Clearly,

$$\hat{\boldsymbol{\theta}}_{\mathbf{IV},LS}(k) = (\mathbf{H}'_{\mathbf{IV}}(k)\mathbf{H}(k))^{-1}\mathbf{H}'_{\mathbf{IV}}(k)\mathbf{Z}(k) \tag{8-44}$$

In general, the instrumental variable matrix should be chosen so that it is uncorrelated with $\mathbf{V}(k)$ and is highly correlated with $\mathbf{H}(k)$. The former is needed so that $\hat{\boldsymbol{\theta}}_{\mathbf{IV},LS}(k)$ is a consistent estimator of $\boldsymbol{\theta}$, whereas the latter is needed so that the asymptotic variance of $\hat{\boldsymbol{\theta}}_{\mathbf{IV},LS}(k)$ does not become infinitely large (see Problem 8-12). In general, $\hat{\boldsymbol{\theta}}_{\mathbf{IV},LS}(k)$ is not an asymptotically efficient estimator of $\boldsymbol{\theta}$ (Fomby et al., 1984). The selection of $\mathbf{H}_{\mathbf{IV}}(k)$ is problem dependent and is not unique. In some applications, lagged variables are used as the instrumental variables. See the references cited previously (especially, Young, 1984) for discussions on how to choose the instrumental variables.

**Theorem 8-6.** *Let* $\mathbf{H}_{\mathbf{IV}}(\mathrm{k})$ *be an instrumental variable matrix for* $\mathbf{H}(\mathrm{k})$. *Then the instrumental variable estimator* $\hat{\boldsymbol{\theta}}_{\mathbf{IV},LS}(\mathrm{k})$, *given by (8-44), is consistent.* □

Because the proof of this theorem is so similar to the proof of Theorem 8-5, we leave it as an exercise (Problem 8-11).

**Theorem 8-7.** *If (8-38) and (8-39) are true,* $\Sigma_{\mathbf{H}}^{-1}$ *exists, and*

$$\operatorname{plim} \frac{\mathbf{V}'(k)\mathbf{V}(k)}{k} = \sigma_{\mathbf{V}}^2 \tag{8-45}$$

*then*

$$\operatorname{plim} \hat{\sigma}_{\mathbf{V}}^2(k) = \sigma_{\mathbf{V}}^2 \tag{8-46}$$

*where* $\hat{\sigma}_{\mathbf{V}}^2(k)$ *is given by (8-29).*

*Proof.* From (8-31), we find that

$$\tilde{\mathbf{Z}}'\tilde{\mathbf{Z}} = \mathbf{V}'\mathbf{V} - \mathbf{V}'\mathbf{H}(\mathbf{H}'\mathbf{H})^{-1}\mathbf{H}'\mathbf{V} \tag{8-47}$$

Consequently,

$$
\begin{aligned}
\text{plim}\,\hat{\sigma}_{\mathbf{V}}^2(k) &= \text{plim}\,\frac{\tilde{\mathbf{Z}}'\tilde{\mathbf{Z}}}{k-n} \\
&= \text{plim}\,\frac{\mathbf{V}'\mathbf{V}}{k-n} - \text{plim}\,\frac{\mathbf{V}'\mathbf{H}(\mathbf{H}'\mathbf{H})^{-1}\mathbf{H}'\mathbf{V}}{k-n} \\
&= \sigma_v^2 - \text{plim}\,\frac{\mathbf{V}'\mathbf{H}}{k-n}\cdot\text{plim}\left[\frac{\mathbf{H}'\mathbf{H}}{k-n}\right]^{-1}\cdot\text{plim}\,\frac{\mathbf{H}'\mathbf{V}}{k-n} \\
&= \sigma_{\mathbf{v}}^2 - \mathbf{0}'\cdot\Sigma_{\mathbf{H}}^{-1}\cdot\mathbf{0} \\
&= \sigma_{\mathbf{v}}^2 \quad \square
\end{aligned}
$$

## Supplementary Material

### EXPANSION OF TOTAL EXPECTATION

Here we provide a proof of the very important expansion of a total expectation given in (8-22), that

$$\mathbf{E}_{\mathbf{H},\mathbf{V}}\{\cdot\} = \mathbf{E}_{\mathbf{H}}\{\mathbf{E}_{\mathbf{V}|\mathbf{H}}\{\cdot \mid \mathbf{H}\}\}$$

Let $\mathbf{G}(\mathbf{H},\mathbf{V})$ denote an arbitrary matrix function of $\mathbf{H}$ and $\mathbf{V}$ [e.g., see (8-21)], and $p(\mathbf{H},\mathbf{V})$ denote the joint probability density function of $\mathbf{H}$ and $\mathbf{V}$. Then

$$
\begin{aligned}
\mathbf{E}_{\mathbf{H},\mathbf{V}}\{\mathbf{G}(\mathbf{H},\mathbf{V})\} &= \int\cdots\int \mathbf{G}(\mathbf{H},\mathbf{V})p(\mathbf{H},\mathbf{V})d\mathbf{H}d\mathbf{V} \\
&= \int\cdots\int \mathbf{G}(\mathbf{H},\mathbf{V})p(\mathbf{V}|\mathbf{H})p(\mathbf{H})d\mathbf{V}d\mathbf{H} \\
&= \int\cdots\int\{\int\cdots\int \mathbf{G}(\mathbf{H},\mathbf{V})p(\mathbf{V}|\mathbf{H})d\mathbf{V}\}p(\mathbf{H})d\mathbf{H} \\
&= \int\cdots\int \mathbf{E}_{\mathbf{V}|\mathbf{H}}\{\mathbf{G}(\mathbf{H},\mathbf{V}) \mid \mathbf{H}\}p(\mathbf{H})d\mathbf{H} \\
&= \mathbf{E}_{\mathbf{H}}\{\mathbf{E}_{\mathbf{V}|\mathbf{H}}\{\mathbf{G}(\mathbf{H},\mathbf{V}) \mid \mathbf{H}\}\}
\end{aligned}
$$

which is the desired result.

## SUMMARY QUESTIONS

1. When $\mathbf{H}(k)$ is random, $\hat{\boldsymbol{\theta}}_{\text{WLS}}(k)$ is an unbiased estimator of $\boldsymbol{\theta}$:
   **(a)** if and only if $\mathbf{V}(k)$ and $\mathbf{H}(k)$ are statistically independent
   **(b)** $\mathbf{V}(k)$ is zero mean or $\mathbf{V}(k)$ and $\mathbf{H}(k)$ are statistically independent
   **(c)** if $\mathbf{V}(k)$ and $\mathbf{H}(k)$ are statistically independent and $\mathbf{V}(k)$ is zero mean
2. If we fail the unbiasedness test given in Theorem 8-1, it means:
   **(a)** $\hat{\boldsymbol{\theta}}_{\text{WLS}}(k)$ must be biased

(b) $\hat{\theta}_{WLS}(k)$ could be biased
(c) $\hat{\theta}_{WLS}(k)$ is asymptotically biased

3. If $\mathbf{H}(k)$ is deterministic, then $\text{cov}\,[\tilde{\theta}_{WLS}(k)]$ equals:
   (a) $(\mathbf{H}'\mathbf{W}\mathbf{H})^{-1}\mathbf{H}'\mathbf{W}\mathbf{R}\mathbf{W}\mathbf{H}(\mathbf{H}'\mathbf{W}\mathbf{H})^{-1}$
   (b) $\mathbf{H}'\mathbf{W}\mathbf{R}\mathbf{W}\mathbf{H}$
   (c) $(\mathbf{H}'\mathbf{W}\mathbf{H})^{-2}\mathbf{H}'\mathbf{W}\mathbf{R}\mathbf{W}\mathbf{H}$
4. The condition $\text{plim}[\mathbf{H}'(k)\mathbf{H}(k)/k] = \mathbf{\Sigma_H}$ means:
   (a) $\mathbf{H}'(k)\mathbf{H}(k)$ is stationary
   (b) $\mathbf{H}'(k)\mathbf{H}(k)/k$ converges in probability to the population covariance matrix $\mathbf{\Sigma_H}$
   (c) the sample covariance of $\mathbf{H}'(k)\mathbf{H}(k)$ is defined
5. When $\mathbf{H}(k)$ can only be measured with errors, i.e., $\mathbf{H}_m(k) = \mathbf{H}(k) + \mathbf{N}(k)$, $\hat{\theta}_{LS}(k)$ is:
   (a) an unbiased estimator of $\theta$
   (b) a consistent estimator of $\theta$
   (c) not a consistent estimator of $\theta$
6. When $\mathbf{H}(k)$ is a deterministic matrix, the WLSE of $\theta$ is:
   (a) never an unbiased estimator of $\theta$
   (b) sometimes an unbiased estimator of $\theta$
   (c) always an unbiased estimator of $\theta$
7. In the application of impulse response identification, WLSEs of impulse response coefficients are unbiased:
   (a) only if the system's input is deterministic
   (b) only if the system's input is random
   (c) regardless of the nature of the system's input
8. When we say that "$\hat{\theta}_{LS}(k)$ is efficient within the class of linear estimators," this means that we have fixed the structure of the estimator to be a linear transformation of the data, and the resulting estimator:
   (a) just misses the Cramer–Rao bound
   (b) achieves the Cramer–Rao bound
   (c) has the smallest error-covariance matrix of all such structures
9. In general, the instrumental variable matrix should be chosen so that it is:
   (a) orthogonal to $\mathbf{V}(k)$ and highly correlated with $\mathbf{H}(k)$
   (b) uncorrelated with $\mathbf{V}(k)$ and highly correlated with $\mathbf{H}(k)$
   (c) mutually uncorrelated with $\mathbf{V}(k)$ and $\mathbf{H}(k)$

## PROBLEMS

**8-1.** Suppose that $\hat{\theta}_{LS}$ is an unbiased estimator of $\theta$. Is $\hat{\theta}^2_{LS}$ an unbiased estimator of $\theta^2$? (*Hint:* Use the least-squares batch algorithm to study this question.)

**8-2.** For $\hat{\theta}_{WLS}(k)$ to be an unbiased estimator of $\theta$ we required $\mathbf{E}\{\mathbf{V}(k)\} = \mathbf{0}$. This problem considers the case when $\mathbf{E}\{\mathbf{V}(k)\} \neq \mathbf{0}$.
   (a) Assume that $\mathbf{E}\{\mathbf{V}(k)\} = \mathbf{V}_0$, where $\mathbf{V}_0$ is known to us. How is the concatenated measurement equation $\mathbf{Z}(k) = \mathbf{H}(k)\theta + \mathbf{V}(k)$ modified in this case so we can use the results derived in Lesson 3 to obtain $\hat{\theta}_{WLS}(k)$ or $\hat{\theta}_{LS}(k)$?
   (b) Assume that $\mathbf{E}\{\mathbf{V}(k)\} = m_V\mathbf{1}$ where $m_V$ is constant but is unknown. How is the concatenated measurement equation $\mathbf{Z}(k) = \mathbf{H}(k)\theta + \mathbf{V}(k)$ modified in this case so that we can obtain least-squares estimates of both $\theta$ and $m_V$?

**8-3.** Consider the stable autoregressive model $y(k) = \theta_1 y(k-1) + \cdots + \theta_k y(k-K) + \epsilon(k)$ in which the $\epsilon(k)$ are identically distributed random variables with mean zero and finite variance $\sigma^2$. Prove that the least-squares estimates of $\theta_1, \ldots, \theta_k$ are consistent (see also, Ljung, 1976).

**8-4.** In this lesson we have assumed that the $\mathbf{H}(k)$ variables have been measured without error. Here we examine the situation when $\mathbf{H}_m(k) = \mathbf{H}(k) + \mathbf{N}(k)$ in which $\mathbf{H}(k)$ denotes a matrix of true values and $\mathbf{N}(k)$ a matrix of measurement errors. The basic linear model is now

$$\mathbf{Z}(k) = \mathbf{H}(k)\boldsymbol{\theta} + \mathbf{V}(k) = \mathbf{H}_m(k)\boldsymbol{\theta} + [\mathbf{V}(k) - \mathbf{N}(k)\boldsymbol{\theta}]$$

Prove that $\hat{\boldsymbol{\theta}}_{LS}(k)$ is not a consistent estimator of $\boldsymbol{\theta}$. See, also, Problem 8-15.

**8-5.** (Geogiang Yue and Sungook Kim, Fall 1991) Given a random sequence $z(k) = a + bk + \nu(k)$, where $\nu(k)$ is zero-mean Gaussian noise with variance $\sigma_\nu^2$, and $\mathbf{E}\{\nu(i)\nu(j)\} = 0$ for all $i \neq j$. Show that the least-squares estimators of $a$ and $b$, $\hat{a}_{LS}(N)$ and $\hat{b}_{LS}(N)$ are unbiased.

**8-6.** (G. Caso, Fall 1991) [Stark and Woods (1986), Prob. 5.14] Let $\mathbf{Z}(N) = \mathbf{H}(N)\boldsymbol{\theta} + \mathbf{V}(N)$, where $\mathbf{H}(N)$ is deterministic, $\mathbf{E}\{\mathbf{V}(N)\} = \mathbf{0}$, and $\boldsymbol{\phi} = \mathbf{D}\boldsymbol{\theta}$, where $\mathbf{D}$ is also deterministic.

**(a)** Show that for $\hat{\boldsymbol{\phi}} = \mathbf{L}\mathbf{Z}(N)$ to be an unbiased estimator of $\boldsymbol{\phi}$ we require that $\mathbf{L}\mathbf{H}(N) = \mathbf{D}$.

**(b)** Show that $\mathbf{L} = \mathbf{D}[\mathbf{H}'(N)\mathbf{H}(N)]^{-1}\mathbf{H}'(N)$ satisfies this requirement. How is $\hat{\boldsymbol{\phi}}$ related to $\hat{\boldsymbol{\theta}}_{LS}(k)$ for this choice of $\mathbf{L}$?

**8-7.** (Keith M. Chugg, Fall 1991) This problem investigates the effects of preprocessing the observations. If $\hat{\boldsymbol{\theta}}_{LS}(k)$ is based on $\mathbf{Y}(k) = \mathbf{G}(k)\mathbf{Z}(k)$ instead of $\mathbf{Z}(k)$, where $\mathbf{G}(k)$ is an $N \times N$ invertible matrix:

**(a)** What is $\hat{\boldsymbol{\theta}}_{LS}(k)$ based on $\mathbf{Y}(k)$?

**(b)** For $\mathbf{H}(k)$ deterministic and $\text{cov}\,[\mathbf{V}(k)] = \mathbf{R}(k)$, determine the covariance matrix of $\hat{\boldsymbol{\theta}}_{LS}(k)$.

**(c)** Show that for a specific choice of $\mathbf{G}(k)$,

$$\text{cov}[\tilde{\boldsymbol{\theta}}_{LS}(k)] = [\mathbf{H}'(k)\mathbf{R}^{-1}\mathbf{H}(k)]^{-1}$$

What is the structure of the estimator for this choice of $\mathbf{G}(k)$? (*Hint:* Any positive definite symmetric matrix has a Cholesky factorization.)

**(d)** What are the implications of part (c) and the fact that $\mathbf{G}(k)$ is invertible?

**8-8.** Regarding the proof of Theorem 8-3:

**(a)** Prove that $\boldsymbol{\theta}^*(k)$ is an unbiased estimator of $\boldsymbol{\theta}$ *if and only if* $\mathbf{CH} = \mathbf{0}$.

**(b)** Derive Eq. (8-28).

**8-9.** Prove Theorem 8-4 when $\mathbf{H}(k)$ is random.

**8-10.** In this problem we examine the truth of (8-39) when $\mathbf{H}(k)$ is either deterministic or random.

**(a)** Show that (8-39) is satisfied when $\mathbf{H}(k)$ is deterministic, if

$$\lim_{k\to\infty}\left[\frac{\mathbf{H}'(k)\mathbf{H}(k)}{k}\right] = \boldsymbol{\Sigma}_{\mathbf{H}}$$

**(b)** Show that (8-39) is satisfied when $\mathbf{H}(k)$ is random if $\mathbf{E}\{\mathbf{V}(k)\} = \mathbf{0}$, $\mathbf{V}(k)$ and $\mathbf{H}(k)$ are statistically independent , and the components of $\mathbf{V}(k)$ are independent and identically distributed with zero mean and constant variance $\sigma_\nu^2$ (Goldberger, 1964, pp. 269–270).

**8-11.** Prove Theorem 8-6.

**8-12.** (Fomby et al., 1984, pp. 258–259) Consider the scalar parameter model $z(k) = h(k)\theta + v(k)$, where $h(k)$ and $v(k)$ are *correlated*. Let $h_{\text{IV}}(k)$ denote the instrumental variable for $h(k)$. The instrumental variable estimator of $\theta$ is

$$\hat{\theta}_{\text{IV,LS}}(k) = \frac{\sum_{k=1}^{N} h_{\text{IV}}(k)z(k)}{\sum_{k=1}^{N} h(k)h_{\text{IV}}(k)}$$

Assume that $v(k)$ is zero mean and has unity variance.

**(a)** Show that the asymptotic variance of $k^{1/2}[\hat{\theta}_{\text{IV,LS}}(k) - \theta]$ is

$$\frac{\text{plim}\left[\sum_{k=1}^{N} h_{\text{IV}}^2/k\right]}{\{\text{plim}\left[\sum_{k=1}^{N} h(k)h_{\text{IV}}(k)/k\right]\}^2}$$

**(b)** Explain, based upon the result in part (a), why $h_{\text{IV}}(k)$ must be correlated with $h(k)$.

**8-13.** Suppose that $\mathbf{E}\{\mathbf{V}(k)\mathbf{V}'(k)\} = \sigma_v^2\mathbf{\Omega}$, where $\mathbf{\Omega}$ is symmetric and positive semidefinite. Prove that the estimator of $\sigma_v^2$ given in (8-35) is inconsistent. Compare these results with those in Theorem 8-7. See, also, Problem 8-14.

**8-14.** Suppose that $\mathbf{E}\{\mathbf{V}(k)\mathbf{V}'(k)\} = \sigma_v^2\mathbf{\Omega}$, where $\mathbf{\Omega}$ is symmetric and positive semidefinite. Here we shall study $\hat{\theta}_{\text{WLS}}^{\Omega}(k)$, where $\mathbf{W}(k) = \mathbf{\Omega}^{-1}/\sigma_v^2$ [in Lesson 9 $\hat{\theta}_{\text{WLS}}^{\Omega}(k)$ will be shown to be a best linear unbiased estimator of $\theta$, $\hat{\theta}_{\text{BLU}}(k)$]. Consequently, $\hat{\theta}_{\text{WLS}}^{\Omega} = (\mathbf{H}'\mathbf{\Omega}^{-1}\mathbf{H})^{-1}\mathbf{H}'\mathbf{\Omega}^{-1}\mathbf{Z}$.

**(a)** What is the covariance matrix for $\hat{\theta}_{\text{WLS}}^{\Omega}$?

**(b)** Let $\hat{\sigma}_v^2 = \tilde{\mathbf{Z}}'\mathbf{\Omega}^{-1}\tilde{\mathbf{Z}}/(k-n)$, where $\tilde{\mathbf{Z}} = \mathbf{Z} - \mathbf{H}\hat{\theta}_{\text{WLS}}^{\Omega}$. Prove that $\hat{\sigma}_v^2$ is an unbiased and consistent estimator of $\sigma_v^2$.

**8-15.** In this problem we consider the effects of additive noise in least squares on both the regressor and observation (Fomby et al, 1984, pp. 268–269). Suppose we have the case of a scalar parameter, such that $y(k) = h(k)\theta$, but we only observe noisy versions of both $y(k)$ and $h(k)$, that is, $z(k) = y(k) + v(k)$ and $h_m(k) = h(k) + u(k)$. Noises $v(k)$ and $u(k)$ are zero mean, have variances $\sigma_v^2$ and $\sigma_u^2$, respectively, and $\mathbf{E}\{v(k)u(k)\} = \mathbf{E}\{y(k)v(k)\} = \mathbf{E}\{h(k)u(k)\} = \mathbf{E}\{y(k)u(k)\} = \mathbf{E}\{h(k)v(k)\} = 0$. In this case $z(k) = h_m(k)\theta + \rho(k)$, where $\rho(k) = v(k) - \theta u(k)$.

**(a)** Show that $\mathbf{E}\{h_m(k)\rho(k)\} = -\theta\sigma_u^2 \neq 0$.

**(b)** Show that $\text{plim}\,\hat{\theta}_{\text{LS}} = \theta - \theta\sigma_u^2/\sigma_h^2$, where $\sigma_h^2 = \text{plim} \sum h^2(k)/k$.

Note that the method of instrumental variables is a way out of this situation.

**8-16.** This is a continuation of Problem 3-14, for *restricted least squares*.

**(a)** Show that $\hat{\theta}_{\text{LS}}^*(k)$ is an unbiased estimator of $\theta$.

**(b)** Let $\mathbf{P}(k) = \mathbf{E}\{\hat{\theta}_{\text{LS}}(k)\hat{\theta}_{\text{LS}}'(k)\}$ and $\mathbf{P}^*(k) = \mathbf{E}\{\hat{\theta}_{\text{LS}}^*(k)\hat{\theta}_{\text{LS}}^{*\prime}(k)\}$. Prove that $\mathbf{P}^*(k) < \mathbf{P}(k)$.

*a New estimator (the other one was called "LSE")*

# Best Linear Unbiased Estimation "BLUE"

## SUMMARY

The main purpose of this lesson is to develop our second estimator. It is both unbiased and efficient *by design* and is a linear function of the measurements $\mathbf{Z}(k)$. It is called a best linear unbiased estimator (BLUE).

As in the derivation of the WLSE, we begin with our generic linear model; but now we make two assumptions about this model: (1) $\mathbf{H}(k)$ must be deterministic, and (2) $\mathbf{V}(k)$ must be zero mean with positive definite known covariance matrix $\mathbf{R}(k)$. The derivation of the BLUE is more complicated than the derivation of the WLSE because of the design constraints; however, its performance analysis is much easier because we build good performance into its design.

A very remarkable connection exists between the BLUE and WLSE: the BLUE of $\theta$ is the special case of the WLSE of $\theta$ when $\mathbf{W}(k) = \mathbf{R}^{-1}(k)$. Consequently, all results obtained in Lessons 3, 4, and 5 for $\hat{\theta}_{\text{WLS}}(k)$ can be applied to $\hat{\theta}_{\text{BLU}}(k)$ by setting $\mathbf{W}(k) = \mathbf{R}^{-1}(k)$.

In recursive WLSs, matrix $\mathbf{P}(k)$ has no special meaning. In recursive BLUE, matrix $\mathbf{P}(k)$ turns out to be the covariance matrix for the error between $\theta$ and $\hat{\theta}_{\text{BLU}}(k)$. Recall that in Lesson 3 we showed that weighted least-squares estimates may change in numerical value under changes in scale. BLUEs are invariant under changes in scale.

The fact that $\mathbf{H}(k)$ must be deterministic limits the applicability of BLUEs.

When you complete this lesson you will be able to (1) explain and demonstrate how to design an estimator that processes the measurements linearly and has built into it the desirable properties of unbiasedness and efficiency; (2) connect this linear estimator (i.e., the BLUE) with the WLSE; (3) explain why the BLUE is insensitive to scale change; and (4) derive and use recursive BLUE algorithms.

## INTRODUCTION

Least-squares estimation, as described in Lessons 3, 4, and 5, is for the linear model

$$\mathbf{Z}(k) = \mathbf{H}(k)\boldsymbol{\theta} + \mathbf{V}(k) \tag{9-1}$$

where $\boldsymbol{\theta}$ is a deterministic, but unknown vector of parameters, $\mathbf{H}(k)$ can be deterministic or random, and we do not know anything about $\mathbf{V}(k)$ ahead of time. By minimizing $\tilde{\mathbf{Z}}'(k)\mathbf{W}(k)\tilde{\mathbf{Z}}(k)$, where $\tilde{\mathbf{Z}}(k) = \tilde{\mathbf{Z}}(k) - \mathbf{H}(k)\hat{\boldsymbol{\theta}}_{\text{WLS}}(k)$, we determined that $\hat{\boldsymbol{\theta}}_{\text{WLS}}(k)$ is a linear transformation of $\mathbf{Z}(k)$, i.e., $\hat{\boldsymbol{\theta}}_{\text{WLS}}(k) = \mathbf{F}_{\text{WLS}}(k)\mathbf{Z}(k)$. After establishing the structure of $\hat{\boldsymbol{\theta}}_{\text{WLS}}(k)$, we studied its small and large-sample properties. Unfortunately, $\hat{\boldsymbol{\theta}}_{\text{WLS}}(k)$ is not always unbiased or efficient. These properties were not built into $\hat{\boldsymbol{\theta}}_{\text{WLS}}(k)$ during its design.

In this lesson we develop our second estimator. It will be both unbiased and efficient, *by design*. In addition, we want the estimator to be a linear function of the measurements $\mathbf{Z}(k)$. This estimator is called a best linear unbiased estimator (BLUE) or an unbiased minimum-variance estimator (UMVE). To keep notation relatively simple, we will use $\hat{\boldsymbol{\theta}}_{\text{BLU}}(k)$ to denote the BLUE of $\boldsymbol{\theta}$.

As in least squares, we begin with the linear model in (9-1), where $\boldsymbol{\theta}$ is deterministic. Now, however, $\mathbf{H}(k)$ *must be deterministic and* $\mathbf{V}(k)$ *is assumed to be zero mean with positive definite known covariance matrix* $\mathbf{R}(k)$. An example of such a covariance matrix occurs for white noise. In the case of scalar measurements, $z(k)$, this means that scalar noise $\nu(k)$ is white, i.e.,

$$\mathbf{E}\{\nu(k)\nu(j)\} = \sigma_\nu^2(k)\delta_{kj} \tag{9-2}$$

where $\delta_{kj}$ is the Kronecker $\delta$ (i.e., $\delta_{kj} = 0$ for $k \neq j$ and $\delta_{kj} = 1$ for $k = j$). Thus,

$$\mathbf{R}(k) = \mathbf{E}\{\mathbf{V}(k)\mathbf{V}'(k)\} = \text{diag}[\sigma_\nu^2(k), \sigma_\nu^2(k-1), \ldots, \sigma_\nu^2(k-N+1)] \tag{9-3}$$

In the case of vector measurements, $\mathbf{z}(k)$, this means that vector noise, $\mathbf{v}(k)$, is white, i.e.,

$$\mathbf{E}\{\mathbf{v}(k)\mathbf{v}'(j)\} = \bar{\mathbf{R}}(k)\delta_{kj} \tag{9-4}$$

Thus,

$$\mathbf{R}(k) = \text{diag}[\bar{\mathbf{R}}(k), \bar{\mathbf{R}}(k-1), \ldots, \bar{\mathbf{R}}(k-N+1)], \tag{9-5}$$

i.e., $\mathbf{R}(k)$ is block diagonal.

## PROBLEM STATEMENT AND OBJECTIVE FUNCTION

We begin by assuming the following linear structure for $\hat{\boldsymbol{\theta}}_{\text{BLU}}(k)$:

$$\hat{\boldsymbol{\theta}}_{\text{BLU}}(k) = \mathbf{F}(k)\mathbf{Z}(k) \tag{9-6}$$

where, for notational simplicity, we have omitted subscripting $\mathbf{F}(k)$ as $\mathbf{F}_{\text{BLU}}(k)$. We shall design $\mathbf{F}(k)$ such that

⚹ **a.** $\hat{\theta}_{BLU}(k)$ is an unbiased estimator of $\theta$, and

⚹ **b.** the error variance for each of the $n$ parameters is minimized. In this way, $\hat{\theta}_{BLU}(k)$ *will be unbiased and efficient, by design.*

Recall, from Theorem 6-1, that *unbiasedness constrains design matrix* $\mathbf{F}(k)$, *such that*

$$\overset{n \times N}{\mathbf{F}(k)} \overset{N \times n}{\mathbf{H}(k)} = \mathbf{I}, \underset{n \times n}{} \quad \text{for all } k \qquad (9\text{-}7)$$

Our objective now is to choose the elements of $\mathbf{F}(k)$, subject to the constraint of (9-7), in such a way that the error variance for each of the $n$ parameters is minimized. In solving for $\mathbf{F}_{BLU}(k)$, it will be convenient to partition matrix $\mathbf{F}(k)$ as

$$\mathbf{F}(k) = \begin{pmatrix} \mathbf{f}'_1(k) \\ \mathbf{f}'_2(k) \\ \vdots \\ \hline \mathbf{f}'_n(k) \end{pmatrix} \qquad (9\text{-}8)$$

Equation (9-7) can now be expressed in terms of the vector components of $\mathbf{F}(k)$. For our purposes, it is easier to work with the transpose of (9-7), $\mathbf{H}'\mathbf{F}' = \mathbf{I}$, which can be expressed as

$$\mathbf{H}'[\mathbf{f}_1|\mathbf{f}_2|\ldots|\mathbf{f}_n] = [\mathbf{e}_1|\mathbf{e}_2|\ldots|\mathbf{e}_n] \qquad (9\text{-}9)$$

where $\mathbf{e}_i$ is the $i$th unit vector,

$$\mathbf{e}_i = \text{col}(0, 0, \ldots, 0, 1, 0, \ldots, 0) \qquad (9\text{-}10)$$

in which the nonzero element occurs in the $i$th position. Equating respective elements on both sides of (9-9), we find that

$$\mathbf{H}'(k)\mathbf{f}_i(k) = \mathbf{e}_i, \quad i = 1, 2, \ldots, n \qquad (9\text{-}11)$$

Our single unbiasedness constraint on matrix $\mathbf{F}(k)$ is now a set of $n$ constraints on the rows of $\mathbf{F}(k)$.

Next, we express $\mathbf{E}\{[\theta_i - \hat{\theta}_{i,BLU}(k)]^2\}$ in terms of $\mathbf{f}_i (i = 1, 2, \ldots, N)$. We shall make use of (9-11), (9-1), and the following equivalent representation of (9-6):

$$\hat{\theta}_{i,BLU}(k) = \mathbf{f}'_i(k)\mathbf{Z}(k), \quad i = 1, 2, \ldots, n \qquad (9\text{-}12)$$

Proceeding, we find that

$$\begin{aligned}
\mathbf{E}\{[\theta_i - \hat{\theta}_{i,BLU}(k)]^2\} &= \mathbf{E}\{(\theta_i - \mathbf{f}'_i\mathbf{Z})^2\} = \mathbf{E}\{(\theta_i - \mathbf{Z}'\mathbf{f}_i)^2\} \\
&= \mathbf{E}\{\theta_i^2 - 2\theta_i\mathbf{Z}'\mathbf{f}_i + (\mathbf{Z}'\mathbf{f}_i)^2\} \\
&= \mathbf{E}\{\theta_i^2 - 2\theta_i(\mathbf{H}\theta + \mathbf{V})'\mathbf{f}_i + [(\mathbf{H}\theta + \mathbf{V})'\mathbf{f}_i]^2\} \\
&= \mathbf{E}\{\theta_i^2 - 2\theta_i\theta'\mathbf{H}'\mathbf{f}_i - 2\theta_i\mathbf{V}'\mathbf{f}_i + [\theta'\mathbf{H}'\mathbf{f}_i + \mathbf{V}'\mathbf{f}_i]^2\} \\
&= \mathbf{E}\{\theta_i^2 - 2\theta_i\theta'\mathbf{e}_i - 2\theta_i\mathbf{V}'\mathbf{f}_i + [\theta'\mathbf{e}_i + \mathbf{V}'\mathbf{f}_i]^2\} \\
&= \mathbf{E}\{\mathbf{f}'_i\mathbf{V}\mathbf{V}'\mathbf{f}_i\} = \mathbf{f}'_i\mathbf{R}\mathbf{f}_i
\end{aligned} \qquad (9\text{-}13)$$

since $\mathbf{E}\{\mathbf{V}\mathbf{V}'\} = \mathbf{R}$

$\theta'\mathbf{e}_i = \theta_i$

$2\theta_i\theta'\mathbf{e}_i = 2\theta_i^2$

$(\mathbf{V}'\mathbf{f}_i)^2 = \mathbf{f}'_i\mathbf{V}\mathbf{V}'\mathbf{f}_i$

Observe that the error variance for the $i$th parameter depends only on the $i$th row of design matrix $\mathbf{F}(k)$. We, therefore, establish the following objective function:

$$\begin{aligned} J_i(\mathbf{f}_i, \boldsymbol{\lambda}_i) &= \mathbf{E}\{[\theta_i - \hat{\theta}_{i,\mathrm{BLU}}(k)]^2\} + \boldsymbol{\lambda}_i'(\mathbf{H}'\mathbf{f}_i - \mathbf{e}_i) \\ &= \mathbf{f}_i'\mathbf{R}\mathbf{f}_i + \boldsymbol{\lambda}_i'(\mathbf{H}'\mathbf{f}_i - \mathbf{e}_i) \end{aligned} \tag{9-14}$$

where $\boldsymbol{\lambda}_i$ is the $i$th vector of Lagrange multipliers (see the Supplementary Material at the end of this lesson for an explanation of Lagrange's method), which is associated with the $i$th unbiasedness constraint. Observe that the first term in (9-14) is associated with mean-squared error, whereas the second term is associated with unbiasedness. Our objective now is to minimize $J_i$ with respect to $\mathbf{f_i}$ and $\boldsymbol{\lambda}_i (i = 1, 2, \ldots, n)$.

Why is it permissible to determine $\mathbf{f}_i$ and $\boldsymbol{\lambda}_i$ for $i = 1, 2, \ldots, n$ by solving $n$ independent optimization problems? Ideally, we want to minimize all $J_i(\mathbf{f}_i, \boldsymbol{\lambda}_i)$'s *simultaneously*. Consider the following objective function:

$$J(\mathbf{f}_1, \mathbf{f}_2, \ldots, \mathbf{f}_n, \boldsymbol{\lambda}_1, \boldsymbol{\lambda}_2, \ldots, \boldsymbol{\lambda}_n) = \sum_{i=1}^{n} \mathbf{f}_i'\mathbf{R}\mathbf{f}_i + \sum_{i=1}^{n} \boldsymbol{\lambda}_i'(\mathbf{H}'\mathbf{f}_i - \mathbf{e}_i) \tag{9-15}$$

Observe from (9-14) that

$$J(\mathbf{f}_1, \mathbf{f}_2, \ldots, \mathbf{f}_n, \boldsymbol{\lambda}_1, \boldsymbol{\lambda}_2, \ldots, \boldsymbol{\lambda}_n) = \sum_{i=1}^{n} J_i(\mathbf{f}_i, \boldsymbol{\lambda}_i) \tag{9-16}$$

Clearly, the minimum of $J(\mathbf{f}_1, \mathbf{f}_2, \ldots, \mathbf{f}_n, \boldsymbol{\lambda}_1, \boldsymbol{\lambda}_2, \ldots, \boldsymbol{\lambda}_n)$ with respect to all $\mathbf{f}_1, \mathbf{f}_2, \ldots, \mathbf{f}_n, \boldsymbol{\lambda}_1, \boldsymbol{\lambda}_2, \ldots, \boldsymbol{\lambda}_n$ is obtained by minimizing each $J_i(\mathbf{f}_i, \boldsymbol{\lambda}_i)$ with respect to each $\mathbf{f}_i, \boldsymbol{\lambda}_i$. This is because the $J_i$'s are uncoupled; i.e., $J_1$ depends only on $\mathbf{f}_1$ and $\boldsymbol{\lambda}_1$, $J_2$ depends only on $\mathbf{f}_2$ and $\boldsymbol{\lambda}_2$, ..., and $J_n$ depends only on $\mathbf{f}_n$ and $\boldsymbol{\lambda}_n$.

## DERIVATION OF ESTIMATOR

A necessary condition for minimizing $J_i(\mathbf{f}_i, \boldsymbol{\lambda}_i)$ is $\partial J_i(\mathbf{f}_i, \boldsymbol{\lambda}_i)/\partial \mathbf{f}_i = \mathbf{0} (i = 1, 2, \ldots, n)$; hence,

$$2\mathbf{R}\mathbf{f}_i + \mathbf{H}\boldsymbol{\lambda}_i = \mathbf{0} \tag{9-17}$$

from which we determine $\mathbf{f}_i$ as

$$\mathbf{f}_i = -\frac{1}{2}\mathbf{R}^{-1}\mathbf{H}\boldsymbol{\lambda}_i \tag{9-18}$$

For (9-18) to be valid, $\mathbf{R}^{-1}$ must exist. Any noise $\mathbf{V}(k)$ whose covariance matrix $\mathbf{R}$ is positive definite qualifies. Of course, if $\mathbf{V}(k)$ is white, then $\mathbf{R}$ is diagonal (or block diagonal) and $\mathbf{R}^{-1}$ exists. This may also be true if $\mathbf{V}(k)$ is not white. A second necessary condition for minimizing $J_i(\mathbf{f}_i, \boldsymbol{\lambda}_i)$ is $\partial J_i(\mathbf{f}_i, \boldsymbol{\lambda}_i)/\partial \boldsymbol{\lambda}_i = \mathbf{0} (i = 1, 2, \ldots, n)$, which gives us the unbiasedness constraints

$$\mathbf{H}'\mathbf{f}_i = \mathbf{e}_i \qquad i = 1, 2, \ldots, n \tag{9-19}$$

To determine $\boldsymbol{\lambda}_i$, substitute (9-18) into (9-19). Doing this, we find that

$$\boldsymbol{\lambda}_i = -2(\mathbf{H}'\mathbf{R}^{-1}\mathbf{H})^{-1}\mathbf{e}_i \tag{9-20}$$

whereupon

$$\mathbf{f}_i = \mathbf{R}^{-1}\mathbf{H}(\mathbf{H}'\mathbf{R}^{-1}\mathbf{H})^{-1}\mathbf{e}_i \tag{9-21}$$

$(i = 1, 2, \ldots, n)$. Matrix $\mathbf{F}(k)$ is reconstructed from $\mathbf{f}_i(k)$, as follows:

$$\begin{aligned}\mathbf{F}'(k) &= (\mathbf{f}_1|\mathbf{f}_2|\ldots|\mathbf{f}_n)\\ &= \mathbf{R}^{-1}\mathbf{H}(\mathbf{H}'\mathbf{R}^{-1}\mathbf{H})^{-1}(\mathbf{e}_1|\mathbf{e}_2|\ldots|\mathbf{e}_n)\\ &= \mathbf{R}^{-1}\mathbf{H}(\mathbf{H}'\mathbf{R}^{-1}\mathbf{H})^{-1}\end{aligned} \tag{9-22}$$

Hence,

$$\mathbf{F}_{\text{BLU}}(k) = [\mathbf{H}'(k)\mathbf{R}^{-1}(k)\mathbf{H}(k)]^{-1}\mathbf{H}'(k)\mathbf{R}^{-1}(k) \tag{9-23}$$

which means that

$$\hat{\theta}_{\text{BLU}}(k) = [\mathbf{H}'(k)\mathbf{R}^{-1}(k)\mathbf{H}(k)]^{-1}\mathbf{H}'(k)\mathbf{R}^{-1}(k)\mathbf{Z}(k) \tag{9-24}$$

## COMPARISON OF $\hat{\theta}_{\text{BLU}}(k)$ AND $\hat{\theta}_{\text{WLS}}(k)$

We are struck by the close similarity between $\hat{\theta}_{\text{BLU}}(k)$ and $\hat{\theta}_{\text{WLS}}(k)$.

**Theorem 9-1.** *The BLUE of $\theta$ is the special case of the WLSE of $\theta$ when*

$$\mathbf{W}(k) = \mathbf{R}^{-1}(k) \tag{9-25}$$

*If $\mathbf{W}$(k) is diagonal, then (9-25) requires $\mathbf{V}$(k) to be white.*

*Proof.* Compare the formulas for $\hat{\theta}_{\text{BLU}}(k)$ in (9-24) and $\hat{\theta}_{\text{WLS}}(k)$ in (3-10). If $\mathbf{W}(k)$ is a diagonal matrix (which is required in the Lesson 5 derivation of the recursive WLSE), then $\mathbf{R}(k)$ is a diagonal matrix only if $\mathbf{V}(k)$ is white. □

Matrix $\mathbf{R}^{-1}(k)$ weights the contributions of precise measurements heavily and de-emphasizes the contributions of imprecise measurements. The best linear unbiased estimation design technique has led to a weighting matrix that is quite sensible.

**Corollary 9-1.** *All results obtained in Lessons 3, 4, and 5 for $\hat{\theta}_{\text{WLS}}$(k) can be applied to $\hat{\theta}_{\text{BLU}}$(k) by setting $\mathbf{W}$(k) $= \mathbf{R}^{-1}$(k).* □

We leave it to the reader to explore the full implications of this important corollary by reexamining the wide range of topics that was discussed in Lessons 3, 4, and 5.

**Theorem 9-2** (Gauss–Markov Theorem). *If $\mathbf{H}$(k) is deterministic and $\mathbf{R}$(k) $=$ $\sigma_v^2\mathbf{I}$, then $\hat{\theta}_{\text{BLU}}$(k) $=$ $\hat{\theta}_{\text{LS}}$(k).*

*Proof.* Using (9-22) and the fact that $\mathbf{R}(k) = \sigma_v^2\mathbf{I}$, we find that

$$\hat{\theta}_{\text{BLU}}(k) = \left(\frac{1}{\sigma_v^2}\mathbf{H}'\mathbf{H}\right)^{-1}\mathbf{H}'\frac{1}{\sigma_v^2}\mathbf{Z} = (\mathbf{H}'\mathbf{H})^{-1}\mathbf{H}'\mathbf{Z} = \hat{\theta}_{\text{LS}}(k) \quad \square$$

Why is this a very important result? We have connected two seemingly different estimators, one of which, $\hat{\theta}_{\text{BLU}}(k)$, has the properties of unbiased and minimum variance by design; hence, in this case $\hat{\theta}_{\text{LS}}(k)$ inherits these properties. Remember though that the derivation of $\hat{\theta}_{\text{BLU}}(k)$ required $\mathbf{H}(k)$ to be deterministic.

## SOME PROPERTIES OF $\hat{\theta}_{\text{BLU}}(k)$

To begin, we direct our attention at the covariance matrix of parameter estimation error $\hat{\theta}_{\text{BLU}}(k)$.

**Theorem 9-3.** *If* $\mathbf{V}(\mathrm{k})$ *is zero mean, then*

$$\operatorname{cov}[\tilde{\theta}_{\text{BLU}}(k)] = [\mathbf{H}'(k)\mathbf{R}^{-1}(k)\mathbf{H}(k)]^{-1} \tag{9-26}$$

*Proof.* We apply Corollary 9-1 to $\operatorname{cov}[\tilde{\theta}_{\text{WLS}}(k)]$ [given in (8-18)] for the case when $\mathbf{H}(k)$ is deterministic, to see that

$$\begin{aligned}\operatorname{cov}[\tilde{\theta}_{\text{BLU}}(k)] &= \operatorname{cov}[\tilde{\theta}_{\text{WLS}}(k)]\Big|_{\mathbf{W}(k)=\mathbf{R}^{-1}(k)} \\ &= (\mathbf{H}'\mathbf{R}^{-1}\mathbf{H})^{-1}\mathbf{H}'\mathbf{R}^{-1}\mathbf{R}\mathbf{R}^{-1}\mathbf{H}(\mathbf{H}'\mathbf{R}^{-1}\mathbf{H})^{-1} \\ &= (\mathbf{H}'\mathbf{R}^{-1}\mathbf{H})^{-1} \quad \square\end{aligned}$$

Observe the great simplification of the expression for $\operatorname{cov}[\tilde{\theta}_{\text{WLS}}(k)]$, when $\mathbf{W}(k) = \mathbf{R}^{-1}(k)$. Note, also, that the error variance of $\hat{\theta}_{i,\text{BLU}}(k)$ is given by the $i$th diagonal element of $\operatorname{cov}[\tilde{\theta}_{\text{BLU}}(k)]$.

**Corollary 9-2.** *When* $\mathbf{W}(\mathrm{k}) = \mathbf{R}^{-1}(\mathrm{k})$, *then matrix* $\mathbf{P}(\mathrm{k})$, *which appears in the recursive WLSE of* $\theta$ *equals* $\operatorname{cov}[\tilde{\theta}_{\text{BLU}}(\mathrm{k})]$, *i.e.,*

$$\mathbf{P}(k) = \operatorname{cov}[\tilde{\theta}_{\text{BLU}}(k)] \tag{9-27}$$

*Proof.* Recall Equation (4-13), that

$$\mathbf{P}(k) = [\mathbf{H}'(k)\mathbf{W}(k)\mathbf{H}(k)]^{-1} \tag{9-28}$$

When $\mathbf{W}(k) = \mathbf{R}^{-1}(k)$, then

$$\mathbf{P}(k) = [\mathbf{H}'(k)\mathbf{R}^{-1}(k)\mathbf{H}(k)]^{-1} \tag{9-29}$$

Hence, $\mathbf{P}(k) = \operatorname{cov}[\tilde{\theta}_{\text{BLU}}(k)]$ because of (9-26). $\square$

Soon we will examine a recursive BLUE. Matrix $\mathbf{P}(k)$ will have to be calculated, just as it has to be calculated for the recursive WLSE. Every time $\mathbf{P}(k)$ is calculated in our recursive BLUE, we obtain a quantitative measure of how well we are estimating $\theta$. Just look at the diagonal elements of $\mathbf{P}(k)$, $k = 1, 2, \ldots$. The same statement cannot be made for the meaning of $\mathbf{P}(k)$ in the recursive WLSE. In the recursive WLSE, $\mathbf{P}(k)$ has no special meaning.

Next, we examine $\text{cov}\,[\tilde{\theta}_{BLU}(k)]$ in more detail.

**Theorem 9-4.** $\hat{\theta}_{BLU}(k)$ *is a most efficient estimator of* $\theta$ *within the class of all unbiased estimators that are linearly related to the measurements* $\mathbf{Z}(k)$.

In the econometric's literature (e.g., Fomby et al., 1984) this theorem is known as *Aitken's Theorem* (Aitken, 1935).

*Proof* (Mendel, 1973, pp. 155–156). According to Definition 6-3, we must show that

$$\mathbf{\Sigma} = \text{cov}\,[\tilde{\theta}_a(k)] - \text{cov}\,[\tilde{\theta}_{BLU}(k)] \tag{9-30}$$

is positive semidefinite. In (9-30), $\tilde{\theta}_a(k)$ is the error associated with an arbitrary linear unbiased estimate of $\theta$. For convenience, we write $\mathbf{\Sigma}$ as

$$\mathbf{\Sigma} = \mathbf{\Sigma}_a - \mathbf{\Sigma}_{BLU} \tag{9-31}$$

To compute $\mathbf{\Sigma}_a$ we use the facts that

$$\hat{\theta}_a(k) = \mathbf{F}_a(k)\mathbf{Z}(k) \tag{9-32}$$

and

$$\mathbf{F}_a(k)\mathbf{H}(k) = \mathbf{I} \tag{9-33}$$

Thus,

$$\begin{aligned}
\mathbf{\Sigma}_a &= \mathbf{E}\{(\theta - \mathbf{F}_a\mathbf{Z})(\theta - \mathbf{F}_a\mathbf{Z})'\}\\
&= \mathbf{E}\{(\theta - \mathbf{F}_a\mathbf{H}\theta - \mathbf{F}_a\mathbf{V})(\theta - \mathbf{F}_a\mathbf{H}\theta - \mathbf{F}_a\mathbf{V})'\}\\
&= \mathbf{E}\{(\mathbf{F}_a\mathbf{V})(\mathbf{F}_a\mathbf{V})'\}\\
&= \mathbf{F}_a\mathbf{R}\mathbf{F}'_a
\end{aligned} \tag{9-34}$$

Because $\hat{\theta}_{BLU}(k) = \mathbf{F}(k)\mathbf{Z}(k)$ and $\mathbf{F}(k)\mathbf{H}(k) = \mathbf{I}$,

$$\mathbf{\Sigma}_{BLU} = \mathbf{FRF}' \tag{9-35}$$

Substituting (9-34) and (9-35) into (9-31) and making repeated use of the unbiasedness constraints $\mathbf{H}'\mathbf{F}' = \mathbf{H}'\mathbf{F}'_a = \mathbf{F}_a\mathbf{H}_a = \mathbf{I}$, we find that

$$\begin{aligned}
\mathbf{\Sigma} &= \mathbf{F}_a\mathbf{R}\mathbf{F}'_a - \mathbf{FRF}'\\
&= \mathbf{F}_a\mathbf{R}\mathbf{F}'_a - \mathbf{FRF}' + 2(\mathbf{H}'\mathbf{R}^{-1}\mathbf{H})^{-1} - (\mathbf{H}'\mathbf{R}^{-1}\mathbf{H})^{-1} - (\mathbf{H}'\mathbf{R}^{-1}\mathbf{H})^{-1}\\
&= \mathbf{F}_a\mathbf{R}\mathbf{F}'_a - \mathbf{FRF}' + 2(\mathbf{H}'\mathbf{R}^{-1}\mathbf{H})^{-1}(\mathbf{H}'\mathbf{R}^{-1}\mathbf{RF}')\\
&\quad -(\mathbf{H}'\mathbf{R}^{-1}\mathbf{H})^{-1}(\mathbf{H}'\mathbf{R}^{-1}\mathbf{RF}'_a) - (\mathbf{F}_a\mathbf{R}\mathbf{R}^{-1}\mathbf{H})(\mathbf{H}'\mathbf{R}^{-1}\mathbf{H})^{-1}
\end{aligned} \tag{9-36}$$

Making use of the structure of $\mathbf{F}(k)$, given in (9-23), we see that $\mathbf{\Sigma}$ can also be written as

$$\begin{aligned}
\mathbf{\Sigma} &= \mathbf{F}_a\mathbf{R}\mathbf{F}'_a - \mathbf{FRF}' + 2\mathbf{FRF}' - \mathbf{FRF}'_a - \mathbf{F}_a\mathbf{RF}'\\
&= (\mathbf{F}_a - \mathbf{F})\mathbf{R}(\mathbf{F}_a - \mathbf{F})'
\end{aligned} \tag{9-37}$$

To investigate the definiteness of $\mathbf{\Sigma}$, consider the definiteness of $\mathbf{a}'\mathbf{\Sigma}\mathbf{a}$, where $\mathbf{a}$ is an arbitrary nonzero vector:

$$\mathbf{a}'\mathbf{\Sigma}\mathbf{a} = [(\mathbf{F}_a - \mathbf{F})'\mathbf{a}]'\mathbf{R}[(\mathbf{F}_a - \mathbf{F})'\mathbf{a}] \tag{9-38}$$

Matrix $\mathbf{F}$ (i.e., $\mathbf{F}_{\text{BLU}}$) is unique; therefore, $(\mathbf{F}_a - \mathbf{F})'\mathbf{a}$ is a nonzero vector, unless $\mathbf{F}_a - \mathbf{F}$ and $\mathbf{a}$ are orthogonal, which is a possibility that cannot be excluded. Because matrix $\mathbf{R}$ is positive definite, $\mathbf{a}'\mathbf{\Sigma}\mathbf{a} \geq 0$, which means that $\mathbf{\Sigma}$ is positive semidefinite. □

These results serve as further confirmation that designing $\mathbf{F}(k)$ as we have done, by minimizing only the diagonal elements of cov $[\tilde{\theta}_{\text{BLU}}(k)]$, is sound.

Theorem 9-4 proves that $\hat{\theta}_{\text{BLU}}(k)$ is most efficient within the class of linear estimators. The cov $[\tilde{\theta}_{\text{BLU}}(k)]$ is given in (9-26) as $[\mathbf{H}'(k)\mathbf{R}^{-1}(k)\mathbf{H}(k)]^{-1}$; hence, it must be true that the Fisher information matrix [see (6-45)] for this situation is $\mathbf{J}(\theta) = \mathbf{H}'(k)\mathbf{R}^{-1}(k)\mathbf{H}(k)$. We have been able to obtain the Fisher information matrix in this lesson *without knowledge of the probability density function for* $\mathbf{Z}(k)$, because our BLU estimator has been *designed* to be of minimum variance within the class of linear estimators.

Next, let us compare our just determined Fisher information matrix with the results in Example 6-4. In that example, $\mathbf{J}(\theta)$ was obtained for the generic linear model in (9-1) under the assumptions that $\mathbf{H}(k)$ and $\theta$ are deterministic and that generic noise $\mathbf{V}(k)$ is zero mean multivariate Gaussian with covariance matrix equal to $\mathbf{R}(k)$. The latter assumption then lets us determine a probability density function for $\mathbf{Z}(k)$, after which we were able to compute the Fisher information matrix, in (6-56), as $\mathbf{J}(\theta) = \mathbf{H}'(k)\mathbf{R}^{-1}(k)\mathbf{H}(k)$.

How is it possible that we have arrived at exactly the same Fisher information matrix for these two different situations? The answer is simple. The Fisher information matrix in (6-56) is not dependent on the specific structure of an estimator. It merely provides a lower bound for cov $[\tilde{\theta}(k)]$. We have shown, in this lesson, that a linear estimator achieves this lower bound. It is conceivable that a nonlinear estimator might also achieve this lower bound; but, because linear processing is computationally simpler than nonlinear processing, there is no need to search for such a nonlinear estimator.

**Corollary 9-3.** *If* $\mathbf{R}(\mathrm{k}) = \sigma_v^2\mathbf{I}$, *then* $\hat{\theta}_{\text{LS}}(\mathrm{k})$ *is a most efficient estimator of* $\theta$.

The proof of this result is a direct consequence of Theorems 9-2 and 9-4. □

Note that we also proved the truth of Corollary 9-3 in Lesson 8 in Theorem 8-3. The proof of Theorem 8-3 is totally within the context of least squares, whereas the proof of Corollary 9-3 is totally within the context of BLUE, which, of course, is linked to least squares by the Gauss–Markov Theorem 9-2.

At the end of Lesson 3 we noted that $\hat{\theta}_{\text{WLS}}(\mathrm{k})$ may not be invariant under scale changes. We demonstrate next that $\hat{\theta}_{\text{BLU}}(k)$ is invariant to such changes.

**Theorem 9-5.** $\hat{\theta}_{\text{BLU}}(\mathrm{k})$ *is invariant under changes of scale.*

*Proof* (Mendel, 1973, pp. 156–157). Assume that observers $A$ and $B$ are observing a process; but observer $A$ reads the measurements in one set of units and $B$ in another. Let $\mathbf{M}$ be a symmetric matrix of scale factors relating $A$ to $B$ (e.g., 5280 ft/mile, 454 g/lb, etc.), and $\mathbf{Z}_A(k)$ and $\mathbf{Z}_B(k)$ denote the total measurement vectors of $A$ and $B$, respectively. Then

$$\mathbf{Z}_B(k) = \mathbf{H}_B(k)\theta + \mathbf{V}_B(k) = \mathbf{M}\mathbf{Z}_A(k) = \mathbf{M}\mathbf{H}_A(k)\theta + \mathbf{M}\mathbf{V}_A(k) \tag{9-39}$$

which means that

$$\mathbf{H}_B(k) = \mathbf{M}\mathbf{H}_A(k) \tag{9-40}$$

$$\mathbf{V}_B(k) = \mathbf{M}\mathbf{V}_A(k) \tag{9-41}$$

and

$$\mathbf{R}_B(k) = \mathbf{M}\mathbf{R}_A(k)\mathbf{M}' = \mathbf{M}\mathbf{R}_A(k)\mathbf{M} \tag{9-42}$$

Let $\hat{\theta}_{\text{A,BLU}}(k)$ and $\hat{\theta}_{\text{B,BLU}}(k)$ denote the BLUEs associated with observers $A$ and $B$, respectively; then,

$$\begin{aligned}\hat{\theta}_{B,\text{BLU}}(k) &= (\mathbf{H}'_B\mathbf{R}_B^{-1}\mathbf{H}_B)^{-1}\mathbf{H}'_B\mathbf{R}_B^{-1}\mathbf{Z}_B \\ &= [\mathbf{H}'_A\mathbf{M}(\mathbf{M}\mathbf{R}_A\mathbf{M})^{-1}\mathbf{M}\mathbf{H}_A]^{-1}\mathbf{H}'_A\mathbf{M}(\mathbf{M}\mathbf{R}_A\mathbf{M})^{-1}\mathbf{M}\mathbf{Z}_A \\ &= (\mathbf{H}'_A\mathbf{M}\mathbf{M}^{-1}\mathbf{R}_A^{-1}\mathbf{M}^{-1}\mathbf{M}\mathbf{H}_A)^{-1}\mathbf{H}'_A\mathbf{M}\mathbf{M}^{-1}\mathbf{R}_A^{-1}\mathbf{M}^{-1}\mathbf{M}\mathbf{Z}_A \\ &= (\mathbf{H}'_A\mathbf{R}_A^{-1}\mathbf{H}_A)^{-1}\mathbf{H}'_A\mathbf{R}_A^{-1}\mathbf{Z}_A \\ &= \hat{\theta}_{A,\text{BLU}}(k) \quad \square\end{aligned} \tag{9-43}$$

Here is an interesting interpretation of Theorem 9-5. Suppose we "whiten" the data by multiplying $\mathbf{Z}(k)$ by $\mathbf{R}^{-1/2}(k)$. Doing this, our generic linear model can be written as

$$\mathbf{Z}_1(k) = \mathbf{H}_1(k)\theta + \mathbf{V}_1(k) \tag{9-44}$$

where $\mathbf{Z}_1(k) = \mathbf{R}^{-1/2}(k)\mathbf{Z}(k)$, $\mathbf{H}_1(k) = \mathbf{R}^{-1/2}(k)\mathbf{H}(k)$, and $\mathbf{V}_1(k) = \mathbf{R}^{-1/2}(k)\mathbf{V}(k)$. Note that, when the data are whitened, $\text{cov}\,\mathbf{V}_1(k) = \mathbf{R}_1 = \mathbf{I}$. If (9-44) is the starting point for BLUE, then

$$\begin{aligned}\hat{\theta}_{\text{BLU}}(k) &= [\mathbf{H}'_1(k)\mathbf{H}_1(k)]^{-1}\mathbf{H}'_1(k)\mathbf{Z}_1(k) \\ &= [\mathbf{H}'(k)\mathbf{R}^{-1/2}(k)\mathbf{R}^{-1/2}(k)\mathbf{H}(k)]^{-1}\mathbf{H}'(k)\mathbf{R}^{-1/2}(k)\mathbf{R}^{-1/2}(k)\mathbf{Z}(k) \\ &= [\mathbf{H}'(k)\mathbf{R}^{-1}(k)\mathbf{H}(k)]^{-1}\mathbf{H}'(k)\mathbf{R}^{-1}(k)\mathbf{Z}(k)\end{aligned} \tag{9-45}$$

which is precisely the correct formula for $\hat{\theta}_{\text{BLU}}(k)$. Creating $\mathbf{R}^{-1/2}(k)\mathbf{Z}(k)$ properly normalizes the data. This normalization is not actually performed on the data in BLUE. *It is performed by the BLUE algorithm automatically.*

This suggests that we should choose the normalization matrix $\mathbf{N}$ in Lesson 3 as $\mathbf{R}^{-1/2}(k)$; but this can only be done if we know the covariance matrix of $\mathbf{V}(k)$, *something that was not assumed known for the method of least squares.* When $\mathbf{R}(k)$ is known and $\mathbf{H}(k)$ is deterministic, then use $\hat{\theta}_{\text{BLU}}(k)$. When $\mathbf{R}(k)$ is known and

$\mathbf{H}(k)$ is not deterministic, use $\hat{\theta}_{WLS}(k)$ with $\mathbf{W}(k) = \mathbf{R}^{-1}(k)$. We cannot use $\hat{\theta}_{BLU}(k)$ in the latter case because the BLUE was derived under the assumption that $\mathbf{H}(k)$ is deterministic.

## RECURSIVE BLUEs

Because of Corollary 9-1, we obtain recursive formulas for the BLUE of $\theta$ by setting $1/w(k+1) = r(k+1)$ in the recursive formulas for the WLSEs of $\theta$, which are given in Lesson 5. In the case of a vector of measurements, we set (see Table 5-1) $\mathbf{w}^{-1}(k+1) = \mathbf{R}(k+1)$.

**Theorem 9-6** (Information form of recursive BLUE). *A recursive structure for $\hat{\theta}_{BLU}(k)$ is*

$$\hat{\theta}_{BLU}(k+1) = \hat{\theta}_{BLU}(k) + \mathbf{K_B}(k+1)[z(k+1) - \mathbf{h}'(k+1)\hat{\theta}_{BLU}(k)] \tag{9-46}$$

*where*

$$\mathbf{K}_B(k+1) = \mathbf{P}(k+1)\mathbf{h}(k+1)r^{-1}(k+1) \tag{9-47}$$

*and*

$$\mathbf{P}^{-1}(k+1) = \mathbf{P}^{-1}(k) + \mathbf{h}(k+1)r^{-1}(k+1)\mathbf{h}'(k+1) \tag{9-48}$$

*These equations are initialized by $\hat{\theta}_{BLU}(n)$ and $\mathbf{P}^{-1}(n)$ [where $\mathbf{P}(k)$ is $\text{cov}\,[\tilde{\theta}_{BLU}(k)]$, given in (9-26)] and are used for $k = n, n+1, \ldots, N-1$. These equations can also be used for $k = 0, 1, \ldots, N-1$ as long as $\hat{\theta}_{BLU}(0)$ and $\mathbf{P}^{-1}(0)$ are chosen using Equations (5-21) and (5-20) in Lesson 5, respectively, in which $w(0)$ is replaced by $r^{-1}(0)$.* □

**Theorem 9-7** (Covariance form of recursive BLUE). *Another recursive structure for $\hat{\theta}_{BLU}(k)$ is (9-46) in which*

$$\mathbf{K_B}(k+1) = \mathbf{P}(k)\mathbf{h}(k+1)[\mathbf{h}'(k+1)\mathbf{P}(k)\mathbf{h}(k+1) + r(k+1)]^{-1} \tag{9-49}$$

*and*

$$\mathbf{P}(k+1) = [\mathbf{I} - \mathbf{K_B}(k+1)\mathbf{h}'(k+1)]\mathbf{P}(k) \tag{9-50}$$

*These equations are initialized by $\hat{\theta}_{BLU}(n)$ and $\mathbf{P}(n)$ and are used for $k = n, n+1, \ldots, N-1$. They can also be used for $k = 0, 1, \ldots, N-1$ as long as $\hat{\theta}_{BLU}(0)$ and $\mathbf{P}(0)$ are chosen using Equations (5-21) and (5-20), respectively, in which $w(0)$ is replaced by $r^{-1}(0)$.* □

Recall that, in best linear unbiased estimation, $\mathbf{P}(k) = \text{cov}\,[\tilde{\theta}_{BLU}(k)]$. Observe, in Theorem 9-7, that we compute $\mathbf{P}(k)$ recursively, and not $\mathbf{P}^{-1}(k)$. This is why the results in Theorem 9-7 (and, subsequently, Theorem 5-2) are referred to as the *covariance form* of recursive BLUE.

See Problem 9-10 for a generalization of the recursive BLUE to a nondiagonal symmetric covariance matrix.

## COMPUTATION

Because BLUE is a special case of WLSE, refer to Lessons 4 and 5 for discussions on computation. Of course, we must now set $\mathbf{W}(k) = \mathbf{R}^{-1}(k)$ in batch algorithms, or $w(k) = 1/r(k)$ in recursive algorithms.

### Supplementary Material

## LAGRANGE'S METHOD

Lagrange's method for handling optimization problems with constraints was devised by the great eighteenth century mathematician Joseph Louis Lagrange. In the case of optimizing $F(\mathbf{x})$ subject to the scalar constraint $G(\mathbf{x}) = g$, Lagrange's method tells us to proceed as follows: form the function

$$J(\mathbf{x}, \lambda) = F(\mathbf{x}) + \lambda G(\mathbf{x}) \tag{9-51}$$

where $\lambda$ is a constant, as yet undetermined in value. Treat the elements of $\mathbf{x}$ as independent variables, and write down the conditions

$$\frac{\partial J(\mathbf{x}, \lambda)}{\partial x_1} = 0, \quad \frac{\partial J(\mathbf{x}, \lambda)}{\partial x_2} = 0, \ldots, \frac{\partial J(\mathbf{x}, \lambda)}{\partial x_n} = 0 \tag{9-52}$$

Solve these $n$ equations along with the equation of the constraint

$$G(\mathbf{x}) = g \tag{9-53}$$

to find the values of the $n + 1$ quantities $x_1, x_2, \ldots, x_n, \lambda$. More than one point $(x_1, x_2, \ldots, x_n, \lambda)$ may be found in this way, but among the points so found will be the points of optimal values of $F(\mathbf{x})$.

Equivalently, form the function

$$J_1(\mathbf{x}, \lambda) = F(\mathbf{x}) + \lambda[G(\mathbf{x}) - g] \tag{9-54}$$

where $\lambda$ is treated as before. In addition to the $n$ equations $\partial J(\mathbf{x}, \lambda)/\partial x_i = 0$, write down the condition

$$\frac{\partial J_1(\mathbf{x}, \lambda)}{\partial \lambda} = G(\mathbf{x}) - g = 0 \tag{9-55}$$

Solve these $n + 1$ equations for $\mathbf{x}$ and $\lambda$.

The parameter $\lambda$ is called a *Lagrange multiplier*. Multiple constraints are handled by introducing one Lagrange multiplier for each constraint. So, for example, the vector of $n_c$ constraints $\mathbf{G}(\mathbf{x}) = \mathbf{g}$ is handled in (9-54) by replacing the term $\lambda[G(\mathbf{x}) - g]$ by the term $\boldsymbol{\lambda}'[\mathbf{G}(\mathbf{x}) - \mathbf{g}]$.

## SUMMARY QUESTIONS

1. For a BLUE:
   (a) $\mathbf{H}(k)$ and $\mathbf{V}(k)$ must be known
   (b) $\mathbf{H}(k)$ may be random
   (c) $\mathbf{H}(k)$ must be deterministic and $\mathbf{V}(k)$ may be zero mean
2. $\hat{\boldsymbol{\theta}}_{\text{BLU}}(k) = \hat{\boldsymbol{\theta}}_{\text{LS}}(k)$ if:
   (a) $\mathbf{R}(k) = \mathbf{W}(k)$
   (b) $\mathbf{R}(k) = w\mathbf{I}$
   (c) $\mathbf{R}(k) = \sigma_\nu^2\mathbf{I}$
3. By design, a BLUE is:
   (a) consistent and unbiased
   (b) efficient and unbiased
   (c) a least-squares estimator
4. $\hat{\boldsymbol{\theta}}_{\text{BLU}}(k)$ is invariant under scale changes because:
   (a) it was designed to be so
   (b) $\mathbf{W}(k) = \mathbf{R}^{-1}(k)$
   (c) $\mathbf{R}_B(k) = \mathbf{M}\mathbf{R}_A(k)\mathbf{M}$
5. The name *covariance form* of the recursive BLUE (and, subsequently, recursive WLSE) arises, because:
   (a) $\mathbf{P}(k) = \text{cov}\,[\tilde{\boldsymbol{\theta}}_{\text{BLU}}(k)]$ and we compute $\mathbf{P}(k)$ recursively
   (b) $\mathbf{P}(k) = \text{cov}\,[\tilde{\boldsymbol{\theta}}_{\text{BLU}}(k)]$ and we compute $\mathbf{P}^{-1}(k)$ recursively
   (c) $\mathbf{P}(k) = \text{cov}\,[\tilde{\boldsymbol{\theta}}_{\text{BLU}}(k)]$
6. The BLUE of $\boldsymbol{\theta}$ is the special case of the WLSE of $\boldsymbol{\theta}$ when $\mathbf{W}(k)$ equals:
   (a) $\mathbf{R}(k)$
   (b) $\mathbf{R}^{-1}(k)$
   (c) $w\mathbf{I}$
7. Check off the two design constraints imposed on the BLUE:
   (a) $\hat{\boldsymbol{\theta}}_{\text{BLU}}(k)$ is an unbiased estimator of $\boldsymbol{\theta}$
   (b) $\hat{\boldsymbol{\theta}}_{\text{BLU}}(k)$ is a linear function of the measurements
   (c) the error variance of each of the $n$ parameters is minimized
8. Recursive formulas for the BLUE of $\boldsymbol{\theta}$ are obtained from the recursive formulas in Lesson 5 for the WLSE of $\boldsymbol{\theta}$ by setting:
   (a) $w(k+1) = \text{constant}$
   (b) $1/w(k+1) = r(k+1)$
   (c) $w(k+1) = r(k+1)$
9. When $\mathbf{H}(k)$ is random, computations in the derivation of $\hat{\boldsymbol{\theta}}_{\text{BLU}}(k)$ break down in:
   (a) the unbiasedness constraint
   (b) computing the Lagrange multipliers
   (c) computing the error variance
10. When $\mathbf{W}(k) = \mathbf{R}^{-1}(k)$, then matrix $\mathbf{P}(k)$, which appears in the recursive WLSE of $\boldsymbol{\theta}$, equals:
    (a) $\text{cov}\,[\tilde{\boldsymbol{\theta}}_{\text{BLU}}(k)]$
    (b) a diagonal matrix
    (c) $\mathbf{W}(k)$

## PROBLEMS

**9-1.** (Mendel, 1973, Exercise 3-2, p. 175). Assume $\mathbf{H}(k)$ is random and that $\hat{\boldsymbol{\theta}}_{\text{BLU}}(k) = \mathbf{F}(k)\mathbf{Z}(k)$.

**(a)** Show that unbiasedness of the estimate is attained when $\mathbf{E}\{\mathbf{F}(k)\mathbf{H}(k)\} = \mathbf{I}$.

**(b)** At what point in the derivation of $\hat{\boldsymbol{\theta}}_{\text{BLU}}(k)$ do the computations break down because $\mathbf{H}(k)$ is random?

**9-2.** Here we examine the situation when $\mathbf{V}(k)$ is colored noise and how to use a model to compute $\mathbf{R}(k)$. Now our linear model is

$$\mathbf{z}(k+1) = \mathbf{H}(k+1)\boldsymbol{\theta} + \mathbf{v}(k+1)$$

where $\mathbf{v}(k)$ is colored noise modeled as

$$\mathbf{v}(k+1) = \mathbf{A}_v\mathbf{v}(k) + \boldsymbol{\xi}(k)$$

We assume that deterministic matrix $\mathbf{A}_v$ is known and that $\boldsymbol{\xi}(k)$ is zero-mean white noise with covariance $\mathbf{R}_{\xi}(k)$. Working with the *measurement difference* $\mathbf{z}^*(k+1) = \mathbf{z}(k+1) - \mathbf{A}_v\mathbf{z}(k)$, write down the formula for $\hat{\boldsymbol{\theta}}_{\text{BLU}}(k)$ in batch form. Be sure to define all concatenated quantities.

**9-3.** (Sorenson, 1980, Exercise 3-15, p. 130). Suppose $\hat{\theta}_1$ and $\hat{\theta}_2$ are unbiased estimators of $\theta$ with var $(\hat{\theta}_1) = \sigma_1^2$ and var $(\hat{\theta}_2) = \sigma_2^2$. Let $\hat{\theta}_3 = \alpha\hat{\theta}_1 + (1-\alpha)\hat{\theta}_2$.

**(a)** Prove that $\hat{\theta}_3$ is unbiased.

**(b)** Assume that $\hat{\theta}_1$ and $\hat{\theta}_2$ are statistically independent, and find the mean-squared error of $\hat{\theta}_3$.

**(c)** What choice of $\alpha$ minimizes the mean-squared error?

**9-4.** (Mendel, 1973, Exercise 3-12, pp. 176–177). A series of measurements $\mathbf{z}(k)$ is made, where $\mathbf{z}(k) = \mathbf{H}\boldsymbol{\theta} + \mathbf{v}(k)$, $\mathbf{H}$ is an $m \times n$ *constant* matrix, $\mathbf{E}\{\mathbf{v}(k)\} = \mathbf{0}$, and $\text{cov}[\mathbf{v}(k)] = \mathbf{R}$ is a constant matrix.

**(a)** Using the two formulations of the recursive BLUE, show that (Ho, 1963, pp. 152–154):

**(i)** $\mathbf{P}(k+1)\mathbf{H}' = \mathbf{P}(k)\mathbf{H}'[\mathbf{H}\mathbf{P}(k)\mathbf{H}' + \mathbf{R}]^{-1}\mathbf{R}$, and

**(ii)** $\mathbf{H}\mathbf{P}(k) = \mathbf{R}[\mathbf{H}\mathbf{P}(k-1)\mathbf{H}' + \mathbf{R}]^{-1}\mathbf{H}\mathbf{P}(k-1)$.

**(b)** Next, show that

**(i)** $\mathbf{P}(k)\mathbf{H}' = \mathbf{P}(k-2)\mathbf{H}'[2\mathbf{H}\mathbf{P}(k-2)\mathbf{H}' + \mathbf{R}]^{-1}\mathbf{R}$;

**(ii)** $\mathbf{P}(k)\mathbf{H}' = \mathbf{P}(k-3)\mathbf{H}'[3\mathbf{H}\mathbf{P}(k-3)\mathbf{H}' + \mathbf{R}]^{-1}\mathbf{R}$; and

**(iii)** $\mathbf{P}(k)\mathbf{H}' = \mathbf{P}(0)\mathbf{H}'[k\mathbf{H}\mathbf{P}(0)\mathbf{H}' + \mathbf{R}]^{-1}\mathbf{R}$.

**(c)** Finally, show that the asymptotic form ($k \to \infty$) for the BLUE of $\boldsymbol{\theta}$ is (Ho, 1963, pp. 152–154)

$$\begin{aligned}\hat{\boldsymbol{\theta}}_{\text{BLU}}(k+1) &= \hat{\boldsymbol{\theta}}_{\text{BLU}}(k)\\ &\quad + \frac{1}{k+1}\mathbf{P}(0)\mathbf{H}'[\mathbf{H}\mathbf{P}(0)\mathbf{H}']^{-1}[\mathbf{z}(k+1) - \mathbf{H}\hat{\boldsymbol{\theta}}_{\text{BLU}}(k)]\end{aligned}$$

This equation, with its $1/(k+1)$ weighting function, represents a form of multidimensional stochastic approximation.

**9-5.** (Iraj Manocheri, Spring 1992) We are given the BLUE $\hat{\boldsymbol{\theta}}_{\text{BLU:1}}(k)$ of $\boldsymbol{\theta}$ that is based on measurements $\mathbf{Z}_1(k)$, where $\mathbf{Z}_1(k) = \mathbf{H}_1(k)\boldsymbol{\theta} + \mathbf{V}_1(k)$, $\mathbf{E}\{\mathbf{V}_1(k)\} = \mathbf{0}$, and $\mathbf{E}\{\mathbf{V}_1(k)\mathbf{V}_1'(k)\} = \mathbf{R}_1(k)$. Additional measurements $\mathbf{Z}_2(k) = \mathbf{H}_2(k)\boldsymbol{\theta} + \mathbf{V}_2(k)$,

$\mathbf{E}\{\mathbf{V}_2(k)\} = \mathbf{0}$, and $\mathbf{E}\{\mathbf{V}_2(k)\mathbf{V}_2'(k)\} = \mathbf{R}_2(k)$ are obtained, where $\mathbf{V}_2(k)$ is independent of $\mathbf{V}_1(k)$.

**(a)** What are the formulas for $\hat{\boldsymbol{\theta}}_{BLU:1}(k)$ and $\hat{\boldsymbol{\theta}}_{BLU:2}(k)$?

**(b)** Find the batch BLUE of $\boldsymbol{\theta}$ given *both* $\mathbf{Z}_1(k)$ and $\mathbf{Z}_2(k)$.

**(c)** Find the recursive BLUE of $\boldsymbol{\theta}$ given $\mathbf{Z}_2(k)$ and $\hat{\boldsymbol{\theta}}_{BLU:1}(k)$.

**9-6.** (Bruce Rebhan, Spring 1992) One of your colleagues at Major Defense Corporation has come to you with an estimation problem.

**(a)** Two sets of sensors are used to provide an altitude measurement for the fighter radar you are developing. These sensors have the following covariance (**R**) matrices associated with their noises:

$$\mathbf{R}_1 = \begin{bmatrix} 3 & 4 \\ 4 & 3 \end{bmatrix} \quad \text{and} \quad \mathbf{R}_2 = \begin{bmatrix} 2 & 0 \\ 0 & 5 \end{bmatrix}$$

Your colleague tells you that he plans to take a BLUE estimate of the altitude from each set of sensors. You tell him that you don't think that's a good idea. Why not?

**(b)** Your colleague returns to you, again seeking your sage advice regarding BLUE estimates. This time he has two different sets of sensors with noise covariance matrices

$$\mathbf{R}_1 = \begin{bmatrix} 5 & 4 \\ 4 & 5 \end{bmatrix} \quad \text{and} \quad \mathbf{R}_2 = \begin{bmatrix} 4 & 5 \\ 5 & 4 \end{bmatrix}$$

He tells you that neither set of sensors is conducive to a BLUE estimate, since neither covariance matrix is "white." Do you agree with him?

**(c)** Your colleague returns once more, this time with covariance matrices associated with another two sets of sensors:

$$\mathbf{R}_1 = \begin{bmatrix} 2 & 0 \\ 0 & 2 \end{bmatrix} \quad \text{and} \quad \mathbf{R}_2 = \begin{bmatrix} 10 & 0 \\ 0 & 10 \end{bmatrix}$$

He tells you that since both of the noise covariance matrices are of the form $\mathbf{R} = \sigma_v^2\mathbf{I}$ he may take BLUEs from both set of sensors. You agree with this much. He then says that, furthermore, since both BLUEs will reduce to the LSE (by the Gauss–Markov theorem), it makes no difference which BLUE is used by the radar processor. You disagree with his conclusion, and tell him to use the BLUE from the sensors associated with $\mathbf{R}_1$. What is your reasoning?

**9-7.** (Richard S. Lee, Jr., Spring 1992) This computer-oriented problem is designed to explore the differences between the LS and BLU estimators.

**(a)** Generate a set of measurement data, $\{z(k), k = 0, 1, \ldots, 309\}$, where

$$z(k) = a + bk + v(k)$$

in which $a = 1.0$ and $b = 0.5$, $v(k)$ is Gaussian zero-mean white noise with variance $r(k)$, where $r(k) = 1.0$ for $k = 0, 1, \ldots, 9$, and $r(k)$ is uniformly distributed between 1 and 100 for $k = 10, 11, \ldots, 309$.

**(b)** Using the generated measurements $\{z(k), k = 0, 1, \ldots, 9\}$, determine $\hat{a}$ and $\hat{b}$ using the batch LS estimator.

**(c)** For each value of $k = 10, 11, \ldots, 309$, determine $\hat{a}$ and $\hat{b}$ using the recursive LS algorithm. Use the values of $\hat{a}(9)$ and $\hat{b}(9)$ determined in part (b) to initialize the recursive algorithm. Plot $\hat{a}(k)$ and $\hat{b}(k)$ versus $k$.

**(d)** Repeat part (b) using the BLUE. Compare your results to those obtained in part (b).

**(e)** Repeat part (c) using the BLUE. Compare your results to those obtained in part (c).

**9-8.** Suppose covariance matrix $\mathbf{R}$ is unknown and we decide to estimate it using least squares so that we can then use this estimate in the formula for the BLUE of $\boldsymbol{\theta}$; i.e., we estimate $\mathbf{R}$ as

$$\hat{\mathbf{R}} = [\mathbf{Z}(k) - \mathbf{H}(k)\hat{\boldsymbol{\theta}}_{\mathrm{LS}}(k)][\mathbf{Z}(k) - \mathbf{H}(k)\hat{\boldsymbol{\theta}}_{\mathrm{LS}}(k)]'$$

Show that $\hat{\mathbf{R}}$ is rank 1 and is therefore singular; consequently, $\hat{\mathbf{R}}^{-1}$ cannot be computed. Note that if $\mathbf{R}$ can be suitably parameterized (e.g., see Problem 9-9) it may be possible to estimate its parameters. Note, also, that if we know that $\mathbf{H}(k)$ is deterministic and $\mathbf{V}(k)$ is multivariate Gaussian (see Lesson 12) we can determine a maximum-likelihood estimator for $\mathbf{R}$.

**9-9.** Consider the generic linear model $\mathbf{Z}(k) = \mathbf{H}(k)\boldsymbol{\theta} + \mathbf{V}(k)$, where the elements of $\mathbf{V}(k)$ are not white but rather are "colored," in the sense that they satisfy the following first-order difference equation (for more extensive discussions about colored noise, see Lesson 22):

$$\nu(k) = \rho\nu(k-1) + n(k), \qquad k = 1, 2, \ldots, N$$

where $\rho$ is a parameter that may or may not be known ahead of time, $|\rho| < 1$, and $n(k)$ is zero mean, independent, and identically distributed with known variance $\sigma_n^2$ (Fomby et al., 1984, pp. 206–208).

**(a)** Show that

$$\mathbf{E}\{\mathbf{V}(N)\mathbf{V}'(N)\} = \sigma_n^2\boldsymbol{\Omega} = \sigma_n^2\frac{1}{1-\rho^2}\begin{bmatrix} 1 & \rho & \cdots & \rho^{N-1} \\ \rho & 1 & \cdots & \rho^{N-2} \\ & & \cdots & \rho \\ \rho^{N-1} & \rho^{N-2} & \cdots & 1 \end{bmatrix}$$

**(b)** Show that $\boldsymbol{\Omega}^{-1}$ is a *tridiagonal matrix* with diagonal elements equal to $(1, (1+\rho^2), (1+\rho^2), \ldots, (1+\rho^2), 1)$, superdiagonal elements all equal to $-\rho$, and subdiagonal elements all equal to $-\rho$.

**(c)** Assume that $\rho$ is known. What is $\hat{\boldsymbol{\theta}}_{\mathrm{BLU}}(k)$?

**9-10.** (Prof. Tom Hebert) In this lesson we derived the recursive BLUE under the assumption that $\mathbf{R}(k)$ is a diagonal matrix. In this problem we generalize the results to a nondiagonal symmetric covariance matrix (e.g., see Problem 9-9). This could be useful if the generic noise vector is colored where the coloring is known.

**(a)** Prove that:

$$\begin{aligned}\hat{\boldsymbol{\theta}}_{\mathrm{BLU}}(k+1) &= [\mathbf{I} - a_{k+1}\mathbf{P}(k+1)\boldsymbol{\beta}(k+1)\boldsymbol{\beta}'(k+1)]\hat{\boldsymbol{\theta}}_{\mathrm{BLU}}(k) \\ &\quad + a_{k+1}[z(k+1) - \mathbf{x}'(k)\mathbf{R}^{-1}(k)\mathbf{Z}(k)]\mathbf{P}(k+1)\boldsymbol{\beta}(k+1)\end{aligned}$$

where

$$\mathbf{R}(k+1) = \left[\begin{array}{c|c} \alpha_{k+1} & \mathbf{x}'(k) \\ \hline \mathbf{x}(k) & \mathbf{R}(k) \end{array}\right], \qquad \mathbf{R}^{-1}(k+1) = \left[\begin{array}{c|c} a_{k+1} & \mathbf{b}'(k) \\ \hline \mathbf{b}(k) & \mathbf{C}(k) \end{array}\right]$$

$$a_{k+1} = \frac{1}{\alpha_{k+1} - \mathbf{x}'(k)\mathbf{R}^{-1}(k)\mathbf{x}(k)}$$

$$\mathbf{b}(k) = -a_{k+1}\mathbf{R}^{-1}(k)\mathbf{x}(k)$$

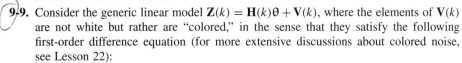

$$\mathbf{C}(k) = \mathbf{R}^{-1}(k) + a_{k+1}\mathbf{R}^{-1}(k)\mathbf{x}(k)\mathbf{x}'(k)\mathbf{R}^{-1}(k)$$

$$\boldsymbol{\beta}(k+1) = \mathbf{h}(k+1) - \mathbf{H}'(k)\mathbf{R}^{-1}(k)\mathbf{x}(k)$$

and

$$\mathbf{P}^{-1}(k+1) = \mathbf{P}^{-1}(k) + a_{k+1}\boldsymbol{\beta}(k+1)\boldsymbol{\beta}'(k+1)$$

**(b)** Prepare a flow chart that shows the correct order of computations for this recursive algorithm.

**(c)** Demonstrate that the algorithm reduces to the usual recursive BLUE when $\mathbf{R}(k)$ is diagonal.

**(d)** What is the recursive WLSE for which $\mathbf{W}(k+1) = \mathbf{R}^{-1}(k+1)$, when $\mathbf{W}(k+1)$ is nondiagonal but symmetric?

# LESSON 10

# Likelihood

## SUMMARY

This lesson provides the background material for the method of maximum likelihood. It explains the relationship of *likelihood* to *probability* and when the terms *likelihood* and *likelihood ratio* can be used interchangeably.

Probability is associated with a forward experiment in which the probability model is specified, including values for the parameters in that model (e.g., mean and variance in a Gaussian density function), and data (i.e., realizations) are generated using this model. In likelihood, on the other hand, the data are given, as well as the nature of the probability model; but the parameters of the probability model are not specified. They must be determined from the given data. Likelihood is, therefore, associated with an inverse experiment.

When you complete this lesson you will be able to (1) describe the relationship between likelihood and probability; (2) explain the notion of likelihood ratio; (3) explain why log-likelihood can be used instead of likelihood; and (4) explain why "likelihood" and "likelihood ratio" can be used interchangeably for parameters that are described by continuous probability models.

## LIKELIHOOD DEFINED

To begin, we define what is meant by an hypothesis, $H$, and results (of an experiment), $R$. The major reference for this lesson is Edwards (1972), a most delightful book. Suppose scalar parameter $\theta$ can assume only two values, 0 or 1; then we say that there are two hypotheses associated with $\theta$, $H_0$ and $H_1$, where, for $H_0, \theta = 0$, and for $H_1, \theta = 1$. This is the situation of a *binary hypothesis*. Suppose next that scalar parameter $\theta$ can assume 10 values, $a, b, c, d, e, f, g, h, i, j$; then we say there are 10 hypotheses associated with $\theta$, $H_1, H_2, H_3, \ldots, H_{10}$, where, for $H_1, \theta = a$, for $H_2, \theta = b, \ldots$, and for $H_{10}, \theta = j$. Parameter $\theta$ may also assume values from an interval, i.e., $a \le \theta \le b$. In this case, we have an infinite, uncountable number of hypotheses about $\theta$, each associated with a real number in the interval $[a, b]$. Finally, we may have a vector of parameters each of whose elements has a collection of hypotheses associated with it. For example, suppose that each of the $n$ elements of $\boldsymbol{\theta}$ is either 0 or 1. Vector $\boldsymbol{\theta}$ is then characterized by $2^n$ hypotheses. Sometimes we use the "null hypothesis" to account for *all* the possibilities that are not already accounted for by the enumerated hypotheses. For example, in the binary hypothesis case, we may also include hypothesis $H_2$ to stand for all $\theta$ not equal to either 0 or 1.

Results, $R$, are the outputs of an experiment. In our work on parameter estimation for the linear model $\mathbf{Z}(k) = \mathbf{H}(k)\boldsymbol{\theta} + \mathbf{V}(k)$, the "results" are the data in $\mathbf{Z}(k)$ and $\mathbf{H}(k)$.

We let $P(R|H)$ denote the probability of obtaining results $R$ given hypothesis $H$ according to some probability model, e.g., $p[z(k)|\boldsymbol{\theta}]$. In probability, $P(R|H)$ is always viewed as a function of $R$ for *fixed* values of $H$. Usually, the explicit dependence of $P$ on $H$ is not shown. To understand the differences between probability and likelihood, it is important to show the explicit dependence of $P$ on $H$.

**EXAMPLE 10-1**

Random number generators are often used to generate a sequence of random numbers that can then be used as the input sequence to a dynamical system, or as an additive measurement noise sequence. To run a random number generator, you must choose a probability model. The Gaussian model is often used; however, it is characterized by two parameters, mean $\mu$ and variance $\sigma^2$. To obtain a stream of Gaussian random numbers from the random number generator, you must fix $\mu$ and $\sigma^2$. Let $\mu_T$ and $\sigma_T^2$ denote (true) values chosen for $\mu$ and $\sigma^2$. The Gaussian probability density function for the generator is $p[z(k)|\mu_T, \sigma_T^2]$, and the numbers we obtain at its output, $z(1), z(2), \ldots$, are of course quite dependent on the hypothesis $H_T = (\mu_T, \sigma_T^2)$. □

For fixed $H$ we can apply the three axioms of probability: (1) probabilities must be nonnegative; (2) the probability of something that occurs every time must be unity; and (3) the probability of the union of mutually exclusive events must be equal to the sum of their individual probabilities (Larson and Shubert, 1979).

**Definition 10-1** (Edwards, 1972, p. 9). *Likelihood*, L(H|R), *of the hypothesis* H *given the results* R *and a specific probability model is proportional to* P(R|H), *the constant of proportionality being arbitrary; i.e.,*

$$L(H|R) = cP(R|H) \quad \square \tag{10-1}$$

For likelihood $R$ is fixed (i.e., given ahead of time) and $H$ is variable. *There are no axioms of likelihood.* Likelihood cannot be compared using different data sets (i.e., different results, say, $R_1$ and $R_2$) unless the data sets are statistically independent.

**EXAMPLE 10-2**

Suppose we are given a sequence of Gaussian random numbers, using the random number generator that was described in Example 10-1, but we do not know $\mu_T$ and $\sigma_T^2$. Is it possible to infer (i.e., estimate) what the values of $\mu$ and $\sigma^2$ were that most likely generated the given sequence? The method of maximum likelihood, which we study in Lesson 11, will show us how to do this. The starting point for the estimation of $\mu$ and $\sigma^2$ will be $p[z(k)|\mu, \sigma^2]$, where now $z(k)$ is fixed and $\mu$ and $\sigma^2$ are treated as variables. □

**EXAMPLE 10-3** (Edwards, 1972, p. 10)

To further illustrate the difference between $P(R|H)$ and $L(H|R)$, we consider the following binomial model, which we assume describes the occurrence of boys and girls in a family of two children:

$$P(R|p) = \frac{(m+f)!}{m!f!} p^m (1-p)^f \tag{10-2}$$

where $p$ denotes the probability of a male child, $m$ equals the number of male children, $f$ equals the number of female children, and, in this example,

$$m + f = 2 \tag{10-3}$$

Our objective is to determine $p$; but to do this we need some results. Knocking on neighbor's doors and conducting a simple survey, we establish two data sets:

$$R_1 = \{1 \text{ boy and } 1 \text{ girl}\} \Rightarrow m = 1 \quad \text{and} \quad f = 1 \tag{10-4}$$

$$R_2 = \{2 \text{ boys}\} \Rightarrow m = 2 \quad \text{and} \quad f = 0 \tag{10-5}$$

To keep the determination of $p$ simple for this meager collection of data, we shall only consider two values for $p$, i.e., two hypotheses,

$$\begin{aligned} H_1 : p &= \tfrac{1}{4} \\ H_2 : p &= \tfrac{1}{2} \end{aligned} \tag{10-6}$$

To begin, we create a *table of probabilities*, in which the entries are $P(R_i|H_j \text{ fixed})$, where this is computed using (10-2). For $H_1$ (i.e., $p = \frac{1}{4}$), $P(R_1|1/4) = 3/8$ and $P(R_2|1/4) = 1/16$; for $H_2$ (i.e., $p = 1/2$), $P(R_1|1/2) = 1/2$ and $P(R_2|1/2) = 1/4$. These results are collected together in Table 10-1.

**TABLE 10-1** $P(R_i|H_j$ FIXED)

| | $R_1$ | $R_2$ |
|---|---|---|
| $H_1$ | $\frac{3}{8}$ | $\frac{1}{16}$ |
| $H_2$ | $\frac{1}{2}$ | $\frac{1}{4}$ |

Next, we create a *table of likelihoods* using (10-1). In this table (Table 10-2) the entries are $L(H_i|R_j \text{ fixed})$. Constants $c_1$ and $c_2$ are arbitrary, and $c_1$, for example, appears in each of the table entries in the $R_1$ column.

**TABLE 10-2**  $L(H_i|R_j$ FIXED)

*R is fixed throughout*

|       | $R_1$      | $R_2$       |
| ----- | ---------- | ----------- |
| $H_1$ | $3/8\,c_1$ | $1/16\,c_2$ |
| $H_2$ | $1/2\,c_1$ | $1/4\,c_2$  |

*since R is fixed, compare things in columns*

*This data set ($R_1$) we see $H_1$ more th* | *This data set =) $H_2 > H_1$*

What can we conclude from the table of likelihoods? First, for data $R_1$, the likelihood of $H_2$ is $\frac{4}{3}$ the likelihood of $H_1$. The number $\frac{4}{3}$ was obtained by taking the ratio of likelihoods $L(H_2|R_1)$ and $L(H_1|R_1)$. Second, on data $R_2$, the likelihood of $H_2$ is four times the likelihood of $H_1$ [note that $4 = 1/4\,c_2\big/1/16\,c_2$]. Finally, we conclude that, even from our two meager results, the value $p = 1/2$ appears to be more likely than the value $p = 1/4$, which, of course, agrees with our intuition. $\square$

## LIKELIHOOD RATIO

In the preceding example we were able to draw conclusions about the likelihood of one hypothesis versus a second hypothesis by comparing ratios of likelihoods, defined, of course, on the same set of data. Forming the ratio of likelihoods, we obtain the *likelihood ratio.*

**Definition 10-2** (Edwards, 1972, p. 10).  *The likelihood ratio of two hypotheses on the same data is the ratio of the likelihoods on the data.  Let* $L(H_1, H_2|R)$ *denote likelihood ratio; then,*

$$L(H_1, H_2|R) = \frac{L(H_1|R)}{L(H_2|R)} = \frac{P(R|H_1)}{P(R|H_2)} \quad \square \tag{10-7}$$

Observe that likelihood ratio statements do not depend on the arbitrary constant $c$ that appears in the definition of likelihood, because $c$ cancels out of the ratio $cP(R|H_1)/cP(R|H_2)$.

**Theorem 10-1** (Edwards, 1972, p. 11).  *Likelihood ratios of two hypotheses on statistically independent sets of data may be multiplied together to form the likelihood ratio of the combined data.*

*Proof.* Let $L(H_1, H_2|R_1 \& R_2)$ denote the likelihood ratio of $H_1$ and $H_2$ on the combined data $R_1 \& R_2$; i.e.,

$$L(H_1, H_2|R_1 \& R_2) = \frac{L(H_1|R_1 \& R_2)}{L(H_2|R_1 \& R_2)} = \frac{P(R_1 \& R_2|H_1)}{P(R_1 \& R_2|H_2)} \tag{10-8}$$

Because $R_1 \& R_2$ are statistically independent data, $P(R_1 \& R_2|H_i) = P(R_1|H_i)P(R_2|H_i)$; hence,

$$L(H_1, H_2|R_1 \& R_2) = \frac{P(R_1|H_1)}{P(R_1|H_2)} \times \frac{P(R_2|H_1)}{P(R_2|H_2)}$$

$$= L(H_1, H_2|R_1) \times L(H_1, H_2|R_2) \quad \square \tag{10-9}$$

**EXAMPLE 10-4**

We can now state the conclusions that are given at the end of Example 10-3 more formally. From Table 10-2, we see that $L(1/4, 1/2|R_1) = 3/4$ and $L(1/4, 1/2|R_2) = 1/4$. Additionally, because $R_1$ and $R_2$ are data from independent experiments, $L(1/4, 1/2|R_1 \,\&\, R_2) = 3/4 \times 1/4 = 3/16$. This reinforces our intuition that $p = 1/2$ is much more likely than $p = 1/4$. □

## RESULTS DESCRIBED BY CONTINUOUS DISTRIBUTIONS

Suppose the results $R$ have a continuous distribution; then we know from probability theory that the probability of obtaining a result that lies in the interval $(R, R + dR)$ is $P(R|H)dR$, as $dR \to 0$. $P(R|H)$ is then a probability density. In this case, $L(H|R) = cP(R|H)dR$; but $cdR$ can be defined as a new arbitrary constant, $c_1$, so that $L(H|R) = c_1 P(R|H)$. In likelihood ratio statements $c_1 = cdR$ disappears entirely.

Recall that a transformation of variables greatly affects probability because of the $dR$ that appears in the probability formula. *Likelihood and likelihood ratio, on the other hand, are unaffected by transformations of variables*, because of the absorption of $dR$ into $c_1$. For additional discussions, see the Supplementary Material at the end of this lesson.

## MULTIPLE HYPOTHESES

Thus far, all our attention has been directed at the case of two hypotheses. To apply likelihood and likelihood ratio concepts to parameter estimation problems, where parameters take on more than two values, we must extend our preceding results to the case of multiple hypotheses. As stated by Edwards (1972, p. 11), "Instead of forming all the pairwise likelihood ratios it is simpler to present the same information in terms of the likelihood ratios for the several hypotheses versus one of their number, which may be chosen quite arbitrarily for this purpose."

Our extensions rely very heavily on the results for the two hypotheses case, because of the convenient introduction of an arbitrary comparison hypothesis, $H^*$. Let $H_i$ denote the $i$th hypothesis; then

$$L(H_i, H^*|R) = \frac{L(H_i|R)}{L(H^*|R)} \tag{10-10}$$

Observe, also, that

$$\frac{L(H_i, H^*|R)}{L(H_j, H^*|R)} = \frac{L(H_i|R)}{L(H_j|R)} = L(H_i, H_j|R) \tag{10-11}$$

which means that we can compute the likelihood ratio between *any* two hypotheses $H_i$ and $H_j$ if we can compute the likelihood-ratio function $L(H_k, H^*|R)$.

Figure 10-1(a) depicts a likelihood function $L(H_k|R)$. Any value of $H_k$ can be chosen as the comparison hypothesis. We choose $H^*$ as the hypothesis associated with the maximum value of $L(H_k|R)$, so that the maximum value of $L(H_k, H^*|R)$ will be

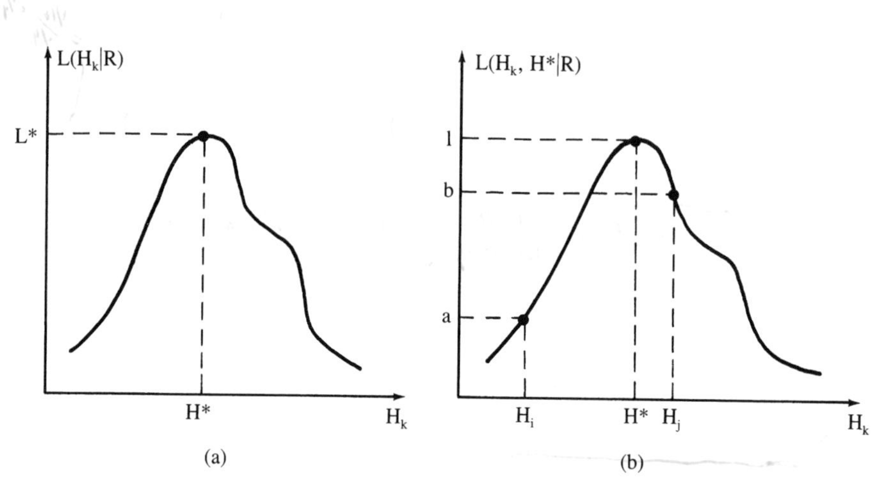


**Figure 10-1** Multiple hypotheses case: (a) likelihood $L(H_k|R)$ versus $H_k$, and (b) likelihood ratio $L(H_k, H^*|R)$ versus $H_k$. Comparison hypothesis $H^*$ has been chosen to be the hypothesis associated with the maximum value of $L(H_k|R)$, $L^*$.

normalized to unity. The likelihood ratio function $L(H_k, H^*|R)$, depicted in Figure 10-1(b), was obtained from Figure 10-1(a) and (10-10). To compute $L(H_i, H_j|R)$, we determine from Figure 10-1(b) that $L(H_i, H^*|R) = a$ and $L(H_j, H^*|R) = b$, so $L(H_i, H_j|R) = a/b$.

Is it really necessary to know $H^*$ in order to carry out the normalization of $L(H_k, H^*|R)$ depicted in Figure 10-1(b)? No, because $H^*$ can be eliminated by a clever "conceptual" choice of constant $c$. We can choose $c$ such that

$$L(H^*|R) = L^* \triangleq 1 \tag{10-12}$$

According to (10-1), this means that

$$c = \frac{1}{P(R|H^*)} \tag{10-13}$$

If $c$ is "chosen" in this manner, then

$$L(H_k, H^*|R) = \frac{L(H_k|R)}{L(H^*|R)} = L(H_k|R) \tag{10-14}$$

which means that, *in the case of multiple hypotheses, likelihood and likelihood ratio can be used interchangeably*. This helps to explain why authors use different names for the function that is the starting point in the method of maximum likelihood, including *likelihood function* and *likelihood-ratio function*.

## DECISION MAKING USING LIKELIHOOD RATIO

Although this book focuses on estimation, some of its concepts are also used in decision making. Likelihood ratio is one such concept (see, also, Lesson 14). In mathematical decision theory, decisions are made based on *tests*. A test is said to be a *likelihood-ratio test* if there is a number $c$, called a threshold, such that this test leads to ($d_i$ refers to the $i$th decision)

$$\begin{aligned} &d_1, && \text{if } L(H_1, H_2|R) > c \\ &d_2, && \text{if } L(H_1, H_2|R) < c \\ &d_1 \text{ or } d_2, && \text{if } L(H_1, H_2|R) = c \end{aligned} \tag{10-15}$$

**EXAMPLE 10-5**

Consider $N$ independent observations $z(1), z(2), \ldots, z(N)$, which are normally distributed with unknown mean $\mu$ and known standard deviation $\sigma$. There are two known possible values for $\mu$, $\mu_1$, and $\mu_2$; i.e., $H_1$: $\mu = \mu_1$ and $H_2$: $\mu = \mu_2$. The sample mean is used as the estimator of $\mu$; i.e.,

$$\hat{\mu}(N) = \frac{1}{N}\sum_{i=1}^{N} z(i) \tag{10-16}$$

To decide between $H_1$ and $H_2$, we shall develop a simple formula for the likelihood ratio and then apply the test in (10-15). Here

$$\begin{aligned} L(H_1, H_2|R) &= \frac{p(z(1), z(2), \ldots, z(N)|H_1)}{p(z(1), z(2), \ldots, z(N)|H_2)} \\ &= \frac{\prod_{i=1}^{N} p(z(i)|H_1)}{\prod_{j=1}^{N} p(z(j)|H_2)} \end{aligned} \tag{10-17}$$

because the measurements are independent. Note that the probability density function for each measurement is given by

$$p(z(i)|H_m) = \frac{1}{\sqrt{2\pi\sigma^2}} \exp\left[\frac{-(z(i) - \mu_m)^2}{2\sigma^2}\right], \qquad m = 1, 2 \tag{10-18}$$

Because of the exponential nature of the probability density function, it is more convenient to alter the test in (10-15) to a *log-likelihood ratio test*; i.e.,

$$\ln L(H_1, H_2|R) = \sum_{i=1}^{N} \ln p(z(i)|H_1) - \sum_{j=1}^{N} \ln p(z(j)|H_2) \square \ln c \tag{10-19}$$

where $\square$ is short for "if left-hand side (lhs) is $>$ than right-hand side (rhs), then decide $H_1$; if lhs is $<$ than rhs, then decide $H_2$; if lhs $=$ rhs, then decide $H_1$ or $H_2$." Substituting (10-18) into (10-19), it is easy to show that the latter equation can be expressed as

$$-\sum_{i=1}^{N}\{[z(i) - \mu_1]^2 - [z(i) - \mu_2]^2\} \square 2\sigma^2 \ln c \tag{10-20}$$

or

$$\hat{\mu}(N) \square \frac{(\mu_1 + \mu_2)}{2} + \frac{\sigma^2 \ln c}{N(\mu_1 - \mu_2)} \triangleq c_1 \tag{10-21}$$

Here is how we use (10-21). First compute $c_1$. To do this, we must fix $c$ and be given $N$, $\mu_1$, and $\mu_2$. Then compute $\hat{\mu}(N)$ using (10-16). Finally, compare $\hat{\mu}(N)$ to $c_1$. If $\hat{\mu}(N) > c_1$, then we decide in favor of $H_1$: $\mu = \mu_1$; if $\hat{\mu}(N) < c_1$, then we decide in favor of $H_2$: $\mu = \mu_2$; or if $\hat{\mu}(N) = c_1$, we decide either $\mu_1$ or $\mu_2$. Finally, observe that if $c = 1$ then the test in (10-21) becomes

$$\hat{\mu}(N) \,\square\, \frac{(\mu_1 + \mu_2)}{2} \tag{10-22}$$

Observe that a difference between estimation and decision making is that if we were estimating $\mu$ we would just use (10-15) as is, whereas in decision making we have additional information available, the two known possible values for $\mu$, $\mu_1$, and $\mu_2$. The likelihood-ratio test uses the estimator, $\hat{\mu}(N)$, as well as the additional information. In short, *estimation can play a very important role in decision making.* $\square$

## Supplementary Material

## TRANSFORMATION OF VARIABLES AND PROBABILITY

How exactly does a transformation of variables affect probability? We state the following result from probability theory (e.g., see Papoulis, 1991, p. 183): suppose $y_1 = g_1(x_1, x_2, \ldots, x_n)$, $y_2 = g_2(x_1, x_2, \ldots, x_n)$, ..., and $y_n = g_n(x_1, x_2, \ldots, x_n)$. Then, to determine the joint density function of $y_1, y_2, \ldots, y_n$ for a given set of numbers $Y_1, Y_2, \ldots, Y_n$, we solve the system $g_1(x_1, x_2, \ldots, x_n) = Y_1$, $g_2(x_1, x_2, \ldots, x_n) = Y_2, \ldots, g_n(x_1, x_2, \ldots, x_n) = Y_n$. Suppose this system has a single real solution $X_1, X_2, \ldots, X_n$; then

$$p_{y_1 y_2 \cdots y_n}(Y_1, Y_2, \ldots, Y_n) = \frac{p_{x_1 x_2 \cdots x_n}(X_1, X_2, \ldots, X_n)}{|J(X_1, X_2, \ldots, X_n)|} \tag{10-23}$$

where $J(X_1, X_2, \ldots, X_n)$ is the following Jacobian matrix:

$$J(X_1, X_2, \ldots, X_n) = \begin{bmatrix} \partial g_1/\partial x_1^* & \cdots & \partial g_1/\partial x_n^* \\ \cdot & \cdots & \cdot \\ \partial g_n/\partial x_1^* & \cdots & \partial g_n/\partial x_n^* \end{bmatrix} \tag{10-24}$$

and

$$\frac{\partial g_i}{\partial x_j^*} = \left.\frac{\partial g_i}{\partial x_j}\right|_{x_1 = X_1, x_2 = X_2, \ldots, x_n = X_n} \tag{10-25}$$

The numbers $X_1, X_2, \ldots, X_n$ are functions of $Y_1, Y_2, \ldots, Y_n$.

If the system of equations $g_1(x_1, x_2, \ldots, x_n) = Y_1$, $g_2(x_1, x_2, \ldots, x_n) = Y_2, \ldots, g_n(x_1, x_2, \ldots, x_n) = Y_n$ has more than one solution, then $p_{y_1 y_2 \ldots y_n}(Y_1, Y_2, \ldots, Y_n)$ is obtained by adding the corresponding expressions [each of which looks like the right-hand side of (10-23)] resulting from all solutions to $g_1(x_1, x_2, \ldots, x_n) = Y_1$, $g_2(x_1, x_2, \ldots, x_n) = Y_2, \ldots, g_n(x_1, x_2, \ldots, x_n) = Y_n$.

The determinant of the Jacobian matrix $J(X_1, X_2, \ldots, X_n)$ is the term that may alter the structure of $p_{x_1 x_2 \cdots x_n}$. This term is absorbed into the constant of proportionality in likelihood.

## SUMMARY QUESTIONS

1. In likelihood, the constant of proportionality:
   **(a)** must be specified
   **(b)** must be positive
   **(c)** is arbitrary
2. In likelihood:
   **(a)** both the hypothesis and results are given
   **(b)** the hypothesis is given and the results are unknown
   **(c)** results are given and the hypothesis is unknown
3. Likelihood ratios of two hypotheses may be multiplied together for:
   **(a)** Gaussian data
   **(b)** statistically independent sets of data
   **(c)** uncorrelated sets of data
4. In the case of multiple hypotheses, the comparison hypothesis $H^*$:
   **(a)** is arbitrary
   **(b)** must occur where the likelihood ratio is a maximum
   **(c)** must be an integer
5. Likelihood cannot be compared using different sets of data unless the latter are:
   **(a)** statistically independent
   **(b)** Gaussian
   **(c)** uncorrelated
6. "Results" are the output of:
   **(a)** a random number generator
   **(b)** an experiment
   **(c)** a probability model

## PROBLEMS

**10-1.** Consider a sequence of $n$ tosses of a coin of which $m$ are heads, i.e., $P(R|p) = p^m(1-p)^{n-m}$. Let $H_1 = p_1$ and $H_2 = p_2$, where $p_1 > p_2$ and $p$ represents the probability of a head. Show that $L(H_1, H_2|R)$ increases as $m$ increases. Then show that $L(H_1, H_2|R)$ increases when $\hat{p} = m/n$ increases, and that the likelihood-ratio test [see (10-15)], which consists of accepting $H_1$ if $L(H_1, H_2|R)$ is larger than some constant, is equivalent to accepting $H_1$ if $\hat{p}$ is larger than some related constant.

**10-2.** Suppose that the data consist of a single observation $z$ with Cauchy density given by

$$p(z|\theta) = \frac{1}{\pi[1 + (z-\theta)^2]}$$

Test $H_1 : \theta = \theta_1 = 1$ versus $H_2 : \theta = \theta_2 = -1$ when $c = 1/2$; i.e., show that, for $c = 1/2$, we accept $H_1$ when $z > -0.35$ *or* when $z < -5.65$.

**10-3.** (G. Caso, Fall 1991) Consider the observation of a Poisson random variable, $R$, for which

$$P(R|\lambda) = \lambda^k e^{-k}/k!$$

where $\lambda$ is unknown. Let $H_1 = \lambda_1$ and $H_2 = \lambda_2$. Determine the likelihood ratio $L(H_1, H_2|R)$ and show that the likelihood-ratio test, which consists of accepting $H$ if

$L(H_1, H_2|R)$ is larger than some constant, reduces to accepting $H_1$ if $k$ is larger than some related constant.

**10-4.** (C. M. Chintall, Fall 1991) (Van Trees, 1968) Given an observation $r$, establish a likelihood-ratio test to decide if

$$H_1 : r = s + n$$

$$H_0 : r = n$$

when $s$ and $n$ are independent random variables whose probability density functions are

$$p_s(S) = a\exp(-aS), \quad s \geq 0 \quad \text{and} \quad p_s(S) = 0, \quad s < 0$$

$$p_n(N) = b\exp(-bN), \quad n \geq 0 \quad \text{and} \quad p_n(N) = 0, \quad n < 0$$

Show that the likelihood-ratio test reduces to a constant.

**10-5.** (Andrew D. Norte, Fall 1991) As part of the training program for an upcoming Ironman triathalon, three college roommates, Dave, John, and Scott, race their bicycles to the top of Mount Crystal Ball. Dave, who eats a lot, weighs 180 lb, while John, a vegetarian, weighs 160 lb, and Scott, who tries to maintain a healthy diet, but occasionally eats a pizza, weighs 175 lb.

The race course is 85 km long. On a given morning, the threesome race their bicycles up the mountain, and a friend at the top of the mountain records only the winning time. We are given information that the winning time was 2 hours. Based on the three hypotheses

$$H_1: \text{ Dave, } 180 \text{ lb}$$

$$H_2: \text{ John, } 160 \text{ lb}$$

$$H_3: \text{ Scott, } 175 \text{ lb}$$

and the probability law

$$P(t = \alpha|W) = \left(\frac{W}{100}\right)^{\alpha}\left(\frac{1}{\alpha!}\right)e^{-W/100}$$

where $W$ =weight in pounds and $t$ =winning time in hours:

**(a)** Use a likelihood-ratio test to determine which bicyclist most likely produced the winning time.

**(b)** For the next two consecutive days, Dave, John, and Scott race up the mountain and the resulting winning times are 2 and 3 hours. Based on the three results, which bicyclist most likely produced the results?

**10-6.** (James Leight, Fall 1991) We are given $N$ statistically independent measurements of a Rician distributed random variable, $x$, where

$$p(x) = (2\pi\sigma_x^2)^{-1/2}\left(\frac{x}{\mu_x}\right)^{1/2}\exp\left[\frac{-\frac{1}{2}(x-\mu_x)}{\sigma_x^2}\right]$$

**(a)** Using the sample mean estimator for $\mu_x$ and the sample variance estimator for $\sigma_x^2$, determine the likelihood ratio for the following two hypotheses: $H_1$: $\sigma_{x_1}^2$ and $H_2$: $\sigma_{x_2}^2$.

**(b)** Show that the likelihood-ratio test leads to a threshold test of the sample variance.

*○ probability: you specify H + provide data*
*○ Likelyhood: you start w/ data + guess H (or parameters in model)*
*(eg. H: $\theta_{Nx1}$*

## LESSON  11

# Maximum-likelihood Estimation

### SUMMARY

The maximum-likelihood method is based on the relatively simple idea that different (statistical) populations generate different samples and that any given sample (i.e., set of data) is more likely to have come from some populations than from others.

To determine $\hat{\theta}_{ML}$, we need to determine a formula for the likelihood function and then maximize that function. Because likelihood is proportional to probability, we need to know the entire joint probability density function of the measurements in order to determine a formula for the likelihood function. This, of course, is much more information about $\mathbf{Z}(k)$ than was required in the derivation of the BLUE. In fact, it is the most information that we can ever expect to know about the measurements. The price we pay for knowing so much information about $\mathbf{Z}(k)$ is complexity in maximizing the likelihood function. Generally, mathematical programming must be used in order to determine $\hat{\theta}_{ML}$.

In one special case it is very easy to compute $\hat{\theta}_{ML}$, i.e., for our generic linear model in which $\mathbf{H}(k)$ is deterministic and $\mathbf{V}(k)$ is Gaussian. In that case, $\hat{\theta}_{ML} = \hat{\theta}_{BLU}$.

Maximum-likelihood estimates are very popular and widely used because they enjoy very good large-sample properties. They are consistent, asymptotically Gaussian, and asymptotically efficient. Functions of maximum-likelihood estimates are themselves maximum-likelihood estimates. This last "invariance" property is usually not enjoyed by WLSEs or BLUEs.

This lesson concludes with the derivation of the log-likelihood function for a linear, time-invariant dynamical system that is excited by a known input, has deterministic initial conditions, and has noisy measurements.

When you complete this lesson, you will be able to (1) describe the two major steps that are needed to obtain a maximum-likelihood estimator: (a) determine a formula for the likelihood function, and (b) maximize the likelihood function; (2) explain the properties of maximum-likelihood estimators that make them so useful; (3) describe the relationships between maximum-likelihood estimators, BLUEs, and WLSEs when the model is our generic linear one; (4) specify a very useful formula for the log-likelihood function of an important dynamical system; and (5) compute maximum-likelihood estimates; however, for many applications you will need commercially available optimization software to actually maximize the likelihood function.

## LIKELIHOOD*

Let us consider a vector of unknown parameters $\theta$ that describes a collection of $N$ independent identically distributed observations $z(k)$, $k = 1, 2, \ldots, N$. We collect these measurements into an $N \times 1$ vector $\mathbf{Z}(N)$, ($\mathbf{Z}$ for short),

$$\mathbf{Z} = \text{col}\,(z(1), z(2), \ldots, z(N)) \tag{11-1}$$

The likelihood of $\theta$, given the observations $\mathbf{Z}$, is defined to be proportional to the value of the probability density function of the observations given the parameters

$$l(\theta|\mathbf{Z}) \propto p(\mathbf{Z}|\theta) \tag{11-2}$$

where $l$ is the likelihood function and $p$ is the conditional joint probability density function. Because $z(i)$ are independent and identically distributed,

$$l(\theta|\mathbf{Z}) \propto p(z(1)|\theta)p(z(2)|\theta)\cdots p(z(N)|\theta) \tag{11-3}$$

In many applications $p(\mathbf{Z}|\theta)$ is exponential (e.g., Gaussian). It is easier then to work with the natural logarithm of $l(\theta|\mathbf{Z})$ than with $l(\theta|\mathbf{Z})$. Let

$$\text{L}(\theta|\mathbf{Z}) = \ln l(\theta|\mathbf{Z}) \tag{11-4}$$

Quantity L is sometimes referred to as the log-likelihood function, the support function (Kmenta, 1971), the likelihood function (Mehra, 1971, Schweppe, 1965, and Stepner and Mehra, 1973, for example), or the conditional likelihood function (Nahi, 1969). We shall use these terms interchangeably.

## MAXIMUM-LIKELIHOOD METHOD AND ESTIMATES†

The *maximum-likelihood method* is based on the relatively simple idea that different populations generate different samples and that any given sample is more likely to have come from some populations than from others. The maximum-likelihood estimate (MLE) $\hat{\theta}_{\text{ML}}$ is the value of $\theta$ that maximizes $l$ or L for a particular set of measurements $\mathbf{Z}$. The logarithm of $l$ is a monotonic transformation of $l$ (i.e., whenever $l$ is decreasing or increasing, $\ln l$ is also decreasing or increasing); therefore, the point corresponding to the maximum of $l$ is also the point corresponding to the maximum of $\ln l = \text{L}$.

* The material in this section is taken from Mendel (1983, pp. 94–95).
† The material in this section is taken from Mendel (1983, pp. 95–98).

→ Obtaining an MLE involves specifying the likelihood function and finding those values of the parameters that give this function its maximum value. It is required that, if L is differentiable, the partial derivatives of L (or $l$) with respect to each of the unknown parameters $\theta_1, \theta_2, \ldots, \theta_n$ equal zero:

*→ to MLE (see example next page)*

$$\left.\frac{\partial L(\boldsymbol{\theta}|\mathbf{Z})}{\partial \theta_i}\right|_{\boldsymbol{\theta}=\hat{\boldsymbol{\theta}}_{\mathrm{ML}}} = 0, \qquad \text{for all } i = 1, 2, \ldots, n \qquad (11\text{-}5)$$

To be sure that the solution of (11-5) gives, in fact, a maximum value of $L(\boldsymbol{\theta}|\mathbf{Z})$, certain second-order conditions must be fulfilled. Consider a Taylor series expansion of $L(\boldsymbol{\theta}|\mathbf{Z})$ about $\hat{\boldsymbol{\theta}}_{\mathrm{ML}}$; i.e.,

$$L(\boldsymbol{\theta}|\mathbf{Z}) = L(\hat{\boldsymbol{\theta}}_{\mathrm{ML}}|\mathbf{Z}) + \sum_{i=1}^{n} \frac{\partial L(\hat{\boldsymbol{\theta}}_{\mathrm{ML}}|\mathbf{Z})}{\partial \theta_i}(\theta_i - \hat{\theta}_{i,\mathrm{ML}})$$

$$+ \frac{1}{2}\sum_{i=1}^{n}\sum_{j=1}^{n} \frac{\partial^2 L(\hat{\boldsymbol{\theta}}_{\mathrm{ML}}|\mathbf{Z})}{\partial \theta_i \partial \theta_j}(\theta_i - \hat{\theta}_{i,\mathrm{ML}})(\theta_j - \hat{\theta}_{j,\mathrm{ML}}) + \cdots \qquad (11\text{-}6)$$

where $\partial L(\hat{\boldsymbol{\theta}}_{\mathrm{ML}}|\mathbf{Z})/\partial \theta_i$, for example, is short for

$$\left[\frac{\partial L(\hat{\boldsymbol{\theta}}|\mathbf{Z})}{\partial \theta_i}\right]\Bigg|_{\boldsymbol{\theta}=\hat{\boldsymbol{\theta}}_{\mathrm{ML}}}$$

Because $\hat{\boldsymbol{\theta}}_{\mathrm{ML}}$ is the MLE of $\boldsymbol{\theta}$, the second term on the right-hand side of (11-6) is zero [by virtue of (11-5)]; hence,

$$L(\boldsymbol{\theta}|\mathbf{Z}) = L(\hat{\boldsymbol{\theta}}_{\mathrm{ML}}|\mathbf{Z})$$

$$+ \frac{1}{2}\sum_{i=1}^{n}\sum_{j=1}^{n} \frac{\partial^2 L(\hat{\boldsymbol{\theta}}_{\mathrm{ML}}|\mathbf{Z})}{\partial \theta_i \partial \theta_j}(\theta_i - \hat{\theta}_{i,\mathrm{ML}})(\theta_j - \hat{\theta}_{j,\mathrm{ML}}) + \cdots \qquad (11\text{-}7)$$

Recognizing that the second term in (11-7) is a quadratic form, we can write it in a more compact notation as $\frac{1}{2}(\boldsymbol{\theta} - \hat{\boldsymbol{\theta}}_{\mathrm{ML}})'\mathbf{J}_o(\hat{\boldsymbol{\theta}}_{\mathrm{ML}}|\mathbf{Z})(\boldsymbol{\theta} - \hat{\boldsymbol{\theta}}_{\mathrm{ML}})$, where $\mathbf{J}_o(\hat{\boldsymbol{\theta}}_{\mathrm{ML}}|\mathbf{Z})$, the *observed Fisher information matrix* [see Equation (6-47)], is

*remember $L(\theta|z) = \ln p(z|\theta)$*

$$\mathbf{J}_o(\hat{\boldsymbol{\theta}}_{\mathrm{ML}}|\mathbf{Z}) = \left(\frac{\partial^2 L(\boldsymbol{\theta}|\mathbf{Z})}{\partial \theta_i \partial \theta_j}\right)\Bigg|_{\boldsymbol{\theta}=\hat{\boldsymbol{\theta}}_{\mathrm{ML}}}, \qquad i, j = 1, 2, \ldots, n \qquad (11\text{-}8)$$

Hence,

$$L(\boldsymbol{\theta}|\mathbf{Z}) = L(\hat{\boldsymbol{\theta}}_{\mathrm{ML}}|\mathbf{Z}) + \frac{1}{2}(\boldsymbol{\theta} - \hat{\boldsymbol{\theta}}_{\mathrm{ML}})'\mathbf{J}_o(\hat{\boldsymbol{\theta}}_{\mathrm{ML}}|\mathbf{Z})(\boldsymbol{\theta} - \hat{\boldsymbol{\theta}}_{\mathrm{ML}}) + \cdots \qquad (11\text{-}9)$$

Now let us examine sufficient conditions for the likelihood function to be maximum. We assume that, close to $\boldsymbol{\theta}$, $L(\boldsymbol{\theta}|\mathbf{Z})$ is approximately quadratic, in which case

$$L(\boldsymbol{\theta}|\mathbf{Z}) \approx L(\hat{\boldsymbol{\theta}}_{\mathrm{ML}}|\mathbf{Z}) + \frac{1}{2}(\boldsymbol{\theta} - \hat{\boldsymbol{\theta}}_{\mathrm{ML}})'\mathbf{J}_o(\hat{\boldsymbol{\theta}}_{\mathrm{ML}}|\mathbf{Z})(\boldsymbol{\theta} - \hat{\boldsymbol{\theta}}_{\mathrm{ML}}) \qquad (11\text{-}10)$$

From vector calculus, it is well known that a sufficient condition for a function of $n$ variables to be maximum is that the matrix of second partial derivatives of that function, evaluated at the extremum, must be negative definite. For $L(\theta|Z)$ in (11-10), this means that *a sufficient condition for* $L(\theta|Z)$ *to be maximized is*

$$\mathbf{J}_o(\hat{\theta}_{ML}|Z) < 0 \qquad \begin{array}{l} \textit{if 11-5 + 11-11} \\ \textit{then the solution is max} \end{array} \qquad (11\text{-}11)$$

### EXAMPLE 11-1

This is a continuation of Example 10-2. We observe a random sample $\{z(1), z(2), \ldots, z(N)\}$ at the output of a Gaussian random number generator and wish to find the maximum-likelihood estimators of $\mu$ and $\sigma^2$.

The Gaussian density function $p(z|\mu, \sigma^2)$ is

$$p(z|\mu, \sigma^2) = (2\pi\sigma^2)^{-1/2} \exp\left\{-\frac{1}{2}\left(\frac{z-\mu}{\sigma}\right)^2\right\} \qquad (11\text{-}12)$$

Its natural logarithm is

$$\ln p(z|\mu, \sigma^2) = -\frac{1}{2}\ln(2\pi\sigma^2) - \frac{1}{2}\left(\frac{z-\mu}{\sigma}\right)^2 \qquad (11\text{-}13)$$

The likelihood function is

$$l(\mu, \sigma^2) = p(z(1)|\mu, \sigma^2)p(z(2)|\mu, \sigma^2)\cdots p(z(N)|\mu, \sigma^2) \qquad (11\text{-}14)$$

and its logarithm is

$$L(\mu, \sigma^2) = \sum_{i=1}^{N} \ln p(z(i)|\mu, \sigma^2) \qquad (11\text{-}15)$$

Substituting for $\ln p(z(i)|\mu, \sigma^2)$ gives

$$\begin{aligned} L(\mu, \sigma^2) &= \sum_{i=1}^{N}\left[-\frac{1}{2}\ln(2\pi\sigma^2) - \frac{1}{2}\left(\frac{z(i)-\mu}{\sigma}\right)^2\right] \\ &= -\frac{N}{2}\ln(2\pi\sigma^2) - \frac{1}{2\sigma^2}\sum_{i=1}^{N}[z(i)-\mu]^2 \end{aligned} \qquad (11\text{-}16)$$

There are two unknown parameters in L, $\mu$ and $\sigma^2$. Differentiating L with respect to each gives

$$\frac{\partial L}{\partial \mu} = \frac{1}{\sigma^2}\sum_{i=1}^{N}[z(i)-\mu] \qquad (11\text{-}17a)$$

$$\frac{\partial L}{\partial(\sigma^2)} = -\frac{N}{2}\frac{1}{\sigma^2} + \frac{1}{2\sigma^4}\sum_{i=1}^{N}[z(i)-\mu]^2 \qquad (11\text{-}17b)$$

Equating these partial derivatives to zero, we obtain

$$\frac{1}{\hat{\sigma}_{ML}^2}\sum_{i=1}^{N}[z(i)-\hat{\mu}_{ML}] = 0 \qquad (11\text{-}18a)$$

$$-\frac{N}{2}\frac{1}{\hat{\sigma}_{ML}^2} + \frac{1}{2\hat{\sigma}_{ML}^4}\sum_{i=1}^{N}[z(i)-\hat{\mu}_{ML}]^2 = 0 \qquad (11\text{-}18b)$$

For $\hat{\sigma}^2_{\text{ML}}$ different from zero, (11-18a) reduces to

$$\sum_{i=1}^{N}[z(i) - \hat{\mu}_{\text{ML}}] = 0$$

giving

$$\hat{\mu}_{\text{ML}} = \frac{1}{N}\sum_{i=1}^{N} z(i) = \bar{z} \tag{11-19}$$

Thus *the MLE of the mean of a Gaussian population is equal to the sample mean* $\bar{z}$. Once again we see that the sample mean is an optimal estimator.

Observe that (11-18b) can now be solved for $\hat{\sigma}^2_{\text{ML}}$, the MLE of $\sigma^2$. Multiplying Equation (11-18b) by $2\hat{\sigma}^4_{\text{ML}}$ leads to

$$-N\hat{\sigma}^2_{\text{ML}} + \sum_{i=1}^{N}[z(i) - \hat{\mu}_{\text{ML}}]^2 = 0$$

Substituting $\bar{z}$ for $\hat{\mu}_{\text{ML}}$ and solving for $\hat{\sigma}^2_{\text{ML}}$ gives

$$\hat{\sigma}^2_{\text{ML}} = \frac{1}{N}\sum_{i=1}^{N}[z(i) - \bar{z}]^2 \tag{11-20}$$

Thus, *the MLE of the variance of a Gaussian population is simply equal to the sample variance.* □

## PROPERTIES OF MAXIMUM-LIKELIHOOD ESTIMATES

The importance of maximum-likelihood estimation is that it produces estimates that have very desirable properties.

**Theorem 11-1.** *Maximum-likelihood estimates are* (1) *consistent,* (2) *asymptotically Gaussian with mean* $\boldsymbol{\theta}$ *and covariance matrix* $\frac{1}{N}\mathbf{J}^{-1}$*, in which* $\mathbf{J}$ *is the Fisher Information Matrix [Equation (6-46)], and* (3) *asymptotically efficient.*

*Proof.* For proofs of consistency, asymptotic normality, and asymptotic efficiency, see Sorenson (1980, pp. 187–190, 190–192, and 192–193, respectively). These proofs, although somewhat heuristic, convey the ideas needed to prove the three parts of this theorem. More detailed analyses can be found in Cramer (1946) and Zacks (1971). See also, Problems 11-18, 11-19, and 11-20. □

**Theorem 11-2** (Invariance Property of MLEs). *Let* $\mathbf{g}(\boldsymbol{\theta})$ *be a vector function mapping* $\boldsymbol{\theta}$ *into an interval in* r*-dimensional Euclidean space. Let* $\hat{\boldsymbol{\theta}}_{\text{ML}}$ *be a MLE of* $\boldsymbol{\theta}$*; then* $\mathbf{g}(\hat{\boldsymbol{\theta}}_{\text{ML}})$ *is a MLE of* $\mathbf{g}(\boldsymbol{\theta})$*; i.e.,*

$$[\widehat{\mathbf{g}(\boldsymbol{\theta})}]_{\text{ML}} = \mathbf{g}(\hat{\boldsymbol{\theta}}_{\text{ML}}) \tag{11-21}$$

*Proof* (see Zacks, 1971). Note that Zacks points out that in many books this theorem is cited only for the case of one-to-one mappings, $\mathbf{g}(\boldsymbol{\theta})$. His proof does not require $\mathbf{g}(\boldsymbol{\theta})$ to be one-to-one. Note, also, that the proof of this theorem is related to the "consistency carry-over" property of a consistent estimator, which was discussed in Lesson 7. □

**EXAMPLE 11-2**

We wish to obtain a MLE of the variance $\sigma_\nu^2$ in our linear model $z(k) = \mathbf{h}'(k)\boldsymbol{\theta} + \nu(k)$. One approach is to let $\theta_1 = \sigma_\nu^2$, establish the log-likelihood function for $\theta_1$, and maximize it to determine $\hat{\theta}_{1,\mathrm{ML}}$. Usually, optimization (i.e., search techniques) must be used to determine $\hat{\theta}_{1,\mathrm{ML}}$. Here is where a difficulty can occur, because $\theta_1$ (a variance) is known to be positive; thus, $\hat{\theta}_{1,\mathrm{ML}}$ must be constrained to be positive. Unfortunately, constrained optimization techniques are more difficult than unconstrained ones.

A second approach is to let $\theta_2 = \sigma_\nu$, establish the log-likelihood function for $\theta_2$ (it will be the same as the one for $\theta_1$, except that $\theta_1$ will be replaced by $\theta_2^2$), and maximize it to determine $\hat{\theta}_{2,\mathrm{ML}}$. Because $\theta_2$ is a standard deviation, which can be positive or negative, unconstrained optimization can be used to determine $\hat{\theta}_{2,\mathrm{ML}}$. Finally, we use the invariance property of MLEs to compute $\hat{\theta}_{1,\mathrm{ML}}$ as

$$\hat{\theta}_{1,\mathrm{ML}} = (\hat{\theta}_{2,\mathrm{ML}})^2 \quad \Box \tag{11-22}$$

## THE LINEAR MODEL [$\mathbf{H}(k)$ Deterministic]

We return now to the linear model

$$\mathbf{Z}(k) = \mathbf{H}(k)\boldsymbol{\theta} + \mathbf{V}(k) \tag{11-23}$$

in which $\boldsymbol{\theta}$ is an $n \times 1$ vector of deterministic parameters, $\mathbf{H}(k)$ *is deterministic*, and $\mathbf{V}(k)$ is zero-mean white noise, with covariance matrix $\mathbf{R}(k)$. This is precisely the same model that was used to derive the BLUE of $\boldsymbol{\theta}$, $\hat{\boldsymbol{\theta}}_{\mathrm{BLU}}(k)$. Our objectives in this paragraph are twofold: (1) to derive the MLE of $\boldsymbol{\theta}$, $\hat{\boldsymbol{\theta}}_{\mathrm{ML}}(k)$, and (2) to relate $\hat{\boldsymbol{\theta}}_{\mathrm{ML}}(k)$ to $\hat{\boldsymbol{\theta}}_{\mathrm{BLU}}$ and $\hat{\boldsymbol{\theta}}_{\mathrm{LS}}(k)$.

To derive the MLE of $\boldsymbol{\theta}$, we need to determine a formula for $p(\mathbf{Z}(k)|\boldsymbol{\theta})$. To proceed, *we assume that* $\mathbf{V}(k)$ *is Gaussian,* with multivariate density function $p(\mathbf{V}(k))$, where

$$p(\mathbf{V}(k)) = \frac{1}{\sqrt{(2\pi)^N|\mathbf{R}(k)|}} \exp\left[-\frac{1}{2}\mathbf{V}'(k)\mathbf{R}^{-1}(k)\mathbf{V}(k)\right] \tag{11-24}$$

Recall (e.g., Papoulis, 1991) that linear transformations on, and linear combinations of, Gaussian random vectors are themselves Gaussian random vectors. For this reason, it is clear that when $\mathbf{V}(k)$ is Gaussian $\mathbf{Z}(k)$ is as well. The multivariate Gaussian density function of $\mathbf{Z}(k)$, derived from $p(\mathbf{V}(k))$, is

$$p(\mathbf{Z}(k)|\boldsymbol{\theta}) = \frac{1}{\sqrt{(2\pi)^N|\mathbf{R}(k)|}} \exp\left\{-\frac{1}{2}[\mathbf{Z}(k) - \mathbf{H}(k)\boldsymbol{\theta}]'\mathbf{R}^{-1}(k)[\mathbf{Z}(k) - \mathbf{H}(k)\boldsymbol{\theta}]\right\} \tag{11-25}$$

**Theorem 11-3.** *For the generic linear model, when* $p(\mathbf{Z}|\theta)$ *is multivariate Gaussian and* $\mathbf{H}(k)$ *is deterministic, then the principle of ML leads to the BLUE of* $\theta$; *i.e.,*

$$\hat{\theta}_{ML}(k) = \hat{\theta}_{BLU}(k) \tag{11-26}$$

*These estimators are unbiased, most efficient (within the class of linear estimators), consistent, and Gaussian.*

*Proof.* We must maximize $p(\mathbf{Z}|\theta)$ with respect to $\theta$. This can be accomplished by minimizing the argument of the exponential in (11-25); hence, $\hat{\theta}_{ML}(k)$ is the solution of

$$\frac{d}{d\theta}(\mathbf{Z} - \mathbf{H}\theta)'\mathbf{R}^{-1}(\mathbf{Z} - \mathbf{H}\theta)\bigg|_{\theta=\hat{\theta}_{ML}} = \mathbf{0} \tag{11-27}$$

This equation can also be expressed as $dJ[\hat{\theta}_{ML}]/d\hat{\theta} = \mathbf{0}$, where $J[\hat{\theta}_{ML}] = \tilde{\mathbf{Z}}'(k)\mathbf{R}^{-1}(k)\tilde{\mathbf{Z}}(k)$ and $\tilde{\mathbf{Z}}(k) = \mathbf{Z}(k) - \mathbf{H}\hat{\theta}_{ML}(k)$. Comparing this version of (11-27) with Equation (3-4) and the subsequent derivation of the WLSE of $\theta$, we conclude that

$$\hat{\theta}_{ML}(k) = \hat{\theta}_{WLS}(k)\bigg|_{\mathbf{W}(k)=\mathbf{R}^{-1}(k)} \tag{11-28}$$

But, we also know, from Lesson 9, that

$$\hat{\theta}_{WLS}(k)\bigg|_{\mathbf{W}(k)=\mathbf{R}^{-1}(k)} = \hat{\theta}_{BLU}(k) \tag{11-29}$$

From (11-28) and (11-29), we conclude that

$$\hat{\theta}_{ML}(k) = \hat{\theta}_{BLU}(k) \tag{11-30}$$

The estimators are

1. unbiased, because $\hat{\theta}_{BLU}(k)$ is unbiased;
2. most efficient, because $\hat{\theta}_{BLU}(k)$ is most efficient;
3. consistent, because $\hat{\theta}_{ML}(k)$ is consistent; and
4. Gaussian, because they depend linearly on $\mathbf{Z}(k)$, which is Gaussian. □

Observe that, when $\mathbf{Z}(k)$ is Gaussian, this theorem permits us to make statements about small-sample properties of MLEs. Usually, we cannot make such statements.

We now suggest a reason why Theorem 9-6 (and, subsequently, Theorem 4-1) is referred to as the "information form" of recursive BLUE. From Theorems 11-1 and 11-3 we know that, when $\mathbf{H}(k)$ is deterministic, $\hat{\theta}_{BLU} \sim N(\hat{\theta}; \theta, \mathbf{J}^{-1}/N)$. This means that $\mathbf{P}(k) = \text{cov}\,[\tilde{\theta}_{BLU}(k)]$ is proportional to $\mathbf{J}^{-1}$. Observe, in Theorem 9-6, that we compute $\mathbf{P}^{-1}$ recursively, and not $\mathbf{P}$. Because $\mathbf{P}^{-1}$ is proportional to the Fisher information matrix $\mathbf{J}$, the results in Theorem 9-6 (and 4-1) are therefore referred to as the "information form" of recursive BLUE.

A second more pragmatic reason is due to the fact that the inverse of any covariance matrix is known as an information matrix (e.g., Anderson and Moore,

1979, p. 138). Consequently, any algorithm that is in terms of information matrices is known as an "information form" algorithm.

**Corollary 11-1.** *For the generic linear model, if* $p[\mathbf{Z}(k)|\theta]$ *is multivariate Gaussian,* $\mathbf{H}(k)$ *is deterministic, and* $\mathbf{R}(k) = \sigma_v^2\mathbf{I}$, *then*

$$\hat{\theta}_{ML}(k) = \hat{\theta}_{BLU}(k) = \hat{\theta}_{LS}(k) \tag{11-31}$$

*These estimators are unbiased, most efficient (within the class of linear estimators), consistent, and Gaussian.*

*Proof.* To obtain (11-31), combine the results in Theorems 11-3 and 9-2. □

## A LOG-LIKELIHOOD FUNCTION FOR AN IMPORTANT DYNAMICAL SYSTEM

In practice, there are two major problems in obtaining MLEs of parameters in models of dynamical systems:

1. Obtaining an expression for $L\{\theta|\mathbf{Z}\}$
2. Maximizing $L\{\theta|\mathbf{Z}\}$ with respect to $\theta$

In this section we direct our attention to the first problem for a linear, time invariant dynamical system that is excited by a known forcing function, has deterministic initial conditions, and has measurements that are corrupted by additive white Gaussian noise. This system is described by the following state-equation model (for information about state-variable models and their relationships to input/output models, see Lesson D):

$$\mathbf{x}(k+1) = \mathbf{\Phi}\mathbf{x}(k) + \mathbf{\Psi}\mathbf{u}(k) \tag{11-32}$$

and

$$\mathbf{z}(k+1) = \mathbf{H}\mathbf{x}(k+1) + \mathbf{v}(k+1), \qquad k = 0, 1, \ldots, N-1 \tag{11-33}$$

In this model, $\mathbf{u}(k)$ is known ahead of time, $\mathbf{x}(0)$ is deterministic, $\mathbf{E}\{\mathbf{v}(k)\} = \mathbf{0}$, $\mathbf{v}(k)$ is Gaussian, and $\mathbf{E}\{\mathbf{v}(k)\mathbf{v}'(j)\} = \bar{\mathbf{R}}\delta_{kj}$.

To begin, we must establish the parameters that constitute $\theta$. In theory, $\theta$ could contain all the elements in $\mathbf{\Phi}$, $\mathbf{\Psi}$, $\mathbf{H}$, and $\bar{\mathbf{R}}$. In practice, however, these matrices are never completely unknown. State equations are either derived from physical principles (e.g., Newton's laws) or associated with a canonical model (e.g., controllability canonical form); hence, we usually know that certain elements in $\mathbf{\Phi}$, $\mathbf{\Psi}$, and $\mathbf{H}$ are identically zero or are known constants.

Even though all the elements in $\mathbf{\Phi}$, $\mathbf{\Psi}$, $\mathbf{H}$, and $\bar{\mathbf{R}}$ will not be unknown in an application, there still may be more unknowns present than can possibly be identified. How many parameters can be identified and which parameters can be identified by maximum-likelihood estimation (or, for that matter, by any type of estimation) are the subject of *identifiability of systems* (Stepner and Mehra, 1973). *We shall assume that* $\theta$ *is identifiable.* Identifiability is akin to "existence". When we assume $\theta$ is identifiable, we assume that it is possible to identify $\theta$ by ML methods. This means that all our statements are predicated by the following statement: If $\theta$ is identifiable, then. . . .

We let

$$\theta = \text{col}\,(\text{elements of } \mathbf{\Phi}, \mathbf{\Psi}, \mathbf{H}, \text{and } \bar{\mathbf{R}}) \tag{11-34}$$

**EXAMPLE 11-3**

The "controllable canonical form" state-variable representation for the discrete-time autoregressive moving average (ARMA) model (see, also, Example D-6)

$$H(z) = \frac{\beta_1 z^{n-1} + \beta_2 z^{n-2} + \cdots + \beta_{n-1} z + \beta_n}{z^n + \alpha_1 z^{n-1} + \cdots + \alpha_{n-1} z + \alpha_n} \tag{11-35}$$

which implies the ARMA difference equation ($z$ denotes the unit advance operator)

$$\begin{aligned} y(k+n) + \alpha_1 y(k+n-1) + \cdots + \alpha_n y(k) \\ = \beta_1 u(k+n-1) + \cdots + \beta_n u(k) \end{aligned} \tag{11-36}$$

is

$$\begin{pmatrix} x_1(k+1) \\ x_2(k+1) \\ \vdots \\ x_n(k+1) \end{pmatrix} = \underbrace{\begin{pmatrix} 0 & 1 & 0 & \cdots & 0 \\ 0 & 0 & 1 & \cdots & 0 \\ \vdots & \vdots & \vdots & \ddots & \vdots \\ -\alpha_n & -\alpha_{n-1} & -\alpha_{n-2} & \cdots & -\alpha_1 \end{pmatrix}}_{\mathbf{\Phi}} \begin{pmatrix} x_1(k) \\ x_2(k) \\ \vdots \\ x_n(k) \end{pmatrix} + \underbrace{\begin{pmatrix} 0 \\ 0 \\ \vdots \\ 1 \end{pmatrix}}_{\mathbf{\Psi}} u(k) \tag{11-37}$$

and

$$y(k) = \underbrace{(\beta_n, \beta_{n-1}, \ldots, \beta_1)}_{\mathbf{H}} \mathbf{x}(k) \tag{11-38}$$

For this model there are no unknown parameters in matrix $\mathbf{\Psi}$, and $\mathbf{\Phi}$ and $\mathbf{H}$ each contain exactly $n$ (unknown) parameters. Matrix $\mathbf{\Phi}$ contains the $n$ $\alpha$ parameters that are associated with the poles of $H(z)$, whereas matrix $\mathbf{H}$ contains the $n$ $\beta$ parameters that are associated with the zeros of $H(z)$. *In general, an* n*th-order, single-input single-output system is completely characterized by* 2n *parameters.* □

Our objective now is to determine $L(\theta|\mathbf{Z})$ for the system in (11-32) and (11-33). To begin, we must determine $p(\mathbf{Z}|\theta) = p(\mathbf{z}(1), \mathbf{z}(2), \ldots, \mathbf{z}(N)|\theta)$. This is easy to do, because of the whiteness of noise $\mathbf{v}(k)$; i.e., $p(\mathbf{z}(1), \mathbf{z}(2), \ldots, \mathbf{z}(N)|\theta = p(\mathbf{z}(1)|\theta)p(\mathbf{z}(2)|\theta)\ldots p(\mathbf{z}(N)|\theta)$. Thus,

$$L(\theta|\mathbf{Z}) = \ln\left[\prod_{i=1}^{N} p(\mathbf{z}(i)|\theta)\right] \tag{11-39}$$

From the Gaussian nature of $\mathbf{v}(k)$ and the linear measurement model in (11-33), we know that

$$p(\mathbf{z}(i)|\theta) = \frac{1}{\sqrt{(2\pi)^m |\bar{\mathbf{R}}|}} \exp\left\{-\frac{1}{2}[\mathbf{z}(i) - \mathbf{H}\mathbf{x}(i)]'\bar{\mathbf{R}}^{-1}[\mathbf{z}(i) - \mathbf{H}\mathbf{x}(i)]\right\} \tag{11-40}$$

Thus,

$$\mathrm{L}(\boldsymbol{\theta}|\mathbf{Z}) = -\frac{1}{2}\sum_{i=1}^{N}[\mathbf{z}(i) - \mathbf{H}\mathbf{x}(i)]'\bar{\mathbf{R}}^{-1}[\mathbf{z}(i) - \mathbf{H}\mathbf{x}(i)] - \frac{N}{2}\ln|\bar{\mathbf{R}}| - \frac{N}{2}m\ln 2\pi \tag{11-41}$$

The log-likelihood function $\mathrm{L}(\boldsymbol{\theta}|\mathbf{Z})$ is a function of $\boldsymbol{\theta}$. To indicate which quantities on the right-hand side of (11-41) may depend on $\boldsymbol{\theta}$, we subscript all such quantities with $\boldsymbol{\theta}$. Additionally, because $\frac{N}{2}m\ln 2\pi$ does not depend on $\boldsymbol{\theta}$, we neglect it in subsequent discussions. Our final log-likelihood function is

$$\mathrm{L}(\boldsymbol{\theta}|\mathbf{Z}) = -\frac{1}{2}\sum_{i=1}^{N}[\mathbf{z}(i) - \mathbf{H}_{\theta}\mathbf{x}_{\theta}(i)]'\bar{\mathbf{R}}_{\theta}^{-1}[\mathbf{z}(i) - \mathbf{H}_{\theta}\mathbf{x}_{\theta}(i)] - \frac{N}{2}\ln|\bar{\mathbf{R}}_{\theta}| \tag{11-42}$$

Observe that $\boldsymbol{\theta}$ occurs explicitly and implicitly in $\mathrm{L}(\boldsymbol{\theta}|\mathbf{Z})$. Matrices $\mathbf{H}_{\theta}$ and $\bar{\mathbf{R}}_{\theta}$ contain the explicit dependence of $\mathrm{L}(\boldsymbol{\theta}|\mathbf{Z})$ on $\boldsymbol{\theta}$, whereas state vector $\mathbf{x}_{\theta}(i)$ contains the implicit dependence of $\mathrm{L}(\boldsymbol{\theta}|\mathbf{Z})$ on $\boldsymbol{\theta}$. To numerically calculate the right-hand side of (11-42), we must solve state equation (11-32). This can be done when values of the unknown parameters that appear in $\boldsymbol{\Phi}$ and $\boldsymbol{\Psi}$ are given specific values; for then (11-32) becomes

$$\mathbf{x}_{\theta}(k+1) = \boldsymbol{\Phi}_{\theta}\mathbf{x}_{\theta}(k) + \boldsymbol{\Psi}_{\theta}\mathbf{u}(k), \qquad \mathbf{x}_{\theta}(0) \text{ known} \tag{11-43}$$

In essence, then, *state equation (11-43) is a constraint that is associated with the computation of the log-likelihood function.*

How do we determine $\hat{\boldsymbol{\theta}}_{\mathrm{ML}}$ for $\mathrm{L}(\boldsymbol{\theta}|\mathbf{Z})$ given in (11-42) [subject to its constraint in (11-43)]? No simple closed-form solution is possible, because $\boldsymbol{\theta}$ enters into $\mathrm{L}(\boldsymbol{\theta}|\mathbf{Z})$ in a complicated *nonlinear* manner. The only way presently known for obtaining $\hat{\boldsymbol{\theta}}_{\mathrm{ML}}$ is by means of optimization.

**Comment.** This completes our studies of methods for estimating unknown deterministic parameters. This is a good place to read Lessons A–C. Lesson A is on the subject of sufficient statistics and statistical estimation of parameters and will provide you with interesting perspectives on both maximum-likelihood estimators and unbiased minimum-variance estimators. Lessons B and C are on the subject of higher-order statistics and will literally open up a whole new world for you, one that is not totally dominated by second-order statistics. These three supplemental lessons are still directed at estimating deterministic parameters. Prior to studying methods for estimating unknown random parameters, we pause to review an important body of material about multivariate Gaussian random variables.

## COMPUTATION

Using the *Optimization Toolbox*, two approaches are possible for obtaining a maximum-likelihood estimator: (1) We are given a likelihood or log-likelihood function and we use an unconstrained optimization M-file to maximize them directly; or, (2) We are given a likelihood or log-likelihood function and we analytically establish the necessary conditions for an extremum, by taking their gradient with respect to the parameter vector and setting the gradient equal to zero. Doing this leads to a nonlinear system of equations (unless you are very fortunate and the resulting system is linear) which must then be solved.

### Unconstrained Optimization

**fminu (fmins)**: Finds the minimum of an unconstrained multivariable function, i.e., **fminu** solves the problem: minimize $f(\mathbf{x})$ with respect to $\mathbf{x}$. To find the maximum of an unconstrained function replace $f(\mathbf{x})$ by $-f(\mathbf{x})$. Gradients are calculated using an adaptive finite-difference method, or they can be supplied by the end-user.

The Algorithms section of the *Optimization Toolbox* Reference Manual gives a very complete description of the algorithm used in **fminu**. **fmins** does not use gradient information and is recommended when $f(\mathbf{x})$ is highly discontinuous.

### Nonlinear Equations

**fsolve**: Solves a system of nonlinear equations, $\mathbf{F}(\mathbf{X}) = \mathbf{0}$ for $\mathbf{X}$, where $\mathbf{F}(\mathbf{X})$ and $\mathbf{X}$ are scalars, vectors, or matrices. In this lesson, we typically have the situation $\mathbf{f}(\mathbf{x}) = \mathbf{0}$, where $\mathbf{f}$ is a vector (i.e., the gradient of the likelihood or log-likelihood function) and $\mathbf{x}$ is a vector.

## SUMMARY QUESTIONS

**1.** Check off two things that must accompany finding $\hat{\theta}_{ML}$:
  **(a)** maximize the likelihood function
  **(b)** differentiate the likelihood function
  **(c)** specify the likelihood function

**2.** The invariance property of MLEs is that:
  **(a)** MLEs are consistent
  **(b)** functions of MLEs are themselves MLEs
  **(c)** MLEs are asymptotically efficient

**3.** Because MLEs are asymptotically Gaussian:
  **(a)** their complete probability density function can be described by $\theta$ and $\mathbf{J}^{-1}/N$
  **(b)** they are invariant
  **(c)** they are asymptotically unbiased

**4.** When we say that $\theta$ is "identifiable," we mean:
  **(a)** $\theta$ has fewer unknown parameters than the total number of measurements
  **(b)** $\theta$ exists
  **(c)** it is possible to identify $\theta$ by ML methods

5. If $\hat{b}$ denotes the ML estimate of $b$ and $f(b) = b!\exp(-b^2 t)$, then $\hat{f}_{\text{ML}} = \hat{b}!\exp(-\hat{b}^2 t)$ is a result of the:
   (a) fundamental theorem of estimation theory
   (b) orthogonality principle
   (c) invariance principle
6. The logarithm of $l$ is a monotonic function of $l$ means that whenever $l$ is:
   (a) decreasing, $\ln l$ is increasing, and vice versa
   (b) decreasing or increasing, $\ln l$ is also decreasing or increasing
   (c) positive or negative, $\ln l$ is also positive or negative
7. The MLE of the mean of a Gaussian population is equal to the:
   (a) population mean
   (b) median
   (c) sample mean
8. When $\mathbf{H}(k)$ is deterministic in our generic linear model and the measurements are jointly Gaussian:
   (a) $\hat{\theta}_{\text{ML}}(k) = \hat{\theta}_{\text{LS}}(k) = \hat{\theta}_{\text{BLU}}(k)$
   (b) $\hat{\theta}_{\text{ML}}(k) = \hat{\theta}_{\text{BLU}}(k)$
   (c) $\hat{\theta}_{\text{BLU}}(k) = \hat{\theta}_{\text{WLS}}(k)$
9. The log-likelihood function for the problem of estimating the parameters in a deterministic state-variable model is constrained by:
   (a) the requirement that the log-likelihood function must be positive
   (b) the measurement equation
   (c) the solution of the deterministic state equation

## PROBLEMS

**11-1.** If $\hat{a}$ is the MLE of $a$, what is the MLE of $a^{1/a}$? Explain your answer.

**11-2.** (Sorenson, 1980, Theorem 5.1, p. 185). If an estimator exists such that equality is satisfied in the Cramer–Rao inequality, prove that it can be determined as the solution of the likelihood equation.

**11-3.** Consider a sequence of independently distributed random variables $x_1, x_2, \ldots, x_N$, having the probability density function $\theta^2 x_i e^{-\theta x_i}$, where $\theta > 0$.
   (a) Derive $\hat{\theta}_{\text{ML}}(N)$.
   (b) You want to study whether $\hat{\theta}_{\text{ML}}(N)$ is an unbiased estimator of $\theta$. Explain (without working out all the details) how you would do this.

**11-4.** Consider a random variable $z$ that can take on the values $z = 0, 1, 2, \ldots$. This variable is Poisson distributed, i.e., its probability function $P(z)$ is $P(z) = \mu^z e^{-\mu}/z!$ Let $z(1), z(2), \ldots, z(N)$ denote $N$ independent observations of $z$.
   (a) Find $\hat{\mu}_{\text{ML}}$.
   (b) Prove that $\hat{\mu}_{\text{ML}}$ is an unbiased estimator of $\mu$.
   (c) Prove that $\hat{\mu}_{\text{ML}}$ is an efficient estimator of $\mu$.
   (d) Prove that $\hat{\mu}_{\text{ML}}$ is a consistent estimator of $\mu$.

**11-5.** If $p(z) = \theta e^{-\theta z}$, $z > 0$ and $p(z) = 0$, otherwise, find $\hat{\theta}_{\text{ML}}$ given a sample of $N$ independent observations.

**11-6.** Find the maximum-likelihood estimator of $\theta$ from a sample of $N$ independent observations that are uniformly distributed over the interval $(0, \theta)$. Is $\hat{\theta}_{\text{ML}}$ an unbiased or an asymptotically unbiased estimator of $\theta$? Obtain the answer to this question from first principles. Verify that $\hat{\theta}_{\mathrm{ML}}$ is a consistent estimator of $\theta$.

**11-7.** Derive the maximum-likelihood estimate of the signal power, defined as $\zeta^2$, in the signal $z(k) = \zeta s(k), k = 0, 1, \ldots, N$, where $s(k)$ is a scalar, stationary, zero-mean Gaussian random sequence with autocorrelation $\mathbf{E}\{s(i)s(j)\} = \phi(j-i)$.

**11-8.** Suppose $x$ is a binary variable that assumes a value of 1 with probability $a$ and a value of 0 with probability $(1-a)$. The probability distribution of $x$ can be described by

$$P(x) = (1-a)^{(1-x)}a^x$$

Suppose we draw a random sample of $N$ values $\{x_1, x_2, \ldots, x_N\}$. Find the MLE of $a$.

**11-9.** (Mendel, 1973, Exercise 3-15, p. 177). Consider the linear model for which $\mathbf{V}(k)$ is Gaussian with zero mean, $\mathbf{H}$ is deterministic, and $\mathbf{R}(k) = \sigma^2\mathbf{I}$.

**(a)** Show that the maximum-likelihood estimator of $\sigma^2$, denoted $\hat{\sigma}^2_{\mathrm{ML}}$, is

$$\hat{\sigma}^2_{\mathrm{ML}} = \frac{(\mathbf{Z} - \mathbf{H}\hat{\boldsymbol{\theta}}_{\mathrm{ML}})'(\mathbf{Z} - \mathbf{H}\hat{\boldsymbol{\theta}}_{\mathrm{ML}})}{N}$$

where $\hat{\boldsymbol{\theta}}_{\mathrm{ML}}$ is the maximum-likelihood estimator of $\boldsymbol{\theta}$.

**(b)** Show that $\hat{\sigma}^2_{\mathrm{ML}}$ is biased, but that it is asymptotically unbiased.

**11-10.** We are given $N$ independent samples $\mathbf{z}(1), \mathbf{z}(2), \ldots, \mathbf{z}(N)$ of the identically distributed two-dimensional random vector $\mathbf{z}(i) = \mathrm{col}\,[z_1(i), z_2(i)]$, with the Gaussian density function

$$p(z_1(i), z_2(i)|\rho) = \frac{1}{2\pi\sqrt{1-\rho^2}} \exp\left\{-\frac{z_1^2(i) - 2\rho z_1(i)z_2(i) + z_2^2(i)}{2(1-\rho^2)}\right\}$$

where $\rho$ is the correlation coefficient between $z_1$ and $z_2$.

**(a)** Determine $\hat{\rho}_{\mathrm{ML}}$. (*Hint*: You will obtain a cubic equation for $\hat{\rho}_{\mathrm{ML}}$ and must show that $\hat{\rho}_{\mathrm{ML}} = r$, where $r$ equals the sample correlation coefficient.)

**(b)** Determine the Cramer–Rao bound for $\hat{\rho}_{\mathrm{ML}}$.

**11-11.** (C. M. Chintall, Fall 1991) (Van Trees, 1968) A secure transmission system is desired to transmit two parameters. The simplest system is given by

$$s_1 = x_{11}A_1 + x_{12}A_2$$

$$s_2 = x_{21}A_1 + x_{22}A_2$$

where the $x_{ij}$'s are given. The received signal is

$$r_1 = s_1 + n_1$$

$$r_2 = s_2 + n_2$$

The noises, $n_1$ and $n_2$, are independent, identically distributed, zero-mean, Gaussian random variables. The $A$ parameters are deterministic.

**(a)** Are the maximum-likelihood estimates of $A_1$ and $A_2$ unbiased?

**(b)** Are the maximum-likelihood estimates of $A_1$ and $A_2$ efficient?

**11-12.** (Noel Wilcoxson, Spring 1992) Random variable $z$ has the following Rayleigh density function:

$$p(z|\alpha) = \alpha z \exp\left(-\frac{1}{2}\alpha z^2\right) \quad \text{for } z > 0 \quad \text{and} \quad p(z|\alpha) = 0 \quad \text{for } z \leq 0$$

Given $N$ statistically independent measurements of $z$, find $\hat{\alpha}_{\mathrm{ML}}$.

**11-13.** (Lisa K. Houle, Spring 1992) The Maxwell probability density function for random variable $x$ is $(A > 0)$

$$p(x|A) = \left(\frac{2}{\pi}\right)^{1/2} A^3 x^2 \exp\left(-\frac{A^2 x^2}{2}\right)$$

Given $N$ statistically independent measurements of $x$, calculate $\hat{A}_{\text{ML}}$.

**11-14.** (Tony Hung-yao Wu, Spring 1992) Given a sequence of i.i.d. random variables, $x(1), x(2), \ldots, x(N)$ having the probability density function

$$p[x(i)] = \left(\frac{1}{\theta_2}\right) \exp\left\{-\frac{[x(i) - \theta_1]}{\theta_2}\right\}, \quad \text{where } \theta_1 \leq x(i) < \infty, \ -\infty < \theta_1 < \infty, \text{ and } 0 < \theta_2 < \infty$$

$$= 0 \text{ elsewhere}$$

Find the ML estimators of $\theta_1$ and $\theta_2$.

**11-15.** (Ryuji Maeda, Spring 1992) Using the ML estimation technique of this lesson, we can estimate the probability of a male child considered in Example 10-3.

**(a)** Find $\hat{p}_{\text{ML}}$, given the observations $R_1$ and $R_2$.

**(b)** Suppose one new independent survey is conducted in which it is found that $R_3 = \{1 \text{ girl and } 1 \text{ boy}\}$. Using the results $R_3$, along with the previous observations $R_1$ and $R_2$, find $\hat{p}_{\text{ML}}$.

**(c)** Suppose that a more extensive independent survey has been conducted, one with $N$ samples of two families with two children. It is found that the populations of boys and girls are exactly the same, so $p$ must equal $\frac{1}{2}$. Show that $\hat{p}_{\text{ML}} = 1/2$.

**11-16.** (Pablo Valle, Spring 1992) When applying statistical notions to speech signals, it is necessary to estimate the probability density function from speech waveforms. Paez and Glison (1972) estimated the probability density function using a histogram of speech amplitudes for a large number of samples. They showed that a good approximation to this density function is the gamma density function

$$p(z) = \left(\frac{\sqrt{3}}{(8\pi\sigma_z|z|}\right)^{1/2} \exp\left(-\frac{\sqrt{3}|z|}{2\sigma_z}\right)$$

Find the maximum-likelihood estimator of $\sigma_z^2$ given a sample, $z(1), z(2), \ldots, z(N)$, of independent observations of the speech waveform.

**11-17.** It is well known, from linear system theory, that the choice of state variables used to describe a dynamical system is not unique. Consider the $n$th-order difference equation

$$y(k+n) + a_1 y(k+n-1) + \cdots + a_n y(k) = b_n^0 m(k+n-1) + b_{n-1}^0 m(k+n-2) + \cdots + b_1^0 m(k)$$

A state-variable model for this system is

$$\begin{pmatrix} x_1(k+1) \\ x_2(k+1) \\ \vdots \\ x_n(k+1) \end{pmatrix} = \begin{pmatrix} 0 & 1 & 0 & \cdots & 0 \\ 0 & 0 & 1 & \cdots & 0 \\ \vdots & \vdots & \vdots & \ddots & \vdots \\ -a_n & -a_{n-1} & -a_{n-2} & \cdots & -a_1 \end{pmatrix} \begin{pmatrix} x_1(k) \\ x_2(k) \\ \vdots \\ x_n(k) \end{pmatrix} + \begin{pmatrix} b_1 \\ b_2 \\ \vdots \\ b_n \end{pmatrix} m(k)$$

and

$$y(k) = (1 \ 0 \ldots 0) x(k)$$

where

$$\begin{pmatrix} b_1 \\ b_2 \\ \vdots \\ b_n \end{pmatrix} = \begin{pmatrix} 1 & 0 & 0 & \dots & 0 & 0 \\ a_1 & 1 & 0 & \dots & 0 & 0 \\ a_2 & a_1 & 1 & \dots & 0 & 0 \\ \vdots & \vdots & \vdots & \ddots & \vdots & \vdots \\ a_{n-2} & a_{n-3} & a_{n-4} & \dots & 1 & 0 \\ a_{n-1} & a_{n-2} & a_{n-3} & \dots & a_1 & 1 \end{pmatrix} \begin{pmatrix} b_1^0 \\ b_2^0 \\ \vdots \\ b_n^0 \end{pmatrix}$$

Suppose maximum-likelihood estimates have heen determined for the $a_i$ and $b_i$ parameters. How do we compute the maximum-likelihood estimates of the $b_i^0$ parameters?

**11-18.** Prove that for a random sample of measurements, a maximum-likelihood estimator is a consistent estimator. [*Hints*: (1) Show that $\mathbf{E}\{\partial \ln p(\mathbf{Z}|\boldsymbol{\theta})/\partial\boldsymbol{\theta}|\boldsymbol{\theta}\} = \mathbf{0}$; (2) expand $\partial \ln p(\mathbf{Z}|\hat{\boldsymbol{\theta}}_{ML})/\partial\boldsymbol{\theta}$ in a Taylor series about $\boldsymbol{\theta}$, and show that $\partial \ln p(\mathbf{Z}|\boldsymbol{\theta})/\partial\boldsymbol{\theta} = -(\hat{\boldsymbol{\theta}}_{ML} - \boldsymbol{\theta})'\partial^2 \ln p(\mathbf{Z}|\boldsymbol{\theta}^*)/\partial\boldsymbol{\theta}^2$, where $\boldsymbol{\theta}^* = \lambda\boldsymbol{\theta} + (1-\lambda)\hat{\boldsymbol{\theta}}_{ML}$ and $0 \le \lambda \le 1$; (3) show that $\partial \ln p(\mathbf{Z}|\boldsymbol{\theta})/\partial\boldsymbol{\theta} = \sum_{i=1}^N \partial \ln p(\mathbf{z}(i)|\boldsymbol{\theta})/\partial\boldsymbol{\theta}$ and $\partial^2 \ln p(\mathbf{Z}|\boldsymbol{\theta})/\partial\boldsymbol{\theta}^2 = \sum_{i=1}^N \partial^2 \ln p(z(i)|\boldsymbol{\theta})/\partial\boldsymbol{\theta}^2$; (4) using the strong law of large numbers to assert that, with probability 1, sample averages converge to ensemble averages, and assuming that $\mathbf{E}\{\partial^2 \ln p(z(i)|\boldsymbol{\theta})/\partial\boldsymbol{\theta}^2\}$ is negative definite, show that $\hat{\boldsymbol{\theta}}_{ML} \to \boldsymbol{\theta}$ with probability 1; thus, $\hat{\boldsymbol{\theta}}_{ML}$ is a consistent estimator of $\boldsymbol{\theta}$. The steps in this proof have been taken from Sorenson, 1980, pp. 187–191.]

**11-19.** Prove that, for a random sample of measurements, a maximum likelihood estimator is asymptotically Gaussian with mean value $\boldsymbol{\theta}$ and covariance matrix $(N\mathbf{J}_1)^{-1}$, where $\mathbf{J}_1$ is the Fisher information matrix for a single measurement $z(i)$; [*Hints*: (1) Expand $\partial \ln p(\mathbf{Z}|\hat{\boldsymbol{\theta}}_{ML})/\partial\boldsymbol{\theta}$ in a Taylor series about $\boldsymbol{\theta}$ and neglect second- and higher-order terms; (2) show that

$$\frac{1}{N}\frac{\partial}{\partial\boldsymbol{\theta}} \ln p(\mathbf{Z}|\boldsymbol{\theta}) = (\hat{\boldsymbol{\theta}}_{ML} - \boldsymbol{\theta})'\left[-\frac{1}{N}\frac{\partial^2}{\partial\boldsymbol{\theta}^2} \ln p(\mathbf{Z}|\boldsymbol{\theta})\right]$$

(3) let $\mathbf{s}(\boldsymbol{\theta}) = \partial \ln p(\mathbf{Z}|\boldsymbol{\theta})/\partial\boldsymbol{\theta}$ and show that $\mathbf{s}(\boldsymbol{\theta}) = \sum_{i=1}^N \mathbf{s}_i(\boldsymbol{\theta})$, where $\mathbf{s}_i(\boldsymbol{\theta}) = \partial \ln p(z(i)|\boldsymbol{\theta})/\partial\boldsymbol{\theta}$; (4) let $\bar{\mathbf{s}}$ denote the sample mean of $\mathbf{s}_i(\boldsymbol{\theta})$, and show that the distribution of $\bar{\mathbf{s}}$ asymptotically converges to a Gaussian distribution having mean zero and covariance $\mathbf{J}_1/N$; (5) using the strong law of large numbers we know that $\frac{1}{N}\partial^2 \ln p(\mathbf{Z}|\boldsymbol{\theta})/\partial\boldsymbol{\theta}^2 \to \mathbf{E}\{\partial^2 \ln p(z(i)|\boldsymbol{\theta})\partial\boldsymbol{\theta}^2|\boldsymbol{\theta}\}$; consequently, show that $(\hat{\boldsymbol{\theta}}_{ML} - \boldsymbol{\theta})'\mathbf{J}_1$ is asymptotically Gaussian with zero mean and covariance $\mathbf{J}_1/N$; (6) complete the proof of the theorem. The steps in this proof have been taken from Sorenson, 1980, p. 192.]

**11-20.** Prove that, for a random sample of measurements, a maximum-likelihood estimator is asymptotically efficient. [*Hints*: (1) Let $\mathbf{J}(\boldsymbol{\theta}) = -\mathbf{E}\{\partial^2 \ln p(\mathbf{Z}|\boldsymbol{\theta})/\partial\boldsymbol{\theta}^2|\boldsymbol{\theta}\}$ and show that $\mathbf{J}(\boldsymbol{\theta}) = N\mathbf{J}_1$, where $\mathbf{J}_1$ is the Fisher information matrix for a single measurement $z(i)$; (2) use the result stated in Problem 11-19 to complete the proof. The steps in this proof have been taken from Sorenson, 1980, p. 193.]

**11-21.** (Mithat Dogan, 1993) Here is an array signal-processing model that occurs when the desired but unknown signal of interest is *time gated*:

$$\mathbf{y}(t) = \mathbf{Lr}(t) = \mathbf{Lb}_d s_d(t) + \mathbf{LBi}(t) + \mathbf{n}_w(t) \tag{11-21.1}$$

where $s_d(t)$ is a time-gated signal; i.e.,

$$s_d(t) = d(t)m(t) = \begin{cases} d(t), & t \in \Omega \\ 0, & t \in \Omega^c \end{cases} \tag{11-21.2}$$

in which $d(t)$ is the *desired signal*, which is available only over time intervals $t \in \Omega$ [$m(t)$ modulates $d(t)$; $m(t) = 1$ for $t \in \Omega$, and $m(t) = 0$ for $t \in \Omega^c$]; $\mathbf{r}(t)$ is a vector of received signals (at an array of sensors) that contains the desired signal (subject to multipaths), interferers, and colored noise; $\mathbf{i}(t)$ are the interferers (jammers); $\mathbf{n}_w(t)$ is white noise, with covariance matrix equal to $\sigma_w^2\mathbf{I}$; $\mathbf{L}$ is a whitening matrix that is prespecified; $\mathbf{b}_d$ is a generalized steering vector for the gated signal $s_d(t)$; and $\mathbf{B}$ is a generalized steering matrix that represents the effective appearance of interference waveforms in the received data. $s_d(t)$ and $\mathbf{i}(t)$ are modeled as deterministic signals, and the number of interferers is assumed known.

Let $\boldsymbol{\xi}$ be a vector of unknown parameters, consisting of samples of waveforms from desired and interference signals and elements of generalized steering vectors.

**(a)** Show that the ML method for estimating $\boldsymbol{\xi}$ is identical to minimizing $J_{\text{ML}}(\boldsymbol{\xi})$, where

$$J_{\text{ML}}(\boldsymbol{\xi}) = \sum_t \|\mathbf{y}(t) - \mathbf{Lb}_d s_d(t) - \mathbf{LBi}(t)\|^2 \tag{11-21.3}$$

In effect, this is a least-squares problem. [*Hint:* Let $\mathbf{Y}(t) = \text{col}(\mathbf{y}(t_1), \mathbf{y}(t_2), \ldots, \mathbf{y}(t_N))$ and determine $L(\boldsymbol{\xi}|\mathbf{Y}(t))$.]

**(b)** Show that the ML estimate of $\mathbf{i}(t)$, given the desired signal $s_d(t)$, is

$$\hat{\mathbf{i}}(t) = (\mathbf{LB})^+[\mathbf{y}(t) - \mathbf{Lb}_d s_d(t)], \qquad t \in \Omega \cup \Omega^c \tag{11-21.4}$$

in which $(\mathbf{LB})^+$ denotes the pseudoinverse of $\mathbf{LB}$.

**(c)** Next, show that, given $\hat{\mathbf{i}}(t)$ in (11-21.4), the ML estimate of $s_d(t)$, $\hat{s}_d(t)$, is

$$\hat{s}_d(t) = \mathbf{c}^+\mathbf{y}(t), \qquad t \in \Omega \tag{11-21.5}$$

where

$$\mathbf{c} = (\mathbf{I} - \mathbf{P}_{\text{LB}})\mathbf{Lb}_d \tag{11-21.6}$$

and

$$\mathbf{P}_{\text{LB}} = (\mathbf{LB})(\mathbf{LB})^+ \tag{11-21.7}$$

[*Hint:* Show that $(\mathbf{I} - \mathbf{P_X})^{\text{H}}(\mathbf{I} - \mathbf{P_X}) = (\mathbf{I} - \mathbf{P_X})$.]

**(d)** Show that, based on the results obtained for $\hat{\mathbf{i}}(t)$ in (11-21.4) and $\hat{s}_d(t)$ in (11-21.5), $J_{\text{ML}}(\boldsymbol{\xi})$ can be expressed as

$$J_{\text{ML}}(\boldsymbol{\xi}) = \sum_{t\in\Omega} \|\mathbf{P}_{\text{LB}}^{\perp}[\mathbf{y}(t) - \mathbf{Lb}_d\mathbf{c}^+\mathbf{y}(t)]\|^2 + \sum_{t\in\Omega^c} \|\mathbf{P}_{\text{LB}}^{\perp}\mathbf{y}(t)\|^2 \tag{11-21.8}$$

where

$$\mathbf{P}_{\text{LB}}^{\perp} = \mathbf{I} - \mathbf{P}_{\text{LB}} \tag{11-21.9}$$

and $\boldsymbol{\xi}$ now only contains elements of the generalized steering vectors.

**(e)** Show that (11-21.8) can be reexpressed as

$$J_{\text{ML}}(\boldsymbol{\xi}) = -\text{tr}\,[(\mathbf{P}_{\text{LB}} + \mathbf{P}_c)\hat{\mathbf{R}}_s] - \text{tr}\,(\mathbf{P}_{\text{LB}}\hat{\mathbf{R}}_I) \tag{11-21.10}$$

where

$$\hat{\mathbf{R}}_s \triangleq \sum_{t \in \Omega} \mathbf{y}(t)\mathbf{y}^H(t) \tag{11-21.11}$$

and

$$\hat{\mathbf{R}}_I \triangleq \sum_{t \in \Omega^c} \mathbf{y}(t)\mathbf{y}^H(t) \tag{11-21.12}$$

[*Hints:* Show that $\mathbf{P}_{\text{LB}}^2 = \mathbf{P}_{\text{LB}}$ and $\mathbf{P}_c^2 = \mathbf{P}_c$ using properties of the pseudoinverse given in Lesson 4. Show that $\mathbf{P}_{\text{LB}}\mathbf{P}_c = \mathbf{0}$. Consequently, $(\mathbf{I} - \mathbf{P}_{\text{LB}} - \mathbf{P}_c)^2 = \mathbf{I} - \mathbf{P}_{\text{LB}} - \mathbf{P}_c$.]

**(f)** Explain how $\xi$ is determined from $J_{\text{ML}}(\xi)$.

# LESSON 12

# Multivariate Gaussian Random Variables


## SUMMARY

This is a transition lesson. Prior to studying the methods for estimating unknown random parameters, we review an important body of material about multivariate (i.e., vector) Gaussian random variables. Most, if not all, of the material in this lesson should be a review for a reader who has had a course in probability theory.

The purpose of this lesson is to collect a wide range of facts about multivariate Gaussian random variables in one place, because they are often needed in the remaining lessons. Two of the most important facts are that (1) the conditional density function for two jointly Gaussian vectors is also Gaussian, and (2) the conditional mean is random.

When you complete this lesson, you will be able to state many useful and important formulas and facts about multivariate Gaussian random variables.


## INTRODUCTION

Gaussian random variables are important and widely used for at least two reasons. First, they sometimes provide a model that is a reasonable approximation to observed random behavior. Second, if the random phenomenon that we observe at the macroscopic level is the superposition of an arbitrarily large number of independent random phenomena, which occur at the microscopic level, the macroscopic description is justifiably Gaussian.

Most (if not all) of the material in this lesson should be a review for a reader who has had a course in probability theory. We collect a wide range of facts about multivariate Gaussian random variables here, in one place, because they are often needed in the remaining lessons.

## UNIVARIATE GAUSSIAN DENSITY FUNCTION

A random variable $y$ is said to be distributed as the *univariate Gaussian distribution* with mean $m_y$ and variance $\sigma_y^2$ [i.e., $y \sim N(y; m_y, \sigma_y^2)$] if the density function of $y$ is given as

$$p(y) = \frac{1}{\sqrt{2\pi\sigma_y^2}} \exp\left[-\frac{(y - m_y)^2}{2\sigma_y^2}\right], \qquad -\infty < y < \infty \qquad (12\text{-}1)$$

For notational simplicity throughout this chapter, we do not condition density functions on their parameters; this conditioning is understood. Density $p(y)$ is the familiar bell-shaped curve, centered at $y = m_y$.

## MULTIVARIATE GAUSSIAN DENSITY FUNCTION   $y = m_x 1$

Let $y_1, y_2, \ldots, y_m$ be random variables, and $\mathbf{y} = \text{col}(y_1, y_2, \ldots, y_m)$. The function

$$p(y_1, y_2, \ldots, y_m) = p(\mathbf{y})$$

$$= \frac{1}{\sqrt{(2\pi)^m |\mathbf{P_y}|}} \exp\left[-\frac{1}{2}(\mathbf{y} - \mathbf{m_y})' \mathbf{P_y}^{-1} (\mathbf{y} - \mathbf{m_y})\right] \qquad (12\text{-}2)$$

is said to be a *multivariate* (m-*variate*) *Gaussian density function* [i.e., $\mathbf{y} \sim N(\mathbf{y}; \mathbf{m_y}, \mathbf{P_y})$]. In (12-2),

$$\mathbf{m_y} = \mathbf{E}\{y\} \quad = m_{1 \times 1} \qquad (12\text{-}3)$$

and

$$\mathbf{P_y} = \mathbf{E}\{(\mathbf{y} - \mathbf{m_y})(\mathbf{y} - \mathbf{m_y})'\} = covariance \ge 0 \qquad (12\text{-}4) \ne 0$$
$$= m \times m$$

*pos def for this course*
*#4,7,9 P,9 70*

Note that, although we refer to $p(\mathbf{y})$ as a density function, it is actually a *joint density function* between the random variables $y_1, y_2, \ldots, y_m$. If $\mathbf{P_y}$ is not positive definite, it is more convenient to define the multivariate Gaussian distribution by its characteristic function. We will not need to do this.

## JOINTLY GAUSSIAN RANDOM VECTORS   $\underline{x} = n \times 1$
$y = m \times 1$

Let $\mathbf{x}$ and $\mathbf{y}$ individually be $n$- and $m$-dimensional Gaussian random vectors, i.e., $\mathbf{x} \sim N(\mathbf{x}; \mathbf{m_x}, \mathbf{P_x})$ and $\mathbf{y} \sim N(\mathbf{y}; \mathbf{m_y}, \mathbf{P_y})$. Let $\mathbf{P_{xy}}$ and $\mathbf{P_{yx}}$ denote the cross covariance matrices between $\mathbf{x}$ and $\mathbf{y}$; i.e.,

$$\mathbf{P_{xy}} = \mathbf{E}\{(\mathbf{x} - \mathbf{m_x})(\mathbf{y} - \mathbf{m_y})'\} \qquad (12\text{-}5)$$

and

$$\mathbf{P_{yx}} = \mathbf{E}\{(\mathbf{y} - \mathbf{m_y})(\mathbf{x} - \mathbf{m_x})'\} \tag{12-6}$$

Clearly, $\mathbf{P_{xy}} = \mathbf{P}'_{\mathbf{yx}}$. We are interested in the joint density between $\mathbf{x}$ and $\mathbf{y}$, $p(\mathbf{x}, \mathbf{y})$. Vectors $\mathbf{x}$ and $\mathbf{y}$ are jointly Gaussian if

$$p(\mathbf{x}, \mathbf{y}) = \frac{1}{\sqrt{(2\pi)^{n+m}|\mathbf{P_z}|}} \exp\left\{-\frac{1}{2}(\mathbf{z} - \mathbf{m_z})'\mathbf{P_z}^{-1}(\mathbf{z} - \mathbf{m_z})\right\} \tag{12-7}$$

where

$$\mathbf{z} = \text{col}\,(\mathbf{x}, \mathbf{y}) \tag{12-8}$$

$$\mathbf{m_z} = \text{col}\,(\mathbf{m_x}, \mathbf{m_y}) \tag{12-9}$$

and

$$\mathbf{P_z} = \begin{pmatrix} \mathbf{P_x} & \mathbf{P_{xy}} \\ \mathbf{P_{yx}} & \mathbf{P_y} \end{pmatrix} \tag{12-10}$$

If $\mathbf{x}$ and $\mathbf{y}$ are marginally Gaussian and mutually uncorrelated, they are jointly Gaussian. Note, also, that if $\mathbf{x}$ and $\mathbf{y}$ are jointly Gaussian then they are marginally (i.e., individually) Gaussian. The converse is true if $\mathbf{x}$ and $\mathbf{y}$ are independent, but it is not necessarily true if they are not independent (Papoulis, 1991, p. 145).

To evaluate $p(\mathbf{x}, \mathbf{y})$ in (12-7), we need $|\mathbf{P_z}|$ and $\mathbf{P_z}^{-1}$. It is straightforward to compute $|\mathbf{P_z}|$ once the given values for $\mathbf{P_x}$, $\mathbf{P_y}$, $\mathbf{P_{xy}}$, and $\mathbf{P_{yx}}$ are substituted into (12-10). It is often useful to be able to express the components of $\mathbf{P_z}^{-1}$ directly in terms of the components of $\mathbf{P_z}$. It is a straightforward exercise in algebra (just form $\mathbf{P_z}\mathbf{P_z}^{-1} = \mathbf{I}$ and equate elements on both sides) to show that

$$\mathbf{P_z}^{-1} = \begin{pmatrix} \mathbf{A} & \mathbf{B} \\ \mathbf{B}' & \mathbf{C} \end{pmatrix} \tag{12-11}$$

where

$$\mathbf{A} = (\mathbf{P_x} - \mathbf{P_{xy}}\mathbf{P_y}^{-1}\mathbf{P_{yx}})^{-1} = \mathbf{P_x}^{-1} + \mathbf{P_x}^{-1}\mathbf{P_{xy}}\mathbf{C}\mathbf{P_{yx}}\mathbf{P_x}^{-1} \tag{12-12}$$

$$\mathbf{B} = -\mathbf{A}\mathbf{P_{xy}}\mathbf{P_y}^{-1} = -\mathbf{P_x}^{-1}\mathbf{P_{xy}}\mathbf{C} \tag{12-13}$$

and

$$\mathbf{C} = (\mathbf{P_y} - \mathbf{P_{yx}}\mathbf{P_x}^{-1}\mathbf{P_{xy}})^{-1} = \mathbf{P_y}^{-1} + \mathbf{P_y}^{-1}\mathbf{P_{yx}}\mathbf{A}\mathbf{P_{xy}}\mathbf{P_y}^{-1} \tag{12-14}$$

## THE CONDITIONAL DENSITY FUNCTION

One of the most important density functions we will be interested in is the conditional density function $p(\mathbf{x}|\mathbf{y})$. Recall, from probability theory (e.g., Papoulis, 1991), that

$$p(\mathbf{x}|\mathbf{y}) = \frac{p(\mathbf{x}, \mathbf{y})}{p(\mathbf{y})} \tag{12-15}$$

**Theorem 12-1.** *Let* $\mathbf{x}$ *and* $\mathbf{y}$ *be* n- *and* m-*dimensional vectors that are jointly Gaussian. Then*

$$p(\mathbf{x}|\mathbf{y}) = \frac{1}{\sqrt{(2\pi)^n |\mathbf{P}_{\mathbf{x}|\mathbf{y}}|}} \exp\left\{-\frac{1}{2}(\mathbf{x}-\mathbf{m})'\mathbf{P}_{\mathbf{x}|\mathbf{y}}^{-1}(\mathbf{x}-\mathbf{m})\right\} \tag{12-16}$$

*where*

$$\mathbf{m} = E\{\mathbf{x}|\mathbf{y}\} = \mathbf{m}_{\mathbf{x}} + \mathbf{P}_{\mathbf{xy}}\mathbf{P}_{\mathbf{y}}^{-1}(\mathbf{y}-\mathbf{m}_{\mathbf{y}}) \tag{12-17}$$

*and*

$$\mathbf{P}_{\mathbf{x}|\mathbf{y}} = \mathbf{A}^{-1} = \mathbf{P}_{\mathbf{x}} - \mathbf{P}_{\mathbf{xy}}\mathbf{P}_{\mathbf{y}}^{-1}\mathbf{P}_{\mathbf{yx}} \quad = covariance \; later \; Kalman \; K_{pf}aj \tag{12-18}$$

*This means* p(x|y) *is also multivariate Gaussian with (conditional) mean* **m** *and covariance* $\mathbf{P}_{\mathbf{x}|\mathbf{y}}$. *Note* p(x|y) *is random! since Y is nature & Y is random (X & Z = measurements whichever random)* (i.e. m is random (since y is this equation 12-17))

*Proof.* From (12-15), (12-7), and (12-2), we find that *explicitly*

$$p(\mathbf{x}|\mathbf{y}) = \frac{1}{\sqrt{(2\pi)^n \frac{|\mathbf{P}_{\mathbf{z}}|}{|\mathbf{P}_{\mathbf{y}}|}}} \exp\left[-\frac{1}{2}(\mathbf{z}-\mathbf{m}_{\mathbf{z}})'\begin{pmatrix}\mathbf{A} & \mathbf{B} \\ \mathbf{B}' & \mathbf{C}-\mathbf{P}_{\mathbf{y}}^{-1}\end{pmatrix}(\mathbf{z}-\mathbf{m}_{\mathbf{z}})\right] \tag{12-19}$$

Taking a closer look at the quadratic exponent, which we denote $E(\mathbf{x}, \mathbf{y})$, we find that

$$\begin{aligned}
E(\mathbf{x}, \mathbf{y}) &= (\mathbf{x}-\mathbf{m}_{\mathbf{x}})'\mathbf{A}(\mathbf{x}-\mathbf{m}_{\mathbf{x}}) + 2(\mathbf{x}-\mathbf{m}_{\mathbf{x}})'\mathbf{B}(\mathbf{y}-\mathbf{m}_{\mathbf{y}}) \\
&\quad + (\mathbf{y}-\mathbf{m}_{\mathbf{y}})'(\mathbf{C}-\mathbf{P}_{\mathbf{y}}^{-1})(\mathbf{y}-\mathbf{m}_{\mathbf{y}}) \\
&= (\mathbf{x}-\mathbf{m}_{\mathbf{x}})'\mathbf{A}(\mathbf{x}-\mathbf{m}_{\mathbf{x}}) - 2(\mathbf{x}-\mathbf{m}_{\mathbf{x}})'\mathbf{A}\mathbf{P}_{\mathbf{xy}}\mathbf{P}_{\mathbf{y}}^{-1}(\mathbf{y}-\mathbf{m}_{\mathbf{y}}) \\
&\quad + (\mathbf{y}-\mathbf{m}_{\mathbf{y}})'\mathbf{P}_{\mathbf{y}}^{-1}\mathbf{P}_{\mathbf{yx}}\mathbf{A}\mathbf{P}_{\mathbf{xy}}\mathbf{P}_{\mathbf{y}}^{-1}(\mathbf{y}-\mathbf{m}_{\mathbf{y}})
\end{aligned} \tag{12-20}$$

In obtaining (12-20) we have used (12-13) and (12-14). We now recognize that (12-20) looks like a quadratic expression in $\mathbf{x}-\mathbf{m}_{\mathbf{x}}$ and $\mathbf{P}_{\mathbf{xy}}\mathbf{P}_{\mathbf{y}}^{-1}(\mathbf{y}-\mathbf{m}_{\mathbf{y}})$ and express it in factored form as

$$E(\mathbf{x}, \mathbf{y}) = [(\mathbf{x}-\mathbf{m}_{\mathbf{x}})-\mathbf{P}_{\mathbf{xy}}\mathbf{P}_{\mathbf{y}}^{-1}(\mathbf{y}-\mathbf{m}_{\mathbf{y}})]'\mathbf{A}[(\mathbf{x}-\mathbf{m}_{\mathbf{x}})-\mathbf{P}_{\mathbf{xy}}\mathbf{P}_{\mathbf{y}}^{-1}(\mathbf{y}-\mathbf{m}_{\mathbf{y}})] \tag{12-21}$$

Defining **m** and $\mathbf{P}_{\mathbf{x}|\mathbf{y}}$ as in (12-17) and (12-18), we see that

$$E(\mathbf{x}, \mathbf{y}) = (\mathbf{x}-\mathbf{m})'\mathbf{P}_{\mathbf{x}|\mathbf{y}}^{-1}(\mathbf{x}-\mathbf{m}) \tag{12-22}$$

If we can show that $|\mathbf{P}_{\mathbf{z}}|/|\mathbf{P}_{\mathbf{y}}| = |\mathbf{P}_{\mathbf{x}|\mathbf{y}}|$, then we will have shown that $p(\mathbf{x}|\mathbf{y})$ is given by (12-16), which will mean that $p(\mathbf{x}|\mathbf{y})$ is multivariate Gaussian with mean **m** and covariance $\mathbf{P}_{\mathbf{x}|\mathbf{y}}$.

It is not at all obvious that $|\mathbf{P}_{\mathbf{x}|\mathbf{y}}| = |\mathbf{P}_{\mathbf{z}}|/|\mathbf{P}_{\mathbf{y}}|$. We shall reexpress matrix $\mathbf{P}_{\mathbf{z}}$ so that $|\mathbf{P}_{\mathbf{z}}|$ can be determined by appealing to the following theorems (Graybill, 1961, p. 6):

**i.** If **V** and **G** are $n \times n$ matrices, then $|\mathbf{VG}| = |\mathbf{V}||\mathbf{G}|$.

**ii.** If **M** is a square matrix such that

$$\mathbf{M} = \begin{pmatrix}\mathbf{M}_{11} & \mathbf{M}_{12} \\ \mathbf{M}_{21} & \mathbf{M}_{22}\end{pmatrix}$$

where $\mathbf{M}_{11}$ and $\mathbf{M}_{22}$ are square matrices, and if $\mathbf{M}_{12} = \mathbf{0}$ or $\mathbf{M}_{21} = \mathbf{0}$, then $|\mathbf{M}| = |\mathbf{M}_{11}||\mathbf{M}_{22}|$.

We now show that two matrices, **L** and **N**, can be found so that

$$\mathbf{P_z} = \begin{pmatrix} \mathbf{P_x} & \mathbf{P_{xy}} \\ \mathbf{P_{yx}} & \mathbf{P_y} \end{pmatrix} = \begin{pmatrix} \mathbf{L} & \mathbf{P_{xy}} \\ \mathbf{0} & \mathbf{P_y} \end{pmatrix} \begin{pmatrix} \mathbf{I_n} & \mathbf{0} \\ \mathbf{N} & \mathbf{I_m} \end{pmatrix} \tag{12-23}$$

This is an L–U decomposition of matrix $\mathbf{P_z}$. Multiplying the two matrices on the right-hand side of (12-23) and equating the 1–1 and 2–1 components from both sides of the resulting equation, we find that

$$\mathbf{P_x} = \mathbf{L} + \mathbf{P_{xy}N} \tag{12-24}$$

and

$$\mathbf{P_{yx}} = \mathbf{P_y N} \tag{12-25}$$

from which it follows that

$$\mathbf{N} = \mathbf{P_y^{-1} P_{yx}} \tag{12-26}$$

and

$$\mathbf{L} = \mathbf{P_x} - \mathbf{P_{xy} P_y^{-1} P_{yx}} \tag{12-27}$$

From (12-23), we see that

$$|\mathbf{P_z}| = |\mathbf{L}||\mathbf{P_y}|$$

or

$$|\mathbf{L}| = \frac{|\mathbf{P_z}|}{|\mathbf{P_y}|} \tag{12-28}$$

Comparing the equations of **L** and $\mathbf{P_{x|y}}$, we find they are the same; thus, we have proven that

$$|\mathbf{P_{x|y}}| = \frac{|\mathbf{P_z}|}{|\mathbf{P_y}|} \tag{12-29}$$

which completes the proof of this theorem. □

## PROPERTIES OF MULTIVARIATE GAUSSIAN RANDOM VARIABLES

From the preceding formulas for $p(\mathbf{y})$, $p(\mathbf{x}, \mathbf{y})$, and $p(\mathbf{x}|\mathbf{y})$, we see that *multivariate Gaussian probability density functions are completely characterized by their first two moments*, i.e., their mean vector and covariance matrix. All other moments can be expressed in terms of their first two moments (see, e.g., Papoulis, 1991).

From probability theory, we also recall the following two important facts about Gaussian random variables:

1. Statistically independent Gaussian random variables are uncorrelated and vice versa; thus, $\mathbf{P_{xy}}$ and $\mathbf{P_{yx}}$ are both zero matrices.

So this can be tested see if $P_{xy} = 0$

⬩ 2. Linear (or affine) transformations on and linear (or affine) combinations of Gaussian random variables are themselves Gaussian random variables; thus, if **x** and **y** are jointly Gaussian, then **z** = **Ax** + **By** + **c** is also Gaussian. We refer to this property as the *linearity property*. $\underline{z} = H\underline{\theta} + \underline{v}$ *so if θ and v are G then z is G*

## PROPERTIES OF CONDITIONAL MEAN

We learned, in Theorem 12-1, that

$$E\{x|y\} = m_x + P_{xy}P_y^{-1}(y - m_y) \quad \text{random} \tag{12-30}$$

Because $E\{x|y\}$ depends on **y**, which is random, it is also random.

**Theorem 12-2.** *When* **x** *and* **y** *are jointly Gaussian,* $E\{x|y\}$ *is multivariate Gaussian and is an affine combination of the elements of* **y**.

*Proof.* That $E\{x|y\}$ is Gaussian follows from the linearity property applied to (12-30). An affine transformation of **y** has the structure **Ty** + **f**. $E\{x|y\}$ has this structure; thus, it is an affine transformation. Note that if $m_x = 0$ and $m_y = 0$ then $E\{x|y\}$ is a linear transformation of **y**. □

**Theorem 12-3.** *Let* **x**, **y**, *and* **z** *be* n × 1, m × 1, *and* r × 1 *jointly Gaussian random vectors. If* **y** *and* **z** *are statistically independent, then*

$$E\{x|y, z\} = E\{x|y\} + E\{x|z\} - m_x \tag{12-31}$$

*Proof.* Let ξ = col(y, z); then

$$E\{x|\xi\} = m_x + P_{x\xi}P_\xi^{-1}(\xi - m_\xi) \tag{12-32}$$

We leave it to the reader to show that

$$P_{x\xi} = (P_{xy}|P_{xz}) \tag{12-33}$$

and

$$P_\xi = \left(\begin{array}{c|c} P_y & 0 \\ \hline 0 & P_z \end{array}\right) \tag{12-34}$$

(the off-diagonal elements in $P_\xi$ are zero if **y** and **z** are statistically independent; this is also true if **y** and **z** are uncorrelated, because **y** and **z** are jointly Gaussian); thus,

$$E\{x|\xi\} = m_x + P_{xy}P_y^{-1}(y - m_y) + P_{xz}P_z^{-1}(z - m_z)$$
$$= E\{x|y\} + E\{x|z\} - m_x \quad \square$$

In our developments of recursive estimators, **y** and **z** (which will be associated with a partitioning of measurement vectors **Z**) are not necessarily independent. The following important generalization of Theorem 12-3 will be needed.

**Theorem 12-4.** *Let* **x**, **y**, *and* **z** *be* n × 1, m × 1, *and* r × 1 *jointly Gaussian random vectors. If* **y** *and* **z** *are not necessarily statistically independent, then*

$$\mathbf{E}\{\mathbf{x}|\mathbf{y}, \mathbf{z}\} = \mathbf{E}\{\mathbf{x}|\mathbf{y}, \tilde{\mathbf{z}}\} \tag{12-35}$$

*where*

$$\tilde{\mathbf{z}} = \mathbf{z} - \mathbf{E}\{\mathbf{z}|\mathbf{y}\} \tag{12-36}$$

*so that*

$$\mathbf{E}\{\mathbf{x}|\mathbf{y}, \mathbf{z}\} = \mathbf{E}\{\mathbf{x}|\mathbf{y}\} + \mathbf{E}\{\mathbf{x}|\tilde{\mathbf{z}}\} - \mathbf{m}_{\mathbf{x}} \tag{12-37}$$

*Proof* (Mendel, 1983, p. 53). The proof proceeds in two stages: (a) assume that (12-35) is true and demonstrate the truth of (12-37), and (b) demonstrate the truth of (12-35).

**a.** If we can show that $\mathbf{y}$ and $\tilde{\mathbf{z}}$ are jointly Gaussian and statistically independent, then (12-37) follows from Theorem 12-3. For Gaussian random vectors, however, uncorrelatedness implies independence.

To begin, we assert that $\mathbf{y}$ and $\tilde{\mathbf{z}}$ are jointly Gaussian, because $\tilde{\mathbf{z}} = \mathbf{z} - \mathbf{E}\{\mathbf{z}|\mathbf{y}\} = \mathbf{z} - \mathbf{m}_{\mathbf{z}} - \mathbf{P}_{\mathbf{zy}}\mathbf{P}_{\mathbf{y}}^{-1}(\mathbf{y} - \mathbf{m}_{\mathbf{y}})$ depends on $\mathbf{y}$ and $\mathbf{z}$, which are jointly Gaussian.

Next, we show that $\tilde{\mathbf{z}}$ is zero mean. This follows from the calculation

$$\mathbf{m}_{\tilde{\mathbf{z}}} = \mathbf{E}\{\mathbf{z} - \mathbf{E}\{\mathbf{z}|\mathbf{y}\}\} = \mathbf{E}\{\mathbf{z}\} - \mathbf{E}\{\mathbf{E}\{\mathbf{z}|\mathbf{y}\}\} \tag{12-38}$$

where the outer expectation in the second term on the right-hand side of (12-38) is with respect to $\mathbf{y}$. From probability theory (Papoulis, 1991, p. 172, or the Supplementary Material at the end of Lesson 8), $\mathbf{E}\{\mathbf{z}\}$ can be expressed as

$$\mathbf{E}\{\mathbf{z}\} = \mathbf{E}\{\mathbf{E}\{\mathbf{z}|\mathbf{y}\}\} \tag{12-39}$$

From (12-38) and (12-39), we see that $\mathbf{m}_{\tilde{\mathbf{z}}} = \mathbf{0}$.

Finally, we show that $\mathbf{y}$ and $\tilde{\mathbf{z}}$ are uncorrelated. This follows from the calculation

$$\begin{aligned}\mathbf{E}\{(\mathbf{y} - \mathbf{m}_{\mathbf{y}})(\tilde{\mathbf{z}} - \mathbf{m}_{\tilde{\mathbf{z}}})'\} &= \mathbf{E}\{(\mathbf{y} - \mathbf{m}_{\mathbf{y}})\tilde{\mathbf{z}}'\} = \mathbf{E}\{\mathbf{y}\tilde{\mathbf{z}}'\} \\ &= \mathbf{E}\{\mathbf{y}\mathbf{z}'\} - \mathbf{E}\{\mathbf{y}\mathbf{E}\{\mathbf{z}'|\mathbf{y}\}\} \\ &= \mathbf{E}\{\mathbf{y}\mathbf{z}'\} - \mathbf{E}\{\mathbf{y}\mathbf{z}'\} = \mathbf{0}\end{aligned} \tag{12-40}$$

**b.** A detailed proof is given by Meditch (1969, pp. 101–102). The idea is to (1) compute $\mathbf{E}\{\mathbf{x}|\mathbf{y}, \mathbf{z}\}$ in expanded form, (2) compute $\mathbf{E}\{\mathbf{x}|\mathbf{y}, \tilde{\mathbf{z}}\}$ in expanded form, using $\tilde{\mathbf{z}}$ given in (12-36), and (3) compare the results from (1) and (2) to prove the truth of (12-35). □

Equation (12-35) is very important. It states that, when $\mathbf{z}$ and $\mathbf{y}$ are dependent, conditioning on $\mathbf{z}$ can always be replaced by conditioning on another Gaussian random vector $\tilde{\mathbf{z}}$, where $\tilde{\mathbf{z}}$ and $\mathbf{y}$ are statistically independent.

The results in Theorems 12-3 and 12-4 depend on all random vectors being jointly Gaussian. Very similar results, which are distribution free, are described in Problem 13-6; however, these results are restricted in yet another way.

## SUMMARY QUESTIONS

1. The notation $\theta \sim N(\theta; \mathbf{m}_\theta, \mathbf{P}_\theta)$ means that $\theta$ is:
   (a) nonsingular
   (b) a linear function of mean $\mathbf{m}_\theta$ and covariance $\mathbf{P}_\theta$
   (c) Gaussian with mean $\mathbf{m}_\theta$ and covariance $\mathbf{P}_\theta$
2. Gaussian random variables are important and widely used for which two reasons?
   (a) Central limit theorem
   (b) they often provide a reasonable model
   (c) they lead to simple calculations
3. The notation for joint density between $\mathbf{x}$ and $\mathbf{y}$ is:
   (a) $p(\mathbf{x}|\mathbf{y})$
   (b) $p(\mathbf{x})p(\mathbf{y})$
   (c) $p(\mathbf{x}, \mathbf{y})$
4. If $\mathbf{x}$ and $\mathbf{y}$ are jointly Gaussian, then:
   (a) $p(\mathbf{x}|\mathbf{y})$ is non-Gaussian
   (b) $p(\mathbf{x}|\mathbf{y})$ is Gaussian with a deterministic mean
   (c) $p(\mathbf{x}|\mathbf{y})$ is Gaussian with a random mean
5. Linear transformations of Gaussian random variables are:
   (a) non-Gaussian
   (b) always Gaussian
   (c) sometimes Gaussian
6. When $\mathbf{z}$ and $\mathbf{y}$ are dependent:
   (a) $\mathbf{E}\{\mathbf{x}|\mathbf{y}, \mathbf{z}\} = \mathbf{E}\{\mathbf{x}|\mathbf{y}\} + \mathbf{E}\{\mathbf{x}|\mathbf{z}\} - \mathbf{m}_\mathbf{x}$
   (b) conditioning on $\mathbf{z}$ can always be replaced by conditioning on another Gaussian random vector $\tilde{\mathbf{z}}$, where $\tilde{\mathbf{z}}$ and $\mathbf{y}$ are again dependent
   (c) conditioning on $\mathbf{z}$ can always be replaced by conditioning on another Gaussian random vector $\tilde{\mathbf{z}}$, where $\tilde{\mathbf{z}}$ and $\mathbf{y}$ are statistically independent
7. If $\mathbf{x}$ and $\mathbf{y}$ are jointly Gaussian, and $\mathbf{m}_\mathbf{x} = \mathbf{m}_\mathbf{y} = \mathbf{0}$, then $\mathbf{E}\{\mathbf{x}|\mathbf{y}\}$ equals:
   (a) $\mathbf{P}_{\mathbf{xy}}\mathbf{P}_\mathbf{y}^{-1}\mathbf{y}$
   (b) $\mathbf{P}_{\mathbf{xy}}\mathbf{P}_\mathbf{x}^{-1}\mathbf{x}$
   (c) $\mathbf{P}_{\mathbf{xy}}\mathbf{P}_\mathbf{y}^{-1}\mathbf{x}$

## PROBLEMS

**12-1.** Let $\mathbf{x}, \mathbf{y}$, and $\mathbf{z}$ be jointly distributed random vectors; $c$ and $h$ fixed constants; and $g(\cdot)$ a scalar-valued function. Assume that $\mathbf{E}\{\mathbf{x}\}, \mathbf{E}\{\mathbf{z}\}$, and $\mathbf{E}\{g(\mathbf{y})\mathbf{x}\}$ exist. Prove the following useful properties of conditional expectation:
   (a) $\mathbf{E}\{\mathbf{x}|\mathbf{y}\} = \mathbf{E}\{\mathbf{x}\}$ if $\mathbf{x}$ and $\mathbf{y}$ are independent
   (b) $\mathbf{E}\{g(\mathbf{y})\mathbf{x}|\mathbf{y}\} = g(\mathbf{y})\mathbf{E}\{\mathbf{x}|\mathbf{y}\}$
   (c) $\mathbf{E}\{c|\mathbf{y}\} = c$
   (d) $\mathbf{E}\{g(\mathbf{y})|\mathbf{y}\} = g(\mathbf{y})$
   (e) $\mathbf{E}\{c\mathbf{x} + h\mathbf{z}|\mathbf{y}\} = c\mathbf{E}\{\mathbf{x}|\mathbf{y}\} + h\mathbf{E}\{\mathbf{z}|\mathbf{y}\}$
   (f) $\mathbf{E}\{g(\mathbf{y})\mathbf{x}\} = \mathbf{E}\{g(\mathbf{y})\mathbf{E}\{\mathbf{x}|\mathbf{y}\}\}$ where the outer expectation is with respect to $\mathbf{y}$

**12-2.** Prove that the cross-covariance matrix of two uncorrelated random vectors is zero.

**12-3.** Prove that if $\mathbf{x}$ and $\mathbf{y}$ are marginally Gaussian and mutually uncorrelated they are jointly Gaussian.

**12-4.** Prove that if $\mathbf{x}$ and $\mathbf{y}$ are jointly Gaussian then they are marginally Gaussian.

**12-5.** (R. T. Okida, Fall 1991) $\mathbf{x}$ and $\mathbf{y}$ are Gaussian random vectors. Prove that if they are uncorrelated they are independent. Under what conditions will these vectors be orthogonal if they are uncorrelated?

**12-6.** (T. A. Parker and M. A. Ranta, Fall 1991) Derive Equations (12-12) through (12-14).

**12-7.** Derive Equations (12-33) and (12-34).

**12-8.** Fill in all of the details required to prove part b of Theorem 12-4.

**12-9.** (C. M. Chintall, Fall 1991) (Meditch, 1969, p. 100) As an extension of Theorem 12-2, show that $\mathbf{x} - \mathbf{E}\{\mathbf{x}|\mathbf{y}\}$ is independent of any linear (or affine) transformation on $\mathbf{y}$.

**12-10.** (G. Caso, Fall 1991) Let $y_1$, $y_2$ be jointly Gaussian random variables with means $m_1$, $m_2$, variances $\sigma_1$, $\sigma_2$, and correlation coefficient $\rho$, where

$$\rho \triangleq \frac{\mathbf{E}\{(y_1 - m_1)(y_2 - m_2)\}}{\sigma_1 \sigma_2}$$

Determine the joint probability density function of $y_1$ and $y_2$ in terms of $\rho$.

**12-11.** (Keith M. Chugg, Fall 1991) For each of the following joint density functions, find $\mathbf{E}\{Z|x\}$ and discuss the structures of these functions of $x$:

**(a)** The random variables $Z$ and $X$ are jointly Gaussian with correlation coefficient $\rho$.

**(b)** $p(z, x) = z \exp[-(1 + x)z]$ for $x > 0$ and $z > 0$.

**(c)** $p(z, x) = 2$ for $x \in [0, 1]$ and $x + z < 1$.

**12-12.** (Andrew D. Norte, Fall 1991) (This is a numerical problem.) Consider the following 4-variate jointly Gaussian random vector: $\mathbf{x} = \text{col}\,(x_1, x_2, x_3, x_4)$, where $\mathbf{m}_\mathbf{x} = \mathbf{E}\{\mathbf{x}\} = \text{col}\,(1, 2, 2.5, 3)$. The random vector $\mathbf{x}$ is obtained from sampling a Gaussian random process whose autocorrelation function is $R_x(\tau) = 10\,\text{sinc}\,(\tau/2)$, where $\text{sinc}\,(y) = \sin \pi y / \pi y$. The random variables $x_1, x_2, x_3, x_4$ are obtained by sampling the random process $x(t)$ at times $t = 0, 1, 3$, and 4 seconds, respectively.

**(a)** Find the covariance matrix $\mathbf{E}\{(\mathbf{x} - \mathbf{m}_\mathbf{x})(\mathbf{x} - \mathbf{m}_\mathbf{x})'\} = \mathbf{P}_\mathbf{x}$.

**(b)** Find the covariance matrices for the following random vectors: $\mathbf{x}_1 = \text{col}\,(x_1, x_2)$, $\mathbf{x}_2 = \text{col}\,(x_2, x_3)$, and $\mathbf{x}_3 = \text{col}\,(x_3, x_4)$.

**(c)** Find the conditional expected value of $\mathbf{x}_1$ given $\mathbf{x}_3$.

# LESSON 13

before (up to BLUE) are used to estimate deterministic stuff.

# Mean-squared Estimation of Random Parameters

## SUMMARY

There are two major types of estimators of random parameters: mean-squared (MS) and maximum a posteriori (MAP). The former does not use statistical information about the random parameters, whereas the latter does. MS estimators are the subject of this lesson; MAP estimators are the subject of Lesson 14.

The MSE is shown to equal a conditional expectation, i.e., $\hat{\theta}_{MS}(k) = \mathbf{E}\{\theta|\mathbf{Z}(k)\}$. This result is referred to as the *fundamental theorem of estimation theory*. When $\theta$ and $\mathbf{Z}(k)$ are jointly Gaussian, then it is possible to compute this expectation, and, in fact, $\hat{\theta}_{MS}(k)$ processes the data, $\mathbf{Z}(k)$, linearly. When $\theta$ and $\mathbf{Z}(k)$ are not jointly Gaussian, then $\mathbf{E}\{\theta|\mathbf{Z}(k)\}$ is, in general, a nonlinear function of the data $\mathbf{Z}(k)$. Supplementary Material at the end of this lesson provides a full-blown nonlinear estimator. In general, it is very difficult to compute $\mathbf{E}\{\theta|\mathbf{Z}(k)\}$; hence, we often use a linear MS estimator. This lesson answers the question "When is the *linear* MS estimator the same as the MS estimator?"

Another important result obtained for MS estimation is the *orthogonality principle*, which is a relationship between the MS error and the data that is used to compute the MS estimate.

The following five properties of MSE's are examined: unbiasedness, minimum variance, linearity, Gaussianity, and uniqueness.

We also compute $\hat{\theta}_{MS}(k) = \mathbf{E}\{\theta|\mathbf{Z}(k)\}$ for our generic linear model in the special case when $\mathbf{H}(k)$ is deterministic, $\mathbf{V}(k)$ is white Gaussian noise with known covariance matrix $\mathbf{R}(k)$, and $\theta$ is multivariate Gaussian with known mean $\mathbf{m}_\theta$ and covariance $\mathbf{P}_\theta$. By examining the structure of $\hat{\theta}_{MS}(k)$ in this case, we learn that there are connections between MS and BLU estimation and, consequently, WLS estimation.

These connections are made very explicit. In fact, we are able to conclude that $\hat{\theta}_{\text{BLU}}(k)$, given in Lesson 9, is applicable to random as well as deterministic parameters, and, because the BLUE of $\theta$ is the special case of the WLSE when $\mathbf{W}(k) = \mathbf{R}^{-1}(k)$, $\hat{\theta}_{\text{WLS}}(k)$, given in Lesson 3, is also applicable to random as well as deterministic parameters—but only for this special case of the generic linear model.

When you complete this lesson, you will be able to (1) derive a general formula for the mean-squared estimator of random parameters, (2) explain the consequences of the "Gaussian assumption" on this general formula, (3) understand what happens to the general formula for the generic linear and Gaussian model, and (4) explain why and when WLSE and BLUE apply also to random parameter estimation.


## INTRODUCTION

In Lesson 2 we showed that state estimation and deconvolution can be viewed as problems in which we are interested in estimating a vector of random parameters. For us, state estimation and deconvolution serve as the primary motivation for studying methods for estimating random parameters; however, the statistics literature is filled with other applications for these methods.

We now view $\theta$ as an $n \times 1$ vector of random unknown parameters. The information available to us are measurements $\mathbf{z}(1), \mathbf{z}(2), \ldots, \mathbf{z}(k)$, which are assumed to depend on $\theta$. In this lesson, we begin by not assuming a specific structural dependency between $\mathbf{z}(i)$ and $\theta$. This is quite different than what we did in WLSE and BLUE. Those methods were studied for the generic linear model $\mathbf{Z}(k) = \mathbf{H}(k)\theta + \mathbf{V}(k)$ and closed-form solutions for $\hat{\theta}_{\text{WLS}}(k)$ and $\hat{\theta}_{\text{BLU}}(k)$ could not have been obtained had we not begun by assuming this linear model. Later in this lesson we shall also focus on the generic linear model.

Many different measures of parameter estimation error can be minimized in order to obtain an estimate of $\theta$ (see Jazwinski, 1970; Meditch 1969; van Trees, 1968; and Sorenson 1980; see, also, the Supplementary Material at the end of this lesson); but, by far the most widely studied measure is the mean-squared error. The resulting estimator is called a *mean-squared estimator* and is denoted $\hat{\theta}_{\text{MS}}(k)$.

*i.e. $z = H\theta + v$ is not assumed*

*no model assumed ∴ This is a general result*

## OBJECTIVE FUNCTION AND PROBLEM STATEMENT

Given measurements $\mathbf{z}(1), \ldots, \mathbf{z}(k)$, we shall determine an estimator of $\theta$,

$$\hat{\theta}_{\text{MS}}(k) = \phi[\mathbf{z}(i), i = 1, 2, \ldots, k] \quad \text{(nonlinear function of measurements)} \tag{13-1}$$

such that the mean-squared error

$$J[\tilde{\theta}_{\text{MS}}(k)] = \mathbf{E}\{\tilde{\theta}'_{\text{MS}}(k)\tilde{\theta}_{\text{MS}}(k)\} \tag{13-2}$$

is minimized. In (13-2), $\tilde{\theta}_{\text{MS}}(k) = \theta - \hat{\theta}_{\text{MS}}(k)$.

The right-hand side of (13-1) means that we have some arbitrary and as yet unknown function of all the measurements. The $n$ components of $\hat{\theta}_{\text{MS}}(k)$ may each depend differently on the measurements. The function $\phi[\mathbf{z}(i), i = 1, 2, \ldots, k]$

may be nonlinear or linear. Its exact structure will be determined by minimizing $J[\tilde{\theta}_{MS}(k)]$. If perchance $\hat{\theta}_{MS}(k)$ is a linear estimator, then

$$\hat{\theta}_{MS}(k) = \sum_{i=1}^{k} \mathbf{A}(i)\mathbf{z}(i) \tag{13-3}$$

We now show that the notion of conditional expectation is central to the calculation of $\hat{\theta}_{MS}(k)$. As usual, we let

$$\mathbf{Z}(k) = \text{col}\,[\mathbf{z}(k), \mathbf{z}(k-1), \ldots, \mathbf{z}(1)] \tag{13-4}$$

The underlying random quantities in our estimation problem are $\theta$ and $\mathbf{Z}(k)$. We assume that their joint density function $p[\theta, \mathbf{Z}(k)]$ exists, so that

$$\mathbf{E}\{\tilde{\theta}'_{MS}(k)\tilde{\theta}_{MS}(k)\} = \int_{-\infty}^{\infty} \cdots \int_{-\infty}^{\infty} \tilde{\theta}'_{MS}(k)\tilde{\theta}_{MS}(k)p[\theta, \mathbf{Z}(k)]\, d\theta\, d\mathbf{Z}(k) \tag{13-5}$$

where $d\theta = d\theta_1 d\theta_2 \ldots d\theta_n$, $d\mathbf{Z}(k) = dz_1(1) \ldots dz_1(k) dz_2(1) \ldots dz_2(k) \ldots dz_m(1) \ldots dz_m(k)$, and there are $n + km$ integrals. Using the fact that

$$p[\theta, \mathbf{Z}(k)] = p[\theta|\mathbf{Z}(k)]p[\mathbf{Z}(k)] \tag{13-6}$$

we rewrite (13-5) as

$$\mathbf{E}\{\tilde{\theta}'_{MS}(k)\tilde{\theta}_{MS}(k)\} = \int_{-\infty}^{\infty} \cdots \int_{-\infty}^{\infty} \left\{ \int_{-\infty}^{\infty} \cdots \int_{-\infty}^{\infty} \tilde{\theta}'_{MS}(k)\tilde{\theta}_{MS}(k)p[\theta|\mathbf{Z}(k)]\, d\theta \right\} p[\mathbf{Z}(k)]\, d\mathbf{Z}(k)$$

$$= \int_{-\infty}^{\infty} \cdots \int_{-\infty}^{\infty} \mathbf{E}\{\tilde{\theta}'_{MS}(k)\tilde{\theta}_{MS}(k)|\mathbf{Z}(k)\}p[\mathbf{Z}(k)]\, d\mathbf{Z}(k) \tag{13-7}$$

From this equation we see that minimizing the conditional expectation $\mathbf{E}\{\tilde{\theta}'_{MS}(k)\tilde{\theta}_{MS}(k)|\mathbf{Z}(k)\}$ with respect to $\hat{\theta}_{MS}(k)$ is equivalent to our original objective of minimizing the total expectation $\mathbf{E}\{\tilde{\theta}'_{MS}(k)\tilde{\theta}_{MS}(k)\}$. Note that the integrals on the right-hand side of (13-7) remove the dependency of the integrand on the data $\mathbf{Z}(k)$.

In summary, we have the following *mean-squared estimation problem: Given the measurements* $\mathbf{z}(1), \mathbf{z}(2), \ldots, \mathbf{z}(k)$, *determine an estimator of* $\theta$,

$$\hat{\theta}_{MS}(k) = \phi[\mathbf{z}(i), i = 1, 2, \ldots, k]$$

*such that the conditional mean-squared error*

$$J_1[\tilde{\theta}_{MS}(k)] = \mathbf{E}\{\tilde{\theta}'_{MS}(k)\tilde{\theta}_{MS}(k)|\mathbf{z}(1), \ldots, \mathbf{z}(k)\} \tag{13-8}$$

*is minimized.*

## DERIVATION OF ESTIMATOR

The solution to the mean-squared estimation problem is given in Theorem 13-1, which is known as the *fundamental theorem of estimation theory.*

*FTET:*

**Theorem 13-1.** *The estimator that minimizes the mean-squared error is*

$$\hat{\theta}_{MS}(k) = E\{\theta|\mathbf{Z}(k)\} \qquad (13\text{-}9)$$

*Proof* (Mendel, 1983). In this proof we omit all functional dependences on $k$, for notational simplicity. Our approach is to substitute $\tilde{\theta}_{MS}(k) = \theta - \hat{\theta}_{MS}(k)$ into (13-8) and to complete the square, as follows:

*E[θ|z] = θ̂ since θ̂ is a function z? ... we need to compute θ̂; we also z to compute θ̂ ∴ θ̂ here is not random (wrt z).*

$$
\begin{aligned}
J_1[\tilde{\theta}_{MS}(k)] &= E\{(\theta - \hat{\theta}_{MS})'(\theta - \hat{\theta}_{MS})|\mathbf{Z}\} \\
&= E\{\theta'\theta - \theta'\hat{\theta}_{MS} - \hat{\theta}_{MS}'\theta + \hat{\theta}_{MS}'\hat{\theta}_{MS}|\mathbf{Z}\} \\
&= E\{\theta'\theta|\mathbf{Z}\} - E\{\theta'|\mathbf{Z}\}\hat{\theta}_{MS} - \hat{\theta}_{MS}'E\{\theta|\mathbf{Z}\} + \hat{\theta}_{MS}'\hat{\theta}_{MS} \qquad (13\text{-}10) \\
&= E\{\theta'\theta|\mathbf{Z}\} + [\hat{\theta}_{MS} - E\{\theta|\mathbf{Z}\}]'[\hat{\theta}_{MS} - E\{\theta|\mathbf{Z}\}] \\
&\quad - E\{\theta'|\mathbf{Z}\}E\{\theta|\mathbf{Z}\}
\end{aligned}
$$

To obtain the third line, we used the fact that $\hat{\theta}_{MS}$, by definition, is a function of $\mathbf{Z}$; hence, $E\{\hat{\theta}_{MS}|\mathbf{Z}\} = \hat{\theta}_{MS}$. The first and last terms in (13-10) do not depend on $\hat{\theta}_{MS}$; hence, the smallest value of $J_1[\tilde{\theta}_{MS}(k)]$ is obviously attained by setting the bracketed terms equal to zero. This means that $\hat{\theta}_{MS}$ must be chosen as in (13-9). $\square$

Let $J_1^*[\tilde{\theta}_{MS}(k)]$ denote the minimum value of $J_1[\tilde{\theta}_{MS}(k)]$. We see, from (13-10) and (13-9), that

$$J_1^*[\tilde{\theta}_{MS}(k)] = E\{\theta'\theta|\mathbf{Z}\} - \hat{\theta}_{MS}'(k)\hat{\theta}_{MS}(k) \qquad (13\text{-}11)$$

As it stands, (13-9) is not terribly useful for computing $\hat{\theta}_{MS}(k)$. In general, we must first compute $p[\theta|\mathbf{Z}(k)]$ and then perform the requisite number of integrations of $\theta p[\theta|\mathbf{Z}(k)]$ to obtain $\hat{\theta}_{MS}(k)$. For pedagogical reasons, it is useful to separate this computation into two major cases: (1) $\theta$ and $\mathbf{Z}(k)$ are jointly Gaussian, and (2) $\theta$ and $\mathbf{Z}(k)$ are not jointly Gaussian. The former is often referred to as the *Gaussian case*, whereas the latter is often referred to as the *non-Gaussian case*. The non-Gaussian case can be further subdivided into three subcases: (2a) a full-blown nonlinear estimator is obtained for $E\{\theta|\mathbf{Z}(k)\}$; (2b) a linear estimator is obtained for $E\{\theta|\mathbf{Z}(k)\}$; and (2c) the mean-squared estimator is abandoned, and estimates are obtained using second-order and some higher-order statistics. Cases 1 and 2b are discussed extensively in the main body of this lesson, case 2a is covered in the Supplementary Material at the end of this lesson, and case 2c is covered in Lessons B and C.

When $\theta$ and $\mathbf{Z}(k)$ are jointly Gaussian, we have a very important and practical corollary to Theorem 13-1.  *$\hat{\theta} = \hat{\theta}_{2k}$*

**Corollary 13-1.** *When $\theta$ and $\mathbf{Z}$(k) are jointly Gaussian, the estimator that minimizes the mean-squared error is*

$$\hat{\theta}_{MS}(k) = \mathbf{m}_\theta + \mathbf{P}_{\theta\mathbf{Z}}(k)\mathbf{P}_\mathbf{Z}^{-1}(k)[\mathbf{Z}(k) - \mathbf{m}_\mathbf{Z}(k)] \qquad (13\text{-}12)$$

*cross Covariance matrix     Covariance matrix*

*So to use 13-12 you must be given 13-12  $P_{\theta z}, P_z$*

*Proof.* When $\theta$ and $\mathbf{Z}(k)$ are jointly Gaussian, then $\mathbf{E}\{\theta|\mathbf{Z}(k)\}$ can be evaluated using (12-17) of Lesson 12. Doing this we obtain (13-12). □

Corollary 13-1 gives us an explicit structure for $\hat{\theta}_{MS}(k)$. We see that $\hat{\theta}_{MS}(k)$ is an affine transformation of $\mathbf{Z}(k)$. If $\mathbf{m}_\theta = \mathbf{0}$ and $\mathbf{m}_\mathbf{Z}(k) = \mathbf{0}$, then $\hat{\theta}_{MS}(k)$ is a linear transformation of $\mathbf{Z}(k)$.

To compute $\hat{\theta}_{MS}(k)$ using (13-12), we must know $\mathbf{m}_\theta$ and $\mathbf{m}_\mathbf{Z}(k)$, and we must first compute $\mathbf{P}_{\theta\mathbf{Z}}(k)$ and $\mathbf{P}_\mathbf{Z}^{-1}(k)$. We perform these computations later in this lesson for the generic linear model, $\mathbf{Z}(k) = \mathbf{H}(k)\theta + \mathbf{V}(k)$.

**Corollary 13-2.** *Suppose $\theta$ and $\mathbf{Z}$(k) are not necessarily jointly Gaussian, and that we know $\mathbf{m}_\theta$, $\mathbf{m}_\mathbf{Z}$(k), $\mathbf{P}_\mathbf{Z}$(k), and $\mathbf{P}_{\theta\mathbf{Z}}$(k). In this case, the estimator that is constrained to be an affine transformation of $\mathbf{Z}$(k) and that minimizes the mean-squared error is also given by (13-12).*

*Proof.* This corollary can be proved in a number of ways. A direct proof begins by assuming that $\hat{\theta}_{MS}(k) = \mathbf{A}(k)\mathbf{Z}(k) + \mathbf{b}(k)$ and choosing $\mathbf{A}(k)$ and $\mathbf{b}(k)$ so that $\hat{\theta}_{MS}(k)$ is an unbiased estimator of $\theta$ and $\mathbf{E}\{\tilde{\theta}'_{MS}(k)\tilde{\theta}_{MS}(k)\} = \text{trace}\ \mathbf{E}\{\tilde{\theta}_{MS}(k)\tilde{\theta}'_{MS}(k)\}$ is minimized. We leave the details of this direct proof to the reader.

A less direct proof is based on the following Gedanken (thought) experiment. Using known first and second moments of $\theta$ and $\mathbf{Z}(k)$, we can conceptualize unique Gaussian random vectors that have these same first and second moments. For these statistically equivalent (through second-order moments) Gaussian vectors, we know, from Corollary 13-1, that the mean-squared estimator is given by the affine transformation of $\mathbf{Z}(k)$ in (13-12). □

Corollaries 13-1 and 13-2, as well as Theorem 13-1, provide us with the answer to the following important question: *When is the linear (affine) mean-squared estimator the same as the mean-squared estimator?* The answer is when $\theta$ and $\mathbf{Z}(k)$ are jointly Gaussian. If $\theta$ and $\mathbf{Z}(k)$ are not jointly Gaussian, then $\hat{\theta}_{MS}(k) = \mathbf{E}\{\theta|\mathbf{Z}(k)\}$, which, in general, is a nonlinear function of measurements $\mathbf{Z}(k)$, i.e., it is a nonlinear estimator.

**Corollary 13-3** (Orthogonality Principle). *Suppose $\mathbf{f}[\mathbf{Z}(k)]$ is any function of the data $\mathbf{Z}$(k). Then the error in the mean-squared estimator is orthogonal to $\mathbf{f}[\mathbf{Z}(k)]$ in the sense that*

$$\mathbf{E}\{[\theta - \hat{\theta}_{MS}(k)]\mathbf{f}'[\mathbf{Z}(k)]\} = \mathbf{0} \tag{13-13}$$

*Proof* (Mendel, 1983, pp. 46–47). We use the following result from probability theory [Papoulis, 1991; see, also, Problem 12-1(f)]. Let $\alpha$ and $\beta$ be jointly distributed random vectors and $g(\beta)$ be a scalar-valued function; then

$$\mathbf{E}\{\alpha g(\beta)\} = \mathbf{E}\{\mathbf{E}\{\alpha|\beta\}g(\beta)\} \tag{13-14}$$

where the outer expectation on the right-hand side is with respect to $\beta$. We proceed as follows (again, omitting the argument $k$):

*E{(θ)=θ̂ₘₛ*

$$\mathbf{E}\{(\theta - \hat{\theta}_{\text{MS}})\mathbf{f}'(\mathbf{Z})\} = \mathop{\mathbf{E}}_{\mathbf{z}}\{\mathop{\mathbf{E}}_{\theta}\{(\theta - \hat{\theta}_{\text{MS}})|\mathbf{Z}\}\mathbf{f}'(\mathbf{Z})\}$$
$$= \mathbf{E}\{(\hat{\theta}_{\text{MS}} - \hat{\theta}_{\text{MS}})\mathbf{f}'(\mathbf{Z})\} = \mathbf{0}$$

where we have used the facts that $\hat{\theta}_{\text{MS}}$ is no longer random when $\mathbf{Z}$ is specified and $\mathbf{E}\{\theta|\mathbf{Z}\} = \hat{\theta}_{\text{MS}}$. □

A frequently encountered special case of (13-13) occurs when $\mathbf{f}[\mathbf{Z}(k)] = \hat{\theta}_{\text{MS}}(k)$; then Corollary 13-3 can be written as *this will allow us to*
*neg cross terms*
*later on*

$$\mathbf{E}\{\tilde{\theta}_{\text{MS}}(k)\hat{\theta}'_{\text{MS}}(k)\} = \mathbf{0} \tag{13-15}$$

## PROPERTIES OF MEAN-SQUARED ESTIMATORS WHEN Θ AND Z(k) ARE JOINTLY GAUSSIAN

In this section we present a collection of important and useful properties associated with $\hat{\theta}_{\text{MS}}(k)$ for the case when $\theta$ and $\mathbf{Z}(k)$ are jointly Gaussian. In this case $\hat{\theta}_{\text{MS}}(k)$ is given by (13-12).

**Property 1** *(Unbiasedness)*. The mean-squared estimator, $\hat{\theta}_{\text{MS}}(k)$ in (13-12), is unbiased.

*Proof.* Taking the expected value of (13-12), we see that $\mathbf{E}\{\hat{\theta}_{\text{MS}}(k)\} = \mathbf{m}_\theta$; thus, $\hat{\theta}_{\text{MS}}(k)$ is an unbiased estimator of $\theta$. □

**Property 2** *(Minimum Variance)*. Dispersion about the mean value of $\hat{\theta}_{i,\text{MS}}(k)$ is measured by the error variance $\sigma^2_{\tilde{\theta}_{i,\text{MS}}}(k)$, where $i = 1, 2, \ldots, n$. An estimator that has the smallest error variance is a minimum-variance estimator (an MVE). The mean-squared estimator in (13-12) is an MVE.

*Proof.* From Property 1 and the definition of error variance, we see that

$$\sigma^2_{\tilde{\theta}_{i,\text{MS}}}(k) = \mathbf{E}\{\tilde{\theta}^2_{i,\text{MS}}(k)\}, \qquad i = 1, 2, \ldots n \tag{13-16}$$

Our mean-squared estimator was obtained by minimizing $J[\tilde{\theta}_{\text{MS}}(k)]$ in (13-2), which can now be expressed as $= E[\tilde{\theta}^{\mathsf{T}}\tilde{\theta}] = 2E[\tilde{\theta}_i]$

$$J[\tilde{\theta}_{\text{MS}}(k)] = \sum_{i=1}^{n} \sigma^2_{\tilde{\theta}_{i,\text{MS}}}(k) \tag{13-17}$$

Because variances are always positive, the minimum value of $J[\tilde{\theta}_{\text{MS}}(k)]$ must be achieved when each of the $n$ variances is minimized; hence, our mean-squared estimator is equivalent to an MVE. □

**Property 3** *(Linearity)*. $\hat{\theta}_{\text{MS}}(k)$ in (13-12) is a "linear" (i.e., affine) estimator.

*Proof.* This is obvious from the form of (13-12). □

Linearity of $\hat{\theta}_{MS}(k)$ permits us to infer the following very important property about both $\hat{\theta}_{MS}(k)$ and $\tilde{\theta}_{MS}(k)$.

**Property 4** *(Gaussianity)*. Both $\hat{\theta}_{MS}(k)$ and $\tilde{\theta}_{MS}(k)$ are multivariate Gaussian.

*Proof.* We use the linearity property of jointly Gaussian random vectors stated in Lesson 12. Estimator $\hat{\theta}_{MS}(k)$ in (13-12) is an affine transformation of Gaussian random vector $\mathbf{Z}(k)$; hence, $\hat{\theta}_{MS}(k)$ is multivariate Gaussian. Estimation error $\tilde{\theta}_{MS}(k) = \theta - \hat{\theta}_{MS}(k)$ is an affine transformation of jointly Gaussian vectors $\theta$ and $\mathbf{Z}(k)$; hence, $\tilde{\theta}_{MS}(k)$ is also Gaussian. □

Estimate $\hat{\theta}_{MS}(k)$ in (13-12) is itself random, because measurements $\mathbf{Z}(k)$ are random. To characterize it completely in a statistical sense, we must specify its probability density function. Generally, this is very difficult to do and often requires that the probability density function of $\hat{\theta}_{MS}(k)$ be approximated using many moments (in theory an infinite number are required). In the Gaussian case, we have just learned that the structure of the probability density function for $\hat{\theta}_{MS}(k)$ [and $\tilde{\theta}_{MS}(k)$] is known. Additionally, we know that a Gaussian density function is completely specified by exactly two moments, its mean and covariance; thus, tremendous simplifications occur when $\theta$ and $\mathbf{Z}(k)$ are jointly Gaussian.

**Property 5** *(Uniqueness)*. Mean-squared estimator $\hat{\theta}_{MS}(k)$, in (13-12), is unique.

The proof of this property is not central to our developments; hence, it is omitted.

## GENERALIZATIONS

Many of the results presented in this section are applicable to objective functions other than the mean-squared objective function in (13-2). See the Supplementary Material at the end of this lesson for discussions on a wide number of objective functions that lead to $\mathbf{E}\{\theta|\mathbf{Z}(k)\}$ as the optimal estimator of $\theta$.

## MEAN-SQUARED ESTIMATOR FOR THE GENERIC LINEAR AND GAUSSIAN MODEL

So far we have required $\theta$ and $\mathbf{Z}(k)$ to be jointly Gaussian. When will this be true? Let us now begin with the generic linear model

$$\mathbf{Z}(k) = \mathbf{H}(k)\theta + \mathbf{V}(k) \tag{13-18}$$

where $\mathbf{H}(k)$ is deterministic, $\mathbf{V}(k)$ is white Gaussian noise with known covariance matrix $\mathbf{R}(k)$, $\theta$ is multivariate Gaussian with known mean $\mathbf{m}_\theta$, and covariance $\mathbf{P}_\theta$, i.e.,

$$\theta \sim N(\theta; \mathbf{m}_\theta, \mathbf{P}_\theta) \qquad (13\text{-}19)$$

and $\theta$ and $\mathbf{V}(k)$ are mutually uncorrelated. Because $\theta$ and $\mathbf{V}(k)$ are Gaussian and mutually uncorrelated, they are jointly Gaussian (see Lesson 12). Consequently, $\mathbf{Z}(k)$ and $\theta$ are also jointly Gaussian; thus, $\hat{\theta}_{MS}(k)$ is given by (13-12).

**Theorem 13-2.** *For the generic linear model in which* $\mathbf{H}(k)$ *is deterministic,* $\mathbf{V}(k)$ *is white Gaussian noise with known covariance matrix* $\mathbf{R}(k)$, $\theta \sim N(\theta; \mathbf{m}_\theta, \mathbf{P}_\theta)$, *and* $\theta$ *and* $\mathbf{V}(k)$ *are ~~mutually uncorrelated~~:* θ&v indy ⇒ θ+v are Jointly Gaussian statistically indys

$$\hat{\theta}_{MS}(k) = \mathbf{m}_\theta + \mathbf{P}_\theta \mathbf{H}'(k)[\mathbf{H}(k)\mathbf{P}_\theta \mathbf{H}'(k) + \mathbf{R}(k)]^{-1}[\mathbf{Z}(k) - \mathbf{H}(k)\mathbf{m}_\theta] \qquad (13\text{-}20)$$

*Let* $\mathbf{P}_{MS}(k)$ *denote the error-covariance matrix associated with* $\hat{\theta}_{MS}(k)$. *Then*

$$\mathbf{P}_{MS}(k) = \mathbf{P}_\theta - \mathbf{P}_\theta \mathbf{H}'(k)[\mathbf{H}(k)\mathbf{P}_\theta \mathbf{H}'(k) + \mathbf{R}(k)]^{-1}\mathbf{H}(k)\mathbf{P}_\theta \qquad (13\text{-}21)$$

*or*

$$\mathbf{P}_{MS}(k) = [\mathbf{P}_\theta^{-1} + \mathbf{H}'(k)\mathbf{R}^{-1}(k)\mathbf{H}(k)]^{-1} \qquad (13\text{-}22)$$

*so that* $\hat{\theta}_{MS}(k)$ *can be reexpressed as*

$$\hat{\theta}_{MS}(k) = \mathbf{m}_\theta + \mathbf{P}_{MS}(k)\mathbf{H}'(k)\mathbf{R}^{-1}(k)[\mathbf{Z}(k) - \mathbf{H}(k)\mathbf{m}_\theta] \qquad (13\text{-}23)$$

*remember θ is Random + you are estimating this Random thing (θ ≠ mθ on any given measurement)*

*crosshair target analogy*

*mean is bulls eye. But if you want an estimate of spot 3 (θ(3)) use Thm 13-2.*

*Proof.* It is straightforward, using (13-18) and (13-19), to show that

$$\mathbf{m}_Z(k) = \mathbf{H}(k)\mathbf{m}_\theta \qquad (13\text{-}24)$$

$$\mathbf{P}_Z(k) = \mathbf{H}(k)\mathbf{P}_\theta \mathbf{H}'(k) + \mathbf{R}(k) \qquad (13\text{-}25)$$

and

$$\mathbf{P}_{\theta Z}(k) = \mathbf{P}_\theta \mathbf{H}'(k) \qquad (13\text{-}26)$$

Equation (13-20) is obtained by substituting (13-24)–(13-26) into (13-12). Because $\hat{\theta}_{MS}(k)$ is unbiased, $\tilde{\theta}_{MS}(k) = \mathbf{0}$ for all $k$; thus,

$$\mathbf{P}_{MS}(k) = \mathbf{E}\{\tilde{\theta}_{MS}(k)\tilde{\theta}'_{MS}(k)\} \qquad (13\text{-}27)$$

From (13-20) and the fact that $\tilde{\theta}_{MS}(k) = \theta - \hat{\theta}_{MS}(k)$, we see that

$$\tilde{\theta}_{MS}(k) = (\theta - \mathbf{m}_\theta) - \mathbf{P}_\theta \mathbf{H}'(\mathbf{H}\mathbf{P}_\theta \mathbf{H}' + \mathbf{R})^{-1}(\mathbf{Z} - \mathbf{H}\mathbf{m}_\theta) \qquad (13\text{-}28)$$

From (13-28), (13-27), and (13-18), it is a straightforward exercise (Problem 13-5) to show that $\mathbf{P}_{MS}(k)$ is given by (13-21). Equation (13-22) is obtained by applying matrix inversion Lemma 5-1 to (13-21).

To express $\hat{\theta}_{MS}(k)$ as an explicit function of $\mathbf{P}_{MS}(k)$, as in (13-23), we note that

$$\mathbf{P}_\theta \mathbf{H}'(\mathbf{H}\mathbf{P}_\theta \mathbf{H}' + \mathbf{R})^{-1} = \mathbf{P}_\theta \mathbf{H}'(\mathbf{H}\mathbf{P}_\theta \mathbf{H}' + \mathbf{R})^{-1}(\mathbf{H}\mathbf{P}_\theta \mathbf{H}' + \mathbf{R} - \mathbf{H}\mathbf{P}_\theta \mathbf{H}')\mathbf{R}^{-1}$$

$$= \mathbf{P}_\theta \mathbf{H}'[\mathbf{I} - (\mathbf{H}\mathbf{P}_\theta \mathbf{H}' + \mathbf{R})^{-1}\mathbf{H}\mathbf{P}_\theta \mathbf{H}']\mathbf{R}^{-1}$$

$$= [\mathbf{P}_\theta - \mathbf{P}_\theta \mathbf{H}'(\mathbf{H}\mathbf{P}_\theta \mathbf{H}' + \mathbf{R})^{-1}\mathbf{H}\mathbf{P}_\theta]\mathbf{H}'\mathbf{R}^{-1}$$

$$= \mathbf{P}_{MS}\mathbf{H}'\mathbf{R}^{-1} \qquad (13\text{-}29)$$

Substituting (13-29) into (13-20), we obtain (13-23). $\square$

Observe that $\hat{\theta}_{MS}(k)$ depends on all the given information: $\mathbf{Z}(k)$, $\mathbf{H}(k)$, $\mathbf{m}_\theta$, $\mathbf{P}_\theta$, and $\mathbf{R}(k)$. Observe, also, that (13-23) has a recursive "feel" to it. Prior to any measurements, we can estimate $\theta$ by $\mathbf{m}_\theta$. After measurements have been made, $\mathbf{m}_\theta$ is updated by the second term in (13-23) in which $\mathbf{H}(k)\mathbf{m}_\theta$ represents the predicted value of $\mathbf{Z}(k)$.

**EXAMPLE 13-1 (Minimum-variance Deconvolution)**

In Example 2-6 we showed that, for the application of deconvolution, our linear model is

$$\mathbf{Z}(N) = \mathbf{H}(N-1)\boldsymbol{\mu} + \mathbf{V}(N) \tag{13-30}$$

We shall assume that $\boldsymbol{\mu}$ and $\mathbf{V}(N)$ are jointly Gaussian and that $\mathbf{m}_\mu = \mathbf{0}$ and $\mathbf{m}_V = \mathbf{0}$; hence, $\mathbf{m}_Z = \mathbf{0}$. Additionally, we assume that $\text{cov}\,[\mathbf{V}(N)] = \rho\mathbf{I}$. From (13-20), we determine the following formula for $\hat{\boldsymbol{\mu}}_{MS}(N)$:

$$\hat{\boldsymbol{\mu}}_{MS}(N) = \mathbf{P}_\mu\mathbf{H}'(N-1)[\mathbf{H}(N-1)\mathbf{P}_\mu\mathbf{H}'(N-1) + \rho\mathbf{I}]^{-1}\mathbf{Z}(N) \tag{13-31}$$

When $\mu(k)$ is described by the product model $\mu(k) = q(k)r(k)$, then

$$\boldsymbol{\mu} = \mathbf{Q}_q\mathbf{r} \tag{13-32}$$

where

$$\mathbf{Q}_q = \text{diag}\,[q(1), q(2), \ldots, q(N)] \tag{13-33}$$

and

$$\mathbf{r} = \text{col}\,(r(1), r(2), \ldots, r(N)) \tag{13-34}$$

In the product model, $r(k)$ is white Gaussian noise with variance $\sigma_r^2$, and $q(k)$ is a Bernoulli sequence. Obviously, if we know $\mathbf{Q}_q$ then $\boldsymbol{\mu}$ is Gaussian, in which case

$$\mathbf{P}_\mu = \mathbf{Q}_q^2\sigma_r^2 = \mathbf{Q}_q\sigma_r^2 \tag{13-35}$$

where we have used the fact that $\mathbf{Q}_q^2 = \mathbf{Q}_q$, because $q(k) = 0$ or 1. When $\mathbf{Q}_q$ is known, (13-31) becomes

$$\hat{\boldsymbol{\mu}}_{MS}(N|\mathbf{Q}_q) = \sigma_r^2\mathbf{Q}_q\mathbf{H}'(N-1)[\sigma_r^2\mathbf{H}(N-1)\mathbf{Q}_q\mathbf{H}'(N-1) + \rho\mathbf{I}]^{-1}\mathbf{Z}(N) \tag{13-36}$$

Although $\hat{\boldsymbol{\mu}}_{MS}$ is a mean-squared estimator, so that it enjoys all the properties of such an estimator (e.g., unbiased and minimum-variance), $\hat{\boldsymbol{\mu}}_{MS}$ is not a consistent estimator of $\boldsymbol{\mu}$. Consistency is a large-sample property of an estimator; however, as $N$ increases, the dimension of $\boldsymbol{\mu}$ increases, because $\boldsymbol{\mu}$ is $N \times 1$. Consequently, we cannot prove consistency of $\hat{\boldsymbol{\mu}}_{MS}$ (recall that, in all other problems, $\theta$ is $n \times 1$, where $n$ is data independent; in these problems we can study consistency of $\hat{\theta}$).

Equations (13-31) and (13-36) are not very practical for actually computing $\hat{\boldsymbol{\mu}}_{MS}$, because both require the inversion of an $N \times N$ matrix, and $N$ can become quite large (it equals the number of measurements). We return to a more practical way for computing $\hat{\boldsymbol{\mu}}_{MS}$ in Lesson 21. □

## BEST LINEAR UNBIASED ESTIMATION, REVISITED

In Lesson 9 we derived the BLUE of $\theta$ for the linear model (13-18) under the following assumptions about this model:

1. $\theta$ is a deterministic but unknown vector of parameters,

2. $\mathbf{H}(k)$ is deterministic, and
3. $\mathbf{V}(k)$ is zero-mean noise with covariance matrix $\mathbf{R}(k)$.

We assumed that $\hat{\boldsymbol{\theta}}_{BLU}(k) = \mathbf{F}(k)\mathbf{Z}(k)$ and chose $\mathbf{F}_{BLU}(k)$ so that $\hat{\boldsymbol{\theta}}_{BLU}(k)$ is an unbiased estimator of $\boldsymbol{\theta}$, and the error variance for each of the $n$ elements of $\boldsymbol{\theta}$ is minimized. The reader should return to the derivation of $\hat{\boldsymbol{\theta}}_{BLU}(k)$ to see that the assumption "$\boldsymbol{\theta}$ is deterministic" is never needed, either in the derivation of the unbiasedness constraint (see the proof of Theorem 6-1), or in the derivation of $J_i(\mathbf{f}_i, \boldsymbol{\lambda}_i)$ in Equation (9-14) (due to some remarkable cancellations); thus, $\hat{\boldsymbol{\theta}}_{BLU}(k)$, *given in (9-24), is applicable to random as well as deterministic parameters in our linear model (13-18); and because the BLUE of* $\boldsymbol{\theta}$ *is the special case of the WLSE when* $\mathbf{W}(k) = \mathbf{R}^{-1}(k)$, $\hat{\boldsymbol{\theta}}_{WLS}(k)$, *given in (3-10), is also applicable to random as well as deterministic parameters in our linear model.* The latter is true even when $\mathbf{H}(k)$ is not deterministic (see Lesson 3). Of course, in this case a performance analysis of $\hat{\boldsymbol{\theta}}_{WLS}(k)$ is very difficult because of the random natures of both $\boldsymbol{\theta}$ and $\mathbf{H}(k)$.

**Theorem 13-3.** *If* $\mathbf{P}_\theta^{-1} = \mathbf{0}$ *and* $\mathbf{H}(\mathrm{k})$ *is deterministic, then*

$$\hat{\boldsymbol{\theta}}_{MS}(k) = \hat{\boldsymbol{\theta}}_{BLU}(k) \tag{13-37}$$

*Proof.* Set $\mathbf{P}_\theta^{-1} = \mathbf{0}$ in (13-22) to see that

$$\mathbf{P}_{MS}(k) = [\mathbf{H}'(k)\mathbf{R}^{-1}(k)\mathbf{H}(k)]^{-1} \tag{13-38}$$

and, therefore,

$$\hat{\boldsymbol{\theta}}_{MS}(k) = [\mathbf{H}'(k)\mathbf{R}^{-1}(k)\mathbf{H}(k)]^{-1}\mathbf{H}'(k)\mathbf{R}^{-1}(k)\mathbf{Z}(k) \tag{13-39}$$

Compare (13-39) and (9-24) to conclude that $\hat{\boldsymbol{\theta}}_{MS}(k) = \hat{\boldsymbol{\theta}}_{BLU}(k)$. □

What does the condition $\mathbf{P}_\theta^{-1} = \mathbf{0}$, given in Theorem 13-3, mean? Suppose, for example, that the elements of $\boldsymbol{\theta}$ are uncorrelated; then $\mathbf{P}_\theta$ is a diagonal matrix, with diagonal elements $\sigma_{i,\theta}^2$. When all these variances are very large, $\mathbf{P}_\theta^{-1} = \mathbf{0}$. A large variance for $\theta_i$ means we have no idea where $\theta_i$ is located about its mean value.

Theorem 13-3 relates $\hat{\boldsymbol{\theta}}_{MS}(k)$ and $\hat{\boldsymbol{\theta}}_{BLU}(k)$ under some very stringent conditions that are needed in order to remove the dependence of $\hat{\boldsymbol{\theta}}_{MS}$ on the a priori statistical information about $\boldsymbol{\theta}$ (i.e., $\mathbf{m}_\theta$ and $\mathbf{P}_\theta$), because this information was never used in the derivation of $\hat{\boldsymbol{\theta}}_{BLU}(k)$.

Next, we derive a different BLUE of $\boldsymbol{\theta}$, one that incorporates the a priori statistical information about $\boldsymbol{\theta}$. To do this (Sorenson, 1980, p. 210), we treat $\mathbf{m}_\theta$ as an additional measurement, which will be augmented to $\mathbf{Z}(k)$. Our additional measurement equation is obtained by adding and subtracting $\boldsymbol{\theta}$ in the identity $\mathbf{m}_\theta = \mathbf{m}_\theta$; i.e.,

$$\mathbf{m}_\theta = \boldsymbol{\theta} + (\mathbf{m}_\theta - \boldsymbol{\theta}) \tag{13-40}$$

Quantity $\mathbf{m}_\theta - \boldsymbol{\theta}$ is now treated as zero-mean noise with covariance matrix $\mathbf{P}_\theta$. Our augmented linear model is

$$\underbrace{\begin{pmatrix} \mathbf{Z}(k) \\ \hline \mathbf{m}_\theta \end{pmatrix}}_{\mathbf{Z}_a} = \underbrace{\begin{pmatrix} \mathbf{H}(k) \\ \hline \mathbf{I} \end{pmatrix}}_{\mathbf{H}_a} \boldsymbol{\theta} + \underbrace{\begin{pmatrix} \mathbf{V}(k) \\ \hline \mathbf{m}_\theta - \boldsymbol{\theta} \end{pmatrix}}_{\mathbf{V}_a} \tag{13-41}$$

which can be written as

$$\mathbf{Z}_a(k) = \mathbf{H}_a(k)\boldsymbol{\theta} + \mathbf{V}_a(k) \tag{13-42}$$

where $\mathbf{Z}_a(k)$, $\mathbf{H}_a(k)$, and $\mathbf{V}_a(k)$ are defined in (13-41). Additionally,

$$\mathbf{E}\{\mathbf{V}_a(k)\mathbf{V}'_a(k)\} \triangleq \mathbf{R}_a(k) = \left(\begin{array}{c|c} \mathbf{R(k)} & \mathbf{0} \\ \hline \mathbf{0} & \mathbf{P}_\theta \end{array}\right) \tag{13-43}$$

We now treat (13-41) as the starting point for derivation of a BLUE of $\boldsymbol{\theta}$, which we denote $\hat{\boldsymbol{\theta}}^a_{\text{BLU}}(k)$. Obviously,

$$\hat{\boldsymbol{\theta}}^a_{\text{BLU}}(k) = [\mathbf{H}'_a(k)\mathbf{R}_a^{-1}(k)\mathbf{H}_a(k)]^{-1}\mathbf{H}'_a(k)\mathbf{R}_a^{-1}(k)\mathbf{Z}_a(k) \tag{13-44}$$

**Theorem 13-4.** *For the linear Gaussian model, when* $\mathbf{H}$(k) *is deterministic, it is always true that*

$$\hat{\boldsymbol{\theta}}_{\text{MS}}(k) = \hat{\boldsymbol{\theta}}^a_{\text{BLU}}(k) \tag{13-45}$$

*Proof.* To begin, substitute the definitions of $\mathbf{H}_a$, $\mathbf{R}_a$, and $\mathbf{Z}_a$ into (13-44) and multiply out all the partitioned matrices to show that

$$\hat{\boldsymbol{\theta}}^a_{\text{BLU}}(k) = (\mathbf{P}_\theta^{-1} + \mathbf{H}'\mathbf{R}^{-1}\mathbf{H})^{-1}(\mathbf{P}_\theta^{-1}\mathbf{m}_\theta + \mathbf{H}'\mathbf{R}^{-1}\mathbf{Z}) \tag{13-46}$$

From (13-22), we recognize $(\mathbf{P}_\theta^{-1} + \mathbf{H}'\mathbf{R}^{-1}\mathbf{H})^{-1}$ as $\mathbf{P}_{\text{MS}}$; hence,

$$\hat{\boldsymbol{\theta}}^a_{\text{BLU}}(k) = \mathbf{P}_{\text{MS}}(k)[\mathbf{P}_\theta^{-1}\mathbf{m}_\theta + \mathbf{H}'(k)\mathbf{R}^{-1}(k)\mathbf{Z}(k)] \tag{13-47}$$

Next, we rewrite $\hat{\boldsymbol{\theta}}_{\text{MS}}(k)$ in (13-23) as

$$\hat{\boldsymbol{\theta}}_{\text{MS}}(k) = [\mathbf{I} - \mathbf{P}_{\text{MS}}\mathbf{H}'\mathbf{R}^{-1}\mathbf{H}]\mathbf{m}_\theta + \mathbf{P}_{\text{MS}}\mathbf{H}'\mathbf{R}^{-1}\mathbf{Z} \tag{13-48}$$

However,

$$\mathbf{I} - \mathbf{P}_{\text{MS}}\mathbf{H}'\mathbf{R}^{-1}\mathbf{H} = \mathbf{P}_{\text{MS}}(\mathbf{P}_{\text{MS}}^{-1} - \mathbf{H}'\mathbf{R}^{-1}\mathbf{H}) = \mathbf{P}_{\text{MS}}\mathbf{P}_\theta^{-1} \tag{13-49}$$

where we have again used (13-22). Substitute (13-49) into (13-42) to see that

$$\hat{\boldsymbol{\theta}}_{\text{MS}}(k) = \mathbf{P}_{\text{MS}}(k)[\mathbf{P}_\theta^{-1}\mathbf{m}_\theta + \mathbf{H}'(k)\mathbf{R}^{-1}(k)\mathbf{Z}(k)] \tag{13-50}$$

Hence, (13-45) follows when we compare (13-47) and (13-50). □

To conclude this section, we note that the weighted least-squares objective function that is associated with $\hat{\boldsymbol{\theta}}^a_{\text{BLU}}(k)$ is

$$\begin{aligned} J_a[\hat{\boldsymbol{\theta}}^a(k)] &= \tilde{\mathbf{Z}}'_a(k)\mathbf{R}_a^{-1}(k)\tilde{\mathbf{Z}}'_a(k) \\ &= (\mathbf{m}_\theta - \hat{\boldsymbol{\theta}}^a(k))'\mathbf{P}_\theta^{-1}(\mathbf{m}_\theta - \hat{\boldsymbol{\theta}}^a(k)) + \tilde{\mathbf{Z}}'(k)\mathbf{R}^{-1}(k)\tilde{\mathbf{Z}}(k) \end{aligned} \tag{13-51}$$

The first term in (13-51) contains all the a priori information about $\theta$. Quantity $\mathbf{m}_\theta - \hat{\theta}^a(k)$ is treated as the difference between "measurement" $\mathbf{m}_\theta$ and its estimate $\hat{\theta}^a(k)$.

## COMPUTATION

There is no M-file to do MSE. In the special but important case of jointly Gaussian random vectors, Equation (13-12) needs to be implemented. The interesting part of the computation is $\mathbf{P}_{\theta\mathbf{Z}}(k)\mathbf{P}_{\mathbf{Z}}{}^{-1}(k)[\mathbf{Z}(k) - \mathbf{m}_{\mathbf{Z}}(k)]$. It is good practice *never to invert a matrix.* Let $\mathbf{a}(k) \triangleq [\mathbf{Z}(k) - \mathbf{m}_{\mathbf{Z}}(k)]$. Compute $\hat{\theta}_{MS}(k)$ as follows: (1) solve $\mathbf{P}_{\mathbf{Z}}(k)\mathbf{b}(k) = \mathbf{a}(k)$ for $\mathbf{b}(k)$; (2) form $\mathbf{P}_{\theta\mathbf{Z}}(k)\mathbf{b}(k) \triangleq \mathbf{c}(k)$; and (3) compute $\hat{\theta}_{MS}(k) = \mathbf{m}_\theta + \mathbf{c}(k)$.

# Supplementary Material

## THE CONDITIONAL MEAN ESTIMATOR

The conditional mean estimator of random $\theta$ is valid in many more situations than described in the main body of this lesson. In this section we summarize these situations. To begin, we need the notion of a *loss function* $L = L[\tilde{\theta}(k)]$, where (Meditch, 1969, pp. 157–158):

1. $L$ is a scalar-valued function of $n$ variables [the $n$ components of $\tilde{\theta}(k)$].
2. $L(\mathbf{0}) = 0$ (i.e., no loss is assigned when the estimate is exact).
3. $L[\tilde{\theta}^a(k)] \geq L[\tilde{\theta}^b(k)]$ whenever $\rho[\tilde{\theta}^a(k)] \geq \rho[\tilde{\theta}^b(k)]$, where $\rho$ is a scalar-valued, nonnegative, convex function of $n$ variables [i.e., $\rho$ is a measure of the distance of $\tilde{\theta}(k)$ from the origin of $n$-dimensional Euclidean space, and $L$ is a nondecreasing function of this distance].
4. $L[\tilde{\theta}(k)] = L[-\tilde{\theta}(k)]$ (i.e., $L$ is symmetric).

A loss function that possesses these four properties is called an *admissible loss function.* Note that $L$ need not be a convex function.

Some examples of admissible loss functions are

$$L[\tilde{\theta}(k)] = \sum_{i=1}^{n} \alpha_i |\tilde{\theta}_i(k)|, \quad \text{where } \alpha_i \geq 0 \text{ but not all are identically zero} \tag{13-52}$$

$$L[\tilde{\theta}(k)] = \alpha \|\tilde{\theta}(k)\|^{2p}, \quad \text{where } \alpha > 0,\ p = \text{positive integer, and } \|\tilde{\theta}(k)\| = \left(\sum_{i=1}^{n} \tilde{\theta}_i^2(k)\right)^{1/2} \tag{13-53}$$

$$L[\tilde{\theta}(k)] = \alpha_1\{1 - \exp[-\alpha_2\|\tilde{\theta}(k)\|^2]\}, \quad \text{where } \alpha_1 \text{ and } \alpha_2 > 0 \tag{13-54}$$

$$L[\tilde{\theta}(k)] = \begin{cases} 0, & \text{for } \|\tilde{\theta}(k)\|^4 < \alpha_1 \\ \mu, & \text{for } \|\tilde{\theta}(k)\|^4 \geq \alpha_1 \end{cases}, \qquad \text{where } \alpha_1 \text{ and } \mu > 0 \tag{13-55}$$

Because $L$ is random, to obtain a useful measure of the loss, we define a *performance measure* $J$ as the mean value of $L$; i.e.,

$$J[\tilde{\theta}(k)] = \mathbf{E}\{L[\tilde{\theta}(k)]\} \tag{13-56}$$

For an admissible loss function, it is clear that

$$J[\tilde{\theta}^a(k)] \geq J[\tilde{\theta}^b(k)] \tag{13-57}$$

whenever $L[\tilde{\theta}^a(k)] \geq L[\tilde{\theta}^b(k)]$; hence, $J$ is a nondecreasing function of the loss. Consequently, the estimate $\hat{\theta}(k)$ that minimizes $J[\tilde{\theta}(k)]$ is said to be the *optimal estimate*. Note that an optimal estimate does not minimize the loss; it minimizes the mean value of the loss.

In addition to the notion of a loss function, we will also need the notions of conditional probability density and distribution functions: $p[\theta|\mathbf{Z}(k)]$ and $F[\theta^*|\mathbf{Z}(k)]$, where $F[\theta^*|\mathbf{Z}(k)] = P[\theta \leq \theta^*|\mathbf{Z}(k)]$.

Here are *seven* situations for which $\hat{\theta}(k) = \mathbf{E}\{\theta|\mathbf{Z}(k)\}$:

1. $L[\tilde{\theta}(k)]$ is admissible and $F[\theta^*|\mathbf{Z}(k)]$ is symmetric about its mean and is convex (this result holds for all combinations of admissible $L$ and symmetric and convex distributions; it is due to Sherman, 1958).
2. $L[\tilde{\theta}(k)]$ is admissible and convex, and $F[\theta^*|\mathbf{Z}(k)]$ is symmetric about its mean (this result holds for all combinations of admissible and convex $L$ and symmetric distributions).
3. $L[\tilde{\theta}(k)]$ is admissible, and $\theta$ and $\mathbf{Z}(k)$ are jointly Gaussian [in this case $\mathbf{E}\{\theta|\mathbf{Z}(k)\}$ is also given by (13-12)].
4. $L[\tilde{\theta}(k)]$ is admissible but not symmetric, and $\theta$ and $\mathbf{Z}(k)$ are jointly Gaussian [in this case $\mathbf{E}\{\theta|\mathbf{Z}(k)\}$ is also given by (13-12)] (Deutsch, 1965).
5. $L[\tilde{\theta}(k)]$ is admissible, and only first and second moments are known for both $\theta$ and $\mathbf{Z}(k)$ (this is a generalization of Corollary 13-2).
6. $L[\tilde{\theta}(k)] = \tilde{\theta}'(k)\tilde{\theta}(k)$ and $p[\theta|\mathbf{Z}(k)]$ is defined and continuous (this is a specialization of situation 2 relative to a specific $L$; but it is also a generalization of situation 2 to rather arbitrary distributions; it is the situation treated in Theorem 13-1).
7. $L[\tilde{\theta}(k)] = \tilde{\theta}'(k)\tilde{\theta}(k)$, and $\theta$ and $\mathbf{Z}(k)$ are jointly Gaussian (this is a specialization of situation 6 to jointly Gaussian $\theta$ and $\mathbf{Z}(k)$; it is also a specialization of situation 3 to a specific $L$; it is the situation treated in Corollary 13-1).

## A NONLINEAR ESTIMATOR

The fundamental theorem of estimation theory states that $\hat{\theta}_{MS}(k) = \mathbf{E}\{\theta|\mathbf{Z}(k)\}$. In the main body of this lesson we developed $\mathbf{E}\{\theta|\mathbf{Z}(k)\}$ for two related situations:

(1) $\boldsymbol{\theta}$ and $\mathbf{Z}(k)$ are jointly Gaussian, and (2) $\boldsymbol{\theta}$ and $\mathbf{Z}(k)$ are jointly Gaussian *and* $\mathbf{Z}(k) = \mathbf{H}(k)\boldsymbol{\theta} + \mathbf{V}(k)$. Here we develop $\mathbf{E}\{\boldsymbol{\theta}|\mathbf{Z}(k)\}$ in yet another direction, one that requires no statistical a priori assumptions about $\boldsymbol{\theta}$ and $\mathbf{Z}(k)$ and is for a *nonparametric* relationship between $\boldsymbol{\theta}$ and $\mathbf{Z}$. In order not to obscure the following derivation with notation, we develop it for a scalar random variable $\theta$. Our development in the rest of this section follows Specht (1991) very closely. Whereas we refer to the final result in (13-62) as a *nonlinear estimator*, Specht has coined the phrase "generalized regression neural network (GRNN)" to describe it.

Using the facts that $p(\theta|\mathbf{Z}) = p(\theta, \mathbf{Z})/p(\mathbf{Z})$ and $p(\mathbf{Z}) = \int_{-\infty}^{\infty} p(\theta, \mathbf{Z})d\theta$, we know that

$$\mathbf{E}\{\theta|\mathbf{Z}\} = \int_{-\infty}^{\infty} \theta p(\theta|\mathbf{Z})d\theta = \int_{-\infty}^{\infty} \frac{\theta p(\theta, \mathbf{Z})}{p(\mathbf{Z})}d\theta = \frac{\displaystyle\int_{-\infty}^{\infty} \theta p(\theta, \mathbf{Z})d\theta}{p(\mathbf{Z})}$$

$$\mathbf{E}\{\theta|\mathbf{Z}\} = \frac{\displaystyle\int_{-\infty}^{\infty} \theta p(\theta, \mathbf{Z})d\theta}{\displaystyle\int_{-\infty}^{\infty} p(\theta, \mathbf{Z})d\theta} \tag{13-58}$$

When the probability density function $p(\theta, \mathbf{Z})$ is unknown, it usually has to be estimated from a given sample of observations of $\theta$ and $\mathbf{Z}$ (i.e., a *training sample*), denoted $\theta^i$ and $\mathbf{Z}^i$, respectively, where $i = 1, 2, \dots, M$. Parzen (1962) proposed estimating probability density functions as the sum of window functions, such as Gaussian functions, centered at each of the observed samples $\mathbf{Z}^i$. Parzen did this for the case of a scalar measurement; however, Cacoullos (1966) extended this result to the multidimensional case when $\mathbf{Z}$ is a vector. The Parzen window estimator for $p(\theta, \mathbf{Z})$ can be written as

$$\hat{p}(\theta, \mathbf{Z}) = \frac{1}{(2\pi)^{(N+1)/2}\sigma^{(N+1)}} \times \frac{1}{M}\sum_{i=1}^{M} \exp\left[-\frac{(\mathbf{Z} - \mathbf{Z}^i)'(\mathbf{Z} - \mathbf{Z}^i)}{2\sigma^2}\right] \times \exp\left[-\frac{(\theta - \theta^i)^2}{2\sigma^2}\right] \tag{13-59}$$

where $N$ is the dimension of $\mathbf{Z}$.

Parzen and Cacoullos have shown that density estimators of the form (13-59) are consistent provided that $\sigma = \sigma(M)$ is chosen as a decreasing function of $M$ such that $\lim_{M\to\infty} \sigma(M) = 0$ and $\lim_{M\to\infty} M\sigma^N(M) = \infty$. Specht (1967) has shown that density estimators of the form (13-59) are a good choice for estimating $p(\theta, \mathbf{Z})$ if it can be assumed that $p(\theta, \mathbf{Z})$ is continuous and that the first partial derivatives of $p(\theta, \mathbf{Z})$ evaluated at any $\mathbf{Z}$ are small. Note, also, that (13-59) is a *Gaussian sum approximation* to $\hat{p}(\theta, \mathbf{Z})$. Gaussian sum approximations to probability density functions were also developed by Sorenson and Alspach (1971), who were apparently unaware of Parzen's work.

Specht (1991) notes that "A physical interpretation of the probability estimate $\hat{p}(\theta, \mathbf{Z})$ is that it assigns sample probability of width $\sigma$ for each sample $\theta^i$ and $\mathbf{Z}^i$, and the probability estimate is the sum of those sample probabilities."

Substituting (13-59) into (13-58) and interchanging the order of integration and summation, we obtain the following conditional mean estimator $\hat{\theta}(\mathbf{Z})$:

$$\hat{\theta}(\mathbf{Z}) = \left\{ \sum_{i=1}^{M} \exp\left[ -\frac{(\mathbf{Z}-\mathbf{Z}^i)'(\mathbf{Z}-\mathbf{Z}^i)}{2\sigma^2} \right] \times \int_{-\infty}^{\infty} \theta \exp\left[ -\frac{(\theta-\theta^i)^2}{2\sigma^2} \right] d\theta \right\} /$$

$$\left\{ \sum_{i=1}^{M} \exp\left[ -\frac{(\mathbf{Z}-\mathbf{Z}^i)'(\mathbf{Z}-\mathbf{Z}^i)}{2\sigma^2} \right] \times \int_{-\infty}^{\infty} \exp\left[ -\frac{(\theta-\theta^i)^2}{2\sigma^2} \right] d\theta \right\} \quad (13\text{-}60)$$

Let

$$d_i^2 \triangleq (\mathbf{Z}-\mathbf{Z}^i)'(\mathbf{Z}-\mathbf{Z}^i) \quad (13\text{-}61)$$

Then it is straightforward to perform the integrations in (13-60) to show that

$$\hat{\theta}(\mathbf{Z}) = \left[ \sum_{i=1}^{M} \theta^i \exp\left( -\frac{d_i^2}{2\sigma^2} \right) \right] / \left[ \sum_{i=1}^{M} \exp\left( -\frac{d_i^2}{2\sigma^2} \right) \right] \quad (13\text{-}62)$$

which is a nonlinear estimator (GRNN) for $\theta$. Observe that $\hat{\theta}(\mathbf{Z})$ depends on $M$, the given number of sample observations (i.e., the training sample).

Specht (1991) notes that

> The estimate $\hat{\theta}(\mathbf{Z})$ can be visualized as a weighted average of all the observed values, $\mathbf{Z}^i$, where each observed value is weighted exponentially according to its Euclidean distance from $\mathbf{Z}$. When the smoothing parameter $\sigma$ is made large, the estimated density is forced to be smooth and in the limit becomes a multivariate Gaussian with covariance $\sigma^2\mathbf{I}$. On the other hand, a smaller value of $\sigma$ allows the estimated density to assume non-Gaussian shapes, but with the hazard that wild points may have too great an effect on the estimate. As $\sigma$ becomes very large, $\hat{\theta}(\mathbf{Z})$ assumes the value of the sample mean of the observed $\mathbf{Z}^i$, and as $\sigma$ goes to 0, $\hat{\theta}(\mathbf{Z})$ assumes the value of the $\mathbf{Z}^i$ associated with the observation closest to $\mathbf{Z}$. For intermediate values of $\sigma$, all values of $\mathbf{Z}^i$ are taken into account, but those corresponding to points closer to $\mathbf{Z}$ are given heavier weight.

For much more discussion about $\hat{\theta}(\mathbf{Z})$, as well as some applications of $\hat{\theta}(\mathbf{Z})$, see Specht (1991).

**EXAMPLE 13-2**

Suppose we have conducted a sample survey and have observed a scalar pair of values $\{\theta^i, Z^i\}$ for $i = 1, 2, \ldots, M$. In this case $\dim \mathbf{Z} = 1$. Equation (13-62) can then be used to provide us with an approximation to $\hat{\theta}_{\text{MS}}(Z)$. Each time a new value of $Z$ becomes available we can compute $\hat{\theta}(Z)$. □

**EXAMPLE 13-3**

Suppose we want to predict $y(k)$ based on some of its previous values, using (13-62). As an illustration, suppose that we believe that $y(k)$ depends mostly on $N$ preceding values, i.e.,

$$y(k) = f[y(k-1), y(k-2), \ldots, y(k-N)] \quad (13\text{-}63)$$

For example, we may wish to predict tomorrow's Dow–Jones index value by means of (13-62), using the past $N = 50$ day's Dow–Jones values [note that use of (13-62) implies that we believe that the Dow–Jones index can be described by a probability model, an assumption that in itself needs to be verified]. Let

$$\mathbf{Z}(k-1) \triangleq \text{col}\,[y(k-1), y(k-2), \ldots, y(k-N)] \tag{13-64}$$

and suppose that we are given $M$ input/output pairs $\{y(k), \mathbf{Z}(k-1)\}$, where $k = N+1, N+2, \ldots, N+M$ [for this choice of $k$, we are using $\mathbf{Z}(k-1)$ values that involve $y(\cdot)$ values all of whose arguments are nonnegative].

Now we come to a crucial assumption if we are to use (13-62) to provide us with $\hat{y}(k)$: *we must assume that $y(k)$ is stationary.* This is necessary because the derivation of (13-62) was for a random variable $\theta$. While it is true that at each frozen time point $y(k)$ is a random variable, it is also true that in this application we do not have an ensemble of sample observations of $y(k)$ at each frozen time point. If, however, $y(k)$ is stationary, then the probability density functions for all observed values of $y(k)$, used to obtain $\hat{y}(k)$ in (13-62), are the same, and in this case we are permitted to use (13-62). This may help to explain why (13-62) would not give a very good prediction of the Dow–Jones index value when a sudden jump or drop in the index occurs. It is precisely at those very interesting time points that the probability density function of the Dow–Jones index changes. The resulting equation for $\hat{y}(k)$ is

$$\hat{y}(k) = \left[\sum_{i=N+1}^{N+M} y(i) \exp\left(-\frac{d_i^2}{2\sigma^2}\right)\right] \Big/ \left[\sum_{i=N+1}^{N+M} \exp\left(-\frac{d_i^2}{2\sigma^2}\right)\right] \tag{13-65}$$

where $k = N+M+1, N+M+2, \ldots,$ and

$$d_i^2 = [\mathbf{Z}(k-1) - \mathbf{Z}(i-1)]'[\mathbf{Z}(k-1) - \mathbf{Z}(i-1)] \tag{13-66}$$

Of course, many other variations of this problem are possible, depending on which measurements are to be used for the prediction of $y(k)$ as well as the use of either a fixed-memory or an expanding memory estimator for $y(k)$. □

## SUMMARY QUESTIONS

1. What is central to the calculation of $\hat{\theta}_{MS}(k)$?
   (a) notion of joint density function
   (b) notion of conditional expectation
   (c) notion of quadratic cost function
2. In general, $\mathbf{E}\{\theta|\mathbf{Z}(k)\}$ is a:
   (a) linear function of the data
   (b) nonlinear function of the data
   (c) matrix function of the data
3. Check the properties of $\hat{\theta}_{MS}(k)$ that are valid. When $\theta$ and $\mathbf{Z}(k)$ are jointly Gaussian, the mean-squared estimator is:
   (a) Gaussian
   (b) consistent
   (c) linear
   (d) unbiased

(e) efficient
(f) unique

4. If $\mathbf{m}_\theta = \mathbf{0}$, then $\hat{\theta}_{MS}(k)$ for the generic linear and Gaussian model is:
(a) $\mathbf{P}_\theta \mathbf{H}'(k)\mathbf{R}^{-1}(k)\mathbf{Z}(k)$
(b) $\mathbf{P}_{MS}\mathbf{H}'(k)\mathbf{R}^{-1}(k)\mathbf{Z}(k)$
(c) $\mathbf{P}_\theta \mathbf{H}'(k)[\mathbf{H}(k)\mathbf{P}_\theta \mathbf{H}'(k)]^{-1}\mathbf{Z}(k)$

5. In the application of deconvolution, where $\hat{\mu}_{MS}$ is a mean-squared estimator of $\mu$:
(a) $\hat{\mu}_{MS}$ is not a consistent estimator of $\mu$
(b) $\hat{\mu}_{MS}$ is a consistent estimator of $\mu$
(c) $\hat{\mu}_{MS}$ converges in mean square to $\mu$

6. Which of the following are valid orthogonality principles?
(a) $\mathbf{E}\{\tilde{\theta}_{MS}(k)\mathbf{W}(k)\hat{\theta}'_{MS}(k)\} = \mathbf{0}$
(b) $\mathbf{E}\{\tilde{\theta}_{MS}(k)\mathbf{f}'(\mathbf{Z}(j)\} = \mathbf{0}$, where $j > k$
(c) $\mathbf{E}\{\tilde{\theta}_{MS}(k)\mathbf{f}'(\mathbf{Z}(j)\} = \mathbf{0}$, where $j < k$

7. The linear mean-squared estimator is the same as the mean-squared estimator:
(a) always
(b) when $\theta$ is Gaussian
(c) when $\theta$ and $\mathbf{Z}(k)$ are jointly Gaussian

8. The orthogonality principle is valid for:
(a) any function of the data
(b) only linear functions of the data
(c) only continuous functions of the data

9. When we say that one sequence is "statistically equivalent" to another Gaussian random sequence, we mean that:
(a) their first two moments are equal
(b) they are asymptotically Gaussian
(c) they are exactly the same

10. It is always true that $\hat{\theta}_{MS}(k) = \hat{\theta}_{BLU}(k)$, if:
(a) $\mathbf{P}_\theta^{-1} = \mathbf{0}$ and $\mathbf{H}(k)$ is deterministic
(b) $\mathbf{P}_\theta^{-1} = \mathbf{0}$, $\mathbf{H}(k)$ is deterministic, and $\theta$ and $\mathbf{V}(k)$ are jointly Gaussian
(c) $\theta$ and $\mathbf{Z}(k)$ are jointly Gaussian

11. The conditional mean estimator of random $\theta$:
(a) is only valid when the loss function equals $\tilde{\theta}'\tilde{\theta}$
(b) is valid for a broad range of *admissible* loss functions
(c) is valid for totally arbitrary loss function as long as the probability distribution function $F[\theta^*|\mathbf{Z}(k)]$ is symmetric about its mean and is convex

12. The nonlinear estimator for $\hat{\theta}(\mathbf{Z})$, given in (13-62), is
(a) independent of the *number* of elements from the training sample
(b) a linear function of the *number* of elements from the training sample
(c) strongly dependent on the *number* of elements from the training sample in both its numerator and denominator

## PROBLEMS

**13-1.** (Todd B. Leach, Spring 1992) Prove the fundamental theorem of estimation theory by differentiating the objective function $J[\tilde{\theta}_{MS}(k)]$.

**13-2.** Prove that $\hat{\theta}_{MS}(k)$, given in (13-12), is unique.

**13-3.** Prove Corollary 13-2 by means of a direct proof.

**13-4.** Let $\theta$ and $\mathbf{Z}(N)$ be zero mean $n \times 1$ and $N \times 1$ random vectors, respectively, with known second-order statistics $\mathbf{P}_\theta$, $\mathbf{P}_\mathbf{Z}$, $\mathbf{P}_{\theta\mathbf{Z}}$, and $\mathbf{P}_{\mathbf{Z}\theta}$. View $\mathbf{Z}(N)$ as a vector of measurements. It is desired to determine a linear estimator of $\theta$,

$$\hat{\theta}(N) = \mathbf{K}_L(N)\mathbf{Z}(N)$$

where $\mathbf{K}_L(N)$ is an $n \times N$ matrix that is chosen to minimize the mean-squared error $\mathbf{E}\{[\theta - \mathbf{K}_L(N)\mathbf{Z}(N)]'[\theta - \mathbf{K}_L(N)\mathbf{Z}(N)]\}$.

**(a)** Show that the gain matrix, which minimizes the mean-squared error, is $\mathbf{K}_L(N) = \mathbf{E}\{\theta\mathbf{Z}'(N)\}[\mathbf{E}\{\mathbf{Z}(N)\mathbf{Z}'(N)\}]^{-1} = \mathbf{P}_{\theta\mathbf{Z}}\mathbf{P}_\mathbf{Z}^{-1}$.

**(b)** Show that the covariance matrix, $\mathbf{P}(N)$, of the estimation error, $\tilde{\theta}(N) = \theta - \hat{\theta}(N)$, is

$$\mathbf{P}(N) = \mathbf{P}_\theta - \mathbf{P}_{\theta\mathbf{Z}}\mathbf{P}_\mathbf{Z}^{-1}\mathbf{P}_{\mathbf{Z}\theta}$$

**(c)** Relate the results obtained in this problem to those in Corollary 13-2.

**13-5.** Derive Equation (13-21) for $\mathbf{P}_{MS}(k)$.

**13-6.** For random vectors $\theta$ and $\mathbf{Z}(k)$, the *linear projection*, $\theta^*(k)$, of $\theta$ on a Hilbert space spanned by $\mathbf{Z}(k)$ is defined as $\theta^*(k) = \mathbf{a} + \mathbf{B}\mathbf{Z}(k)$, where $\mathbf{E}\{\theta^*(k)\} = \mathbf{E}\{\theta\}$ and $\mathbf{E}\{[\theta - \theta^*(k)]\mathbf{Z}'(k)\} = \mathbf{0}$. We denote the linear projection, $\theta^*(k)$, as $\hat{\mathbf{E}}\{\theta|\mathbf{Z}(k)\}$.

**(a)** Prove that the linear (i.e., affine), unbiased mean-squared estimator of $\theta$, $\hat{\theta}_{MS}(k)$, is the linear projection of $\theta$ on $\mathbf{Z}(k)$.

**(b)** Prove that the linear projection, $\theta^*(k)$, of $\theta$ on the Hilbert space spanned by $\mathbf{Z}(k)$ is uniquely equal to the linear (i.e., affine), unbiased mean-squared estimator of $\theta$, $\hat{\theta}_{MS}(k)$.

**(c)** For random vectors $\mathbf{x}$, $\mathbf{y}$, $\mathbf{z}$, where $\mathbf{y}$ and $\mathbf{z}$ are uncorrelated, prove that

$$\hat{\mathbf{E}}\{\mathbf{x}|\mathbf{y},\mathbf{z}\} = \hat{\mathbf{E}}\{\mathbf{x}|\mathbf{y}\} + \hat{\mathbf{E}}\{\mathbf{x}|\mathbf{z}\} - \mathbf{m}_\mathbf{x}$$

**(d)** For random vectors $\mathbf{x}$, $\mathbf{y}$, $\mathbf{z}$, where $\mathbf{y}$ and $\mathbf{z}$ are correlated, prove that

$$\hat{\mathbf{E}}\{\mathbf{x}|\mathbf{y},\mathbf{z}\} = \hat{\mathbf{E}}\{\mathbf{x}|\mathbf{y},\tilde{\mathbf{z}}\}$$

where

$$\tilde{\mathbf{z}} = \mathbf{z} - \hat{\mathbf{E}}\{\mathbf{z}|\mathbf{y}\}$$

so that

$$\hat{\mathbf{E}}\{\mathbf{x}|\mathbf{y},\mathbf{z}\} = \hat{\mathbf{E}}\{\mathbf{x}|\mathbf{y}\} + \hat{\mathbf{E}}\{\mathbf{x}|\tilde{\mathbf{z}}\} - \mathbf{m}_\mathbf{x}$$

Parts (c) and (d) show that the results given in Theorems 12-3 and 12-4 are "distribution free" within the class of linear (i.e., affine), unbiased, mean-squared estimators.

**13-7.** (David Adams, Spring 1992) Find the mean-squared estimator of positive scalar $\theta$ when $N$ independent observations $\{z(1), z(2), \ldots, z(N)\}$ are obtained. The joint probability density function between $\theta$ and the measurements is

$$p(z(i), \theta) = \theta e^{-\theta z(i)}$$

**13-8.** (Andrew D. Norte, Fall 1991) (This is a numerical problem.) Consider the following linear model for a particular group of measurements:

$$\begin{bmatrix} 3 \\ 4 \end{bmatrix} = \begin{bmatrix} 1 & 2 \\ 2 & 3 \end{bmatrix} \begin{bmatrix} \theta_1 \\ \theta_2 \end{bmatrix} + \begin{bmatrix} \nu_1 \\ \nu_2 \end{bmatrix}, \qquad \text{i.e., } \mathbf{Z}(2) = \mathbf{H}(2)\boldsymbol{\theta} + \mathbf{V}(2)$$

$\theta_1$ and $\theta_2$ are themselves related to two jointly Gaussian random variables $x_1$ and $x_2$ in the following manner:

$$\theta_1 = x_1 + 2x_2 \quad \text{and} \quad \theta_2 = x_1 + 3x_2$$

where $\mathbf{x} = \text{col}\,(x_1, x_2)$, $\mathbf{m_x} = \text{col}\,(0, 0)$, and

$$\mathbf{P_x} = \begin{bmatrix} 1 & -1 \\ -1 & 2 \end{bmatrix}$$

Noise vector $\mathbf{V}(2)$ is multivariate Gaussian with covariance matrix $\mathbf{R}_\nu(2) = \text{diag}\,(3, 3)$; $\boldsymbol{\theta}$ and $\mathbf{V}(2)$ are uncorrelated.

**(a)** Find the mean-square estimator $\hat{\boldsymbol{\theta}}_{\text{MS}}(2)$.

**(b)** Find $\mathbf{P}_{\text{MS}}(2)$, the error covariance matrix associated with $\hat{\boldsymbol{\theta}}_{\text{MS}}(2)$.

**13-9.** (Li-Chien Lin, Fall 1991) Consider a mean-square estimator for the following linear model:

$$\begin{bmatrix} z(2) \\ z(1) \end{bmatrix} = \begin{bmatrix} 1 & 0 \\ 1 & 4 \end{bmatrix} \begin{bmatrix} \theta_1 \\ \theta_2 \end{bmatrix} + \begin{bmatrix} \nu(2) \\ \nu(1) \end{bmatrix}, \qquad \text{i.e., } \mathbf{Z}(2) = \mathbf{H}(2)\boldsymbol{\theta} + \mathbf{V}(2)$$

where $\mathbf{V}$ and $\boldsymbol{\theta}$ are Gaussian and mutually uncorrelated, $\mathbf{V}$ is white noise with unit power $\mathbf{m}_{\boldsymbol{\theta}} = \text{col}\,(4, 4)$, and

$$\mathbf{P}_{\boldsymbol{\theta}} = \begin{bmatrix} 4 & 1 \\ 1 & 2 \end{bmatrix}$$

**(a)** Is $\mathbf{Z}(2)$ Gaussian?

**(b)** Find $\hat{\boldsymbol{\theta}}_{\text{MS}}(k)$.

**13-10.** (Keith M. Chugg, Fall 1991) Consider the problem of estimating $\mathbf{s}$ from the following observation:

$$\mathbf{r} = h\mathbf{s} + \mathbf{n}$$

where $h$ is a real random variable with $\Pr(h = 1) = \Pr(h = 0) = 1/2$, and $\mathbf{s}$ and $\mathbf{n}$ are $n \times 1$ independent zero-mean Gaussian random vectors with covariance matrices $\mathbf{P_s}$ and $\mathbf{P_n}$, respectively. What is the mean-squared estimate of $\mathbf{s}$ based on $\mathbf{r}$?

LESSON 14

# Maximum a Posteriori Estimation of Random Parameters

(MAP)

## SUMMARY

MAP estimation is also known as *Bayesian estimation*. Obtaining a MAP estimate involves specifying both $p[\mathbf{Z}(k)|\theta]$ and $p(\theta)$ and finding the value of $\theta$ that maximizes $p(\theta|\mathbf{Z}(k))$. Generally, optimization must be used to compute $\hat{\theta}_{MAP}(k)$. When $\mathbf{Z}(k)$ and $\theta$ are jointly Gaussian, then we show that $\hat{\theta}_{MAP}(k) = \hat{\theta}_{MS}(k)$.

We also examine $\hat{\theta}_{MAP}(k)$ for the generic linear and Gaussian model when $\mathbf{H}(k)$ is deterministic, $\mathbf{V}(k)$ is white Gaussian noise with known covariance matrix $\mathbf{R}(k)$, and $\theta$ is multivariate Gaussian with known mean $\mathbf{m}_\theta$ and covariance $\mathbf{P}_\theta$. We learn that $\hat{\theta}_{MAP}(k)$ is the same as a BLUE; hence, for the generic linear Gaussian model, MS, MAP, and BLUE estimates of $\theta$ are all the same.

Finally, this lesson examines the important applications of deconvolution and state estimation from a MAP point of view.

When you have completed this lesson, you will be able to (1) derive a general formula for a maximum a posteriori estimator of random parameters, (2) explain the consequences of the "Gaussian assumption" on this general formula, (3) understand what happens to the general formula for the generic linear and Gaussian model, and (4) explain when MS, MAP, and BLU estimates of $\theta$ are all the same.

## INTRODUCTION

now we are given $p(\theta)$

As in Lesson 13, we view $\theta$ as an $n \times 1$ vector of random unknown parameters. The information available to us are the measurements $\mathbf{z}(1), \mathbf{z}(2), \ldots, \mathbf{z}(k)$, which are assumed to depend on $\theta$ and an a priori probability model for $\theta$, that is $p(\theta)$.

The latter information is what distinguishes the problem formulation for maximum a posteriori (MAP) estimation from MS estimation.

We begin by not assuming a specific structural dependency between $\mathbf{z}(i)$ and $\theta$. Bayes's rule is the starting point for our development. It leads to the problem of maximizing an unconditional likelihood function, one that not only requires knowledge of $p(\mathbf{Z}|\theta)$, as in Lesson 11, but also of $p(\theta)$. The resulting estimator is called a *maximum a posteriori estimator*, and is denoted $\hat{\theta}_{\text{MAP}}(k)$.

We then determine $\hat{\theta}_{\text{MAP}}(k)$ for the linear generic model

$$\mathbf{Z}(k) = \mathbf{H}(k)\theta + \mathbf{V}(k) \tag{14-1}$$

where $\mathbf{H}(k)$ is deterministic, $\mathbf{V}(k)$ is white Gaussian noise with known covariance matrix $\mathbf{R}(k)$, $\theta$ is multivariate Gaussian with known mean, $\mathbf{m}_\theta$ and covariance, $\mathbf{P}_\theta$, i.e.,

$$\theta \sim N(\theta; \mathbf{m}_\theta, \mathbf{P}_\theta) \tag{14-2}$$

and $\theta$ and $\mathbf{V}(k)$ are jointly Gaussian and uncorrelated. One of our main objectives for studying this linear Gaussian model is to compare $\hat{\theta}_{\text{MAP}}(k)$ and $\hat{\theta}_{\text{MS}}(k)$ to see how these estimators are related for it.

We shall also illustrate many of our results for the deconvolution and state estimation examples that were described in Lesson 2.

## GENERAL RESULTS

Recall Bayes's rule (Papoulis, 1991, p. 30):

$$p(\theta|\mathbf{Z}(k)) = p(\mathbf{Z}(k)|\theta)p(\theta)/p(\mathbf{Z}(k)) \quad = posteriori \, pdf \tag{14-3}$$

in which density function $p(\theta|\mathbf{Z}(k))$ is known as the a posteriori (or posterior) conditional density function, and $p(\theta)$ is the prior probability density function for $\theta$. Observe that $p(\theta|\mathbf{Z}(k))$ is related to likelihood function $l\{\theta|\mathbf{Z}(k)\}$, because $l\{\theta|\mathbf{Z}(k)\} \propto p(\mathbf{Z}(k)|\theta)$. Additionally, because $p(\mathbf{Z}(k))$ does not depend on $\theta$,

$$p(\theta|\mathbf{Z}(k)) \propto p(\mathbf{Z}(k)|\theta)p(\theta) \quad (ie \, we \, will \, find \, \hat{\theta} \, that \, maximizes \, the \, numerator \, wrt \, \theta) \tag{14-4}$$

In maximum a posteriori (MAP) estimation, values of $\theta$ are found that maximize $p(\theta|\mathbf{Z}(k))$ in (14-4).

If $\theta_1, \theta_2, \ldots, \theta_n$ are uniformly distributed, then $p(\theta|\mathbf{Z}(k)) \propto p(\mathbf{Z}(k)|\theta)$, and the MAP estimator of $\theta$ equals the ML estimator of $\theta$. Generally, MAP estimates are quite different from ML estimates. For example, the invariance property of MLEs usually does not carry over to MAP estimates. One reason for this can be seen from (14-4). Suppose, for example, that $\phi = \mathbf{g}(\theta)$ and we want to determine $\hat{\phi}_{\text{MAP}}$ by first computing $\hat{\theta}_{\text{MAP}}$. Because $p(\theta)$ depends on the Jacobian matrix of $\mathbf{g}^{-1}(\phi)$, $\hat{\phi}_{\text{MAP}} \neq \mathbf{g}(\hat{\theta}_{\text{MAP}})$. See the Supplementary Material in the section entitled "Transformation of Variables and Probability" at the end of Lesson 10. Kashyap and Rao (1976, p. 137) note "the two estimates are usually asymptotically identical to one another since in the large sample case the knowledge of the observations *swamps that of the prior distribution*." For additional discussions on the asymptotic properties of MAP estimators, see Zacks (1971).

Quantity $p(\theta|\mathbf{Z}(k))$ in (14-4) is sometimes called an *unconditional likelihood function*, because the random nature of $\theta$ has been accounted for by $p(\theta)$. Density $p(\mathbf{Z}(k)|\theta)$ is the called a *conditional likelihood function* (Nahi, 1969).

Obtaining a MAP estimate involves specifying both $p(\mathbf{Z}(k)|\theta)$ and $p(\theta)$ and finding those values of $\theta$ that maximize $p(\theta|\mathbf{Z}(k))$, or $\ln p(\theta|\mathbf{Z}(k))$. Generally, optimization must be used to compute $\hat{\theta}_{MAP}(k)$. When $\mathbf{Z}(k)$ is related to $\theta$ by our linear model, $\mathbf{Z}(k) = \mathbf{H}(k)\theta + \mathbf{V}(k)$, then it may be possible to obtain $\hat{\theta}_{MAP}(k)$ in closed form. We examine this situation in some detail in the next section.

*[margin note: the algorithm → (Just like ML) (Not obviously linear model yet)]*

### EXAMPLE 14-1

This is a continuation of Example 11-1. We observe a random sample $\{z(1), z(2), \ldots, z(N)\}$ at the output of a Gaussian random number generator, i.e., $z(i) \sim N(z(i); \mu, \sigma_z^2)$. Now, however, $\mu$ is a random variable with prior distribution $N(\mu; 0, \sigma_\mu^2)$. Both $\sigma_z^2$ and $\sigma_\mu^2$ are assumed known, and we wish to determine the MAP estimator of $\mu$.

We can view this random number generator as a cascade of two random number generators. The first is characterized by $N(\mu; 0, \sigma_\mu^2)$ and provides at its output a single realization for $\mu$, say $\mu_R$. The second is characterized by $N(z(i); \mu_R, \sigma_z^2)$. Observe that $\mu_R$, which is unknown to us in this example, is transferred from the first random number generator to the second one before we can obtain the given random sample $\{z(1), z(2), \ldots, z(N)\}$.

Using the facts that

*[margin note: $p(a,b) = p(a|b)p(b) = p(b|a)p(a)$ So $p(b|a) = \dfrac{p(a|b)p(b)}{p(a)}$]*

$$p(z(i)|\mu) = (2\pi\sigma_z^2)^{-1/2} \exp\left\{-\frac{[z(i) - \mu]^2}{2\sigma_z^2}\right\} \tag{14-5}$$

$$p(\mu) = (2\pi\sigma_\mu^2)^{-1/2} \exp\left\{-\frac{\mu^2}{2\sigma_\mu^2}\right\} \tag{14-6}$$

and

$$p(\mathbf{Z}(N)|\mu) = \prod_{i=1}^{N} p(z(i)|\mu) \tag{14-7}$$

we find

*[margin note: $p(\mu|z) = \dfrac{p(z|\mu)p(\mu)}{p(z)}$]*

$$p(\mu|\mathbf{Z}(N)) \propto (2\pi\sigma_z^2)^{-N/2} \exp\left\{-\frac{1}{2}\sum_{i=1}^{N}\frac{[z(i) - \mu]^2}{\sigma_z^2}\right\}$$
$$\cdot (2\pi\sigma_\mu^2)^{-1/2} \exp\left\{-\frac{\mu^2}{2\sigma_\mu^2}\right\} \tag{14-8}$$

Taking the logarithm of (14-8) and neglecting the terms that do not depend $\mu$, we obtain

$$L_{MAP}(\mu|\mathbf{Z}(N)) = -\frac{1}{2}\sum_{i=1}^{N}\left\{\frac{[z(i) - \mu]^2}{\sigma_z^2}\right\} - \frac{\mu^2}{2\sigma_\mu^2} \tag{14-9}$$

Setting $\partial L_{MAP}/\partial \mu = 0$ and solving for $\hat{\mu}_{MAP}(N)$, we find that

$$\hat{\mu}_{\mathrm{MAP}}(N) = \frac{\sigma_\mu^2}{\sigma_z^2 + N\sigma_\mu^2} \sum_{i=1}^{N} z(i) \qquad (14\text{-}10)$$

Next, we compare $\hat{\mu}_{\mathrm{MAP}}(N)$ and $\hat{\mu}_{\mathrm{ML}}(N)$, where [see (11-19]

$$\hat{\mu}_{\mathrm{ML}}(N) = \frac{1}{N} \sum_{i=1}^{N} z(i) \qquad (14\text{-}11)$$

In general, $\hat{\mu}_{\mathrm{MAP}}(N) \neq \hat{\mu}_{\mathrm{ML}}(N)$. If, however, no a priori information about $\mu$ is available, then we let $\sigma_\mu^2 \to \infty$, in which case $\hat{\mu}_{\mathrm{MAP}}(N) = \hat{\mu}_{\mathrm{ML}}(N)$. Observe also that, as $N \to \infty$, $\hat{\mu}_{\mathrm{MAP}}(N) = \hat{\mu}_{\mathrm{ML}}(N)$, which implies (Sorenson, 1980) that the influence of the prior knowledge about $\mu$ [i.e., $\mu \sim N(\mu; 0, \sigma_\mu^2)$] diminishes as the number of measurements increase. $\square$

*no model assumed. Z is what you measure, $\Theta$ is what you estimate.*

**Theorem 14-1.** *If $\mathbf{Z}(k)$ and $\theta$ are jointly Gaussian, then $\hat{\theta}_{\mathrm{MAP}}(k) = \hat{\theta}_{\mathrm{MS}}(k)$ .*

*Proof.* If $\mathbf{Z}(k)$ and $\theta$ are jointly Gaussian, then (see Theorem 12-1)

$$p(\theta|\mathbf{Z}(k)) = \frac{1}{\sqrt{(2\pi)^n |\mathbf{P}_{\theta|z}(k)|}} \exp\left\{ -\frac{1}{2}[\theta - \mathbf{m}(k)]'\mathbf{P}_{\theta|z}^{-1}(k)[\theta - \mathbf{m}(k)] \right\} \qquad (14\text{-}12)$$

where

$$\mathbf{m}(k) = \mathbf{E}\{\theta|\mathbf{Z}(k)\} \qquad (14\text{-}13)$$

$\hat{\theta}_{\mathrm{MAP}}(k)$ is found by maximizing $p(\theta|\mathbf{Z}(k))$ or, equivalently, by minimizing the argument of the exponential in (14-12). The minimum value of $[\theta - \mathbf{m}(k)]'\mathbf{P}_{\theta|z}^{-1}(k)[\theta - \mathbf{m}(k)]$ is zero, and this occurs when

$$[\theta - \mathbf{m}(k)]|_{\theta = \hat{\theta}_{\mathrm{MAP}}(k)} = \mathbf{0} \qquad (14\text{-}14)$$

i.e.,

$$\hat{\theta}_{\mathrm{MAP}}(k) = \mathbf{E}\{\theta|\mathbf{Z}(k)\} \qquad (14\text{-}15)$$

Comparing (14-15) and (13-9), we conclude that $\hat{\theta}_{\mathrm{MAP}}(k) = \hat{\theta}_{\mathrm{MS}}(k)$. $\square$

The result in Theorem 14-1 is true regardless of the nature of the model relating $\theta$ to $\mathbf{Z}(k)$. Of course, in order to use it, we must first establish that $\mathbf{Z}(k)$ and $\theta$ are jointly Gaussian. Except for the linear model, which we examine next, this is very difficult to do.

## THE GENERIC LINEAR AND GAUSSIAN MODEL

To determine $\hat{\theta}_{\mathrm{MAP}}(k)$ for the linear model in (14-1), we first need to determine $p(\theta|\mathbf{Z}(k))$. Using the facts that $\theta \sim N(\theta; \mathbf{m}_\theta, \mathbf{P}_\theta)$ and $\mathbf{V}(k) \sim N(\mathbf{V}(k); \mathbf{0}, \mathbf{R}(k))$, it follows that

$$p(\theta) = \frac{1}{\sqrt{(2\pi)^n |\mathbf{P}_\theta|}} \exp\left\{ -\frac{1}{2}(\theta - \mathbf{m}_\theta)'\mathbf{P}_\theta^{-1}(\theta - \mathbf{m}_\theta) \right\} \qquad (14\text{-}16)$$

and

$$p(\mathbf{Z}(k)|\boldsymbol{\theta}) = \frac{1}{\sqrt{(2\pi)^N |\mathbf{R}(k)|}}$$

$$\exp\left\{-\frac{1}{2}[\mathbf{Z}(k) - \mathbf{H}(k)\boldsymbol{\theta}]'\mathbf{R}^{-1}(k)[\mathbf{Z}(k) - \mathbf{H}(k)\boldsymbol{\theta}]\right\} \tag{14-17}$$

Hence,

$$\ln p(\boldsymbol{\theta}|\mathbf{Z}) \propto -\frac{1}{2}(\boldsymbol{\theta} - \mathbf{m}_\theta)'\mathbf{P}_\theta^{-1}(\boldsymbol{\theta} - \mathbf{m}_\theta) - \frac{1}{2}(\mathbf{Z} - \mathbf{H}\boldsymbol{\theta})'\mathbf{R}^{-1}(\mathbf{Z} - \mathbf{H}\boldsymbol{\theta}) \tag{14-18}$$

To find $\hat{\boldsymbol{\theta}}_{\text{MAP}}(k)$, we must maximize $\ln p(\boldsymbol{\theta}|\mathbf{Z})$ in (14-18).

Note that to find $\hat{\boldsymbol{\theta}}^a_{\text{BLU}}(k)$, in Lesson 13, we had to minimize $J_a[\hat{\boldsymbol{\theta}}^a(k)]$ in (13-51), which is repeated here for the convenience of the reader:

$$\begin{aligned} J_a[\hat{\boldsymbol{\theta}}^a(k)] &= \tilde{\mathbf{Z}}'_a(k)\mathbf{R}_a^{-1}(k)\tilde{\mathbf{Z}}_a(k) \\ &= (\mathbf{m}_\theta - \hat{\boldsymbol{\theta}}^a(k))'\mathbf{P}_\theta^{-1}(\mathbf{m}_\theta - \hat{\boldsymbol{\theta}}^a(k)) + \tilde{\mathbf{Z}}'(k)\mathbf{R}^{-1}(k)\tilde{\mathbf{Z}}(k) \end{aligned} \tag{14-19}$$

But

$$\ln p(\boldsymbol{\theta}|\mathbf{Z}) \propto -J_a[\hat{\boldsymbol{\theta}}^a(k)] \tag{14-20}$$

This means that maximizing $\ln p(\boldsymbol{\theta}|\mathbf{Z})$ leads to the same value of $\hat{\boldsymbol{\theta}}$ as does minimizing $J_a[\hat{\boldsymbol{\theta}}^a(k)]$. We have, therefore, shown that $\hat{\boldsymbol{\theta}}_{\text{MAP}}(k) = \hat{\boldsymbol{\theta}}^a_{\text{BLU}}(k)$.

**Theorem 14-2.** *For the linear Gaussian model, when* $\mathbf{H}$(k) *is deterministic it is always true that*

$$\hat{\boldsymbol{\theta}}_{\text{MAP}}(k) = \hat{\boldsymbol{\theta}}^a_{\text{BLU}}(k) \ \square \tag{14-21}$$

Combining the results in Theorems 13-4 and 14-2, we have the very important result that *for the linear Gaussian model*

$$\hat{\boldsymbol{\theta}}_{\text{MS}}(k) = \hat{\boldsymbol{\theta}}^a_{\text{BLU}}(k) = \hat{\boldsymbol{\theta}}_{\text{MAP}}(k) \tag{14-22}$$

Put another way, for the linear Gaussian model, all roads lead to the same estimator.

Of course, the fact that $\hat{\boldsymbol{\theta}}_{\text{MS}}(k) = \hat{\boldsymbol{\theta}}_{\text{MAP}}(k)$ should not come as any surprise, because we already established it (in a model-free environment) in Theorem 14-1. To use Theorem 14-1, we made use of the fact that $\mathbf{Z}(k)$ and $\boldsymbol{\theta}$ are jointly Gaussian for the generic linear model. This was proved just after Equation (13-19).

**EXAMPLE 14-2 (Maximum-likelihood Deconvolution)**

As in Example 13-1, we begin with the deconvolution linear model

$$\mathbf{Z}(N) = \mathbf{H}(N-1)\boldsymbol{\mu} + \mathbf{V}(N) \tag{14-23}$$

Now, however, we use the product model for $\boldsymbol{\mu}$, given in (13-32), to express $\mathbf{Z}(N)$ as

$$\mathbf{Z}(N) = \mathbf{H}(N-1)\mathbf{Q}_\mathbf{q}\mathbf{r} + \mathbf{V}(N) \tag{14-24}$$

For notational convenience, let

$$\mathbf{q} = \text{col}\,(q(1), q(2), \ldots, q(N)) \tag{14-25}$$

Our objectives in this example are to obtain MAP estimators for both $\mathbf{q}$ and $\mathbf{r}$. In the literature on maximum-likelihood deconvolution (e.g., Mendel, 1983, 1990) these estimators are referred to as unconditional ML estimators and are denoted $\hat{\mathbf{r}}$ and $\hat{\mathbf{q}}$. We denote these estimators as $\hat{\mathbf{r}}_{\text{MAP}}$ and $\hat{\mathbf{q}}_{\text{MAP}}$ in order to be consistent with this book's notation.

The starting point for determining $\hat{\mathbf{q}}_{\text{MAP}}$ and $\hat{\mathbf{r}}_{\text{MAP}}$ is the unconditional likelihood function (see, also, Mendel, 1991)

$$l(\mathbf{q}, \mathbf{r}|\mathbf{Z}) = p(\mathbf{Z}, \mathbf{q}, \mathbf{r}) = p(\mathbf{Z}|\mathbf{q}, \mathbf{r})p(\mathbf{q}, \mathbf{r}) = p(\mathbf{Z}|\mathbf{q}, \mathbf{r})\,\Pr(\mathbf{q})p(\mathbf{r}) \tag{14-26}$$

where we have used the facts that $\mathbf{q}$ and $\mathbf{r}$ are independent and $\mathbf{q}$ is a vector of *discrete* random variables. Although this looks like a very formidable expression, it is relatively easy to evaluate each of the three probability functions on its right-hand side. The keys to doing this are the vector measurement equation (14-24), the Bernoulli nature of the elements of $\mathbf{q}$, and the Gaussian natures of the elements of $\mathbf{r}$ and $\mathbf{V}(N)$. From these facts, it is easy to establish that [note that $\text{cov}\,[\mathbf{V}(N)] = \rho\mathbf{I}$]

$$p(\mathbf{Z}|\mathbf{q}, \mathbf{r}) = (2\pi\rho)^{-N/2} \exp\left[-\frac{(\mathbf{Z} - \mathbf{H}\mathbf{Q}_{\mathbf{q}}\mathbf{r})'(\mathbf{Z} - \mathbf{H}\mathbf{Q}_{\mathbf{q}}\mathbf{r})}{2\rho}\right] \tag{14-27}$$

$$\Pr(\mathbf{q}) = \prod_{k=1}^{N} \Pr[q(k)] = \lambda^{m(q)}(1-\lambda)^{[N-m(q)]} \tag{14-28}$$

in which

$$m(q) = \sum_{k=1}^{N} q(k) \tag{14-29}$$

and

$$p(\mathbf{r}) = (2\pi\sigma_r^2)^{-N/2} \exp\left(-\frac{\mathbf{r}'\mathbf{r}}{2\sigma_r^2}\right) \tag{14-30}$$

Putting all this together, we obtain the final expression for $l(\cdot)$:

$$\begin{aligned} l(\mathbf{q}, \mathbf{r}|\mathbf{Z}) &= (2\pi)^{-N}(\rho\sigma_r^2)^{-N/2} \exp\left[-\frac{(\mathbf{Z} - \mathbf{H}\mathbf{Q}_{\mathbf{q}}\mathbf{r})'(\mathbf{Z} - \mathbf{H}\mathbf{Q}_{\mathbf{q}}\mathbf{r})}{2\rho} - \frac{\mathbf{r}'\mathbf{r}}{2\sigma_r^2}\right] \\ &\quad \times \lambda^{m(q)}(1-\lambda)^{[N-m(q)]} \end{aligned} \tag{14-31}$$

Because of the strong exponential dependence on the right-hand side of (14-31), it is easier to work with the log-likelihood function $L(\cdot)$, i.e., with

$$\begin{aligned} L(\mathbf{q}, \mathbf{r}|\mathbf{Z}) &= -\frac{(\mathbf{Z} - \mathbf{H}\mathbf{Q}_{\mathbf{q}}\mathbf{r})'(\mathbf{Z} - \mathbf{H}\mathbf{Q}_{\mathbf{q}}\mathbf{r})}{2\rho} - \frac{\mathbf{r}'\mathbf{r}}{2\sigma_r^2} \\ &\quad + m(q)\ln\lambda + [N - m(q)]\ln(1-\lambda) \end{aligned} \tag{14-32}$$

where we have dropped the terms that do not depend on $\mathbf{q}$ or $\mathbf{r}$. Values of $\mathbf{q}$ and $\mathbf{r}$ that maximize $L(\mathbf{q}, \mathbf{r}|\mathbf{Z})$ are $\hat{\mathbf{q}}_{\text{MAP}}(N)$ and $\hat{\mathbf{r}}_{\text{MAP}}(N)$.

Observe from (14-32) that $L(\mathbf{q}, \mathbf{r}|\mathbf{Z})$ is quadratic in $\mathbf{r}$; hence, we expect $\hat{\mathbf{r}}_{\text{MAP}}$ to be a linear function of the measurements in $\mathbf{Z}$. Indeed, a direct minimization of $L(\mathbf{q}, \mathbf{r}|\mathbf{Z})$ with respect to $\mathbf{r}$ leads to the following equation (Problem 14-1) for $\hat{\mathbf{r}}_{\text{MAP}}$:

$$\hat{\mathbf{r}}_{\text{MAP}}(N) = \sigma_r^2 \mathbf{Q_q} \mathbf{H}'[\sigma_r^2 \mathbf{H} \mathbf{Q_q} \mathbf{H}' + \rho \mathbf{I}]^{-1} \mathbf{Z} \tag{14-33}$$

Of course, we can only compute the right-hand side of this equation if we are given $\mathbf{Q_q}$; hence, $\hat{\mathbf{r}}_{\text{MAP}}(N)$ is actually conditioned on $\mathbf{Q_q}$, i.e., $\hat{\mathbf{r}}_{\text{MAP}}(N) = \hat{\mathbf{r}}_{\text{MAP}}(N|\mathbf{Q_q})$. Observe, also, that because $\mathbf{Q_q}^2 = \mathbf{Q_q}$ and $\boldsymbol{\mu} = \mathbf{Q_q r}$ then

$$\begin{aligned} \hat{\boldsymbol{\mu}}_{\text{MAP}}(N|\mathbf{Q_q}) = \mathbf{Q_q} \hat{\mathbf{r}}_{\text{MAP}}(N|\mathbf{Q_q}) &= \sigma_r^2 \mathbf{Q_q}^2 \mathbf{H}'[\sigma_r^2 \mathbf{H} \mathbf{Q_q} \mathbf{H}' + \rho \mathbf{I}]^{-1} \mathbf{Z} \\ &= \sigma_r^2 \mathbf{Q_q} \mathbf{H}'[\sigma_r^2 \mathbf{H} \mathbf{Q_q} \mathbf{H}' + \rho \mathbf{I}]^{-1} \mathbf{Z} \\ &= \hat{\mathbf{r}}_{\text{MAP}}(N|\mathbf{Q_q}) \end{aligned} \tag{14-34}$$

Additionally, comparing the second line of (14-34) and (13-36), we see that $\hat{\boldsymbol{\mu}}_{\text{MAP}}(N|\mathbf{Q_q}) = \hat{\boldsymbol{\mu}}_{\text{MS}}(N|\mathbf{Q_q})$. Of course, when $\mathbf{Q_q}$ is given, (14-24) becomes a generic linear and Gaussian model, so Theorem 14-1 applies, and the fact that $\hat{\boldsymbol{\mu}}_{\text{MAP}}(N|\mathbf{Q_q}) = \hat{\boldsymbol{\mu}}_{\text{MS}}(N|\mathbf{Q_q})$ is no big surprise.

Substituting (14-33) into $L(\mathbf{q}, \mathbf{r}|\mathbf{Z})$ as given by (14-32), we obtain $L(\mathbf{q}, \hat{\mathbf{r}}_{\text{MAP}}(N|\mathbf{Q_q})|\mathbf{Z})$, which is just a function of $\mathbf{q}$. We let $M(\mathbf{q}|\mathbf{Z})$ denote $L(\mathbf{q}, \hat{\mathbf{r}}_{\text{MAP}}(N|\mathbf{Q_q})|\mathbf{Z})$. With a little bit of algebra, it is possible to show that (Problem 14-1)

$$M(\mathbf{q}|\mathbf{Z}) = -\frac{1}{2} \mathbf{Z}' \boldsymbol{\Omega}^{-1} \mathbf{Z} + m(q) \ln \lambda + [N - m(q)] \ln(1 - \lambda) \tag{14-35}$$

where

$$\boldsymbol{\Omega} = \mathbf{E}\{\mathbf{Z}\mathbf{Z}'|\mathbf{Q_q}) = \sigma_r^2 \mathbf{H} \mathbf{Q_q} \mathbf{H}' + \rho \mathbf{I} \tag{14-36}$$

Note that $M(\mathbf{q}|\mathbf{Z})$ is not a log-likelihood function; it is an *objective function* that can be maximized to determine $\hat{\mathbf{q}}_{\text{MAP}}(N)$. Unfortunately, there is no simple solution for determining $\hat{\mathbf{q}}_{\text{MAP}}(N)$. Because the elements of $\mathbf{q}$ in $M(\mathbf{q}|\mathbf{Z})$ have nonlinear interactions and because the elements of $\mathbf{q}$ are constrained to take on binary values, it is necessary to evaluate $M(\mathbf{q}|\mathbf{Z})$ for every possible $\mathbf{q}$ sequence to find the $\mathbf{q}$ for which $M(\mathbf{q}|\mathbf{Z})$ is a global maximum. Because $q(k)$ can take on one of two possible values, there are $2^N$ possible sequences where $N$ is the number of elements in $\mathbf{q}$. For reasonable values of $N$ (such as $N = 400$), finding the global maximum of $M(\mathbf{q}|\mathbf{Z})$ would require several centuries of computer time.

We can always design a method for *detecting* significant values of $\mu(k)$ so that the resulting $\hat{\mathbf{q}}$ will be nearly as likely as the unconditional maximum-likelihood estimate, $\hat{\mathbf{q}}_{\text{MAP}}$. Two MAP detectors for accomplishing this are described in Mendel (1990, pp. 143–150). Some elementary aspects of binary detection are contained in the Supplementary Material at the end of this lesson. □

In summary, we have shown that a *separation principle* exists for determination of $\hat{\mathbf{r}}_{\text{MAP}}$ and $\hat{\mathbf{q}}_{\text{MAP}}$; i.e., first we must determine $\hat{\mathbf{q}}_{\text{MAP}}$, after which $\hat{\mathbf{r}}_{\text{MAP}}$ can be computed using (14-34).

Our last example in this lesson is the very important one of *state estimation*, which, as we showed in Example 2-4, is equivalent to estimating a vector of random parameters.

**EXAMPLE 14-3 (State Estimation)** (See lesson D)

In Example 2-4 we showed that, for the application of state estimation, our linear model is

$$\mathbf{Z}(N) = \mathbf{H}(N, k_1)\mathbf{x}(k_1) + \mathbf{V}(N, k_1) \tag{14-37}$$

From (2-17), we see that $x_{k+1} = \phi x_k + \Gamma u(k)$

$z_{k+1} = h^T x_{k+1} + v_{k+1}$

$$\mathbf{x}(k_1) = \mathbf{\Phi}^{k_1}\mathbf{x}(0) + \sum_{i=1}^{k_1} \mathbf{\Phi}^{k_1-i}\gamma u(i-1) = \mathbf{\Phi}^{k_1}\mathbf{x}(0) + \mathbf{L}\mathbf{u} \tag{14-38}$$

where

$$\mathbf{u} = \text{col}\,(u(0), u(1), \ldots, u(N)) \tag{14-39}$$

and $\mathbf{L}$ is an $n \times (N+1)$ matrix, the exact structure of which is not important for this example. Additionally, from (2-21), we see that

$$\mathbf{V}(N, k_1) = \mathbf{M}(N, k_1)\mathbf{u} + \mathbf{v} \tag{14-40}$$

where

$$\mathbf{v} = \text{col}\,(\nu(1), \nu(2), \ldots, \nu(N)) \tag{14-41}$$

Observe that both $\mathbf{x}(k_1)$ and $\mathbf{V}(N, k_1)$ can be viewed as linear functions of $\mathbf{x}(0)$, $\mathbf{u}$, and $\mathbf{v}$.

We now assume that $\mathbf{x}(0)$, $\mathbf{u}$, and $\mathbf{v}$ are jointly Gaussian. Reasons for doing this are discussed in Lesson 15. Then, because $\mathbf{x}(k_1)$ and $\mathbf{V}(N, k_1)$ are linear functions of $\mathbf{x}(0)$, $\mathbf{u}$, and $\mathbf{v}$, $\mathbf{x}(k_1)$ and $\mathbf{V}(N, k_1)$ are jointly Gaussian (Papoulis, 1991, p. 197); consequently, $\mathbf{Z}(N)$ and $\mathbf{x}(k_1)$ are jointly Gaussian. Hence,

$$\begin{aligned}\hat{\mathbf{x}}_{\text{MS}}(k_1|N) = \mathbf{m}_{\mathbf{x}(\mathbf{k}_1)} + \mathbf{P}_{\mathbf{x}(\mathbf{k}_1)}\mathbf{H}'(N, k_1)[\mathbf{H}(N, k_1)\mathbf{P}_{\mathbf{x}(\mathbf{k}_1)}\mathbf{H}'(N, k_1) \\ + \mathbf{R}(N, k_1)]^{-1}[\mathbf{Z}(N) - \mathbf{H}(N, k_1)\mathbf{m}_{\mathbf{x}(k_1)}]\end{aligned} \tag{14-42}$$

and

$$\hat{\mathbf{x}}_{\text{MAP}}(k_1|N) = \hat{\mathbf{x}}_{\text{MS}}(k_1|N) \tag{14-43}$$

To evaluate $\hat{\mathbf{x}}_{\text{MS}}(k_1|N)$, we must first compute $\mathbf{m}_{\mathbf{x}(k_1)}$ and $\mathbf{P}_{\mathbf{x}(k_1)}$. We show how to do this in Lesson 15.

Formula (14-42) is very cumbersome. It appears that its right-hand side changes as a function of $k_1$ (and $N$). *We conjecture, however, that it ought to be possible to express* $\hat{\mathbf{x}}_{\text{MS}}(k_1|N)$ *as an affine transformation of* $\hat{\mathbf{x}}_{\text{MS}}(k_1 - 1|N)$, because $\mathbf{x}(k_1)$ is an affine transformation of $\mathbf{x}(k_1 - 1)$, i.e., $\mathbf{x}(k_1) = \mathbf{\Phi}\mathbf{x}(k_1 - 1) + \gamma u(k_1 - 1)$.

Because of the importance of state estimation in many different fields (e.g., control theory, communication theory, and signal processing), we shall examine it in great detail in many of our succeeding lessons. □

**Comment.** This completes our studies of methods for estimating random parameters.

## COMPUTATION

In the general MAP case, we reach an unconditional likelihood function that has to be maximized. See the discussions in Lesson 11 on how to do this.

In the case of jointly Gaussian random vectors, $\hat{\theta}_{\text{MAP}}(k) = \hat{\theta}_{\text{MS}}(k)$, so that computation of $\hat{\theta}_{\text{MAP}}(k)$ reverts to the computation of $\hat{\theta}_{\text{MS}}(k)$, which is discussed in Lesson 13.

## Supplementary Material

### ELEMENTS OF BINARY DETECTION THEORY*

A detection problem is composed of four elements (Melsa and Cohn, 1978): message, signal, observation, and decision (see Fig. 14-1). Here we shall focus on the binary situation when there are only two possible cases at each $t_k$: a message is present or a message is not present. Our *message space* $\mathcal{M}$ therefore contains only two elements: $m_1$ (there is a message) and $m_2$ (there is no message). We assume that the occurrence of a message is a random event and that a priori probabilities $\Pr(m_1)$ and $\Pr(m_2)$ are specified.

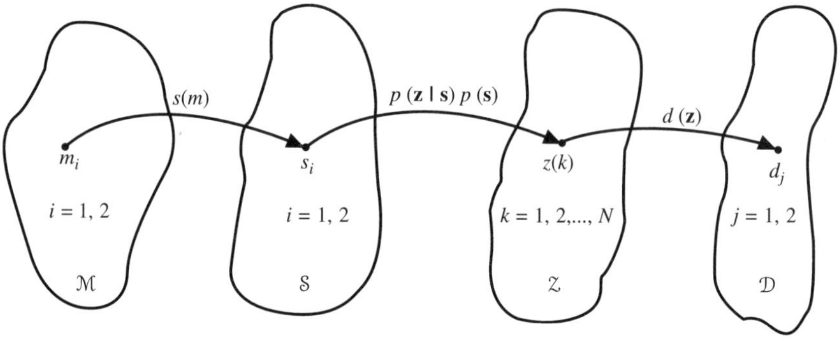


**Figure 14-1** Elements of a detection problem: $\mathcal{M}$ = message space, $\mathcal{S}$ = signal space, $\mathcal{Z}$ = observation space, $\mathcal{D}$ = decision space; $s(m)$ = signaling mechanism, $p(\mathbf{z}|\mathbf{s})p(\mathbf{s})$ = probabilistic transition mechanism (e.g., convolutional model), $d(\mathbf{z})$ = detection rule. (Adapted from Fig. 1.1-2 of Melsa and Cohn, 1978.)

A *signal space* isolates the portion of the problem where information is generated from the portion where that information is transmitted. Our *signal space* $\mathcal{S}$ also contains only two elements: $s_1 \triangleq q(k) = 1$ and $s_2 \triangleq q(k) = 0$. We assume that there is a unique and invertible mapping between elements of our message and signal spaces:

$$\begin{aligned} q(k) &= 1 \Leftrightarrow m_1 \text{ (there is a message)} \\ q(k) &= 0 \Leftrightarrow m_2 \text{ (there is no message)} \end{aligned} \tag{14-44}$$

Our observations $\mathbf{Z} = \text{col}(z(1), \ldots, z(N))$ constitute the *observation space* $\mathcal{Z}$. We assume that a given mathematical model connects the signal and observation spaces, e.g., convolutional model or state-variable model. For example, (14-23) is the concatenated convolutional model that is the starting point for deconvolution. From (14-23)–(14-25), observe that $\mathbf{Z}$ depends on $N$ elements from $\mathcal{S}$: $\mathbf{s} \triangleq \mathbf{q} = \text{col}(q(1), q(2), \ldots, q(N))$. We refer to $\mathbf{s}$ as a vector point in $\mathcal{S}$. Signals are mapped into the observation space by means of a probabilistic transition mechanism $p(\mathbf{Z}, \mathbf{s}) = p(\mathbf{Z}|\mathbf{s})p(\mathbf{s})$. This mechanism is our mathematical model.

*The material in this section is taken from Mendel, 1983, pp. 123-127.

Finally, a decision must be reached. In our problem we must decide, at each $t_k$, whether or not a message is present; thus, our *decision space* $\mathcal{D}$ consists of the same set of points as our message space. More specifically,

$$d_1 \Rightarrow m_1, \qquad d_2 \Rightarrow m_2 \tag{14-45}$$

Because $\mathcal{D}$ contains only two elements, it is a *binary decision space.*

The relationship between the observation space and decision space is called the *detection rule* $d(\mathbf{Z})$. Because there are only two decisions, we must divide $\mathcal{Z}$ into two decision regions $\mathcal{Z}_1$ and $\mathcal{Z}_2$ such that $d(\mathbf{Z}) = d_1$ if $\mathbf{Z} \in \mathcal{Z}_1$ and $d(\mathbf{Z}) = d_2$ if $\mathbf{Z} \in \mathcal{Z}_2$. As pointed out by Melsa and Cohn (1978), "the regions $\mathcal{Z}_1$ and $\mathcal{Z}_2$ must be disjoint ($\mathcal{Z}_1 \bigcap \mathcal{Z}_2 = \emptyset$) in order that each point in $\mathcal{Z}$ will yield a unique decision. The regions $\mathcal{Z}_1$ and $\mathcal{Z}_2$ must cover $\mathcal{Z}(\mathcal{Z}_1 \bigcup \mathcal{Z}_2 = \mathcal{Z})$ in order that each point in $\mathcal{Z}$ will have a decision associated with it."

Many detection rules are possible (see Melsa and Cohn, 1978, for example). We direct our attention to the unconditional maximum-likelihood (MAP) detection rule for two reasons: (1) it makes use of a priori knowledge about random quantity $q(k)$, and (2) it minimizes the probability of error, as described later.

The MAP decision criterion is: given an observation $\mathbf{Z}$, select $d_1$ if $m_1$ is more likely than $m_2$. The *MAP detection rule* is

$$d(\mathbf{Z}) = \begin{cases} d_1, & \text{if } \Pr(m_1|\mathbf{Z}) > \Pr(m_2|\mathbf{Z}) \\ d_2, & \text{if } \Pr(m_1|\mathbf{Z}) < \Pr(m_2|\mathbf{Z}) \end{cases} \tag{14-46}$$

An equivalent method for representing detection rule (14-46) is to define the decision regions $\mathcal{Z}_1$ and $\mathcal{Z}_2$ as

$$\begin{aligned} \mathcal{Z}_1 &= \{\mathbf{Z} : \Pr(m_1|\mathbf{Z}) > \Pr(m_2|\mathbf{Z})\} \\ \mathcal{Z}_2 &= \{\mathbf{Z} : \Pr(m_1|\mathbf{Z}) < \Pr(m_2|\mathbf{Z})\} \end{aligned} \tag{14-47}$$

The values of $\mathbf{Z}$ for which $\Pr(m_1|\mathbf{Z}) = \Pr(m_2|\mathbf{Z})$ may be arbitrarily assigned to either $\mathcal{Z}_1$ or $\mathcal{Z}_2$. The following shorthand notation is often used to represent Eq. (14-47).

$$\frac{\Pr(m_1|\mathbf{Z})}{\Pr(m_2|\mathbf{Z})} \underset{d_2}{\overset{d_1}{\gtrless}} 1 \tag{14-48}$$

Next we show how to express the left-hand side of Eq. (14-48) in terms of unconditional likelihoods. From Bayes's rule, it follows that

$$\frac{\Pr(m_1|\mathbf{Z})}{\Pr(m_2|\mathbf{Z})} = \frac{\Pr(m_1|\mathbf{Z})p(\mathbf{Z})}{\Pr(m_2|\mathbf{Z})p(\mathbf{Z})} = \frac{p(\mathbf{Z}|m_1)\Pr(m_1)}{p(\mathbf{Z}|m_2)\Pr(m_2)} \tag{14-49}$$

We now define the *unconditional likelihood ratio* $\Lambda(\mathbf{Z})$ as

$$\Lambda(\mathbf{Z}) \triangleq \frac{p(\mathbf{Z}|m_1)\Pr(m_1)}{p(\mathbf{Z}|m_2)\Pr(m_2)} \tag{14-50}$$

so our MAP detection rule in Eq. (14-48) can be written as

$$\Lambda(\mathbf{Z}) \underset{d_2}{\overset{d_1}{\gtrless}} 1 \tag{14-51}$$

Quantity $\Lambda(\mathbf{Z})$ is called an *unconditional likelihood ratio* because its numerator and denominator are each unconditional likelihoods. Compare (14-51) with (10-15) to see the difference between an unconditional likelihood ratio test and a likelihood ratio test.

The expression for the unconditional likelihood ratio can be operated on by any operator that uniquely retains the ordering of $\Lambda(\mathbf{Z})$ relative to unity. The natural logarithm is often a useful operator, especially for Gaussian problems. In that case our decision rule can be written as

$$\ln \Lambda(\mathbf{Z}) \underset{d_2}{\overset{d_1}{\gtrless}} 0 \tag{14-52}$$

As stated by Melsa and Cohn,

> there are two types of errors that we can make in a binary decision problem. First, we may decide $d_2$ when $m_1$ is true, and second we may decide $d_1$ when $m_2$ is true. Each of these errors has a probability associated with it which depends on the decision rule and conditional densities. The following notation will be employed:
>
> $\Pr\{d_2|m_1\}$ = probability of making decision $d_2$ when $m_1$ is true
>
> $\Pr\{d_1|m_2\}$ = probability of making decision $d_1$ when $m_2$ is true

We shall refer to $\Pr\{d_2|m_1\}$ as the *miss probability* and to $\Pr\{d_1|m_2\}$ as the *false-alarm probability*. When $m_1$ is true, $q(k) = 1$, which means there is a message; but when we then decide that there is no message present [i.e., $d(\mathbf{Z}) = d_2$], we are missing the true message. This explains why we refer to $\Pr\{d_2|m_1\}$ as the miss probability. On the other hand, when $m_2$ is true, $q(k) = 0$, which means there is no message; but when we then decide that there is a message present [i.e., $d(\mathbf{Z}) = d_1$], we are inserting a false message. This explains why we refer to $\Pr\{d_1|m_2\}$ as the false-alarm probability.

Melsa and Cohn also note that

> in addition to two errors, there are also two correct decisions that we can make in the binary decision problem. We may decide $d_1$ when $m_1$ is true and we may decide $d_2$ when $m_2$ is true. Again these correct decisions have associated probabilities represented by
>
> $\Pr\{d_1|m_1\}$ = probability of making decision $d_1$ when $m_1$ is true
>
> $\Pr\{d_2|m_2\}$ = probability of making decision $d_2$ when $m_2$ is true

Another important detection rule is based on minimizing the total probability of error $P_e$, defined as

$$\begin{aligned} P_e &\triangleq \text{Pr}\{\text{make an incorrect decision}\} \\ &= \text{Pr}\{\text{decide } d_2 \text{ when } m_1 \text{ is true or decide } d_1 \text{ when } m_2 \text{ is true}\} \\ &= \text{Pr}\{(d_2 \text{ and } m_1) \text{ or } (d_1 \text{ and } m_2)\} \end{aligned} \tag{14-53}$$

Because messages $m_1$ and $m_2$ are mutually exclusive, $P_e$ can be written as

$$P_e = \text{Pr}\{d_2, m_1\} + \text{Pr}\{d_1, m_2\} \tag{14-54}$$

and, by use of conditional probabilities (Papoulis, 1991), this becomes

$$P_e = \text{Pr}\{d_2|m_1\}\,\text{Pr}\{m_1\} + \text{Pr}\{d_1|m_2\}\,\text{Pr}\{m_2\} \tag{14-55}$$

Observe that $P_e$ depends on the two error probabilities.

**Theorem 14-3.** *The probability-of-error detection rule, obtained by minimizing Eq. (14-55), is identical to the unconditional maximum-likelihood detection rule, given in Eq. (14-51).*

*Proof.* See Melsa and Cohn (1978, pp. 38–42). □

The importance of Theorem 14-3 to us is that we can be assured that a MAP detector that is based on (14-51) will give rise to very few false alarms and missed detections.

**EXAMPLE 14-4 (Square Law Detector)**

Here we consider the following *signal plus noise model*:

$$z(k) = s(k) + e(k) = c(k)u(k) + e(k), \qquad k = 1, 2, \ldots, N \tag{14-56}$$

where $s(k)$ is the "signal" and $e(k)$ is the "noise"; $c(k)$ is a time-varying gain; $u(k)$ is described by the following product model:

$$u(k) = r(k)q(k) \tag{14-57}$$

in which $r(k) \sim N(r(k); 0, \sigma_r^2)$, $q(k)$ is a Bernoulli sequence for which $\text{Pr}[q(k) = 1] = \lambda$ and $\text{Pr}[q(k) = 0] = 1 - \lambda$, and $r(k)$ and $q(k)$ are statistically independent; $e(k) \sim N(e(k); 0, \sigma_e^2)$; and $e(k)$ and $u(k)$ are statistically independent. Our goal is to evaluate (14-51) for this model, but at each value of $t_k$; hence, $\Lambda(\mathbf{Z}) \rightarrow \Lambda(k)$. Additionally, $m_1 = (q(k) = 1)$ and $m_2 = (q(k) = 0)$. From (14-50), we see that

$$\Lambda(k) = \frac{p[z(k)|q(k) = 1]\,\text{Pr}[q(k) = 1]}{p[z(k)|q(k) = 0]\,\text{Pr}[q(k) = 0]} \tag{14-58}$$

Hence,

$$\ln \Lambda(k) = \ln\left\{\frac{p[z(k)|q(k) = 1]}{p[z(k)|q(k) = 0]}\right\} + \ln\left\{\frac{\text{Pr}[q(k) = 1]}{\text{Pr}[q(k) = 0]}\right\} \tag{14-59}$$

When $q(k)$ is fixed, $z(k)$ is the sum of two Gaussian random variables; hence, it is also Gaussian, and therefore

$$p[z(k)|q(k)] = [2\pi A_q(k)]^{-1/2} \exp\left[-\frac{z^2(k)}{2A_q(k)}\right] \tag{14-60}$$

where

$$A_q(k) = \mathbf{E}\{z^2(k)|q(k) = q\}, \qquad q = 0 \text{ or } 1 \tag{14-61}$$

We return to the calculation of $A_q(k)$ later. From (14-60) we find that

$$\ln\left\{\frac{p[z(k)|q(k)=1]}{p[z(k)|q(k)=0]}\right\} = \frac{1}{2}\ln\left[\frac{A_0(k)}{A_1(k)}\right] + \frac{1}{2}z^2(k)\left[\frac{1}{A_0(k)} - \frac{1}{A_1(k)}\right] \tag{14-62}$$

in which $A_0$ and $A_1$ are short for $A_{q=0}$ and $A_{q=1}$, respectively. Additionally, because of the Bernoulli nature of $q(k)$,

$$\ln\left\{\frac{\Pr[q(k)=1]}{\Pr[q(k)=0]}\right\} = \ln\left[\frac{\lambda}{1-\lambda}\right] \tag{14-63}$$

Substituting (14-62) and (14-63) into (14-59), we find that the latter can be expressed as

$$\ln\Lambda(k) = \frac{1}{2}z^2(k)\left[\frac{1}{A_0(k)} - \frac{1}{A_1(k)}\right] + \frac{1}{2}\ln\left[\frac{A_0(k)}{A_1(k)}\right] + \ln\left[\frac{\lambda}{1-\lambda}\right] \tag{14-64}$$

Substituting (14-64) into the left-hand side of (14-52), our MAP detection rule can be reexpressed as

$$\begin{aligned}&\text{If } z^2(k) \geq \left\{\frac{A_0(k)A_1(k)}{A_1(k)-A_0(k)}\right\}\left\{\ln\left[\frac{A_1(k)}{A_0(k)}\right] - 2\ln\left[\frac{\lambda}{1-\lambda}\right]\right\} \quad \text{decide } q(k)=1\\ &\text{If } z^2(k) \leq \left\{\frac{A_0(k)A_1(k)}{A_1(k)-A_0(k)}\right\}\left\{\ln\left[\frac{A_1(k)}{A_0(k)}\right] - 2\ln\left[\frac{\lambda}{1-\lambda}\right]\right\} \quad \text{decide } q(k)=0\end{aligned} \tag{14-65}$$

From (14-61) and (14-56) and making use of the facts that $e(k)$ and $u(k)$ are statistically independent and $r(k)$ and $e(k)$ do not depend on $q(k)$, it follows that

$$\begin{aligned}A_q(k) &= \mathbf{E}\{z^2(k)|q(k)=q\} = c^2(k)q^2\mathbf{E}\{r^2(k)\} + \mathbf{E}\{e^2(k)\}\\ &= c^2(k)q^2\sigma_r^2 + \sigma_e^2\end{aligned} \tag{14-66}$$

Observe, therefore, that $A_0(k) = \sigma_e^2$ and $A_1(k) = \sigma_e^2 + c^2(k)\sigma_r^2$. Use these two values in (14-65) to perform the test.

Because (14-65) involves squaring up the measurements, it is often referred to as a *square-law detector*. For an illustration of how estimators can be used to obtain the quantities needed on the right-hand side of (14-65), see Problem 14-13. □

## SUMMARY QUESTIONS

1. The difference between *conditional* and *unconditional* likelihood functions is:
   **(a)** the constant of proportionality
   **(b)** the random nature of $\theta$ is accounted for in the latter
   **(c)** an exponential density function
2. For the linear Gaussian model, when $\mathbf{H}(k)$ is deterministic:
   **(a)** $\hat{\theta}_{MS}(k) = \hat{\theta}_{BLU}(k) = \hat{\theta}_{MAP}(k)$

(b) $\hat{\theta}_{MS}(k) = \hat{\theta}_{WLS}(k) = \hat{\theta}_{MAP}(k)$
(c) $\hat{\theta}_{MS}(k) = \hat{\theta}^{a}_{BLU}(k) = \hat{\theta}_{MAP}(k)$

3. The a priori information about $\boldsymbol{\theta}$ can be included in a weighted least-squares objective function by adding which term to the usual objective function?
(a) $\tilde{\mathbf{Z}}'(k)\mathbf{R}^{-1}(k)\tilde{\mathbf{Z}}(k)$
(b) $(\mathbf{m}_\theta - \boldsymbol{\theta})'\mathbf{R}^{-1}(k)(\mathbf{m}_\theta - \boldsymbol{\theta})$
(c) $(\mathbf{m}_\theta - \boldsymbol{\theta})'\mathbf{P}_\theta^{-1}(k)(\mathbf{m}_\theta - \boldsymbol{\theta})$

4. $\hat{\theta}_{MAP}(k) = \hat{\theta}_{MS}(k)$ if:
(a) the model is arbitrary, and $\mathbf{Z}(k)$ and $\boldsymbol{\theta}$ are jointly Gaussian
(b) $\mathbf{Z}(k)$ is Gaussian
(c) $\theta_1, \theta_2, \ldots, \theta_n$ are uniformly distributed

5. In state estimation, $\hat{\mathbf{x}}_{MAP}(k_1|N) = \hat{\mathbf{x}}_{MS}(k_1|N)$, if $\mathbf{x}(0)$, $\mathbf{u}$, and $\mathbf{v}$ are:
(a) random
(b) jointly Gaussian
(c) statistically independent

6. In MLD, $\mathbf{r}$ can be estimated:
(a) before $\hat{\mathbf{q}}_{MAP}$ is found, using MVD
(b) while $\mathbf{q}$ is being estimated
(c) after $\hat{\mathbf{q}}_{MAP}$ is found, using MVD

7. A detection problem is composed of which elements?
(a) message space
(b) transformation space
(c) observation space
(d) resolution space
(e) decision space
(f) signal space

8. The MAP decision criterion is:
(a) select decision 1 if message 1 is more likely than message 2, regardless of the measurements
(b) given an observation, select decision 1 if message 1 is less likely than message 2
(c) given an observation, select decision 1 if message 1 is more likely than message 2

9. Which are the two types of error that can be made in a binary decision problem?
(a) miss probability
(b) capture probability
(c) failure probability
(d) false-alarm probability

10. The unconditional maximum-likelihood detection rule is powerful, because it is identical to the:
(a) probability of miss detection rule
(b) probability of error detection rule
(c) probability of false-alarm detection rule

## PROBLEMS

**14-1.** Show that $\hat{\mathbf{r}}_{MAP}(N)$ is given by Eq. (14-33) and that $\hat{\mathbf{q}}_{MAP}$ can be found by maximizing (14-35).

**14-2.** For the linear Gaussian model in which $\mathbf{H}(k)$ is deterministic, prove that $\hat{\boldsymbol{\theta}}_{\text{MAP}}(k)$ is a most efficient estimator of $\boldsymbol{\theta}$. Do this in two different ways. Is $\hat{\boldsymbol{\theta}}_{\text{MS}}(k)$ a most efficient estimator of $\boldsymbol{\theta}$?

**14-3.** $x$ and $\nu$ are independent Gaussian random variables with zero means and variances $\sigma_x^2$ and $\sigma_\nu^2$, respectively. We observe the single measurement $z = x + \nu = 1$.
**(a)** Find $\hat{x}_{\text{ML}}$.
**(b)** Find $\hat{x}_{\text{MAP}}$.

**14-4.** Consider the linear model $z(k) = 2\theta + n(k)$, where

$$p[n(k)] = \begin{cases} 1 - |n(k)|, & |n(k)| \leq 1 \\ 0, & \text{otherwise} \end{cases}$$

and

$$p(\theta) = \begin{cases} \frac{1}{2}, & |\theta| \leq 1 \\ 0, & \text{otherwise} \end{cases}$$

A random sample of $N$ measurements is available. Explain how to find the ML and MAP estimators of $\theta$, and be sure to list all the assumptions needed to obtain a solution (they are not all given).

**14-5.** (Charles Pickman, Spring 1992) When comparing ML and MAP estimators, there are various trade-offs; e.g., (1) the ML approach is sometimes selected because it does not require arbitrary prior probability density function assumptions; (2) the MAP method gives a quantitative measure of the cost–benefit of additional measurements; (3) the MAP estimator is more costly because it requires an "a priori" measurement system (Kashyap and Rao, 1976, pp. 136–137). This problem looks at all costs, treating estimation error as only one type of "cost." Initially, more estimation error is accepted in order to gain (assumed) faster convergence of the MAP estimator.

Consider the quantitative precision of the ML versus the MAP estimators. Assume that a zero-mean Gaussian source $\mu$ with variance $\sigma_\mu^2$ generates the mean value input to a second cascade Gaussian source $z$ with variance $\sigma_z^2$. Also, assume there is a relatively expensive way to determine the a priori probability density function of $\mu$, $p(\mu)$ as a function of data length $N$. As $N$ increases, the cost of determining $p(\mu)$ increases proportionately.

Our goal is to design an estimator for the parameter $\mu$ that uses a combination of MAP and ML estimators throughout the entire experiment and $\mu$-estimation process.
**(a)** Generate a formula for the mean estimator error, $\hat{\mu}_{\text{ML}} - \hat{\mu}_{\text{MAP}}$, as a function of $N$, $\sigma_\mu^2$, and $\sigma_z^2$ (assume that the MLE is unbiased). Also, determine the percentage error of $\hat{\mu}_{\text{MAP}}$ versus $\hat{\mu}_{\text{ML}}$ as a function of $N$, $\sigma_\mu^2$, and $\sigma_z^2$.
**(b)** The MAP estimator is useful at the beginning of an estimation procedure in order to quantify various measurement costs versus other costs; but, after a specific number of measurements and related $p(\mu)$ calculations, the cheaper MLE can be used for the remainder of the estimation procedure. Assuming the following relative variance values and percentage error switchover point, determine the number of MAP estimations ($N$) required before switching over to MLE: (1) $\sigma_\mu^2 = \frac{1}{4}\sigma_z^2$, percent switchover error= 50%, 10%, and 1% (i.e., determine three values of $N$); and, (2) $\sigma_\mu^2 = \sigma_z^2$, percent switchover error= 50%, 10%, and 1% (i.e., determine three values of $N$).
**(c)** Comment on the results from parts (b1) and (b2). As the source gets noisier (i.e., $\sigma_\mu^2$ increases), which estimator (ML or MAP) is best, based on the switchover point and equivalent estimation error?

**14-6.** (Chanil Chung, Spring 1992) A compound random number generator has the following density functions associated with it:

$$p(z(i)|a) = \frac{1}{a} \times \exp\left[-\frac{z(i)-a}{a}\right]$$

$$p(a) = \frac{1}{a} \times \exp(1)$$

The mean of $z(i)$ is $a$ and the variance of $z(i)$ is $a^2$. $N$ independent measurements $\{z(1), z(2), \ldots, z(N)\}$ are made at the output of the random number generator. Compute the MAP estimator of $a$.

**14-7.** (Michiel van Nieustadt, Spring 1992) Consider a series of $N$ observations $x_i$ with an exponential density function $p(x) = ae^{-ax}$. The decay parameter $a$ has the Rayleigh distribution with parameter $\sigma$; i.e.,

$$p(a) = 2a\sigma \exp(-\sigma a^2), \qquad \text{where } a \in (0, \infty)$$

**(a)** Find the MLE of $a$.
**(b)** Find the MAP estimator of $a$, and verify that the effect of $\sigma$ is intuitively correct.

**14-8.** (Brad Verona, Spring 1992) We observe a random sample $z(1), z(2), \ldots, z(N)$ of an i.i.d. random variable $z$ that is exponentially distributed with parameter $\theta$; i.e.,

$$p(z(i)|\theta) = \theta \exp[-z(i)\theta], \qquad \text{where } \theta \geq 0 \text{ and } z(i) \geq 0$$

Additionally, $\theta$ is a random variable with prior probability density function

$$p(\theta) = \theta \exp(-2\theta), \qquad \text{where } \theta \geq 0$$

Assume that the means and variances of $z(i)$ and $\theta$ are known.
**(a)** Determine $\hat{\theta}_{\text{MAP}}(N)$.
**(b)** Consider the sample mean, which is given as

$$\bar{s} = \frac{1}{N}\sum_{i=1}^{N} z(i)$$

As $N$ gets large, how does $\hat{\theta}_{\text{MAP}}(N)$ compare with $\bar{s}$?

**14-9.** (G. Caso, Fall 1991) In communications, the channel output is often modeled as

$$z(k) = \sum_{i=0}^{L-1} h(i)d(k-i) + \nu(k)$$

where $\{d(k)\}$ is the data symbol sequence, $\{h(i)\}$ is the channel impulse response (assumed to be limited to $L$ samples), and $\{\nu(k)\}$ is additive white Gaussian noise (AWGN) with variance $\sigma_\nu^2$. The sequence $\{d(k)\}$ is formed by independent selections from a finite alphabet of equally likely symbols. The optimal detector (with respect to minimizing the probability of error) assigns the estimated sequence $\{\hat{d}(k)\}$ as the one that maximizes the a posteriori probability; i.e., it is the MAP estimator of $\{d(k)\}$. Since $\{\nu(k)\}$ is Gaussian and $\{h(i)\}$ is assumed to be deterministic and known, $p(\mathbf{Z}(N)|\boldsymbol{\theta}(N))$ is Gaussian with mean given by $\mathbf{m}(N) = \mathbf{H}(N)\boldsymbol{\theta}(N)$ and covariance given by $\mathbf{R}(N) = \sigma_\nu^2\mathbf{I}$, where $\mathbf{Z}(N) = \text{col}\,(z(N), \ldots, z(1))$, $\boldsymbol{\theta}(N) = \text{col}\,(d(N), \ldots, d(1))$, and

$$\mathbf{H}(N) = \begin{bmatrix} h(0) & h(1) & h(2) & \cdots & h(L-1) & 0 & 0 & \cdots & 0 \\ 0 & h(0) & h(1) & \cdots & h(L-2) & h(L-1) & 0 & \cdots & 0 \\ 0 & 0 & h(0) & \cdots & h(L-3) & h(L-2) & h(L-1) & \cdots & 0 \\ \cdot & \cdot & \cdot & \cdots & \cdot & \cdot & \cdot & \cdots & \cdot \\ \cdot & \cdot & \cdot & \cdots & \cdot & \cdot & \cdot & \cdots & \cdot \\ \cdot & \cdot & \cdot & \cdots & \cdot & \cdot & \cdot & \cdots & \cdot \\ 0 & 0 & 0 & \cdots & 0 & 0 & 0 & \cdots & h(0) \end{bmatrix}$$

Thus,

$$p(\mathbf{Z}(N)|\boldsymbol{\theta}(N)) = [(2\pi)^N N\sigma_v^2]^{-1/2} \exp\left\{-\frac{[\mathbf{Z}(N) - \mathbf{H}(N)\boldsymbol{\theta}(N)]'[\mathbf{Z}(N) - \mathbf{H}(N)\boldsymbol{\theta}(N)]}{2\sigma_v^2}\right\}$$

and

$$p(\boldsymbol{\theta}(N)) = \frac{1}{K^N}$$

for all sequences, where $K$ is the number of symbols in the alphabet of symbols.

Assuming that $K = 2$ with $d(k) = \pm 1$, determine the MAP estimate of $\boldsymbol{\theta}(4) = \text{col}\,(d(4), d(3), d(2), d(1))$ when $\mathbf{Z}(4) = \text{col}\,(-0.5, 0.9, -1.7, -0.5)$ and $L = 4$ with $h(0) = 1, h(1) = 0.2, h(2) = 0.04$, and $h(3) = 0.0008$. [*Hint:* Determine which of the possible $2^4$ sequences maximizes $p(\mathbf{Z}(4))$ by brute force; i.e., evaluate $p(\mathbf{Z}(4))$ for each sequence.]

**14-10.** The signal-to-noise-ratio (SNR) for the model in (14-56) is defined as the ratio of the variance of the signal to the variance of the additive measurement noise.

**(a)** Show that $\text{SNR} = c^2(k)\lambda\sigma_r^2/\sigma_e^2$.

**(b)** Show that in the square-law detector $A_1/A_0 = 1 + \text{SNR}/\lambda$ and $A_1A_0/(A_1 - A_0) = \sigma_e^2(1 + \lambda/\text{SNR})$.

**(c)** What happens to the decision rule in (14-65) as $\text{SNR} \to \infty$?

**14-11.** Suppose $q(k) = \pm 1$ in the square-law detector. What happens to the decision rule (14-65)? Interpret the meaning of this result. What has gone wrong and can it be fixed?

**14-12.** Beginning with the second line of (14-34) and using the fact that $\tilde{\boldsymbol{\mu}}_{\text{MAP}}(N|\mathbf{Q_q}) = \boldsymbol{\mu} - \hat{\boldsymbol{\mu}}_{\text{MAP}}(N|\mathbf{Q_q})$, show that:

**(a)** $\text{Covar}\,[\tilde{\boldsymbol{\mu}}_{\text{MAP}}(N|\mathbf{Q_q})] = \sigma_r^2\mathbf{Q_q} - \sigma_r^4\mathbf{Q_q}\mathbf{H}'[\sigma_r^2\mathbf{H}\mathbf{Q_q}\mathbf{H}' + \rho\mathbf{I}]^{-1}\mathbf{H}\mathbf{Q_q}$

**(b)** $\text{Var}\,[\tilde{\mu}_{\text{MAP}}(k|N)] = \sigma_r^2 q(k) - \sigma_r^4 q(k)\mathbf{h}_k'[\sigma_r^2\mathbf{H}\mathbf{Q_q}\mathbf{H}' + \rho\mathbf{I}]^{-1}\mathbf{h}_k$, where $\mathbf{h}_k$ is the $k$th column of $\mathbf{H}$.

**14-13.** Here we study square-law detection for seismic deconvolution when $z(k)$ is replaced by $\hat{\mu}_{\text{MAP}}(k|N)$. This is permissible because $\hat{\mu}_{\text{MAP}}(k|N)$ is, indeed, a function of the measurements. To begin, refer to (14-34), in which $\hat{\boldsymbol{\mu}}_{\text{MAP}}(N|\mathbf{Q_q}) = \text{col}\,[\hat{\mu}_{\text{MAP}}(1|N), \hat{\mu}_{\text{MAP}}(2|N), \ldots, \hat{\mu}_{\text{MAP}}(N|N)]$, and where, for notational convenience, we omit the explicit dependence of the components of $\hat{\boldsymbol{\mu}}_{\text{MAP}}(N|\mathbf{Q_q})$ on $\mathbf{Q_q}$.

**(a)** Show that $\hat{\mu}_{\text{MAP}}(k|N) = \sigma_r^2 q(k)\mathbf{h}_k'[\sigma_r^2\mathbf{H}\mathbf{Q_q}\mathbf{H}' + \rho\mathbf{I}]^{-1}\mathbf{Z}$, where $\mathbf{h}_k$ is the $k$th column of $\mathbf{H}$.

**(b)** Show that $\hat{\mu}_{\text{MAP}}(k|N)$ can be expressed as in (14-56). What are $c(k)$ and $e(k)$? Is $e(k)$ zero mean?

**(c)** Noise $e(k)$ is not Gaussian; hence, how are we interpreting $e(k)$ when we treat it as Gaussian, as required by our derivation of the square-law detector?

**(d)** Using the fact that $\tilde{\mu}_{\text{MAP}}(k|N) = \mu(k) - \hat{\mu}_{\text{MAP}}(k|N)$, show that

$$\sigma_e^2(k) = \sigma_{\tilde{\mu}}^2(k|N) - [1 - c(k)]^2\sigma_r^2\lambda$$

**(e)** From your formula for $c(k)$, obtained in part (b), and the formula for $\text{Var}[\tilde{\mu}_{\text{MAP}}(k|N)] = \sigma_{\tilde{\mu}}^2(k|N)$, given in part (b) of Problem 14-12, show that $c(k) = 1 - \sigma_{\tilde{\mu}}^2(k|N)/\sigma_r^2\lambda$.

**(f)** Express $A_q(k)$ in terms of known statistics and $\sigma_{\tilde{\mu}}^2(k|N)$.

**(g)** Explain why we can do better than a square-law detector for this problem. What exactly does "do better" mean?

**14-14.** Here we develop another detector for seismic deconvolution (see Example 14-2), a *single most likely replacement detector* (SMLR detector) that is recursive (Kormylo, 1979; Kormylo and Mendel, 1982; Mendel, 1983, 1990). It starts where the square-law detector leaves off. Consider the following experiment. We are given a reference $\mathbf{q}$, which we denote $\mathbf{q}_r$; the first choice for $\mathbf{q}_r$ could be the value of $\mathbf{q}$ obtained from the square-law detector, $\mathbf{q}_{\text{SL}}$, or from other a priori knowledge. Recall that $\mathbf{q}_r$ is a vector with $N$ elements. We now create $N$ test vectors each of which differs from $\mathbf{q}_r$ in exactly one location. These test vectors are denoted $\mathbf{q}_{t,1}, \mathbf{q}_{t,2}, \ldots, \mathbf{q}_{t,N}$. Note that the $k$th test vector, $\mathbf{q}_{t,k}$, differs from $\mathbf{q}_r$ only at the $k$th time point. This means that $q_{t,k}(i) = q_r(i)$ for all $i \neq k$ and $q_{t,k}(i) = 1 - q_r(i)$ for $i = k$. By evaluating a likelihood ratio between $\mathbf{q}_{t,k}$ and $\mathbf{q}_r$, we obtain a decision function $D(\mathbf{Z}; k)$. As we shall see, it processes the measurements in $\mathbf{Z}$ in a very nonlinear manner. The SMLR detector decision strategy is as follows: Find the value of $k$ at which the likelihood-ratio decision function $D(\mathbf{Z}; k)$ is maximum. Call this time point $k'$. This will be the single time point at which a change is made in our reference sequence. The winning test sequence is $\mathbf{q}_{t,k'}$. Once $\mathbf{q}_{t,k'}$ has been found, this event vector replaces the original reference vector, and the procedure is repeated either for a fixed number of iterations or until no test sequence can be found that is better than the most recent reference sequence. The latter occurs when the entire likelihood ratio decision function is negative.

It would appear that we will need $N$ independent calculations to calculate the likelihood decision function $D(\mathbf{Z}; k)$. In this problem you will show that only one calculation is necessary, and it involves just the reference sequence $\mathbf{q}_r$.

Here we shall use $M(\mathbf{q}|\mathbf{Z})$ given in (14-35) as the starting point for derivation of the SMLR detector. More specifically, we shall use $M(\mathbf{q}_{t,k}|\mathbf{Z})$ and $M(\mathbf{q}_r|\mathbf{Z})$ for choosing between the test and reference $\mathbf{q}$, $\mathbf{q}_{t,k}$, and $\mathbf{q}_r$, respectively. We define $\ln D(\mathbf{Z}; k) = \ln\{M(\mathbf{q}_{t,k}|\mathbf{Z})/M(\mathbf{q}_r|\mathbf{Z})\}$.

**(a)** Show that

$$2\ln D(\mathbf{Z}; k) = \frac{(\mathbf{h}_k'\mathbf{P}_z^{-1}\mathbf{Z})^2}{[\sigma_r^2(q_{t,k}(k) - q_r(k))]^{-1} + \mathbf{h}_k'\mathbf{P}_z^{-1}\mathbf{h}_k} + 2[q_{t,k}(k) - q_r(k)]\ln\left[\frac{\lambda}{1-\lambda}\right]$$

where $\mathbf{h}_k$ is the $k$th column of $\mathbf{H}$ and $\mathbf{P}_z = \mathbf{E}\{\mathbf{ZZ}'\} = \sigma_r^2\lambda\mathbf{HH}' + \rho\mathbf{I}$.

**(b)** Show that $\mathbf{h}_k'\mathbf{P}_z^{-1}\mathbf{Z}$ can be expressed in terms of the $k$th component of $\hat{\boldsymbol{\mu}}_{\text{MS}}(N)$, $\hat{\mu}_{\text{MS}}(k|N)$, by starting with (13-31) and calculating $\mathbf{P}_\mu = \lambda\sigma_r^2\mathbf{I}$, and that

$$\mathbf{h}_k'\mathbf{P}_z^{-1}\mathbf{Z} = \frac{\hat{\mu}_{\text{MS}}(k|N)}{\lambda\sigma_r^2}$$

**(c)** Similarly, use results from Problem 14-12b in which $q(k)$ is replaced by $\lambda$ to show that

$$\mathbf{h}_k'\mathbf{P}_z^{-1}\mathbf{h}_k = \frac{\lambda\sigma_r^2 - \sigma_{\hat{\mu}}^2(k|N)}{\lambda^2\sigma_r^4}$$

**(d)** Parts (b) and (c) show that minimum-variance deconvolution is an essential ingredient to SMLR detection. Explain how MVD is actually used only one time to obtain $\ln D(\mathbf{Z}; k)$.

**(e)** Draw a figure of $D(\mathbf{Z}; k)$ versus $k$ and explain the SMLR detector decision strategy.

**14-15.** Deterministic *cost functions*, $C(\epsilon)$, which are different from the quadratic cost function (loss function) that was used in Lesson 13, are depicted in Figure P14-15 (Van Trees, 1968; Kay, 1993). The absolute error cost function penalizes errors proportionally, whereas the uniform (or hit and miss) cost function assigns zero cost for small errors and unity cost for all errors in excess of the threshold error $\delta$. An error less than $|\delta|$ is as good as no error, and all errors larger than $|\delta|$ are penalized by the same amount.

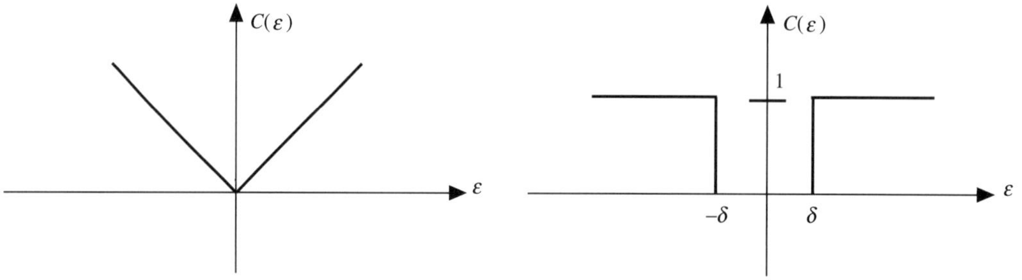


(a) Absolute error

(b) Uniform error

**Figure P14-15**

The average cost, $\mathbf{E}\{C(\epsilon)\}$, is called the *Bayes risk* $R$, i.e., $R = \mathbf{E}\{C(\epsilon)\}$, and provides a measure of the performance of an estimator. Note that

$$R = \mathbf{E}\{C(\epsilon)\} = \int\int C(\theta - \hat{\theta})p(\mathbf{Z}, \theta)d\mathbf{Z}\,d\theta = \int [\int C(\theta - \hat{\theta})p(\theta|\mathbf{Z})d\theta]p(\mathbf{Z})\,d\mathbf{Z}$$

**(a)** Prove that the estimator that minimizes the Bayes risk for the absolute error cost function is the *median* of the posterior probability density function. (*Hint:* Use Leibnitz's rule of differentiation.)

**(b)** Prove that the estimator that minimizes the Bayes risk for the uniform error cost function is the *mode* of the posterior probability density function, i.e., the location of the maximum $p(\theta|\mathbf{Z})$; this is the MAP estimator.

**(c)** Show that when $p(\theta|\mathbf{Z})$ is Gaussian the mean-squared estimator (which is the *mean* of the posterior density function), the median estimator, and the mode estimator are all the same.

# Elements of Discrete-time Gauss–Markov Random Sequences

*This is State analysis no estimation (see next 2 chaps for estimation)*

## SUMMARY

This is another transition lesson. Prior to studying recursive state estimation, we first review an important body of material on discrete-time Gauss–Markov random sequences. Most if not all of this material should be a review for a reader who has had courses in random processes and linear systems.

A *first-order Markov sequence* is one whose probability law depends only on the immediate past value of the random sequence; hence, the infinite past does not have to be remembered for such a sequence.

It is in this lesson that we provide a formal definition of *Gaussian white noise*. We also introduce the *basic state-variable model*. It consists of a state equation and a measurement equation and can be used to describe time-varying nonstationary systems. Of course, it is a simple matter to specialize this model to time-invariant and stationary systems, simply by making all the time-varying matrices that appear in the state-variable model constant matrices. The basic state-variable model is not the most general state-variable model; however, it is a simple one to use in deriving recursive state estimators. Lesson 22 treats more general state-variable models.

Many statistical properties are given for the basic state-variable model. They require that *all* sources of uncertainty in that model must be jointly Gaussian. One important result from this lesson is a procedure for computing the mean vector and covariance matrix of a state vector using recursive formulas.

Finally, we define the important concept of *signal-to-noise ratio* and show how it can be computed from a state-variable model.

When you complete this lesson, you will be able to (1) state many useful definitions and facts about Gauss–Markov random sequences; (2) define Gaussian white noise; (3) explain the *basic state-variable model* and specify its properties; (4) use recursive algorithms to compute the mean vector and covariance matrix for the state vector for our basic state-variable model; and (5) define signal-to-noise ratio and explain how it can be computed.

## INTRODUCTION

Lesson 13 and 14 have demonstrated the importance of Gaussian random variables in estimation theory. In this lesson we extend some of the basic concepts that were introduced in Lesson 12, for Gaussian random variables, to indexed random variables, that is, random sequences. These extensions are needed in order to develop state estimators.

## DEFINITION AND PROPERTIES OF DISCRETE-TIME GAUSS–MARKOV RANDOM SEQUENCES

Recall that a random process is a collection of random variables in which the notion of time plays a role.

**Definition 15-1** (Meditch, 1969, p. 106). *A vector random process is a family of random vectors* $\{\mathbf{s}(\mathrm{t}), \mathrm{t} \in \mathcal{I}\}$ *indexed by a parameter* t *all of whose values lie in some appropriate index set* $\mathcal{I}$. *When* $\mathcal{I} = \{\mathrm{k} : \mathrm{k} = 0, 1, \ldots\}$ *we have a discrete-time random sequence.* □

**Definition 15-2** (Meditch, 1969, p. 117). *A vector random sequence* $\{\mathbf{s}(\mathrm{t}), \mathrm{t} \in \mathcal{I}\}$ *is defined to be multivariate Gaussian if, for* **any** $\ell$ *time points* $\mathrm{t}_1, \mathrm{t}_2, \ldots, \mathrm{t}_\ell$ *in* $\mathcal{I}$, *where* $\ell$ *is an integer, the set of* $\ell$ *random* n *vectors* $\mathbf{s}(\mathrm{t}_1), \mathbf{s}(\mathrm{t}_2), \ldots, \mathbf{s}(\mathrm{t}_\ell)$, *is jointly Gaussian distributed.* □

Let $\mathcal{S}(l)$ be defined as

$$\mathcal{S}(l) = \mathrm{col}\,(\mathbf{s}(t_1), \mathbf{s}(t_2), \ldots, \mathbf{s}(t_l)) \tag{15-1}$$

Then Definition 15-2 means that

$$p[\mathcal{S}(l)] = (2\pi)^{-nl/2}|\mathbf{P}_\mathcal{S}(l)|^{-1/2} \exp\left\{-\frac{1}{2}[\mathcal{S}(l) - \mathbf{m}_\mathcal{S}(l)]'\mathbf{P}_\mathcal{S}^{-1}(l)[\mathcal{S}(l) - \mathbf{m}_\mathcal{S}(l)]\right\} \tag{15-2}$$

in which

$$\mathbf{m}_\mathcal{S}(l) = \mathbf{E}\{\mathcal{S}(l)\} \tag{15-3}$$

and $\mathbf{P}_\mathcal{S}(l)$ is the $nl \times nl$ matrix $\mathbf{E}\{[\mathcal{S}(l) - \mathbf{m}_\mathcal{S}(l)][\mathcal{S}(l) - \mathbf{m}_\mathcal{S}(l)]'\}$ with elements $\mathbf{P}_\mathcal{S}(i, j)$, where

$$\mathbf{P}_\mathcal{S}(i, j) = \mathbf{E}\{[\mathbf{s}(t_i) - \mathbf{m}_s(t_i)][\mathbf{s}(t_j) - \mathbf{m}_s(t_j)]'\} \tag{15-4}$$

$i, j = 1, 2, \ldots, l$.

**Definition 15-3** (Meditch, 1969, p. 118). *A vector random sequence* $\{\mathbf{s}(t), t \in \mathcal{J}\}$ *is a Markov sequence, if, for* **any** $m$ *time points* $t_1 < t_2 < \cdots < t_m$ *in* $\mathcal{J}$, *where* $m$ *is any integer, it is true that*

$$P[\mathbf{s}(t_m) \leq \mathbf{S}(t_m)|\mathbf{s}(t_{m-1}) = \mathbf{S}(t_{m-1}), \ldots, \mathbf{s}(t_1) = \mathbf{S}(t_1)]$$
$$= P[\mathbf{s}(t_m) \leq \mathbf{S}(t_m)|\mathbf{s}(t_{m-1}) = \mathbf{S}(t_{m-1})] \qquad (15\text{-}5)$$

*For continuous random variables, this means that*

$$p[\mathbf{s}(t_m)|\mathbf{s}(t_{m-1}), \ldots, \mathbf{s}(t_1)] = p[\mathbf{s}(t_m)|\mathbf{s}(t_{m-1})] \ \square \qquad (15\text{-}6)$$

Note that, in (15-5), $\mathbf{s}(t_m) \leq \mathbf{S}(t_m)$ means $s_i(t_m) \leq S_i(t_m)$ for $i = 1, 2, \ldots, n$. If we view time point $t_m$ as the present time and time points $t_{m-1}, \ldots, t_1$ as the past, then a Markov sequence is one whose probability law (e.g., probability density function) depends only on the immediate past value, $t_{m-1}$. This is often referred to as the *Markov property* for a vector random sequence. Because the probability law depends only on the immediate past value, we often refer to such a process as a *first-order Markov sequence* (if it depended on the immediate two past values it would be a second-order Markov sequence).

**Theorem 15-1.** *Let* $\{\mathbf{s}(t), t \in \mathcal{J}\}$ *be a first-order Markov sequence, and* $t_1 < t_2 < \cdots < t_m$ *be any time points in* $\mathcal{J}$, *where* $m$ *is an integer. Then*

$$p[\mathbf{s}(t_m), \mathbf{s}(t_{m-1}), \ldots, \mathbf{s}(t_1)] = p[\mathbf{s}(t_m)|\mathbf{s}(t_{m-1})]p[\mathbf{s}(t_{m-1})|\mathbf{s}(t_{m-2})]\ldots$$
$$p[\mathbf{s}(t_2)|\mathbf{s}(t_1)]p[\mathbf{s}(t_1)] \qquad (15\text{-}7)$$

*Proof.* From probability theory (e.g., Papoulis, 1991) and the Markov property of $\mathbf{s}(t)$, we know that

$$\begin{aligned} p[\mathbf{s}(t_m), \mathbf{s}(t_{m-1}), \ldots, \mathbf{s}(t_1)] &= p[\mathbf{s}(t_m)|\mathbf{s}(t_{m-1}), \ldots, \mathbf{s}(t_1)]p[\mathbf{s}(t_{m-1}), \ldots, \mathbf{s}(t_1)] \\ &= p[\mathbf{s}(t_{m)})|\mathbf{s}(t_{m-1})]p[\mathbf{s}(t_{m-1}), \ldots, \mathbf{s}(t_1)] \end{aligned} \qquad (15\text{-}8)$$

In a similar manner, we find that

$$\left.\begin{aligned} p[\mathbf{s}(t_{m-1}), \ldots, \mathbf{s}(t_1)] &= p[\mathbf{s}(t_{m-1})|\mathbf{s}(t_{m-2})]p[\mathbf{s}(t_{m-2}), \ldots, \mathbf{s}(t_1)] \\ p[\mathbf{s}(t_{m-2}), \ldots, \mathbf{s}(t_1)] &= p[\mathbf{s}(t_{m-2})|\mathbf{s}(t_{m-3})]p[\mathbf{s}(t_{m-3}), \ldots, \mathbf{s}(t_1)] \\ &\ldots \\ p[\mathbf{s}(t_{2)}, \mathbf{s}(t_1)] &= p[\mathbf{s}(t_2)|\mathbf{s}(t_1)]p[\mathbf{s}(t_1)] \end{aligned}\right\} \qquad (15\text{-}9)$$

Equation (15-7) is obtained by successively substituting each of the equations in (15-9) into (15-8). $\square$

Theorem 15-1 demonstrates that a first-order Markov sequence is completely characterized by two probability density functions; the *transition probability density function* $p[\mathbf{s}(t_i)|\mathbf{s}(t_{i-1})]$ and the initial (prior) *probability density function* $p[\mathbf{s}(t_1)]$. Note that generally the transition probability density functions can all be different, in which case they should be subscripted [e.g., $p_m[\mathbf{s}(t_m)|\mathbf{s}(t_{m-1})]$ and $p_{m-1}[\mathbf{s}(t_{m-1})|\mathbf{s}(t_{m-2})]$].

**Theorem 15-2.** *For a first-order Markov sequence,*

$$\mathbf{E}\{\mathbf{s}(t_m)|\mathbf{s}(t_{m-1}), \ldots, \mathbf{s}(t_1)\} = \mathbf{E}\{\mathbf{s}(t_m)|\mathbf{s}(t_{m-1})\} \quad \square \tag{15-10}$$

We leave the proof of this useful result as an exercise.

A vector random sequence that is both Gaussian and a first-order Markov sequence will be referred to in the sequel as a Gauss–Markov sequence.

**Definition 15-4.** *A vector random sequence* $\{\mathbf{s}(t), t \in \mathcal{J}\}$ *is said to be a (discrete-time) Gaussian white sequence if, for* **any** $m$ *time points* $t_1, t_2, \ldots, t_m$ *in* $\mathcal{J}$, *where* $m$ *is any integer, the* $m$ *random vectors* $\mathbf{s}(t_1), \mathbf{s}(t_2), \ldots, \mathbf{s}(t_m)$ *are uncorrelated Gaussian random vectors.* $\square$

White noise is zero mean, or else it cannot have a flat spectrum. For white noise

$$\mathbf{E}\{\mathbf{s}(t_i)\mathbf{s}'(t_j)\} = \mathbf{0}, \qquad \text{for all } i \neq j \tag{15-11}$$

Additionally, for Gaussian white noise

$$p[\mathcal{S}(l)] = p[\mathbf{s}(t_1)]p[\mathbf{s}(t_2)], \ldots, p[\mathbf{s}(t_l)] \tag{15-12}$$

[because uncorrelatedness implies statistical independence (see Lesson 12)], where $p[\mathbf{s}(t_i)]$ is a multivariate Gaussian probability density function.

**Theorem 15-3.** *A vector Gaussian white sequence* $\mathbf{s}(t)$ *can be viewed as a first-order Gauss-Markov sequence for which*

$$p[\mathbf{s}(t)|\mathbf{s}(\tau)] = p[\mathbf{s}(t)] \tag{15-13}$$

*for all* $t, \tau \in \mathcal{J}$ *and* $t \neq \tau$.

*Proof.* For a Gaussian white sequence, we know from (15-12), that

$$p[\mathbf{s}(t), \mathbf{s}(\tau)] = p[\mathbf{s}(t)]p[\mathbf{s}(\tau)] \tag{15-14}$$

But we also know that

$$p[\mathbf{s}(t), \mathbf{s}(\tau)] = p[\mathbf{s}(t)|\mathbf{s}(\tau)]p[\mathbf{s}(\tau)] \tag{15-15}$$

Equating (15-14) and (15-15), we obtain (15-13). $\square$

Theorem 15-3 means that past and future values of $\mathbf{s}(t)$ in no way help determine present values of $\mathbf{s}(t)$. For Gaussian white sequences, the transition probability density function equals the marginal density function, $p[\mathbf{s}(t)]$, which is multivariate Gaussian. Additionally (Problem 15-1),

$$\mathbf{E}\{\mathbf{s}(t_m)|\mathbf{s}(t_{m-1}), \ldots, \mathbf{s}(t_1)\} = \mathbf{E}\{\mathbf{s}(t_m)\} \tag{15-16}$$

**Definition 15-5.** *A vector random sequence* $\{\mathbf{s}(t), t \in \mathcal{J}\}$ *is strictly stationary if its probability density function is the same for all values of time. It is wide-sense stationary if its first- and second-order statistics do not depend on time.* $\square$

## THE BASIC STATE-VARIABLE MODEL

In succeeding lessons we shall develop a variety of state estimators for the following basic linear, (possibly) time-varying, discrete-time, dynamical system (our *basic state-variable model*), which is characterized by $n \times 1$ state vector $\mathbf{x}(k)$ and $m \times 1$ measurement vector $\mathbf{z}(k)$:

$$\mathbf{x}(k+1) = \mathbf{\Phi}(k+1,k)\mathbf{x}(k) + \mathbf{\Gamma}(k+1,k)\mathbf{w}(k) + \mathbf{\Psi}(k+1,k)\mathbf{u}(k) \tag{15-17}$$

and

$$\mathbf{z}(k+1) = \mathbf{H}(k+1)\mathbf{x}(k+1) + \mathbf{v}(k+1) \tag{15-18}$$

where $k = 0, 1, \ldots$. In this model $\mathbf{w}(k)$ and $\mathbf{v}(k)$ are $p \times 1$ and $m \times 1$ mutually uncorrelated (possibly nonstationary) jointly Gaussian white noise sequences; i.e.,

$$\mathbf{E}\{\mathbf{w}(i)\mathbf{w}'(j)\} = \mathbf{Q}(i)\delta_{ij} \tag{15-19}$$

$$\mathbf{E}\{\mathbf{v}(i)\mathbf{v}'(j)\} = \mathbf{R}(i)\delta_{ij} \tag{15-20}$$

and

$$\mathbf{E}\{\mathbf{w}(i)\mathbf{v}'(j)\} = \mathbf{S} = \mathbf{0}, \qquad \text{for all } i \text{ and } j \tag{15-21}$$

Covariance matrix $\mathbf{Q}(i)$ is positive semidefinite and $\mathbf{R}(i)$ is positive definite [so that $\mathbf{R}^{-1}(i)$ exists]. Additionally, $\mathbf{u}(k)$ is an $l \times 1$ vector of known system inputs, and initial state vector $\mathbf{x}(0)$ is multivariate Gaussian, with mean $\mathbf{m}_x(0)$ and covariance $\mathbf{P}_x(0)$; i.e.,

$$\mathbf{x}(0) \sim N(\mathbf{x}(0); \mathbf{m}_\mathbf{x}(0), \mathbf{P}_\mathbf{x}(0)) \tag{15-22}$$

and $\mathbf{x}(0)$ is not correlated with $\mathbf{w}(k)$ and $\mathbf{v}(k)$. The dimensions of matrices $\mathbf{\Phi}$, $\mathbf{\Gamma}$, $\mathbf{\Psi}$, $\mathbf{H}$, $\mathbf{Q}$, and $\mathbf{R}$ are $n \times n$, $n \times p$, $n \times l$, $m \times n$, $p \times p$, and $m \times m$, respectively.

### Comments

1. The double arguments in matrices $\mathbf{\Phi}$, $\mathbf{\Gamma}$, and $\mathbf{\Psi}$ may not always be necessary, in which case we can replace $(k+1,k)$ by $(k)$. This will be true if the underlying system, whose state-variable model is described by (15-17) and (15-18), is discrete time in nature. When the underlying system is continuous time in nature, it must first be discretized, a process that is described in careful detail in Lesson 23 (in the section entitled "Discretization of Linear Time-varying State-variable Model"), where you will learn that, as a result of discretization, the matrices $\mathbf{\Phi}$, $\mathbf{\Gamma}$, and $\mathbf{\Psi}$ do have the double arguments $(k+1,k)$. Because so many engineering systems are continuous time in nature and will therefore have to first be discretized, we adopt the double argument notation in this book.
2. In Lesson 11 we used the notation $\bar{\mathbf{R}}$ for $\mathbf{E}\{\mathbf{v}(k)\mathbf{v}'(k)\}$; here we use $\mathbf{R}$. In the earlier lessons, the symbol $\mathbf{R}$ was reserved for the covariance of the noise vector in our generic linear model. Because state estimation, which is the major subject in the rest of this book, is based on the state-variable model and

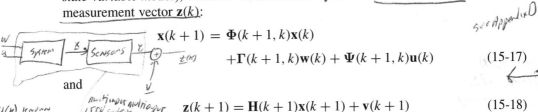

not on the generic linear model, there should be no confusion between our use of symbol $\mathbf{R}$ in state estimation and its use in the earlier lessons.

**3.** I have found that many students of estimation theory are not as familiar with state-variable models as they are with input/output models; hence, I have included a very extensive introduction to state-variable models and methods as a supplementary lesson (Lesson D).

**4.** Disturbance $\mathbf{w}(k)$ is often used to model the following types of uncertainty:

   **a.** disturbance forces acting on the system (e.g., wind that buffets an airplane);

   **b.** errors in modeling the system (e.g., neglected effects); and

   **c.** errors, due to actuators, in the traslation of the known input, $\mathbf{u}(k)$, into physical signals.

**5.** Vector $\mathbf{v}(k)$ is often used to model the following types of uncertainty:

   **a.** errors in measurements made by sensing instruments;

   **b.** unavoidable disturbances that act directly on the sensors; and

   **c.** errors in the realization of feedback compensators using physical components [this is valid only when the measurement equation contains a direct throughput of the input $\mathbf{u}(k)$, i.e., when $\mathbf{z}(k+1) = \mathbf{H}(k+1)\mathbf{x}(k+1) + \mathbf{G}(k+1)\mathbf{u}(k+1) + \mathbf{v}(k+1)$; we shall examine this situation in Lesson 22].

**6.** Of course, not all dynamical systems are described by this basic model. In general, $\mathbf{w}(k)$ and $\mathbf{v}(k)$ may be correlated, some measurements may be made so accurate that, for all practical purposes, they are "perfect" (i.e., no measurement noise is associated with them), and either $\mathbf{w}(k)$ or $\mathbf{v}(k)$, or both, may be colored noise processes. We shall consider the modification of our basic state-variable model for each of these important situations in Lesson 22.

## PROPERTIES OF THE BASIC STATE-VARIABLE MODEL

In this section we state and prove a number of important statistical properties for our basic state-variable model.

**Theorem 15-4.** *When* $\mathbf{x}(0)$ *and* $\{\mathbf{w}(\mathrm{k}), \mathrm{k} = 0, 1, \ldots\}$ *are jointly Gaussian, then* $\mathbf{x}(\mathrm{k}), \mathrm{k} = 0, 1 \ldots\}$ *is a Gauss–Markov sequence.*

Note that if $\mathbf{x}(0)$ and $\mathbf{w}(k)$ are individually Gaussian and statistically independent (or uncorrelated), they will be jointly Gaussian (Papoulis, 1991).

*Proof*

**a.** *Gaussian property* [assuming $\mathbf{u}(k)$ nonrandom]. Because $\mathbf{u}(k)$ is nonrandom, it has no effect on determining whether $\mathbf{x}(k)$ is Gaussian; hence, for this part of the proof, we assume $\mathbf{u}(k) = \mathbf{0}$. The solution to (15-17) is (Problem 15-2)

$$x(k) = \Phi(k, 0)x(0) + \sum_{i=1}^{k} \Phi(k, i)\Gamma(i, i-1)w(i-1) \qquad (15\text{-}23)$$

where

$+ \sum \delta(ki) \, \Psi(i,i-1)u(i-1)$    but this is zero

$$\Phi(k, i) = \Phi(k, k-1)\Phi(k-1, k-2)\ldots\Phi(i+1, i) \qquad (15\text{-}24)$$

Observe that $x(k)$ is a linear transformation of jointly Gaussian random vectors $x(0), w(0), w(1)\ldots, w(k-1)$; hence, $x(k)$ is Gaussian.

**b.** *Markov property.* This property does not require $x(k)$ or $w(k)$ to be Gaussian. Because $x$ satisfies state equation (15-17), we see that $x(k)$ depends only on its immediate past value; hence, $x(k)$ is Markov. $\square$

We have been able to show that our dynamical system is Markov because we specified a model for it. Without such a specification, it would be quite difficult (or impossible) to test for the Markov nature of a random sequence.

By stacking up $x(1), x(2)\ldots$, into a supervector, it is easily seen that this super-vector is just a linear transformation of jointly Gaussian quatities $x(0), w(0), w(1), \ldots$ (Problem 15-3); hence, $x(1), x(2),\ldots$ are themselves *jointly Gaussian*.

A Gauss–Markov sequence can be completely characterized in two ways:

**1.** Specify the marginal density of the initial state vector, $p[x(0)]$, and the transition density $p[x(k+1)|x(k)]$.

*OR* **2.** Specify the mean and covariance of the state vector sequence. full set into about density function is known (since it is Gaussian)

The second characterization is a complete one because Gaussian random vectors are completely characterized by their means and covariances (Lesson 12). We shall find the second characterization more useful than the first.

The Gaussian density function for state vector $x(k)$ is

$$p[x(k)] = [(2\pi)^n |P_x(k)|]^{-1/2}$$
$$\exp\left\{-\frac{1}{2}[x(k) - m_x(k)]'P_x^{-1}(k)[x(k) - m_x(k)]\right\} \qquad (15\text{-}25)$$

where

$$m_x(k) = E\{x(k)\} \qquad (15\text{-}26)$$

and

$$P_x(k) = E\{[x(k) - m_x(k)][x(k) - m_x(k)]'\} \qquad (15\text{-}27)$$

We now demonstrate that $m_x(k)$ and $P_x(k)$ can be computed by means of recursive equations. this => if $x(0) + P_x(0)$ is know, then we know it all. $p(x(k))$

**Theorem 15-5.** *For our basic state-variable model:* since $E\{u\} = 0$ $m_x = E[cy \, (15.17)]$ using (15.17)

**a.** $m_x(k)$ *can be computed from the vector recursive equation* since $E\{u\} = 0$

$$m_x(k+1) = \Phi(k+1, k)m_x(k) + \Psi(k+1, k)u(k) \qquad (15\text{-}28)$$

Properties of the Basic State-variable Model      **217**

*where* $k = 0, 1, \ldots,$ *and* $\mathbf{m_x}(0)$ *initializes (15-28).*

**b.** $\mathbf{P_x}(k)$ *can be computed from the matrix recursive equation*

$$\begin{aligned}\mathbf{P_x}(k+1) &= \mathbf{\Phi}(k+1,k)\mathbf{P_x}(k)\mathbf{\Phi}'(k+1,k)\\ &\quad + \mathbf{\Gamma}(k+1,k)\mathbf{Q}(k)\mathbf{\Gamma}'(k+1,k)\end{aligned} \tag{15-29}$$

*where* $k = 0, 1, \ldots,$ *and* $\mathbf{P_x}(0)$ *initializes (15-29).*

**c.** $\mathbf{E}\{[\mathbf{x}(i) - \mathbf{m_x}(i)][\mathbf{x}(j) - \mathbf{m_x}(j)]'\} \triangleq \mathbf{P_x}(i, j)$ *can be computed from*

$$\mathbf{P_x}(i,j) = \begin{cases} \mathbf{\Phi}(i,j)\mathbf{P_x}(j), & \text{when } i > j \\ \mathbf{P_x}(i)\mathbf{\Phi}'(j,i), & \text{when } i < j \end{cases} \tag{15-30}$$

*Proof*

**a.** Take the expected value of both sides of (15-17), using the facts that expectation is a linear operation (Papoulis, 1991) and $\mathbf{w}(k)$ is zero mean, to obtain (15-28).

**b.** For notational simplicity, we omit the temporal arguments of $\mathbf{\Phi}$ and $\mathbf{\Gamma}$ in this part of the proof. Using (15-17) and (15-28), we obtain

$$\begin{aligned}\mathbf{P_x}(k+1) &= \mathbf{E}\{[\mathbf{x}(k+1) - \mathbf{m_x}(k+1][\mathbf{x}(k+1) - \mathbf{m_x}(k+1)]'\}\\ &= \mathbf{E}\{[\mathbf{\Phi}[\mathbf{x}(k) - \mathbf{m_x}(k)] + \mathbf{\Gamma}\mathbf{w}(k)]\\ &\quad [\mathbf{\Phi}[\mathbf{x}(k) - \mathbf{m_x}(k)] + \mathbf{\Gamma}\mathbf{w}(k)]'\}\\ &= \mathbf{\Phi}\mathbf{P_x}(k)\mathbf{\Phi}' + \mathbf{\Gamma}\mathbf{Q}(k)\mathbf{\Gamma}' + \mathbf{\Phi}\mathbf{E}\{[\mathbf{x}(k) - \mathbf{m_x}(k)]\mathbf{w}'(k)\}\mathbf{\Gamma}'\\ &\quad + \mathbf{\Gamma}\mathbf{E}\{\mathbf{w}(k)[\mathbf{x}(k) - \mathbf{m_x}(k)]'\}\mathbf{\Phi}'\end{aligned} \tag{15-31}$$

Because $\mathbf{m_x}(k)$ is not random and $\mathbf{w}(k)$ is zero mean, $\mathbf{E}\{\mathbf{m_x}(k)\mathbf{w}'(k)\} = \mathbf{m_x}\mathbf{E}\{\mathbf{w}'(k)\} = \mathbf{0}$, and $\mathbf{E}\{\mathbf{w}(k)\mathbf{m_x}'(k)\} = \mathbf{0}$. State vector $\mathbf{x}(k)$ depends at most on random input $\mathbf{w}(k-1)$ [see (15-17)]; hence,

$$\mathbf{E}\{\mathbf{x}(k)\mathbf{w}'(k)\} = \mathbf{E}\{\mathbf{x}(k)\}\mathbf{E}\{\mathbf{w}'(k)\} = \mathbf{0} \tag{15-32}$$

and $\mathbf{E}\{\mathbf{w}(k)\mathbf{x}'(k)\} = \mathbf{0}$ as well. The last two terms in (15-31) are therefore equal to zero, and the equation reduces to (15-29).

**c.** We leave to proof of (15-30) as an exercise. Observe that once we know covariance matrix $\mathbf{P_x}(k)$ it is an easy matter to determine any cross-covariance matrix between state $\mathbf{x}(k)$ and $\mathbf{x}(i)(i \neq k)$. The Markov nature of our basic state-variable model is responsible for this. □

Observe that mean vector $\mathbf{m_x}(k)$ satisfies a deterministic vector state equation, (15-28), covariance matrix $\mathbf{P_x}(k)$ satisfies a deterministic matrix state equation, (15-29), and (15-28) and (15-29) are easily programmed for digital computation.

Next we direct our attention to the statistics of measurement vector $\mathbf{z}(k)$.

**Theorem 15-6.** *For our basic state-variable model, when* $\mathbf{x}(0)$, $\mathbf{w}(k)$, *and* $\mathbf{v}(k)$ *are jointly Gaussian, then* $\{\mathbf{z}(k), k = 1, 2, \ldots\}$ *is Gaussian, and*

$$\mathbf{m_z}(k+1) = \mathbf{H}(k+1)\mathbf{m_x}(k+1) \tag{15-33}$$

*and*

$$\mathbf{P_z}(k+1) = \mathbf{H}(k+1)\mathbf{P_x}(k+1)\mathbf{H}'(k+1) + \mathbf{R}(k+1) \tag{15-34}$$

*where* $\mathbf{m_x}$(k+1) *and* $\mathbf{P_x}$(k+1) *are computed from (15-28) and (15-29), respectively.* □

We leave the proof as an exercise for the reader. Note that if $\mathbf{x}(0)$, $\mathbf{w}(k)$, and $\mathbf{v}(k)$ are statistically independent and Gaussian they will be jointly Gaussian.

**EXAMPLE 15-1**

Consider the simple single-input, single-output first-order system

$$x(k+1) = \frac{1}{2}x(k) + w(k) \tag{15-35}$$

$$z(k+1) = x(k+1) + \nu(k+1) \tag{15-36}$$

where $w(k)$ and $\nu(k)$ are wide-sense stationary white noise processes, for which $q = 20$ and $r = 5$. Additionally, $m_x(0) = 4$ and $p_x(0) = 10$.

The mean of $x(k)$ is computed from the following homogeneous equation:

$$m_x(k+1) = \frac{1}{2}m_x(k), \qquad m_x(0) = 4 \tag{15-37}$$

and the variance of $x(k)$ is computed from Equation (15-29), which in this case simplifies to

$$p_x(k+1) = \frac{1}{4}p_x(k) + 20, \qquad p_x(0) = 10 \tag{15-38}$$

Additionally, the mean and variance of $z(k)$ are computed from

$$m_z(k+1) = m_x(k+1) \tag{15-39}$$

and

$$p_z(k+1) = p_x(k+1) + 5 \tag{15-40}$$

Figure 15-1 depicts $m_x(k)$ and $p_x^{1/2}(k)$. Observe that $m_x(k)$ decays to zero very rapidly and that $p_x^{1/2}(k)$ approaches a steady-state value $\bar{p}_x^{1/2} = 5.163$. This steady-state value can be computed from Equation (15-38) by setting $p_x(k) = p_x(k+1) = \bar{p}_x$. The existence of $\bar{p}_x$ is guaranteed by our first-order system being stable.

Although $m_x(k) \to 0$, there is a lot of uncertainty about $x(k)$, as evidenced by the large value of $\bar{p}_x$. There will be an even larger uncertainty about $z(k)$, because $\bar{p}_z \to 31.66$. These large values for $\bar{p}_z$ and $\bar{p}_x$ are due to the large values of $q$ and $r$. In many practical applications, both $q$ and $r$ will be much less than unity, in which case $\bar{p}_z$ and $\bar{p}_x$ will be quite small. □

For our basic state-variable model to be *stationary*, it must be time invariant, and the probability density functions of $\mathbf{w}(k)$ and $\mathbf{v}(k)$ must be the same for all values of time (Definition 15-5). Because $\mathbf{w}(k)$ and $\mathbf{v}(k)$ are Gaussian, this means that $\mathbf{Q}(k)$ must equal the constant matrix $\mathbf{Q}$ and $\mathbf{R}(k)$ must equal the constant matrix $\mathbf{R}$. Additionally, either $\mathbf{x}(0) = \mathbf{0}$ or $\mathbf{\Phi}(k, 0)\mathbf{x}(0) \approx \mathbf{0}$ when $k > k_0$ [see (15-23)]. In both cases $\mathbf{x}(k)$ will be in its "steady-state" regime, so stationarity is possible.

If our basic state-variable model is time invariant and stationary and if $\mathbf{\Phi}$ is associated with an asymptotically stable system (i.e., one whose poles all lie within the

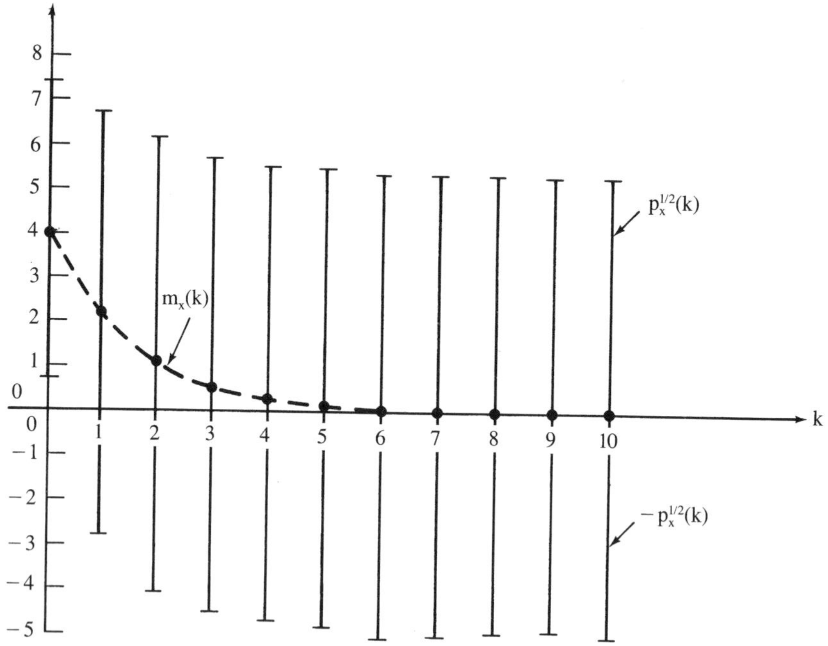

**Figure 15-1**  Mean (dashed) and standard deviation (bars) for first-order system (15-35) and (15-36).

unit circle), then (Anderson and Moore, 1979) matrix $\mathbf{P_x}(k)$ reaches a limiting (steady-state) solution $\bar{\mathbf{P}}_\mathbf{x}$; i.e.,

*only if we stationary + Time invariant (prev. pg)*

$$\lim_{k \to \infty} \mathbf{P_x}(k) = \bar{\mathbf{P}}_\mathbf{x} \tag{15-41}$$

Matrix $\bar{\mathbf{P}}_\mathbf{x}$ is the solution of the following steady-state version of (15-29):

$$\bar{\mathbf{P}}_\mathbf{x} = \mathbf{\Phi} \bar{\mathbf{P}}_\mathbf{x} \mathbf{\Phi}' + \mathbf{\Gamma Q \Gamma}' \tag{15-42}$$

This equation is called a *discrete-time Lyapunov equation*. See Laub (1979) for an excellent numerical method that can be used to solve (15-42) for $\bar{\mathbf{P}}_\mathbf{x}$.

## SIGNAL-TO-NOISE RATIO

In this section we simplify our basic state-variable model (15-17) and (15-18) to a time-invariant, stationary, two-input, single-output model:

$$\mathbf{x}(k+1) = \mathbf{\Phi x}(k) + \boldsymbol{\gamma} w(k) + \mathbf{\Psi} u(k) \tag{15-43}$$

$$z(k+1) = \mathbf{h}'\mathbf{x}(k+1) + \nu(k+1) \tag{15-44}$$

Measurement $z(k)$ is of the classical form of signal plus noise, where "signal" $s(k) = \mathbf{h}'\mathbf{x}(k)$.

The *signal-to-noise ratio* is an often-used measure of quality of measurement $z(k)$. Here we define that ratio, denoted by $\text{SNR}(k)$, as

$$\text{SNR}(k) = \frac{\sigma_s^2(k)}{r} \qquad r = \sigma_v^2 = E[v\dot{v}]$$
$$\sigma_s^2 = E[ss] \qquad (15\text{-}45)$$

From preceding analyses, we see that

$$\text{SNR}(k) = \frac{\mathbf{h}'\mathbf{P_x}(k)\mathbf{h}}{r} \qquad (15\text{-}46)$$

Because $\mathbf{P_x}(k)$ is in general a function of time, $\text{SNR}(k)$ is also a function of time. If, however, $\boldsymbol{\Phi}$ is associated with an asymptotically stable system, then (15-41) is true. In this case we can use $\bar{\mathbf{P}}_x$ in (15-46) to provide us with a single number, $\overline{\text{SNR}}$, for the signal-to-noise ratio; i.e.,

$$\overline{\text{SNR}} = \frac{\mathbf{h}'\bar{\mathbf{P}}_x\mathbf{h}}{r} \qquad (15\text{-}47)$$

Finally, we demonstrate that $\text{SNR}(k)$ (or $\overline{\text{SNR}}$) can be computed without knowing $q$ and $r$ explicitly; all that is needed is the ratio $q/r$. Multiplying and dividing the right-hand side of (15-46) by $q$, we find that

$$\text{SNR}(k) = \left[ \mathbf{h}' \frac{\mathbf{P_x}(k)}{q} \mathbf{h} \right] \frac{q}{r} \qquad q \text{ is } Q = E[ww] \qquad (15\text{-}48)$$

Scaled covariance matrix $\mathbf{P_x}(k)/q$ is computed from the following version of (15-29):

$$\frac{\mathbf{P_x}(k+1)}{q} = \boldsymbol{\Phi} \frac{\mathbf{P_x}(k)}{q} \boldsymbol{\Phi}' + \boldsymbol{\gamma}\boldsymbol{\gamma}' \qquad (15\text{-}49)$$

One of the most useful ways for using (15-48) is to compute $q/r$ for a given signal-to-noise ratio $\overline{\text{SNR}}$; i.e.,

$$\frac{q}{r} = \frac{\overline{\text{SNR}}}{\mathbf{h}' \dfrac{\bar{\mathbf{P}}_x}{q} \mathbf{h}} \qquad (15\text{-}50)$$

In Lesson 18 we show that $q/r$ can be viewed as an estimator tuning parameter; hence, signal-to-noise ratio, $\overline{\text{SNR}}$, can also be treated as such a parameter.

**EXAMPLE 15-2**   (Mendel, 1981)

Consider the first-order system

$$x(k+1) = \phi x(k) + \gamma w(k) \qquad (15\text{-}51)$$

$$z(k+1) = hx(k+1) + v(k+1) \qquad (15\text{-}52)$$

In this case, it is easy to solve (15-49) to show that

$$\frac{\bar{P}_x}{q} = \frac{\gamma^2}{1 - \phi^2} \qquad (15\text{-}53)$$

Hence,

$$\overline{\text{SNR}} = \frac{h^2\gamma^2}{1-\phi^2}\frac{q}{r} \tag{15-54}$$

Observe that if $h^2\gamma^2 = 1 - \phi^2$ then $\overline{\text{SNR}} = q/r$. The condition $h^2\gamma^2 = 1 - \phi^2$ is satisfied if, for example, $\gamma = 1$, $\phi = 1/\sqrt{2}$, and $h = 1/\sqrt{2}$. □

## COMPUTATION

This lesson as well as its companion lesson, Lesson E, are loaded with computational possibilities. The *Control System Toolbox* contains all the M-files listed next except **eig**, which is a MATLAB M-file.

Models come in different guises, such as transfer functions, zeros and poles, and state variable. Sometimes it is useful to be able to go from one type of model to another. The following Model Conversion M-files will let you do this:

**ss2tf**: State space to transfer function conversion.
**ss2zp**: State space to zero pole conversion.
**tf2ss**: Transfer function to state space conversion.
**tf2zp**: Transfer function to zero pole conversion.
**zp2tf**: Zero pole to transfer function conversion.
**zp2ss**: Zero pole to state space conversion.

Once a model has been established, it can be used to generate time responses. The following time response M-files will let you do this:

**dlsim**: Discrete-time simulation to arbitrary input.
**dimpulse**: Discrete-time unit sample response.
**dinitial**: Discrete-time initial condition response.
**dstep**: Discrete-time step response.

Frequency responses can be obtained from:

**dbode**: Discrete Bode plot. It computes the magnitude and phase response of discrete-time LTI systems.

Eigenvalues can be obtained from:

**ddamp**: Discrete damping factors and natural frequencies.
**eig**: eigenvalues and eigenvectors.

There is no M-file that lets us implement the full-blown time-varying state-vector covariance equation in (15-29). If, however, your system is time invariant and stationary, so that all the matrices associated with the basic state-variable model are constant, then the steady-state covariance matrix of $\mathbf{x}(k)$ [i.e., the solution of (15-42)] can be computed using:

**dcovar**: Discrete (steady-state) covariance response to white noise.

## SUMMARY QUESTIONS

1. A first-order Markov sequence is one whose probability law depends:
   **(a)** on its first future value
   **(b)** only on its immediate past value
   **(c)** on a few past values
2. Gaussian white noise is:
   **(a)** biased
   **(b)** uncorrelated
   **(c)** colored
3. The basic state-variable model is applicable:
   **(a)** to time-varying and nonstationary systems
   **(b)** only to time-invariant and nonstationary systems
   **(c)** only to time-varying and stationary systems
4. If no disturbance is present in the state equation, then:
   **(a)** $\mathbf{m}_\mathbf{x}(k+1) = \mathbf{\Phi m}_\mathbf{x}(k)$
   **(b)** $\mathbf{P}_\mathbf{x}(i, j) = \mathbf{I}$ for all $i$ and $j$
   **(c)** $\mathbf{P}_\mathbf{x}(k+1) = \mathbf{\Phi P}_\mathbf{x}(k)\mathbf{\Phi}'$
5. A steady-state value of $\mathbf{P}_\mathbf{x}(k)$ exists if our system is:
   **(a)** time invariant and has all its poles inside the unit circle
   **(b)** time varying and stable
   **(c)** time invariant and has some of its poles outside the unit circle
6. In general, signal-to-noise ratio is:
   **(a)** always a constant
   **(b)** a function of time
   **(c)** a polynomial in time
7. If measurements are perfect, then:
   **(a)** $\mathbf{P}_\mathbf{z}(k+1) = \mathbf{HP}_\mathbf{x}(k+1)\mathbf{H}'$
   **(b)** $\mathbf{Q} = \mathbf{0}$
   **(c)** $\mathbf{m}_\mathbf{z}(k+1) = \mathbf{0}$
8. Which statistical statements apply to the description of the basic state-variable model?
   **(a)** $\mathbf{w}(k)$ and $\mathbf{v}(k)$ are jointly Gaussian
   **(b)** $\mathbf{w}(k)$ and $\mathbf{v}(k)$ are correlated
   **(c)** $\mathbf{w}(k)$ and $\mathbf{v}(k)$ are zero mean
   **(d)** $\mathbf{Q}(k) > 0$
   **(e)** $\mathbf{R}(k) > 0$
   **(f)** $\mathbf{x}(0)$ is non-Gaussian
9. $\{\mathbf{x}(k), k = 0, 1, \ldots\}$ is a Gauss–Markov sequence if $\mathbf{x}(0)$ and $\mathbf{w}(k)$ are:
   **(a)** independent
   **(b)** jointly Gaussian
   **(c)** uncorrelated

10. A random sequence is a (an) ______________ family of random variables:
(a) large
(b) Markov
(c) indexed

11. To compute signal-to-noise ratio for a two-input single-output model, we need to know:
(a) both $q$ and $r$
(b) $q$ but not $r$
(c) just $q/r$

12. A first-order Markov sequence is completely characterized by which probability density functions?
(a) $p[\mathbf{s}(t_1)]$
(b) $p[\mathbf{s}(t_i|t_1)]$
(c) $p[\mathbf{s}(t_i)|\mathbf{s}(t_{i-1})]$

## PROBLEMS

**15-1.** Prove Theorem 15-2, and then show that for Gaussian white noise $\mathbf{E}\{\mathbf{s}(t_m)|\mathbf{s}(t_{m-1}), \ldots, \mathbf{s}(t_1)\} = \mathbf{E}\{\mathbf{s}(t_m)\}$.

**15-2.** (a) Derive the solution to the time-varying state equation (15-17) that is given in (15-23).

(b) The solution in (15-23) expresses $\mathbf{x}(k)$ as a function of $\mathbf{x}(0)$; derive the following solution that expresses $\mathbf{x}(k)$ as a function of $\mathbf{x}(j)$, where $j < k$:

$$\mathbf{x}(k) = \mathbf{\Phi}(k, j)\mathbf{x}(j) + \sum_{m=j+1}^{k} \mathbf{\Phi}(k, m)[\mathbf{\Gamma}(m, m-1)\mathbf{w}(m-1) + \mathbf{\Psi}(m, m-1)\mathbf{u}(m-1)]$$

**15-3.** In the text it is stated that "By stacking up $\mathbf{x}(1), \mathbf{x}(2), \ldots$ into a supervector it is easily seen that this supervector is just a linear transformation of jointly Gaussian quantities $\mathbf{x}(0), \mathbf{w}(0), \mathbf{w}(1), \ldots$; hence, $\mathbf{x}(1), \mathbf{x}(2), \ldots$ are themselves *jointly Gaussian*." Demonstrate the details of this "stacking" procedure for $\mathbf{x}(1), \mathbf{x}(2)$, and $\mathbf{x}(3)$.

**15-4.** Derive the formula for the cross-covariance of $\mathbf{x}(k)$, $\mathbf{P}_\mathbf{x}(i, j)$ given in (15-30). [*Hint:* Use the *solution* to the state equation that is given in part (b) of Problem 15-2.]

**15-5.** Derive the first- and second-order statistics of measurement vector $\mathbf{z}(k)$, that are summarized in Theorem 15-6.

**15-6.** Reconsider the basic state-variable model when $\mathbf{x}(0)$ is correlated with $\mathbf{w}(0)$, and $\mathbf{w}(k)$ and $\mathbf{v}(k)$ are correlated $[\mathbf{E}\{\mathbf{w}(k)\mathbf{v}'(k)\} = \mathbf{S}(k)]$.
(a) Show that the covariance equation for $\mathbf{z}(k)$ remains unchanged.
(b) Show that the covariance equation for $\mathbf{x}(k)$ is changed, but only at $k = 1$.
(c) Compute $\mathbf{E}\{\mathbf{z}(k+1)\mathbf{z}'(k)\}$.

**15-7.** In this problem, assume that $u(k)$ and $\nu(k)$ are individually Gaussian and uncorrelated. Impulse response $h$ depends on parameter $a$, where $a$ is a Gaussian random variable that is statistically independent of $u(k)$ and $\nu(k)$.
(a) Evaluate $\mathbf{E}\{z(k)\}$.
(b) Explain whether or not $y(k)$ is Gaussian.

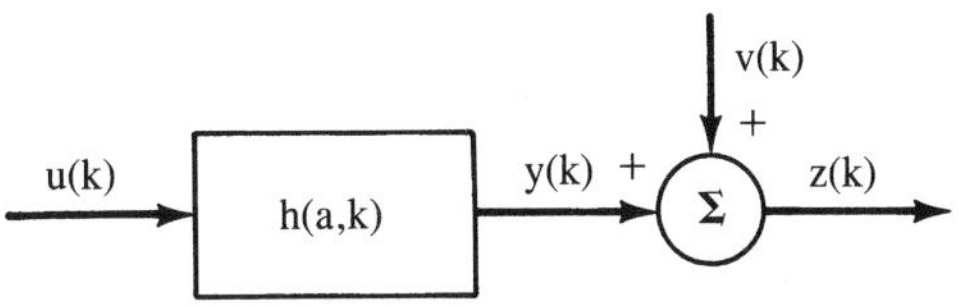


**Figure P15-7**

**15-8.** (Kiet D. Ngo, Spring, 1992) White noise is very useful in theoretical studies; however, it usually does not exist in real physical systems. The typical noise in physical systems is colored. A first-order Gauss–Markov process $x(k)$ has an autocorrelation function $r_x(k) = \sigma_w^2 a^{|k|}$ and spectral density $\Phi_x(\omega) = \sigma_w^2(1-a^2)/(1+a^2-2a\cos\omega)$. Find the transfer function of the filter that will shape a unity-variance Gaussian white noise process into a first-order Gauss–Markov process whose autocorrelation and spectral density are as given. Draw the block diagram of the filter.

**15-9.** (Mike Walter, Spring 1992) Consider the following system:

$$x(k+2) = \frac{1}{2}x(k+1) + \frac{1}{4}x(k) + w(k)$$

$$z(k+2) = x(k+2) + v(k+2)$$

where $w(k)$ and $v(k)$ are wide-sense stationary white noise processes, $m_x(0) = 0$, $m_x(1) = 0$, and $q = 25$.

**(a)** Find a basic state-variable model for this system.

**(b)** Find the steady-state values of $\mathbf{E}\{x^2(k)\}$ and $\mathbf{E}\{x(k+1)x(k)\}$.

**15-10.** (Richard S. Lee, Jr., Spring 1992) Consider the following measurement equation for the basic state-variable model:

$$\mathbf{z}(k+1) = \mathbf{H}(k+1)\mathbf{x}(k+1) + \mathbf{v}(k+1)$$

where $\mathbf{x}(k+1) = \text{col}\,(x_1(k+1), x_2(k+1))$:

$$x_1(k+1) = x_1(k) + e^{-k}x_2(k) + w(k)$$

$$x_2(k+1) = \frac{1}{2}x_2(k) + u(k)$$

$w(k)$ and $\mathbf{v}(k)$ are mutually uncorrelated, zero-mean, jointly Gaussian white noise sequences, and $\mathbf{E}\{w(i)w(j)\} = \delta_{ij}$.

**(a)** Determine the state equation of the basic state-variable model.

**(b)** Given that $\mathbf{E}\{x_1(0)\} = \mathbf{E}\{x_2(0)\} = 0$, $u(0) = 1$, $u(1) = 0.5$, and $\mathbf{P}_\mathbf{x}(0) = \mathbf{I}$, find $\mathbf{m}_\mathbf{x}(1)$, $\mathbf{m}_\mathbf{x}(2)$, $\mathbf{P}_\mathbf{x}(1)$, and $\mathbf{P}_\mathbf{x}(2)$.

**15-11.** (Lance Kaplan, Spring 1992) Suppose the basic state-variable model is modified so that $\mathbf{x}(0)$ is not multivariate Gaussian but has finite moments [in this problem, $\mathbf{u}(k)$ is deterministic].

**(a)** Is $\mathbf{x}(k)$, in general, multivariate Gaussian for any $k \geq 0$?

**(b)** Now let $\mathbf{\Phi}(k+1, k) = \mathbf{\Phi}$, where all the eigenvalues of $\mathbf{\Phi}$ have magnitude less than unity. What can be said about the multivariate probability density function of $\mathbf{x}(k)$ as $k$ becomes large?

**(c)** What are the mean and covariance of $\mathbf{x}(k)$ as $k$ approaches infinity?

**(d)** Let $\bar{\mathbf{x}} = \lim \mathbf{x}(k)$ as $k \to \infty$. Does the distribution of $\bar{\mathbf{x}}$ depend on the initial condition $\mathbf{x}(0)$?

**15-12.** (Iraj Manocheri, Spring 1992) Consider the state equation

$$x(k+1) = \frac{1}{2}x(k) + u(k)$$

where $x(0) = 2$, $\mathbf{E}\{u(k)\} = 1$, and $\text{cov}\,[u(k)] = \delta(k_2 - k_1)$.

**(a)** Find the recursions for the mean and covariance of $x(k)$.

**(b)** Find the steady-state mean and covariance of $x(k)$.

# LESSON 16

# State Estimation: Prediction


## SUMMARY

Prediction, filtering, and smoothing are three types of mean-squared state estimation that have been developed during the past 35 years. The purpose of this lesson is to study *prediction*.

A predicted estimate of a state vector $\mathbf{x}(k)$ uses measurements that occur earlier than $t_k$ and a model to make the transition from the last time point, say $t_j$, at which a measurement is available to $t_k$. The success of prediction depends on the quality of the model. In state estimation we use the state equation model. Without a model, prediction is dubious at best.

Filtered and predicted state estimates are very tightly coupled together; hence, most of the results from this lesson cannot be implemented until we have developed results for mean-squared filtering. This is done in Lesson 17.

A *single-stage predicted estimate* of $\mathbf{x}(k)$ is denoted $\hat{\mathbf{x}}(k|k-1)$. It is the mean-squared estimate of $\mathbf{x}(k)$ that uses all the measurements up to and including the one made at time $t_{k-1}$; hence, a single-stage predicted estimate looks exactly one time point into the future. This estimate is needed by the Kalman filter in Lesson 17.

The difference between measurement $\mathbf{z}(k)$ and its single-stage predicted value, $\hat{\mathbf{z}}(k|k-1)$, is called the *innovations process*. This process plays a very important role in mean-squared filtering and smoothing. Interestingly enough, the innovations process for our basic state-variable model is a zero-mean Gaussian white noise sequence.

When you complete this lesson, you will be able to (1) derive and use the important single-stage predictor; (2) derive and use more general state predictors; and (3) explain the innovations process and state its important properties.

## INTRODUCTION

We have mentioned, a number if times in this book, that in state estimation three situations are possible depending on the relative relationship of total number of available measurements, $N$, and the time point, $k$, at which we estimate state vector $\mathbf{x}(k)$: prediction ($N < k$), filtering ($N = k$), and smoothing ($N > k$). In this lesson we develop algorithms for mean-squared predicted estimates, $\hat{\mathbf{x}}_{MS}(k|j)$ of state $\mathbf{x}(k)$. To simplify our notation, we shall abbreviate $\hat{\mathbf{x}}_{MS}(k|j)$ as $\hat{\mathbf{x}}(k|j)$. [Just in case you have forgotten what the notation $\hat{\mathbf{x}}(k|j)$ stands for, see Lesson 2]. Note that, in prediction, $k > j$.

## SINGLE-STAGE PREDICTOR

The most important predictor of $\mathbf{x}(k)$ for our future work on filtering and smoothing is the *single-stage predictor* $\hat{\mathbf{x}}(k|k-1)$. From the fundamental theorem of estimation theory (Theorem 13-1), we know that

$$\hat{\mathbf{x}}(k|k-1) = \mathbf{E}\{\mathbf{x}(k)|\mathbf{Z}(k-1)\} \tag{16-1}$$

where

$$\mathbf{Z}(k-1) = \text{col}\,(\mathbf{z}(1), \mathbf{z}(2), \ldots, \mathbf{z}(k-1)) \tag{16-2}$$

It is very easy to derive a formula for $\hat{\mathbf{x}}(k|k-1)$ by operating on both sides of the state equation

$$\mathbf{x}(k) = \mathbf{\Phi}(k, k-1)\mathbf{x}(k-1) + \mathbf{\Gamma}(k, k-1)\mathbf{w}(k-1) + \mathbf{\Psi}(k, k-1)\mathbf{u}(k-1) \tag{16-3}$$

with the linear expectation operator $\mathbf{E}\{\cdot|\mathbf{Z}(\text{k-1})\}$. Doing this, we find that

$$\hat{\mathbf{x}}(k|k-1) = \mathbf{\Phi}(k, k-1)\hat{\mathbf{x}}(k-1|k-1) + \mathbf{\Psi}(k, k-1)\mathbf{u}(k-1) \tag{16-4}$$

where $k = 1, 2, \ldots$. To obtain (16-4), we have used the facts that $\mathbf{E}\{\mathbf{w}(k-1)\} = \mathbf{0}$ and $\mathbf{u}(k-1)$ is deterministic.

Observe, from (16-4), that the single-stage predicted estimate $\hat{\mathbf{x}}(k|k-1)$ depends on the filtered estimate $\hat{\mathbf{x}}(k-1|k-1)$ of the preceding state vector $\mathbf{x}(k-1)$. At this point, (16-4) is an interesting theoretical result; but there is nothing much we can do with it, because we do not as yet know how to compute filtered state estimates. In Lesson 17 we shall begin our study into filtered state estimates and shall learn that such estimates of $\mathbf{x}(k)$ depend on predicted estimates of $\mathbf{x}(k)$, just as predicted estimates of $\mathbf{x}(k)$ depend on filtered estimates of $\mathbf{x}(k-1)$; thus, *filtered and predicted state estimates are very tightly coupled together.*

Let $\mathbf{P}(k|k-1)$ denote the error-covariance matrix that is associated with $\hat{\mathbf{x}}(k|k-1)$; i.e.,

$$\mathbf{P}(k|k-1) = \mathbf{E}\{[\tilde{\mathbf{x}}(k|k-1) - \mathbf{m}_{\tilde{\mathbf{x}}}(k|k-1][\tilde{\mathbf{x}}(k|k-1) - \mathbf{m}_{\tilde{\mathbf{x}}}(k|k-1)]'\} \tag{16-5}$$

where

$$\tilde{\mathbf{x}}(k|k-1) = \mathbf{x}(k) - \hat{\mathbf{x}}(k|k-1) \tag{16-6}$$

Additionally, let $\mathbf{P}(k-1|k-1)$ denote the error-covariance matrix that is associated with $\hat{\mathbf{x}}(k-1|k-1)$; i.e., ← *filtering cov matrix*

$$\mathbf{P}(k-1|k-1) = \mathbf{E}\{[\tilde{\mathbf{x}}(k-1|k-1)$$
$$-\mathbf{m}_{\tilde{\mathbf{x}}}(k-1|k-1)][\tilde{\mathbf{x}}(k-1|k-1) - \mathbf{m}_{\tilde{\mathbf{x}}}(k-1|k-1)]'\} \tag{16-7}$$

For our basic state-variable model (see Property 1 of Lesson 13), $\mathbf{m}_{\tilde{\mathbf{x}}}(k|k-1) = \mathbf{0}$ and $\mathbf{m}_{\tilde{\mathbf{x}}}(k-1|k-1) = \mathbf{0}$, so *Jointly Gaussian*

$$\mathbf{P}(k|k-1) = \mathbf{E}\{\tilde{\mathbf{x}}(k|k-1)\tilde{\mathbf{x}}'(k|k-1)\} \tag{16-8}$$

and

$$\mathbf{P}(k-1|k-1) = \mathbf{E}\{\tilde{\mathbf{x}}(k-1|k-1)\tilde{\mathbf{x}}'(k-1|k-1)\} \tag{16-9}$$

Combining (16-3) and (16-4), we see that

$$\tilde{\mathbf{x}}(k|k-1) = \mathbf{\Phi}(k,k-1)\tilde{\mathbf{x}}(k-1|k-1) + \mathbf{\Gamma}(k,k-1)\mathbf{w}(k-1) \tag{16-10}$$

A straightforward calculation leads to the following formula for $\mathbf{P}(k|k-1)$:

$$\mathbf{P}(k|k-1) = \mathbf{\Phi}(k,k-1)\mathbf{P}(k-1|k-1)\mathbf{\Phi}'(k,k-1)$$
$$+\mathbf{\Gamma}(k,k-1)\mathbf{Q}(k-1)\mathbf{\Gamma}'(k,k-1) \tag{16-11}$$

where $k = 1, 2, \ldots$. ← *covariance of estimate of state, NOT of (15-24) which is covariance of state.*

Observe, from (16-4) and (16-11), that $\hat{\mathbf{x}}(0|0)$ and $\mathbf{P}(0|0)$ initialize the single-stage predictor and its error covariance. Additionally,

$$\hat{\mathbf{x}}(0|0) = \mathbf{E}\{\mathbf{x}(0)|\text{no measurements}\} = \mathbf{m}_{\mathbf{x}}(0) \tag{16-12}$$

and

$$\mathbf{P}(0|0) = \mathbf{E}\{\tilde{\mathbf{x}}(0|0)\tilde{\mathbf{x}}'(0|0)\} = \mathbf{E}\{[\mathbf{x}(0) - \mathbf{m}_{\mathbf{x}}(0)][\mathbf{x}(0) - \mathbf{m}_{\mathbf{x}}(0)]'\} = \mathbf{P}(0) \tag{16-13}$$

Finally, recall (Property 4 of Lesson 13) that both $\hat{\mathbf{x}}(k|k-1)$ and $\tilde{\mathbf{x}}(k|k-1)$ are Gaussian.

## A GENERAL STATE PREDICTOR

In this section we generalize the results of the preceding section so as to obtain predicted values of $\mathbf{x}(k)$ that look further into the future than just one step. We shall determine $\hat{\mathbf{x}}(k|j)$ where $k > j$ under the assumption that filtered state estimate $\hat{\mathbf{x}}(j|j)$ and its error-covariance matrix $\mathbf{E}\{\tilde{\mathbf{x}}(j|j)\tilde{\mathbf{x}}'(j|j)\} = \mathbf{P}(j|j)$ are known for some $j = 0, 1, \ldots$.

### Theorem 16-1

**a.** *If input* $\mathbf{u}(k)$ *is deterministic, or does not depend on any measurements, then the mean-squared predicted estimator of* $\mathbf{x}(k)$, $\hat{\mathbf{x}}(k|j)$, *is given by the expression*

$$\hat{\mathbf{x}}(k|j) = \mathbf{\Phi}(k, j)\hat{\mathbf{x}}(j|j)$$

$$+ \sum_{i=j+1}^{k} \mathbf{\Phi}(k, i)\mathbf{\Psi}(i, i-1)\mathbf{u}(i-1), \qquad k > j \qquad (16\text{-}14)$$

**b.** *The vector random sequence* $\{\tilde{\mathbf{x}}(k|j), k = j+1, j+2, \ldots\}$ *is:*

**i.** *zero mean*

**ii.** *Gaussian, and*

**iii.** *first-order Markov, and*

**iv.** *its covariance matrix is governed by*

$$\mathbf{P}(k|j) = \mathbf{\Phi}(k, k-1)\mathbf{P}(k-1|j)\mathbf{\Phi}'(k, k-1)$$
$$+\mathbf{\Gamma}(k, k-1)\mathbf{Q}(k-1)\mathbf{\Gamma}'(k, k-1) \qquad (16\text{-}15)$$

Before proving this theorem, let us observe that the prediction formula (16-14) is intuitively what we would expect. Why is this so? Suppose we have processed all the measurements $\mathbf{z}(1), \mathbf{z}(2), \ldots, \mathbf{z}(j)$ to obtain $\hat{\mathbf{x}}(j|j)$ and are asked to predict the value of $\mathbf{x}(k)$, where $k > j$. No additional measurements can be used during prediction. All that we can therefore use is our dynamical state equation. When that equation is used for purposes of prediction, we neglect the random disturbance term, because the disturbances cannot be measured. We can only use measured quantities to assist our prediction efforts. The simplified state equation is

$$\mathbf{x}(k+1) = \mathbf{\Phi}(k+1, k)\mathbf{x}(k) + \mathbf{\Psi}(k+1, k)\mathbf{u}(k) \qquad (16\text{-}16)$$

a solution of which is (see Problem 15-2b)

$$\mathbf{x}(k) = \mathbf{\Phi}(k, j)\mathbf{x}(j) + \sum_{i=j+1}^{k} \mathbf{\Phi}(k, i)\mathbf{\Psi}(i, i-1)\mathbf{u}(i-1) \qquad (16\text{-}17)$$

Substituting $\hat{\mathbf{x}}(j|j)$ for $\mathbf{x}(j)$, we obtain the predictor in (16-14). In our proof of Theorem 16-1, we establish (16-14) in a more rigorous manner.

*Proof*

a. The solution to state equation (16-3), for $\mathbf{x}(k)$, can be expressed in terms of $\mathbf{x}(j)$, where $j < k$, as (see Problem 15-2b)

$$\mathbf{x}(k) = \mathbf{\Phi}(k, j)\mathbf{x}(j) + \sum_{i=j+1}^{k} \mathbf{\Phi}(k, i)[\mathbf{\Gamma}(i, i-1)\mathbf{w}(i-1)$$
$$+\mathbf{\Psi}(i, i-1)\mathbf{u}(i-1)] \qquad (16\text{-}18)$$

We apply the fundamental theorem of estimation theory to (16-18) by taking the conditional expectation with respect to $\mathbf{Z}(j)$ on both sides of it. Doing this, we find that

$$\hat{\mathbf{x}}(k|j) = \mathbf{\Phi}(k, j)\hat{\mathbf{x}}(j|j) + \sum_{i=j+1}^{k} \mathbf{\Phi}(k, i)[\mathbf{\Gamma}(i, i-1)\mathbf{E}\{\mathbf{w}(i-1)|\mathbf{Z}(j)\}$$
$$+\mathbf{\Psi}(i, i-1)\mathbf{E}\{\mathbf{u}(i-1)|\mathbf{Z}(j)\}] \tag{16-19}$$

Note that $\mathbf{Z}(j)$ depends at most on $\mathbf{x}(j)$, which, in turn, depends at most on $\mathbf{w}(j-1)$. Consequently,

$$\mathbf{E}\{\mathbf{w}(i-1)|\mathbf{Z}(j)\} = \mathbf{E}\{\mathbf{w}(i-1)|\mathbf{w}(0), \mathbf{w}(1), \ldots, \mathbf{w}(j-1)\} \tag{16-20}$$

where $i = j+1, j+2, \ldots, k$. Because of this range of values on argument $i$, $\mathbf{w}(i-1)$ is never included in the conditioning set of values $\mathbf{w}(0), \mathbf{w}(1), \ldots, \mathbf{w}(j-1)$; hence,

$$\mathbf{E}\{\mathbf{w}(i-1)|\mathbf{Z}(j)\} = \mathbf{E}\{\mathbf{w}(i-1)\} = \mathbf{0} \tag{16-21}$$

for all $i = j+1, j+2, \ldots, k$.

Note, also, that

$$\mathbf{E}\{\mathbf{u}(i-1)|\mathbf{Z}(j)\} = \mathbf{E}\{\mathbf{u}(i-1)\} = \mathbf{u}(i-1) \tag{16-22}$$

because we have assumed that $\mathbf{u}(i-1)$ does not depend on any of the measurements. Substituting (16-21) and (16-22) into (16-19), we obtain the prediction formula (16-14).

**b(i) and b(ii).** Have already been proved in Properties 1 and 4 of Lesson 13.

**b(iii).** Starting with $\tilde{\mathbf{x}}(k|j) = \mathbf{x}(k) - \hat{\mathbf{x}}(k|j)$ and substituting (16-18) and (16-14) into this relation, we find that

$$\tilde{\mathbf{x}}(k|j) = \mathbf{\Phi}(k, j)\tilde{\mathbf{x}}(j|j) + \sum_{i=j+1}^{k} \mathbf{\Phi}(k, i)\mathbf{\Gamma}(i, i-1)\mathbf{w}(i-1) \tag{16-23}$$

This equation looks quite similar to the solution of state equation (16-3), when $\mathbf{u}(k) = \mathbf{0}$ (for all $k$), e.g., see (16-18); hence, $\tilde{\mathbf{x}}(k|j)$ also satisfies the state equation

$$\tilde{\mathbf{x}}(k|j) = \mathbf{\Phi}(k, k-1)\tilde{\mathbf{x}}(k-1|j) + \mathbf{\Gamma}(k, k-1)\mathbf{w}(k-1) \tag{16-24}$$

Because $\tilde{\mathbf{x}}(k|j)$ depends only on its previous value, $\tilde{\mathbf{x}}(k-1|j)$, it is first-order Markov.

**b(iv).** We derived a recursive covariance equation for $\mathbf{x}(k)$ in Theorem 15-5. That equation is (15-29). Because $\tilde{\mathbf{x}}(k|j)$ satisfies the state equation for $\mathbf{x}(k)$, its covariance $\mathbf{P}(k|j)$ is also given by (15-29). We have rewritten this equation as in (16-15). □

Observe that by setting $j = k-1$ in (16-14) we obtain our previously derived single-stage predictor $\hat{\mathbf{x}}(k|k-1)$.

Theorem 16-1 is quite limited because presently the only values of $\hat{\mathbf{x}}(j|j)$ and $\mathbf{P}(j|j)$ that we know are those at $j = 0$. For $j = 0$, (16-14) becomes

$$\hat{\mathbf{x}}(k|0) = \mathbf{\Phi}(k, 0)\mathbf{m}_x(0) + \sum_{i=1}^{k} \mathbf{\Phi}(k, i)\mathbf{\Psi}(i, i-1)\mathbf{u}(i-1) \tag{16-25}$$

The reader might feel that this predictor of $\mathbf{x}(k)$ becomes poorer and poorer as $k$ gets farther and farther away from zero. The following example demonstrates that this is not necessarily true.

**EXAMPLE 16-1**

Let us examine prediction performance, as measured by $p(k|0)$, for the first-order system

$$x(k+1) = \frac{1}{\sqrt{2}}x(k) + w(k) \tag{16-26}$$

where $q = 25$ and $p_x(0)$ is variable. Quantity $p(k|0)$, which in the case of a scalar state vector is a variance, is easily computed from the recursive equation

$$p(k|0) = \frac{1}{2}p(k-1|0) + 25 \tag{16-27}$$

for $k = 1, 2, \ldots$. Two cases are summarized in Figure 16-1. When $p_x(0) = 6$, we have relatively small uncertainty about $\hat{x}(0|0)$, and, as we expected, our predictions of $x(k)$ for $k \geq 1$ do become worse, because $p(k|0) > 6$ for all $k \geq 1$. After a while, $p(k|0)$ reaches a limiting value equal to 50. When this occurs, we are estimating $\hat{x}(k|0)$ by a number that is very close to zero, because $\hat{x}(k|0) = (1/\sqrt{2})^k\hat{x}(0|0)$, and $(1/\sqrt{2})^k$ approaches zero for large values of $k$.

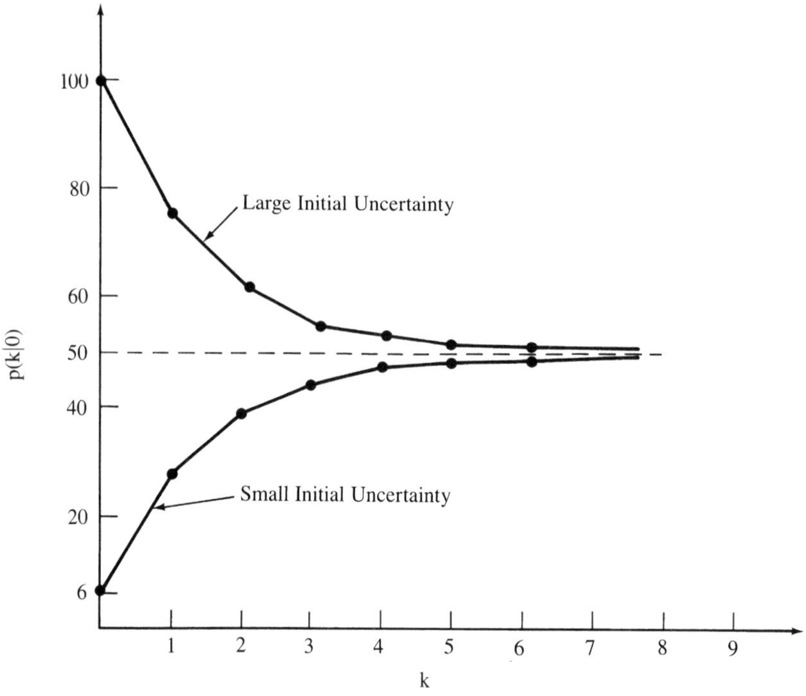


**Figure 16-1** Prediction error variance $p(k|0)$.

When $p_x(0) = 100$, we have large uncertainty about $\hat{x}(0|0)$, and, perhaps to our surprise, our predictions of $x(k)$ for $k \geq 1$, i.e., $\hat{x}(k|0) = (1/\sqrt{2})^k\hat{x}(0|0)$, improve in performance, because $p(k|0) < 100$ for all $k \geq 1$. In this case the predictor discounts the large initial uncertainty; however, as in the former case, $p(k|0)$ again reaches the limiting value of 50.

For suitably large values of $k$, the predictor is completely insensitive to $p_x(0)$. It reaches a *steady-state* level of performance equal to 50, which can be predetermined by setting $p(k|0)$ and $p(k-1|0)$ equal $\bar{p}$, in (16-27), and solving the resulting equation for $\bar{p}$. □

Prediction is possible only because we have a known dynamical model, our state-variable model. Without a model, prediction is dubious at best (e.g., try predicting tomorrow's price of a stock listed on any stock exchange using today's closing price).

## THE INNOVATIONS PROCESS

Suppose we have just computed the single-stage predicted estimate of $\mathbf{x}(k+1)$, $\hat{\mathbf{x}}(k+1|k)$. Then the single-stage predicted estimate of $\mathbf{z}(k+1)$, $\hat{\mathbf{z}}(k+1|k)$, is

$$\hat{\mathbf{z}}(k+1|k) = \mathbf{E}\{\mathbf{z}(k+1)|\mathbf{Z}(k)\} = \mathbf{E}\{[\mathbf{H}(k+1)\mathbf{x}(k+1) + \mathbf{v}(k+1)]|\mathbf{Z}(k)\}$$

or

$$\hat{\mathbf{z}}(k+1|k) = \mathbf{H}(k+1)\hat{\mathbf{x}}(k+1|k) \tag{16-28}$$

The error between $\mathbf{z}(k+1)$ and $\hat{\mathbf{z}}(k+1|k)$ is $\tilde{\mathbf{z}}(k+1|k)$; i.e.,

$$\tilde{\mathbf{z}}(k+1|k) = \mathbf{z}(k+1) - \hat{\mathbf{z}}(k+1|k) \tag{16-29}$$

Signal $\tilde{\mathbf{z}}(k+1|k)$ is often referred to either as the *innovations process, prediction error process*, or *measurement residual process*. We shall refer to it as the innovations process, because this is most commonly done in the estimation theory literature (e.g., Kailath, 1968). The innovations process plays a very important role in mean-squared filtering and smoothing. We summarize important facts about it in the following:

**Theorem 16-2** (Innovations)

**a.** *The following representations of the innovations process* $\tilde{\mathbf{z}}(k+1|k)$ *are equivalent*:

$$\tilde{\mathbf{z}}(k+1|k) = \mathbf{z}(k+1) - \hat{\mathbf{z}}(k+1|k) \tag{16-30}$$

$$\tilde{\mathbf{z}}(k+1|k) = \mathbf{z}(k+1) - \mathbf{H}(k+1)\hat{\mathbf{x}}(k+1|k) \tag{16-31}$$

*or*

$$\tilde{\mathbf{z}}(k+1|k) = \mathbf{H}(k+1)\tilde{\mathbf{x}}(k+1|k) + \mathbf{v}(k+1) \tag{16-32}$$

**b.** *The innovations is a zero-mean Gaussian white noise sequence, with*

$$\begin{aligned}\mathbf{E}\{\tilde{\mathbf{z}}(k+1|k)\tilde{\mathbf{z}}'(k+1|k)\} &= \mathbf{P}_{\tilde{z}\tilde{z}}(k+1|k)\\ &= \mathbf{H}(k+1)\mathbf{P}(k+1|k)\mathbf{H}'(k+1) + \mathbf{R}(k+1)\end{aligned} \tag{16-33}$$

*Proof* (Mendel, 1983).

**a.** Substitute (16-28) into (16-29) to obtain (16-31). Next, substitute the measurement equation $\mathbf{z}(k+1) = \mathbf{H}(k+1)\mathbf{x}(k+1) + \mathbf{v}(k+1)$ into (16-31), and use the fact that $\tilde{\mathbf{x}}(k+1|k) = \mathbf{x}(k+1) - \hat{\mathbf{x}}(k+1|k)$ to obtain (16-32).

**b.** Because $\tilde{\mathbf{x}}(k+1|k)$ and $\mathbf{v}(k+1)$ are both zero mean, $\mathbf{E}\{\tilde{\mathbf{z}}(k+1|k)\} = \mathbf{0}$. The innovations is Gaussian because $\mathbf{z}(k+1)$ and $\hat{\mathbf{x}}(k+1|k)$ are Gaussian, and therefore $\tilde{\mathbf{z}}(k+1|k)$ is a linear transformation of Gaussian random vectors. To prove that $\tilde{\mathbf{z}}(k+1|k)$ is white noise, we must show that

$$\mathbf{E}\{\tilde{\mathbf{z}}(i+1|i)\tilde{\mathbf{z}}'(j+1|j)\} = \mathbf{P}_{\tilde{z}\tilde{z}}(i+1|i)\delta_{ij} \tag{16-34}$$

We shall consider the cases $i > j$ and $i = j$ leaving the case $i < j$ as an exercise for the reader. When $i > j$,

$$\begin{aligned}\mathbf{E}\{\tilde{\mathbf{z}}(i+1|i)\tilde{\mathbf{z}}'(j+1|j)\} &= \mathbf{E}\{[\mathbf{H}(i+1)\tilde{\mathbf{x}}(i+1|i) + \mathbf{v}(i+1)]\\ &\qquad [\mathbf{H}(j+1)\tilde{\mathbf{x}}(j+1|j) + \mathbf{v}(j+1)]'\}\\ &= \mathbf{E}\{\mathbf{H}(i+1)\tilde{\mathbf{x}}(i+1|i)\\ &\qquad [\mathbf{H}(j+1)\tilde{\mathbf{x}}(j+1|i) + \mathbf{v}(j+1)]'\}\end{aligned}$$

because $\mathbf{E}\{\mathbf{v}(i+1)\mathbf{v}'(j+1)\} = \mathbf{0}$ and $\mathbf{E}\{\mathbf{v}(i+1)\tilde{\mathbf{x}}'(j+1|j)\} = \mathbf{0}$. The latter is true because, for $i > j$, $\tilde{\mathbf{x}}(j+1|j)$ does not depend on measurement $\mathbf{z}(i+1)$; hence, for $i > j$, $\mathbf{v}(i+1)$ and $\tilde{\mathbf{x}}(j+1|j)$ are independent, so $\mathbf{E}\{\mathbf{v}(i+1)\tilde{\mathbf{x}}'(j+1|j)\} = \mathbf{E}\{\mathbf{v}(i+1)\}\mathbf{E}\{\tilde{\mathbf{x}}'(j+1|j)\} = \mathbf{0}$. We continue as follows:

$$\begin{aligned}&\mathbf{E}\{\tilde{\mathbf{z}}(i+1|i)\tilde{\mathbf{z}}'(j+1|j)\}\\ &\quad = \mathbf{H}(i+1)\mathbf{E}\{\tilde{\mathbf{x}}(i+1|i)[\mathbf{z}(j+1) - \mathbf{H}(j+1)\hat{\mathbf{x}}(j+1|j)]'\} = \mathbf{0}\end{aligned}$$

by repeated application of the orthogonality principle (Corollary 13-3).

When $i = j$,

$$\begin{aligned}\mathbf{P}_{\tilde{z}\tilde{z}}(i+1|i) &= \mathbf{E}\{[\mathbf{H}(i+1)\tilde{\mathbf{x}}(i+1|i) + \mathbf{v}(i+1)][\mathbf{H}(i+1)\tilde{\mathbf{x}}(i+1|i)\\ &\qquad + \mathbf{v}(i+1)]'\} = \mathbf{H}(i+1)\mathbf{P}(i+1|i)\mathbf{H}'(i+1) + \mathbf{R}(i+1)\end{aligned}$$

because, once again, $\mathbf{E}\{\mathbf{v}(i+1)\tilde{\mathbf{x}}'(i+1|i)\} = \mathbf{0}$, and $\mathbf{P}(i+1) = \mathbf{E}\{\tilde{\mathbf{x}}(i+1|i)\tilde{\mathbf{x}}'(i+1|i)\}$. □

In Lesson 17 the inverse of $\mathbf{P}_{\tilde{z}\tilde{z}}(k+1|k)$ is needed; hence, we shall assume that $\mathbf{H}(k+1)\mathbf{P}(k+1|k)\mathbf{H}'(k+1) + \mathbf{R}(k+1)$ is nonsingular. This is usually true and will always be true if, as in our basic state-variable model, $\mathbf{R}(k+1)$ is positive definite.

**Comment.** This is a good place to mention a very important historical paper by Tom Kailath, "A View of Three Decades of Linear Filtering Theory" (Kailath, 1974). More than 400 references are given in this paper. Kailath outlines the developments in the theory of linear least-squares estimation and pays particular attention to early mathematical works in the field and to more modern developments. He shows some of the many connections between least-squares filtering and other fields. The paper includes a very good historical account of the innovations process. His use of the term "least-squares filtering" is all-inclusive of just about everything covered in this text.

## COMPUTATION

Because prediction and filtering are inexorably intertwined, we defer discussions on computation of predicted state estimates until Lessons 17 and 19.

## Supplementary Material

## LINEAR PREDICTION

We will see, in Lesson 17, that the filtered state estimate $\hat{\mathbf{x}}(k|k)$ is a linear transformation of the measurements $\mathbf{Z}(k)$; hence, the predicted estimates of $\mathbf{x}(k)$ are also linear transformations of the measurements. The linearity of these "digital filters" is due to the fact that all sources of uncertainty in our basic state-variable model are Gaussian, so we can expand the conditional expectation as in (13-12).

A well-studied problem in digital signal processing (e.g., Haykin, 1991), is the *linear prediction problem*, in which the structure of the predictor is fixed ahead of time to be a linear transformation of the data. The "forward" linear prediction problem is to predict a future value of a *stationary* discrete-time random sequence $\{y(k), k = 1, 2, \ldots\}$ using a set of past samples of the sequence. Let $\hat{y}(k)$ denote the predicted value of $y(k)$ that uses $M$ past measurements; i.e.,

$$\hat{y}(k) = \sum_{i=1}^{M} a_{M,i} y(k-i) \tag{16-35}$$

The forward prediction error filter (PEF) coefficients, $a_{M,1}, \ldots, a_{M,M}$, are chosen so that either the mean-squared or least-squares forward prediction error (FPE), $f_M(k)$, is minimized, where

$$f_M(k) = y(k) - \hat{y}(k) \tag{16-36}$$

Note that in this filter design problem the length of the filter, $M$, is treated as a design variable, which is why the PEF coefficients are argumented by $M$. Note, also, that the PEF coefficients do not depend on $t_k$; i.e., the PEF is a constant coefficent predictor. We will see, in Lesson 17, that our state predictor and filter are time-varying digital filters.

Predictor $\hat{y}(k)$ uses a finite window of past measurements: $y(k-1), y(k-2), \ldots, y(k-M)$. This window of measurements is different for different values of $t_k$. This use of the measurements is quite different than our use of the measurements in state prediction, filtering, and smoothing. The latter are based on an expanding memory, whereas the former is based on a fixed memory (see Lesson 3).

We return to the minimization of $\mathbf{E}\{f_M^2(k)\}$ in Lesson 19.

Digital signal-processing specialists have invented a related type of linear prediction named *backward linear prediction* in which the objective is to predict a past value of a stationary discrete-time random sequence using a set of future values of the sequence. Of course, backward linear prediction is not prediction at all; it is smoothing (interpolation). But the term backward linear prediction is firmly entrenched in the DSP literature.

Let $\hat{y}(k-M)$ denote the predicted value of $y(k-M)$ that uses $M$ future measurements; i.e.,

$$\hat{y}(k-M) = \sum_{i=1}^{M} c_{M,i}\, y(k-i+1) \tag{16-37}$$

The backward PEF coefficients $c_{M,1}, \ldots, c_{M,M}$ are chosen so that either the mean-squared or least-squares backward prediction error (BPE), $b_M(k)$, is minimized, where

$$b_M(k) = y(k-M) - \hat{y}(k-M) \tag{16-38}$$

Both the forward and backward PEFs have a filter architecture associated with them that is known as a *tapped delay line* (Figure 16-2). Remarkably, when the two filter design problems are considered simultaneously, their solutions can be shown to be coupled. This is the main reason for introducing the backward prediction error filter design problem. Coupling occurs between the forward and backward prediction errors. For example, if the forward and backward PEF coefficients are chosen to minimize $\mathbf{E}\{f_M^2(k)\}$ and $\mathbf{E}\{b_M^2(k)\}$, respectively, then we can show that the following two equations are true:

$$f_m(k) = f_{m-1}(k) + \Gamma_m^* b_{m-1}(k-1) \tag{16-39}$$

$$b_m(k) = b_{m-1}(k-1) + \Gamma_m f_{m-1}(k) \tag{16-40}$$

where $m = 1, 2, \ldots, M$, $k = 1, 2, \ldots,$ and coefficent $\Gamma_m$ is called a *reflection coefficient*. The forward and backward PEF coefficients can be calculated from the reflection coefficients.

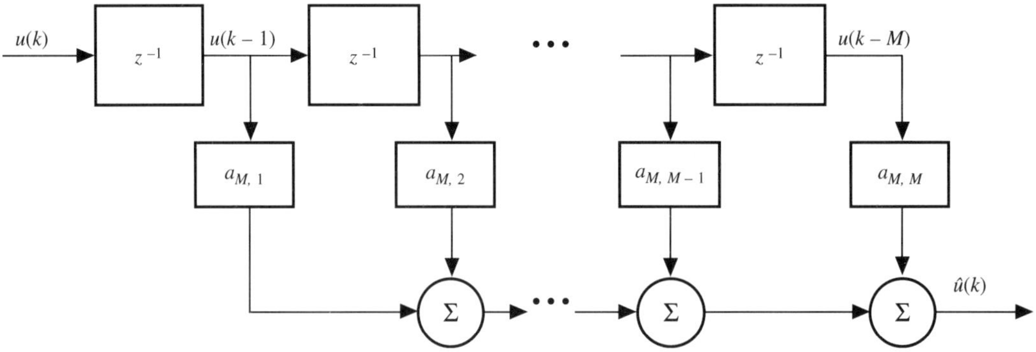


**Figure 16-2** Tapped delay line architecture for the forward prediction error filter. A similar structure exists for the backward prediction error filter.

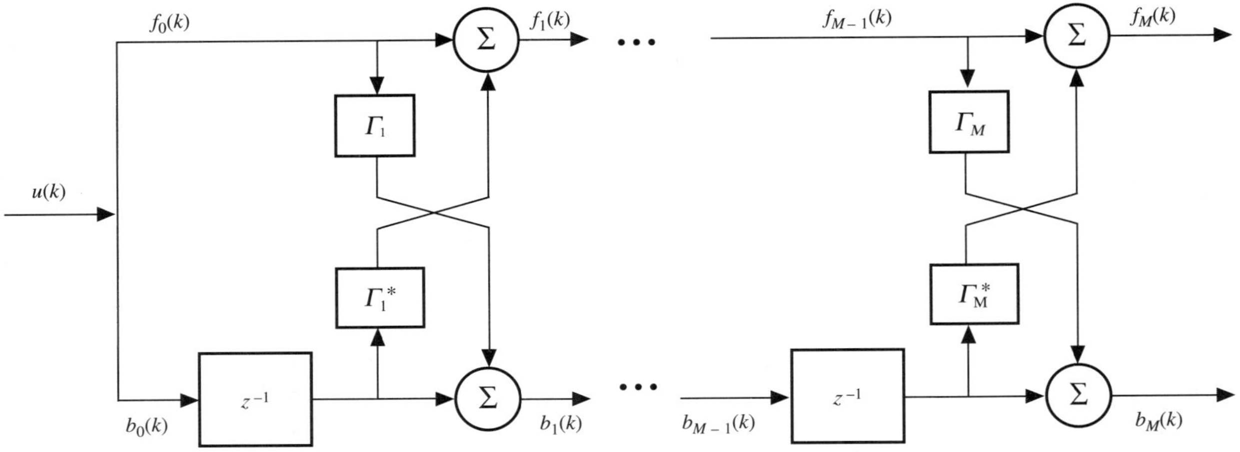


**Figure 16-3** Lattice prediction error filter of order $M$.

A block diagram for (16-39) and (16-40) is depicted in Figure 16-3. The resulting filter architecture is called a *lattice*. Observe that the lattice filter is doubly recursive in both time, $k$, and filter order, $m$. The tapped delay line is only recursive in time. Changing its filter length leads to a completely new set of filter coefficients. Adding another stage to the lattice filter does not affect the earlier reflection coefficients. Consequently, the lattice filter is a very powerful architecture.

No such lattice architecture is known for mean-squared state estimators. The derivation of the lattice architecture depends to a large extent on the assumption that the process to be "predicted" is stationary. Our state-variable model is almost always associated with a nonstationary process. Lesson 19 treats the very special case where the state vector becomes stationary.

## SUMMARY QUESTIONS

1. The proper initial conditions for state estimation are:
   **(a)** $\hat{\mathbf{x}}(0|0) = \mathbf{0}$ and $\mathbf{P}(0|0) = \mathbf{P}_{\mathbf{x}}(0)$
   **(b)** $\hat{\mathbf{x}}(0|0) = \mathbf{m}_{\mathbf{x}}(0)$ and $\mathbf{P}(-1|0) = \mathbf{0}$
   **(c)** $\hat{\mathbf{x}}(0|0) = \mathbf{m}_{\mathbf{x}}(0)$ and $\mathbf{P}(0|0) = \mathbf{P}_{\mathbf{x}}(0)$
2. The statement $\mathbf{P}(k+1|k) = \mathbf{P}_{\mathbf{x}}(k+1)$ is:
   **(a)** always true
   **(b)** never true
   **(c)** obvious
3. Prediction is possible only because we have:
   **(a)** measurements
   **(b)** a known measurement model
   **(c)** a known state equation model
4. If $\tilde{\mathbf{z}}(k+1|k)$ is an innovations process, then:
   **(a)** it is white
   **(b)** $\tilde{\mathbf{x}}(k+1|k)$ is also an innovations process
   **(c)** it is numerically the same as $\mathbf{z}(k+1)$
5. The single-stage predictor is:
   **(a)** $\hat{\mathbf{x}}(k|k+1)$
   **(b)** $\hat{\mathbf{x}}(k|k-1)$
   **(c)** $\hat{\mathbf{x}}(k|k)$
6. If $\mathbf{u}(k) = \mathbf{0}$ for all $k$, then $\hat{\mathbf{x}}(k|j)$, where $j < k$, equals:
   **(a)** $\mathbf{\Phi}(k, j)\hat{\mathbf{x}}(j|j)$
   **(b)** $\hat{\mathbf{x}}(k|k-1)$
   **(c)** $\mathbf{\Phi}(j, k)\hat{\mathbf{x}}(k|k)$
7. In forward linear prediction, we:
   **(a)** predict a past value of a stationary discrete-time random sequence using a past set of measurements
   **(b)** predict a future value of a stationary discrete-time random sequence using a past set of measurements
   **(c)** predict a future value of a stationary discrete-time random sequence using a future set of measurements
8. Which of the following distinguish the forward linear predictor from our state predictor?
   **(a)** its length is treated as a design variable

(b) it is a linear transformation of innovations
(c) it uses a finite window of measurements, whereas the state predictor uses an expanding memory
(d) it is noncausal
(e) its coefficients don't depend on time

9. Two architectures associated with linear prediction are:
(a) tapped delay line
(b) pi-network
(c) lattice
(d) systolic array
(e) neural network

## PROBLEMS

**16-1.** Develop the counterpart to Theorem 16-1 for the case when input $\mathbf{u}(k)$ is random and independent of $\mathbf{Z}(j)$. What happens if $\mathbf{u}(k)$ is random and dependent on $\mathbf{Z}(j)$?

**16-2.** For the innovations process $\tilde{\mathbf{z}}(k+1|k)$, prove that $\mathbf{E}\{\tilde{\mathbf{z}}(i+1|i)\tilde{\mathbf{z}}'(j+1|j)\} = \mathbf{0}$ when $i < j$.

**16-3.** In the proof of part (b) of Theorem 16-2, we make repeated use of the orthogonality principle, stated in Corollary 13-3. In the latter corollary $\mathbf{f}'[\mathbf{Z}(k)]$ appears to be a function of *all* the measurements used in $\hat{\theta}_{MS}(k)$. In the expression $\mathbf{E}\{\tilde{\mathbf{x}}(i+1|i)\mathbf{z}'(j+1)\}, i > j, \mathbf{z}'(j+1)$ certainly is not a function of all the measurements used in $\tilde{\mathbf{x}}(i+1|i)$. What is $\mathbf{f}[\cdot]$ when we apply the orthogonality principle to $\mathbf{E}\{\tilde{\mathbf{x}}(i+1|i)\mathbf{z}'(j+1)\}, i > j$, to conclude that this expectation is zero?

**16-4.** Refer to Problem 15-7. Assume that $u(k)$ can be measured [e.g., $u(k)$ might be the output of a random number generator] and that $a = a[z(1), z(2), \ldots, z(k-1)]$. What is $\hat{z}(k|k-1)$?

**16-5.** Consider the following autoregressive model: $z(k+n) = a_1 z(k+n-1) - a_2 z(k+n-2) - \cdots - a_n z(k) + w(k+n)$ in which $w(k)$ is white noise. Measurements $z(k), z(k+1), \ldots, z(k+n-1)$ are available.
(a) Compute $\hat{z}(k+n|k+n-1)$.
(b) Explain why the result in part (a) is the overall mean-squared prediction of $z(k+n)$ even if $w(k+n)$ is non-Gaussian.

**16-6.** The state vector covariance matrix can be determined from (see Lesson 15)

$$\mathbf{P}_{\mathbf{x}}(k+1) = \mathbf{\Phi}(k+1,k)\mathbf{P}_{\mathbf{x}}(k)\mathbf{\Phi}'(k+1,k) + \mathbf{\Gamma}(k+1,k)\mathbf{Q}(k)\mathbf{\Gamma}'(k+1,k)$$

whereas the single-stage predicted state estimation error covariance can be determined from

$$\mathbf{P}(k+1|k) = \mathbf{\Phi}(k+1,k)\mathbf{P}(k|k)\mathbf{\Phi}'(k+1,k) + \mathbf{\Gamma}(k+1,k)\mathbf{Q}(k)\mathbf{\Gamma}'(k+1,k)$$

The similarity of these equations has led many a student to conclude that $\mathbf{P}_{\mathbf{x}}(k) = \mathbf{P}(k|k-1)$. Explain why this is *incorrect.*

**16-7.** (Keith M. Chugg, Fall 1991) In this lesson we developed the framework for mean-squared predicted state estimates for the basic state-variable model. Develop the corresponding framework for the MAP estimator [i.e., the counterparts to Equations (16-4) and (16-11)].

**16-8.** (Andrew David Norte, Fall 1991) We are given $n$ measurements $z(1), z(2), \ldots, z(n)$ that are zero mean and jointly Gaussian. Consequently, the innovations is a linear combination of these measurements; i.e.,

$$\tilde{z}(k|k-1) = z(k) - \sum_{i=1}^{k-1} \alpha(k,i)z(i), \qquad k = 1, 2, \ldots, n \tag{16-8.1}$$

**(a)** Express (16-8.1) in vector-matrix form as $\tilde{\mathbf{Z}}(n) = \mathbf{\Gamma Z}(n)$.
**(b)** Demonstrate, by an *explicit calculation*, that $\mathbf{E}\{\tilde{z}(2|1)\tilde{z}(1|0)\} = 0$.
**(c)** Given the three measurements $z(1)$, $z(2)$, and $z(3)$ and their associated covariance matrix

$$\mathbf{P}_{\mathbf{Z}(3)} = \begin{bmatrix} 1 & \frac{1}{4} & \frac{1}{2} \\ \frac{1}{4} & 1 & \frac{1}{4} \\ \frac{1}{2} & \frac{1}{4} & 1 \end{bmatrix}$$

find the numerical elements of matrix $\mathbf{\Gamma}$ that relate the innovations to the measurements.
**(d)** Using the results obtained in part (c), find the numerical values of the innovations covariance matrix $\mathbf{P}_{\tilde{\mathbf{Z}}(3)}$.
**(e)** Show that $\mathbf{Z}(3)$ can be obtained from $\tilde{\mathbf{Z}}(3)$, thereby showing that $\mathbf{Z}(3)$ and $\tilde{\mathbf{Z}}(3)$ are causally invertible.

**16-9.** (Gregg Isara and Loan Bui, Fall 1991) Curly, Moe, and Larry have just been hired by the Department of Secret Military Products to monitor measurements of the test setup shown in Figure P16-9.

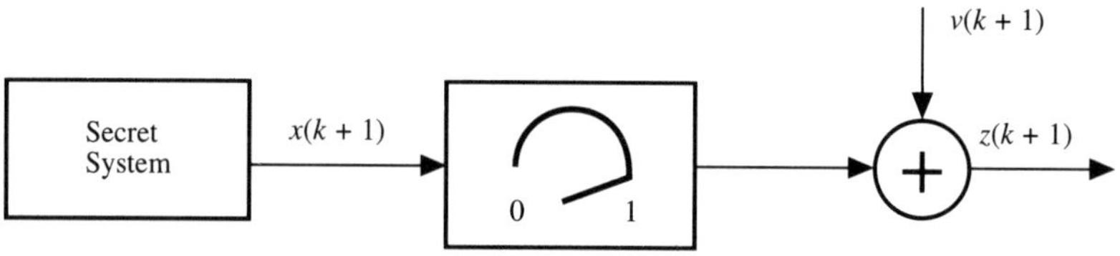


**Figure P16-9**

A set of measurements had been recorded earlier with the dial setting of the gain box set to unity. Curly, Moe, and Larry are to monitor a second set of measurements to verify the first. Since the supervisor left no instructions regarding the gain box setting, Curly decides it should be 1/3 because there are 3 people monitoring the measurements. Now

$$x(k+1) = 0.5x(k) + 1.25w(k)$$

and the measurement equations are

$$z_f(k+1) = x(k+1) + \nu(k+1)$$

and

$$z_s(k+1) = 0.333x(k+1) + \nu(k+1)$$

where $z_f(k+1)$ and $z_s(k+1)$ correspond to the first and second sets of measurements, respectively.

**(a)** Compute and compare $\hat{x}(1|0)$ for the two sets of measurements, assuming $\hat{x}(0|0) = 0.99$ for both systems.

**(b)** Given that $\hat{x}(1|1) = \hat{x}(1|0) + K(1)\tilde{z}(1|0)$ and $K(1) = P(1|0)h[hP(1|0)h + r]^{-1}$, find the equations for $\hat{x}(2|1)$ for the two sets of measurements assuming that $P(0|0) = 0.2$, $q = 10$, and $r = 5$.

**(c)** For $z_f(1) = 0.8$ and $z_s(1) = \frac{1}{3}z_f(1)$, compare the predicted estimates $\hat{x}_f(2|1)$ and $\hat{x}_s(2|1)$.

**16-10.** (Todd A. Parker and Mark A. Ranta, Fall 1991) Using single-stage prediction techniques, design a tool that could be used to estimate the output of the FIR system depicted in Figure P16-10. Note that $u(k)$ is the system input, which is deterministic, $w_i(k)$ are Gaussian white noise sequences representing errors with the register delay circuitry (i.e., the $z^{-1}$), and $v(k)$ is a Gaussian white measurement noise sequence. What is the error covariance of the predicted output value?

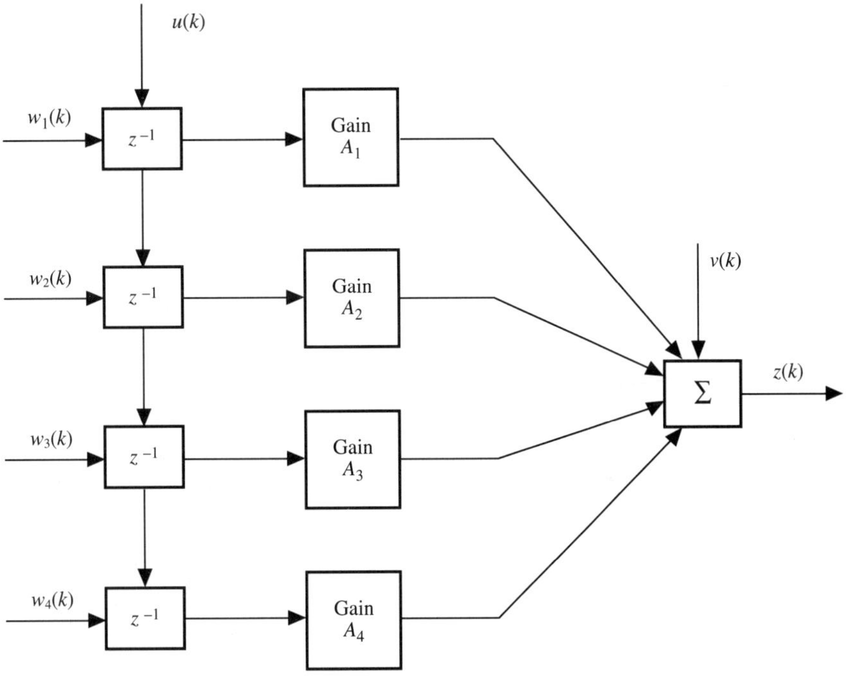


**Figure P16-10**

**16-11.** (Gregory Caso, Fall 1991) For the AR model suggested in Problem 16-5, determine the mean-squared $p$-step predictor $\hat{z}(k + m - 1 + p|k + m - 1)$, where $p \geq 1$, when the measurements $z(k), z(k + 1), \ldots, z(k + m - 1)$ are available.

# LESSON 17

# State Estimation: Filtering (the Kalman Filter)

## SUMMARY

A recursive mean-squared *state filter* is called a Kalman filter, because it was developed by Kalman around 1959. Although it was originally developed within a community of control theorists, and is regarded as the most widely used result of so-called "modern control theory," it is no longer viewed as a control theory result. It is a result within estimation theory; consequently, we now prefer to view it as a signal-processing result.

A filtered estimate of state vector $\mathbf{x}(k)$ uses all of the measurements up to and including the one made at time $t_k$. The main purpose of this lesson is to derive the Kalman filter. Another purpose is to view the Kalman filter from many perspectives. Lessons 18 and 19 will continue our study of Kalman filtering.

Some of the things we learn about the Kalman filter are: (1) it is a time-varying digital filter; (2) it uses information from both the state and measurement equation; (3) embedded within the Kalman filter is a set of recursive matrix equations which permit us to perform a *performance analysis* of the filter before any data is processed, i.e., we are able to precompute state estimation error-covariance matrices prior to data processing; and (4) it also gives MAP estimates of state vector $\mathbf{x}(k)$ for our basic state-variable model.

Finally, we explain why $\mathbf{z}(k)$ and $\tilde{\mathbf{z}}(k|k-1)$ are said to be *causally invertible*, i.e., why it is possible to compute $\mathbf{z}(k)$ from $\tilde{\mathbf{z}}(k|k-1)$ and vice-versa using a causal (i.e., physically realizable) system. The importance of the relationship between $\mathbf{z}(k)$ and $\tilde{\mathbf{z}}(k|k-1)$ is that it permits us to replace the measurements by their informationally equivalent innovations. This simplifies certain statistical calculations, because, as we learned in Lesson 16, the innovations are white (i.e., uncorrelated from one time point to another).

When you complete this lesson, you will be able to (1) derive and use the Kalman filter; (2) explain the Kalman filter from many perspectives; (3) state recursive filter and predictor algorithms, which are time-varying digital filters; (4) explain what we mean when we say the innovations process and the measurement process are causally invertible; and, (5) implement a Kalman filter on a digital computer, although commercial software is available for doing this in a numerically sound way.

## INTRODUCTION

In this lesson we shall develop the Kalman filter, which is a recursive mean-squared error filter for computing $\hat{\mathbf{x}}(k+1|k+1)$, $k = 0, 1, 2, \ldots$. As its name implies, this filter was developed by Kalman [circa 1959 (Kalman, 1960)].

From the fundamental theorem of estimation theory, Theorem 13-1, we know that

$$\hat{\mathbf{x}}(k+1|k+1) = \mathbf{E}\{\mathbf{x}(k+1)|\mathbf{Z}(k+1)\} \tag{17-1}$$

Our approach to developing the Kalman filter is to partition $\mathbf{Z}(k+1)$ into two sets of measurements, $\mathbf{Z}(k)$ and $\mathbf{z}(k+1)$, and to then expand the conditional expectation in terms of data sets $\mathbf{Z}(k)$ and $\mathbf{z}(k+1)$; i.e.,

$$\hat{\mathbf{x}}(k+1|k+1) = \mathbf{E}\{\mathbf{x}(k+1)|\mathbf{Z}(k), \mathbf{z}(k+1)\} \tag{17-2}$$

What complicates this expansion is the fact that $\mathbf{Z}(k)$ and $\mathbf{z}(k+1)$ are statistically dependent. Measurement vector $\mathbf{Z}(k)$ depends on state vectors $\mathbf{x}(1), \mathbf{x}(2), \ldots, \mathbf{x}(k)$, because $\mathbf{z}(j) = \mathbf{H}(j)\mathbf{x}(j) + \mathbf{v}(j) (j = 1, 2, \ldots, k)$. Measurement vector $\mathbf{z}(k+1)$ also depends on state vector $\mathbf{x}(k)$, because $\mathbf{z}(k+1) = \mathbf{H}(k+1)\mathbf{x}(k+1) + \mathbf{v}(k+1)$ and $\mathbf{x}(k+1) = \mathbf{\Phi}(k+1,k)\mathbf{x}(k) + \mathbf{\Gamma}(k+1,k)\mathbf{w}(k) + \mathbf{\Psi}(k+1,k)\mathbf{u}(k)$. Hence $\mathbf{Z}(k)$ and $\mathbf{z}(k+1)$ both depend on $\mathbf{x}(k)$ and are, therefore, dependent.

Recall that $\mathbf{x}(k+1)$, $\mathbf{Z}(k)$, and $\mathbf{z}(k+1)$ are jointly Gaussian random vectors; hence, we can use Theorem 12-4 to express (17-2) as

$$\hat{\mathbf{x}}(k+1|k+1) = \mathbf{E}\{\mathbf{x}(k+1)|\mathbf{Z}(k), \tilde{\mathbf{z}}\} \tag{17-3}$$

where

$$\tilde{\mathbf{z}} = \mathbf{z}(k+1) - \mathbf{E}\{\mathbf{z}(k+1)|\mathbf{Z}(k)\} \tag{17-4}$$

We immediately recognize $\tilde{\mathbf{z}}$ as the innovations process $\tilde{\mathbf{z}}(k+1|k)$ [see (16-29)]; thus, we rewrite (17-3) as

$$\hat{\mathbf{x}}(k+1|k+1) = \mathbf{E}\{\mathbf{x}(k+1)|\mathbf{Z}(k), \tilde{\mathbf{z}}(k+1|k)\} \tag{17-5}$$

Applying (12-37) to (17-5), we find that

$$\hat{\mathbf{x}}(k+1|k+1) = \mathbf{E}\{\mathbf{x}(k+1)|\mathbf{Z}(k)\} + \mathbf{E}\{\mathbf{x}(k+1)|\tilde{\mathbf{z}}(k+1|k)\} - \mathbf{m}_{\mathbf{x}}(k+1) \tag{17-6}$$

We recognize the first term on the right-hand side of (17-6) as the single-stage predicted estimator of $\mathbf{x}(k+1)$, $\hat{\mathbf{x}}(k+1|k)$; hence,

$$\hat{\mathbf{x}}(k+1|k+1) = \hat{\mathbf{x}}(k+1|k) + \mathbf{E}\{\mathbf{x}(k+1)|\tilde{\mathbf{z}}(k+1|k)\} - \mathbf{m}_{\mathbf{x}}(k+1) \tag{17-7}$$

This equation is the starting point for our derivation of the Kalman filter.

Before proceeding further, we observe, upon comparison of (17-2) and (17-5), that our original conditioning on $\mathbf{z}(k+1)$ has been replaced by conditioning on the innovations process $\tilde{\mathbf{z}}(k+1|k)$. We can show that $\tilde{\mathbf{z}}(k+1|k)$ is computable from $\mathbf{z}(k+1)$ and that $\mathbf{z}(k+1)$ is computable from $\tilde{\mathbf{z}}(k+1|k)$; hence, it is said that $\mathbf{z}(k+1)$ and $\tilde{\mathbf{z}}(k+1|k)$ are *causally invertible* (Anderson and Moore, 1979). We explain this statement more carefully at the end of this lesson.

## Kalman: A Short Biography

Rudolph Emil Kalman was born in Budapest, Hungary on May 19, 1930. The son of an electrical engineer, he decided to follow in his father's footsteps. He immigrated to the United States and obtained a bachelor's and master's degree in Electrical Engineering from the Massachusetts Institute of Technology in 1953 and 1954, respectively. He left M.I.T. and continued his studies at Columbia University, where he received his ScD. in 1957 under the direction of John R. Ragazzini.

His early interest in control systems was evident by his research at M.I.T. and especially at Columbia. His early research, which was based on the notion of state-variable representations, was mathematically advanced, but was motivated by practical problems.

From 1957 to 1958 he was employed as a staff engineer at the IBM Research Laboratory in Poughkeepsie, New York. During that period of time, he made important contributions to the design of linear sampled-data control systems using quadratic performance criteria, as well as in the use of Lyapunov stability theory for the analysis and design of control systems.

In 1958 he joined the Research Institute for Advanced Study (RIAS), which was started by the late Solomon Lefschetz. It was during the period of time from 1958 to 1964 that Kalman made some of his truly pioneering contributions to modern control theory. His research in fundamental systems concepts, such as controllability and observability, helped put on a solid theoretical basis some of the most important engineering systems structural aspects. He unified, in both discrete time and continuous time, the theory and design of linear systems with respect to quadratic criteria. He was instrumental in introducing the work of Caratheodory in optimal control theory and clarifying the interrelations between Pontryagin's maximum principle and the Hamilton–Jacobi–Bellman equation, as well as the variational calculus in general. His research not only stressed mathematical generality, but in addition it was guided by the use of the digital computer as an integral part of the design process and of the control system implementations.

It was also during his stay at RIAS that Kalman developed what is perhaps his most well known contribution, the *Kalman filter*. He obtained results on the discrete-time version of the mean-squared state-estimation problem in late 1958 and early 1959. He blended together fundamental work in filtering by Wiener, Kolmogorov, Bode, Shannon, Pugachev, and others with the modern state-space approach. His solution to the discrete-time problem led him to the continuous-time version of the problem, and in 1960–1961 he developed, in collaboration with Richard S. Bucy, the continuous-time version of the Kalman filter, which has become to be known as the Kalman–Bucy filter.

The Kalman filter and its later extensions to nonlinear problems represent the most widely applied by-product of modern control theory. From a theoretical point of view, the filter brought under a common roof related concepts of filtering and control and the duality between these two problems.

In 1964, Kalman went to Stanford University. During his stay at Stanford his research efforts shifted toward the fundamental issues associated with realization theory and algebraic system theory. Once more he opened up new research avenues in a new and basic area, modern system theory.

In 1971 Kalman was appointed graduate research professor at the University of Florida, Gainesville, Florida, where he is also director of the Center for Mathematical Systems Theory.

Kalman not only shaped the field of modern control theory, but he has also been instrumental in promoting its wide usage. *remember* $P(k+1|k) = f(\hat{x}(k+1|k))$

*we need* $P(k+1|k+1) = f(\hat{x}(k+1|k+1))$

## A PRELIMINARY RESULT

In our derivation of the Kalman filter, we shall determine that

$$\hat{\mathbf{x}}(k+1|k+1) = \hat{\mathbf{x}}(k+1|k) + \mathbf{K}(k+1)\tilde{\mathbf{z}}(k+1|k) \qquad (17\text{-}8)$$

where $\mathbf{K}(k+1)$ is an $n \times m$ (Kalman) gain matrix. We will calculate the optimal gain matrix in the next section.

Here let us view (17-8) as the structure of an arbitrary recursive linear filter, which is written in *predictor–corrector* format; i.e., the filtered estimate of $\mathbf{x}(k+1)$ is obtained by a predictor step, $\hat{\mathbf{x}}(k+1|k)$, and a corrector step, $\mathbf{K}(k+1)\tilde{\mathbf{z}}(k+1|k)$. The predictor step uses information from the state equation, because $\hat{\mathbf{x}}(k+1|k) = \mathbf{\Phi}(k+1,k)\hat{\mathbf{x}}(k|k) + \mathbf{\Psi}(k+1,k)\mathbf{u}(k)$. The corrector step uses the new measurement available at $t_{k+1}$. The correction is proportional to the difference between that measurement and its best predicted value, $\hat{\mathbf{z}}(k+1|k)$. The following result provides us with the means for evaluating $\hat{\mathbf{x}}(k+1|k+1)$ in terms of its error-covariance matrix $\mathbf{P}(k+1|k+1)$.

**Preliminary Result.** *Filtering error-covariance matrix* $\mathbf{P}(k+1|k+1)$ *for the arbitrary linear recursive filter (17-8) is computed from the following equation:*

$$\mathbf{P}(k+1|k+1) = [\mathbf{I} - \mathbf{K}(k+1)\mathbf{H}(k+1)]\mathbf{P}(k+1|k)$$

$$[\mathbf{I} - \mathbf{K}(k+1)\mathbf{H}(k+1)]' + \mathbf{K}(k+1)\mathbf{R}(k+1)\mathbf{K}'(k+1) \qquad (17\text{-}9)$$

*Proof.* Substitute (16-32) into (17-8) and then subtract the resulting equation from $\mathbf{x}(k+1)$ in order to obtain $x_{k+1} - \hat{x}_{k+1|k+1} = [x_{k+1} - \hat{x}_{k+1|k} - k_{k+1} \tilde{z}_{k+1|k}] =$

$$\tilde{\mathbf{x}}(k+1|k+1) = [\mathbf{I} - \mathbf{K}(k+1)\mathbf{H}(k+1)]\tilde{\mathbf{x}}(k+1|k)$$

$$-\mathbf{K}(k+1)\mathbf{v}(k+1) \qquad (17\text{-}10)$$

Substitute this equation into $\mathbf{P}(k+1|k+1) = \mathbf{E}\{\tilde{\mathbf{x}}(k+1|k+1)\tilde{\mathbf{x}}'(k+1|k+1)\}$ to obtain Equation (17-9). As in the proof of Theorem 16-2, we have used the fact that $\tilde{\mathbf{x}}(k+1|k)$ and $\mathbf{v}(k+1)$ are independent to show that $\mathbf{E}\{\tilde{\mathbf{x}}(k+1|k)\mathbf{v}'(k+1)\} = \mathbf{0}$. □

The state prediction-error covariance matrix $\mathbf{P}(k+1|k)$ is given by Equation (16-11). Observe that (17-9) and (16-11) can be computed recursively; once gain matrix $\mathbf{K}(k+1)$ is specified, as follows: $\mathbf{P}(0|0) \to \mathbf{P}(1|0) \to \mathbf{P}(1|1) \to \mathbf{P}(2|1) \to \mathbf{P}(2|2) \to \ldots$, etc.

It is important to reiterate the fact that (17-9) is true for *any* gain matrix, including the optimal gain matrix given next in Theorem 17-1.

## THE KALMAN FILTER

### Theorem 17-1

**a.** *The mean-squared filtered estimator of* $\mathbf{x}(k+1)$, $\hat{\mathbf{x}}(k+1|k+1)$, *written in predictor–corrector format, is*

$$\hat{\mathbf{x}}(k+1|k+1) = \hat{\mathbf{x}}(k+1|k) + \mathbf{K}(k+1)\tilde{\mathbf{z}}(k+1|k) \tag{17-11}$$

*for* $k = 0, 1, \ldots$, *where* $\hat{\mathbf{x}}(0|0) = \mathbf{m}_\mathbf{x}(0)$, *and* $\tilde{\mathbf{z}}(k+1|k)$ *is the innovations process* $[\tilde{\mathbf{z}}(k+1|k) = \mathbf{z}(k+1) - \mathbf{H}(k+1)\hat{\mathbf{x}}(k+1|k)]$.

**b.** $\mathbf{K}(k+1)$ *is an* $n \times m$ *matrix (commonly referred to as the Kalman gain matrix, or weighting matrix), which is specified by the set of relations*

$$\mathbf{K}(k+1) = \mathbf{P}(k+1|k)\mathbf{H}'(k+1)[\mathbf{H}(k+1)\mathbf{P}(k+1|k)\mathbf{H}'(k+1) + \mathbf{R}(k+1)]^{-1} \tag{17-12}$$

$$\mathbf{P}(k+1|k) = \mathbf{\Phi}(k+1,k)\mathbf{P}(k|k)\mathbf{\Phi}'(k+1,k) + \mathbf{\Gamma}(k+1,k)\mathbf{Q}(k)\mathbf{\Gamma}'(k+1,k) \tag{17-13}$$

*and*

$$\mathbf{P}(k+1|k+1) = [\mathbf{I} - \mathbf{K}(k+1)\mathbf{H}(k+1)]\mathbf{P}(k+1|k) \tag{17-14}$$

*for* $k = 0, 1, \ldots$, *where* $\mathbf{I}$ *is the* $n \times n$ *identity matrix, and* $\mathbf{P}(0|0) = \mathbf{P}_\mathbf{x}(0)$.

**c.** *The stochastic sequence* $\{\tilde{\mathbf{x}}(k+1|k+1), k = 0, 1, \ldots\}$, *which is defined by the filtering error relation*

$$\tilde{\mathbf{x}}(k+1|k+1) = \mathbf{x}(k+1) - \hat{\mathbf{x}}(k+1|k+1) \tag{17-15}$$

$k = 0, 1, \ldots$, *is a zero-mean Gauss–Markov sequence whose covariance matrix is given by (17-14).*

*Proof* (Mendel 1983, pp. 56–57)

**a.** We begin with the formula for $\hat{\mathbf{x}}(k+1|k+1)$ in (17-7). Recall that $\mathbf{x}(k+1)$ and $\mathbf{z}(k+1)$ are jointly Gaussian. Because $\mathbf{z}(k+1)$ and $\tilde{\mathbf{z}}(k+1|k)$ are causally invertible, $\mathbf{x}(k+1)$ and $\tilde{\mathbf{z}}(k+1|k)$ are also jointly Gaussian. Additionally,

$P_{x\tilde{z}}(k+1, k+1|k)$ note $x = f(k+1)$, $\tilde{z} = f(k+1|k)$

$\mathbf{E}\{\tilde{\mathbf{z}}(k+1|k)\} = \mathbf{0}$; hence,

$$\mathbf{E}\{\mathbf{x}(k+1)|\tilde{\mathbf{z}}(k+1|k)\} = \mathbf{m}_{\mathbf{x}}(k+1) + \mathbf{P}_{\mathbf{x}\tilde{\mathbf{z}}}(k+1, k+1|k)\mathbf{P}_{\tilde{\mathbf{z}}\tilde{\mathbf{z}}}^{-1}(k+1|k)\tilde{\mathbf{z}}(k+1|k) \tag{17-16}$$

We define gain matrix $\mathbf{K}(k+1)$ as

$$\mathbf{K}(k+1) = \mathbf{P}_{\mathbf{x}\tilde{\mathbf{z}}}(k+1, k+1|k)\mathbf{P}_{\tilde{\mathbf{z}}\tilde{\mathbf{z}}}^{-1}(k+1|k) \tag{17-17}$$

Substituting (17-16) and (17-17) into (17-7), we obtain the Kalman filter equation (17-11). Because $\hat{\mathbf{x}}(k+1|k) = \mathbf{\Phi}(k+1, k)\hat{\mathbf{x}}(k|k) + \mathbf{\Psi}(k+1, k)\mathbf{u}(k)$, Equation (17-11) must be initialized by $\hat{\mathbf{x}}(0|0)$, which we have shown must equal $\mathbf{m}_{\mathbf{x}}(0)$ [see Equation (16-12)].

**b.** To evaluate $\mathbf{K}(k+1)$, we must evaluate $\mathbf{P}_{\mathbf{x}\tilde{\mathbf{z}}}$ and $\mathbf{P}_{\tilde{\mathbf{z}}\tilde{\mathbf{z}}}^{-1}$. Matrix $\mathbf{P}_{\tilde{\mathbf{z}}\tilde{\mathbf{z}}}$ has been computed in (16-33). By definition of cross-covariance,

$$\begin{aligned}\mathbf{P}_{\mathbf{x}\tilde{\mathbf{z}}} &= \mathbf{E}\{[\mathbf{x}(k+1) - \mathbf{m}_{\mathbf{x}}(k+1)]\tilde{\mathbf{z}}'(k+1|k)\} \\ &= \mathbf{E}\{\mathbf{x}(k+1)\tilde{\mathbf{z}}'(k+1|k)\}\end{aligned} \tag{17-18}$$

because $\tilde{\mathbf{z}}(k+1|k)$ is zero mean. Substituting (16-32) into this expression, we find that

$$\mathbf{P}_{\mathbf{x}\tilde{\mathbf{z}}} = \mathbf{E}\{\mathbf{x}(k+1)\tilde{\mathbf{x}}'(k+1|k)\}\mathbf{H}'(k+1) \tag{17-19}$$

because $\mathbf{E}\{\mathbf{x}(k+1)\mathbf{v}'(k+1)\} = \mathbf{0}$. Finally, expressing $\mathbf{x}(k+1)$ as $\tilde{\mathbf{x}}(k+1|k) + \hat{\mathbf{x}}(k+1|k)$ and applying the orthogonality principle (13-15), we find that

$$\mathbf{P}_{\mathbf{x}\tilde{\mathbf{z}}} = \mathbf{P}(k+1|k)\mathbf{H}'(k+1) \tag{17-20}$$

Combining Equations (17-20) and (16-33) into (17-17), we obtain Equation (17-12) for the Kalman gain matrix.

State prediction-error covariance matrix $\mathbf{P}(k+1|k)$ was derived in Lesson 16.

State filtering-error covariance matrix $\mathbf{P}(k+1|k+1)$ is obtained by substituting (17-12) for $\mathbf{K}(k+1)$ into (17-9), as follows:

$$\begin{aligned}\mathbf{P}(k+1|k+1) &= (\mathbf{I} - \mathbf{KH})\mathbf{P}(\mathbf{I} - \mathbf{KH})' + \mathbf{KRK}' \\ &= (\mathbf{I} - \mathbf{KH})\mathbf{P} - \mathbf{PH}'\mathbf{K}' + \mathbf{KHPH}'\mathbf{K}' + \mathbf{KRK}' \\ &= (\mathbf{I} - \mathbf{KH})\mathbf{P} - \mathbf{PH}'\mathbf{K}' + \mathbf{K}(\mathbf{HPH}' + \mathbf{R})\mathbf{K}' \\ &= (\mathbf{I} - \mathbf{KH})\mathbf{P} - \mathbf{PH}'\mathbf{K}' + \mathbf{PH}'\mathbf{K}' \\ &= (\mathbf{I} - \mathbf{KH})\mathbf{P}\end{aligned} \tag{17-21}$$

**c.** The proof that $\tilde{\mathbf{x}}(k+1|k+1)$ is zero mean, Gaussian, and Markov is so similar to the proof of part b of Theorem 16-1 that we omit its details [see Meditch, 1969, pp. 181–182]. □

## OBSERVATIONS ABOUT THE KALMAN FILTER

1. Figure 17-1 depicts the interconnection of our basic dynamical system [Equations (15-17) and (15-18)] and Kalman filter system. The feedback nature of the Kalman filter is quite evident. Observe, also, that the Kalman filter contains within its structure a model of the plant.

   The feedback nature of the Kalman filter manifests itself in *two* different ways: in the calculation of $\hat{\mathbf{x}}(k+1|k+1)$ and also in the calculation of the matrix of gains, $\mathbf{K}(k+1)$, both of which we shall explore later.

2. The predictor–corrector form of the Kalman filter is illuminating from an information-usage viewpoint. Observe that the predictor equations, which compute $\hat{\mathbf{x}}(k+1|k)$ and $\mathbf{P}(k+1|k)$, use information only from the state equation, whereas the corrector equations, which compute $\mathbf{K}(k+1)$, $\hat{\mathbf{x}}(k+1|k+1)$ and $\mathbf{P}(k+1|k+1)$, use information only from the measurement equation.

3. Once the gain matrix is computed, then (17-11) represents a *time-varying recursive digital filter*. This is seen more clearly when Equations (16-4) and (16-31) are substituted into (17-11). The resulting equation can be rewritten as

$$\begin{aligned}\hat{\mathbf{x}}(k+1|k+1) = {} & [\mathbf{I} - \mathbf{K}(k+1)\mathbf{H}(k+1)]\mathbf{\Phi}(k+1,k)\hat{\mathbf{x}}(k|k) \\ & + \mathbf{K}(k+1)\mathbf{z}(k+1) \\ & + [\mathbf{I} - \mathbf{K}(k+1)\mathbf{H}(k+1)]\mathbf{\Psi}(k+1,k)\mathbf{u}(k)\end{aligned} \tag{17-22}$$

for $k = 0, 1, \ldots$. This is a state equation for state vector $\hat{\mathbf{x}}$, whose time-varying plant matrix is $[\mathbf{I} - \mathbf{K}(k+1)\mathbf{H}(k+1)]\mathbf{\Phi}(k+1,k)$. Equation (17-22) is time varying even if our dynamical system in Equations (15-17) and (15-18) is time invariant and stationary, because gain matrix $\mathbf{K}(k+1)$ is still time varying in that case. It is possible, however, for $\mathbf{K}(k+1)$ to reach a limiting value (i.e., a steady-state value, $\bar{\mathbf{K}}$), in which case (17-22) reduces to a recursive constant coefficient filter. We will have more to say about this important steady-state case in Lesson 19.

Equation (17-22) is in a *recursive filter form*, in that it relates the filtered estimate of $\mathbf{x}(k+1)$, $\hat{\mathbf{x}}(k+1|k+1)$, to the filtered estimate of $\mathbf{x}(k)$, $\hat{\mathbf{x}}(k|k)$. Using substitutions similar to those used in the derivation of (17-22), we can also obtain the following *recursive predictor form* of the Kalman filter (Problem 17-2):

$$\begin{aligned}\hat{\mathbf{x}}(k+1|k) = {} & \mathbf{\Phi}(k+1,k)[\mathbf{I} - \mathbf{K}(k)\mathbf{H}(k)]\hat{\mathbf{x}}(k|k-1) \\ & + \mathbf{\Phi}(k+1,k)\mathbf{K}(k)\mathbf{z}(k) + \mathbf{\Psi}(k+1,k)\mathbf{u}(k)\end{aligned} \tag{17-23}$$

Observe that in (17-23) the predicted estimate of $\mathbf{x}(k+1)$, $\hat{\mathbf{x}}(k+1|k)$, is related to the predicted estimate of $\mathbf{x}(k)$, $\hat{\mathbf{x}}(k|k-1)$. Interestingly enough, the recursive predictor (17-23), and not the recursive filter (17-22), plays an important role in mean-squared smoothing, as we shall see in Lesson 21.

The structures of (17-22) and (17-23) are summarized in Figure 17-2. This figure supports the claim made in Lesson 1 that our recursive estimators

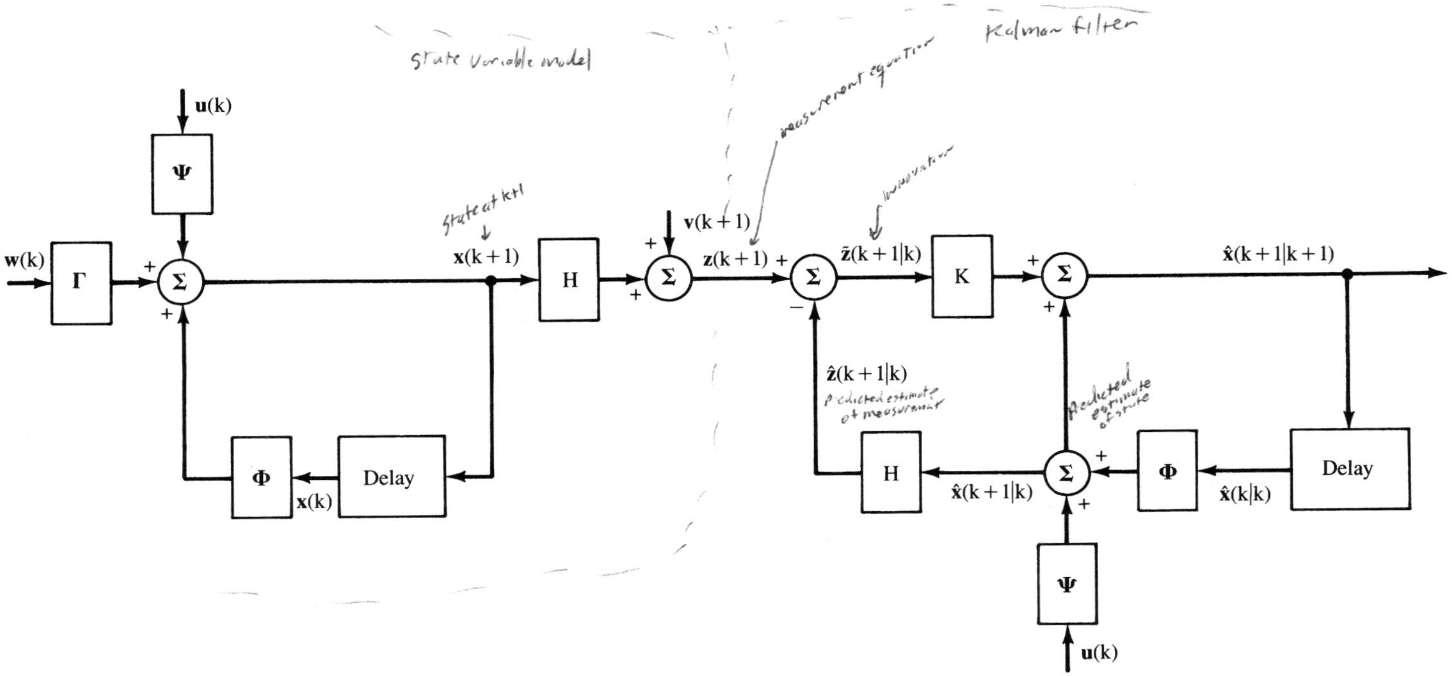


**Figure 17-1** Interconnection of system and Kalman filter. (Mendel, 1983, © 1983, Academic Press, Inc.)

are nothing more than time-varying digital filters that operate on random (and also deterministic) inputs.

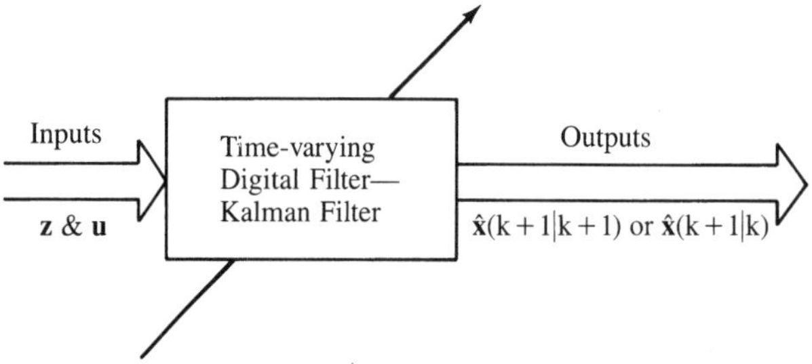


**Figure 17-2** Input/output interpretation of the Kalman filter.

4. Embedded within the recursive Kalman filter equations is another set of recursive equations: (17-12), (17-13), and (17-14). Because $\mathbf{P}(0|0)$ initializes these calculations, these equations must be ordered as follows: $\mathbf{P}(k|k) \rightarrow \mathbf{P}(k+1|k) \rightarrow \mathbf{K}(k+1) \rightarrow \mathbf{P}(k+1|k+1) \rightarrow$, etc.

   By combining these three equations, it is possible to get a matrix recursive equation for $\mathbf{P}(k+1|k)$ as a function of $\mathbf{P}(k|k-1)$ or a similar equation for $\mathbf{P}(k+1|k+1)$ as a function of $\mathbf{P}(k|k)$. These equations are nonlinear and are known as *matrix Riccati equations*. For example, the matrix Riccati equation for $\mathbf{P}(k+1|k)$ is

$$\mathbf{P}(k+1|k) = \mathbf{\Phi P}(k|k-1)\{\mathbf{I} - \mathbf{H}'[\mathbf{HP}(k|k-1)\mathbf{H}' + \mathbf{R}]^{-1} \mathbf{HP}(k|k-1)\}\mathbf{\Phi} + \mathbf{\Gamma Q \Gamma}' \quad (17\text{-}24)$$

   where we have omitted the temporal arguments on $\mathbf{\Phi}, \mathbf{\Gamma}, \mathbf{H}, \mathbf{Q}$, and $\mathbf{R}$ for notational simplicity [their correct arguments are $\mathbf{\Phi}(k+1,k), \mathbf{\Gamma}(k+1,k), \mathbf{H}(k), \mathbf{Q}(k)$, and $\mathbf{R}(k)$]. We leave the derivations of the matrix Riccati equations for $\mathbf{P}(k+1|k)$ and $\mathbf{P}(k+1|k+1)$ as an exercise (Problem 17-3).

5. A measure of recursive predictor performance is provided by matrix $\mathbf{P}(k+1|k)$. This covariance matrix can be calculated prior to any processing of real data, using its matrix Riccati equation (17-24) or Equations (17-13), (17-14), and (17-12). A measure of recursive filter performance is provided by matrix $\mathbf{P}(k+1|k+1)$, and this covariance matrix can also be calculated prior to any processing of real data. Note that $\mathbf{P}(k+1|k+1) \neq \mathbf{P}(k+1|k)$. These calculations are often referred to as *performance analyses*. It is indeed interesting that the Kalman filter utilizes a measure of its mean-squared error during its real-time operation.

6. Two formulas are available for computing $\mathbf{P}(k+1|k+1)$, i.e., (17-14), which is known as the *standard form*, and (17-9), which is known as the *stabilized form*. Although the stabilized form requires more computations than the standard form, it is much less sensitive to numerical errors from the prior calculation of gain matrix $\mathbf{K}(k+1)$ than is the standard form. In fact, we can show that first-order errors in the calculation of $\mathbf{K}(k+1)$ propagate as first-order errors in the calculation of $\mathbf{P}(k+1|k+1)$ when the standard form is used, but only as second-order errors in the calculation of $\mathbf{P}(k+1|k+1)$ when the stabilized

form is used. This is why (17-9) is called the stabilized form (for detailed derivations, see Aoki, 1967; Jazwinski, 1970; or Mendel, 1973). We caution the reader not to believe that this type of analysis is a complete numerical analysis of the Kalman filter. It is not! See Observation 8.

**7.** On the subject of computation, the calculation of $\mathbf{P}(k+1|k)$ is the most costly for the Kalman filter because of the term $\mathbf{\Phi}(k+1,k)\mathbf{P}(k|k)\mathbf{\Phi}'(k+1,k)$, which entails two multiplications of two $n \times n$ matrices [i.e., $\mathbf{P}(k|k)\mathbf{\Phi}(k+1,k)$ and $\mathbf{\Phi}(k+1,k)(\mathbf{P}(k|k)\mathbf{\Phi}'(k+1,k))$]. Total computation for the two matrix multiplications is on the order of $2n^3$ multiplications and $2n^3$ additions (for more detailed computation counts, including storage requirements, for all the Kalman filter equations, see Mendel, 1971; Gura and Bierman, 1971; and Bierman, 1973a).

We must be very careful to code the standard or stabilized forms for $\mathbf{P}(k+1|k+1)$ so that $n \times n$ matrices are never multiplied. We leave it to the reader to show that the standard algorithm can be coded in such a manner that it only requires on the order of $\frac{1}{2}mn^2$ multiplications, whereas the stabilized algorithm can be coded in such a manner that it only requires on the order of $\frac{5}{2}mn^2$ multiplications (Problem 17-9). In many applications, system order $n$ is larger than number of measurements $m$, so $mn^2 << n^3$. Usually, computation is most sensitive to system order; so, *whenever possible*, use low-order (but adequate) models.

**8.** Because of the equivalence between mean-squared, best-linear unbiased, and weighted least-squares filtered estimates of our state vector $\mathbf{x}(k)$ (see Lesson 14), we must realize that our Kalman filter equations are just a recursive solution to a system of normal equations (see Lesson 3). Other implementations of the Kalman filter that solve the normal equations using stable algorithms from numerical linear algebra (see, e.g., Bierman, 1977) and involve orthogonal transformations have better numerical properties than (17-11)–(17-14). Grewal and Andrews (1993) have an excellent comprehensive discussion on implementation methods for the Kalman filter. They state (p. 262), "It would be difficult to overemphasize the importance of good numerical methods in Kalman filtering. Limited to finite precision, computers will always make approximation errors. They are not infallible. One must always take this into account in problem analysis. The effect of roundoff may be thought to be minor, but overlooking them could be a major blunder." Finally, Verhaegen and Van Dooren (1986) present a theoretical analysis of the error propagation due to numerical round-off for four different Kalman filter implementations. They recommend the square-root covariance filter, and its observer–Hessenberg implementation in the time-invariant case as the optimal choice of Kalman filter implementation, because of its good balance of reliability and efficiency.

**9.** In Lesson 5 we developed two forms for a recursive least-squares estimator, the covariance and information forms. Compare $\mathbf{K}(k+1)$ in (17-12) with $\mathbf{K}_w(k+1)$ in (5-25) to see they have the same structure; hence, our formulation of the Kalman filter is often known as the *covariance formulation*. We leave it to the reader (Problem 17-5) to show that $\mathbf{K}(k+1)$ can also be computed as

$$\mathbf{K}(k+1) = \mathbf{P}(k+1|k+1)\mathbf{H}'(k+1)\mathbf{R}^{-1}(k+1) \qquad (17\text{-}25)$$

where $\mathbf{P}(k+1|k+1)$ is computed as

$$\mathbf{P}^{-1}(k+1|k+1) = \mathbf{P}^{-1}(k+1|k)$$

$$+\mathbf{H}'(k+1)\mathbf{R}^{-1}(k+1)\mathbf{H}(k+1) \quad (17\text{-}26)$$

When these equations are used along with (17-11) and (17-13), we have the *information formulation* of the Kalman filter. Of course, the orderings of the computations in these two formulations of the Kalman filter are different. See Lessons 5 and 9 for related discussions.

10. In Lesson 14 we showed that $\hat{\mathbf{x}}_{\text{MAP}}(k_1|N) = \hat{\mathbf{x}}_{\text{MS}}(k_1|N)$; hence,

$$\hat{\mathbf{x}}_{\text{MAP}}(k|k) = \hat{\mathbf{x}}_{\text{MS}}(k|k) \qquad (17\text{-}27)$$

This means that the Kalman filter also gives MAP estimates of the state vector $\mathbf{x}(k)$ for our basic state-variable model. See the Supplementary Material at the end of this lesson for a MAP derivation of the Kalman filter.

11. At the end of the introduction section in this lesson, we mentioned that $\mathbf{z}(k+1)$ and $\tilde{\mathbf{z}}(k+1|k)$ are *causally invertible*. This means that we can compute one from the other using a causal (i.e., realizable) system. For example, when the measurements are available, then $\tilde{\mathbf{z}}(k+1|k)$ can be obtained from Equations (17-23) and (16-31), which we repeat here for the convenience of the reader:

$$\hat{\mathbf{x}}(k+1|k) = \boldsymbol{\Phi}(k+1,k)[\mathbf{I} - \mathbf{K}(k)\mathbf{H}(k)]\hat{\mathbf{x}}(k|k-1)$$

$$+\boldsymbol{\Psi}(k+1,k)\mathbf{u}(k) + \boldsymbol{\Phi}(k+1,k)\mathbf{K}(k)\mathbf{z}(k) \qquad (17\text{-}28)$$

and

$$\tilde{\mathbf{z}}(k+1|k) = -\mathbf{H}(k+1)\hat{\mathbf{x}}(k+1|k) + \mathbf{z}(k+1), \qquad k = 0, 1, \ldots \quad (17\text{-}29)$$

We refer to (17-28) and (17-29) in the rest of this book as the *Kalman innovations system*. It is initialized by $\hat{\mathbf{x}}(0|-1) = \mathbf{0}$. Because the innovations sequence is known to be white (Lesson 16), the Kalman innovations system can be viewed as a *whitening* filter.

On the other hand, if the innovations are given a priori, then $\mathbf{z}(k+1)$ can be obtained from

$$\hat{\mathbf{x}}(k+1|k) = \boldsymbol{\Phi}(k+1,k)\hat{\mathbf{x}}(k|k-1) + \boldsymbol{\Psi}(k+1,k)\mathbf{u}(k)$$

$$+\boldsymbol{\Phi}(k+1,k)\mathbf{K}(k)\tilde{\mathbf{z}}(k|k-1) \qquad (17\text{-}30)$$

and

$$\mathbf{z}(k+1) = \mathbf{H}(k+1)\hat{\mathbf{x}}(k+1|k) + \tilde{\mathbf{z}}(k+1|k), \qquad k = 0, 1, \ldots \quad (17\text{-}31)$$

Equation (17-30) was obtained by rearranging the terms in (17-28), and (17-31) was obtained by solving (17-29) for $\mathbf{z}(k+1)$. Equation (17-30) is again initialized by $\hat{\mathbf{x}}(0|-1) = \mathbf{0}$.

Note that (17-30) and (17-31) are equivalent to our basic state-variable model in (15-17) and (15-18) from an input/output point of view. Consequently,

Remember innovations are white
$\hat{x}^2 = f(k)$
↑ may not be white!
white
∴ may not be white!

model (17-30) and (17-31) is often the starting point for important problems such as the stochastic realization problem (e.g., Faurre, 1976).

## COMPUTATION

We could find no existing M-file for a full-blown Kalman filter in any toolbox; hence, we prepared our own. In fact, we have prepared four M-files, as follows:

**kf**: a recursive Kalman filter that can be used to obtain mean-squared *filtered* estimates of the states of our basic state-variable model. This model can be time varying or nonstationary.

**kp**: a recursive Kalman predictor that can be used to obtain mean-squared *predicted* estimates of the states of our basic state-variable model. This model can be time varying or nonstationary.

**sof**: a recursive *suboptimal filter* in which the gain matrix must be prespecified. It implements all the Kalman filter equations except (17-12) and uses (17-9) to compute $\mathbf{P}(k+1|k+1)$.

**sop**: a recursive *suboptimal predictor* in which the gain matrix must be prespecified. It implements all the Kalman filter equations except (17-12) and uses (17-9) to compute $\mathbf{P}(k+1|k+1)$.

A complete listing of these M-files can be found in Appendix B. We caution the reader that these M-files may not work well for moderate- to high-order systems, or even for low-order systems that have rank problems in the observation matrix $\mathbf{H}(k)$. More sophisticated M-files need to be written. See Observation 8 in the section "Observations about the Kalman Filter."

### Supplementary Material

## MAP DERIVATION OF THE KALMAN FILTER

The Kalman filter can be derived in many different ways. Because the orthogonality principle derives from the fundamental theorem of estimation theory, the Kalman filter can be derived using the orthogonality principle. On the other hand, we know, from Example 14-3, that the MAP state estimate equals the MS state estimate; hence, as we have pointed out in Observation 10 in the main body of this lesson, $\hat{\mathbf{x}}_{MAP}(k|k) = \hat{\mathbf{x}}_{MS}(k|k)$. Consequently, the Kalman filter can also be derived from a MAP point of view. We show how to do this here (Sage and Melsa, 1971; Melsa and Cohn, 1978).

The MAP estimate of $\mathbf{x}(k)$ given $\mathbf{Z}(k)$ is given by the value of $\mathbf{x}(k)$ that maximizes $p[\mathbf{x}(k)|\mathbf{Z}(k)]$. Consequently, we must determine $\hat{\mathbf{x}}_{MAP}(k|k)$ by solving the equation

$$\left.\frac{\partial p[\mathbf{x}(k)|\mathbf{Z}(k)]}{\partial \mathbf{x}(k)}\right|_{\mathbf{x}(k)=\hat{\mathbf{x}}_{MAP}(k|k)} = \mathbf{0} \tag{17-32}$$

and also show that

$$\left.\frac{\partial^2 p[\mathbf{x}(k)|\mathbf{Z}(k)]}{\partial \mathbf{x}^2(k)}\right|_{\mathbf{x}(k)=\hat{\mathbf{x}}_{MAP}(k|k)} < \mathbf{0} \tag{17-33}$$

To begin, we need an expression for $p[\mathbf{x}(k)|\mathbf{Z}(k)]$. This conditional density function cannot be computed directly because in our state-variable model it is $\mathbf{z}(k)$ that depends on $\mathbf{x}(k)$ and not vice versa. Recall, however, that

$$p[\mathbf{x}(k)|\mathbf{Z}(k)] = \frac{p[\mathbf{x}(k), \mathbf{Z}(k)]}{p[\mathbf{Z}(k)]} = \frac{p[\mathbf{x}(k), \mathbf{z}(k), \mathbf{Z}(k-1)]}{p[\mathbf{z}(k), \mathbf{Z}(k-1)]} \tag{17-34}$$

where

$$\begin{aligned} p[\mathbf{x}(k), \mathbf{z}(k), \mathbf{Z}(k-1)] &= p[\mathbf{z}(k)|\mathbf{x}(k), \mathbf{Z}(k-1)]p[\mathbf{x}(k), \mathbf{Z}(k-1)] \\ &= p[\mathbf{z}(k)|\mathbf{x}(k), \mathbf{Z}(k-1)]p[\mathbf{x}(k)|\mathbf{Z}(k-1)]p[\mathbf{Z}(k-1)] \\ &= p[\mathbf{z}(k)|\mathbf{x}(k)]p[\mathbf{x}(k)|\mathbf{Z}(k-1)]p[\mathbf{Z}(k-1)] \end{aligned} \tag{17-35}$$

because, when $\mathbf{x}(k)$ is given, then $\mathbf{z}(k) = \mathbf{H}(k)\mathbf{x}(k) + \mathbf{v}(k)$ depends only on the random quantity $\mathbf{v}(k)$, and $\mathbf{v}(k)$ does not depend on $\mathbf{Z}(k-1)$ (why?); consequently, $p[\mathbf{z}(k)|\mathbf{x}(k), \mathbf{Z}(k-1)] = p[\mathbf{z}(k)|\mathbf{x}(k)]$. Substituting (17-35) into (17-34), the latter equation becomes

$$\begin{aligned} p[\mathbf{x}(k)|\mathbf{Z}(k)] &= \frac{p[\mathbf{z}(k)|\mathbf{x}(k)]p[\mathbf{x}(k)|\mathbf{Z}(k-1)]p[\mathbf{Z}(k-1)]}{p[\mathbf{z}(k), \mathbf{Z}(k-1)]} \\ &= \frac{p[\mathbf{z}(k)|\mathbf{x}(k)]p[\mathbf{x}(k)|\mathbf{Z}(k-1)]p[\mathbf{Z}(k-1)]}{p[\mathbf{z}(k)|\mathbf{Z}(k-1)]p[\mathbf{Z}(k-1)]} \\ &= \frac{p[\mathbf{z}(k)|\mathbf{x}(k)]p[\mathbf{x}(k)|\mathbf{Z}(k-1)]}{p[\mathbf{z}(k)|\mathbf{Z}(k-1)]} \end{aligned} \tag{17-36}$$

The two conditional densities in the numerator of this equation can be evaluated because they are Gaussian; i.e.,

$$\begin{aligned} p[\mathbf{z}(k)|\mathbf{x}(k)] = \{(2\pi)^m|\mathbf{R}(k)|\}^{-1/2}\exp\{-\frac{1}{2}[\mathbf{z}(k) - \mathbf{H}(k)\mathbf{x}(k)]' \\ \mathbf{R}^{-1}(k)[\mathbf{z}(k) - \mathbf{H}(k)\mathbf{x}(k)]\} \end{aligned} \tag{17-37}$$

$$\begin{aligned} p[\mathbf{x}(k)|\mathbf{Z}(k-1)] = \{(2\pi)^n|\mathbf{P}(k|k-1)|\}^{-1/2}\exp\{-\frac{1}{2}[\mathbf{x}(k) - \hat{\mathbf{x}}(k|k-1)]' \\ \mathbf{P}^{-1}(k|k-1)[\mathbf{x}(k) - \hat{\mathbf{x}}(k|k-1)]\} \end{aligned} \tag{17-38}$$

The latter result follows from Lesson 16. It is not necessary to evaluate $p[\mathbf{z}(k)|\mathbf{Z}(k-1)]$ in the denominator of (17-36) because this term does not depend on $\mathbf{x}(k)$, which means that $p[\mathbf{z}(k)|\mathbf{Z}(k-1)]$ acts like a constant in the denominator of (17-36) (Problem 17-9). Substituting (17-37) and (17-38) into (17-36), we find that

$$p[\mathbf{x}(k)|\mathbf{Z}(k)] = C\exp\{-\frac{1}{2}[\mathbf{z}(k) - \mathbf{H}(k)\mathbf{x}(k)]'\mathbf{R}^{-1}(k)[\mathbf{z}(k) - \mathbf{H}(k)\mathbf{x}(k)]\}$$
$$\times \exp\{-\frac{1}{2}[\mathbf{x}(k) - \hat{\mathbf{x}}(k|k-1)]'\mathbf{P}^{-1}(k|k-1)$$
$$[\mathbf{x}(k) - \hat{\mathbf{x}}(k|k-1)]\} \quad (17\text{-}39)$$

where $C = \{(2\pi)^m|\mathbf{R}(k)|\}^{-1/2}\{(2\pi)^n|\mathbf{P}(k|k-1)|\}^{-1/2}/p[\mathbf{z}(k)|\mathbf{Z}(k-1)]$.

We are now ready to maximize $p[\mathbf{x}(k)|\mathbf{Z}(k)]$ with respect to $\mathbf{x}(k)$; however, because of the exponential nature of the right-hand side, it is easier to maximize $\ln p[\mathbf{x}(k)|\mathbf{Z}(k)]$ with respect to $\mathbf{x}(k)$; i.e., we now evaluate

$$\left.\frac{\partial \ln p[\mathbf{x}(k)|\mathbf{Z}(k)]}{\partial \mathbf{x}(k)}\right|_{\mathbf{x}(k)=\hat{\mathbf{x}}_{\mathrm{MAP}}(k|k)} = \mathbf{0} \quad (17\text{-}40)$$

We leave the rest of the calculations to the reader, because they are routine (Problem 17-10). The end result will be an equation for $\hat{\mathbf{x}}_{\mathrm{MAP}}(k|k)$. Finally, we leave it to the reader (Problem 17-10) to show that

$$\left.\frac{\partial^2 \ln p[\mathbf{x}(k)|\mathbf{Z}(k)]}{\partial \mathbf{x}^2(k)}\right|_{\mathbf{x}(k)=\hat{\mathbf{x}}_{\mathrm{MAP}}(k|k)} = -[\mathbf{H}'(k)\mathbf{R}^{-1}(k)\mathbf{H}(k) + \mathbf{P}^{-1}(k|k)] < 0 \quad (17\text{-}41)$$

because both $\mathbf{H}'(k)\mathbf{R}^{-1}(k)\mathbf{H}(k)$ and $\mathbf{P}^{-1}(k|k)$ are positive definite; hence, $\hat{\mathbf{x}}_{\mathrm{MAP}}(k|k)$ does indeed maximize $p[\mathbf{x}(k)|\mathbf{Z}(k)]$.

## SUMMARY QUESTIONS

1. When we say that $\mathbf{z}(k+1)$ and $\tilde{\mathbf{z}}(k+1|k)$ are causally invertible, we mean that:
   **(a)** $\mathbf{z}'\mathbf{z} = \tilde{\mathbf{z}}'\tilde{\mathbf{z}}$
   **(b)** one is computable from the other
   **(c)** $\mathbf{M}^{-1}(\mathbf{z}) = \mathbf{M}^{-1}(\tilde{\mathbf{z}})$, where $\mathbf{M}$ is a square matrix
2. Formula (17-9) for $\mathbf{P}(k+1|k+1)$ is valid for:
   **(a)** any linear estimator
   **(b)** only the Kalman filter
   **(c)** any estimator
3. The Kalman filter has how many feedback loops?
   **(a)** one
   **(b)** two
   **(c)** three
4. The matrix equations within the KF must be solved in the following order:
   **(a)** $\mathbf{P}(k+1|k) \to \mathbf{P}(k|k) \to \mathbf{K}(k+1) \to \mathbf{P}(k+1|k+1) \to$, etc.
   **(b)** $\mathbf{P}(k|k) \to \mathbf{P}(k+1|k) \to \mathbf{P}(k+1|k+1) \to \mathbf{K}(k+1) \to$, etc.
   **(c)** $\mathbf{P}(k|k) \to \mathbf{P}(k+1|k) \to \mathbf{K}(k+1) \to \mathbf{P}(k+1|k+1) \to$, etc.
5. In a performance analysis, we:
   **(a)** run the KF to observe its output
   **(b)** compute $\mathbf{P}(k+1|k)$ or $\mathbf{P}(k+1|k+1)$ prior to processing data
   **(c)** evaluate the innovations to see if it is indeed white

**6.** The most costly computation in the KF is:
   **(a)** $(\mathbf{I} - \mathbf{KH})\mathbf{P}(k+1|k)$
   **(b)** $\mathbf{\Phi P}(k|k)\mathbf{\Phi}'$
   **(c)** $\mathbf{P}(k+1|k+1)$

**7.** First-order errors made in calculating $\mathbf{K}(k+1)$ propagate into the:
   **(a)** stabilized form equation for $\mathbf{P}(k|k)$ as first-order errors
   **(b)** stabilized form equation for $\mathbf{P}(k|k)$ as second-order errors
   **(c)** standard form equation for $\mathbf{P}(k|k)$ as second-order errors

**8.** The KF uses information:
   **(a)** just from the measurement equation
   **(b)** just from the state equation
   **(c)** from state and measurement equations

**9.** The filtered estimate of $\mathbf{x}(k)$ is:
   **(a)** $\hat{\mathbf{x}}(k|k+1)$
   **(b)** $\hat{\mathbf{x}}(k|k-1)$
   **(c)** $\hat{\mathbf{x}}(k|k)$

**10.** The predictor–corrector form of the KF demonstrates the:
   **(a)** independence between prediction and filtering
   **(b)** coupling between prediction and filtering
   **(c)** noncausal nature of the KF

**11.** The recursive filter and predictor are:
   **(a)** time varying
   **(b)** time invariant
   **(c)** unforced

**12.** A matrix Riccati equation is:
   **(a)** nonlinear
   **(b)** linear
   **(c)** homogeneous

**13.** The stabilized form of the filtering error-covariance equation:
   **(a)** always requires on the order of $n^3$ flops for its implementation
   **(b)** requires fewer flops than the standard form
   **(c)** requires about five times as many flops as the standard form

**14.** The plant matrices of the recursive filter and predictor are:
   **(a)** the same
   **(b)** different
   **(c)** the transposes of one another

## PROBLEMS

**17-1.** Prove that $\tilde{\mathbf{x}}(k+1|k+1)$ is zero mean, Gaussian, and first-order Markov.

**17-2.** Derive the recursive predictor form of the Kalman filter given in (17-23).

**17-3.** Derive the matrix Riccati equations for $\mathbf{P}(k+1|k)$ and $\mathbf{P}(k+1|k+1)$.

**17-4.** (Atindra Mitra, Spring 1992) Derive the expression for the Kalman gain matrix in Theorem 17-1 by minimizing

$$\mathbf{P}(k+1|k+1) = [\mathbf{I} - \mathbf{K}(k+1)\mathbf{H}(k+1)]\mathbf{P}(k+1|k)[\mathbf{I} - \mathbf{K}(k+1)\mathbf{H}(k+1)]' + \mathbf{K}(k+1)\mathbf{R}(k+1)\mathbf{K}'(k+1)$$

with respect to $\mathbf{K}(k+1)$. (*Hint:* Complete the square.)

**17-5.** Show that gain matrix $\mathbf{K}(k+1)$ can also be computed using (17-25).

**17-6.** Suppose a small error $\delta\mathbf{K}$ is made in the computation of the Kalman filter gain $\mathbf{K}(k+1)$.

**(a)** Show that when $\mathbf{P}(k+1|k+1)$ is computed from the "standard form" equation then, to first-order terms,

$$\delta\mathbf{P}(k+1|k+1) = -\delta\mathbf{K}(k+1)\mathbf{H}(k+1)\mathbf{P}(k+1|k)$$

**(b)** Show that when $\mathbf{P}(k+1|k+1)$ is computed from the "stabilized form" equation then, to first-order terms,

$$\delta\mathbf{P}(k+1|k+1) \simeq \mathbf{0}$$

**17-7.** Derive the Kalman filter using the orthogonality principle.

**17-8.** In Lesson 16 we studied the innovations process $\tilde{\mathbf{z}}(k+1|k)$. Here we study the process $\tilde{\mathbf{z}}(k+1|k+1)$. Obtain the value of $\tilde{\mathbf{z}}(k+1|k+1)$.

**17-9.** Referring to the derivation of the Kalman filter using MAP, show that it is not necessary to evaluate $p[\mathbf{z}(k)|\mathbf{Z}(k-1)]$ because it does not depend on $\mathbf{x}(k)$, which means that $p[\mathbf{z}(k)|\mathbf{Z}(k-1)]$ acts like a constant in the denominator of (17-36).

**17-10.** Referring to the derivation of the Kalman filter using MAP, beginning with Equations (17-39) and (17-40), complete the derivation of the Kalman filter. Note that you will need the matrix inversion Lemma 5-1. Finally, demonstrate the truth of (17-41).

**17-11.** Consider the basic scalar system $x(k+1) = \phi x(k) + w(k)$ and $z(k+1) = x(k+1) + \nu(k+1)$.

**(a)** Show that $p(k+1|k) \geq q$, which means that the variance of the system disturbance sets the performance limit on prediction accuracy.

**(b)** Show that $0 \leq K(k+1) \leq 1$.

**(c)** Show that $0 \leq p(k+1|k+1) \leq r$.

**17-12.** An RC filter with time constant $\tau$ is excited by Gaussian white noise, and the output is measured every $T$ seconds. The output at the sample times obeys the equation $x(k) = e^{-T/\tau}x(k-1) + w(k-1)$, where $\mathbf{E}\{x(0)\} = 1$, $\mathbf{E}\{x^2(0)\} = 2$, $q = 2$, and $T = \tau = 0.1$ sec. The measurements are described by $z(k) = x(k) + \nu(k)$, where $\nu(k)$ is a white but *non-Gaussian* noise sequence for which $\mathbf{E}\{\nu(k)\} = 0$ and $\mathbf{E}\{\nu^2(k)\} = 4$. Additionally, $w(k)$ and $\nu(k)$ are uncorrelated. Measurements $z(1) = 1.5$ and $z(2) = 3.0$.

**(a)** Find the best linear estimate of $x(1)$ based on $z(1)$.

**(b)** Find the best linear estimate of $x(2)$ based on $z(1)$ and $z(2)$.

**17-13.** (Brad Verona, Spring 1992) Consider the second-order tracking system

$$\mathbf{x}(k+1) = \begin{pmatrix} 0 & 1 \\ 1 & 1 \end{pmatrix} \mathbf{x}(k) + \mathbf{w}(k)$$

$$z(k+1) = (1 \quad 0)\mathbf{x}(k) + \nu(k)$$

where $\mathbf{w}(k)$ and $\nu(k)$ are zero-mean white noise sequences for which $\mathbf{Q} = \mathbf{I}$ and $r = 1$. Additionally, $\mathbf{P}(0|0) = \text{diag}(5, 5)$.

**(a)** Compute and plot the first 10 iterations of the Kalman gain vector $\mathbf{K}(k+1)$ and $p_1(k|k)$.

**(b)** What is the approximate steady-state filter equation for $\hat{\mathbf{x}}(k+1|k+1)$?

**(c)** Given the first 10 measurements $\{-2, -1, 0, 1, 2, 3, 4, 5, 6, 7\}$, find the state estimate for $\hat{x}_1(k+1|k+1)$. Plot $\hat{x}_1(k+1|k+1)$ and $z(k)$ versus $k$ to compare the two quantities.

**17-14.** (Tony Hung-yao Wu, Spring 1992) In this problem, you will see how nonstationary noise affects the Kalman gain vector. Consider the following second-order system:

$$\mathbf{x}(k+1) = \begin{pmatrix} 1 & 0 \\ 0 & 1 \end{pmatrix} \mathbf{x}(k) + \mathbf{w}(k)$$

$$z(k) = x_1(k) + \nu(k)$$

The disturbance is stationary, with $\mathbf{Q}(k) = \text{diag}\,(0, 1)$; but the measurement noise is nonstationary, with $r(k) = 2 + (-1)^k$. Additionally, $\mathbf{P}(0|0) = 10\mathbf{I}$.

**(a)** Calculate and plot $\mathbf{K}(k)$ for $k = 1, 2, \ldots, 10$.

**(b)** From your results in part (a), explain why the Kalman gains increase for odd values of $k$ and decrease for even values of $k$, and explain the nature of the "steady-state" values for the two gains.

**17-15.** Consider the following first-order basic state-variable model:

$$x(k+1) = \frac{1}{\sqrt{2}} x(k) + \frac{1}{4} w(k)$$

$$z(k+1) = \begin{cases} x(k+1), & k = 0, 2, 4, \ldots \\ \frac{1}{3} x(k+1) + \nu(k+1), & k = 1, 3, 5, 7, \ldots \end{cases}$$

where $\mathbf{E}\{x(0)\} = \sqrt{2}, \sigma_x^2 = 1, q = 4, r = \frac{1}{9}$, and $s = 0$. Give the formula for the recursive MSE filtered estimate of $x(k+1)$.

**17-16.** Table 17-1 lists multiplications and additions for all the basic matrix operations used in a Kalman filter. Using the formulas for the Kalman filter given in Theorem 17-1, establish the number of multiplications and additions required to compute $\hat{\mathbf{x}}(k+1|k)$, $\mathbf{P}(k+1|k)$, $\mathbf{K}(k+1)$, $\hat{\mathbf{x}}(k+1|k+1)$, and $\mathbf{P}(k+1|k+1)$.

**TABLE 17-1 OPERATION CHARACTERISTICS**

| Name | Function | Multiplications | Additions |
|---|---|---|---|
| Matrix addition | $\mathbf{C}_{MN} = \mathbf{A}_{MN} + \mathbf{B}_{MN}$ | — | $MN$ |
| Matrix subtraction | $\mathbf{C}_{MN} = \mathbf{A}_{MN} - \mathbf{B}_{MN}$ | — | $MN$ |
| Matrix multiply | $\mathbf{C}_{ML} = \mathbf{A}_{MN}\mathbf{B}_{NL}$ | $MNL$ | $ML(N-1)$ |
| Matrix transpose multiply | $\mathbf{C}_{MN} = \mathbf{A}_{ML}(\mathbf{B}'_{NL})$ | $MNL$ | $ML(N-1)$ |
| Matrix inversion | $\mathbf{A}_{NN} \rightarrow \mathbf{A}_{NN}^{-1}$ | $\alpha N^3$ | $\beta N^3$ |
| Scalar-vector product | $\mathbf{C}_{N1} = \rho \mathbf{A}_{N1}$ | $N$ | — |

**17-17.** Show that the standard algorithm for computing $\mathbf{P}(k+1|k+1)$ only requires on the order of $\frac{1}{2}mn^2$ multiplications, whereas the stabilized algorithm requires on the order of $\frac{5}{2}mn^2$ multiplications (use Table 17-1). Note that this last result requires a very clever coding of the stabilized algorithm.

# LESSON 18

# State Estimation: Filtering Examples

## SUMMARY

In this lesson we present seven examples that illustrate some interesting numerical and theoretical aspects of Kalman filtering. Example 18-1 demonstrates what is meant by a *performance analysis*. Example 18-2 introduces the notion of *sensitivity* and what is meant by a *Kalman filter sensitivity system*. Using this system it is possible to study (by simulations) how robust the Kalman filter is to small or large variations in the parameters of the basic state-variable model. Example 18-3 treats the single-input, single-output model, i.e., the single-channel case. It shows that signal-to-noise ratio can be viewed as a Kalman filter tuning parameter. Example 18-4 explains how the Kalman filter equations can be used to obtain BLUEs of a constant random parameter vector $\boldsymbol{\theta}$. It links state estimation to parameter estimation. Example 18-5 illustrates the *divergence phenomenon*, which often occurs when either the process noise or the measurement noise or both are small. In essence, the Kalman filter locks onto the wrong values for the state, but believes them to be the true values, i.e., it "learns" the wrong state too well. A number of different remedies are described for controlling divergence effects. Example 18-6 demonstrates a pitfall when using Kalman filtering in an ad hoc manner. Example 18-7 shows how a Kalman filter can adversely affect the performance of a feedback control system when the feedback control law is one that is proportional to the output of the Kalman filter. In essence, state estimation errors act as additional plant disturbances, which must be properly accounted for at the front end of a control system design.

When you complete this lesson, you will be able to (1) describe the transient and steady-state behaviors of a Kalman filter; (2) describe the sensitivity of the Kalman filter to variations of some system parameters; (3) explain how a BLUE can be obtained from the Kalman filter equations; (4) explain how signal-to-noise ratio can be used as a Kalman filter tuning parameter; (5) explain the divergence phenomenon; (6) explain how not to use the Kalman filter; and (7) explain the effect of a Kalman filter when it is used as part of a feedback control law.

## INTRODUCTION

In this lesson (which is an excellent one for self-study) we present seven examples, which illustrate some interesting numerical and theoretical aspects of Kalman filtering.

## EXAMPLES

### EXAMPLE 18-1 *this is time invariant & stationary system*

In Lesson 17 we learned that the Kalman filter is a dynamical feedback system. Its gain matrix and predicted- and filtering-error covariance matrices comprise a matrix feedback system operating within the Kalman filter. Of course, these matrices can be calculated prior to processing of data, and such calculations constitute a *performance analysis* of the Kalman filter. Here we examine the results of these calculations for two second-order systems, $H_1(z) = 1/(z^2 - 1.32z + 0.875)$ and $H_2(z) = 1/(z^2 - 1.36z + 0.923)$. The second system is less damped than the first. Impulse responses of both systems are depicted in Figure 18-1.

In Figure 18-1 we also depict $p_{11}(k|k)$, $p_{22}(k|k)$, $K_1(k)$, and $K_2(k)$ versus $k$ for both systems. In both cases $\mathbf{P}(0|0)$ was set equal to the zero matrix. For system 1, $q = 1$ and $r = 5$, whereas for system 2, $q = 1$ and $r = 20$; hence, there is more measurement noise in system 2 than there is in system 1. Observe that the error variances and Kalman gains exhibit a transient response as well as a steady-state response, i.e., after a certain value of $k(k \cong 10$ for system 1 and $k \cong 15$ for system 2), $p_{11}(k|k)$, $p_{22}(k|k)$, $K_1(k)$, and $K_2(k)$ reaching limiting values. These limiting values do not depend on $\mathbf{P}(0|0)$, as can be seen from Figure 18-2. The Kalman filter is initially influenced by its initial conditions, but eventually ignores them, paying much greater attention to model parameters and the measurements. The relatively large steady-state values for $p_{11}$ and $p_{22}$ are due to the large value of $r$. $\square$

### EXAMPLE 18-2

A state estimate is implicitly conditioned on knowing the true values for all system parameters (i.e., $\boldsymbol{\Phi}, \boldsymbol{\Gamma}, \boldsymbol{\Psi}, \mathbf{H}, \mathbf{Q}$, and $\mathbf{R}$). Sometimes we do not know these values exactly; hence, it is important to learn how *sensitive* (i.e., robust) the Kalman filter is to parameter errors. Many references that treat Kalman filter sensitivity issues can be found in Mendel and Gieseking (1971) under category $2f$, "State Estimation: Sensitivity Considerations."

Let $\theta$ denote any parameter that may appear in $\boldsymbol{\Phi}, \boldsymbol{\Gamma}, \boldsymbol{\Psi}, \mathbf{H}, \mathbf{Q}$, or $\mathbf{R}$. To determine the sensitivity of the Kalman filter to *small* variations in $\theta$, we compute $\partial\hat{\mathbf{x}}(k + 1|k + 1)/\partial\theta$, whereas, for large variations in $\theta$, we compute $\Delta\hat{\mathbf{x}}(k + 1|k + 1)/\Delta\theta$. An analysis of $\Delta\hat{\mathbf{x}}(k+1|k+1)/\Delta\theta$, for example, reveals the interesting chain of events that $\Delta\hat{\mathbf{x}}(k+1|k+1)/\Delta\theta$ depends on $\Delta\mathbf{K}(k+1)/\Delta\theta$, which in turn depends on $\Delta\mathbf{P}(k+1|k)/\Delta\theta$, which in turn depends on $\Delta\mathbf{P}(k|k)/\Delta\theta$. Hence, for each variable parameter $\theta$, we have a *Kalman filter sensitivity system* comprised of equations for which we compute $\Delta\hat{\mathbf{x}}(k + 1|k + 1)/\Delta\theta$ (see also, Lesson 25). An alternative to using these equations is to perform a computer perturbation study. For example,

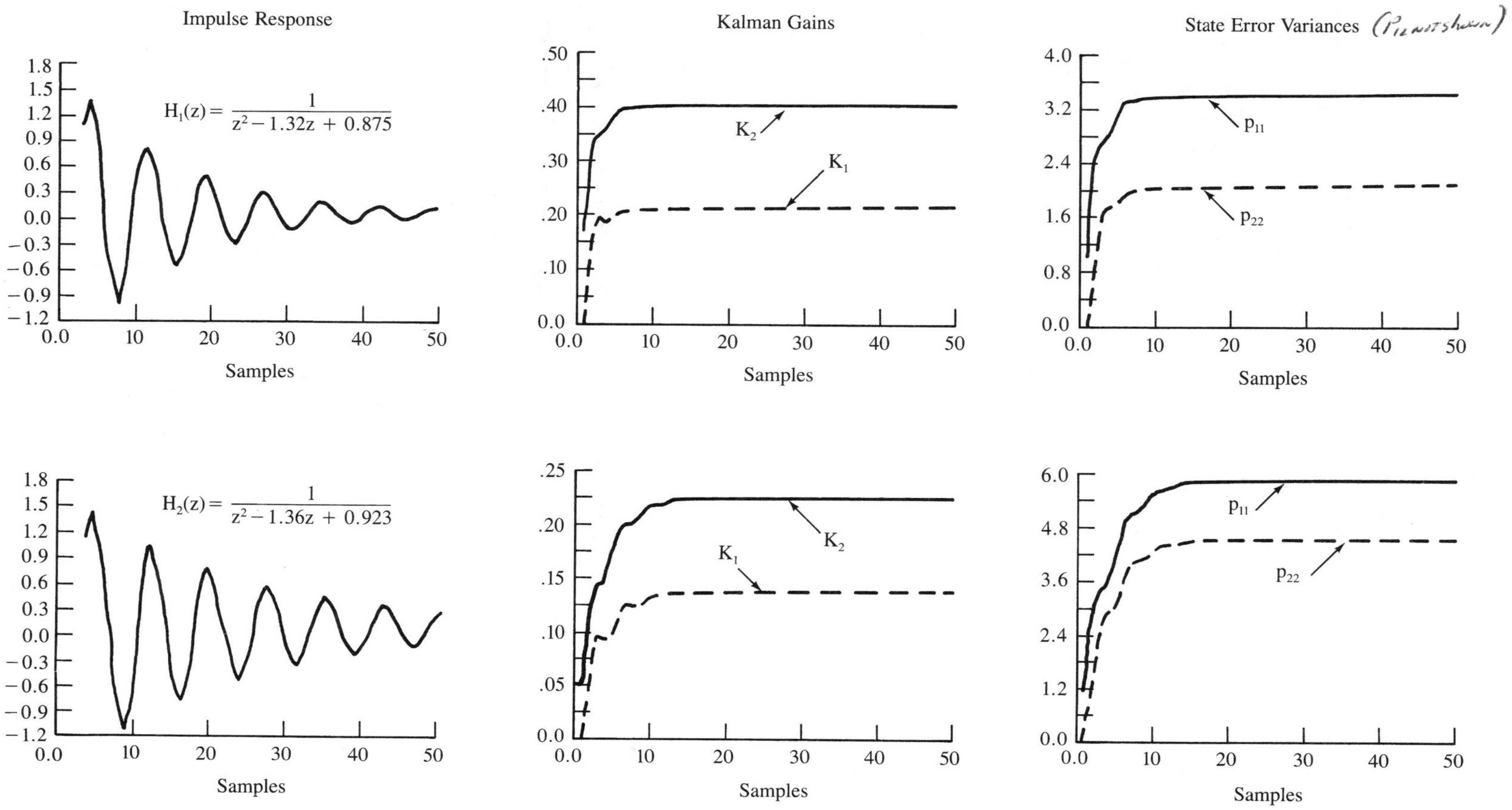


**Figure 18-1** Kalman gains and filtering error variances for two second-order systems.

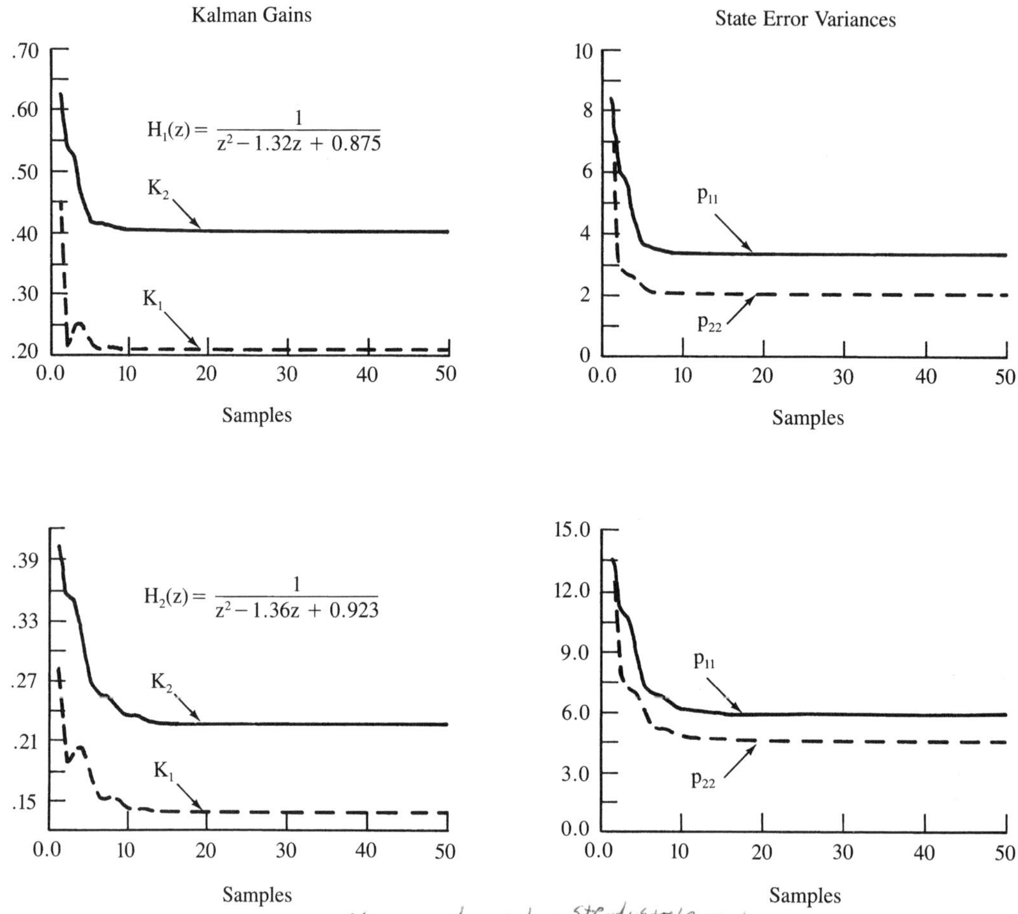


diff init conditions, but steady state result was the same.

**Figure 18-2** Same as Figure 18-1, except for a different choice of $\mathbf{P}(0|0)$.

$$\frac{\Delta \mathbf{K}(k+1)}{\Delta \theta} = \frac{\mathbf{K}(k+1)\,\big|_{\theta=\theta_N+\Delta\theta} - \mathbf{K}(k+1)\big|_{\theta=\theta_N}}{\Delta \theta} \tag{18-1}$$

where $\theta_N$ denotes a nominal valute of $\theta$.

We define *sensitivity* as the ratio of the percentage change in a function [e.g., $\mathbf{K}(k+1)$] to the percentage change in a parameter, $\theta$. For example,

$$S_\theta^{K_{ij}(k+1)} = \frac{\dfrac{\Delta K_{ij}(k+1)}{K_{ij}(k+1)}}{\dfrac{\Delta \theta}{\theta}} \tag{18-2}$$

denotes the sensitivity of the *ij*th element of matrix $\mathbf{K}(k+1)$ with respect to parameter $\theta$. All other sensitivities, such as $S_\theta^{P_{ij}(k|k)}$, are defined similarly.

Here we present some numerical sensitivity results for the simple first-order system

$$x(k+1) = ax(k) + bw(k) \tag{18-3}$$

$$z(k) = hx(k) + n(k) \tag{18-4}$$

where $a_N = 0.7, b_N = 1.0, h_N = 0.5, q_N = 0.2$, and $r_N = 0.1$. Figure 18-3 depicts $S_a^{K(k+1)}$, $S_b^{K(k+1)}$, $S_h^{K(k+1)}$, $S_q^{K(k+1)}$, and $S_r^{K(k+1)}$ for parameter variations of $\pm 5\%, \pm 10\%, \pm 20\%$, and $\pm 50\%$ about nominal values $a_N, b_N, h_N, q_N, r_N$. Similar sets of curves can be computed for sensitivities of the predicted or filtered error variances with respect to each of the system's parameters.

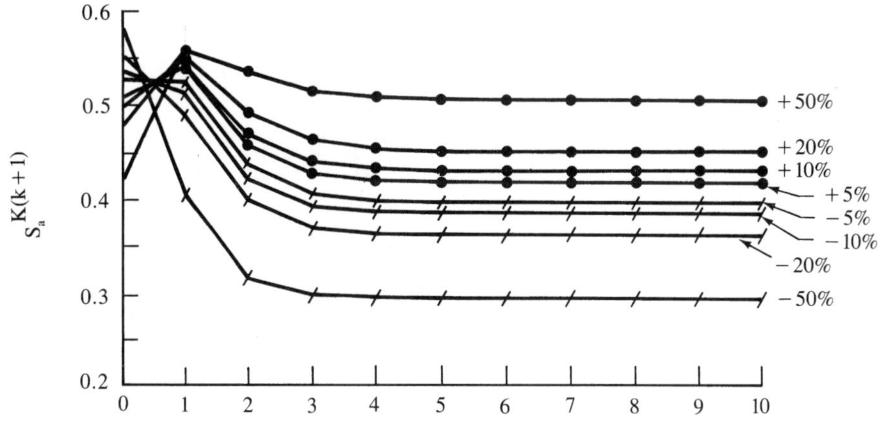


(a)

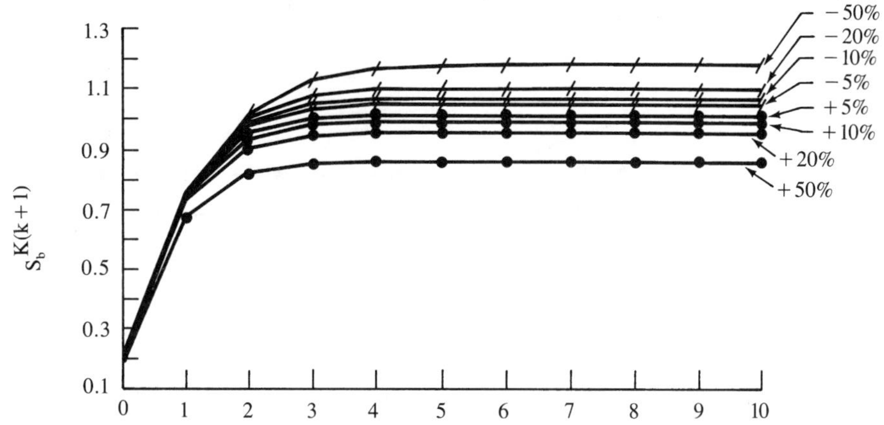


(b)

**Figure 18-3** Sensitivity plots.

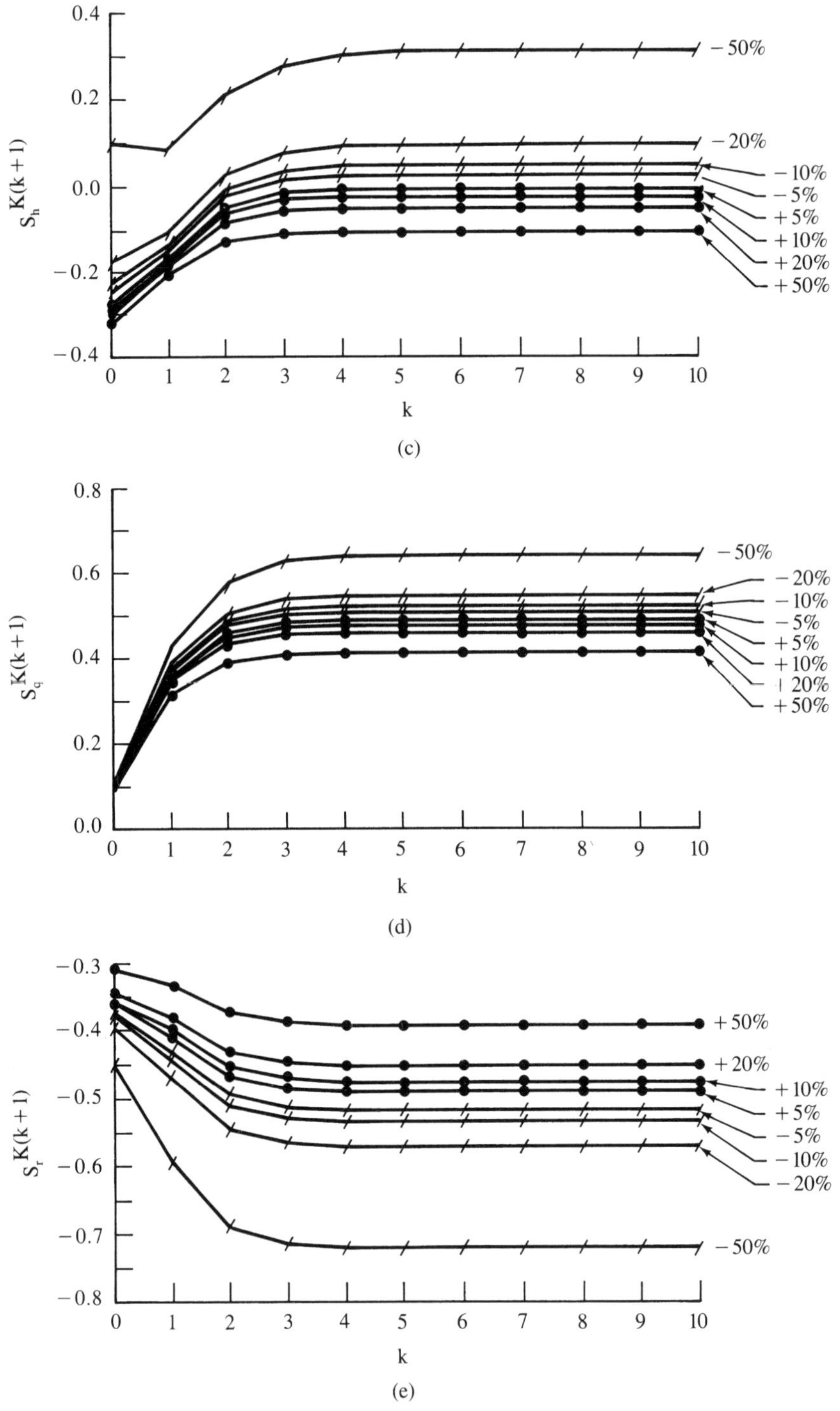


**Figure 18-3** *(continued)*

Observe that the sensitivity functions vary with time and that they all reach steady-state values. Table 18-1 summarizes the steady-state sensitivity coefficients $\bar{S}_\theta^{K(k+1)}$, $\bar{S}_\theta^{P(k|k)}$, and $\bar{S}_\theta^{P(k+1|k)}$.

Some conclusions that can be drawn from these numerical results are (1) $K(k+1)$, $P(k|k)$, and $P(k+1|k)$ are most sensitive to changes in parameter $b$, and (2) $\bar{S}_\theta^{K(k+1)} = \bar{S}_\theta^{P(k|k)}$ for $\theta = a, b$, and $q$. This last observation could have been foreseen, because of our alternative equation for $K(k+1)$ [Equation (17-25)].

$$\mathbf{K}(k+1) = \mathbf{P}(k+1|k+1)\mathbf{H}(k+1)\mathbf{R}^{-1}(k+1) \tag{18-5}$$

This expression shows that if $h$ and $r$ are fixed then $\mathbf{K}(k+1)$ varies exactly the same way as $\mathbf{P}(k+1|k+1)$.

**TABLE 18-1** STEADY-STATE SENSITIVITY COEFFICIENTS

| (a) $\bar{S}_\theta^{K(k+1)}$ | Percentage change in $\theta$ | | | | | | | |
|---|---|---|---|---|---|---|---|---|
| $\theta$ | +50 | +20 | +10 | +5 | −5 | −10 | −20 | −50 |
| $a$ | 0.506 | 0.451 | 0.430 | 0.419 | 0.396 | 0.384 | 0.361 | 0.294 |
| $b$ | 0.838 | 0.937 | 0.972 | 0.989 | 1.025 | 1.042 | 1.077 | 1.158 |
| $h$ | −0.108 | −0.053 | −0.026 | −0.010 | 0.026 | 0.047 | 0.096 | 0.317 |
| $q$ | 0.417 | 0.465 | 0.483 | 0.493 | 0.515 | 0.526 | 0.551 | 0.648 |
| $r$ | −0.393 | −0.452 | −0.476 | −0.490 | −0.518 | −0.534 | −0.570 | −0.717 |
| (b) $\bar{S}_\theta^{P(k\|k)}$ | Percentage change in $\theta$ | | | | | | | |
| $\theta$ | +50 | +20 | +10 | +5 | −5 | −10 | −20 | −50 |
| $a$ | 0.506 | 0.451 | 0.430 | 0.419 | 0.396 | 0.384 | 0.361 | 0.294 |
| $b$ | 0.838 | 0.937 | 0.972 | 0.989 | 1.025 | 1.042 | 1.077 | 1.158 |
| $h$ | −0.739 | −0.877 | −0.932 | −0.962 | −1.025 | −1.059 | −1.130 | −1.367 |
| $q$ | 0.417 | 0.465 | 0.483 | 0.493 | 0.515 | 0.526 | 0.551 | 0.648 |
| $r$ | 0.410 | 0.457 | 0.476 | 0.486 | 0.507 | 0.519 | 0.544 | 0.642 |
| (c) $\bar{S}_\theta^{P(k+1\|k)}$ | Percentage change in $\theta$ | | | | | | | |
| $\theta$ | +50 | +20 | +10 | +5 | −5 | −10 | −20 | −50 |
| $a$ | 1.047 | 0.820 | 0.754 | 0.723 | 0.664 | 0.637 | 0.585 | 0.453 |
| $b$ | 2.022 | 1.836 | 1.775 | 1.745 | 1.684 | 1.653 | 1.591 | 1.402 |
| $h$ | −0.213 | −0.253 | −0.268 | −0.277 | −0.295 | −0.305 | −0.325 | −0.393 |
| $q$ | 0.832 | 0.846 | 0.851 | 0.854 | 0.860 | 0.864 | 0.871 | 0.899 |
| $r$ | 0.118 | 0.131 | 0.137 | 0.140 | 0.146 | 0.149 | 0.157 | 0.185 |

Here is how to use the results in Table 18-1. From Equation (18-2), for example, we see that

$$\frac{\Delta K_{ij}(k+1)}{K_{ij}(k+1)} = \left(\frac{\Delta\theta}{\theta}\right) S_\theta^{K_{ij}(k+1)} \tag{18-6}$$

or

$$\% \text{ change in } K_{ij}(k+1) = (\% \text{ change in } \theta) \times S_\theta^{K_{ij}(k+1)} \tag{18-7}$$

From Table 18-1 and this formula we see, for example, that a 20% change in $a$ produces a $(20)(0.451) = 9.02\%$ steady-state change in $K$, whereas a 20% change in $h$ only produces a $(20)(-0.053) = -1.06\%$ steady-state change in $K$, etc. □

## EXAMPLE 18-3

In the single-channel case, when $\mathbf{w}(k)$, $\mathbf{z}(k)$, and $\mathbf{v}(k)$ are scalars, then $\mathbf{K}(k+1)$, $\mathbf{P}(k+1|k)$, and $\mathbf{P}(k+1|k+1)$ do not depend on $q$ and $r$ separately. Instead, as we demonstrate next, they depend only on the ratio $q/r$. In this case, $\mathbf{H} = \mathbf{h}'$, $\mathbf{\Gamma} = \gamma$, $\mathbf{Q} = q$, and $\mathbf{R} = r$, and Equations (17-12), 17-13), and (17-14) can be expressed as

$$\mathbf{K}(k+1) = \frac{\mathbf{P}(k+1|k)}{r}\mathbf{h}\left[\mathbf{h}'\frac{\mathbf{P}(k+1|k)}{r}\mathbf{h}+1\right]^{-1} \tag{18-8}$$

$$\frac{\mathbf{P}(k+1|k)}{r} = \mathbf{\Phi}\frac{\mathbf{P}(k|k)}{r}\mathbf{\Phi}' + \frac{q}{r}\gamma\gamma' \tag{18-9}$$

and

$$\frac{\mathbf{P}(k+1|k+1)}{r} = [\mathbf{I} - \mathbf{K}(k+1)\mathbf{h}']\frac{\mathbf{P}(k+1|k)}{r} \tag{18-10}$$

Observe that, given ratio $q/r$, we can compute $\mathbf{P}(k+1|k)/r$, $\mathbf{K}(k+1)$, and $\mathbf{P}(k+1|k+1)/r$. We refer to Equations (18-8), (18-9), and (18-10) as the *scaled* Kalman filter equations.

Ratio $q/r$ can be viewed as a *filter tuning parameter*. Recall that, in Lesson 15, we showed $q/r$ is related to signal-to-noise ratio; thus, using (15-50) for example, *we can also view signal-to-noise ratio as a (single-channel) Kalman filter tuning parameter*. Suppose, for example, that data quality is quite poor, in which case $r$ is very large, so that signal-to-noise ratio, as measured by $\overline{\text{SNR}}$, is very small. Then $q/r$ will also be very small, because $q/r = \overline{\text{SNR}}/\mathbf{h}'(\bar{\mathbf{P}}_{\mathbf{x}}/q)\mathbf{h}$. In this case the Kalman filter rejects the low-quality data; i.e., the Kalman gain matrix approaches the zero matrix, because (Problem 18-3)

$$\frac{\mathbf{P}(k+1|k)}{r} \simeq \mathbf{\Phi}\frac{\mathbf{P}(k|k)}{r}\mathbf{\Phi}' \to \mathbf{0} \tag{18-11}$$

The Kalman filter is therefore quite sensitive to signal-to-noise ratio, as indeed are most digital filters. Its dependence on signal-to-noise ratio is complicated and nonlinear. Although signal-to-noise ratio (or $q/r$) enters quite simply into the equation for $\mathbf{P}(k+1|k)/r$, it is transformed in a nonlinear manner by (18-8) and (18-10). □

## EXAMPLE 18-4

A recursive unbiased minimum-variance estimator (BLUE) of a random parameter vector $\theta$ can be obtained from the Kalman filter equations in Theorem 17-1 by setting $\mathbf{x}(k) = \theta$, $\mathbf{\Phi}(k+1,k) = \mathbf{I}$, $\mathbf{\Gamma}(k+1,k) = \mathbf{0}$, $\mathbf{\Psi}(k+1,k) = \mathbf{0}$, and $\mathbf{Q}(k) = \mathbf{0}$. Under these conditions we see that $\mathbf{w}(k) = \mathbf{0}$ for all $k$, and

$$\mathbf{x}(k+1) = \mathbf{x}(k)$$

which means, of course, that $\mathbf{x}(k)$ is a vector of constants, $\theta$. The Kalman filter equations reduce to

$$\hat{\theta}(k+1|k+1) = \hat{\theta}(k|k) + \mathbf{K}(k+1)[\mathbf{z}(k+1) - \mathbf{H}(k+1)\hat{\theta}(k|k)] \tag{18-12}$$

$$\mathbf{P}(k+1|k) = \mathbf{P}(k|k)$$

$$\mathbf{K}(k+1) = \mathbf{P}(k|k)\mathbf{H}'(k+1)[\mathbf{H}(k+1)\mathbf{P}(k|k)\mathbf{H}'(k+1) + \mathbf{R}(k+1)]^{-1} \tag{18-13}$$

and

$$\mathbf{P}(k+1|k+1) = [\mathbf{I} - \mathbf{K}(k+1)\mathbf{H}(k+1)]\mathbf{P}(k|k) \tag{18-14}$$

Note that it is no longer necessary to distinguish between filtered and predicted quantities, because $\hat{\theta}(k+1|k) = \hat{\theta}(k|k)$ and $\mathbf{P}(k+1|k) = \mathbf{P}(k|k)$; hence, the notation $\hat{\theta}(k|k)$

can be simplified to $\hat{\theta}(k)$, for example. Equations (18-12), 18-13), and (18-14) were obtained earlier in Lesson 9 (see Theorem 9-7) for the case of scalar measurements.

We described the *forward linear prediction problem*, which is very well studied in digital signal processing (Haykin, 1991), in the Supplementary Material to Lesson 16. Referring to Equation (16-35) and its accompanying discussion, we note that one approach to determining the forward prediction error filter (FPE) coefficients, $a_{M,1}, \ldots, a_{M,M}$, is to treat them as unknown constants and to then minimize either the forward mean-squared or least-squares prediction error.

In another approach, the constraint that the FPE coeffients are constant is transformed into the state equations:

$$\begin{aligned} a_{M,1}(k+1) &= a_{M,1}(k) \\ a_{M,2}(k+1) &= a_{M,2}(k) \\ &\cdots \\ a_{M,M}(k+1) &= a_{M,M}(k) \end{aligned}$$

Equation (16-35) then plays the role of the observation equation in our basic state-variable model and is one in which the observation matrix is time varying. The resulting mean-squared error design is then referred to as the *Kalman filter solution for the PEF coefficients*. Of course, we have just learned that this solution is a very special case of the Kalman filter, the BLUE; hence, it is a bit of a misnomer to refer to the solution as a Kalman filter.

In yet a third approach, the PEF coefficients are modeled as

$$\begin{aligned} a_{M,1}(k+1) &= a_{M,1}(k) + w_1(k) \\ a_{M,2}(k+1) &= a_{M,2}(k) + w_2(k) \\ &\cdots \\ a_{M,M}(k+1) &= a_{M,M}(k) + w_M(k) \end{aligned}$$

where $w_i(k)$ are white noises with variances $q_i$. Equation (16-35) again plays the role of the observation equation in our basic state-variable model and is one in which the observation matrix is time varying. The resulting mean-squared error design is now a full-blown Kalman filter. □

**EXAMPLE 18-5**

This example illustrates the *divergence phenomenon*, which often occurs when either process noise or measurement noise or both are small. We shall see that the Kalman filter locks onto wrong values for the state, but believes them to be the true values; i.e., it "learns" the wrong state too well.

Our example is adapted from Jazwinski (1970, pp. 302-303). We begin with the simple first-order system

$$x(k+1) = x(k) + b \tag{18-15}$$

$$z(k+1) = x(k+1) + \nu(k+1) \tag{18-16}$$

where $b$ is a very small bias, so small that, when we design our Kalman filter, we choose to neglect it. Our Kalman filter is based on the following model;

$$x_m(k+1) = x_m(k) \tag{18-17}$$

$$z(k+1) = x_m(k+1) + \nu(k+1) \tag{18-18}$$

Using this model, we estimate $x(k)$ as $\hat{x}_m(k|k)$, where it is straightforward to show (Problem 18-4) that

$$\hat{x}_m(k+1|k+1) = \hat{x}_m(k|k) + \overbrace{\frac{p(0)}{(k+1)p(0)+r}}^{K(k+1)}[z(k+1) - \hat{x}_m(k|k)] \tag{18-19}$$

Observe that, as $k \to \infty$, $K(k+1) \to 0$ so that $\hat{x}_m(k+1|k+1) \to \hat{x}_m(k|k)$. The Kalman filter is rejecting the new measurements because it believes (18-17) to be the true model for $x(k)$; but, of course, it is not the true model.

The Kalman filter computes the error variance, $p_m(k|k)$, between $\hat{x}_m(k|k)$ and $x_m(k)$. The true error variance is associated with $\tilde{x}(k|k)$, where

$$\tilde{x}(k|k) = x(k) - \hat{x}_m(k|k) \tag{18-20}$$

We leave it to the reader to show (Problem 18-4) that

$$\tilde{x}(k|k) = \frac{r}{kp(0)+r}\tilde{x}(0|0) - \frac{p(0)}{kp(0)+r}\sum_{i=1}^{k} \nu(i) + \frac{[(k-1)k/2]p(0)+kr}{kp(0)+r}b \tag{18-21}$$

As $k \to \infty$, $\tilde{x}(k|k) \to \infty$ because the third term on the right-hand side of (18-21) diverges to infinity. This term contains the bias $b$ that was neglected in the model used by the Kalman filter. Note also that $\tilde{x}_m(k|k) = x_m(k) - \hat{x}_m(k|k) \to 0$ as $k \to \infty$; thus, the Kalman filter has locked onto the wrong state and is unaware that the true error variance is diverging.

A number of different remedies have been proposed for controlling divergence effects, including:

1. Adding fictitious process noise
2. Finite-memory filtering
3. Fading memory filtering

Fictitious process noise, which appears in the state equation, can be used to account for neglected modeling effects that enter into the state equation (e.g., truncation of second- and higher-order effects when a nonlinear state equation is linearized, as described in Lesson 23). This process noise introduces $\mathbf{Q}$ into the Kalman filter equations. Observe, in our first-order example, that $\mathbf{Q}$ does not appear in the equations for $\hat{x}_m(k+1|k+1)$ or $\tilde{x}(k|k)$, because state equation (18-17) contains no process noise.

Divergence is a large-sample property of the Kalman filter. Finite-memory and fading-memory filtering control divergence by not letting the Kalman filter get into its "large-sample" regime. Finite-memory filtering (Jazwinski, 1970) uses a finite window of measurements (of fixed length $W$) to estimate $x(k)$. As we move from $t = k_1$ to $t = k_1 + 1$, we must account for two effects, the new measurement at $t = k_1 + 1$ and a discarded measurement at $t = k_1 - W$. Fading-memory filtering, due to Sorenson and Sacks (1971), exponentially ages the measurements, weighting the recent measurement most heavily and past measurements much less heavily. It is analogous to weighted least squares, as described in Lesson 3. The only change in the Kalman filter equations, when exponential fading is used, is that the prediction covariance matrix is now updated using the equation

$$\mathbf{P}(k+1|k) = \mathbf{\Phi}\mathbf{P}(k|k)\mathbf{\Phi}' \exp(c_k) + \mathbf{\Gamma}(k+1,k)\mathbf{Q}(k)\mathbf{\Gamma}'(k+1,k) \tag{18-22}$$

where $c_k \geq 0$ for all $k$ is a possibly time-varying fading factor. Because $c_k \geq 0$, $\exp(c_k) \geq 1$; hence, $\exp(c_k)$ has the effect of boosting the filter error covariance matrix, $\mathbf{P}(k|k)$, before $\mathbf{P}(k+1|k)$ is updated.

Fading-memory filtering seems to be the most successful and popular way to control divergence effects. □

**EXAMPLE 18-6** — *how not to use kalman filter*

Here we examine a *prediction error filter* (PEF) that was developed in 1972 by Ott and Meder. It uses a scaled version of the innovations process as an estimator of white noise system input $\mathbf{w}(k)$. Obviously, the Ott and Meder PEF is performing *deconvolution* (*which case estimate wk*).

The starting point for the derivation of the Ott and Meder PEF is our basic state-variable model in which $\mathbf{u}(k) = \mathbf{0}$. By combining (17-11) and (16-4), we obtain the following recursive filter equation that is driven by the white innovations process (note that this is *not* a fully recursive filter equation, because the innovations is computed from the predicted estimate)

$x_{k+1} = \phi x_k + \Gamma w_k$ (A) 15-17 — *innovation, is also white noise.*

$$\hat{\mathbf{x}}(k+1|k+1) = \boldsymbol{\Phi}(k+1, k)\hat{\mathbf{x}}(k|k) + \mathbf{K}(k+1)\tilde{\mathbf{z}}(k+1|k) \tag{18-23}$$

Observing the *structural similarity* between Equations (18-23) and (15-17), Ott and Meder equated the white noise terms to establish their PEF for $\hat{\mathbf{w}}(k)$; i.e.,

$$\boldsymbol{\Gamma}(k+1, k)\hat{\mathbf{w}}(k) \triangleq \mathbf{K}(k+1)\tilde{\mathbf{z}}(k+1|k) \tag{18-24}$$

from which it follows that

$$\hat{\mathbf{w}}(k) = \boldsymbol{\Gamma}^+(k+1, k)\mathbf{K}(k+1)\tilde{\mathbf{z}}(k+1|k) \tag{18-25}$$

*we know w is white z is white so w will be white*

where $\boldsymbol{\Gamma}^+(k+1, k)$ is the pseudoinverse of $\boldsymbol{\Gamma}(k+1, k)$ (see Lesson 4). Interestingly enough, the PEF is actually a *smoother*, because $\hat{\mathbf{w}}(k)$ uses the innovations at $t_{k+1}$. (*eg this is not filtered, it uses data from future*)

Observe that $\hat{\mathbf{w}}(k)$ enjoys the property of being a white estimator of the white input $\mathbf{w}(k)$; however, $\hat{\mathbf{w}}(k)$ is not an optimal estimator of $\mathbf{w}(k)$. It has merely been *defined* as in (18-24). As such, the Ott and Meder PEF is an *ad hoc estimator*. The performance of the Ott and Meder PEF can be established by determining the error covariance matrix between $\hat{\mathbf{w}}(k)$ and $\mathbf{w}(k)$. Let

$$\boldsymbol{\Psi}_\mathbf{w}(k) \triangleq \mathbf{E}\{[\tilde{\mathbf{w}}(k) - \mathbf{E}\{\tilde{\mathbf{w}}(k)\}][\tilde{\mathbf{w}}(k) - \mathbf{E}\{\tilde{\mathbf{w}}(k)\}]'\} \tag{18-26}$$

Because $\mathbf{w}(k)$ and $\tilde{\mathbf{z}}(k+1|k)$ are both zero mean,

$$\boldsymbol{\Psi}_\mathbf{w}(k) = \mathbf{E}\{\tilde{\mathbf{w}}(k)\tilde{\mathbf{w}}'(k)\} \tag{18-27}$$

We leave it to the reader to show (Problem 18-5) (Mendel, 1977b) that

$$\boldsymbol{\Psi}_\mathbf{w}(k) = \mathbf{Q} + \boldsymbol{\Gamma}^+\mathbf{PH}'(\mathbf{HPH}' + \mathbf{R})^{-1}\mathbf{HP}\boldsymbol{\Gamma}'^+ - \mathbf{Q}\boldsymbol{\Gamma}'\mathbf{H}'(\mathbf{HPH}' + \mathbf{R})^{-1}\mathbf{HP}\boldsymbol{\Gamma}'^+$$

$$-\boldsymbol{\Gamma}^+\mathbf{PH}'(\mathbf{HPH}' + \mathbf{R})^{-1}\mathbf{H}\boldsymbol{\Gamma}\mathbf{Q} \tag{18-28}$$

where $\mathbf{Q} = \mathbf{Q}(k)$, $\boldsymbol{\Gamma}^+ = \boldsymbol{\Gamma}^+(k+1, k)$, $\mathbf{P} = \mathbf{P}(k+1|k)$, $\mathbf{H} = \mathbf{H}(k+1)$, $\mathbf{R} = \mathbf{R}(k+1)$, and $\boldsymbol{\Gamma} = \boldsymbol{\Gamma}(k+1, k)$.

Because the last two terms in (18-28) are negative, it would appear to be true that for some systems $\boldsymbol{\Psi}_\mathbf{w}(k)$ will be smaller than $\mathbf{Q}(k)$. This is important since the error-covariance matrix of the *zero estimator* of $\mathbf{w}(k)$ [i.e., the estimator that estimates $\mathbf{w}(k)$ as zero for all $k$] is $\mathbf{Q}(k)$; and obviously any useful estimator of $\mathbf{w}(k)$ should perform better than the zero estimator.

Consider the following example of a one-dimensional damped harmonic oscillator that is excited by impulses of random intensity at random time instances:

$$\begin{bmatrix} x_1(k+1) \\ x_2(k+1) \end{bmatrix} = \begin{bmatrix} 1 & T \\ -\beta T & 1 - \alpha T \end{bmatrix} \begin{bmatrix} x_1(k) \\ x_2(k) \end{bmatrix} + \begin{bmatrix} 0 \\ \omega_E T \end{bmatrix} w(k) \tag{18-29}$$

$$z(k) = (1 \quad 0)\text{col}\,[x_1(k), x_2(k)] + v(k) \tag{18-30}$$

Equation (18-29) was obtained by discretizing the equation $d^2x(t)/dt^2 + \alpha \, dx(t)/dt + \beta x(t) = \omega_E w(t)$, assuming $t_{k+1} - t_k = T$ is small (see Lesson 23 for exact details on how to do this). When we proceed to calculate $\boldsymbol{\Psi}_w(k)$, by means of (18-28), we find that $\mathbf{H}\boldsymbol{\Gamma} = \mathbf{0}$ because $\mathbf{H}(k) = (1, 0)$ and $\boldsymbol{\Gamma}(k+1, k) = \mathrm{col}\,(0, \omega_E T)$, so that *↳ those neg. terms are gone*

$$\boldsymbol{\Psi}_w(k) = \mathbf{Q} + \boldsymbol{\Gamma}^+ \mathbf{P} \mathbf{H}' (\mathbf{H}\mathbf{P}\mathbf{H}' + \mathbf{R})^{-1} \mathbf{H}\mathbf{P}\boldsymbol{\Gamma}'^+ \tag{18-31}$$

The second term in (18-31) is positive semidefinite; hence, we see that $\boldsymbol{\Psi}_w(k) \geq \mathbf{Q}$! This means that, for this example, the Ott and Meder PEF performs no better and probably worse than the zero estimator of $\mathbf{w}(k)$. This behavior is due to the ad hoc nature of the Ott and Meder PEF. In Lesson 20 we shall use mean-squared smoothing to obtain an optimal estimator for $\mathbf{w}(k)$ and shall demonstrate that its performance for this example is always better than that of the Ott and Meder PEF. $\square$

## EXAMPLE 18-7

Consider the basic state-variable model, but now the control has been designed (using, e.g., quadratic optimal control) so that

$$\hat{\mathbf{u}}(k) = \mathbf{C}(k)\hat{\mathbf{x}}(k|k) \quad \textit{the control} = f(\text{estimate of state vector}) \tag{18-32}$$

where $\mathbf{C}(k)$ is a control gain matrix and $\hat{\mathbf{x}}(k|k)$ is the output of a Kalman filter. In this case the basic state-variable model can be expressed as

$$\mathbf{x}(k+1) = \boldsymbol{\Phi}_c(k+1, k)\mathbf{x}(k) + \boldsymbol{\Gamma}(k+1, k)\mathbf{w}(k) - \boldsymbol{\Psi}(k+1, k)\mathbf{C}(k)\tilde{\mathbf{x}}(k|k) \tag{18-33}$$

where

$$\boldsymbol{\Phi}_c(k+1, k) = \boldsymbol{\Phi}(k+1, k) + \boldsymbol{\Psi}(k+1, k)\mathbf{C}(k) \tag{18-34}$$

Proceeding as we did in Lesson 15, where we derived the equation for the covariance matrix of $\mathbf{x}(k)$, that is (15-29), we can show (Problem 18-7) that the equation for the covariance matrix of $\mathbf{x}(k)$ for the system in (18-33) is (Mendel, 1971a)

$$
\begin{aligned}
\mathbf{P}(k+1) = {} & \boldsymbol{\Phi}_c(k+1, k)\mathbf{P}(k)\boldsymbol{\Phi}_c'(k+1, k) + \boldsymbol{\Gamma}(k+1, k)\mathbf{Q}(k)\boldsymbol{\Gamma}'(k+1, k) \\
& + \boldsymbol{\Psi}(k+1, k)\mathbf{C}(k)\mathbf{P}(k|k)\mathbf{C}'(k)\boldsymbol{\Psi}'(k+1, k) \\
& - \boldsymbol{\Phi}_c(k+1, k)\mathbf{P}(k|k)\mathbf{C}'(k)\boldsymbol{\Psi}'(k+1, k) \\
& - \boldsymbol{\Psi}(k+1, k)\mathbf{C}(k)\mathbf{P}(k|k)\boldsymbol{\Phi}_c'(k+1, k)
\end{aligned}
\tag{18-35}
$$

What we observe by comparing (18-35) and (15-29) is that, by using the feedback control law in (18-32), *state estimation errors act as an additional plant disturbance*. Even if $\mathbf{w}(k) = \mathbf{0}$, the Kalman filter will "disturb" the system. This is not necessarily bad; it is just a fact. It does suggest that the effects of the Kalman filter must be taken into account at the front end of an optimal control design procedure; i.e., if a controller is designed to achieve a low variance in some of its states assuming that perfect measurements of the states are available, then, when the Kalman filter is used to actually estimate the states, performance, as measured by variance of the states, will usually be worse because of the last three terms in (18-35). $\square$

# APPLICATIONS OF KALMAN FILTERING

Kalman filtering has been widely applied. In 1983 the IEEE Control Systems Society published a special issue of its *Transactions on Automatic Control* on "Applications of Kalman Filtering." This special issue has been entirely reprinted in the IEEE Press volume *Kalman Filtering: Theory and Applications* (Sorenson, 1985). The special issue consists of 19 papers collected together under the following nine categories: orbit determination, tracking, navigation, ship motion, remote sensing, geophysical exploration, industrial processes, power systems, and demography. In this section we describe some of the papers in the special issue, using in many cases the words of the papers' authors.

In the paper "Voyager Orbit Determination at Jupiter" (Campbell et al., 1983) we learn that on March 5, 1979, after a journey of 546 days and slightly more than 1 billion kilometers, the Voyager 1 spacecraft passed very close to the innermost Galilean satellite Io. Four months later, on July 8, Voyager 2 flew close by a third Galilean satellite and then by Jupiter the next day. These Voyager encounters with Jupiter were both spectacular and historic. Each mission returned huge amounts of data from 11 scientific instruments, including some 15,000 high-resolution pictures of Jupiter and five of its satellites. Where does Kalman filtering come into all this?

Voyager's flight path during the Jupiter near-encounter phase had to be controlled very accurately to achieve its scientific objectives. Since Voyager's final trajectory correction maneuver was very small, getting Voyager into its proper Jupiter near-encounter orbit was determined by orbit estimation accuracies that were available at the time of maneuver specification. Accurate instrument sequences, which were required to obtain near-encounter science data, were highly dependent on accurate knowledge of the spacecraft trajectory and satellite orbits. Postflight reconstructed orbits indicated that the near-encounter spacecraft orbits were predicted to within 50 kilometers and that the spacecraft satellite pointing was predicted to within 3 milliradians, even for the close Voyager 1 Io flyby.

A Kalman filter was used to process the Voyager tracking data for two reasons: (1) modeling nongravitational accelerations (for example, thruster plumes that actually hit the spacecraft) and (2) modeling the optical data to account for pointing errors.

Outer space certainly requires high technology, so it's not surprising that a Kalman filter was needed in the United States' Space Program; but what about something more down to earth?

Let's turn to the classical problem of antiaircraft gun fire control, which is described in the paper "Estimation and Prediction for Maneuvering Target Trajectories" (Berg, 1983). This problem has to do with the accurate prediction of the future position of a given target at the time of projectile intercept. Obtaining this information permits us to determine the correct gun pointing angles. Norbert Wiener (see his biography in Lesson 19) worked on the fire control problem; but current approaches to this problem use Kalman filtering to estimate target velocity and acceleration on the basis of target position measurements. Berg's paper describes a numerically efficient scheme for estimating and predicting the future position of maneuvering fixed-wing aircraft. The scheme was implemented in the fire control computer of a radar tracking gun system, which was successfully tested and evaluated against a variety of fixed-wing aircraft.

Next, let's consider the paper "Application of Multiple Model Estimation to a Recursive Terrain Height Correlation System," (Mealy and Tang, 1983). By comparing measured terrain height with stored terrain height data, it is possible to estimate aircraft position. The well-known Terrain Contour Matching System (TERCOM) is one implementation of this concept. TERCOM considers ground clearance as the fundamental system measurement from which vehicle position must be determined. Measured terrain clearance is compared to the value predicted by an onboard navigation system and the measurement model to obtain an error signal that is proportional to navigation errors. Using a system model, which describes how these errors propagate, a state estimator then implements the equations in order to calculate an estimate of the reference-state errors. The state estimator may be any linearized estimator, such as the extended Kalman filter (described in detail in Lesson 24). The error estimate is subtracted from the reference to give a corrected estimate of the vehicle state.

So, a Kalman filter is very useful in airborne applications.

The special issue also has a few papers devoted to shipboard applications of Kalman filtering. To begin, let's consider the paper "Design and Analysis of a Dynamic Positioning System Based on Kalman Filtering and Optimal Control" (Saelid et al., 1983) in which we learn that dynamic positioning systems for drilling vessels and platforms and for support vessels in the offshore oil industry have been manufactured since the early 1960s. In the beginning, these systems were designed using conventional control principles. Since 1974, new systems have been developed based on "modern control theory," in which state and parameter estimation is employed in the form of extended Kalman filters and multivariable control designs that are based on linear quadratic control theory. A Norwegian company has marketed such systems since 1976 under the name "Albatross." These systems are based on extensive mathematical modeling of the dynamic behavior of the drilling vessel, which is subject to forces exerted by the controllable thrusters, i.e. propellors, as well as environmental forces from wind, currents, and waves.

A good system for dynamic positioning should keep the drilling vessel within specified position and heading limits, with a minimum of fuel consumption and a minimum of wear on the propulsion equipment, and it must also tolerate transient failures in the measurement and propulsion systems. To accomplish this, the behavior of the drilling vessel was modeled as the combination of a "low-frequency drift" and a "high-frequency oscillation." The slow motion was caused by the propellors and the forces from the wind and water currents. Since the wind could be measured fairly accurately, only the current forces had to be estimated. It was impossible to counteract wave oscillations, so the control actions of the propellors were derived from the state of the low-frequency model, the state of the current model, and a feed forward from the wind measurement. State estimates were obtained from an extended Kalman filter.

Another interesting shipboard application is described in the paper "On the Feasibility of Real-time Prediction of Aircraft Carrier Motion at Sea" (Sidar and

Doolin, 1983). It deals with the landing phase of an aircraft aboard an aircraft carrier, which represents a complex operation and a demanding task. The last 10 to 15 seconds before aircraft touchdown are critical and involve terminal guidance and difficult control problems, because, not only is the aircraft affected by several kinds of random disturbances, for example, wind, but the touchdown point on the ship is also being moved randomly. Despite the wind disturbances and the final point random motion, the landing accuracy specified for carrier operations is very high, a few tens of feet. The terminal point problem is solved in a very natural way, by assuming that the ship's position can be predicted for several seconds ahead so that the airplane is moved toward the future position of the touchdown point. The scope of this paper's study was to establish to what extent a stochastic process, like the ship's motion, is predictable over moderate periods of time. Being able to predict the ship's motion to within acceptable bounds of error can also lead to improvement of the landing signal officer's decision policy for waveoffs. Another potential application is the incorporation of the predictor algorithm into a control loop used to stabilize the carrier's Fresnel lens system. This provides improved glide path information to the pilot for pilot-controlled landings. This can be accomplished by making use of the ship's motion characteristics as measured by instrumentation existing onboard the ship. In this way it is possible to predict the ship's motion using a mean-squared predictor. The results obtained showed that a maximum achievable prediction time of up to 15 seconds could be reached within reasonable acceptable errors.

So far, we have seen how Kalman filtering has been used by spacecraft, aircraft, and ships. It has also been used in remote sensing applications as described, for example, in the paper "An Integrated Multisensor Aircraft Track Recovery System for Remote Sensing" (Gesing and Reid, 1983). In this paper, an aerial hydrography system was used that employed an aerial survey camera, a high-powered pulsed laser, and some navigation sensors that measured the orientation of the camera and laser in flight. The data obtained in the air were processed on the ground. The laser measures water depth by transmitting pulses of green light 10 times per second. The pulses are reflected from the water surface and from the bottom. The time difference between the surface and bottom reflections provides an accurate measure of water depth. The camera takes overlapping stereo photographs of the survey area. After the flight is over, an analytical plotter forms stereo images of the bottom and near-shore topography. When these images are combined, they produce shallow-water bathymetric plots, which are then merged with plots of deeper waters to produce nautical charts. The deep-water plots are acquired by conventional echo sounding. Conventional methods for aerial photo mapping rely on ground control detail points to establish the orientation of the aerial photographs with respect to Earth. In coastal mapping applications, photo orientation is difficult because overwater photography may contain little or no land for bridging or photo control. One approach for overcoming this problem is to obtain the position and attitude of the camera at the time each photo is taken. This has to be done with an absolute precision sufficient to satisfy requirements for charting of spot depths on nautical charts and a relative precision sufficient enough to form stereo models in the analytical plotter from parts of adjacent photographs. The data collected by individual navigation sensors are not accurate enough to meet these demands and are therefore used by a Kalman filter to estimate sensor errors. The error estimates are combined with the original data to yield high-accuracy position and attitude information.

A second remote sensing application is described in the paper "Bathymetric and Oceanographic Applications of Kalman Filtering Techniques" (Brammer et al., 1983), in which Kalman filtering is applied to the analysis of radar altimeter data for mapping ocean currents and for detecting seamounts. A seamount is an underwater mountain, primarily of volcanic origin. In this application, a satellite is tracked by ground-based tracking stations that determine its orbit and height relative to a reference ellipsoid. After applying various corrections for atmospheric refraction and instrumental bias, the height of the sea surface can be estimated accurately with respect to the reference ellipsoid. These height data can be used to infer gravity anomalies, to estimate ocean tides, and to detect major ocean currents and seamounts. Kalman filtering and Kalman smoothing algorithms were used in this paper to estimate the sea-surface slope.

Kalman filtering has also been used in petroleum exploration to process seismic data and, in analyzing oil wells, in well logging. This textbook explains how estimation theory can be used to perform deconvolution. Deconvolution is very widely used in seismic data processing for petroleum exploration. High-resolution deconvolution leads to nonlinear signal processing and consists of combined estimation and detection (see Lesson 14). The paper "A Kalman Filtering Approach to Natural Gamma Ray Spectroscopy in Well Logging" (Ruckebusch, 1983) describes a well-logging device known as a natural gamma ray spectroscopy tool, which was commercially introduced by Schlumberger several years ago. It provides a direct means to determine the thorium, uranium, and potassium concentrations in a geological formation. These concentrations are useful in computer reservoir evaluations. A petroleum reservoir is a natural earth formation in which hydrocarbons, i.e., oil and gas, are trapped. The natural gamma ray spectroscopy tool is designed to detect natural gamma rays of various energies that are emitted from a formation of radioactive nuclei of potassium and the thorium and uranium–radium series. The thorium, uranium, and potassium concentrations along the borehole of the well have to be estimated on-line from the detection of gamma rays in five energy windows. This is done using a microcomputer at the surface of the well. The standard technique in the logging industry is to compute the elemental concentrations at a given depth using only the observed counting rates at the same depth. Unfortunately, the resulting estimates have pretty large statistical errors, which has limited the application of the natural gamma ray spectroscopy tool. This paper shows that a Kalman filter that is based on a dynamical model of thorium, uranium, and potassium vertical variations can produce real-time estimates that are very useful in computer reservoir evaluation. These real-time estimates can be used to assist a drilling crew in determining whether or not drilling should be continued in the well.

Kalman filtering has also been used in industrial processes. An excellent example of this is given in the paper "Estimation and Prediction of Unmeasureable Variables in the Steel Mill Soaking Pit Control System" (Lumelsky, 1983). In one version of the steel-making process, steel ingots typically pass through the soaking pit operation before they are rolled at the rolling mill. The purpose of this operation is to equalize the temperature throughout the ingot masses at some prespecified level. Accurate description of the ingot temperature distribution requires knowledge of analytical space–temperature relationships, or temperature values at many points along the ingot's side, top, and bottom surfaces, as well as throughout its mass. For practical purposes, it is enough to know some average estimates of the ingot surface and center temperatures. When the ingots come from the stripping yard, their surface temperature is anywhere from 1300° to 1900° F and their center temperature is anywhere from 2000° to 2800° F. Sometimes, the ingots come from cold storage, in which case their temperature will be that of ambient. To equalize the temperatures throughout the ingot, a control system has to continuously estimate the current ingot temperatures, at least on the surface and at the center, and to stop the operation when these temperatures arrive at prespecified levels. Today (circa 1983), this is done by a human operator. Underheating of ingots results in poor rolling mill performance or in returning the ingots to the pit for additional heat-up; and overheating, which is often the case, results in a waste of energy. According to industry estimates, up to 15% to 20% fuel savings could be realized in this energy-intensive operation if an efficient ingot temperature estimation system was put into operation. But the ingot temperature is not directly measurable. It must be estimated using some indirect measurements, such as pit-wall temperature or fuel flow. Another piece of information that is important for the whole operation, especially for operation planning purposes, is the predicted moment of time at which the ingots will arrive at the said prespecified temperature. This paper describes a Kalman filter approach and some results of a joint project between the General Electric Company and the U.S. Steel Company on the estimation and prediction of ingot temperature in the soaking pit operation.

Kalman filters have also played an important role in power systems, ranging from conventional coal-fired to more modern nuclear plants. Their use in nuclear power plants is the subject of the paper "On-line Failure Detection in Nuclear Power Plant Instrumentation" (Tylee, 1983). In nuclear power plants it is very important that instrument failures be detected and accommodated before significant performance degradation results. If such detection can't be made, it may be necessary to shut down the plant. Doing this results in lost power production and lost revenues to the operating utility. One technique for detecting failed instruments in a power plant is hardware redundancy, where a two-out-of-three voting logic is used to eliminate faulty measurements. A more common failure-detection technique is operator reliance, where the human operator uses his or her knowledge of the plant operating state and dynamics to determine if a certain measurement is unreasonable. With the hundreds of measurements made in a nuclear facility, the cost of redundant sensors for each measurement is too high, and it is impossible for the plant operator to continually monitor each instrument for possible failures. Incidents at Three Mile Island, Chernobyl, and other nuclear facilities have demonstrated a real need for better failure-detection capabilities. Digital computers are now being brought into the control room in an attempt to provide the operator with information needed to properly establish plant status. If these computers could also somehow process the measurements available to them and make a decision as to whether a certain instrument has failed or not, improvements in plant safety and operation would be possible. This approach to failure detection, which is described in this paper, is known as *functional redundancy*. In functional redundancy, inputs and available measurements are used to drive a bank of Kalman filters, one for each measurement. Each Kalman filter is designed to be sensitive to failure in just one instrument. Checks are performed on redundant state estimates in the instrument failure-detection logic. If these tests show one estimate to be inconsistent with the other estimates or inconsistent with the current plant operating point, a failure is noted for the instrument associated with the filter that has generated the anomalous estimate. Estimation theory now makes it possible to automatically monitor hundreds of measurements to protect us from a nuclear disaster.

The last application described here is most unusual; it demonstrates the tremendous potential for Kalman filtering in nontraditional areas. The paper "Application of Kalman Filtering to Demographic Models" (Leibundgut et al., 1983) describes simulation models that are used in agricultural economics to make predictions and to help decision makers by analyzing the effects of alternative policies. One example of such an agroeconomic system is a beef cattle herd. In general, such systems are characterized by four items. The first is that only limited data are generally available. New experiments cannot be performed, so identification is always a passive process. The second is that data are often not observed directly, but through a polling system, such as a sample survey, which introduces large errors in comparison with technological sensors. The third is that the simulation horizon is generally long, for example, 10 to 20 years; hence, stationarity assumptions are inappropriate because of structural evolutions. The fourth is that many uncontrolled factors, such as people, weather, or epidemics, can perturb the system so that it is hard to define the dynamical structure. The problem described in this paper is to estimate the number of male and female beef cattle per age class, as accurately as possible, using information available from a variety of statistical services. A methodology was actually developed to do this and it was applied to the French cattle herd, based on Kalman filtering. This method provided results that improve our knowledge of the French cattle herd, as well as the confidence on male and female populations per age classes over the time horizon. Predictions, using state estimation, were performed through slaughter scenarios in order to help decision makers control the cattle herd. This implementation gave very satisfactory results. The authors conclude that "The present methodology is general and could be used in every field of human sciences insofar as the dynamic evolution of a system can be modeled by a state equation."

This completes our short excursion into applications of Kalman filtering. It should provide you with a glimpse into the tremendous applicability of Kalman filtering. This is the reason that Kalman filtering has withstood the test of time.

## SUMMARY QUESTIONS

**1.** Steady-state values of Kalman filter quantities:
  **(a)** always depend on $\mathbf{P}(0|0)$
  **(b)** sometimes depend on $\mathbf{P}(0|0)$
  **(c)** never depend on $\mathbf{P}(0|0)$

2. A Kalman filter sensitivity system is:
   (a) a single equation for computing $\delta\hat{\mathbf{x}}(k|k)/\delta\theta$
   (b) a system of equations for computing $\delta\hat{\mathbf{x}}(k|k)/\delta\theta$
   (c) available only for small parameter variations
3. In the single-channel case, signal-to-noise ratio:
   (a) can be viewed as a KF tuning parameter
   (b) equals $q/r$
   (c) depends on both $q$ and $r$
4. Divergence is:
   (a) always present in a KF
   (b) a large-sample property of the KF
   (c) a small-sample property of the KF
5. Divergence of a KF occurs when:
   (a) either process noise or measurement noise or both are small
   (b) the process model is unstable
   (c) the KF becomes unstable
6. Which of the following can be used to handle divergence effects?
   (a) finite-memory filtering
   (b) smoothing
   (c) adding fictitious process noise
   (d) adding fictitious measurement noise
   (e) fading-memory filtering
7. A KF reduces to a BLUE of a constant $\boldsymbol{\theta}$ when we set:
   (a) $w(k) = 0$, for all $k$
   (b) $\boldsymbol{\Gamma} = \mathbf{0}$, $\mathbf{Q} = \mathbf{0}$, and $\boldsymbol{\Phi} = \mathbf{I}$
   (c) $\boldsymbol{\Phi} = \mathbf{0}$, $\boldsymbol{\Gamma} = \mathbf{0}$, $\boldsymbol{\Phi} = \mathbf{0}$, and $\mathbf{Q} = \mathbf{0}$
8. Which of the following represent "morals" for the Ott and Meder *prediction error filter*?
   (a) look before you leap
   (b) estimation theory is so powerful it can be used in an ad hoc manner with good results
   (c) estimation theory must be used properly or else it can lead to terrible results
9. When a control system is designed in which the controller is a signal that is proportional to the output of a Kalman filter:
   (a) we can optimize the control gains by assuming that there are no state estimation errors
   (b) state estimation errors act as an additional plant disturbance, and these disturbances must be accounted for during the overall design
   (c) state estimation errors act as an additional plant disturbance, but these disturbances do not have to be accounted for during the overall design

## PROBLEMS

**18-1.** Derive a general system of Kalman filter sensitivity equations for small variations in $\theta$. See, also, Lesson 25 and Problem 25-1.

**18-2.** (Computation Project) Choose a second-order system and perform a thorough sensitivity study of its associated Kalman filter. Do this for various nominal values and for both small and large variations of the system's parameters. *You will need a computer for this project.* Present the results both graphically and tabularly, as in Example 18-2. Draw as many conclusions as possible.

**18-3.** (Sam Heidari, 1991) Prove for a stable first-order state-variable system that

$$\lim_{k\to\infty} K(k+1) = \bar{K} \to 0$$

as $q/r \to 0$. Explain how this proof generalizes to an $n$th-order state-variable model.

**18-4.** Derive the equations for $\hat{x}_m(k+1|k+1)$ and $\tilde{x}(k|k)$ in (18-19) and (18-21), respectively.

**18-5.** Derive Equation (18-28).

**18-6.** In Problem 5-7 we described cross-sectional processing for weighted least-squares estimates. Cross-sectional (also known as sequential) processing can be performed in Kalman filtering. Suppose $\mathbf{z}(k+1) = \text{col}\,(\mathbf{z}_1(k+1), \mathbf{z}_2(k+1), \ldots, \mathbf{z}_q(k+1))$, where $\mathbf{z}_i(k+1) = \mathbf{H}_i\mathbf{x}(k+1) + \mathbf{v}_i(k+1)$, $\mathbf{v}_i(k+1)$, are mutually uncorrelated for $i = 1, 2, \ldots, \mathbf{z}_i(k+1)$ is $m_i \times 1$, and $m_1 + m_2 + \ldots + m_q = m$. Let $\hat{\mathbf{x}}_i(k+1|k+1)$ be a "corrected" estimate of $\mathbf{x}(k+1)$ that is associated with processing $\mathbf{z}_i(k+1)$.

**(a)** Using the fundamental theorem of estimation theory, prove that a cross-sectional structure for the corrector equation of the Kalman filter is

$$\begin{aligned}
\hat{\mathbf{x}}_1(k+1|k+1) &= \hat{\mathbf{x}}(k+1|k) + \mathbf{E}\{\mathbf{x}(k+1)|\tilde{\mathbf{z}}_1(k+1|k)\} \\
\hat{\mathbf{x}}_2(k+1|k+1) &= \hat{\mathbf{x}}_1(k+1|k+1) + \mathbf{E}\{\mathbf{x}(k+1)|\tilde{\mathbf{z}}_2(k+1|k)\} \\
&\;\;\vdots \\
\hat{\mathbf{x}}_q(k+1|k+1) &= \hat{\mathbf{x}}_{q-1}(k+1|k+1) + \mathbf{E}\{\mathbf{x}(k+1)|\tilde{\mathbf{z}}_q(k+1|k)\} \\
&= \hat{\mathbf{x}}(k+1|k+1)
\end{aligned}$$

**(b)** Provide equations for computing $\mathbf{E}\{\mathbf{x}(k+1)|\tilde{\mathbf{z}}_i(k+1|k)\}$.

**18-7.** Derive Equation (18-35).

# LESSON 19

# State Estimation: Steady-state Kalman Filter and Its Relationship to a Digital Wiener Filter

## SUMMARY

In this lesson we study the steady-state Kalman filter from different points of view, and we then show how it is related to a digital Wiener filter.

What is a steady-state Kalman filter? We have already noted that a Kalman filter is a time-varying recursive digital filter. It is time-varying because the Kalman gain matrix, $\mathbf{K}(k)$, varies with time. If our basic state-variable model is time invariant and stationary, then (under some additional system-theoretic conditions), after an initial interval of time during which $\mathbf{K}(k)$ does vary with time, $\mathbf{K}(k)$ reaches a steady state in which all its elements are constants. The steady-state Kalman filter uses the steady-state values of the Kalman gain matrix.

If the basic state-variable model is time invariant, stationary, and asymptotically stable, then the steady-state Kalman filter is also asymptotically stable.

The single-channel steady-state Kalman filter is a recursive digital filter with filter coefficients equal to $h_f(j), j = 0, 1, \ldots$. Quite often $h_f(j) \approx 0$ for $j \geq J$, so the filter can be truncated. The *truncated steady-state Kalman filter* can then be implemented as a finite impulse response (FIR) digital filter.

There is a more direct way, which is described in this lesson, for deriving a FIR minimum mean-squared error filter, i.e., a *digital Wiener filter*. The coefficients of this filter can be computed very quickly using the Levinson algorithm, which is described in the Supplementary Material at the end of this lesson. We shall see that the steady-state Kalman filter is an infinite-length digital Wiener filter and that the truncated steady-state Kalman filter is a FIR digital Wiener filter. These observations connect results from two seemingly different fields, estimation theory and digital signal processing.

There are, of course, differences between Kalman and Wiener filters, and these are described in the last section of this lesson.

When you complete this lesson, you will be able to (1) explain what is meant by a steady-state Kalman filter and describe its properties; (2) describe the steady-state Kalman filter as a digital filter; (3) explain what is meant by a digital Wiener filter; and (4) identify the equivalences between a steady-state Kalman filter and a digital Wiener filter.

## INTRODUCTION

In this lesson we study the steady-state Kalman filter from different points of view, and we then show how it is related to a digital Wiener filter.

## STEADY-STATE KALMAN FILTER

For time-invariant and stationary systems, if $\lim_{k\to\infty} \mathbf{P}(k+1|k) = \mathbf{P}_p$ exists, then $\lim_{k\to\infty} \mathbf{K}(k) \to \bar{\mathbf{K}}$ and the Kalman filter (17-11) becomes a constant coefficient filter. Because $\mathbf{P}(k+1|k)$ and $\mathbf{P}(k|k)$ are intimately related, then, if $\mathbf{P}_p$ exists, $\lim_{k\to\infty} \mathbf{P}(k|k) \triangleq \mathbf{P}_f$ also exists. We have already observed limiting behaviors for $\mathbf{K}(k+1)$ and $\mathbf{P}(k|k)$ in Example 18-1. The following theorem, which is adopted from Anderson and Moore (1979, p. 77), tells us when $\mathbf{P}_p$ exists, and assures us that the steady-state Kalman filter will be asymptotically stable.

**Theorem 19-1** (Steady-state Kalman Filter). *If our dynamical model in Equations (15-17) and (15-18) is time invariant, stationary, and asymptotically stable (i.e., all the eigenvalues of $\mathbf{\Phi}$ lie inside the unit circle), then:*

**a.** *For any nonnegative-definite symmetric initial condition $\mathbf{P}(0|-1)$, we have $\lim_{k\to\infty} \mathbf{P}(k+1|k) = \mathbf{P}_p$ with $\mathbf{P}_p$ independent of $\mathbf{P}(0|-1)$ and satisfying the following steady-state version of Equation (17-24):*

$$\mathbf{P}_p = \mathbf{\Phi}\mathbf{P}_p[\mathbf{I} - \mathbf{H}'(\mathbf{H}\mathbf{P}_p\mathbf{H}' + \mathbf{R})^{-1}\mathbf{H}\mathbf{P}_p]\mathbf{\Phi}' + \mathbf{\Gamma}\mathbf{Q}\mathbf{\Gamma}' \tag{19-1}$$

*Equation (19-1) is often referred to either as a steady-state or algebraic Riccati equation.*

**b.** *The eigenvalues of the steady-state Kalman filter, $\boldsymbol{\lambda}[\mathbf{\Phi} - \bar{\mathbf{K}}\mathbf{H}\mathbf{\Phi}]$, all lie within the unit circle, so the filter is asymptotically stable i.e.,*

$$|\boldsymbol{\lambda}[\mathbf{\Phi} - \bar{\mathbf{K}}\mathbf{H}\mathbf{\Phi}]| < 1 \tag{19-2}$$

*If our dynamical model in Equations (15-17) and (15-18) is time invariant and stationary, but is not necessarily asymptotically stable, then points (a) and (b) still hold as long as the system is completely stabilizable and detectable.* □

A proof of this theorem is beyond the scope of this textbook. It can be found in Anderson and Moore, pp. 78–82 (1979). For definitions of the system-theoretic terms *stabilizable* and *detectable*, the reader should consult a textbook on linear systems, such as Kailath (1980) or Chen (1984). See, also, the Supplementary Material at the end of this lesson in the section entitled "Some Linear System Concepts." By *completely detectable* and *completely stabilizable*, we mean that $(\mathbf{\Phi}, \mathbf{H})$ is completely detectable and $(\mathbf{\Phi}, \mathbf{\Gamma Q}_1)$ is completely stabilizable, where $\mathbf{Q} = \mathbf{Q}_1\mathbf{Q}_1'$. Additionally, any asymptotically stable model is always completely stabilizable and detectable.

Probably the most interesting case of a system that is not asymptotically stable, for which we want to design a steady-state Kalman filter, is one that has a pole on the unit circle.

**EXAMPLE 19-1**

In this example (which is similar to Example 5.4 in Meditch, 1969, pp. 189–190), we consider the scalar system

$$x(k+1) = x(k) + w(k) \tag{19-3}$$

$$z(k+1) = x(k+1) + v(k+1) \tag{19-4}$$

It has a pole on the unit circle. When $q = 20$ and $r = 5$, Equations (17-13), (17-12), and (17-14) reduce to

$$p(k+1)|k) = p(k|k) + 20 \tag{19-5}$$

$$K(k+1) = \frac{p(k|k) + 20}{p(k|k) + 25} \tag{19-6}$$

and

$$p(k+1|k+1) = \frac{5[p(k|k) + 20]}{p(k|k) + 25} = 5K(k+1) \tag{19-7}$$

Starting with $p(0|0) = p_x(0) \triangleq 50$, it is a relatively simple matter to compute $p(k+1|k)$, $K(k+1)$, and $p(k+1|k+1)$ for $k = 0, 1, \ldots$. The results for the first few iterations are given in Table 19-1.

**TABLE 19-1** KALMAN FILTER QUANTITIES

| $k$ | $p(k\|k-1)$ | $K(k)$ | $p(k\|k)$ |
|---|---|---|---|
| 0 | ... | ... | 50 |
| 1 | 70 | 0.933 | 4.67 |
| 2 | 24.67 | 0.831 | 4.16 |
| 3 | 24.16 | 0.829 | 4.14 |
| 4 | 24.14 | 0.828 | 4.14 |

The steady-state value of $p(k|k)$ is obtained by setting $p(k+1|k+1) = p(k|k) = p_f$ in the last of the above three relations to obtain

$$p_f^2 + 20p_f - 100 = 0 \tag{19-8}$$

Because $p_f$ is a variance it must be nonnegative; hence, only the solution $p_f = 4.14$ is valid. Comparing this result with $p(3|3)$, we see that the filter is in the steady state to within the indicated computational accuracy after processing just three measurements. In steady state, the Kalman filter equation is

$$\hat{x}(k+1|k+1) = \hat{x}(k|k) + 0.828[z(k+1) - \hat{x}(k|k)]$$
$$= 0.172\hat{x}(k|k) + 0.828z(k+1) \tag{19-9}$$

Observe that the filter's pole at 0.172 lies inside the unit circle. □

Many ways have been reported for solving the algebraic Riccati equation (19-1) [see Laub (1979), for example], ranging from direct iteration of the matrix Riccati equation (17-24) until $\mathbf{P}(k+1|k)$ does not change appreciably from $\mathbf{P}(k|k-1)$, to solving the nonlinear algebraic Riccati equation by an iterative Newton–Raphson procedure, to solving that equation in one shot by the Schur method. Iterative methods are quite sensitive to error accumulation. The one-shot Schur method possesses a high degree of numerical integrity and appears to be one of the most successful ways for obtaining $\mathbf{P}_p$. For details about this method, see Laub (1979).

In summary, then, to design a steady-state Kalman filter:

1. Given $(\mathbf{\Phi}, \mathbf{\Gamma}, \mathbf{\Psi}, \mathbf{H}, \mathbf{Q}, \mathbf{R})$, compute $\mathbf{P}_p$, the positive definite solution of (19-1).
2. Compute $\bar{\mathbf{K}}$, as

$$\bar{\mathbf{K}} = \mathbf{P}_p\mathbf{H}'(\mathbf{H}\mathbf{P}_p\mathbf{H}' + \mathbf{R})^{-1} \tag{19-10}$$

3. Use (19-10) in

$$\hat{\mathbf{x}}(k+1|k+1) = \mathbf{\Phi}\hat{\mathbf{x}}(k|k) + \mathbf{\Psi}\mathbf{u}(k) + \bar{\mathbf{K}}\tilde{\mathbf{z}}(k+1|k)$$
$$= (\mathbf{I} - \bar{\mathbf{K}}\mathbf{H})\mathbf{\Phi}\hat{\mathbf{x}}(k|k) + \bar{\mathbf{K}}\mathbf{z}(k+1) \tag{19-11}$$
$$+(\mathbf{I} - \bar{\mathbf{K}}\mathbf{H})\mathbf{\Psi}\mathbf{u}(k)$$

Equation (19-11) is a *steady-state filter state equation.*

## SINGLE-CHANNEL STEADY-STATE KALMAN FILTER

The steady-state Kalman filter can be viewed outside the context of estimation theory as a recursive digital filter. As such, it is sometimes useful to be able to compute its impulse response, transfer function, and frequency response. In this section we restrict our attention to the single-channel steady-state Kalman filter. From (19-11) we observe that this filter is excited by two inputs, $\mathbf{z}(k+1)$ and $\mathbf{u}(k)$ [in the single-channel case, $z(k+1)$ and $u(k)$]. In this section we shall only be interested in transfer functions that are associated with the effect of $z(k+1)$ on the filter; hence, we set $u(k) = 0$. Additionally, we shall view the signal component of measurement $z(k)$ as our desired output. The signal component of $z(k)$ is $\mathbf{h}'\mathbf{x}(k)$.

Let $H_f(z)$ denote the $z$-transform of the impulse response of the steady-state filter; i.e.,

$$H_f(z) = \mathcal{Z}\{\hat{z}(k|k) \text{ when } z(k) = \delta(k)\} \tag{19-12}$$

This transfer function is found from the following *steady-state filter system* [which is obtained from (19-11), where $k+1 \to k$ and $z(k) \to \delta(k)$, and the fact that, if

$z(k) = \mathbf{h}'\mathbf{x}(k) + \nu(k)$, then $\hat{z}(k|k) = \mathbf{h}'\hat{\mathbf{x}}(k|k)$]:

$$\hat{\mathbf{x}}(k|k) = [\mathbf{I} - \bar{\mathbf{K}}\mathbf{h}']\mathbf{\Phi}\hat{\mathbf{x}}(k-1|k-1) + \bar{\mathbf{K}}\delta(k) \tag{19-13}$$

$$\hat{\mathbf{z}}(k|k) = \mathbf{h}'\hat{\mathbf{x}}(k|k) \tag{19-14}$$

Taking the $z$-transform of (19-13) and (19-14), it follows that ($z$ is the unit advance operator and $z^{-1}$ is the unit delay operator)

$$H_f(z) = \mathbf{h}'(\mathbf{I} - \mathbf{\Phi}_f z^{-1})^{-1}\bar{\mathbf{K}} \tag{19-15}$$

where

$$\mathbf{\Phi}_f = (\mathbf{I} - \bar{\mathbf{K}}\mathbf{h}')\mathbf{\Phi} \tag{19-16}$$

Equation (19-15) can also be written as [because $(\mathbf{I} - \mathbf{\Phi}_f z^{-1})^{-1}$ can be expressed as an infinite series in $z^{-1}$ with a nonzero coefficient for the zero-degree term]

$$H_f(z) = h_f(0) + h_f(1)z^{-1} + h_f(2)z^{-2} + \cdots \tag{19-17}$$

where the filter's coefficients (i.e., Markov parameters), $h_f(j)$, are

$$h_f(j) = \mathbf{h}'\mathbf{\Phi}_f^j\bar{\mathbf{K}} \tag{19-18}$$

for $j = 0, 1, \ldots$ .

In our study of mean-squared smoothing, in Lessons 20 and 21, we will see that the steady-state predictor system plays an important role. The *steady-state predictor*, obtained from (17-23), is given by

$$\hat{\mathbf{x}}(k+1|k) = \mathbf{\Phi}_p\hat{\mathbf{x}}(k|k-1) + \boldsymbol{\gamma}_p z(k) \tag{19-19}$$

where

$$\mathbf{\Phi}_p = \mathbf{\Phi}(\mathbf{I} - \bar{\mathbf{K}}\mathbf{h}') \tag{19-20}$$

and

$$\boldsymbol{\gamma}_p = \mathbf{\Phi}\bar{\mathbf{K}} \tag{19-21}$$

Let $H_p(z)$ denote the $z$-transform of the impulse response of the steady-state predictor; i.e.,

$$H_p(z) = \mathcal{Z}\{\hat{z}(k|k-1) \text{ when } z(k) = \delta(k)\} \tag{19-22}$$

This transfer function is found from the following *steady-state predictor system*:

$$\hat{\mathbf{x}}(k+1|k) = \mathbf{\Phi}_p\hat{\mathbf{x}}(k|k-1) + \boldsymbol{\gamma}_p\delta(k) \tag{19-23}$$

$$\hat{z}(k|k-1) = \mathbf{h}'\hat{\mathbf{x}}(k|k-1) \tag{19-24}$$

Taking the $z$-transform of these equations, we find that

$$H_p(z) = \mathbf{h}'(\mathbf{I} - \mathbf{\Phi}_p z^{-1})^{-1}z^{-1}\boldsymbol{\gamma}_p \tag{19-25}$$

which can also be written as

$$H_p(z) = h_p(1)z^{-1} + h_p(2)z^{-2} + \cdots \tag{19-26}$$

where the predictor's coefficients, $h_p(j)$, are

$$h_p(j) = \mathbf{h}'\mathbf{\Phi}_p^{j-1}\gamma_p \tag{19-27}$$

for $j = 1, 2, \ldots$ .

Although $H_f(z)$ and $H_p(z)$ contain an infinite number of terms, they can usually be truncated, because both are associated with asymptotically stable filters.

**EXAMPLE 19-2**

In this example we examine the impulse response of the steady-state predictor for the first-order system

$$x(k+1) = \frac{1}{\sqrt{2}}x(k) + w(k) \tag{19-28}$$

$$z(k+1) = \frac{1}{\sqrt{2}}x(k+1) + \nu(k+1) \tag{19-29}$$

In Example 15-2, we showed that ratio $q/r$ is proportional to signal-to-noise ratio $\overline{\text{SNR}}$. For this example, $\overline{\text{SNR}} = q/r$. Our numerical results, given below, are for $\overline{\text{SNR}} = 20, 5, 1$ and use the scaled steady-state prediction-error variance $p_p/r$, which can be solved for from (19-1) when it is written as

$$\frac{p_p}{r} = \frac{1}{2}\frac{p_p}{r}\left[1 - \frac{1}{2}\left(\frac{1}{2}\frac{p_p}{r} + 1\right)^{-1}\frac{p_p}{r}\right] + \overline{\text{SNR}} \tag{19-30}$$

Note that we could just as well have solved (19-1) for $p_p/q$, but the structure of its equation is more complicated than (19-30). The positive solution of (19-30) is

$$\frac{p_p}{r} = \frac{1}{2}(\overline{\text{SNR}} - 1) + \frac{1}{2}\sqrt{(\overline{\text{SNR}} - 1)^2 + 8\,\overline{\text{SNR}}} \tag{19-31}$$

Additionally,

$$\bar{K} = \frac{1}{\sqrt{2}}\frac{p_p}{r}\left(\frac{1}{2}\frac{p_p}{r} + 1\right)^{-1} \tag{19-32}$$

Table 19-2 summarizes $p_p/r$, $\bar{K}$, $\phi_p$, and $\gamma_p$, quantities that are needed to compute the impulse response $h_p(k)$ for $k \geq 0$, which is depicted in Figure 19-1. Observe that all three responses peak at $j = 1$; however, the decay time for $\overline{\text{SNR}} = 20$ is quicker than the decay times for lower $\overline{\text{SNR}}$ values, and the peak amplitude is larger for higher $\overline{\text{SNR}}$ values. The steady-state predictor tends to reject measurements that have a low signal-to-noise ratio (the same is true for the steady-state filter). □

**TABLE 19-2** STEADY-STATE PREDICTOR QUANTITIES

| $\overline{\text{SNR}}$ | $p_p/r$ | $\bar{K}$ | $\phi_p$ | $\gamma_p$ |
|---|---|---|---|---|
| 20 | 20.913 | 1.287 | 0.064 | 0.910 |
| 5 | 5.742 | 1.049 | 0.183 | 0.742 |
| 1 | 1.414 | 0.586 | 0.414 | 0.414 |

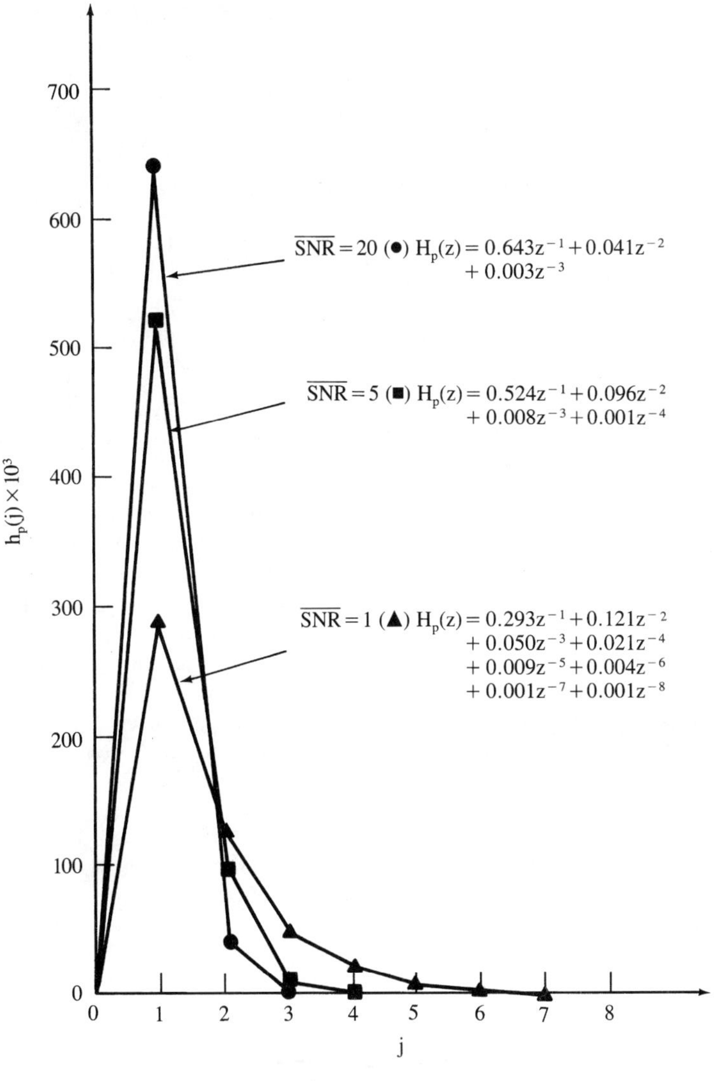


**Figure 19-1** Impulse response of steady-state predictor for $\overline{SNR} = 20$, 5, and 1. $H_p(z)$ is shown to three significant figures (Mendel, 1981, © 1981, IEEE).

**EXAMPLE 19-3**

In this example we present impulse response and frequency response plots for the steady-state predictors associated with the systems

$$H_1(z) = \frac{1}{z^2 - z + 0.5} \tag{19-33}$$

and

$$H_2(z) = \frac{-0.688z^3 + 1.651z^2 - 1.221z + 0.25}{z^4 - 2.586z^3 + 2.489z^2 - 1.033z + 0.168} \tag{19-34}$$

Matrices $\mathbf{\Phi}$, $\boldsymbol{\gamma}$, and $\mathbf{h}$ for these systems are

$$\mathbf{\Phi}_1 = \begin{pmatrix} 0 & 1 \\ -1/2 & 1 \end{pmatrix}, \quad \gamma_1 = \begin{pmatrix} 0 \\ 1 \end{pmatrix}, \quad \mathbf{h}_1 = \begin{pmatrix} 1 \\ 0 \end{pmatrix}$$

and

$$\mathbf{\Phi}_2 = \begin{pmatrix} 0 & 1 & 0 & 0 \\ 0 & 0 & 1 & 0 \\ 0 & 0 & 0 & 1 \\ -0.168 & +1.033 & -2.489 & +2.586 \end{pmatrix}, \quad \gamma_2 = \begin{pmatrix} 0 \\ 0 \\ 0 \\ 1 \end{pmatrix}, \quad \mathbf{h}_2 = \begin{pmatrix} 0.250 \\ -1.221 \\ 1.651 \\ -0.688 \end{pmatrix}$$

Figure 19-2 depicts $h_1(k)$, $|H_1(j\omega)|$ (in decibels) and $\angle H_1(j\omega)$, as well as $h_{p1}(k)$, $|H_{p1}(j\omega)|$, and $\angle H_{p1}(j\omega)$ for $\overline{\text{SNR}} = 1, 5$, and 20. Figure 19-3 depicts comparable quantities for the second system. Observe that, as signal-to-noise ratio decreases, the steady-state predictor rejects the measurements; for the amplitudes of $h_{p1}(k)$ and $h_{p2}(k)$ become smaller as $\overline{\text{SNR}}$ becomes smaller. It also appears, from examination of $|H_{p1}(j\omega)|$ and $|H_{p2}(j\omega)|$, that at high signal-to-noise ratios the steady-state predictor behaves like a high-pass filter for system 1 and a band-pass filter for system 2. On the other hand, at low signal-to-noise ratios it appears to behave like a band-pass filter.

The steady-state predictor appears to be quite dependent on system dynamics at high signal-to-noise ratios, but is much less dependent on the system dynamics at low signal-to-noise ratios. At low signal-to-noise ratios, the predictor is rejecting the measurements regardless of system dynamics.

Just because the steady-state predictor behaves like a high-pass filter at high signal-to-noise ratios does not mean that it passes a lot of noise through it, because *high signal-to-noise ratio* means that measurement noise level is quite low. Of course, a spurious burst of noise would pass through this filter quite easily. □

## RELATIONSHIPS BETWEEN THE STEADY-STATE KALMAN FILTER AND A FINITE IMPULSE RESPONSE DIGITAL WIENER FILTER

The steady-state Kalman filter is a recursive digital filter with filter coefficients equal to $h_f(j)$, $j = 0, 1, \ldots$ [see (19-18)]. Quite often $h_f(j) \approx 0$ for $j > J$, so $H_f(z)$ can be truncated; i.e.,

$$H_f(z) \simeq h_f(0) + h_f(1)z^{-1} + \cdots + h_f(J)z^{-J} \tag{19-35}$$

The *truncated steady-state Kalman filter* can then be implemented as a finite-impulse response (FIR) digital filter.

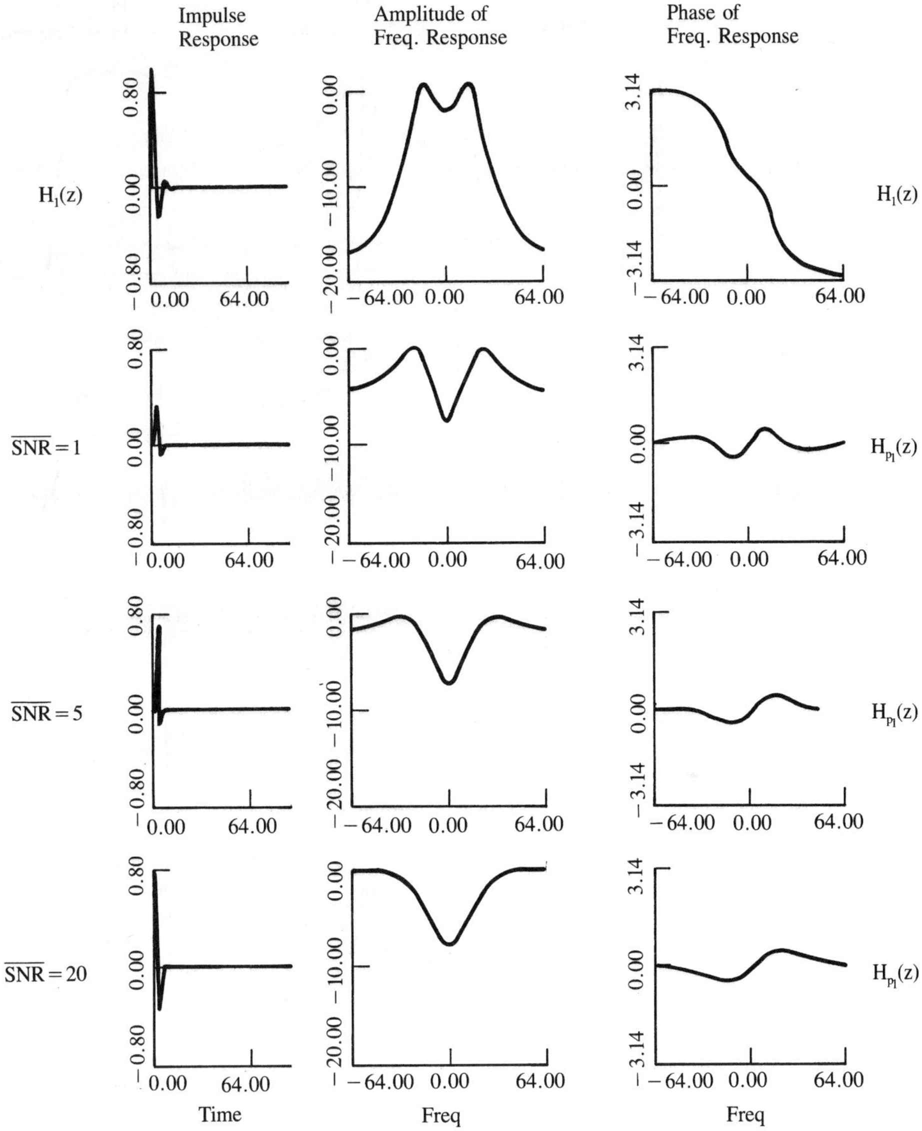


**Figure 19-2** Impulse and frequency responses for system $H_1(z)$ and its associated steady-state predictor.

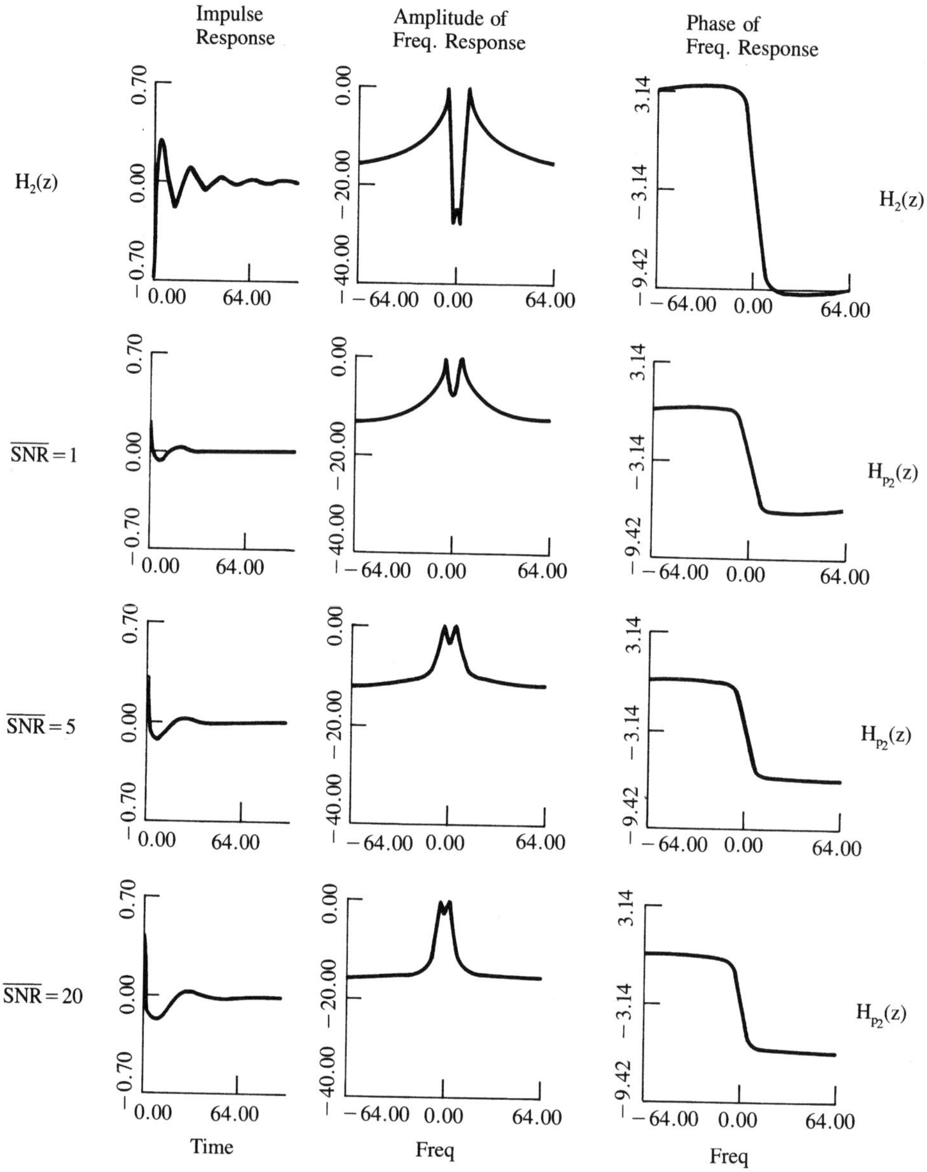


**Figure 19-3** Impulse and frequency responses for system $H_2(z)$ and its associated steady-state predictor.

*some desired signal*

FIR Filter

z(k) → | F(z) | → y(k) → (−)(+) Σ → e(k)

d(k)

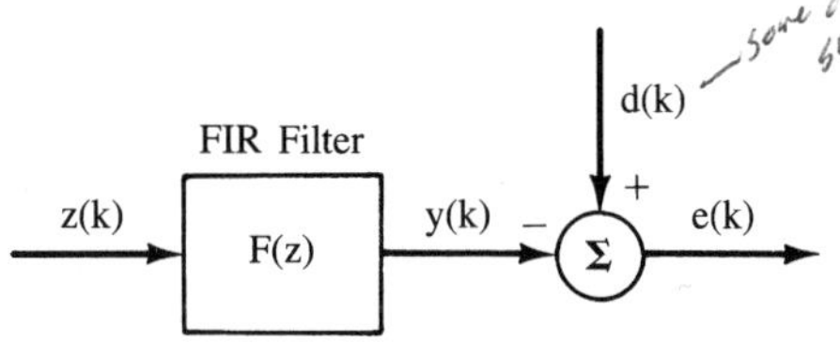

**Figure 19-4** Starting point for design of FIR digital Wiener filter.

There is a more direct way for designing a FIR minimum mean-squared error filter, i.e., a *digital Wiener filter*, as we describe next.

Consider the situation depicted in Figure 19-4. We wish to design digital filter $F(z)$'s coefficients $f(0), f(1), \ldots, f(\eta)$ so that the filter's output, $y(k)$, is close, in some sense, to a desired signal $d(k)$. In a digital Wiener filter design, $f(0), f(1), \ldots, f(\eta)$ are obtained by minimizing the following mean-squared error:

$$I(\mathbf{f}) = \mathbf{E}\{[d(k) - y(k)]^2\} = \mathbf{E}\{e^2(k)\} \qquad (19\text{-}36)$$

*choose f this so $e(k)$ is minimized*

Using the fact that

$$y(k) = f(k) * z(k) = \sum_{i=0}^{\eta} f(i)z(k - i) \qquad (19\text{-}37)$$

*filter coeff* ↑ ↑ *data*

we see that

$$I(\mathbf{f}) = \mathbf{E}\{[d(k) - \sum_{i=0}^{\eta} f(i)z(k - i)]^2\} \qquad (19\text{-}38)$$

which can also be written as

$$I(\mathbf{f}) = \phi_d(0) - 2\sum_{i=1}^{\eta} f(i)\phi_{zd}(i) + \sum_{i=0}^{\eta}\sum_{m=0}^{\eta} f(i)f(m)\phi_{zz}(i - m) \qquad (19\text{-}39)$$

where

$$\phi_d(0) = \mathbf{E}\{d^2(k)\} \qquad (19\text{-}40)$$

*desired*

$$\phi_{zd}(i) = \mathbf{E}\{d(k)z(k - i)\} \quad \text{cross correlation of data + signal} \qquad (19\text{-}41)$$

and

$$\phi_{zz}(i - m) = \mathbf{E}\{z(k - i)z(k - m)\} \quad \text{auto correlation data} \qquad (19\text{-}42)$$

The filter coefficients that minimize $I(\mathbf{f})$ are found by setting $\partial I(\mathbf{f})/\partial f(j) = 0$ for $j = 0, 1, \ldots, \eta$. Clearly,

$$\frac{\partial I(\mathbf{f})}{\partial f(j)} = -2\phi_{zd}(j) + 2\sum_{i=0}^{\eta} f(i)\phi_{zz}(i - j) = 0 \qquad (19\text{-}43)$$

which leads to the following system of $\eta + 1$ equations in the $\eta + 1$ unknown filter coefficients:

*auto correlation of data*

$$\sum_{i=0}^{\eta} f(i)\phi_{zz}(i - j) = \phi_{zd}(j), \quad j = 0, 1, \ldots, \eta \qquad (19\text{-}44)$$

*cross correlation between data + desired signal*

Relationships Between the Kalman Filter and Wiener Filter **289**

*$\phi_{zz}$ + $\phi_{zd}$ (can be computed) so solve for f(i)*

Equations (19-44) are known as the *discrete-time Wiener–Hopf equations*. Observe, from (19-43), that $\partial^2 I(\mathbf{f})/\partial f^2(j) = 2\phi_{zz}(0) > 0$; hence, the filter coefficients found as a solution to the discrete-time Wiener–Hopf equations do indeed minimize $I(\mathbf{f})$. The discrete-time Wiener–Hopf equations are a system of normal equations and can be solved in many different ways, the fastest of which is by the Levinson algorithm (Treitel and Robinson, 1966). See the Supplementary Material at the end of this lesson for a derivation of the Levinson algorithm and a discussion on its applicability to solving (19-44).

The minimum mean-squared error, $I^*(\mathbf{f})$, can be shown to be given by

$$I^*(\mathbf{f}) = \phi_{dd}(0) - \sum_{i=0}^{\eta} f(i)\phi_{zd}(i) \tag{19-45}$$

One property of the digital Wiener filter is that $I^*(\mathbf{f})$ becomes smaller as $\eta$, the number of filter coefficients, increases. In general, $I^*(\mathbf{f})$ approaches a nonzero limiting value, a value that is often reached for modest values of $\eta$.

To relate this FIR Wiener filter to the truncated steady-state Kalman filter, we must first assume a signal-plus-noise model for $z(k)$ (because a Kalman filter uses a system model); i.e., (see Figure 19-5)

$$z(k) = s(k) + \nu(k) = h(k) * w(k) + \nu(k) \tag{19-46}$$

where $h(k)$ is the impulse response of a linear time-invariant system and, as in our basic state-variable model (Lesson 15), $w(k)$ and $\nu(k)$ are mutually uncorrelated (stationary) white noise processes with variances $q$ and $r$, respectively. We must also specify an explicit form for "desired signal" $d(k)$. We shall require that

$$d(k) = s(k) = h(k) * w(k) \tag{19-47}$$

which means, of course, that we want the output of the FIR digital Wiener filter to be as close as possible to signal $s(k)$.

Using (19-46) and (19-47), it is a straightforward exercise to compute $\phi_{zz}(i-j)$ and $\phi_{zd}(j)$ (Problem 19-3). The resulting discrete-time Wiener–Hopf equations are

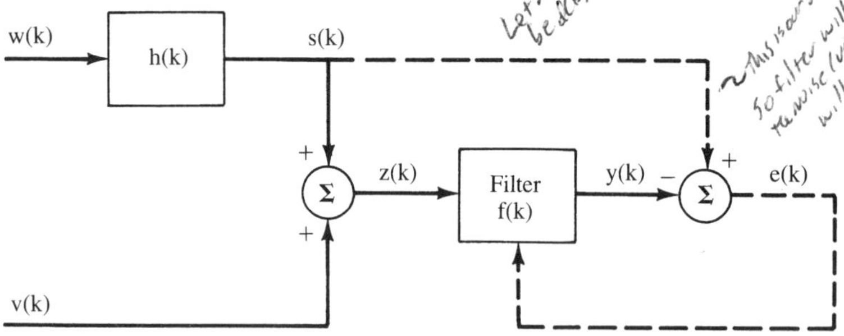


**Figure 19-5** Signal-plus-noise model incorporated into the design of a FIR digital Wiener filter. The dashed lines denote paths that exist only during the design stage; these paths do not exist in the implementation of the filter.

$$\sum_{i=0}^{\eta} f(i)\left[\frac{q}{r}\phi_{hh}(j-i)+\delta(j-i)\right]=\frac{q}{r}\phi_{hh}(j), \qquad j=0,1,\ldots,\eta \tag{19-48}$$

where

$$\phi_{hh}(i)=\sum_{l=0}^{\infty} h(l)h(l+i) \tag{19-49}$$

Observe, that, as in the case of a single-channel Kalman filter, our single-channel Wiener filter only depends on the ratio $q/r$ (and subsequently on $\overline{\text{SNR}}$).

**Theorem 19-2.** *The steady-state Kalman filter is an infinite-length digital Wiener filter. The truncated steady-state Kalman filter is a FIR digital Wiener filter.*

*Proof* (heuristic). The digital Wiener filter has constant coefficients, as does the steady-state Kalman filter. Both filters minimize error variances. The infinite-length Wiener filter has the smallest error variance of all Wiener filters, as does the steady-state Kalman filter; hence, the steady-state Kalman filter is an infinite-length digital Wiener filter.

The second part of the theorem is proved in a similar manner. □

Using Theorem 19-2, we suggest the following *procedure for designing a recursive Wiener filter* (i.e., a steady-state Kalman filter):

1. Obtain a state-variable representation of $h(k)$.
2. Determine the steady-state Kalman gain matrix, $\bar{\mathbf{K}}$, as described above.
3. Implement the steady-state Kalman filter (19-11).
4. Compute the estimate of desired signal $s(k)$, as

$$\hat{s}(k|k)=\mathbf{h}'\hat{\mathbf{x}}(k|k) \tag{19-50}$$

## Wiener: A Short Biography

Norbert Wiener was born in Columbia, Missouri, November 26, 1894. He entered Tufts College at age 11. Believe it or not, he was *not* a child prodigy in mathematics. At 15, he enrolled in the Harvard Graduate School to work in zoology. Levinson (1966) notes that "Unfortunately for Norbert he was a failure at laboratory work which meant that he could not continue in biology. He lacked manipulative skill, good eyesight, and the patience required for meticulous work." At Harvard he studied the border between philosophy and mathematics. His Ph. D. thesis, done at Harvard, was on mathematical logic. At 19 he studied with Bertrand Russell at Cambridge University in England, where he took his first real mathematics courses. He also studied at Göttingen with Hilbert.

Wiener's position in the Mathematics Department at Massachusetts Institute of Technology began in 1919. Early areas where he made profound contributions are generalized harmonic analysis, Tauberian theorems, and absolutely convergent series. He was promoted to professorship in 1932. During that year he spent time at Cambridge, England, where he collaborated with E. Hopf on the solution of the integral equation

$$f(x) = \int_0^{\infty} K(x - y) f(y) dy$$

for $f(x)$. This equation occurs commonly in many applications and has come to be known as the *Wiener–Hopf equation*, although Wiener referred to it as the Hopf–Wiener equation.

During 1932–1933, back at M.I.T., Wiener collaborated with the young British mathematician R. E. A. C. Paley, who was visiting him. They collaborated on a number of theorems involving the Fourier transform in the complex plane. The most famous of these is the (Root, 1966) "Paley–Wiener Theorem which gives a necessary and sufficient condition that a given gain function be possible for a linear, time-invariant system that is realizable, that is, that operates only on the past."

According to Doob (1966), "Wiener's most important contributions to probability theory were centered about the Brownian motion process, now sometimes called the 'Wiener process.' He constructed this process rigorously more than a decade before probabilists had made their subject respectable and he applied the process both inside and outside mathematics in many important problems."

According to Levinson, when World War II

> finally came to Europe the United States began to mobilize its scientists. Wiener obtained a small grant to work on fire-control. The problem was to design apparatus that would direct anti-aircraft guns effectively. With the increase in speed of planes this was a difficult problem. It was necessary to determine the position and direction of flight of the aeroplane and then extrapolate over the flight time of the projectile to determine where the plane would be so that the projectile could be aimed so as to reach it. This problem stimulated Wiener to produce his prediction theory and also eventually brought him to cybernetics.

Wiener's prediction theory has had an important influence in engineering. Levinson states "His [Wiener's] introduction of the statistical element in the question of existence and significance of the auto- and cross-correlation functions proved to be very important, and his generalized harmonic analysis finally became a tool of applied mathematics."

Most of Wiener's important work was inspired by physics or engineering, and in this sense he was very much an applied mathematician. On the other hand, he formulated his theories in the framework of rigorous mathematics, and as a consequence his impact on engineering was very much delayed. His book *Extrapolation, Interpolation, and Smoothing of Stationary Time Series*, which was published in 1949, was originally written as a wartime report to the National Defense Research Committee. As a report it was known as the "yellow peril" because of its yellow covers and its difficulty, especially to engineers.

Wiener was also interested in nonlinear problems in engineering; he wanted to provide a rather general structure of theory to support problems in the synthesis and analysis of nonlinear electric networks and other nonlinear systems.

Levinson notes that

> Wiener had wanted very much to be a biologist but had failed in his first year of graduate school because he lacked the ability to do satisfactory laboratory work. Some thirty-five years later, he was able to make his mark in biology by the way of his theory of control and communications as applied to man and machine in the theory he called cybernetics. . . . His role in cybernetics was not only that of innovator but also of publicist, synthesizer, unifier, popularizer, prophet and philosopher. He has been called the philosophizer of the age of automation.

## COMPARISONS OF KALMAN AND WIENER FILTERS

We conclude this lesson with a brief comparison of Kalman and Wiener filters. First, the filter designs use different types of modeling information. Autocorrelation information is used for FIR digital Wiener filtering, whereas a combination of autocorrelation and difference equation information is used in Kalman filtering. To compute $\phi_{hh}(l)$, $h(k)$ must be known; it is always possible to go directly from $h(k)$ to the parameters in a state-variable model [i.e., $(\mathbf{\Phi}, \boldsymbol{\gamma}, \mathbf{h}]$ using an approximate realization procedure (e.g., Kung, 1978).

Because the Kalman filter is recursive, it is an infinite-length filter; hence, unlike the FIR Wiener filter, where filter length is a design variable, the filter's length is not a design variable in the Kalman filter.

Our derivation of the digital Wiener filter was for a single-channel model, whereas the derivation of the Kalman filter was for a more general multichannel model (i.e., the case of vector inputs and outputs). For a derivation of a multichannel digital Wiener filter, see Treitel (1970).

There is a conceptual difference between Kalman and FIR Wiener filtering. A Wiener filter acts directly on the measurements to provide a signal $\hat{s}(k|k)$ that is "close" to $s(k)$. The Kalman filter does not do this directly. It provides us with such a signal in two steps. In the first step, a signal is obtained that is "close" to $\mathbf{x}(k)$, and in the second step, a signal is obtained that is "close" to $s(k)$. The first step provides the optimal filtered value of $\mathbf{x}(k)$, $\hat{\mathbf{x}}(k|k)$; the second step provides the optimal filtered value of $s(k)$, because $\hat{s}(k|k) = \mathbf{h}'\hat{\mathbf{x}}(k|k)$.

Finally, we can picture a diagram similar to the one in Figure 19-5 for the Kalman filter, except that the dashed lines are solid ones. In the Kalman filter the filter coefficients are usually time varying and are affected in real time by a measure of error $e(k)$ (i.e., by error-covariance matrices), whereas Wiener filter coefficients are constant and have only been affected by a measure of $e(k)$ during their design.

In short, a Kalman filter is a generalization of the FIR digital Wiener filter to include time-varying filter coefficients that incorporate the effects of error in an active manner. Additionally, a Kalman filter is applicable to either time-varying or nonstationary systems or both, whereas the digital Wiener filter is not.

## COMPUTATION

Using the *Control System Toolbox*, you must use two M-files. One (**dlqe**) provides gain, covariance, and eigenvalue information about the steady-state KF; the other (**destim**) computes both the steady-state **predicted** and **filtered** values for $\mathbf{x}(k)$.

**dlqe**: Discrete linear quadratic estimator design. Computes the steady-state Kalman gain matrix, predictor, and filter steady-state covariance matrices and closed-loop eigenvalues of the predictor. Does this for the basic state-variable model.

**destim**: Forms steady-state Kalman filter. The steady-state Kalman gain matrix, computed by **dlqe**, must be passed on to **destim**. Outputs steady-state predicted and filtered values for $\mathbf{x}(k)$, as well as the predicted value of the measurement vector, all as functions of time. For correlated noise case, see Lesson 22.

M-file **destim** can also be used to provide suboptimal filtered state estimates simply by providing it with a gain matrix other than the steady-state Kalman gain matrix.

### Supplementary Material

## SOME LINEAR SYSTEM CONCEPTS

Friedland (1986, pp. 204, 207, and 220) states that

> A system is said to be controllable if and only if it is possible, by means of the input, to transfer the system from *any* initial state $\mathbf{x}(t) = \mathbf{x}_t$ to any other state $\mathbf{x}_T = \mathbf{x}(T)$ in a finite time $T - t \geq 0 \ldots$. If a system is not controllable, it can be divided into two subsystems, one of which (if it exists) is controllable and the other is uncontrollable. If the uncontrollable subsystem is stable, the entire system is said to be *stabilizable*. The set of stabilizable systems thus includes the controllable systems as a subset: every controllable system is stabilizable, but not every stabilizable system is controllable.... An unforced system is said to be observable if and only if it is possible to determine an (arbitrary initial) state $\mathbf{x}(t) = \mathbf{x}_t$ by using only a finite record, $y(\tau)$ for $t \leq \tau \leq T$, of the output. A system that is not observable ... can be divided into two subsystems, one of which (if it exists) is observable and the other is not. If the unobservable subsystem is stable, the entire system is said to be *detectable*. Thus the observable systems are a subset of the detectable systems.

Kalman (1963) showed that a dynamical system, such as our basic state-variable model, can be transformed into an interconnection of four subsystems. The first subsystem is both controllable and observable; the second is uncontrollable but observable; the third is controllable but unobservable; and the fourth is uncontrollable and unobservable.

Kalman showed that a perfect pole–zero cancellation in an unstable system can produce an unstable system with a stable transfer function (i.e., the unstable poles are canceled by zeros situated at the locations of the unstable poles). Obviously, the resulting transfer function is of lower order than the original system, and the unstable modes are either invisible at the system's output (i.e., unobservable) or cannot be affected by the system's input (uncontrollable). The transfer function of the system is determined only by the controllable and observable portion of the system. Many real-world systems can be either uncontrollable or unobservable. Friedland (1986) shows that if a system has redundant state variables, or is one in which the only forces and torques that are present are internal to the system, or has too much symmetry (e.g., an electrical bridge network), it can be uncontrollable. Symmetry can also lead to a system being unobservable. If a system is the cascade of two subsystems and only the states of the first subsystem can be measured, then the system is unobservable because the states of the second subsystem cannot be measured.

Any asymptotically stable time-invariant system is stabilizable and detectable (see, for example, Kwakernaak and Sivan, 1972). Any controllable system is stabilizable, and any controllable system is detectable. Mathematical tests have been developed to check whether or not a system is controllable and observable (Friedland, 1986; Kwakernaak and Sivan, 1972). For many examples that illustrate controllable and uncontrollable systems as well as some very interesting historical notes about controllability and observability, see Friedland (1986, Chapter 5).

## THE LEVINSON ALGORITHM

### Overview

The normal equations are associated with many different problems, such as Wiener filtering, spectral analysis, and seismic inverse problems. We shall express them here as

$$\begin{bmatrix} r_0 & r_1 & r_2 & \cdots & r_m \\ r_1 & r_0 & r_1 & \cdots & r_{m-1} \\ r_2 & r_1 & r_0 & \cdots & r_{m-2} \\ & & \cdot \quad \cdot \quad \cdot & & \\ r_m & r_{m-1} & r_{m-2} & \cdots & r_0 \end{bmatrix} \begin{bmatrix} f_0 \\ f_1 \\ f_2 \\ \cdots \\ f_m \end{bmatrix} = \begin{bmatrix} g_0 \\ g_1 \\ g_2 \\ \cdots \\ g_m \end{bmatrix} \tag{19-51}$$

or as

$$\mathbf{Rf} = \mathbf{g} \tag{19-52}$$

The Levinson algorithm (LA) finds the solution for **f** in a recursive manner; it makes use of the high degree of symmetry that this system of equations possesses, the Toeplitz structure of **R**. Its advantage over other more traditional linear equation-solving methods (e.g., Gaussian elimination) is that its computational effort grows as $m^2$, whereas the computational efforts of traditional methods grow as $m^3$. This speedup in computation time is due to the explicit use of the Toeplitz nature of **R**. Additionally, the LA requires computer storage proportional to $m$ versus $m^2$ for traditional methods.

The LA consists of two recursions: first, the recursion for finding solutions of the simpler system of equations

$$\begin{bmatrix} r_0 & r_1 & r_2 & \cdots & r_i \\ r_1 & r_0 & r_1 & \cdots & r_{i-1} \\ r_2 & r_1 & r_0 & \cdots & r_{i-2} \\ & & \cdot\ \cdot\ \cdot & & \\ r_i & r_{i-1} & r_{i-2} & \cdots & r_0 \end{bmatrix} \begin{bmatrix} a_{i0} \\ a_{i1} \\ a_{i2} \\ \cdots \\ a_{ii} \end{bmatrix} = \begin{bmatrix} \alpha_i \\ 0 \\ 0 \\ \cdots \\ 0 \end{bmatrix} \tag{19-53}$$

for $i = 0, 1, \ldots, m$, and then with the knowledge of $(a_{i0}, a_{i1}, \ldots, a_{ii})$, a second recursion solves the equation

$$\begin{bmatrix} r_0 & r_1 & r_2 & \cdots & r_i \\ r_1 & r_0 & r_1 & \cdots & r_{i-1} \\ r_2 & r_1 & r_0 & \cdots & r_{i-2} \\ & & \cdot\ \cdot\ \cdot & & \\ r_i & r_{i-1} & r_{i-2} & \cdots & r_0 \end{bmatrix} \begin{bmatrix} f_{i0} \\ f_{i1} \\ f_{i2} \\ \cdots \\ f_{ii} \end{bmatrix} = \begin{bmatrix} g_0 \\ g_1 \\ g_2 \\ \cdots \\ g_i \end{bmatrix} \tag{19-54}$$

for $i = 0, 1, \ldots, m$. In (19-53) and (19-54) the matrix on the left-hand side is a principal minor of matrix $\mathbf{R}$ in (19-52) and is also Toeplitz. At the last step of this recursive procedure, i.e., when $i = m$, the final values obtained for $\mathbf{f}_i$, the $\text{col}\,(f_{m0}, f_{m1}, \ldots, f_{mm})$, represent the desired solution $\mathbf{f} = \text{col}\,(f_0, f_1, \ldots, f_m)$ of (19-52).

### Relation between Equations (19-53) and (19-54)

Here we show that the solution of (19-54) reduces to the solution of (19-53). Suppose we have just solved (19-54) for $i = n$ and are going to solve it for $i = n + 1$, but we want to make use of the result obtained for $i = n$; i.e., we have obtained $\text{col}\,(f_{n0}, f_{n1}, \ldots, f_{nn})$ as the solution of

$$\begin{bmatrix} r_0 & r_1 & r_2 & \cdots & r_n \\ r_1 & r_0 & r_1 & \cdots & r_{n-1} \\ r_2 & r_1 & r_0 & \cdots & r_{n-2} \\ & & \cdot\ \cdot\ \cdot & & \\ r_i & r_{n-1} & r_{n-2} & \cdots & r_0 \end{bmatrix} \begin{bmatrix} f_{n0} \\ f_{n1} \\ f_{n2} \\ \cdots \\ f_{nn} \end{bmatrix} = \begin{bmatrix} g_0 \\ g_1 \\ g_2 \\ \cdots \\ g_n \end{bmatrix} \tag{19-55}$$

and from this solution we are to solve the equation

$$\begin{bmatrix} r_0 & r_1 & \cdots & r_n & r_{n+1} \\ r_1 & r_0 & \cdots & r_{n-1} & r_n \\ & & \cdot\ \cdot\ \cdot & & \\ r_n & r_{n-1} & \cdots & r_0 & r_1 \\ r_{n+1} & r_n & \cdots & r_1 & r_0 \end{bmatrix} \begin{bmatrix} f_{n+1,0} \\ f_{n+1,1} \\ \cdots \\ f_{n+1,n} \\ f_{n+1,n+1} \end{bmatrix} = \begin{bmatrix} g_0 \\ g_1 \\ \cdots \\ g_n \\ g_{n+1} \end{bmatrix} \tag{19-56}$$

To begin, we relate (19-55) to (19-56). To do this, we add another row and column to the Toeplitz matrix in (19-55) so that the augmented Toeplitz matrix looks just like the one in (19-56). Of course, this must be done in such a way that we don't alter (19-55). Doing this, we obtain

$$\left[\begin{array}{cccc|c} r_0 & r_1 & \cdots & r_n & r_{n+1} \\ r_1 & r_0 & \cdots & r_{n-1} & r_n \\ & & \cdots & & \\ r_n & r_{n-1} & \cdots & r_0 & r_1 \\ \hline r_{n+1} & r_n & \cdots & r_1 & r_0 \end{array}\right]\left[\begin{array}{c} f_{n0} \\ f_{n1} \\ \cdots \\ f_{nn} \\ \hline 0 \end{array}\right]=\left[\begin{array}{c} g_0 \\ g_1 \\ \cdots \\ g_n \\ \hline \gamma_n \end{array}\right] \tag{19-57}$$

From (19-57), we see that

$$\gamma_n = f_{n0}r_{n+1} + f_{n1}r_n + \cdots + f_{nn}r_1 = \sum_{i=0}^{n} f_{ni}r_{n+1-i} \tag{19-58}$$

Observe that the first $n+1$ equations in (19-57) are precisely those given by (19-55). By expanding matrix $\mathbf{R}_n$ [we call the $n+1 \times n+1$ Toeplitz matrix in (19-55) $\mathbf{R}_n$] to $\mathbf{R}_{n+1}$ [we call the $n+2 \times n+2$ Toeplitz matrix in (19-56) $\mathbf{R}_{n+1}$] as indicated in (19-57), we have introduced a design degree of freedom $\gamma_n$, which can be computed from (19-58) once col $(f_{n0}, f_{n1}, \ldots, f_{nn})$ is known.

Next, subtract (19-57) from (19-56), to obtain

$$\begin{aligned} \mathbf{R}_{n+1}\text{col}\,(f_{n+1,0} - f_{n0}, f_{n+1,1} - f_{n1}, \ldots, f_{n+1,n} - f_{nn}, f_{n+1,n+1}) \\ = \text{col}\,(0, 0, \ldots, 0, g_{n+1} - \gamma_n) \end{aligned} \tag{19-59}$$

Let

$$\begin{aligned} \text{col}\,(f_{n+1,0} - f_{n0}, f_{n+1,1} - f_{n1}, \ldots, f_{n+1,n} - f_{nn}, f_{n+1,n+1}) \\ \stackrel{\triangle}{=} q_n\text{col}\,(a_{n+1,n+1}, a_{n+1,n}, \ldots, a_{n+1,1}, a_{n+1,0}) \end{aligned} \tag{19-60}$$

where $q_n$ is a *normalizing factor* chosen to make

$$a_{n+1,0} \stackrel{\triangle}{=} 1 \tag{19-61}$$

We can, therefore, write (19-59) as

$$\mathbf{R}_{n+1}\text{col}\,(a_{n+1,n+1}, a_{n+1,n}, \ldots, a_{n+1,1}, a_{n+1,0}) = \text{col}\,(0, 0, \ldots, 0, \alpha_{n+1}) \tag{19-62}$$

where

$$\alpha_{n+1} = \frac{g_{n+1} - \gamma_n}{q_n} \tag{19-63}$$

Now comes the crucial step. Because of the Toeplitz structure of matrix $\mathbf{R}_{n+1}$, (19-62) can be manipulated into the following equivalent form:

$$\mathbf{R}_{n+1}\text{col}\,(a_{n+1,0}, a_{n+1,1}, \ldots, a_{n+1,n}, a_{n+1,n+1}) = \text{col}\,(\alpha_{n+1}, 0, \ldots, 0, 0) \tag{19-64}$$

Equations (19-63), (19-60), (19-61), and (19-64) are the key results so far. We write (19-60) as

$$\begin{aligned} f_{n+1,0} &= f_{n0} + q_n a_{n+1,n+1} \\ f_{n+1,1} &= f_{n1} + q_n a_{n+1,n} \\ &\cdots \\ f_{n+1,n} &= f_{nn} + q_n a_{n+1,1} \\ f_{n+1,n+1} &= q_n a_{n+1,0} \end{aligned} \tag{19-65}$$

What we have done so far is to reduce the problem of finding the solution of (19-56) to the problem of finding the solution of the simpler equation (19-64): $\text{col}\,(a_{n+1,0}, a_{n+1,1}, \ldots, a_{n+1,n}, a_{n+1,n+1})$. Then, using (19-65), together with knowledge of $\text{col}\,(f_{n0}, f_{n1}, \ldots, f_{nn})$, we can compute $\text{col}\,(f_{n+1,0}, f_{n+1,1}, \ldots, f_{n+1,n}, f_{n+1,n+1})$.

### Recursive Method for Solving Equation (19-53)

Suppose we have just completed step $i = n$ of the recursion and have evaluated $\text{col}\,(a_{n0}, a_{n1}, \ldots, a_{nn})$ using $\alpha_n$. These quantities satisfy the matrix equation

$$\mathbf{R}_n \text{col}\,(a_{n0}, a_{n1}, \ldots, a_{nn}) = \text{col}\,(\alpha_n, 0, \ldots, 0) \tag{19-66}$$

where

$$\mathbf{R}_n = \begin{bmatrix} r_0 & r_1 & \cdots & r_n \\ r_1 & r_0 & \cdots & r_{n-1} \\ & \cdot & \cdot & \cdot \\ r_n & r_{n-1} & \cdots & r_0 \end{bmatrix} \tag{19-67}$$

We now augment $\mathbf{R}_n$ in (19-66) to $\mathbf{R}_{n+1}$, as we did in (19-57):

$$\mathbf{R}_{n+1} \text{col}\,(a_{n0}, a_{n1}, \ldots, a_{nn}|0) = \text{col}\,(\alpha_n, 0, \ldots, 0|\beta_n) \tag{19-68}$$

where

$$\beta_n = a_{n0} r_{n+1} + a_{n1} r_n + \ldots + a_{nn} r_1 \tag{19-69}$$

Just as we were able to write (19-62) as in (19-64), by making use of the Toeplitz nature of $\mathbf{R}_{n+1}$, we are able to write (19-68) as

$$\mathbf{R}_{n+1} \text{col}\,(0|a_{nn}, \ldots, a_{n1}, a_{n0}) = \text{col}\,(\beta_n|0, \ldots, 0, \alpha_n) \tag{19-70}$$

Next write (19-66) for $n + 1$; i.e.,

$$\mathbf{R}_{n+1} \text{col}\,(a_{n+1,0}, a_{n+1,1}, \ldots, a_{n+1,n+1}) = \text{col}\,(\alpha_{n+1}, 0, \ldots, 0) \tag{19-71}$$

We wish to solve (19-71) for $\text{col}\,(a_{n+1,0}, a_{n+1,1}, \ldots, a_{n+1,n+1})$ in terms of $\text{col}\,(a_{n0}, a_{n1}, \ldots, a_{nn})$ and $\alpha_n$ and $\beta_n$, which are known once $\text{col}\,(a_{n0}, a_{n1}, \ldots, a_{nn})$ is known.

We take a linear combination of (19-68) and (19-70), by multiplying (19-70) by a constant $k_n$, as yet undetermined, and then adding the result to (19-68) to obtain

$$\mathbf{R}_{n+1} \text{col}\,(a_{n0}, a_{n1}+k_n a_{nn}, \ldots, a_{nn}+k_n a_{n1}, k_n a_{n0}) = \text{col}\,(\alpha_n+k_n\beta_n, 0, \ldots, 0, \beta_n+k_n\alpha_n) \tag{19-72}$$

Now compare (19-72) with (19-71). Clearly, (19-72) reduces to (19-71) if

$$\beta_n + k_n \alpha_n = 0 \Rightarrow k_n = -\frac{\beta_n}{\alpha_n} \tag{19-73}$$

$$\alpha_{n+1} = \alpha_n + k_n \beta_n \tag{19-74}$$

and

$$
\begin{aligned}
a_{n+1,0} &= a_{n0} \stackrel{\triangle}{=} 1 \\
a_{n+1,1} &= a_{n1} + k_n a_{nn} \\
&\cdots \\
a_{n+1,n} &= a_{nn} + k_n a_{n1} \\
a_{n+1,n+1} &= k_n a_{n0} = k_n
\end{aligned}
\tag{19-75}
$$

Given $\alpha_n$, $\beta_n$, and $\text{col}\,(a_{n0}, a_{n1}, \ldots, a_{nn})$, we can solve for $k_n$ from (19-73) and then solve for $\text{col}\,(a_{n+1,0}, a_{n+1,1}, \ldots, a_{n+1,n+1})$ using (19-75). This is the recursive procedure for solving (19-53).

Initial conditions for this recursion are

$$a_{00} = 1 \tag{19-76a}$$

$$\alpha_0 = r_0 \tag{19-76b}$$

$$\beta_0 = r_1 \tag{19-76c}$$

Equation (19-76a) is our normalization requirement in (19-61). For $i = 0$, (19-53) is $r_0 a_{00} = \alpha_0$; hence, $r_0 = \alpha_0$, which is (19-76b). Additionally, from (19-69), $\beta_0 = a_{00} r_1 = r_1$, which is (19-76c).

### Recursive Method for Solving Equation (19-53)

Once we compute $\text{col}\,(a_{i0}, a_{i1}, \ldots, a_{ii})$ for $i = 1, 2, \ldots, m$ using (19-73)–(19-76), we can compute $\text{col}\,(f_{i0}, f_{i1}, \ldots, f_{ii})$ for $i = 1, 2, \ldots, m$ using (19-65) where, from (19-63),

$$q_n = \frac{g_{n+1} - \gamma_n}{\alpha_{n+1}} \tag{19-77}$$

Observe that, when $\text{col}\,(f_{n0}, f_{n1}, \ldots, f_{nn})$ is known, then $\gamma_n$ can be computed using (19-58); $g_{n+1}$ is given by the data [see (19-51)]; and $\alpha_{n+1}$ is computed as part of the calculation of $\text{col}\,(a_{n0}, a_{n1}, \ldots, a_{nn})$ using (19-74). Hence, the right-hand side of (19-77) is a computable number.

The initial conditions for solving (19-53) using (19-65) are

$$f_{00} = \frac{g_0}{r_0} \tag{19-78a}$$

$$\gamma_0 = f_{00} r_1 \tag{19-78b}$$

When $i = 0$, (19-54) becomes $r_0 f_{00} = g_0$, which leads to (19-78a), and (19-58) becomes $\gamma_0 = f_{00} r_1$, which is (19-78b).

### Summary of the Levinson Algorithm

Here we collect all the previous results together to obtain the order in which calculations are performed in the LA: (1) initializations, using (19-76) and (19-78); (2) $k_n$ using (19-73); (3) $\text{col}\,(a_{n+1,0}, a_{n+1,1}, \ldots, a_{n+1,n+1})$ using (19-75); (4) $\alpha_{n+1}$ using (19-74); (5) $\beta_{n+1}$ using (19-69); (6) $q_n$ using (19-77); (7)

col $(f_{n+1,0}, f_{n+1,1}, \ldots, f_{n+1,n+1})$ using (19-65); and (8) $\gamma_n$ using (19-58). Steps (2)–(8) are iterated for $n = 0, 1, \ldots, m-1$. At the final step, when $n = m-1$, the solution col $(f_{m0}, f_{m1}, \ldots, f_{mm}) \triangleq$ col $(f_0, f_1, \ldots, f_m)$ is obtained.

In some filter designs or other applications, we do not begin with (19-51); instead, we begin with the simpler system in (19-53). In those cases, the LA reduces to steps (1)–(5) in the above procedure. At the final step, when $n = m - 1$, the solution col $(a_{m0}, a_{m1}, \ldots, a_{mm}) \triangleq$ col $(a_0, a_1, \ldots, a_m)$ is obtained.

## SUMMARY QUESTIONS

1. The steady-state Kalman filter:
   **(a)** is always asymptotically stable
   **(b)** is asymptotically stable under certain conditions
   **(c)** depends on $\mathbf{P}(0|-1)$
2. Which of the following are true? The steady-state KF is:
   **(a)** an all-pass filter
   **(b)** a finite impulse response filter
   **(c)** an infinite impulse response filter
   **(d)** an FIR digital WF
   **(e)** an infinite-length digital WF
   **(f)** a low-pass filter
3. The discrete-time Wiener–Hopf equations are:
   **(a)** singular
   **(b)** nonlinear
   **(c)** linear
4. A Kalman filter is:
   **(a)** the same as a Wiener filter
   **(b)** more general than a Wiener filter
   **(c)** a special case of a Wiener filter
5. A digital Wiener filter is:
   **(a)** an all-pass filter
   **(b)** a finite impulse response filter
   **(c)** an infinite impulse response filter
6. The steady-state Riccati equations for the predicted and filtered error-covariance matrices are:
   **(a)** always different
   **(b)** sometimes different
   **(c)** the same
7. The Markov parameters for a single-channel steady-state predictor are:
   **(a)** $\mathbf{h}'\mathbf{\Phi}_f^j\bar{\mathbf{K}}, j = 0, 1, \ldots$
   **(b)** $\mathbf{h}'\mathbf{\Phi}_p^{j-1}\gamma_p, j = 1, 2, \ldots$
   **(c)** $\mathbf{h}'\mathbf{\Phi}_p^j\bar{\mathbf{K}}, j = 1, 2, \ldots$
8. The Levinson algorithm is:
   **(a)** a fast way to solve the normal equations $\mathbf{Ax} = \mathbf{b}$, where $\mathbf{A}$ is a Toeplitz matrix
   **(b)** a fast way to solve the normal equations $\mathbf{Ax} = \mathbf{b}$, where $\mathbf{A}$ is a circulant matrix
   **(c)** a fast Wiener filter

## PROBLEMS

**19-1.** Prove that the steady-state filter and predictor have the same eigenvalues.

**19-2.** Prove that the minimum mean-squared error $I^*(\mathbf{f})$, given in (19-45), becomes smaller as $\eta$, the number of filter coefficients, increases.

**19-3.** Derive formulas for $\phi_{zz}(i-j)$ and $\phi_{zd}(j)$, which are needed to obtain (19-48).

**19-4.** Consider the basic state-variable model $x(k+1) = \frac{2}{\sqrt{3}}x(k) + w(k)$ and $z(k+1) = x(k+1) + \nu(k+1)$, where $q = 2, r = 1, m_x(0) = 0$ and $\mathbf{E}\{x^2(0)\} = 2$.

**(a)** Specify $\hat{x}(0|0)$ and $p(0|0)$.

**(b)** Give the recursive predictor for this system.

**(c)** Obtain the steady-state predictor.

**(d)** Suppose $z(5)$ is not provided (i.e., there is a gap in the measurements at $k = 5$). How does this affect $p(6|5)$ and $p(100|99)$?

**19-5.** Consider the basic scalar system $x(k+1) = \phi x(k) + w(k)$ and $z(k+1) = x(k+1) + \nu(k+1)$. Assume that $q = 0$ and let $\lim_{k\to\infty} p(k|k) = p_f$.

**(a)** Show that $p_f = 0$ and $p_f = (\phi^2 - 1)r/\phi^2$.

**(b)** Which of the two solutions in part (a) is the correct one when $\phi^2 < 1$?

**19-6.** (Richard W. Baylor, Spring 1992) (Computational Problem) Given the following basic state-variable model:

$$x(k+1) = \frac{1}{\sqrt{3}}x(k) + w(k)$$

$$z(k+1) = x(k+1) + \nu(k+1)$$

where $q = 10, r = 2, p(0|0) = 4$, and $\mathbf{E}\{x(0)\} = 0$.

**(a)** Determine the steady-state Kalman gain, prediction-error variance, and filtering-error variance by direct iterations for $k = 0, 1, \ldots, 8$.

**(b)** Redo part (a) using the appropriate steady-state Riccati equations.

**(c)** Compare the results in part (a) and (b). How does $p(0|0)$ enter into the respective calculations?

**(d)** Where are the poles of the steady-state filter and predictor located?

**19-7.** (Todd B. Leach, Spring 1992) (Computational Problem) A certain system is defined as follows:

$$x(k+1) = (\phi + d\phi)x(k) + w(k)$$

$$z(k+1) = x(k+1) + \nu(k+1)$$

where $w(k) \sim N[w(k); 0, q], \nu(k) \sim N[\nu(k); 0, r], w(k)$ and $\nu(k)$ are mutually uncorrelated, and $d\phi$ is the uncertainty in the plant parameter $\phi$.

**(a)** Determine expressions for the steady-state KF gain and error variance.

**(b)** Using $q = 5$ and $r = 1$, plot $p_f$ and $\bar{K}$ for values of $d\phi$ ranging from 0% to 40% of a nominal value for $\phi$.

**19-8.** (David Adams, Spring 1992) (Computational Problem) Along lines similar to Example 19-3, find the impulse response and frequency response for the following system, as well as for its steady-state KF, when $q/r = 10$:

$$\mathbf{\Phi} = \begin{bmatrix} 0 & 1 & 0 & 0 \\ 0 & 0 & 1 & 0 \\ 0 & 0 & 0 & 1 \\ -21 & -0.072 & -0.4 & -0.1 \end{bmatrix}$$

$\boldsymbol{\gamma} = \text{col}\,(0, 0, 0, 1)$ and $\mathbf{h} = \text{col}\,(0.3, 1.4, -1.1, -0.5)$. Where are the poles and zeros of this system located?

**19-9.** (Michiel van Nieustadt, Spring 1992) (Computational Problem) Consider the dynamical system

$$x(k+1) = 0.5x(k) + w(k)$$

where $w(k)$ is *nonstationary* white Gaussian noise with time-varying mean $\mathbf{E}\{w(k)\} = 10 \times \sin[2\pi/20 \times (k+1)]$ and constant variance $q = 1$. Additionally, $x(0) = 0$. The time-varying mean acts like a deterministic input (see Lesson 22). The observation equation for our system is

$$s(k) = 0.7x(k)$$

$$z(k) = s(k) + v(k)$$

where $v(k)$ is zero-mean stationary Gaussian noise with $r = 1$. Signal $s(k)$ is the *desired output.*

**(a)** Implement a steady-state KF for this system and plot 100 estimated output values, as well as the desired output $s(k)$.

**(b)** Compare the desired output $s(k)$ of the dynamical system with the KF. Plot the transfer function of the steady-state KF. Verify that the gain for the frequency of our input accounts for the difference between the obtained $s(k)$ and the output of the KF.

**(c)** Implement a FIR WF for this system and again plot its output for 100 samples. Set the filter length to 5. Compare with the KF.

**(d)** Plot the summed-squared error between the steady-state KF and the FIR WF as a function of filter length. Use filter lengths $1, 2, \ldots, 10$.

**19-10.** (Computational Problem) A very simple *target tracking model* (e.g., Blair and Kazakos, 1993) is one in which the motion of the target is defined in a single coordinate, and it is assumed that the target is moving with constant velocity but is subject to zero-mean white Gaussian acceleration errors. Measurements are made of the position of the target. The state-variable model for this target tracking scenario is $2 \times 1$, where $\mathbf{x}(k) = \text{col}\,(\text{position, velocity})$,

$$\mathbf{\Phi} = \begin{bmatrix} 1 & T \\ 0 & 1 \end{bmatrix}$$

$\mathbf{\Gamma} = \text{col}\,(T^2/2, T)$, $\mathbf{H} = (1, 0)$, and $T$ is sampling time.

The resulting steady-state Kalman filter is known as the *$\alpha$–$\beta$ filter.* In this filter, $\bar{\mathbf{K}} = \text{col}\,(\alpha, \beta/T)$.

**(a)** Show that $\alpha$ and $\beta$ are determined by solving the simultaneous equations

$$\frac{\beta^2}{1-\alpha} = \frac{T^4 q}{r}$$

$$\beta = 2(2-\alpha) - 4(1-\alpha)^{1/2}$$

(b) Show that

$$\mathbf{\Phi}_f = \begin{bmatrix} 1-\alpha & (1-\alpha)T \\ -\dfrac{\beta}{T} & 1-\beta \end{bmatrix}$$

(c) Show that

$$\mathbf{P}_f = \frac{r}{\alpha d_1}\begin{bmatrix} 2\alpha^2+\beta(2-3\alpha) & \dfrac{(2\alpha-\beta)\beta}{T} \\ \dfrac{(2\alpha-\beta)\beta}{T} & \dfrac{2\beta^2}{T^2} \end{bmatrix}$$

where $d_1 = 4 - 2\alpha - \beta$.

(d) Determine numerical values of $\alpha$ and $\beta$ for a wide range of $q/r$. Relate $q/r$ to steady-state SNR. Locate the eigenvalues of the $\alpha$–$\beta$ filter for this range of $q/r$. Determine the steady-state error variances in tracking the position and velocity of the target for this range of $q/r$.

**19-11.** In the Supplementary Material to Lesson 16, we introduced a forward linear predictor [see (16-35)]. Relate the problem of determining the forward PEF coefficients, $a_{M,1}, \ldots, a_{M,M}$, to the design of the FIR digital Wiener filter that is described in this lesson.

**19-12.** Consider the following first-order basic state-variable model:

$$x(k+1) = \frac{1}{\sqrt{2}}x(k) + \frac{1}{4}w(k)$$

$$z(k+1) = \frac{1}{3}x(k+1) + v(k+1)$$

where $\mathbf{E}\{x(0)\} = \sqrt{2}$, $\sigma_x^2 = 1$, $q = 4$, $r = \frac{1}{9}$, and $s = 0$.

(a) What is the steady-state value of the variance of $z(k+1)$?
(b) What is the steady-state SNR?
(c) What is the steady-state Kalman gain?
(d) What is the location of the pole of the steady-state recursive Kalman filter?
(e) What is the location of the pole of the steady-state recursive Kalman predictor?
(f) What are the steady-state predictor and filter state-estimation error variances?
(g) What is the steady-state variance of the innovations process?
(h) Demonstrate that filtering is better than prediction.

*3rd & final types of state estimation;*

# State Estimation: Smoothing

## SUMMARY

This is the first of two lessons on smoothing. A smoothed estimate of state vector $\mathbf{x}(k)$ not only uses measurements that occur earlier than $t_k$, plus the one at $t_k$, but also uses measurements to the right of $t_k$. Consequently, smoothing can never be carried out in real time, because we have to collect "future" measurements before we can compute a smoothed estimate. If we don't look too far into the future, then smoothing can be performed subject to a delay of $LT$ sec, where $T$ is our data sampling time and $L$ is a fixed positive integer that describes how many sample points to the right of $t_k$ are to be used in smoothing.

Depending on how many future measurements are used and how they are used, it is possible to create *three* types of smoothers: (1) the fixed-interval smoother, $\hat{\mathbf{x}}(k|N)$, $k = 0, 1, \ldots, N - 1$, where $N$ is a fixed positive integer; (2) the fixed-point smoother, $\hat{\mathbf{x}}(k|j)$, $j = k + 1, k + 2, \ldots$, where $k$ is a fixed positive integer; and (3) the fixed-lag smoother, $\hat{\mathbf{x}}(k|k + L)$, $k = 0, 1, \ldots$, where $L$ is a fixed positive integer. One main purpose of this lesson is to explain the meanings of these three smoothers.

Rather than jump right into derivations of these three smoothers, this lesson first derives single- and double-stage smoothers, $\hat{\mathbf{x}}(k|k + 1)$ and $\hat{\mathbf{x}}(k|k + 2)$, respectively. A table at the end of the lesson connects the single- and double-stage smoothers to fixed-interval, fixed-point, and fixed-lag smoothers.

From the derivations of $\hat{\mathbf{x}}(k|k + 1)$ and $\hat{\mathbf{x}}(k|k + 2)$, we shall see one of the major differences between the structures of smoothers and earlier state estimators: whereas the Kalman gain matrix depends explicitly on the error-covariance matrix $\mathbf{P}(k + 1|k)$ [or $\mathbf{P}(k|k)$], smoother gain matrices do not depend explicitly on smoothing error-covariance matrices. In fact, smoothing error-covariance matrices do not appear at all in smoothing equations and must be computed separately.

When you complete this lesson, you will be able to (1) distinguish between fixed-interval, fixed-point, and fixed-lag smoothing; (2) derive and use single- and double-stage smoothers; and (3) relate single- and double-stage smoothers to fixed-interval, fixed-point, and fixed-lag smoothers.

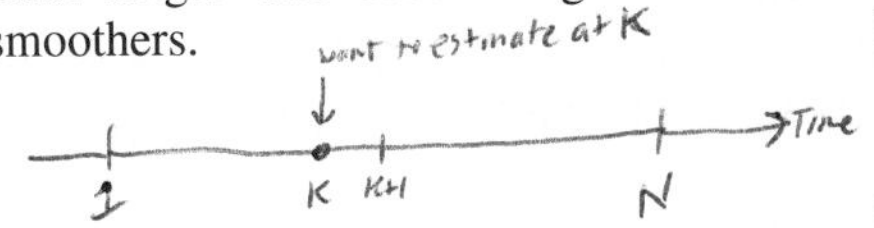

## THREE TYPES OF SMOOTHERS

Recall that in smoothing we obtain mean-squared estimates of state vector $\mathbf{x}(k)$, $\hat{\mathbf{x}}(k|j)$ for which $k < j$. From the fundamental theorem of estimation theory, Theorem 13-1, we know that the structure of the mean-squared smoother is

$$\hat{\mathbf{x}}(k|j) = \mathbf{E}\{\mathbf{x}(k)|\mathbf{Z}(j)\}, \qquad \text{where } k < j \tag{20-1}$$

In this lesson we shall develop recursive smoothing algorithms that are comparable to our recursive Kalman filter algorithm.

Smoothing is much more complicated than filtering, primarily because we are using future measurements to obtain estimates at earlier time points. Because there is some flexibility associated with how we choose to process future measurements, it is convenient to distinguish between three types of smoothing (Meditch, 1969, pp. 204-208): fixed interval, fixed point, and fixed lag.

The *fixed-interval smoother* is $\hat{\mathbf{x}}(k|N)$, $k = 0, 1, \ldots, N-1$, where $N$ is a fixed positive integer. The situation here is as follows: with an experiment completed, we have measurements available over the fixed interval $1 \le k \le N$. For each time point within this interval we wish to obtain the optimal estimate of state vector $\mathbf{x}(k)$, which is based on *all* the available measurement data $\{\mathbf{z}(j), j = 1, 2, \ldots, N\}$. Fixed-interval smoothing is very useful in signal processing, where the processing is done after all the data are collected. It cannot be carried out on-line during an experiment, as filtering can be. Because all the available data are used, we cannot hope to do better (by other forms of smoothing) than by fixed-interval smoothing.

The *fixed-point* smoothed estimate is $\hat{\mathbf{x}}(k|j)$, $j = k+1, k+2, \ldots$, where $k$ is a fixed positive integer. Suppose we want to improve our estimate of a state at a specific time by making use of future measurements. Let this time be $\bar{k}$. Then fixed-point smoothed estimates of $\mathbf{x}(\bar{k})$ will be $\hat{\mathbf{x}}(\bar{k}|\bar{k}+1)$, $\hat{\mathbf{x}}(\bar{k}|\bar{k}+2), \ldots$, etc. The last possible fixed-point estimate is $\hat{\mathbf{x}}(\bar{k}|N)$, which is the same as the fixed-interval estimate of $\mathbf{x}(\bar{k})$. Fixed-point smoothing can be carried out on-line, if desired, but the calculation of $\hat{\mathbf{x}}(\bar{k}|\bar{k}+d)$ is subject to a delay of $dT$ sec.

The *fixed-lag* smoothed estimate is $\hat{\mathbf{x}}(k|k+L)$, $k = 0, 1, \ldots$, where $L$ is a fixed positive integer. In this case, the point at which we seek the estimate of the system's state lags the time point of the most recent measurement by a fixed interval of time, $L$; i.e., $t_{k+L} - t_k = L$, which is a positive constant for all $k = 0, 1, \ldots$. This type of estimator can be used where a constant lag betwen measurements and state estimates is permissible. Fixed-lag smoothing can be carried out on-line, if desired, but the calculation of $\hat{\mathbf{x}}(k|k+L)$ is subject to an $LT$ sec delay.

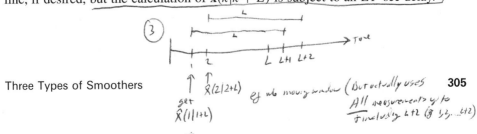

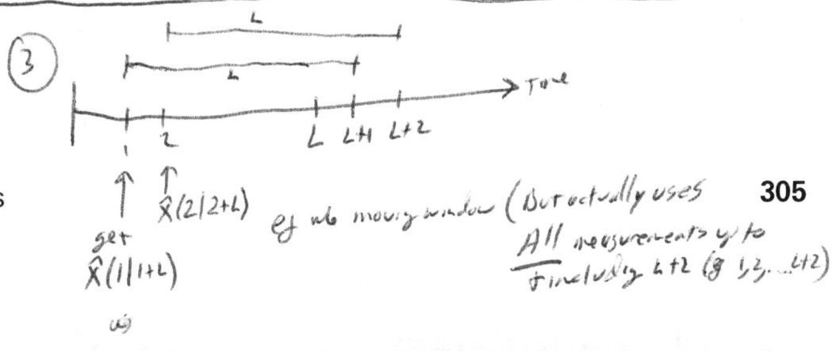

## APPROACHES FOR DERIVING SMOOTHERS

The literature on smoothing is filled with many different approaches for deriving recursive smoothers. By augmenting suitably defined states to (15-17) and (15-18), we can reduce the derivation of smoothing formulas to a Kalman filter for the augmented state-variable model (Anderson and Moore, 1979). The "filtered" estimates of the newly introduced states turn out to be equivalent to smoothed values of $\mathbf{x}(k)$. We shall examine this augmentation approach in Lesson 21. A second approach is to use the orthogonality principle to derive a discrete-time Wiener–Hopf equation, which can then be used to establish the smoothing formulas. We do not treat this approach in this book.

A third approach, the one we shall follow in this lesson, is based on the causal invertibility between the innovations process $\tilde{\mathbf{z}}(k+j|k+j-1)$ and measurement $\mathbf{z}(k+j)$ and repeated applications of Theorem 12-4.

## A SUMMARY OF IMPORTANT FORMULAS

The following formulas, which have been derived in earlier lessons, are used so frequently in this lesson, as well as in Lesson 21, that we collect them here for the convenience of the reader:

$$\tilde{\mathbf{z}}(k+1|k) = \mathbf{z}(k+1) - \mathbf{H}(k+1)\hat{\mathbf{x}}(k+1|k) \tag{20-2}$$

$$\tilde{\mathbf{z}}(k+1|k) = \mathbf{H}(k+1)\tilde{\mathbf{x}}(k+1|k) + \mathbf{v}(k+1) \tag{20-3}$$

$$\mathbf{P}_{\tilde{\mathbf{z}}\tilde{\mathbf{z}}}(k+1|k) = \mathbf{H}(k+1)\mathbf{P}(k+1|k)\mathbf{H}'(k+1) + \mathbf{R}(k+1) \tag{20-4}$$

$$\tilde{\mathbf{x}}(k+1|k) = \mathbf{\Phi}(k+1,k)\tilde{\mathbf{x}}(k|k) + \mathbf{\Gamma}(k+1,k)\mathbf{w}(k) \tag{20-5}$$

and

$$\tilde{\mathbf{x}}(k+1|k+1) = [\mathbf{I} - \mathbf{K}(k+1)\mathbf{H}(k+1)]\tilde{\mathbf{x}}(k+1|k) - \mathbf{K}(k+1)\mathbf{v}(k+1) \tag{20-6}$$

## SINGLE-STAGE SMOOTHER

As in our study of prediction, it is useful first to develop a single-stage smoother and then to obtain more general smoothers.

**Theorem 20-1.** *The single-stage mean-squared smoothed estimator of* $\mathbf{x}(\mathrm{k})$, $\hat{\mathbf{x}}(\mathrm{k}|\mathrm{k}+1)$, *is given by the expression*

$$\hat{\mathbf{x}}(k|k+1) = \hat{\mathbf{x}}(k|k) + \mathbf{M}(k|k+1)\tilde{\mathbf{z}}(k+1|k) \tag{20-7}$$

*where single-stage smoother gain matrix,* $\mathbf{M}(\mathrm{k}|\mathrm{k}+1)$, *is*

$$\mathbf{M}(k|k+1) = \mathbf{P}(k|k)\mathbf{\Phi}'(k+1,k)\mathbf{H}'(k+1)[\mathbf{H}(k+1)\mathbf{P}(k+1|k)\mathbf{H}'(k+1) + \mathbf{R}(k+1)]^{-1} \tag{20-8}$$

*Proof.* From the fundamental theorem of estimation theory and Theorem 12-4, we know that

$$\begin{aligned}\hat{\mathbf{x}}(k|k+1) &= \mathbf{E}\{\mathbf{x}(k)|\mathbf{Z}(k+1)\}\\ &= \mathbf{E}\{\mathbf{x}(k)|\mathbf{Z}(k), \mathbf{z}(k+1)\}\\ &= \mathbf{E}\{\mathbf{x}(k)|\mathbf{Z}(k), \tilde{\mathbf{z}}(k+1|k)\}\\ &= \mathbf{E}\{\mathbf{x}(k)|\mathbf{Z}(k)\} + \mathbf{E}\{\mathbf{x}(k)|\tilde{\mathbf{z}}(k+1|k)\} - \mathbf{m}_{\mathbf{x}}(k)\end{aligned} \tag{20-9}$$

which can also be expressed as (see Corollary 13-1)

$$\hat{\mathbf{x}}(k|k+1) = \hat{\mathbf{x}}(k|k) + \mathbf{P}_{\mathbf{x}\tilde{\mathbf{z}}}(k, k+1|k)\mathbf{P}_{\tilde{\mathbf{z}}\tilde{\mathbf{z}}}^{-1}(k+1|k)\tilde{\mathbf{z}}(k+1|k) \tag{20-10}$$

Defining single-stage smoother gain matrix $\mathbf{M}(k|k+1)$ as

$$\mathbf{M}(k|k+1) = \mathbf{P}_{\mathbf{x}\tilde{\mathbf{z}}}(k, k+1|k)\mathbf{P}_{\tilde{\mathbf{z}}\tilde{\mathbf{z}}}^{-1}(k+1|k) \tag{20-11}$$

(20-10) reduces to (20-7).

Next, we must show that $\mathbf{M}(k|k+1)$ can also be expressed as in (20-8). We already have expression (20-4) for $\mathbf{P}_{\tilde{\mathbf{z}}\tilde{\mathbf{z}}}(k+1|k)$; hence, we must compute $\mathbf{P}_{\mathbf{x}\tilde{\mathbf{z}}}(k, k+1|k)$. To do this, we make use of (20-3), (20-5), and the orthogonality principle, as follow:

$$\begin{aligned}\mathbf{P}_{\mathbf{x}\tilde{\mathbf{z}}}(k, k+1|k) &= \mathbf{E}\{\mathbf{x}(k)\tilde{\mathbf{z}}'(k+1|k)\}\\ &= \mathbf{E}\{\mathbf{x}(k)[\mathbf{H}(k+1)\tilde{\mathbf{x}}(k+1|k) + \mathbf{v}(k+1)]'\}\\ &= \mathbf{E}\{\mathbf{x}(k)\tilde{\mathbf{x}}'(k+1|k)\}\mathbf{H}'(k+1)\\ &= \mathbf{E}\{\mathbf{x}(k)[\mathbf{\Phi}(k+1, k)\tilde{\mathbf{x}}(k|k)\\ &\quad +\mathbf{\Gamma}(k+1, k)\mathbf{w}(k)]'\}\mathbf{H}'(k+1)\\ &= \mathbf{E}\{\mathbf{x}(k)\tilde{\mathbf{x}}'(k|k)\}\mathbf{\Phi}'(k+1, k)\mathbf{H}'(k+1)\\ &= \mathbf{E}\{[\hat{\mathbf{x}}(k|k) + \tilde{\mathbf{x}}(k|k)]\tilde{\mathbf{x}}'(k|k)\}\mathbf{\Phi}'(k+1, k)\mathbf{H}'(k+1)\\ &= \mathbf{P}(k|k)\mathbf{\Phi}'(k+1, k)\mathbf{H}'(k+1)\end{aligned} \tag{20-12}$$

Substituting (20-12) and (20-4) into (20-11), we obtain (20-8). □

For future reference, we record the following fact:

$$\mathbf{E}\{\mathbf{x}(k)\tilde{\mathbf{x}}'(k+1|k)\} = \mathbf{P}(k|k)\mathbf{\Phi}'(k+1, k) \tag{20-13}$$

This is obtained by comparing the third and last lines of (20-12).

Observe that the structure of the single-stage smoother is quite similar to that of the Kalman filter. The Kalman filter obtains $\hat{\mathbf{x}}(k+1|k+1)$ by adding a correction that depends on the most recent innovations, $\tilde{\mathbf{z}}(k+1|k)$, to the *predicted* value of $\mathbf{x}(k+1)$. The single-stage smoother, on the other hand, obtains $\hat{\mathbf{x}}(k|k+1)$, the estimate of $\mathbf{x}$ at $t_k$, by adding a correction that also depends on $\tilde{\mathbf{z}}(k+1|k)$, the innovations at $t_{k+1}$, to the *filtered* value of $\mathbf{x}(k)$. We see that filtered estimates are required to obtain smoothed estimates.

*theoretical result*
*don't implement on computer*
↳**Corollary 20-1.** *Kalman gain matrix* $\mathbf{K}(k+1)$ *is a factor in* $\mathbf{M}(k|k+1)$, *i.e.*,

$$\mathbf{M}(k|k+1) = \mathbf{A}(k)\mathbf{K}(k+1) \tag{20-14}$$

*where*

$$\mathbf{A}(k) \stackrel{\triangle}{=} \mathbf{P}(k|k)\mathbf{\Phi}'(k+1,k)\mathbf{P}^{-1}(k+1|k) \tag{20-15}$$

*Proof.* Using the fact that [Equation (17-12)]

$$\mathbf{K}(k+1) = \mathbf{P}(k+1|k)\mathbf{H}'(k+1)$$
$$[\mathbf{H}(k+1)\mathbf{P}(k+1|k)\mathbf{H}'(k+1) + \mathbf{R}(k+1)]^{-1} \tag{20-16}$$

we see that

$$\mathbf{H}'(k+1)[\mathbf{H}(k+1)\mathbf{P}(k+1|k)\mathbf{H}'(k+1) + \mathbf{R}(k+1)]^{-1}$$
$$= \mathbf{P}^{-1}(k+1|k)\mathbf{K}(k+1) \tag{20-17}$$

When (20-17) is substituted into (20-8), we obtain (20-14). □

**Corollary 20-2.** *Another way to express* $\hat{\mathbf{x}}(k|k+1)$ *is*

$$\hat{\mathbf{x}}(k|k+1) = \hat{\mathbf{x}}(k|k) + \mathbf{A}(k)[\hat{\mathbf{x}}(k+1|k+1) - \hat{\mathbf{x}}(k+1|k)] \tag{20-18}$$

*Proof.* Substitute (20-14) into (20-7) to see that

$$\hat{\mathbf{x}}(k|k+1) = \hat{\mathbf{x}}(k|k) + \mathbf{A}(k)\mathbf{K}(k+1)\tilde{\mathbf{z}}(k+1|k) \tag{20-19}$$

but [see (17-11)]

$$\mathbf{K}(k+1)\tilde{\mathbf{z}}(k+1|k) = \hat{\mathbf{x}}(k+1|k+1) - \hat{\mathbf{x}}(k+1|k) \tag{20-20}$$

Substitute (20-20) into (20-19) to obtain the desired result in (20-18). □

*implementation* →  Formula (20-7) is useful for computational purposes, because it can be programmed with no multiplications of $n \times n$ matrices, whereas (20-18) is most useful for theoretical purposes [matrix $\mathbf{A}(k)$ requires the multiplication of two $n \times n$ matrices]. These facts will become more clear when we examine double-stage smoothing in our next section.

Whereas the structure of the single-stage smoother is similar to that of the Kalman filter, we see that $\mathbf{M}(k|k+1)$ does not depend on single-stage smoothing error-covariance matrix $\mathbf{P}(k|k+1)$. Kalman gain $\mathbf{K}(k+1)$, of course, does depend on $\mathbf{P}(k+1|k)$ [or $\mathbf{P}(k|k)$]. In fact, $\mathbf{P}(k|k+1)$ does not appear at all in the smoothing equations and must be computed (if we desire to do so) separately. We address this calculation in Lesson 21.

**EXAMPLE 20-1**

Once again we return to the important application of *deconvolution*. The first meaningful MS estimator of input $\mathbf{w}(k)$ is *not* $E\{\mathbf{w}(k)|\mathbf{Z}(k)\}$, because $E\{\mathbf{w}(k)|\mathbf{Z}(k)\} = \mathbf{0}$ (by now you ought to be able to explain why this is true in your sleep). To obtain the first meaningful MS estimator of input $\mathbf{w}(k)$, operate on both sides of state equation (15-17) [for simplicity, we set $\mathbf{u}(k) = \mathbf{0}$] with the operator $E\{\cdot|\mathbf{Z}(k+1)\}$:

$$\mathbf{E}\{\mathbf{x}(k+1)|\mathbf{Z}(k+1)\} = \mathbf{\Phi}'\mathbf{E}\{\mathbf{x}(k)|\mathbf{Z}(k+1)\} + \mathbf{\Gamma}\mathbf{E}\{\mathbf{w}(k)|\mathbf{Z}(k+1)\} \tag{20-21}$$

The last term in (20-21) will not be zero because $\mathbf{Z}(k+1)$ depends on $\mathbf{w}(k)$; hence, we obtain the following equation for an estimator of $\mathbf{w}(k)$:

$$\mathbf{\Gamma}\hat{\mathbf{w}}(k|k+1) = \hat{\mathbf{x}}(k+1|k+1) - \mathbf{\Phi}\hat{\mathbf{x}}(k|k+1) \tag{20-22}$$

Next, we show that (Mendel, 1977b)

$$\hat{\mathbf{w}}(k|k+1) = \mathbf{Q}\mathbf{\Gamma}'\mathbf{P}^{-1}(k+1|k)\mathbf{K}(k+1)\tilde{\mathbf{z}}(k+1|k) \tag{20-23a}$$

or, alternatively,

$$\hat{\mathbf{w}}(k|k+1) = \mathbf{Q}(\mathbf{H}\mathbf{\Gamma})'[\mathbf{H}\mathbf{P}(k+1|k)\mathbf{H}' + \mathbf{R}]^{-1}\tilde{\mathbf{z}}(k+1|k) \tag{20-23b}$$

Additionally, letting

$$\mathbf{\Psi}_{\mathbf{w}}(k|k+1) \triangleq \mathbf{E}\{[\tilde{\mathbf{w}}(k|k+1) - \mathbf{E}\{\tilde{\mathbf{w}}(k|k+1)\}][\tilde{\mathbf{w}}(k|k+1) - \mathbf{E}\{\tilde{\mathbf{w}}(k|k+1)\}]'\}, \tag{20-24}$$

it follows that

$$\mathbf{\Psi}_{\mathbf{w}}(k|k+1) = \mathbf{Q} - \mathbf{Q}\mathbf{\Gamma}'\mathbf{H}'[\mathbf{H}\mathbf{P}(k+1|k)\mathbf{H}' + \mathbf{R}]^{-1}\mathbf{H}\mathbf{\Gamma}\mathbf{Q} \tag{20-25}$$

*Derivation of (20-23)*: We begin with (20-22), substitute (20-19) into it, and make use of the fact that $\hat{\mathbf{x}}(k+1|k) = \mathbf{\Phi}\hat{\mathbf{x}}(k|k)$ and (17-11) to show that

$$\mathbf{\Gamma}\hat{\mathbf{w}}(k|k+1) = [\mathbf{I} - \mathbf{\Phi}\mathbf{A}(k)]\mathbf{K}(k+1)\tilde{\mathbf{z}}(k+1|k) \tag{20-26}$$

Matrix $\mathbf{A}(k)$ is defined in (20-15). From that equation and (17-13) (assuming that matrix $\mathbf{\Gamma}$ is of full column rank), it follows that

$$\mathbf{I} - \mathbf{\Phi}\mathbf{A}(k) = \mathbf{\Gamma}\mathbf{Q}\mathbf{\Gamma}'\mathbf{P}^{-1}(k+1|k) \tag{20-27}$$

Substitute (20-27) into (20-26) to obtain the desired expression for $\hat{\mathbf{w}}(k|k+1)$ in (20-23a). Substitute (17-12) into (20-23a) to obtain (20-23b).

We leave the derivation of $\mathbf{\Psi}_{\mathbf{w}}(k|k+1)$ for the reader (Problem 20-1).

Let's compare the optimal estimator $\hat{\mathbf{w}}(k|k+1)$ with Ott and Meder's PEF, which is given in (18-25) and repeated here for the convenience of the reader:

$$\hat{\mathbf{w}}(k) = \mathbf{\Gamma}^{+}\mathbf{K}(k+1)\tilde{\mathbf{z}}(k+1|k) \tag{20-28}$$

Clearly, the two estimators look quite similar. Both are linear transformations of the innovations $\tilde{\mathbf{z}}(k+1|k)$. Both are themselves white because of this. The only difference between the two estimators is the factor of $\mathbf{Q}\mathbf{\Gamma}'\mathbf{P}^{-1}(k+1|k)$ in $\hat{\mathbf{w}}(k|k+1)$ versus the factor $\mathbf{\Gamma}^{+}$ in $\hat{\mathbf{w}}(k)$; but oh what a difference these two factors make! In Example 18-6 we derived an expression for $\mathbf{\Psi}_{\mathbf{w}}(k)$ and showed, by way of an example, that $\mathbf{\Psi}_{\mathbf{w}}(k)$ could be larger than $\mathbf{Q}$ (in a matrix sense). Examining (20-25), we see that $\mathbf{\Psi}_{\mathbf{w}}(k|k+1)$ can be no larger than $\mathbf{Q}$.

If $\mathbf{H}\mathbf{\Gamma} = \mathbf{0}$, as in the case of the harmonic oscillator, then, from (20-23b), we see that $\hat{\mathbf{w}}(k|k+1) = \mathbf{0}$, which means that in this case even $\hat{\mathbf{w}}(k|k+1)$ is not the first meaningful estimator of $\mathbf{w}(k)$. This motivates a need for a double-staged smoothed estimator, $\hat{\mathbf{w}}(k|k+2)$. We return to the harmonic oscillator in Example 20-2. □

## DOUBLE-STAGE SMOOTHER

Instead of immediately generalizing the single-stage smoother to an $N$ stage smoother, we first present results for the double-stage smoother. We will then be able to write down the general results (almost) by inspection of the single- and double-stage results.

**Theorem 20-2.** *The double-stage mean-squared smoothed estimator of* $\mathbf{x}(k)$, $\hat{\mathbf{x}}(k|k+2)$, *is given by the expression*

$$\hat{\mathbf{x}}(k|k+2) = \hat{\mathbf{x}}(k|k+1) + \mathbf{M}(k|k+2)\tilde{\mathbf{z}}(k+2|k+1) \tag{20-29}$$

*where double-stage smoother gain matrix,* $\mathbf{M}(k|k+2)$, *is*

$$\begin{aligned}\mathbf{M}(k|k+2) = \; & \mathbf{P}(k|k)\mathbf{\Phi}'(k+1,k)[\mathbf{I} - \mathbf{K}(k+1) \\ & \mathbf{H}(k+1)]'\mathbf{\Phi}'(k+2,k+1) \\ & \mathbf{H}'(k+2)[\mathbf{H}(k+2)\mathbf{P}(k+2|k+1) \\ & \mathbf{H}'(k+2) + \mathbf{R}(k+2)]^{-1}\end{aligned} \tag{20-30}$$

*Proof.* From the fundamental theorem of estimation theory, Theorem 12-4, and Corollary 13-1, we know that

$$\begin{aligned}\hat{\mathbf{x}}(k|k+2) &= \mathbf{E}\{\mathbf{x}(k)|\mathbf{Z}(k+2)\} \\ &= \mathbf{E}\{\mathbf{x}(k)|\mathbf{Z}(k+1), \mathbf{z}(k+2)\} \\ &= \mathbf{E}\{\mathbf{x}(k)|\mathbf{Z}(k+1), \tilde{\mathbf{z}}(k+2|k+1)\} \\ &= \mathbf{E}\{\mathbf{x}(k)|\mathbf{Z}(k+1)\} + \mathbf{E}\{\mathbf{x}(k)|\tilde{\mathbf{z}}(k+2|k+1)\} - \mathbf{m}_{\mathbf{x}}(k)\end{aligned} \tag{20-31}$$

which can also be expressed as

$$\begin{aligned}\hat{\mathbf{x}}(k|k+2) = \; & \hat{\mathbf{x}}(k|k+1) \\ & + \mathbf{P}_{\mathbf{x}\tilde{\mathbf{z}}}(k, k+2|k+1)\mathbf{P}_{\tilde{\mathbf{z}}\tilde{\mathbf{z}}}^{-1}(k+2|k+1)\tilde{\mathbf{z}}(k+2|k+1)\end{aligned} \tag{20-32}$$

Defining double-stage smoother gain matrix $\mathbf{M}(k|k+2)$ as

$$\mathbf{M}(k|k+2) = \mathbf{P}_{\mathbf{x}\tilde{\mathbf{z}}}(k, k+2|k+1)\mathbf{P}_{\tilde{\mathbf{z}}\tilde{\mathbf{z}}}^{-1}(k+2|k+1) \tag{20-33}$$

(20-32) reduces to (20-29).

To show that $\mathbf{M}(k|k+2)$ in (20-33) can be expressed as in (20-30), we proceed as in our derivation of $\mathbf{M}(k|k+1)$ in (20-12); however, the details are lengthier because $\tilde{\mathbf{z}}(k+2|k+1)$ involves quantities that are two time units away from $\mathbf{x}(k)$, whereas $\tilde{\mathbf{z}}(k+1|k)$ involves quantities that are only one time unit away from $\mathbf{x}(k)$. Equation (20-13) is used during the derivation. We leave the detailed derivation of (20-30) as an exercise for the reader. □

Whereas (20-30) is a computationally useful formula, it is not useful from a theoretical viewpoint; i.e., when we examine $\mathbf{M}(k|k+1)$ in (20-8) and $\mathbf{M}(k|k+2)$ in (20-30), it is not at all obvious how to generalize these formulas to $\mathbf{M}(k|k+N)$, or even to $\mathbf{M}(k|k+3)$. The following result for $\mathbf{M}(k|k+2)$ is easily generalized.

**Corollary 20-3.** *Kalman gain matrix* $\mathbf{K}(k+2)$ *is a factor in* $\mathbf{M}(k|k+2)$*; i.e.,*

$$\mathbf{M}(k|k+2) = \mathbf{A}(k)\mathbf{A}(k+1)\mathbf{K}(k+2) \tag{20-34}$$

*where* $\mathbf{A}(k)$ *is defined in (20-15).*

*Proof.* Increment $k$ to $k+1$ in (20-17) to see that

$$\mathbf{H}'(k+2)[\mathbf{H}(k+2)\mathbf{P}(k+2|k+1)\mathbf{H}'(k+2)+\mathbf{R}(k+2)]^{-1} = \mathbf{P}^{-1}(k+2|k+1)\mathbf{K}(k+2) \quad (20\text{-}35)$$

Next, note from (17-14), that

$$\mathbf{I}-\mathbf{K}(k+1)\mathbf{H}(k+1) = \mathbf{P}(k+1|k+1)\mathbf{P}^{-1}(k+1|k) \quad (20\text{-}36)$$

Hence,

$$[\mathbf{I}-\mathbf{K}(k+1)\mathbf{H}(k+1)]' = \mathbf{P}^{-1}(k+1|k)\mathbf{P}(k+1|k+1) \quad (20\text{-}37)$$

Substitute (20-35) and (20-37) into (20-30) to see that

$$\begin{aligned}\mathbf{M}(k|k+2) = \; & \mathbf{P}(k|k)\boldsymbol{\Phi}'(k+1,k)\mathbf{P}^{-1}(k+1|k) \\ & \mathbf{P}(k+1|k+1)\boldsymbol{\Phi}'(k+2,k+1) \\ & \mathbf{P}^{-1}(k+2|k+1)\mathbf{K}(k+2)\end{aligned} \quad (20\text{-}38)$$

Using the definition of matrix $\mathbf{A}(k)$ in (20-15), we see that (20-38) can be expressed as in (20-34). □

**Corollary 20-4.** *Two other ways to express* $\hat{\mathbf{x}}(k|k+2)$ *are*

$$\begin{aligned}\hat{\mathbf{x}}(k|k+2) = \; & \hat{\mathbf{x}}(k|k+1) \\ & +\mathbf{A}(k)\mathbf{A}(k+1)[\hat{\mathbf{x}}(k+2|k+2)-\hat{\mathbf{x}}(k+2|k+1)]\end{aligned} \quad (20\text{-}39)$$

*and*

$$\hat{\mathbf{x}}(k|k+2) = \hat{\mathbf{x}}(k|k) + \mathbf{A}(k)[\hat{\mathbf{x}}(k+1|k+2)-\hat{\mathbf{x}}(k+1|k)] \quad (20\text{-}40)$$

*Proof.* The derivation of (20-39) follows exactly the same path as the derivation of (20-18) and is therefore left as an exercise for the reader. Equation (20-39) is the starting place for the derivation of (20-40). Observe, from (20-18), that

$$\mathbf{A}(k+1)[\hat{\mathbf{x}}(k+2|k+2)-\hat{\mathbf{x}}(k+2|k+1)] = \hat{\mathbf{x}}(k+1|k+2)-\hat{\mathbf{x}}(k+1|k+1) \quad (20\text{-}41)$$

Thus, (20-39) can be written as

$$\begin{aligned}\hat{\mathbf{x}}(k|k+2) = \; & \hat{\mathbf{x}}(k|k+1) \\ & +\mathbf{A}(k)[\hat{\mathbf{x}}(k+1|k+2)-\hat{\mathbf{x}}(k+1|k+1)]\end{aligned} \quad (20\text{-}42)$$

Substituting (20-18) into (20-42), we obtain the desired result in (20-40). □

The alternative forms we have obtained for both $\hat{\mathbf{x}}(k|k+1)$ and $\hat{\mathbf{x}}(k|k+2)$ will suggest how we can generalize our single- and double-stage smoothers to $N$-stage smoothers.

**EXAMPLE 20-2**

To obtain the second meaningful MS estimator of input $\mathbf{w}(k)$, operate on both sides of state equation (15-17) [for simplicity, we set $\mathbf{u}(k)=\mathbf{0}$] with the operator $\mathbf{E}\{\cdot|\mathbf{Z}(k+2)\}$. We leave it to the reader to show that

$$\hat{\mathbf{w}}(k|k+2) = \hat{\mathbf{w}}(k|k+1) + \mathbf{N_W}(k|k+2)\tilde{\mathbf{z}}(k+2|k+1) \tag{20-43}$$

which can also be expressed as

$$\hat{\mathbf{w}}(k|k+2) = \mathbf{N_w}(k|k+1)\tilde{\mathbf{z}}(k+1|k) + \mathbf{N_w}(k|k+2)\tilde{\mathbf{z}}(k+2|k+1) \tag{20-44}$$

where

$$\mathbf{N_w}(k|k+1) = \mathbf{Q\Gamma}'\mathbf{P}^{-1}(k+1|k)\mathbf{K}(k+1) \tag{20-45}$$

and

$$\mathbf{N_w}(k|k+2) = \mathbf{Q\Gamma}'\mathbf{P}^{-1}(k+1|k)\mathbf{A}(k+1)\mathbf{K}(k+2) \tag{20-46}$$

Additionally, letting

$$\mathbf{\Psi_w}(k|k+2) \triangleq \mathbf{E}\{[\tilde{\mathbf{w}}(k|k+2) - \mathbf{E}\{\tilde{\mathbf{w}}(k|k+2)\}][\tilde{\mathbf{w}}(k|k+2) - \mathbf{E}\{\tilde{\mathbf{w}}(k|k+2)\}]'\} \tag{20-47}$$

it follows that (Mendel, 1977a)

$$\mathbf{\Psi_w}(k|k+2) = \mathbf{\Psi_w}(k|k+1) - \mathbf{N_w}(k|k+2)[\mathbf{HP}(k+2|k+1)\mathbf{H}' + \mathbf{R}]\mathbf{N}'_\mathbf{w}(k|k+2) \tag{20-48}$$

Let us return to the harmonic oscillator (see Examples 18-6 and 20-1), for which $\mathbf{H\Gamma} = \mathbf{0}$ so that $\mathbf{N_w}(k|k+1) = \mathbf{0}$. Now, however, $\mathbf{N_w}(k|k+2) \neq \mathbf{0}$, so that

$$\hat{\mathbf{w}}(k|k+2) = \mathbf{N_w}(k|k+2)\tilde{\mathbf{z}}(k+2|k+1) \tag{20-49}$$

and

$$\mathbf{\Psi_w}(k|k+2) = \mathbf{Q} - \mathbf{N_w}(k|k+2)[\mathbf{HP}(k+2|k+1)\mathbf{H}' + \mathbf{R}]\mathbf{N}'_\mathbf{w}(k|k+2) < \mathbf{Q} \tag{20-50}$$

$\hat{\mathbf{w}}(k|k+2)$ is the first MS meaningful estimator for the white input of an harmonic oscillator and is substantially different from Ott and Meder's PEF stated in (20-28). □

## SINGLE- AND DOUBLE-STAGE SMOOTHERS AS GENERAL SMOOTHERS

At the beginning of this lesson we described three types of smoothers: fixed-interval, fixed-point, and fixed-lag smoothers. Table 20-1 shows how our single- and double-stage smoothers fit into these three categories.

To obtain the fixed-interval smoother formulas, given for both the single- and double-stage smoothers, set $k+1 = N$ in (20-18) and $k+2 = N$ in (20-40), respectively. Doing this forces the left-hand side of both equations to be conditioned on data length $N$. Observe that, before we can compute $\hat{\mathbf{x}}(N-1|N)$ or $\hat{\mathbf{x}}(N-2|N)$, we must run a Kalman filter on all the data in order to obtain $\hat{\mathbf{x}}(N|N)$. This last filtered state estimate initializes the backward running fixed-interval smoother. Observe, also, that we must compute $\hat{\mathbf{x}}(N-1|N)$ before we can compute $\hat{\mathbf{x}}(N-2|N)$. Clearly, the limitation of our results so far is that we can only perform fixed-interval smoothing for $k = N-1$ and $N-2$. More general results that will permit us to perform fixed-interval smoothing for any $k < N$ are described in Lesson 21.

To obtain the fixed-point smoother formulas, given for both the single- and double-stage smoothers, set $k = \bar{k}$ in (20-18) and (20-39), respectively. Time point $\bar{k}$

**TABLE 20-1** SMOOTHING INTERRELATIONSHIPS

<table>
<tr><th></th><th>Single-Stage</th><th>Double-Stage</th></tr>
<tr><td>Fixed-Interval</td><td>$\hat{\mathbf{x}}(N-1|N) = \hat{\mathbf{x}}(N-1|N-1) + \mathbf{A}(N-1)[\hat{\mathbf{x}}(N|N) - \hat{\mathbf{x}}(N|N-1)]$<br>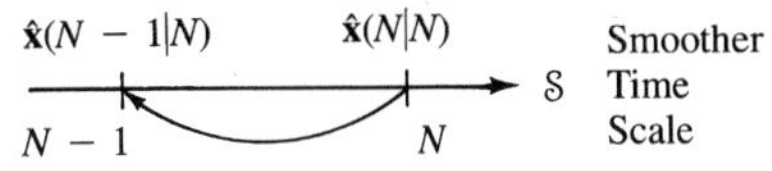<br><br>Solution proceeds in reverse time, from $N$ to $N-1$, where $N$ is fixed.</td><td>$\hat{\mathbf{x}}(N-2|N) = \hat{\mathbf{x}}(N-2|N-2) + \mathbf{A}(N-2)[\hat{\mathbf{x}}(N-1|N) - \hat{\mathbf{x}}(N-1|N-2)]$<br>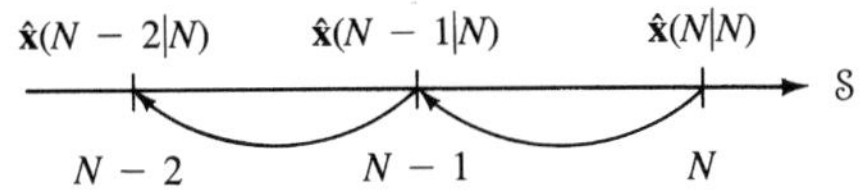<br><br>Solution proceeds in reverse time, from $N$ to $N-1$, to $N-2$, where $N$ is fixed.</td></tr>
<tr><td>Fixed-Point</td><td>$\hat{\mathbf{x}}(\bar{k}|\bar{k}+1) = \hat{\mathbf{x}}(\bar{k}|\bar{k}) + \mathbf{A}(\bar{k})[\hat{\mathbf{x}}(\bar{k}+1|\bar{k}+1) - \hat{\mathbf{x}}(\bar{k}+1|\bar{k})]$<br>$k$ fixed at $\bar{k}$<br>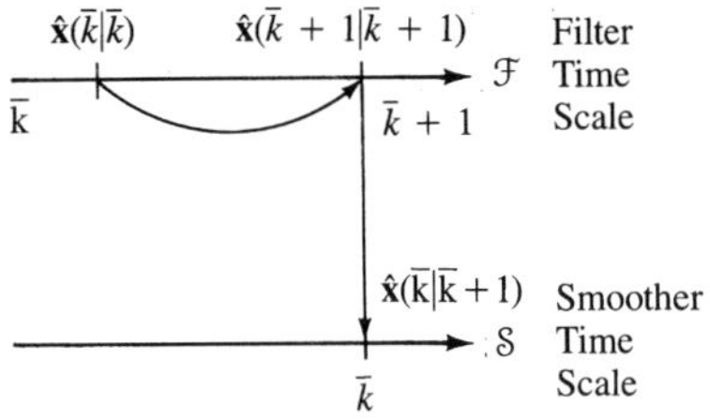<br><br>Solution proceeds in forward time from $\bar{k}$ to $\bar{k}+1$ on the filtering time scale and then back to $\bar{k}$ on the smoothing time scale. A one-unit time delay is present.</td><td>$\hat{\mathbf{x}}(\bar{k}|\bar{k}+2) = \hat{\mathbf{x}}(\bar{k}|\bar{k}+1) + \mathbf{A}(\bar{k})\mathbf{A}(\bar{k}+1)[\hat{\mathbf{x}}(\bar{k}+2|\bar{k}+2) - \hat{\mathbf{x}}(\bar{k}+2|\bar{k}+1)]$<br>$k$ fixed at $\bar{k}$<br>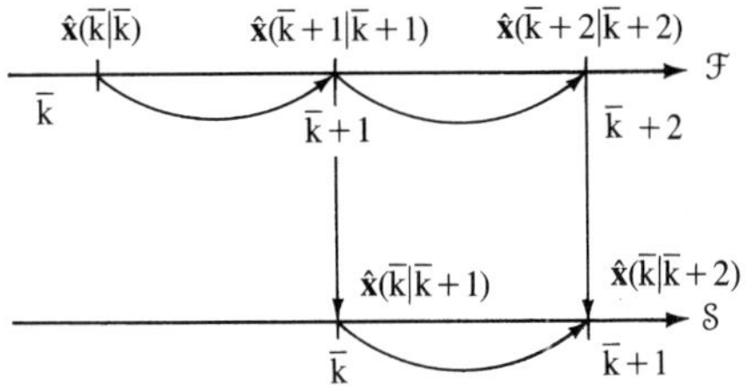<br><br>Solution proceeds in forward time. Results from single-stage smoother as well as optimal filter are required at $\mathcal{S}=\bar{k}+1$, whereas at $\mathcal{S}=\bar{k}$ only results from optimal filter are required. A two-unit time delay is present.</td></tr>
<tr><th></th><th>Single-Stage</th><th>Double-Stage</th></tr>
<tr><td>Fixed-Lag</td><td>$\hat{\mathbf{x}}(k|k+1) = \hat{\mathbf{x}}(k|k) + \mathbf{A}(k)[\hat{\mathbf{x}}(k+1|k+1) - \hat{\mathbf{x}}(k+1|k)]$<br>$k$ variable<br>Picture and discussion same as in Fixed-Point case, replacing $\bar{k}$ by $k$.</td><td>$\hat{\mathbf{x}}(k|k+2) = \hat{\mathbf{x}}(k|k+1) + \mathbf{A}(k)\mathbf{A}(k+1)[\hat{\mathbf{x}}(k+2|k+2) - \hat{\mathbf{x}}(k+2|k+1)]$<br>$k$ variable<br>Picture and discussion same as in Fixed-Point case, replacing $\bar{k}$ by $k$.</td></tr>
</table>

represents the fixed point in our smoother formulas. As noted at the beginning of this lesson, fixed-point smoothing can be carried out on line, but it is subject to a delay. From an information availability point of view, a one-unit time delay is present in the single-stage fixed-point smoother, because our smoothed estimate at $k = \bar{k}$ uses the filtered estimate computed at $k = \bar{k} + 1$. A two-unit time delay is present in the double-stage fixed-point smoother, because our smoothed estimate at $\mathcal{S} = \bar{k} + 1$ uses the filtered estimate computed at $k = \bar{k} + 1$ and $\bar{k} + 2$.

The fixed-lag smoother formulas look just like the fixed-point formulas, except that $k$ is a variable in the former and is fixed in the latter. As in the case of fixed-point smoothing, fixed-lag smoothing can be carried out on-line, subject, of course, to a fixed delay equal to the "lag" of the smoother.

## COMPUTATION

No M-files exist in any toolbox for single- or double-stage smoothing. Using the results in Table 20-1, it is straightforward to specialize the general fixed-point or fixed-lag smoothers, which are described in the Computation section in Lesson 21, to single- or double-stage smoothers. Our M-file **fis**, for fixed-interval smoothing, is described in Lesson 21.

## SUMMARY QUESTIONS

**1.** Of the three types of smoothing, which always gives the best performance?
  **(a)** fixed interval
  **(b)** fixed point
  **(c)** fixed lag

**2.** Which of the following denotes a *smoothed* estimate of state $\mathbf{x}(k)$:
  **(a)** $\hat{\mathbf{x}}(k|k)$
  **(b)** $\hat{\mathbf{x}}(k|j)$, where $j > k$
  **(c)** $\hat{\mathbf{x}}(k|j)$, where $j < k$

**3.** The single-stage smoothed estimate of $\mathbf{x}(k)$ is:
  **(a)** $\hat{\mathbf{x}}(k|k-1)$
  **(b)** $\hat{\mathbf{x}}(k|k+1)$
  **(c)** $\hat{\mathbf{x}}(k+1|k+1)$

**4.** Matrix $\mathbf{A}(k)$, in (20-15), requires on the order of how many flops for its calculation?
  **(a)** $2n^3$
  **(b)** $3n^3$
  **(c)** $2.5n^3$

**5.** Smoother $\hat{\mathbf{x}}(k|k+3)$:
  **(a)** has a 3-unit time delay
  **(b)** can be calculated in real time
  **(c)** has a 3-unit time advance

**6.** Which of the following is a fixed-point smoother?
  **(a)** $\hat{\mathbf{x}}(k|k+L), k = 1, 2, \ldots$

(b) $\hat{\mathbf{x}}(k|N), k = 0, 1, \ldots$
(c) $\hat{\mathbf{x}}(k|k+j), j = 1, 2, \ldots$

7. Based on the formulas that have been derived for $\hat{\mathbf{x}}(k|k+1)$ and $\hat{\mathbf{x}}(k|k+2)$, which is the correct formula for $\hat{\mathbf{x}}(k|k+3)$?
(a) $\hat{\mathbf{x}}(k|k+3) = \hat{\mathbf{x}}(k|k+1) + \mathbf{M}(k|k+3)\tilde{\mathbf{z}}(k+3|k+2)$
(b) $\hat{\mathbf{x}}(k|k+3) = \hat{\mathbf{x}}(k|k) + \mathbf{A}(k)[\hat{\mathbf{x}}(k+2|k+2) - \hat{\mathbf{x}}(k+1|k)]$
(c) $\hat{\mathbf{x}}(k|k+3) = \hat{\mathbf{x}}(k|k+2) + \mathbf{M}(k|k+3)\tilde{\mathbf{z}}(k+3|k+2)$

8. The first meaningful MS estimator of white process noise $\mathbf{w}(k)$ is:
(a) $\mathbf{E}\{\mathbf{w}(k)|\mathbf{Z}(k)\}$
(b) $\mathbf{E}\{\mathbf{w}(k)|\mathbf{Z}(k-1)\}$
(c) $\mathbf{E}\{\mathbf{w}(k)|\mathbf{Z}(k+1)\}$

9. If the first $l$ Markov parameters equal zero, then the first meaningful MS estimator of white process noise $\mathbf{w}(k)$ is:
(a) $\hat{\mathbf{w}}(k|k+1)$
(b) $\hat{\mathbf{w}}(k|k+l+1)$
(c) $\hat{\mathbf{w}}(k|k+l+2)$
(d) $\hat{\mathbf{w}}(k|k+l)$

## PROBLEMS

**20-1.** Derive the equation for $\boldsymbol{\Psi}_{\mathbf{w}}(k|k+1)$ given in (20-25).

**20-2.** Derive the formula for the double-stage smoother gain matrix $\mathbf{M}(k|k+2)$, given in (20-30).

**20-3.** Derive the alternative expression for $\hat{\mathbf{x}}(k|k+2)$ given in (20-39).

**20-4.** Derive the equations for $\hat{\mathbf{w}}(k|k+2)$ and $\boldsymbol{\Psi}_{\mathbf{w}}(k|k+2)$, which are given in (20-43) and (20-48), respectively.

**20-5.** Prove that, even though the innovations $\tilde{\mathbf{z}}(k+1|k)$ and $\tilde{\mathbf{z}}(k+2|k+1)$ are white, $\hat{\mathbf{w}}(k|k+2)$ given in (20-44) is *not* white. Explain in words why this is true.

**20-6.** (Gregg Isara and Loan Bui, Fall 1991) (Computation Problem) A skeptical student doesn't believe that single- and double-stage smoothers can perform any better than filtering.
(a) Write a program that simulates measurements for the following first-order system and compare results for the filtered, single-stage smoothed, and double-stage smoothed estimates of $x(k)$ (of course, you will also have to write programs to do these estimations):

$$x(k+1) = 0.85x(k) + w(k), \quad \text{where } q = 1, m_x(0) = 35, \text{ and } p_x(0) = 1$$
$$z(k+1) = 0.3x(k+1) + \nu(k+1), \quad \text{where } r = 4$$

What conclusions can you draw by running the estimation programs?
(b) How do you create a meaningful simulation experiment that accounts for the random nature of the data?
(c) While writing the program, you realize that a better way to make the comparison would be to compare the error variances. Formulas for these variances are given in Lesson 21. Program the appropriate variances and plot them to see whether the results you obtained in part (a) make sense.

**20-7.** (Todd A. Parker and Mark A. Ranta, Fall 1991) Given the FIR system depicted in Figure P20-7 in which $s(k)$ is a deterministic input signal, $w(k)$ is Gaussian white noise that represents errors in the values stored in the shift registers (i.e., the $z^{-1}$), and $\nu(k)$ is an additive Gaussian white measurement noise sequence. Determine a single-stage smoother that could be used to smooth the state of this "machine," which in turn could be used to obtain a smoothed estimate of the system's output.

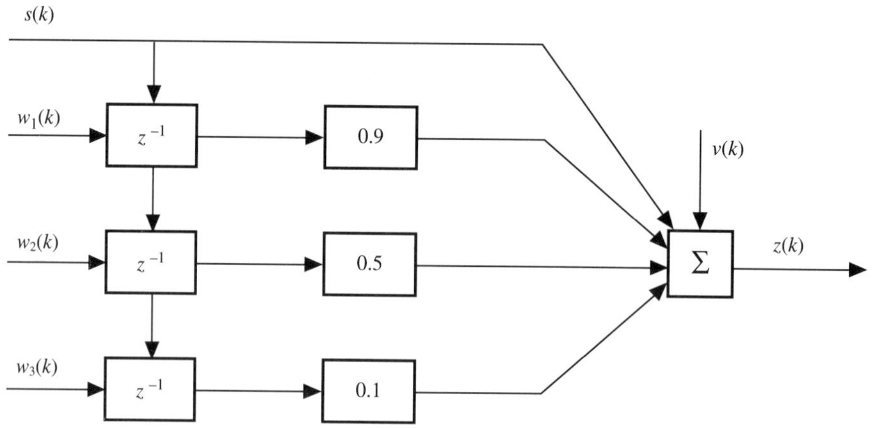


**Figure P20-7**

**20-8.** (Terry L. Bates, Fall 1991) (Computation Problem) We are given the first-order system

$$x(k+1) = 0.95x(k) + w(k)$$

$$z(k+1) = 0.9x(k+1) + \nu(k+1)$$

where $p(0|0) = 10, q = 25$, and $r = 15$.

**(a)** Graphically compare the Kalman filter gain with the single-stage smoother gain.
**(b)** Repeat part (a) when $z(k+1) = 0.35x(k+1) + \nu(k+1)$.
**(c)** Repeat part (a) when $z(k+1) = 0.10x(k+1) + \nu(k+1)$.
**(d)** Draw conclusions from parts (a)–(c).

**20-9.** (Gregory Caso, Fall 1991) Consider the first-order system

$$x(k+1) = ax(k) + w(k)$$

$$z(k+1) = x(k+1) + \nu(k+1)$$

where $Q(k) = q$ and $R(k) = r$.

**(a)** Determine the single-stage smoother for this system.
**(b)** Determine the steady-state single-stage smoother for $a = 0.9, q = 10$, and $r = 4$.

# LESSON 21

# State Estimation: Smoothing (General Results)

## SUMMARY

The purpose of this lesson is to develop general formulas for fixed-interval, fixed-point, and fixed-lag smoothers. The structure of our "most useful" fixed-interval smoother is very interesting, in that it is a two-pass signal processor. First, data are processed in the forward direction by a Kalman filter to produce the sequence of innovations $\tilde{\mathbf{z}}(k|k-1), k = 1, 2, \ldots, N$. Then these innovations are processed in a backward direction by a time-varying recursive digital filter that resembles a backward-running recursive predictor. The noncausal nature of the fixed-interval smoother is very apparent from this two-pass structure.

One of the easiest ways to derive both fixed-point and fixed-lag smoothers is by a *state-augmentation procedure*. In this procedure, a clever choice of additional state variables is made and then a new "augmented" basic state-variable model is created by combining the original basic state-variable model with the state-variable model that is associated with these new states. A Kalman filter is then obtained for the augmented model. By partitioning this Kalman filter's equations, we obtain the filtered estimates of the augmented states. These filtered estimates turn out to be smoothed estimates of the original states, because of the "clever" choice of these new states.

When you complete this lesson, you will be able to (1) describe a computationally efficient fixed-interval smoother algorithm, and (2) explain how to obtain fixed-point and fixed-lag smoothers by a state augmentation procedure.

## INTRODUCTION

In Lesson 20 we introduced three general types of smoothers: fixed-interval, fixed-point, and fixed-lag smoothers. We also developed formulas for single- and double-stage smoothers and showed how these specialized smoothers fit into the three general categories. In this lesson we shall develop general formulas for fixed-interval, fixed-point, and fixed-lag smoothers.

## FIXED-INTERVAL SMOOTHING

In this section we develop a number of algorithms for the fixed-interval smoothed estimator of $\mathbf{x}(k)$, $\hat{\mathbf{x}}(k|N)$. Not all of them will be useful, because we will not be able to compute with some. Only those with which we can compute will be considered useful.

Recall that

$$\hat{\mathbf{x}}(k|N) = \mathbf{E}\{\mathbf{x}(k)|\mathbf{Z}(N)\} \tag{21-1}$$

It is straightforward to proceed as we did in Lesson 20, by showing first that

$$\begin{aligned} \hat{\mathbf{x}}(k|N) &= \mathbf{E}\{\mathbf{x}(k)|\mathbf{Z}(N-1), \tilde{\mathbf{z}}(N|N-1)\} \\ &= \mathbf{E}\{\mathbf{x}(k)|\mathbf{Z}(N-1)\} + \mathbf{E}\{\mathbf{x}(k)|\tilde{\mathbf{z}}(N|N-1)\} - \mathbf{m}_{\mathbf{x}}(k) \\ &= \hat{\mathbf{x}}(k|N-1) + \mathbf{M}(k|N)\tilde{\mathbf{z}}(N|N-1) \end{aligned} \tag{21-2}$$

where

$$\mathbf{M}(k|N) = \mathbf{P}_{\mathbf{x}\tilde{\mathbf{z}}}(k, N|N-1)\mathbf{P}_{\tilde{\mathbf{z}}\tilde{\mathbf{z}}}^{-1}(N|N-1) \tag{21-3}$$

and then that

$$\mathbf{M}(k|N) = \mathbf{A}(k)\mathbf{A}(k+1)\ldots\mathbf{A}(N-1)\mathbf{K}(N) \tag{21-4}$$

and, finally, that other algorithms for $\hat{\mathbf{x}}(k|N)$ are

$$\hat{\mathbf{x}}(k|N) = \hat{\mathbf{x}}(k|N-1) + \left[\prod_{i=k}^{N-1}\mathbf{A}(i)\right][\hat{\mathbf{x}}(N|N) - \hat{\mathbf{x}}(N|N-1)] \tag{21-5}$$

and

$$\hat{\mathbf{x}}(k|N) = \hat{\mathbf{x}}(k|k) + \mathbf{A}(k)[\hat{\mathbf{x}}(k+1|N) - \hat{\mathbf{x}}(k+1|k)] \tag{21-6}$$

A detailed derivation of these results that uses a mathematical induction proof can be found in Meditch (1969, pp. 216–220). The proof relies, in part, on our previously derived results in Lesson 20, for the single- and double-stage smoothers. The reader, however, ought to be able to obtain (21-4), (21-5), and (21-6) directly from the rules of formation that can be inferred from the Lesson 20 formulas for $\mathbf{M}(k|k+1)$ and $\mathbf{M}(k|k+2)$ and $\hat{\mathbf{x}}(k|k+1)$ and $\hat{\mathbf{x}}(k|k+2)$.

To compute $\hat{\mathbf{x}}(k|N)$ for specific values of $k$, all the terms that appear on the right-hand side of an algorithm must be available. We have three possible algorithms for $\hat{\mathbf{x}}(k|N)$; however, as we explain next, only one is "useful."

The first value of $k$ for which the right-hand side of (21-2) is fully available is $k = N - 1$. By running a Kalman filter over all the measurements, we are able to compute $\hat{\mathbf{x}}(N-1|N-1)$ and $\tilde{\mathbf{z}}(N|N-1)$, so that we can compute $\hat{\mathbf{x}}(N-1|N)$. We can now try to iterate (21-2) in the backward direction to see if we can compute $\hat{\mathbf{x}}(N-2|N)$, $\hat{\mathbf{x}}(N-3|N)$, etc. Setting $k = N-2$ in (21-2), we obtain

$$\hat{\mathbf{x}}(N-2|N) = \hat{\mathbf{x}}(N-2|N-1) + \mathbf{M}(N-2|N)\tilde{\mathbf{z}}(N|N-1) \tag{21-7}$$

Unfortunately, $\hat{\mathbf{x}}(N-2|N-1)$, which is a single-stage smoothed estimate of $\mathbf{x}(N-2)$, has not been computed; hence, (21-7) is not useful for computing $\hat{\mathbf{x}}(N-2|N)$. *We, therefore, reject (21-2) as a useful fixed-interval smoother.*

A similar argument can be made against (21-5); thus, *we also reject (21-5) as a useful fixed-interval smoother.*

*Equation (21-6) is a useful fixed-interval smoother*, because all its terms on its right-hand side are available when we iterate it in the backward direction. For example,

$$\hat{\mathbf{x}}(N-1|N) = \hat{\mathbf{x}}(N-1|N-1) + \mathbf{A}(N-1)[\hat{\mathbf{x}}(N|N) - \hat{\mathbf{x}}(N|N-1)] \tag{21-8}$$

and

$$\hat{\mathbf{x}}(N-2|N) = \hat{\mathbf{x}}(N-2|N-2) + \mathbf{A}(N-2)[\hat{\mathbf{x}}(N-1|N) - \hat{\mathbf{x}}(N-1|N-2)] \tag{21-9}$$

Observe how $\hat{\mathbf{x}}(N-1|N)$ is used in the calculation of $\hat{\mathbf{x}}(N-2|N)$.

Equation (21-6) was developed by Rauch et al. (1965).

**Theorem 21-1.** *A useful mean-squared fixed-interval smoothed estimator of* $\mathbf{x}(\mathrm{k})$, $\hat{\mathbf{x}}(\mathrm{k|N})$, *is given by the expression*

$$\hat{\mathbf{x}}(k|N) = \hat{\mathbf{x}}(k|k) + \mathbf{A}(k)[\hat{\mathbf{x}}(k+1|N) - \hat{\mathbf{x}}(k+1|k)] \tag{21-10}$$

*where matrix* $\mathbf{A}(\mathrm{k})$ *is defined in (20-15), and* $\mathrm{k} = \mathrm{N}-1, \mathrm{N}-2, \ldots, 0$. *Additionally, the error-covariance matrix associated with* $\hat{\mathbf{x}}(\mathrm{k|N})$, $\mathbf{P}(\mathrm{k|N})$, *is given by*

$$\mathbf{P}(k|N) = \mathbf{P}(k|k) + \mathbf{A}(k)[\mathbf{P}(k+1|N) - \mathbf{P}(k+1|k)]\mathbf{A}'(k) \tag{21-11}$$

*where* $\mathrm{k} = \mathrm{N}-1, \mathrm{N}-2, \ldots, 0$.

*Proof.* We have already derived (21-10); hence, we direct our attention to the derivation of the algorithm for $\mathbf{P}(k|N)$. Our derivation of (21-11) follows the one given in Meditch, 1969, pp. 222–224. To begin, we know that

$$\tilde{\mathbf{x}}(k|N) = \mathbf{x}(k) - \hat{\mathbf{x}}(k|N) \tag{21-12}$$

Substitute (21-10) into (21-12) to see that

$$\tilde{\mathbf{x}}(k|N) = \tilde{\mathbf{x}}(k|k) - \mathbf{A}(k)[\hat{\mathbf{x}}(k+1|N) - \hat{\mathbf{x}}(k+1|k)] \tag{21-13}$$

Collecting terms conditioned on $N$ on the left-hand side and terms conditioned on $k$ on the right-hand side, (21-13) becomes

$$\tilde{\mathbf{x}}(k|N) + \mathbf{A}(k)\hat{\mathbf{x}}(k+1|N) = \tilde{\mathbf{x}}(k|k) + \mathbf{A}(k)\hat{\mathbf{x}}(k+1|k) \tag{21-14}$$

Treating (21-14) as an identity, we see that the covariance of its left-hand side must equal the covariance of its right-hand side; hence [in order to obtain (21-15) we use the orthogonality principle to eliminate all the cross-product terms],

$$\mathbf{P}(k|N) + \mathbf{A}(k)\mathbf{P}_{\hat{\mathbf{x}}\hat{\mathbf{x}}}(k+1|N)\mathbf{A}'(k) = \mathbf{P}(k|k) + \mathbf{A}(k)\mathbf{P}_{\hat{\mathbf{x}}\hat{\mathbf{x}}}(k+1|k)\mathbf{A}'(k) \tag{21-15}$$

or

$$\mathbf{P}(k|N) = \mathbf{P}(k|k) + \mathbf{A}(k)[\mathbf{P}_{\hat{\mathbf{x}}\hat{\mathbf{x}}}(k+1|k) - \mathbf{P}_{\hat{\mathbf{x}}\hat{\mathbf{x}}}(k+1|N)]\mathbf{A}'(k) \tag{21-16}$$

where $\mathbf{P}_{\hat{\mathbf{x}}\hat{\mathbf{x}}}(k+1|k) \triangleq \mathbf{E}\{\hat{\mathbf{x}}(k+1|k)\hat{\mathbf{x}}'(k+1|k)\}$. Note that $\mathbf{P}_{\hat{\mathbf{x}}\hat{\mathbf{x}}}(k+1|k)$ is *not* a covariance matrix, because $\mathbf{E}\{\hat{\mathbf{x}}(k+1|k)\} \neq \mathbf{0}$. We must now evaluate $\mathbf{P}_{\hat{\mathbf{x}}\hat{\mathbf{x}}}(k+1|k)$ and $\mathbf{P}_{\hat{\mathbf{x}}\hat{\mathbf{x}}}(k+1|N)$.

Recall that $\mathbf{x}(k+1) = \hat{\mathbf{x}}(k+1|k) + \tilde{\mathbf{x}}(k+1|k)$; thus,

$$\mathbf{P}_{\mathbf{x}}(k+1) = \mathbf{P}_{\hat{\mathbf{x}}\hat{\mathbf{x}}}(k+1|k) + \mathbf{P}(k+1|k) - \mathbf{m}_{\mathbf{x}}(k+1)\mathbf{m}'_{\mathbf{x}}(k+1) \tag{21-17}$$

Additionally, $\mathbf{x}(k+1) = \hat{\mathbf{x}}(k+1|N) + \tilde{\mathbf{x}}(k+1|N)$, so

$$\mathbf{P}_{\mathbf{x}}(k+1) = \mathbf{P}_{\hat{\mathbf{x}}\hat{\mathbf{x}}}(k+1|N) + \mathbf{P}(k+1|N) - \mathbf{m}_{\mathbf{x}}(k+1)\mathbf{m}'_{\mathbf{x}}(k+1) \tag{21-18}$$

Equating (21-17) and (21-18), we find that

$$\mathbf{P}_{\hat{\mathbf{x}}\hat{\mathbf{x}}}(k+1|k) - \mathbf{P}_{\hat{\mathbf{x}}\hat{\mathbf{x}}}(k+1|N) = \mathbf{P}(k+1|N) - \mathbf{P}(k+1|k) \tag{21-19}$$

Finally, substituting (21-19) into (21-16), we obtain the desired expression for $\mathbf{P}(k|N)$, (21-11). □

A proof of the fact that $\{\tilde{\mathbf{x}}(k|N), k = N, N-1, \ldots, 0\}$, is a zero-mean *second-order* Gauss–Markov sequence is given in the Supplementary Material at the end of this lesson.

**EXAMPLE 21-1**

To illustrate fixed-interval smoothing and obtain at the same time a comparison of the relative accuracies of smoothing and filtering, we return to Example 19-1. To review briefly, we have the scalar system $x(k+1) = x(k) + w(k)$ with the scalar measurement $z(k+1) = x(k+1) + v(k+1)$ and $p(0) = 50$, $q = 20$, and $r = 5$. In this example (which is similar to Example 6.1 in Meditch, 1969, p. 225), we choose $N = 4$ and compute quantities that are associated with $\hat{x}(k|4)$, where from (21-10)

$$\hat{x}(k|4) = \hat{x}(k|k) + A(k)[\hat{x}(k+1|4) - \hat{x}(k+1|k)] \tag{21-20}$$

$k = 3, 2, 1, 0$. Because $\Phi = 1$ and $p(k+1|k) = p(k|k) + 20$,

$$A(k) = p(k|k)\Phi' p^{-1}(k+1|k) = \frac{p(k|k)}{p(k|k)+20} \tag{21-21}$$

and, therefore,

$$p(k|4) = p(k|k) + \left[\frac{p(k|k)}{p(k|k)+20}\right]^2 [p(k+1|4) - p(k+1|k)] \tag{21-22}$$

Utilizing these last two expressions, we compute $A(k)$ and $p(k|4)$ for $k = 3, 2, 1, 0$

and present them, along with the results summarized in Table 19-1, in Table 21-1. The three estimation error variances are given in adjacent columns for ease in comparison.

**TABLE 21-1** KALMAN FILTER AND FIXED-INTERVAL SMOOTHER QUANTITIES

| $k$ | $p(k\|k-1)$ | $p(k\|k)$ | $p(k\|4)$ | $K(k)$ | $A(k)$ |
|---|---|---|---|---|---|
| 0 | ... | 50 | 16.31 | ... | 0.714 |
| 1 | 70 | 4.67 | 3.92 | 0.933 | 0.189 |
| 2 | 24.67 | 4.16 | 3.56 | 0.831 | 0.172 |
| 3 | 24.16 | 4.14 | 3.55 | 0.829 | 0.171 |
| 4 | 24.14 | 4.14 | 4.14 | 0.828 | 0.171 |

Observe the large improvement (percentage wise) of $p(k|4)$ over $p(k|k)$. Improvement seems to get larger the farther away we get from the end of our data; thus, $p(0|4)$ is more than three times as small as $p(0|0)$. Of course, it should be, because $p(0|0)$ is an *initial condition* that is data independent, whereas $p(0|4)$ is a result of processing $z(1), z(2), z(3)$, and $z(4)$. In essence, fixed-interval smoothing has let us look into the future and reflect the future back to time zero.

Finally, note that, for large values of $k$, $A(k)$ reaches a steady-state value, $\bar{A}$, where in this example

$$\bar{A} = \frac{\bar{p}_f}{\bar{p}_f + 20} = 0.171 \tag{21-23}$$

This steady-state value is achieved for $k = 3$. □

Equation (21-10) requires two multiplications of $n \times n$ matrices, as well as a matrix inversion at each iteration; hence, it is somewhat limited for practical computing purposes. The following results, which are due to Bryson and Frazier (1963) and Bierman (1973b), represent the most practical way for computing $\hat{\mathbf{x}}(k|N)$ and also $\mathbf{P}(k|N)$.

**Theorem 21-2.** (a) *A most useful mean-squared fixed-interval smoothed estimator of* $\mathbf{x}(k)$, $\hat{\mathbf{x}}(k|N)$, *is*

$$\hat{\mathbf{x}}(k|N) = \hat{\mathbf{x}}(k|k-1) + \mathbf{P}(k|k-1)\mathbf{r}(k|N) \tag{21-24}$$

*where* $k = N-1, N-2, \ldots, 1$, *and* $n \times 1$ *vector* $\mathbf{r}$ *satisfies the backward-recursive equation*

$$\mathbf{r}(j|N) = \boldsymbol{\Phi}'_p(j+1, j)\mathbf{r}(j+1|N) + \mathbf{H}'(j)[\mathbf{H}(j)\mathbf{P}(j|j-1)\mathbf{H}'(j) + \mathbf{R}(j)]^{-1}\tilde{\mathbf{z}}(j|j-1) \tag{21-25}$$

*where* $j = N, N-1, \ldots, 1$ *and* $\mathbf{r}(N+1|N) = \mathbf{0}$. *Matrix* $\boldsymbol{\Phi}_p$ *is defined in (21-33).*
(b) *The smoothing error-covariance matrix,* $\mathbf{P}(k|N)$, *is*

$$\mathbf{P}(k|N) = \mathbf{P}(k|k-1) - \mathbf{P}(k|k-1)\mathbf{S}(k|N)\mathbf{P}(k|k-1) \tag{21-26}$$

*where* $\mathrm{k = N-1, N-2, \ldots, 1}$, *and* $\mathrm{n \times n}$ *matrix* $\mathbf{S}(\mathrm{j|N})$, *which is the covariance matrix of* $\mathbf{r}(\mathrm{j|N})$, *satisfies the backward-recursive equation*

$$\mathbf{S}(j|N) = \mathbf{\Phi}'_p(j+1, j)\mathbf{S}(j+1|N)\mathbf{\Phi}_p(j+1, j) + \mathbf{H}'(j)[\mathbf{H}(j)\mathbf{P}(j|j-1)\mathbf{H}'(j) + \mathbf{R}(j)]^{-1}\mathbf{H}(j) \tag{21-27}$$

*where* $\mathrm{j = N, N-1, \ldots, 1}$ *and* $\mathbf{S}(\mathrm{N+1|N}) = \mathbf{0}$.

*Proof* (Mendel, 1983, pp. 64–65). (a) Substitute the Kalman filter equation (17-11) for $\hat{\mathbf{x}}(k|k)$ into Equation (21-10), to show that

$$\hat{\mathbf{x}}(k|N) = \hat{\mathbf{x}}(k|k-1) + \mathbf{K}(k)\tilde{\mathbf{z}}(k|k-1) + \mathbf{A}(k)[\hat{\mathbf{x}}(k+1|N) - \hat{\mathbf{x}}(k+1|k)] \tag{21-28}$$

*Residual state vector* $\mathbf{r}(k|N)$ is defined as

$$\mathbf{r}(k|N) = \mathbf{P}^{-1}(k|k-1)[\hat{\mathbf{x}}(k|N) - \hat{\mathbf{x}}(k|k-1)] \tag{21-29}$$

Next, substitute $\mathbf{r}(k|N)$ and $\mathbf{r}(k+1|N)$, using (21-29), into (21-28) to show that

$$\mathbf{r}(k|N) = \mathbf{P}^{-1}(k|k-1)[\mathbf{K}(k)\tilde{\mathbf{z}}(k|k-1) + \mathbf{P}(k|k)\mathbf{\Phi}'(k+1, k)\mathbf{r}(k+1|N)] \tag{21-30}$$

From (17-12) and (17-14) and the symmetry of covariance matrices, we find that

$$\mathbf{P}^{-1}(k|k-1)\mathbf{K}(k) = \mathbf{H}'(k)[\mathbf{H}(k)\mathbf{P}(k|k-1)\mathbf{H}'(k) + \mathbf{R}(k)]^{-1} \tag{21-31}$$

and

$$\mathbf{P}^{-1}(k|k-1)\mathbf{P}(k|k) = [\mathbf{I} - \mathbf{K}(k)\mathbf{H}(k)]' \tag{21-32}$$

Substituting (21-31) and (21-32) into Equation (21-30) and defining

$$\mathbf{\Phi}_p(k+1, k) = \mathbf{\Phi}(k+1, k)[\mathbf{I} - \mathbf{K}(k)\mathbf{H}(k)] \tag{21-33}$$

we obtain Equation (21-25). Setting $k = N+1$ in (21-29), we establish $\mathbf{r}(N+1|N) = \mathbf{0}$. Finally, solving (21-29) for $\hat{\mathbf{x}}(k|N)$, we obtain Equation (21-24).
(b) The orthogonality principle in Corollary 13-3 leads us to conclude that

$$\mathbf{E}\{\tilde{\mathbf{x}}(k|N)\mathbf{r}'(k|N)\} = \mathbf{0} \tag{21-34}$$

because $\mathbf{r}(k|N)$ is simply a linear combination of all the observations $\mathbf{z}(1), \mathbf{z}(2), \ldots, \mathbf{z}(N)$. From (21-24) we find that

$$\tilde{\mathbf{x}}(k|k-1) = \tilde{\mathbf{x}}(k|N) - \mathbf{P}(k|k-1)\mathbf{r}(k|N) \tag{21-35}$$

and, therefore, using (21-34), we find that

$$\mathbf{P}(k|k-1) = \mathbf{P}(k|N) + \mathbf{P}(k|k-1)\mathbf{S}(k|N)\mathbf{P}(k|k-1) \tag{21-36}$$

where

$$\mathbf{S}(k|N) = \mathbf{E}\{\mathbf{r}(k|N)\mathbf{r}'(k|N)\} \tag{21-37}$$

is the covariance-matrix of $\mathbf{r}(k|N)$ [note that $\mathbf{r}(k|N)$ is zero mean]. Equation (21-36) is solved for $\mathbf{P}(k|N)$ to give the desired result in Equation (21-26).

Because the innovations process is uncorrelated, (21-27) follows from substitution of (21-25) into (21-37). Finally, $\mathbf{S}(N+1|N) = \mathbf{0}$ because $\mathbf{r}(N+1|N) = \mathbf{0}$. $\square$

Equations (21-24) and (21-25) are very efficient; they require no matrix inversions or multiplications of $n \times n$ matrices. The calculation of $\mathbf{P}(k|N)$ does require two multiplications of $n \times n$ matrices.

Matrix $\mathbf{\Phi}_p(k+1, k)$ in (21-33) is the plant matrix of the recursive predictor (Lesson 16). It is interesting that the recursive predictor and not the recursive filter plays the predominant role in fixed-interval smoothing. This is further borne out by the appearance of predictor quantities on the right-hand side of (21-24). Observe that (21-25) looks quite similar to a recursive predictor [see (17-23)] that is excited by the innovations—one that is running in a backward direction.

Finally, note that (21-24) can also be used for $k = N$, in which case its right-hand side reduces to $\hat{\mathbf{x}}(N|N-1) + \mathbf{K}(N)\tilde{\mathbf{z}}(N|N-1)$, which, of course, is $\hat{\mathbf{x}}(N|N)$

For an application of the results in Theorem 21-2 to deconvolution, see the Supplementary Material at the end of this lesson.

## FIXED-POINT SMOOTHING

A fixed-point smoother, $\hat{\mathbf{x}}(k|j)$, where $j = k+1, k+2, \ldots$, can be obtained in exactly the same manner as we obtained fixed-interval smoother (21-5). It is obtained from this equation by setting $N$ equal to $j$ and then letting $j = k+1, k+2, \ldots$; thus,

$$\hat{\mathbf{x}}(k|j) = \hat{\mathbf{x}}(k|j-1) + \mathbf{B}(j)[\hat{\mathbf{x}}(j|j) - \hat{\mathbf{x}}(j|j-1)] \tag{21-38}$$

where

$$\mathbf{B}(j) = \prod_{i=k}^{j-1} \mathbf{A}(i) \tag{21-39}$$

and $j = k+1, k+2, \ldots$. Additionally, we can show that the fixed-point smoothing error-covariance matrix, $\mathbf{P}(k|j)$, is computed from

$$\mathbf{P}(k|j) = \mathbf{P}(k|j-1) + \mathbf{B}(j)[\mathbf{P}(j|j) - \mathbf{P}(j|j-1)]\mathbf{B}'(j) \tag{21-40}$$

where $j = k+1, k+2, \ldots$.

Equation (21-38) is impractical from a computational viewpoint, because of the many multiplications of $n \times n$ matrices required first to form the $\mathbf{A}(i)$ matrices and then to form the $\mathbf{B}(j)$ matrices. Additionally, the inverse of matrix $\mathbf{P}(i+1|i)$ is needed in order to compute matrix $\mathbf{A}(i)$. The following results present a "fast" algorithm for computing $\hat{\mathbf{x}}(k|j)$. It is fast in the sense that no multiplications of $n \times n$ matrices are needed to implement it.

**Theorem 21-3.** *A most useful mean-squared fixed-point smoothed estimator of* $\mathbf{x}(k)$, $\hat{\mathbf{x}}(k|k+l)$, *where* $l = 1, 2, \ldots$, *is given by the expression*

$$\hat{\mathbf{x}}(k|k+l) = \hat{\mathbf{x}}(k|k+l-1) + \mathbf{N}_{\mathbf{x}}(k|k+l)[\mathbf{z}(k+l) - \mathbf{H}(k+l)\hat{\mathbf{x}}(k+l|k+l-1)] \tag{21-41}$$

*where*

$$\mathbf{N}_{\mathbf{x}}(k|k+l) = \mathbf{D}_{\mathbf{x}}(k,l)\mathbf{H}'(k+l)[\mathbf{H}(k+l)\mathbf{P}(k+l|k+l-1)\mathbf{H}'(k+l) + \mathbf{R}(k+l)]^{-1} \tag{21-42}$$

*and*

$$\mathbf{D}_{\mathbf{x}}(k,l) = \mathbf{D}_{\mathbf{x}}(k,l-1)[\mathbf{I} - \mathbf{K}(k+l-1)\mathbf{H}(k+l-1)]'\mathbf{\Phi}'(k+l,k+l-1) \tag{21-43}$$

*Equations (21-41) and (21-43) are initialized by* $\hat{\mathbf{x}}(k|k)$ *and* $\mathbf{D}_{\mathbf{x}}(k,1) = \mathbf{P}(k|k)\mathbf{\Phi}'(k+1,k)$, *respectively. Additionally,*

$$\mathbf{P}(k|k+l) = \mathbf{P}(k|k+l-1) - \mathbf{N}_{\mathbf{x}}(k|k+l)[\mathbf{H}(k+l)\mathbf{P}(k+l|k+l-1)\mathbf{H}'(k+l) + \mathbf{R}(k+l)]\mathbf{N}_{\mathbf{x}}'(k|k+l) \tag{21-44}$$

*which is initialized by* $\mathbf{P}(k|k)$. □

We leave the proof of this useful theorem, which is similar to a result given by Fraser (1967), as an exercise for the reader.

**EXAMPLE 21-2**

Here we consider the problem of fixed-point smoothing to obtain a refined estimate of the initial condition for the system described in Example 21-1. Recall that $p(0|0) = 50$ and that by fixed-interval smoothing we had obtained the result $p(0|4) = 16.31$, which is a significant reduction in the uncertainty associated with the initial condition.

Using Equation (21-40) or (21-44), we compute $p(0|1)$, $p(0|2)$, and $p(0|3)$ to be 16.69, 16.32, and 16.31, respectively. Observe that a major reduction in the smoothing error variance occurs as soon as the first measurement is incorporated and that the improvement in accuracy thereafter is relatively modest. This seems to be a general trait of fixed-point smoothing. □

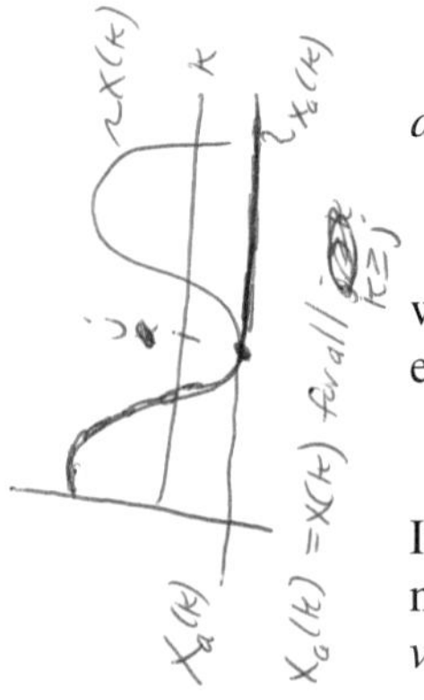

Another way to derive fixed-point smoothing formulas is by the following *state augmentation procedure* (Anderson and Moore, 1979). We assume that for $k \geq j$

$$\mathbf{x}_a(k) \triangleq \mathbf{x}(j) \tag{21-45}$$

which means that $\mathbf{x}(k)$ is held constant at the value $\mathbf{x}(j)$ for all $k \geq j$. The state equation for state vector $\mathbf{x}_a(k)$ is

$$\mathbf{x}_a(k+1) = \mathbf{x}_a(k), \qquad k \geq j \tag{21-46}$$

It is initialized at $k = j$ by (21-45). Augmenting (21-46) to our basic state-variable model in (15-17) and (15-18), we obtain the following *augmented basic state-variable model*:

$$
\begin{pmatrix} \mathbf{x}(k+1) \\ \mathbf{x}_a(k+1) \end{pmatrix} = \begin{pmatrix} \mathbf{\Phi}(k+1,k) & \mathbf{0} \\ \mathbf{0} & \mathbf{I} \end{pmatrix} \begin{pmatrix} \mathbf{x}(k) \\ \mathbf{x}_a(k) \end{pmatrix}
$$

$$
+ \begin{pmatrix} \mathbf{\Psi}(k+1,k) \\ \mathbf{0} \end{pmatrix} \mathbf{u}(k) + \begin{pmatrix} \mathbf{\Gamma}(k+1,k) \\ \mathbf{0} \end{pmatrix} \mathbf{w}(k) \qquad (21\text{-}47)
$$

and

$$
\mathbf{z}(k+1) = (\mathbf{H}(k+1)\,\mathbf{0}) \begin{pmatrix} \mathbf{x}(k) \\ \mathbf{x}_a(k) \end{pmatrix} + \mathbf{v}(k+1) \qquad (21\text{-}48)
$$

The following two-step procedure can be used to obtain an algorithm for $\hat{\mathbf{x}}(j|k)$:

1. Write down the Kalman filter equations for the augmented basic state-variable model. Anderson and Moore (1979) give these equations for the recursive predictor; i.e., they find

$$
\mathbf{E}\left\{ \begin{pmatrix} \mathbf{x}(k+1) \\ \mathbf{x}_a(k+1) \end{pmatrix} \middle| \mathbf{Z}(k) \right\} = \begin{pmatrix} \hat{\mathbf{x}}(k+1|k) \\ \hat{\mathbf{x}}_a(k+1|k) \end{pmatrix} = \begin{pmatrix} \hat{\mathbf{x}}(k+1|k) \\ \hat{\mathbf{x}}(j|k) \end{pmatrix} \qquad (21\text{-}49)
$$

   where $k \geq j$. The last equality in (21-49) makes use of (21-46) and (21-45), i.e., $\hat{\mathbf{x}}_a(k+1|k) = \hat{\mathbf{x}}_a(k|k) = \hat{\mathbf{x}}(j|k)$. Observe that $\hat{\mathbf{x}}(j|k)$, the fixed-point smoother of $\mathbf{x}(j)$, has been found as the second component of the recursive predictor for the augmented model.

2. The Kalman filter (or recursive predictor) equations are partitioned in order to obtain the explicit structure of the algorithm for $\hat{\mathbf{x}}(j|k)$. We leave the details of this two-step procedure as an exercise for the reader.

## FIXED-LAG SMOOTHING

The earliest attempts to obtain a fixed-lag smoother $\hat{\mathbf{x}}(k|k+L)$ led to an algorithm (e.g., Meditch, 1969) that was later shown to be unstable (Kelly and Anderson, 1971). The following state augmentation procedure leads to a stable fixed-interval smoother for $\hat{\mathbf{x}}(k-L|k)$

We introduce $L+1$ state vectors, as follows: $\mathbf{x}_1(k+1) = \mathbf{x}(k)$, $\mathbf{x}_2(k+1) = \mathbf{x}(k-1)$, $\mathbf{x}_3(k+1) = \mathbf{x}(k-2), \ldots, \mathbf{x}_{L+1}(k+1) = \mathbf{x}(k-L)$ [i.e., $\mathbf{x}_i(k+1) = \mathbf{x}(k+1-i)$, $i = 1, 2, \ldots, L+1$]. The state equations for these $L+1$ state vectors are

$$
\left.\begin{aligned}
\mathbf{x}_1(k+1) &= \mathbf{x}(k) \\
\mathbf{x}_2(k+1) &= \mathbf{x}_1(k) \\
\mathbf{x}_3(k+1) &= \mathbf{x}_2(k) \\
&\cdots \\
\mathbf{x}_{L+1}(k+1) &= \mathbf{x}_L(k)
\end{aligned}\right\} \qquad (21\text{-}50)
$$

Augmenting (21-50) to our basic state-variable model in (15-17) and (15-18), we obtain yet another *augmented basic state-variable model*:

$$
\begin{pmatrix} \mathbf{x}(k+1) \\ \mathbf{x}_1(k+1) \\ \mathbf{x}_2(k+1) \\ \vdots \\ \mathbf{x}_{L+1}(k+1) \end{pmatrix} = \begin{pmatrix} \mathbf{\Phi}(k+1,k) & \mathbf{0} & \cdots & \mathbf{0} & \mathbf{0} \\ \mathbf{I} & \mathbf{0} & \cdots & \mathbf{0} & \mathbf{0} \\ \mathbf{0} & \mathbf{I} & \cdots & \mathbf{0} & \mathbf{0} \\ \vdots & \vdots & \ddots & \vdots & \vdots \\ \mathbf{0} & \mathbf{0} & \cdots & \mathbf{I} & \mathbf{0} \end{pmatrix} \begin{pmatrix} \mathbf{x}(k) \\ \mathbf{x}_1(k) \\ \mathbf{x}_2(k) \\ \vdots \\ \mathbf{x}_{L+1}(k) \end{pmatrix}
+ \begin{pmatrix} \mathbf{\Psi}(k+1,k) \\ \mathbf{0} \\ \mathbf{0} \\ \vdots \\ \mathbf{0} \end{pmatrix} \mathbf{u}(k) + \begin{pmatrix} \mathbf{\Gamma}(k+1,k) \\ \mathbf{0} \\ \mathbf{0} \\ \vdots \\ \mathbf{0} \end{pmatrix} \mathbf{w}(k) \quad (21\text{-}51)
$$

Using (21-51), it is easy to obtain the following Kalman filter:

$$
\begin{bmatrix} \hat{\mathbf{x}}(k+1|k+1) \\ \hat{\mathbf{x}}_1(k+1|k+1) \\ \hat{\mathbf{x}}_2(k+1|k+1) \\ \vdots \\ \hat{\mathbf{x}}_{L+1}(k+1|k+1) \end{bmatrix} = \begin{bmatrix} \mathbf{\Phi}(k+1,k) & \mathbf{0} & \cdots & \mathbf{0} & \mathbf{0} \\ \mathbf{I} & \mathbf{0} & \cdots & \mathbf{0} & \mathbf{0} \\ \mathbf{0} & \mathbf{I} & \cdots & \mathbf{0} & \mathbf{0} \\ \vdots & \vdots & \ddots & \vdots & \vdots \\ \mathbf{0} & \mathbf{0} & \cdots & \mathbf{I} & \mathbf{0} \end{bmatrix} \begin{bmatrix} \hat{\mathbf{x}}(k|k) \\ \hat{\mathbf{x}}_1(k|k) \\ \hat{\mathbf{x}}_2(k|k) \\ \vdots \\ \hat{\mathbf{x}}_{L+1}(k|k) \end{bmatrix}
+ \begin{bmatrix} \mathbf{K}_0(k+1) \\ \mathbf{K}_1(k+1) \\ \mathbf{K}_2(k+1) \\ \vdots \\ \mathbf{K}_{L+1}(k+1) \end{bmatrix} \tilde{\mathbf{z}}(k+1|k) \quad (21\text{-}52)
$$

where, for the sake of simplicity, we have assumed that $\mathbf{u}(k) = \mathbf{0}$. Using the facts that $\mathbf{x}_1(k+1) = \mathbf{x}(k)$, $\mathbf{x}_2(k+1) = \mathbf{x}(k-1), \ldots, \mathbf{x}_{L+1}(k+1) = \mathbf{x}(k-L)$, we see that

$$
\begin{aligned}
\hat{\mathbf{x}}_1(k+1|k+1) &= \hat{\mathbf{x}}(k|k+1) \\
\hat{\mathbf{x}}_2(k+1|k+1) &= \hat{\mathbf{x}}(k-1|k+1) \\
&\cdots \\
\hat{\mathbf{x}}_{L+1}(k+1|k+1) &= \hat{\mathbf{x}}(k-L|k+1) \qquad (21\text{-}53)
\end{aligned}
$$

Consequently, (21-52) can be written out as

$$
\begin{aligned}
\hat{\mathbf{x}}(k+1|k+1) &= \mathbf{\Phi}(k+1,k)\hat{\mathbf{x}}(k|k) + \mathbf{K}_0(k+1)\tilde{\mathbf{z}}(k+1|k) \\
\hat{\mathbf{x}}(k|k+1) &= \hat{\mathbf{x}}(k|k) + \mathbf{K}_1(k+1)\tilde{\mathbf{z}}(k+1|k) \\
\hat{\mathbf{x}}(k-1|k+1) &= \hat{\mathbf{x}}(k-1|k) + \mathbf{K}_2(k+1)\tilde{\mathbf{z}}(k+1|k) \\
&\cdots \\
\hat{\mathbf{x}}(k-L|k+1) &= \hat{\mathbf{x}}(k-L+1|k) + \mathbf{K}_{L+1}(k+1)\tilde{\mathbf{z}}(k+1|k) \qquad (21\text{-}54)
\end{aligned}
$$

Values for $\mathbf{K}_0, \mathbf{K}_1, \ldots, \mathbf{K}_{L+1}$ are obtained by partitioning the Kalman gain matrix equation (17-12). The block diagram given in Figure 21-1 demonstrates that this implementation of the fixed-lag smoother introduces no new feedback loops other than the one that appears in the Kalman filter; hence, the eigenvalues of this smoother are those of the Kalman filter. Because the Kalman filter is stable [this is not only true for the steady-state Kalman filter, but is also true for the non-steady-state Kalman filter; see Kalman (1960) for a proof of this that uses Lyapunov stability theory], this fixed-lag smoother is also stable.

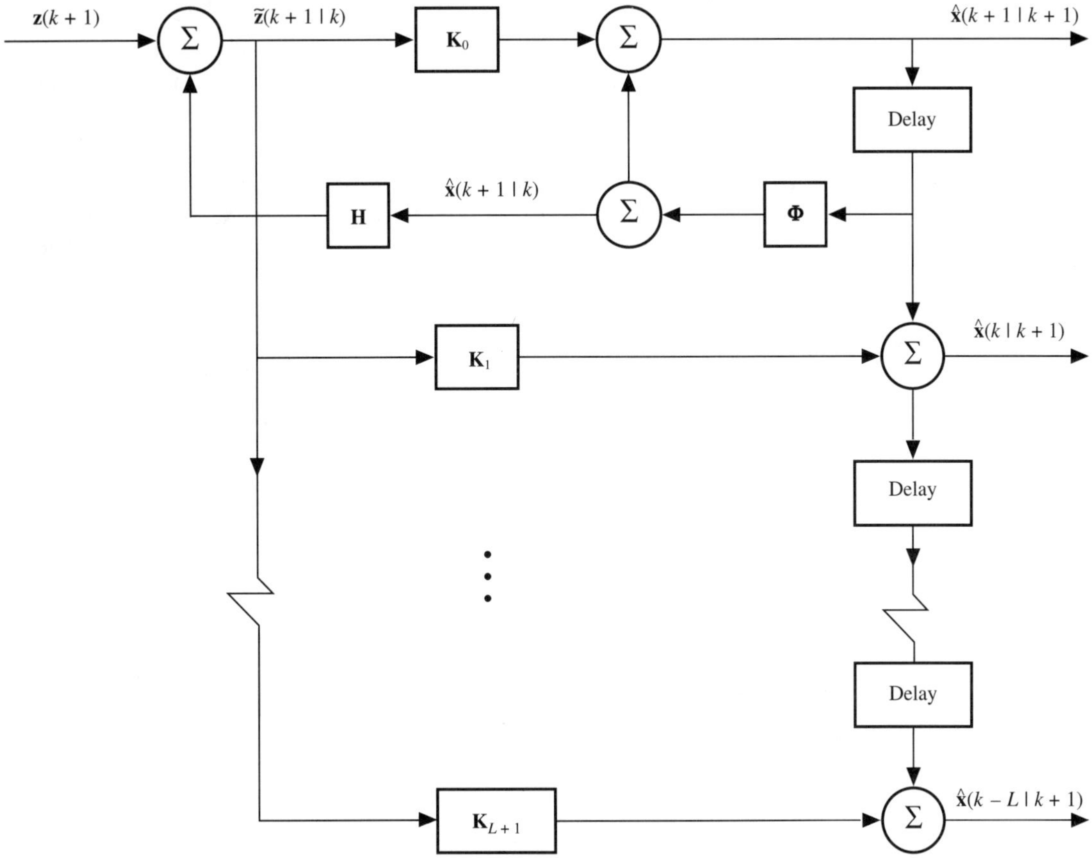


**Figure 21-1** Fixed-lag smoother block diagram.

Some additional aspects of this fixed-lag smoother are:

1. To compute $\hat{\mathbf{x}}(k-L|k)$, we must also compute the $L-1$ fixed-lag estimates, $\hat{\mathbf{x}}(k-1|k), \hat{\mathbf{x}}(k-2|k), \ldots, \hat{\mathbf{x}}(k-L+1|k)$; this may be costly to do from a computational point of view.
2. Computation can be reduced by careful coding of the partitioned recursive predictor equations.

## COMPUTATION

No M-file exists in any toolbox for fixed-interval smoothers; hence, we have prepared the following M-file to do it:

> **fis**: obtains the mean-squared fixed-interval smoothed estimate of the state vector of our basic state-variable model. This model can be time varying or nonstationary.

A complete listing of this M-file can be found in Appendix B.

In this lesson we have also shown how both fixed-point and fixed-lag smoothers can be obtained from a Kalman filter by using state augmentation techniques. The augmented basic state-variable model for fixed-point smoothing is given in (21-47), whereas the basic state-variable model for fixed-lag smoothing is given in (21-51). We can use the M-file **kp** described in Lesson 17 and listed in Appendix B to compute fixed-point and fixed-lag smoothers using the respective augmented models, or we can use existing M-files in the *Control System Toolbox* to generate steady-state fixed-point or fixed-lag smoothers, again using the respective augmented models. See Lesson 19 for discussions on the latter.

### Supplementary Material

## SECOND-ORDER GAUSS–MARKOV RANDOM SEQUENCES

Suppose we have a stochastic sequence $\{\mathbf{x}(k), k = 0, 1, \ldots\}$ that is defined by the second-order linear vector difference equation

$$\mathbf{x}(k+2) = \mathbf{\Phi}(k+2, k+1)\mathbf{x}(k+1) + \mathbf{\Lambda}(k+2, k)\mathbf{x}(k) + \mathbf{\Gamma}(k+2, k)\mathbf{w}(k) \tag{21-55}$$

where $\mathbf{x}(0)$ and $\mathbf{x}(1)$ are jointly Gaussian and $\{\mathbf{w}(k), k = 0, 1, \ldots\}$ is a Gaussian white sequence that is independent of $\mathbf{x}(0)$ and $\mathbf{x}(1)$.

Because all sources of uncertainty are Gaussian and (21-55) is linear, $\mathbf{x}(k)$ must be Gaussian. Additionally, because (21-55) is second order, the conditional probability density function of $\mathbf{x}$ at any time point, given the values of $\mathbf{x}$ at an arbitrary number of previous time points, depends only on its two previous values; hence, $\mathbf{x}(k)$ is a second-order Markov sequence. Note, also, that to go from input $\mathbf{w}(k)$ to the state at time $k+2$ requires two delays.

Another situation that also leads to a second-order Gauss–Markov sequence is one in which our system is described by the two coupled first-order difference equations:

$$\mathbf{x}(k+1) = \mathbf{\Phi}_{\mathbf{x}}(k+1, k)\mathbf{x}(k) + \mathbf{\Gamma}(k+1, k)\mathbf{w}(k) \tag{21-56}$$

$$\mathbf{w}(k+1) = \mathbf{\Phi}_{\mathbf{w}}(k+1, k)\mathbf{w}(k) + \mathbf{\Delta}(k+1, k)\boldsymbol{\xi}(k) \tag{21-57}$$

where, again, all sources of uncertainty are jointly Gaussian. In this case, $\mathbf{w}(k)$ is *colored noise*, a case that is covered more fully in Lesson 22. There are two delays in this system between the input $\xi(k)$ and "output" $\mathbf{x}(k+1)$; hence, the system is second order and is therefore Markov 2.

The state augmentation procedure that is described in Lesson 22 permits us to treat a Markov 2 sequence as an augmented Markov 1 sequence. For additional details, see Meditch (1969).

Next we explain why $\{\tilde{\mathbf{x}}(k|N), k = N, N-1, \ldots, 0\}$, is a second-order Markov sequence. We could present a number of detailed analyses to prove that $\tilde{\mathbf{x}}(k+1|N)$ can be expressed in terms of $\tilde{\mathbf{x}}(k|N)$ and $\tilde{\mathbf{x}}(k-1|N)$; but this is tedious and not very enlightening. Instead, we give a heuristic explanation.

First, we see from (21-24) that $\tilde{\mathbf{x}}(k|N)$ depends on both $\hat{\mathbf{x}}(k|k-1)$ and $\mathbf{r}(k|N)$. The recursive predictor, which is given in (17-23), is a first-order Markov sequence, because $\hat{\mathbf{x}}(k+1|k)$ depends only on $\hat{\mathbf{x}}(k|k-1)$. The recursive equation for residual state vector $\mathbf{r}(k|N)$, given in (21-25), is a first-order backward-Markovian sequence, because $\mathbf{r}(j|N)$ depends only on $\mathbf{r}(j+1|N)$. Finally, as we just noted above, the interrelationship between two first-order coupled Markov sequences is that of a second-order Markov sequence. Note that, in this case, the coupling occurs in the equation for $\hat{\mathbf{x}}(k|N)$, which plays the role of an output equation for the $\hat{\mathbf{x}}(k|k-1)$ and $\mathbf{r}(k|N)$ systems.

## MINIMUM-VARIANCE DECONVOLUTION (MVD)

Here, as in Example 2-6 and 13-1, we begin with the convolutional model

$$z(k) = \sum_{i=1}^{k} \mu(i)h(k-i) + \nu(k), \qquad k = 1, 2, \ldots, N \tag{21-58}$$

Recall that deconvolution is the signal-processing procedure for removing the effects of $h(j)$ and $\nu(j)$ from the measurements so that we are left with an estimate of $\mu(j)$. Here we shall obtain a very useful algorithm for a mean-squared fixed-interval estimator of $\mu(j)$.

To begin, we must convert (21-58) into an equivalent state-variable model.

**Theorem 21-4** (Mendel, 1983, pp. 13–14). *The single-channel state-variable model*

$$\mathbf{x}(k+1) = \mathbf{\Phi}\mathbf{x}(k) + \boldsymbol{\gamma}\mu(k) \tag{21-59}$$

$$z(k) = \mathbf{h}'\mathbf{x}(k) + \nu(k) \tag{21-60}$$

*is equivalent to the convolutional sum model in (21-58) when* $\mathbf{x}(0) = \mathbf{0}$, $\mu(0) = 0$, $h(0) = 0$, *and*

$$h(l) = \mathbf{h}'\mathbf{\Phi}^{l-1}\boldsymbol{\gamma}, \qquad l = 1, 2, \ldots \tag{21-61}$$

*Proof.* Iterate (21-59) (see Equation D-49) and substitute the results into (21-60). Compare the resulting equation with (21-58) to see that, under the conditions $\mathbf{x}(0) = \mathbf{0}$, $\mu(0) = 0$, and $h(0) = 0$, they are the same. □

The condition $\mathbf{x}(0) = \mathbf{0}$ merely initializes our state-variable model. The condition $\mu(0) = 0$ means there is no input at time zero. The coefficients in (21-61) represent sampled values of the impulse response. If we are given impulse response data $\{h(1), h(2), \ldots, h(L)\}$, then we can determine matrices $\mathbf{\Phi}$, $\boldsymbol{\gamma}$, and $\mathbf{h}$ as well as system order $n$ by applying an approximate realization procedure, such as Kung's (1978), to $\{h(1), h(2), \ldots, h(L)\}$. Additionally, if $h(0) \neq 0$, it is simple to modify Theorem 21-4.

In Example 13-1 we obtained a rather unwieldy formula for $\hat{\boldsymbol{\mu}}_{MS}(N)$. Note that, in terms of our conditioning notation, the elements of $\hat{\boldsymbol{\mu}}_{MS}(N)$ are $\hat{\mu}_{MS}(k|N), k = 1, 2, \ldots, N$. We now obtain a very useful algorithm for $\hat{\mu}_{MS}(k|N)$. For notational convenience, we shorten $\hat{\mu}_{MS}$ to $\hat{\mu}$.

**Theorem 21-5** (Mendel, 1983, pp. 68–70)

**a.** *A two-pass fixed-interval smoother for* $\mu(k)$ *is*

$$\hat{\mu}(k|N) = q(k)\boldsymbol{\gamma}'\mathbf{r}(k+1|N) \tag{21-62}$$

*where* $k = N-1, N-2, \ldots, 1$.

**b.** *The smoothing error variance,* $\sigma_{\mu}^2(k|N)$, *is*

$$\sigma_{\mu}^2(k|N) = q(k) - q(k)\boldsymbol{\gamma}'\mathbf{S}(k+1|N)\boldsymbol{\gamma}q(k) \tag{21-63}$$

*where* $k = N-1, N-2, \ldots, 1$. *In these formulas* $\mathbf{r}(k|N)$ *and* $\mathbf{S}(k|N)$ *are computed using (21-25) and (21-27), respectively, and* $\mathbf{E}\{\mu^2(k)\} = q(k)$ [here $q(k)$ denotes the variance of $\mu(k)$ and should not be confused with the event sequence, which appears in the product model for $\mu(k)$].

*Proof*

**a.** To begin, we apply the fundamental theorem of estimation theory, Theorem 13-1, to (21-59). We operate on both sides of that equation with $\mathbf{E}\{\cdot|\mathbf{Z}(N)\}$ to show that

$$\boldsymbol{\gamma}\hat{\mu}(k|N) = \hat{\mathbf{x}}(k+1|N) - \mathbf{\Phi}\hat{\mathbf{x}}(k|N) \tag{21-64}$$

By performing appropriate manipulations on this equation, we can derive (21-62) as follows. Substitute $\hat{\mathbf{x}}(k|N)$ and $\hat{\mathbf{x}}(k+1|N)$ from Equation (21-24) into Equation (21-64), to see that

$$\begin{aligned}\boldsymbol{\gamma}\hat{\mu}(k|N) &= \hat{\mathbf{x}}(k+1|k) + \mathbf{P}(k+1|k)\mathbf{r}(k+1|N) \\ &\quad - \mathbf{\Phi}[\hat{\mathbf{x}}(k|k-1) + \mathbf{P}(k|k-1)\mathbf{r}(k|N)] \\ &= \hat{\mathbf{x}}(k+1|k) - \mathbf{\Phi}\hat{\mathbf{x}}(k|k-1) \\ &\quad + \mathbf{P}(k+1|k)\mathbf{r}(k+1|N) - \mathbf{\Phi}\mathbf{P}(k|k-1)\mathbf{r}(k|N)\end{aligned} \tag{21-65}$$

Applying (17-11) and (16-4) to the state-variable model in (21-59) and (21-60), it is straightforward to show that

$$\hat{\mathbf{x}}(k+1|k) = \mathbf{\Phi}\hat{\mathbf{x}}(k|k-1) + \mathbf{\Phi}\mathbf{K}(k)\tilde{z}(k|k-1) \tag{21-66}$$

Hence, (21-65) reduces to

$$\boldsymbol{\gamma}\hat{\mu}(k|N) = \boldsymbol{\Phi}\mathbf{K}(k)\tilde{z}(k|k-1) + \mathbf{P}(k+1|k)\mathbf{r}(k+1|N) - \boldsymbol{\Phi}\mathbf{P}(k|k-1)\mathbf{r}(k|N) \tag{21-67}$$

Next, substitute (21-25) into (21-67), to show that

$$\begin{aligned}\boldsymbol{\gamma}\hat{\mu}(k|N) = {} & \boldsymbol{\Phi}\mathbf{K}(k)\tilde{z}(k|k-1) + \mathbf{P}(k+1|k)\mathbf{r}(k+1|N) \\ & - \boldsymbol{\Phi}\mathbf{P}(k|k-1)\boldsymbol{\Phi}'_p(k+1,k)\mathbf{r}(k+1|N) \\ & - \boldsymbol{\Phi}\mathbf{P}(k|k-1)\mathbf{h}'[\mathbf{h}'\mathbf{P}(k|k-1)\mathbf{h} + r]^{-1}\tilde{z}(k|k-1)\end{aligned} \tag{21-68}$$

Making use of Equation (17-12) for $\mathbf{K}(k)$, we find that the first and last terms in Equation (21-68) are identical; hence,

$$\boldsymbol{\gamma}\hat{\mu}(k|N) = \mathbf{P}(k+1|k)\mathbf{r}(k+1|N) - \boldsymbol{\Phi}\mathbf{P}(k|k-1) \boldsymbol{\Phi}'_p(k+1,k)\mathbf{r}(k+1|N) \tag{21-69}$$

Combine Equations (17-13) and (17-14) to see that

$$\mathbf{P}(k+1|k) = \boldsymbol{\Phi}\mathbf{P}(k|k-1)\boldsymbol{\Phi}'_p(k+1,k) + \boldsymbol{\gamma}q(k)\boldsymbol{\gamma}' \tag{21-70}$$

Finally, substitute (21-70) into Equation (21-69) to observe that

$$\boldsymbol{\gamma}\hat{\mu}(k|N) = \boldsymbol{\gamma}q(k)\boldsymbol{\gamma}'\mathbf{r}(k+1|N)$$

which has the unique solution given by

$$\hat{\mu}(k|N) = q(k)\boldsymbol{\gamma}'\mathbf{r}(k+1|N) \tag{21-71}$$

which is Equation (21-62).

**b.** To derive Equation (21-63), we use (21-62) and the definition of estimation error $\tilde{\mu}(k|N)$,

$$\tilde{\mu}(k|N) = \mu(k) - \hat{\mu}(k|N) \tag{21-72}$$

to form

$$\mu(k) = \tilde{\mu}(k|N) + q(k)\boldsymbol{\gamma}'\mathbf{r}(k+1|N) \tag{21-73}$$

Taking the variance of both sides of (21-73) and using the orthogonality condition

$$\mathbf{E}\{\tilde{\mu}(k|N)\mathbf{r}(k+1|N)\} = \mathbf{0} \tag{21-74}$$

we see that

$$q(k) = \sigma^2_{\tilde{\mu}}(k|N) + q(k)\boldsymbol{\gamma}'\mathbf{S}(k+1|N)\boldsymbol{\gamma}q(k) \tag{21-75}$$

which is equivalent to (21-63). □

Observe, from (21-62) and (21-63), that $\hat{\mu}(k|N)$ and $\sigma^2_{\tilde{\mu}}(k|N)$ are easily computed once $\mathbf{r}(k|N)$ and $\mathbf{S}(k|N)$ have been computed.

The extension of Theorem 21-5 to a time-varying state-variable model is straightforward and can be found in Mendel (1983).

**EXAMPLE 21-3**

In this example we compute $\hat{\mu}(k|N)$, first for a broadband channel IR, $h_1(k)$, and then for a narrower-band channel IR, $h_2(k)$. The transfer functions of these channel models are

$$H_1(z) = \frac{-0.76286z^3 + 1.5884z^2 - 0.82356z + 0.000222419}{z^4 - 2.2633z^3 + 1.77734z^2 - 0.49803z + 0.045546} \tag{21-76}$$

and

$$H_2(z) = \frac{0.0378417z^3 - 0.0306517z}{z^4 - 3.4016497z^3 + 4.5113732z^2 - 2.7553363z + 0.6561} \tag{21-77}$$

respectively. Plots of these IRs and their squared amplitude spectra are depicted in Figures 21-2 and 21-3.

Measurements, $z(k)(k = 1, 2, \ldots, 250$, where $T = 3$ msec), were generated by convolving each of these IRs with a sparse and different spike train (i.e., a Bernoulli–Gaussian sequence) and then adding measurement noise to the results. These measurements, which, of course, represent the starting point for deconvolution, are depicted in Figure 21-4.

Figure 21-5 depicts $\hat{\mu}(k|N)$. Observe that much better results are obtained for the broadband channel than for the narrower-band channel, even though data quality, as measured by $\overline{\text{SNR}}$, is much lower in the former case. The MVD results for the narrower-band channel appear "smeared out," whereas the MVD results for the broadband channel are quite sharp. We provide a theoretical explanation for this effect below.

Observe, also, that $\hat{\mu}(k|N)$ tends to undershoot $\mu(k)$. See Chi and Mendel (1984) for a theoretical explanation about why this occurs. □

## STEADY-STATE MVD FILTER

For a time-invariant IR and stationary noises, the Kalman gain matrix, as well as the error-covariance matrices, will reach steady-state values. When this occurs, both the Kalman innovations filter and anticausal $\mu$-filter [(21-62) and (21-25)] become time invariant, and we then refer to the MVD filter as a *steady-state MVD filter*. In this section we examine an important property of this steady-state filter.

Let $h_i(k)$ and $h_\mu(k)$ denote the IRs of the steady-state Kalman innovations and anticausal $\mu$-filters, respectively. Then

$$\begin{aligned}\hat{\mu}(k|N) &= h_\mu(k) * \tilde{z}(k|k-1)\\ &= h_\mu(k) * h_i(k) * z(k)\\ &= h_\mu(k) * h_i(k) * h(k) * \mu(k)\\ &\quad + h_\mu(k) * h_i(k) * \nu(k)\end{aligned} \tag{21-78}$$

which can also be expressed as

$$\hat{\mu}(k|N) = \hat{\mu}_s(k|N) + n(k|N) \tag{21-79}$$

where the *signal component* of $\hat{\mu}(k|N)$, $\hat{\mu}_s(k|N)$, is

$$\hat{\mu}_s(k|N) = h_\mu(k) * h_i(k) * h(k) * \mu(k) \tag{21-80}$$

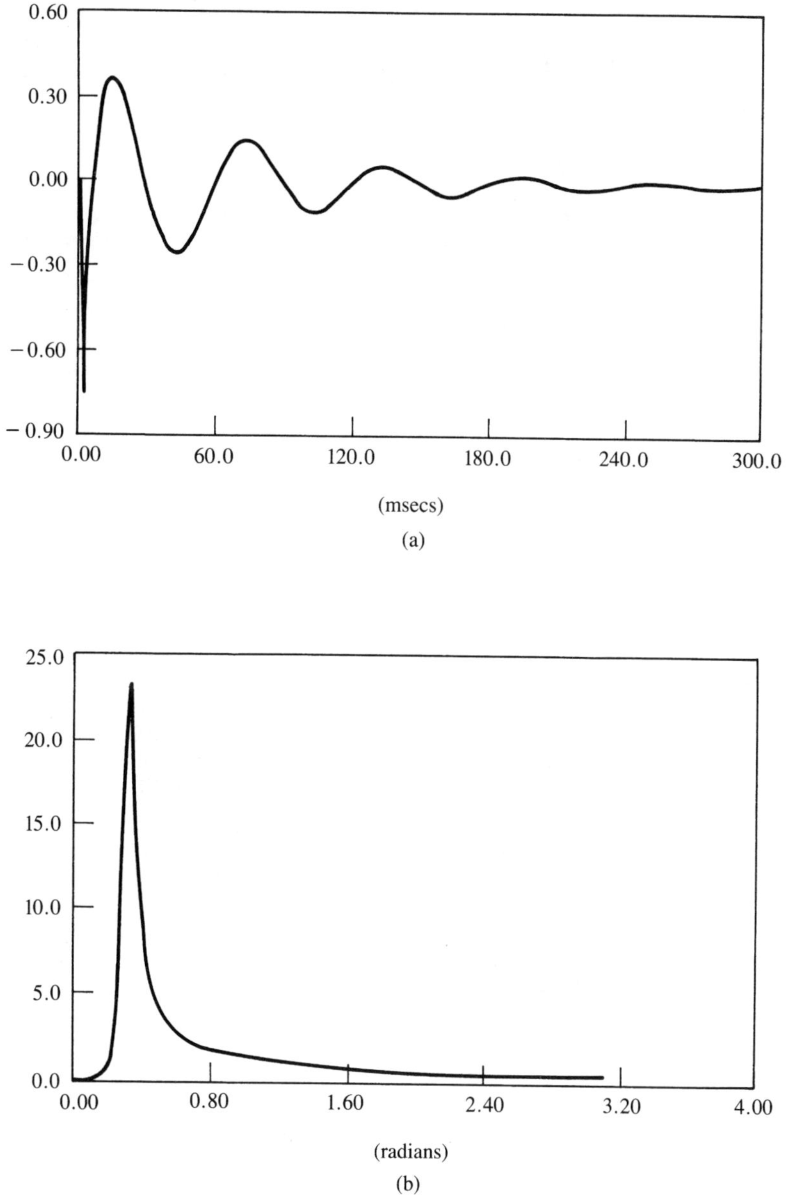


**Figure 21-2** (a) Fourth-order broadband channel IR, and (b) its squared amplitude spectrum (Chi, 1983).

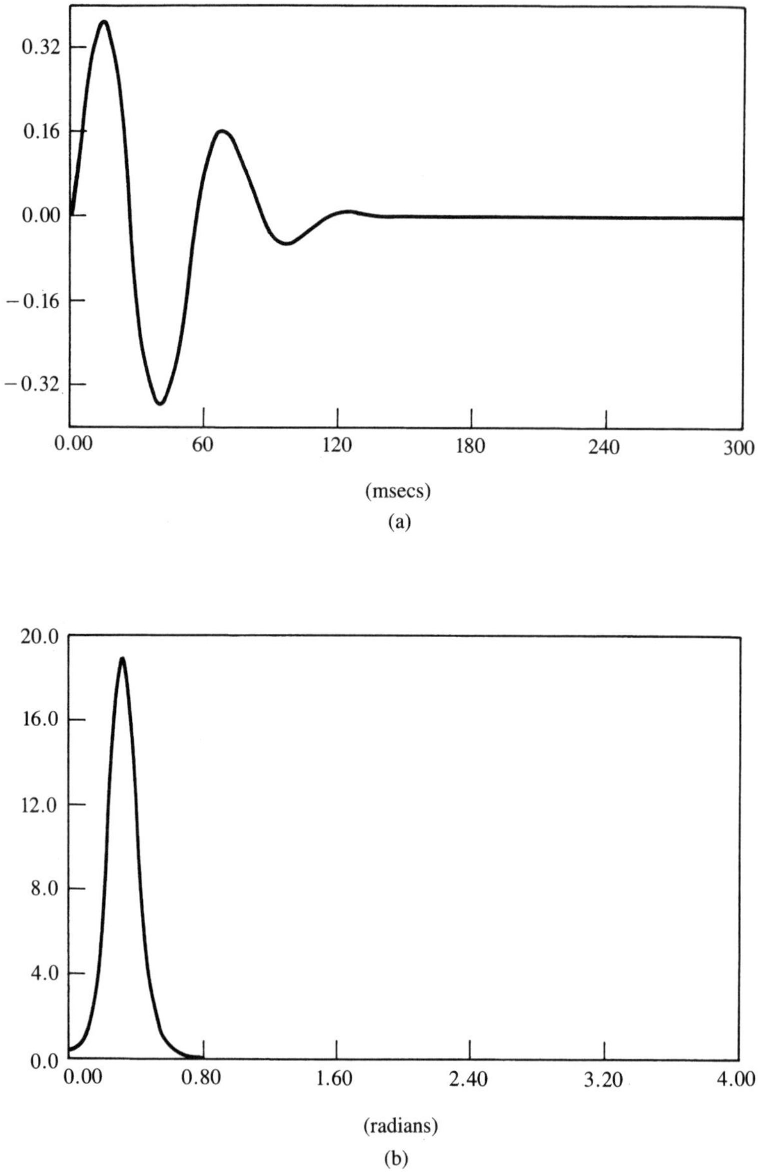


**Figure 21-3** (a) Fourth-order narrower-band channel IR, and (b) its squared amplitude spectrum (Chi and Mendel, 1984, © 1984, IEEE).

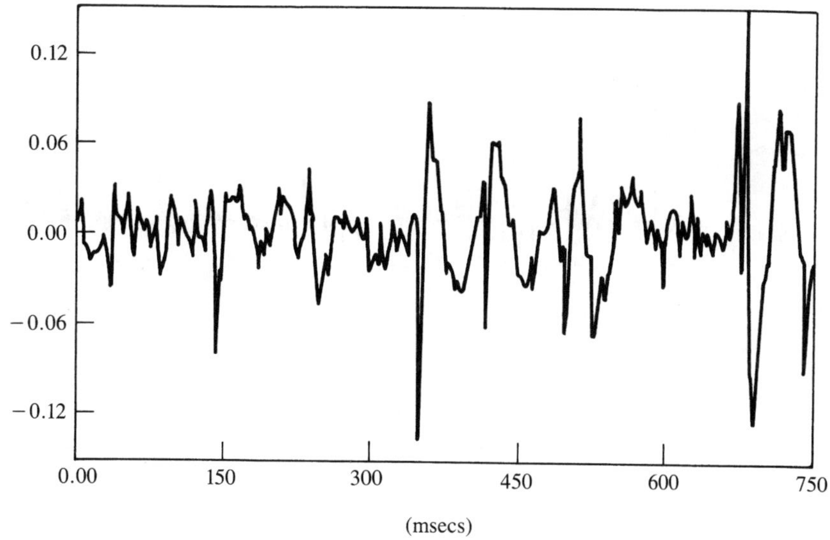


(a)

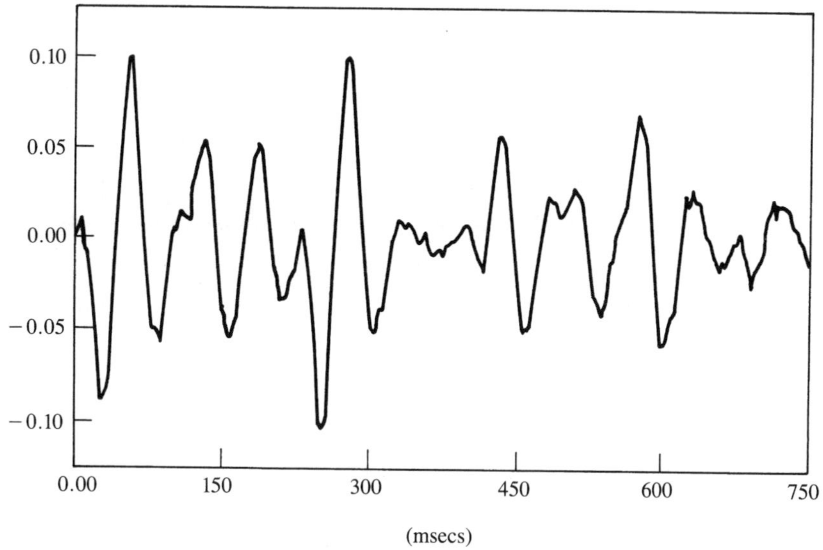


(b)

**Figure 21-4** Measurements associated with (a) broadband channel ($\overline{\text{SNR}} = 10$) and (b) narrower-band channel ($\overline{\text{SNR}} = 100$) (Chi and Mendel, 1984, © 1984, IEEE).

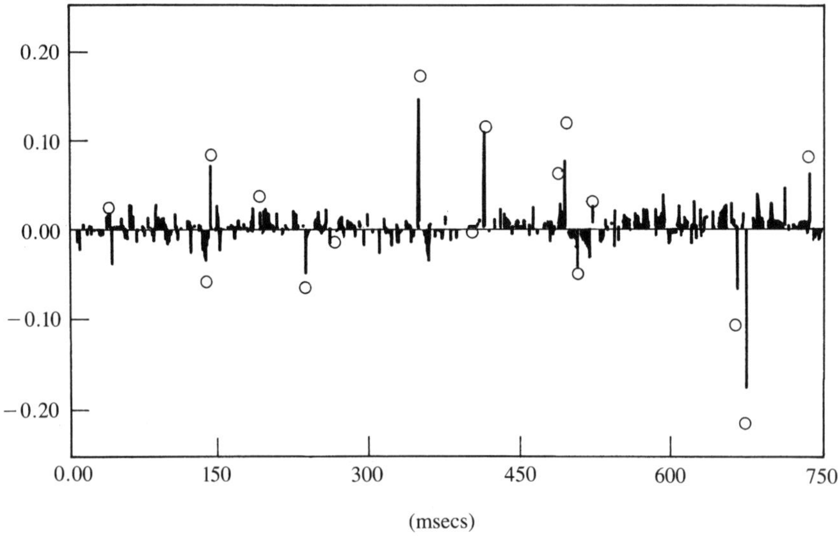


(a)

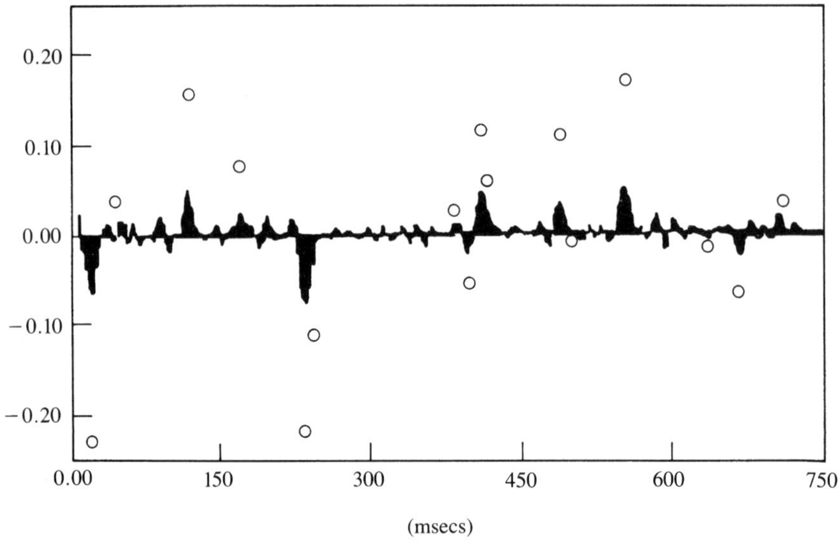


(b)

**Figure 21-5** $\hat{\mu}(k|N)$ for (a) broadband channel ($\overline{\text{SNR}} = 10$) and (b) narrower-band channel ($\overline{\text{SNR}} = 100$). Circles depict true $\mu(k)$ and bars depict estimate of $\mu(k)$ (Chi and Mendel, 1984, © 1984, IEEE).

and the *noise component* of $\hat{\mu}(k|N)$, $n(k|N)$, is

$$n(k|N) = h_\mu(k) * h_i(k) * \nu(k) \tag{21-81}$$

We shall refer to $h_\mu(k) * h_i(k)$ as the IR of the MVD filter, $h_{\text{MV}}(k)$; i.e.,

$$h_{\text{MV}}(k) = h_\mu(k) * h_i(k) \tag{21-82}$$

The following result has been proved by Chi and Mendel (1984) for the slightly modified model $\mathbf{x}(k+1) = \mathbf{\Phi x}(k) + \boldsymbol{\gamma}\mu(k+1)$ and $z(k) = \mathbf{h}'\mathbf{x}(k) + \nu(k)$ [because of the $\mu(k+1)$ input instead of the $\mu(k)$ input, $h(0) \neq 0$].

**Theorem 21-6.** *In the stationary case:*

**a.** *The Fourier transform of* $h_{\text{MV}}(k)$ *is*

$$H_{\text{MV}}(\omega) = \frac{qH^*(\omega)}{q|H(\omega)|^2 + r} \tag{21-83}$$

*where* $H^*(\omega)$ *denotes the complex conjugate of* $H(\omega)$; *and*

**b.** *the signal component of* $\hat{\mu}(k|N)$, $\hat{\mu}_s(k|N)$, *is given by*

$$\hat{\mu}_s(k|N) = R(k) * \mu(k) \tag{21-84}$$

*where* $R(k)$ *is the autocorrelation function*

$$R(k) = \frac{q}{\eta}[h(k) * h_i(k)] * [h(-k) * h_i(-k)] \tag{21-85}$$

*in which*

$$\eta = \mathbf{h}'\mathbf{P}_p\mathbf{h} + r \tag{21-86}$$

*Additionally,*

$$R(\omega) = \frac{q|H(\omega)|^2}{q|H(\omega)|^2 + r} \quad \square \tag{21-87}$$

We leave the proof of this theorem as an exercise for the reader. Observe that part (b) of the theorem means that $\hat{\mu}_s(k|N)$ *is a zero-phase wave-shaped version of* $\mu(k)$. Observe, also, that $R(\omega)$ can be written as

$$R(\omega) = \frac{|H(\omega)|^2 q/r}{1 + |H(\omega)|^2 q/r} \tag{21-88}$$

which demonstrates that $q/r$, and subsequently $\overline{\text{SNR}}$, is an MVD filter tuning parameter. As $q/r \to \infty$, $R(\omega) \to 1$ so that $R(k) \to \delta(k)$; thus, for high signal-to-noise ratios, $\hat{\mu}_s(k|N) \to \mu(k)$. Additionally, when $|H(\omega)|^2 q/r \gg 1$, $R(\omega) \to 1$, and once again $R(k) \to \delta(k)$. Broadband IRs often satisfy this condition. In general, however, $\hat{\mu}_s(k|N)$ is a smeared-out version of $\mu(k)$; however, the nature of the smearing is quite dependent on the bandwidth of $h(k)$ and $\overline{\text{SNR}}$.

**EXAMPLE 21-4**

This example is a continuation of Example 21-3. Figure 21-6 depicts $R(k)$ for both the broadband and narrower-band IRs, $h_1(k)$ and $h_2(k)$, respectively. As predicted by (21-88), $R_1(k)$ is much spikier than $R_2(k)$, which explains why the MVD results for the broadband IR are quite sharp, whereas the MVD results for the narrower-band IR are smeared out. Note, also, the difference in peak amplitudes for $R_1(k)$ and $R_2(k)$. This explains why $\hat{\mu}(k|N)$ underestimates the true values of $\mu(k)$ by such large amounts in the narrower-band case (see Figs. 21-5a and b). □

## RELATIONSHIP BETWEEN STEADY-STATE MVD FILTER AND AN INFINITE IMPULSE RESPONSE DIGITAL WIENER DECONVOLUTION FILTER

We have seen that an MVD filter is a cascade of a causal Kalman innovations filter and an anticausal $\mu$-filter; hence, it is a noncausal filter. Its impulse response extends from $k = -\infty$ to $k = +\infty$, and the IR of the steady-state MVD filter is given in the time domain by $h_{\text{MV}}(k)$ in (21-82) or in the frequency domain by $H_{\text{MV}}(\omega)$ in (21-83).

There is a more direct way for designing an IIR minimum mean-squared error deconvolution filter, i.e., an *IIR digital Wiener deconvolution filter*, as we describe next.

We return to the situation depicted in Figure 19-4, but now we assume that filter $F(z)$ is an IIR filter, with coefficients $\{f(j), j = 0, \pm 1, \pm 2, \ldots\}$;

$$d(k) = \mu(k) \tag{21-89}$$

where $\mu(k)$ is a white noise sequence; $\mu(k)$, $\nu(k)$, and $n(k)$ are stationary; and $\mu(k)$ and $\nu(k)$ are uncorrelated. In this case, (19-44) becomes

$$\sum_{i=-\infty}^{\infty} f(i)\phi_{zz}(i-j) = \phi_{z\mu}(j), \qquad j = 0, \pm 1, \pm 2, \ldots \tag{21-90}$$

Using (21-58), the whiteness of $\mu(k)$, and the assumptions that $\mu(k)$ and $\nu(k)$ are uncorrelated and stationary, it is straightforward to show that (Problem 21-13)

$$\phi_{z\mu}(j) = qh(-j) \tag{21-91}$$

Substituting (21-91) into (21-90), we have

$$\sum_{i=-\infty}^{\infty} f(i)\phi_{zz}(i-j) = qh(-j), \qquad j = 0, \pm 1, \pm 2, \ldots \tag{21-92}$$

We are now ready to solve (21-92) for the deconvolution filter's coefficients. We cannot do this by solving a linear system of equations because there are a doubly infinite number of them. Instead, we take the discrete-time Fourier transform of (21-92), i.e.,

$$F(\omega)\Phi_{zz}(\omega) = qH^*(\omega) \tag{21-93}$$

but, from (21-58), we also know that (Problem 21-13)

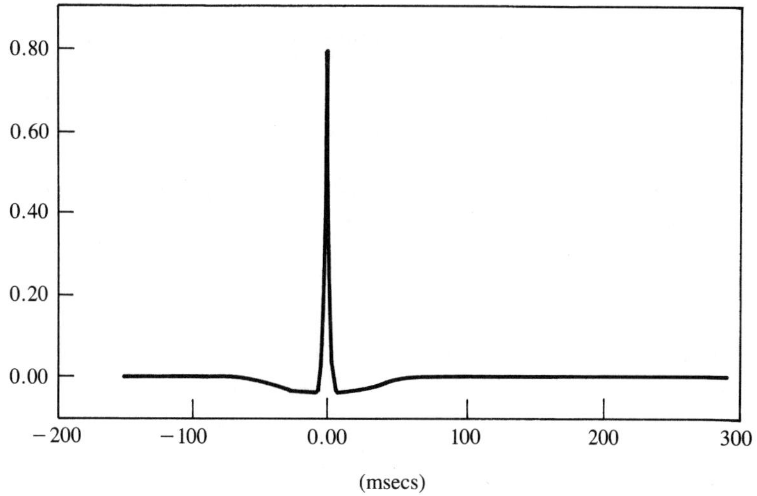


(a)

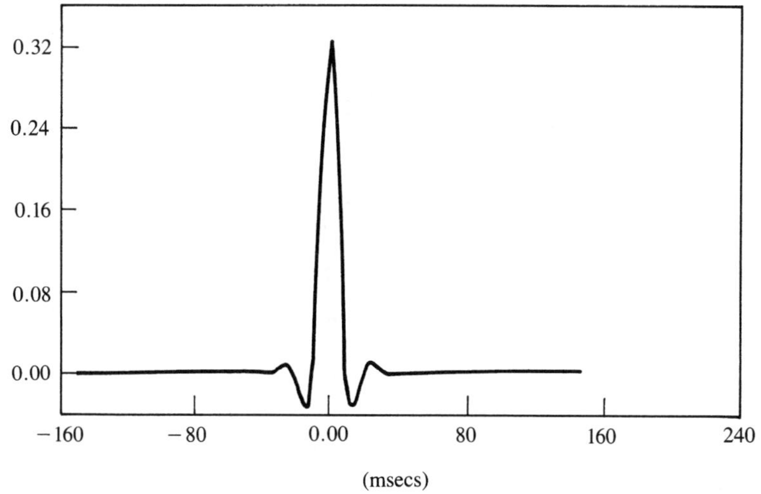


(b)

**Figure 21-6** $R(k)$ for (a) broadband channel ($\overline{\text{SNR}} = 10$) and (b) narrower-band channel ($\overline{\text{SNR}} = 100$) (Chi and Mendel, 1984, © 1984, IEEE).

$$\Phi_{zz}(\omega) = q|H(\omega)|^2 + r \tag{21-94}$$

Substituting (21-94) into (21-93), we determine $F(\omega)$, as

$$F(\omega) = \frac{qH^*(\omega)}{q|H(\omega)|^2 + r} \tag{21-95}$$

This IIR digital Wiener deconvolution filter (i.e., two-sided least-squares inverse filter) was, to the best of our knowledge, first derived by Berkhout (1977).

**Theorem 21-7** (Chi and Mendel, 1984). *The steady-state MVD filter, whose IR is given by* $h_{MV}(k)$, *is exactly the same as Berkhout's IIR digital Wiener deconvolution filter.* □

The steady-state MVD filter is a recursive implementation of Berkhout's infinite-length filter. Of course, the MVD filter is also applicable to time-varying and nonstationary systems, whereas his filter is not.

## MAXIMUM-LIKELIHOOD DECONVOLUTION

In Example 14-2 we began with the deconvolution linear model $\mathbf{Z}(N) = \mathbf{H}(N-1)\boldsymbol{\mu} + \mathbf{V}(N)$, used the product model for $\boldsymbol{\mu}$ (i.e., $\boldsymbol{\mu} = \mathbf{Q_q r}$), and showed that a separation principle exists for the determination of $\hat{\mathbf{r}}_{MAP}$ and $\hat{\mathbf{q}}_{MAP}$. We showed that first we must determine $\hat{\mathbf{q}}_{MAP}$, after which $\hat{\mathbf{r}}_{MAP}$ can be computed using (14-34). Equation (14-34) is terribly unwieldy because of the $N \times N$ matrix, $\sigma_r^2\mathbf{HQ_qH}' + \rho\mathbf{I}$, that must be inverted. The following theorem provides a more practical way to compute the elements of $\hat{\mathbf{r}}_{MAP}(N|\mathbf{Q}_{\hat{\mathbf{q}}})$, where $\hat{\mathbf{q}}$ is short for $\hat{\mathbf{q}}_{MAP}$. These elements are $\hat{\mathbf{r}}_{MAP}(k|N;\hat{\mathbf{q}}), k = 1, 2, \ldots, N$.

**Theorem 21-8** (Mendel, 1983). *Unconditional maximum-likelihood (i.e., MAP) estimates of* $\mathbf{r}$ *can be obtained by applying MVD formulas to the state-variable model*

$$\mathbf{x}(k+1) = \mathbf{\Phi x}(k) + \boldsymbol{\gamma}\hat{q}_{MAP}(k)r(k) \tag{21-96}$$

$$z(k) = \mathbf{h}'\mathbf{x}(k) + \nu(k) \tag{21-97}$$

*where* $\hat{q}_{MAP}(k)$ *is a MAP estimate of* $q(k)$.

*Proof.* Example 14-2 showed that a MAP estimate of $\mathbf{q}$ can be obtained prior to finding a MAP estimate of $\mathbf{r}$. By using the product model for $\mu(k)$ and $\hat{\mathbf{q}}_{MAP}$, our state-variable model in (21-59) and (21-60) can be expressed as in (21-96) and (21-97). Applying (14-22) to this system, we see that

$$\hat{r}_{MAP}(k|N) = \hat{r}_{MS}(k|N) \tag{21-98}$$

But, by comparing (21-96) and (21-59) and (21-97) and (21-60), we see that $\hat{r}_{MS}(k|N)$ can be found from the MVD algorithm in Theorem 21-5 in which we replace $\mu(k)$ by $r(k)$, $\boldsymbol{\gamma}$ by $\boldsymbol{\gamma}\hat{\mathbf{q}}_{MAP}(k)$, and set $q(k) = \sigma_r^2$. □

## SUMMARY QUESTIONS

1. By a "useful" fixed-interval smoother, we mean one that:
   **(a)** runs efficiently
   **(b)** can be computed
   **(c)** is stable
2. By a "most useful" smoother, we mean one that:
   **(a)** can be computed
   **(b)** can be computed with no multiplications of $n \times n$ matrices
   **(c)** is stable
3. Smoother error-covariance matrix $\mathbf{P}(k|N)$:
   **(a)** must be computed prior to processing data
   **(b)** satisfies a forward-running equation
   **(c)** does not have to be calculated at all in order to process data
4. The plant matrix in the backward-running equation for the residual state vector is:
   **(a)** that of the recursive predictor
   **(b)** that of the recursive filter
   **(c)** time invariant
5. In deriving the fixed-lag smoother, we augmented ________________ versions of $\mathbf{x}(k)$ to $\mathbf{x}(k)$?
   **(a)** advanced and delayed
   **(b)** advanced
   **(c)** delayed
6. Deconvolution is the signal-processing procedure for removing the effects of:
   **(a)** $h(j)$ and $\mu(j)$ from the measurements so that we obtain an estimate of $\nu(j)$
   **(b)** $h(j)$ and $\nu(j)$ from the measurements so that we obtain an estimate of $\mu(j)$
   **(c)** $\mu(j)$ and $\nu(j)$ from the measurements so that we obtain an estimate of $h(j)$
7. Smoothing error variance $\sigma_{\mu}^{2}(k|N)$ shows that MVD:
   **(a)** may not reduce the initial uncertainty $q$
   **(b)** may sometimes increase initial uncertainty
   **(c)** reduces initial uncertainty
8. The coefficients of a FIR WF are found by solving the normal equations. The coefficients of an IIR WF cannot be found by solving the normal equations, because they:
   **(a)** are a doubly infinite system of equations
   **(b)** are no longer linear
   **(c)** cannot be expressed in matrix form
9. The eigenvalues of the fixed-lag smoother:
   **(a)** are those of the recursive predictor
   **(b)** may lie outside the unit circle
   **(c)** include new stable elements in addition to those of the recursive predictor
10. A second-order Gauss–Markov random sequence is associated with a system that has:
    **(a)** two states
    **(b)** two delays
    **(c)** two moments
11. The IIR digital Wiener deconvolution filter is:
    **(a)** zero phase
    **(b)** causal
    **(c)** all-pass

## PROBLEMS

**21-1.** Derive the formula for $\hat{\mathbf{x}}(k|N)$ in (21-5) using mathematical induction. Then derive $\hat{\mathbf{x}}(k|N)$ in (21-6).

**21-2.** Derive the formula for the fixed-point smoothing error-covariance matrix $\mathbf{P}(k|j)$ given in (21-40).

**21-3.** Prove Theorem 21-3, which gives formulas for a most useful mean-squared fixed-point smoother $\mathbf{x}(k)$, $\hat{\mathbf{x}}(k|k+l), l = 1, 2, \ldots$.

**21-4.** Using the two-step procedure described at the end of the section entitled Fixed-point Smoothing, derive the resulting fixed-point smoother equations. Be sure to explain how $\hat{\mathbf{x}}(j|k)$ and $\mathbf{P}(j|k)$ are initialized.

**21-5.** (Kwan Tjia, Spring 1992) Show that, for $\ell = 2$, the general result for the mean-squared fixed-point smoothed estimator of $\mathbf{x}(k)$, given in Theorem 21-3, reduces to the expression for the double-stage smoother given in Theorem 20-2.

**21-6.** (Meditch, 1969, Exercise 6-13, p. 245). Consider the scalar system $x(k+1) = 2^{-k}x(k) + w(k)$, $z(k+1) = x(k+1)$, $k = 0, 1, \ldots$, where $x(0)$ has mean zero and variance $\sigma_0^2$ and $w(k)$, $k = 0, 1, \ldots$ is a zero mean Gaussian white sequence that is independent of $x(0)$ and has a variance equal to $q$.

**(a)** Assuming that optimal fixed-point smoothing is to be employed to determine $x(0|j)$, $j = 1, 2, \ldots$, what is the equation for the appropriate smoothing filter?

**(b)** What is the limiting value of $p(0|j)$ as $j \to \infty$?

**(c)** How does this value compare with $p(0|0)$?

**21-7.** (William A. Emanuelsen, Spring 1992) (Computation Problem) The state equation for the $g$ accelerations experienced by a fighter pilot in a certain maneuver is

$$g(k+1) = 0.11(k+1)g(k) + w(k)$$

where $w(k)$ is a noise term to account for turbulence experienced during the maneuver. It is possible to measure $g(k)$ during the maneuver with an accelerometer:

$$z(k+1) = g(k+1) + \nu(k+1)$$

where $\nu(k)$ is measurement noise in the accelerometer. Given the test data in Table P21-7, and that all noises are zero mean, white, and Gaussian, with variances $\sigma_w^2 = 0.04$ and $\sigma_\nu^2 = 0.01$, determine the fixed-interval smoothed mean-squared estimate of $g(k)$. The mean initial $g$ acceleration is $1.0g$ and the variance about the mean is unknown.

**TABLE P21-7** ACCELEROMETER DATA $\{z(k)\}$

| Time (sec) | Acceleration ($g$'s) | Time (sec) | Acceleration ($g$'s) |
|---|---|---|---|
| 0.00 | 1.16290 | 2.75 | 0.13574 |
| 0.25 | 0.32518 | 3.00 | 0.18210 |
| 0.50 | 0.00123 | 3.25 | −0.10527 |
| 0.75 | 0.25095 | 3.50 | −0.06163 |
| 1.00 | 0.12807 | 3.75 | −0.02912 |
| 1.25 | 0.27190 | 4.00 | 0.23072 |
| 1.50 | 0.09897 | 4.25 | 0.36021 |
| 1.75 | −0.19914 | 4.50 | 0.71272 |
| 2.00 | 0.10350 | 4.75 | 1.30116 |
| 2.25 | 0.17581 | 5.00 | 3.64263 |
| 2.50 | −0.09106 | | |

**21-8.** (Mike Walter, Spring 1992) (Computation Problem) Consider the following model:

$$x_1(k+1) = x_1(k) + x_2(k)$$

$$x_2(k+1) = x_2(k)$$

$$z(k+1) = x_1(k+1) + \nu(k+1)$$

where $\nu(k)$ is zero-mean white Gaussian noise with $\sigma_\nu^2(k) = 20$, $x_1(0)$ and $x_2(0)$ are uncorrelated, $\mathbf{E}\{x_1(0)\} = \mathbf{E}\{x_2(0)\} = 0$, and $\mathbf{E}\{x_1^2(0)\} = \mathbf{E}\{x_2^2(0)\} = 20$.

**(a)** Give equations for $\mathbf{P}(k+1|k), \mathbf{P}(k|k), \mathbf{P}(k|N)$, and $\hat{\mathbf{x}}(k|N)$.

**(b)** Use a computer to simulate this model for $k = 0$ to $k = 50$. Initialize your model with $x_1(0) = 1$ and $x_2(0) = 1$. Produce the following plots: (1)$x_1(k)$ and $z(k)$; (2) $\mathbf{P}_{11}(k|k)$ and $\mathbf{P}_{11}(k|50)$ for $x_1$; (3) $\mathbf{P}_{22}(k|k)$ and $\mathbf{P}_{22}(k|50)$ for $x_2$; (4) $\hat{x}_1(k|k)$ and $\hat{x}_1(k|50)$; and (5) $\hat{x}_2(k|k)$ and $\hat{x}_2(k|50)$.

**(c)** Discuss your results.

**21-9.** (Richard S. Lee, Jr., Spring 1992) Consider the following linear model that represents a first-order polynomial: $\mathbf{Z} = \mathbf{H}\boldsymbol{\theta} + \mathbf{V}$, where $\mathbf{Z} = \text{col}\,(z(0), z(1), \ldots, z(N))$, $\mathbf{V} = \text{col}\,(\nu(0), \nu(1), \ldots, \nu(N))$, in which $\nu(k)$ is zero-mean Gaussian white noise, $\boldsymbol{\theta} = \text{col}\,(a, b)$ and

$$\mathbf{H} = \begin{bmatrix} 1 & 0 \\ 1 & 2 \\ & \cdots \\ 1 & N \end{bmatrix}$$

The values $a$ and $b$ correspond to the parameters in the equation of the line $y = a + bx$.

**(a)** Reexpress this model in terms of the basic state-variable model.

**(b)** Show that the mean-squared fixed-interval smoothed estimator of $\mathbf{x}(k)$, $\hat{\mathbf{x}}(k|N)$, is equal to the mean-squared filtered estimator of $\mathbf{x}(N)$, $\hat{\mathbf{x}}(N|N)$, for all $k = 0, 1, \ldots, N$. Explain why this is so.

**21-10.** (Lance Kaplan, Spring 1992) Consider the system

$$\mathbf{x}(k+1) = \boldsymbol{\Phi}(k+1, k)\mathbf{x}(k)$$

$$\mathbf{z}(k+1) = \mathbf{H}(k+1)\mathbf{x}(k+1) + \mathbf{v}(k+1)$$

where $\boldsymbol{\Phi}(k+1, k)$ is full rank for all $k$.

**(a)** Show that, when calculating $\hat{\mathbf{x}}(k|k+\ell)$ for $\ell = 1, 2, \ldots$, then Equation (21-41) yields $\mathbf{N}_\mathbf{x}(k|k+\ell) = \boldsymbol{\Phi}^{-1}(k+\ell, k)\mathbf{K}(k+\ell)$, where $\boldsymbol{\Phi}(k+\ell, k) = \boldsymbol{\Phi}(k+\ell, k+\ell-1)\boldsymbol{\Phi}(k+\ell-1, k+\ell-2)\cdots\boldsymbol{\Phi}(k+1, k)$.

**(b)** Let $\mathbf{y}(k) = \boldsymbol{\Phi}^{-1}(k, 0)\mathbf{x}(k)$ for $\ell = 1, 2, \ldots$. Rewrite the state equations in terms of the new state vector $\mathbf{y}(k)$.

**(c)** Show that $\mathbf{N}_\mathbf{y}(k|k+1) = \mathbf{K}(k+1)$.

**(d)** How do the smoothed estimates $\hat{\mathbf{y}}(k|k+\ell)$ relate to the filtered estimates $\hat{\mathbf{y}}(k+\ell|k+\ell)$?

**(e)** How do the smoothed estimates $\hat{\mathbf{x}}(k|k+\ell)$ relate to the filtered estimates $\hat{\mathbf{x}}(k+\ell|k+\ell)$?

**21-11.** Prove Theorem 21-6. Explain why part (b) of the theorem means that $\hat{\mu}_s(k|N)$ is a zero-phase wave-shaped version of $\mu(k)$.

**21-12.** This problem is a memory refresher. You probably have either seen or carried out the calculations asked for in a course on random processes.

**(a)** Derive Equation (21-92).

**(b)** Derive Equation (21-95).

**21-13.** (James E. Leight, Fall 1991) Show the details of the proof of Theorem 21-4.

**21-14.** (Constance M. Chintall, Fall 1991) Determine $H_{\text{MV}}(\omega)$ and $R(\omega)$ when $H(\omega) = 1/(\alpha + j\omega)$. Do this for $\alpha = 10$ and SNR $= 1$ and $\alpha = 100$ and SNR $= 0.01$. Discuss your results.

**21-15.** (Gregg Isara and Loan Bui, Fall 1991) Write the equations for $\hat{r}_{\text{MAP}}(k|N)$ for the following single-channel state-variable model:

$$x(k+1) = 0.667x(k) + 3r(k)$$

$$z(k+1) = 0.25x(k+1) + v(k+1)$$

where the variance of $r(k)$ is 4, $m_v(k) = 0$, and $\sigma_v^2(k) = 2$. Draw a flow chart for the implementation of $\hat{r}_{\text{MAP}}(k|N)$.

# LESSON 22

# State Estimation for the Not-so-basic State-variable Model


## SUMMARY

In deriving all our state estimators we worked with our basic state-variable model. In many practical applications we begin with a model that differs from the basic state-variable model. The purpose of this lesson is to explain how to use or modify our previous state estimators when there are:

1. Either nonzero-mean noise processes or known bias functions or both in the state or measurement equations; or
2. Correlated noise processes in the state and measurement equations; or
3. Colored disturbances or measurement noises; or
4. Some perfect measurements.

No new concepts are needed to handle items 1 and 2. A *state augmentation procedure* is needed to handle item 3. Colored noises are modeled as low-order Markov sequences, i.e., as low-order difference equations. The states associated with these colored noise models must be augmented to the original state-variable model prior to application of a recursive state estimator to the augmented system.

A set of *perfect* measurements reduces the number of states that have to be estimated. So, for example, if there are $l$ perfect measurements, then we show how to estimate $\mathbf{x}(k)$ by a Kalman filter whose dimension is no greater than $n - l$. Such an estimator is called a reduced-order estimator of $\mathbf{x}(k)$. The payoff for using a reduced-order estimator is fewer computations and less storage.

In actual practice, some or all of the four special cases can occur simultaneously. Just merge the methods learned for treating each case separately to handle more complicated situations.

When you complete this lesson, you will be able to (1) use state estimation algorithms when the starting state-variable model differs from the basic state-variable model; and (2) apply state estimation to any linear, time-varying, discrete-time, nonstationary dynamical system.

## INTRODUCTION

In deriving all our state estimators, we assumed that our dynamical system could be modeled as in Lesson 15, i.e., as our *basic state-variable model.* The results so obtained are applicable only for systems that satisfy all the conditions of that model: the noise processes $\mathbf{w}(k)$ and $\mathbf{v}(k)$ are both zero mean, white, and mutually uncorrelated, no known bias functions appear in the state or measurement equations, and no measurements are noise-free (i.e., perfect). The following cases frequently occur in practice:

1. Either nonzero-mean noise processes or known bias functions or both in the state or measurement equations
2. Correlated noise processes
3. Colored noise processes
4. Some perfect measurements

In this lesson we show how to modify some of our earlier results in order to treat these important special cases. To see the forest from the trees, we consider each of these four cases separately. In practice, some or all of them may occur together.

## BIASES

Here we assume that our basic state-variable model, given in (15-17) and (15-18), has been modified to

$$\mathbf{x}(k+1) = \mathbf{\Phi}(k+1,k)\mathbf{x}(k) + \mathbf{\Gamma}(k+1,k)\mathbf{w}_1(k) + \mathbf{\Psi}(k+1,k)\mathbf{u}(k) \tag{22-1}$$

$$\mathbf{z}(k+1) = \mathbf{H}(k+1)\mathbf{x}(k+1) + \mathbf{G}(k+1)\mathbf{u}(k+1) + \mathbf{v}_1(k+1) \tag{22-2}$$

where $\mathbf{w}_1(k)$ and $\mathbf{v}_1(k)$ are nonzero mean, individually and mutually uncorrelated Gaussian noise sequences; i.e.,

$$E\{\mathbf{w}_1(k)\} = \mathbf{m}_{\mathbf{w}_1}(k) \neq \mathbf{0}, \qquad \mathbf{m}_{\mathbf{w}_1}(k) \text{ known} \tag{22-3}$$

$$E\{\mathbf{v}_1(k)\} = \mathbf{m}_{\mathbf{v}_1}(k) \neq \mathbf{0}, \qquad \mathbf{m}_{\mathbf{v}_1}(k) \text{ known} \tag{22-4}$$

$E\{[\mathbf{w}_1(i)-\mathbf{m}_{\mathbf{w}_1}(i)][\mathbf{w}_1(j)-\mathbf{m}_{\mathbf{w}_1}(j)]'\} = \mathbf{Q}(i)\delta_{ij}$, $E\{[\mathbf{v}_1(i)-\mathbf{m}_{\mathbf{v}_1}(i)][\mathbf{v}_1(j)-\mathbf{m}_{\mathbf{v}_1}(j)]'\} = \mathbf{R}(i)\delta_{ij}$, and $E\{[\mathbf{w}_1(i) - \mathbf{m}_{\mathbf{w}_1}(i)][\mathbf{v}_1(j) - \mathbf{m}_{\mathbf{v}_1}(j)]'\} = \mathbf{0}$. Note that the term $\mathbf{G}(k+1)\mathbf{u}(k+1)$, which appears in the measurement equation (22-2), is known and can be thought of as contributing to the bias of $\mathbf{v}_1(k+1)$.

This case is handled by reducing (22-1) and (22-2) to our previous basic state-variable model, using the following simple transformations. Let

$$\mathbf{w}(k) \triangleq \mathbf{w}_1(k) - \mathbf{m}_{\mathbf{w}_1}(k) \tag{22-5}$$

and

$$\mathbf{v}(k) \triangleq \mathbf{v}_1(k) - \mathbf{m}_{\mathbf{v}_1}(k) \tag{22-6}$$

Observe that both $\mathbf{w}(k)$ and $\mathbf{v}(k)$ are zero-mean white noise processes, with covariances $\mathbf{Q}(k)$ and $\mathbf{R}(k)$, respectively. Adding and subtracting $\Gamma(k + 1, k)\mathbf{m}_{\mathbf{w}_1}(k)$ in state equation (22-1) and $\mathbf{m}_{\mathbf{v}_1}(k + 1)$ in measurement equation (22-2), these equations can be expressed as

$$\mathbf{x}(k + 1) = \Phi(k + 1, k)\mathbf{x}(k) + \Gamma(k + 1, k)\mathbf{w}(k) + \mathbf{u}_1(k) \tag{22-7}$$

and

*a new known $u(k)$.*

$$\mathbf{z}_1(k + 1) = \mathbf{H}(k + 1)\mathbf{x}(k + 1) + \mathbf{v}(k + 1) \tag{22-8}$$

where

$$\mathbf{u}_1(k) = \Psi(k + 1, k)\mathbf{u}(k) + \Gamma(k + 1, k)\mathbf{m}_{\mathbf{w}_1}(k) \tag{22-9}$$

and

*new measurement to use in 22-7, 22-8, which is our normal state eqs.*

$$\mathbf{z}_1(k + 1) = \mathbf{z}(k + 1) - \mathbf{G}(k + 1)\mathbf{u}(k + 1) - \mathbf{m}_{\mathbf{v}_1}(k + 1) \tag{22-10}$$

*known*

Clearly, (22-7) and (22-8) are once again a basic state-variable model, one in which $\mathbf{u}_1(k)$ plays the role of $\Psi(k + 1, k)\mathbf{u}(k)$ and $\mathbf{z}_1(k + 1)$ plays the role of $\mathbf{z}(k + 1)$.

**Theorem 22-1.** *When biases are present in a state-variable model, then that model can always be reduced to a basic state-variable model [e.g., (22-7) to (22-10)]. All our previous state estimators can be applied to this basic state-variable model by replacing $\mathbf{z}(k)$ by $\mathbf{z}_1(k)$ and $\Psi(k + 1, k)\mathbf{u}(k)$ by $\mathbf{u}_1(k)$.* $\square$

*example* $\mathbf{z}(t) = \ddot{y}(r) + w(t)$ ———— *measure acceleration*
*random*
$\ddot{y}(t) + a_1 \dot{y} + a_2 y = w(t)$ $\quad \dot{x}_2 = -a_1 \dot{y} - a_2 y + w \implies \begin{pmatrix} \dot{x}_1 \\ \dot{x}_2 \end{pmatrix} = \begin{pmatrix} 0 & 1 \\ -a_2 & -a_1 \end{pmatrix}\begin{pmatrix} x_1 \\ x_2 \end{pmatrix} + \begin{pmatrix} 0 \\ 1 \end{pmatrix}$

**CORRELATED NOISES** *Let $x_1 = y$.* $\quad so \implies \dot{x}_1 = \dot{y} = x_2$ $\quad = -a_1 x_2 - a_2 x_1 + w$
$\quad x_2 = \dot{y}$ $\qquad\qquad \dot{x}_2 = \ddot{y}$ $\implies$
*$w(t)$ random*

Here we assume that our basic state-variable model is given by (15-17) and (15-18), except that now $\mathbf{w}(k)$ and $\mathbf{v}(k)$ are correlated, i.e.,

$$E\{\mathbf{w}(k)\mathbf{v}'(k)\} = \mathbf{S}(k) \neq \mathbf{0} \tag{22-11}$$

There are many approaches for treating correlated process and measurement noises, some leading to a recursive predictor, some to a recursive filter, and others to a filter in predictor–corrector form, as in the following:

**Theorem 22-2.** *When $\mathbf{w}(k)$ and $\mathbf{v}(k)$ are correlated, then a predictor–corrector form of the Kalman filter is*

$$\hat{\mathbf{x}}(k + 1|k) = \Phi(k + 1, k)\hat{\mathbf{x}}(k|k) + \Psi(k + 1, k)\mathbf{u}(k)$$
$$+ \Gamma(k + 1, k)\mathbf{S}(k)[\mathbf{H}(k)\mathbf{P}(k|k - 1)\mathbf{H}'(k) + \mathbf{R}(k)]^{-1}\tilde{\mathbf{z}}(k|k - 1) \tag{22-12}$$

$z = \ddot{y} + w = -a_1 \dot{y} - a_2 y + w + w$ *(note disturbance is here)*
$= -a_1 x_2 - a_2 x_1 + w + w$
*(ACT)*
$\implies z(t) = (-a_2 \quad -a_1)\begin{pmatrix} x_1 \\ x_2 \end{pmatrix} + (1 \quad 1)\begin{pmatrix} w(t) \\ w(t) \end{pmatrix}$
*$z(t)$ is correlated w/ $x_2$*
*Random stuff = $v_1(t)$*

$$E\left\{\mathbf{w}_{\phi\nu}\mathbf{v}_c'\right\} = E\left(\begin{matrix}\mathbf{u}_\phi\mathbf{w}_c\\ \mathbf{w}_r\,\mathbf{v}_c\end{matrix}\right) = \left(\begin{matrix}Q(t)\,\delta(t-\tau)\\ 0\end{matrix}\right) \neq \phi$$

*since v&w were uncorrelated.*

*and*

$$\hat{\mathbf{x}}(k+1|k+1) = \hat{\mathbf{x}}(k+1|k) + \mathbf{K}(k+1)\tilde{\mathbf{z}}(k+1|k) \qquad (22\text{-}13)$$

*where Kalman gain matrix* $\mathbf{K}(k+1)$ *is given by (17-12), filtering-error covariance matrix* $\mathbf{P}(k+1|k+1)$ *is given by (17-14), and predication-error covariance matrix* $\mathbf{P}(k+1|k)$ *is given by*

$$\mathbf{P}(k+1|k) = \mathbf{\Phi}_1(k+1,k)\mathbf{P}(k|k)\mathbf{\Phi}_1'(k+1,k) + \mathbf{Q}_1(k) \qquad (22\text{-}14)$$

*in which*

$$\mathbf{\Phi}_1(k+1,k) = \mathbf{\Phi}(k+1,k) - \mathbf{\Gamma}(k+1,k)\mathbf{S}(k)\mathbf{R}^{-1}(k)\mathbf{H}(k) \qquad (22\text{-}15)$$

*and*

$$\begin{aligned}\mathbf{Q}_1(k) = {} & \mathbf{\Gamma}(k+1,k)\mathbf{Q}(k)\mathbf{\Gamma}'(k+1,k)\\ & - \mathbf{\Gamma}(k+1,k)\mathbf{S}(k)\mathbf{R}^{-1}(k)\mathbf{S}'(k)\mathbf{\Gamma}'(k+1,k) \qquad (22\text{-}16)\end{aligned}$$

Observe that, if $\mathbf{S}(k) = \mathbf{0}$, then (22-12) reduces to the more familiar predictor equation (16-4), and (22-14) reduces to the more familiar (17-13).

*Proof.* The derivation of correction equation (22-13) is exactly the same, when $\mathbf{w}(k)$ and $\mathbf{v}(k)$ are correlated, as it was when $\mathbf{w}(k)$ and $\mathbf{v}(k)$ were assumed uncorrelated. See the proof of part (a) of Theorem 17-1 for the details so as to convince yourself of the truth of this.

To derive predictor equation (22-12), we begin with the fundamental theorem of estimation theory; i.e.,

$$\hat{\mathbf{x}}(k+1|k) = E\{\mathbf{x}(k+1)|\mathbf{Z}(k)\} \qquad (22\text{-}17)$$

Substitute state equation (15-17) into (22-17) to show that

$$\begin{aligned}\hat{\mathbf{x}}(k+1|k) = {} & \mathbf{\Phi}(k+1,k)\hat{\mathbf{x}}(k|k) + \mathbf{\Psi}(k+1,k)\mathbf{u}(k)\\ & + \mathbf{\Gamma}(k+1,k)\underbrace{E\{\mathbf{w}(k)|\mathbf{Z}(k)\}}_{\text{no longer zero}} \qquad (22\text{-}18)\end{aligned}$$

Next, we develop an expression for $E\{\mathbf{w}(k)|\mathbf{Z}(k)\}$.

Let $\mathbf{Z}(k) = \text{col}\,(\mathbf{Z}(k-1), \mathbf{z}(k))$; then

$$\begin{aligned}E\{\mathbf{w}(k)|\mathbf{Z}(k)\} &= E\{\mathbf{w}(k)|\mathbf{Z}(k-1), \mathbf{z}(k)\}\\ &= E\{\mathbf{w}(k)|\mathbf{Z}(k-1), \tilde{\mathbf{z}}(k|k-1)\}\\ &= E\{\mathbf{w}(k)|\mathbf{Z}(k-1)\} + E\{\mathbf{w}(k)|\tilde{\mathbf{z}}(k|k-1)\} \qquad (22\text{-}19)\\ & \quad - E\{\mathbf{w}(k)\}\\ &= E\{\mathbf{w}(k)|\tilde{\mathbf{z}}(k|k-1)\}\end{aligned}$$

In deriving (22-19) we used the facts that $\mathbf{w}(k)$ is zero mean, and $\mathbf{w}(k)$ and $\mathbf{Z}(k-1)$ are statistically independent. Because $\mathbf{w}(k)$ and $\tilde{\mathbf{z}}(k|k-1)$ are jointly Gaussian,

$$E\{\mathbf{w}(k)|\tilde{\mathbf{z}}(k|k-1)\} = \mathbf{P}_{\mathbf{w}\tilde{\mathbf{z}}}(k, k|k-1)\mathbf{P}_{\tilde{\mathbf{z}}\tilde{\mathbf{z}}}^{-1}(k|k-1)\tilde{\mathbf{z}}(k|k-1) \qquad (22\text{-}20)$$

where $\mathbf{P}_{\tilde{\mathbf{z}}\tilde{\mathbf{z}}}$ is given by (16-33), and

$$\begin{aligned}\mathbf{P}_{\mathbf{w}\tilde{\mathbf{z}}}(k, k|k-1) &= \mathbf{E}\{\mathbf{w}(k)\tilde{\mathbf{z}}'(k|k-1)\} \\ &= \mathbf{E}\{\mathbf{w}(k)[\mathbf{H}(k)\tilde{\mathbf{x}}(k|k-1) + \mathbf{v}(k)]'\} \\ &= \mathbf{S}(k)\end{aligned} \tag{22-21}$$

In deriving (22-21) we used the facts that $\tilde{\mathbf{x}}(k|k-1)$ and $\mathbf{w}(k)$ are statistically independent, and $\mathbf{w}(k)$ is zero mean. Substituting (22-21) and (16-33) into (22-20), we find that

$$\mathbf{E}\{\mathbf{w}(k)|\tilde{\mathbf{z}}(k|k-1)\} = \mathbf{S}(k)[\mathbf{H}(k)\mathbf{P}(k|k-1)\mathbf{H}'(k) + \mathbf{R}(k)]^{-1}\tilde{\mathbf{z}}(k|k-1) \tag{22-22}$$

Substituting (22-22) into (22-19) and the resulting equation into (22-18) completes our derivation of the recursive predictor equation (22-12).

Equation (22-14) can be derived, along the lines of the derivation of (16-11), in a straightforward manner, although it is a bit tedious (Problem 22-1). A more interesting derivation of (22-14) is given in the Supplementary Material at the end of this lesson. □

Recall that the recursive predictor plays the predominant role in smoothing; hence, we present the following corollary:

**Corollary 22-1.** *When* $\mathbf{w}(\mathrm{k})$ *and* $\mathbf{v}(\mathrm{k})$ *are correlated, then a recursive predictor for* $\mathbf{x}(\mathrm{k}+1)$ *is*

$$\begin{aligned}\hat{\mathbf{x}}(k+1|k) = \mathbf{\Phi}(k+1, k)\hat{\mathbf{x}}(k|k-1) \\ + \mathbf{\Psi}(k+1, k)\mathbf{u}(k) + \mathbf{L}(k)\tilde{\mathbf{z}}(k|k-1)\end{aligned} \tag{22-23}$$

*where*

$$\begin{aligned}\mathbf{L}(k) = [\mathbf{\Phi}(k+1, k)\mathbf{P}(k|k-1)\mathbf{H}'(k) \\ + \mathbf{\Gamma}(k+1, k)\mathbf{S}(k)][\mathbf{H}(k)\mathbf{P}(k|k-1)\mathbf{H}'(k) + \mathbf{R}(k)]^{-1}\end{aligned} \tag{22-24}$$

*and*

$$\begin{aligned}\mathbf{P}(k+1|k) = &[\mathbf{\Phi}(k+1, k) - \mathbf{L}(k)\mathbf{H}(k)]\mathbf{P}(k|k-1) \\ &[\mathbf{\Phi}(k+1, k) - \mathbf{L}(k)\mathbf{H}(k)]' \\ &+ \mathbf{\Gamma}(k+1, k)\mathbf{Q}(k)\mathbf{\Gamma}'(k+1, k) \\ &- \mathbf{\Gamma}(k+1, k)\mathbf{S}(k)\mathbf{L}'(k) \\ &- \mathbf{L}(k)\mathbf{S}'(k)\mathbf{\Gamma}'(k+1, k) \\ &+ \mathbf{L}(k)\mathbf{R}(k)\mathbf{L}'(k)\end{aligned} \tag{22-25}$$

*Proof.* These results follow directly from Theorem 22-2; or they can be derived in an independent manner, as explained in Problem 22-1. □

**Corollary 22-2.** *When* $\mathbf{w}(\mathrm{k})$ *and* $\mathbf{v}(\mathrm{k})$ *are correlated, then a recursive filter for* $\mathbf{x}(\mathrm{k}+1)$ *is*

$$\hat{\mathbf{x}}(k+1|k+1) = \mathbf{\Phi}_1(k+1,k)\hat{\mathbf{x}}(k|k) + \mathbf{\Psi}(k+1,k)\mathbf{u}(k) + \mathbf{D}(k)\mathbf{z}(k) + \mathbf{K}(k+1)\tilde{\mathbf{z}}(k+1|k) \tag{22-26}$$

*where*

$$\mathbf{D}(k) = \mathbf{\Gamma}(k+1,k)\mathbf{S}(k)\mathbf{R}^{-1}(k) \tag{22-27}$$

*and all other quantities have been defined above.*

*Proof.* Again, these results follow directly from Theorem 22-2; however, they can also be derived, in a much more elegant and independent manner, as described in Problem 22-2. □

## COLORED NOISES

Quite often, some or all of the elements of either $\mathbf{v}(k)$ or $\mathbf{w}(k)$ or both are colored (i.e., have finite bandwidth). The following three-step procedure is used in these cases:

1. Model each colored noise by a low-order difference equation that is excited by white Gaussian noise.
2. Augment the states associated with the step 1 colored noise models to the original state-variable model.
3. Apply the recursive filter or predictor to the augmented system.

We try to model colored noise processes by low-order Markov sequences, i.e., low-order difference equations. Usually, first- or second-order models are quite adequate. Consider the following first-order model for colored noise process $\omega(k)$,

$$\omega(k+1) = \alpha\omega(k) + n(k) \tag{22-28}$$

In this model $n(k)$ is white noise with variance $\sigma_n^2$; thus, this model contains two parameters, $\alpha$ and $\sigma_n^2$, which must be determined from a priori knowledge about $\omega(k)$. We may know the amplitude spectrum of $\omega(k)$, correlation information about $\omega(k)$, steady-state variance of $\omega(k)$, etc. Two independent pieces of information are needed in order to uniquely identify $\alpha$ and $\sigma_n^2$.

**EXAMPLE 22-1**

We are given the facts that scalar noise $w(k)$ is stationary with the properties $\mathbf{E}\{w(k)\} = 0$ and $\mathbf{E}\{w(i)w(j)\} = e^{-2|j-i|}$. A first-order Markov model for $w(k)$ can easily be obtained as

$$\xi(k+1) = e^{-2}\xi(k) + \sqrt{1-e^{-4}}\,n(k) \tag{22-29}$$

$$w(k) = \xi(k) \tag{22-30}$$

where $\mathbf{E}\{\xi(0)\} = 0$, $\mathbf{E}\{\xi^2(0)\} = 1$, $\mathbf{E}\{n(k)\} = 0$, and $\mathbf{E}\{n(i)n(j)\} = \delta_{ij}$. □

**EXAMPLE 22-2**

Here we illustrate the state augmentation procedure for the first-order system

$$\chi(k+1) = a_1\chi(k) + \omega(k) \tag{22-31}$$

$$z(k+1) = h\chi(k+1) + \nu(k+1) \tag{22-32}$$

where $\omega(k)$ is a first-order Markov sequence, i.e.,

$$\omega(k+1) = a_2\omega(k) + n(k) \tag{22-33}$$

and $\nu(k)$ and $n(k)$ are white noise processes. We *augment* (22-33) to (22-31), as follows. Let

$$\mathbf{x}(k) = \text{col}\,(\chi(k), \omega(k)) \tag{22-34}$$

Then (22-31) and (22-33) can be combined to give

$$\underbrace{\begin{pmatrix} \chi(k+1) \\ \omega(k+1) \end{pmatrix}}_{\mathbf{x}(k+1)} = \underbrace{\begin{pmatrix} a_1 & 1 \\ 0 & a_2 \end{pmatrix}}_{\mathbf{\Phi}} \underbrace{\begin{pmatrix} \chi(k) \\ \omega(k) \end{pmatrix}}_{\mathbf{x}(k)} + \underbrace{\begin{pmatrix} 0 \\ 1 \end{pmatrix}}_{\mathbf{\Upsilon}} n(k) \tag{22-35}$$

Equation (22-25) is our *augmented state equation*. Observe that it is once again excited by a white noise process, just as our basic state equation (15-17) is.

To complete the description of the augmented state-variable model, we must express measurement $z(k+1)$ in terms of the augmented state vector, $\mathbf{x}(k+1)$; i.e.,

$$z(k+1) = \underbrace{(h \quad 0)}_{\mathbf{H}} \underbrace{\begin{pmatrix} \chi(k+1) \\ \omega(k+1) \end{pmatrix}}_{\mathbf{x}(k+1)} + \nu(k+1) \tag{22-36}$$

Equations (22-35) and (22-36) constitute the augmented state-variable model. We observe that, when the original process noise is colored and the measurement noise is white, the state augmentation procedure leads us once again to a basic (augmented) state-variable model, one that is of higher dimension than the original model because of the modeled colored process noise. Hence, in this case we can apply all our state estimation algorithms to the augmented state-variable model. □

**EXAMPLE 22-3**

Here we consider the situation where the process noise is white but the measurement noise is colored, again for a first-order system:

$$\chi(k+1) = a_1\chi(k) + \omega(k) \tag{22-37}$$

$$z(k+1) = h\chi(k+1) + \nu(k+1) \tag{22-38}$$

As in the preceding example, we model $\nu(k)$ by the following first-order Markov sequence

$$\nu(k+1) = a_2\nu(k) + n(k) \tag{22-39}$$

where $n(k)$ is white noise. Augmenting (22-39) to (22-37) and reexpressing (22-38) in terms of the augmented state vector $\mathbf{x}(k)$, where

$$\mathbf{x}(k) = \text{col}\,(\chi(k), \nu(k)) \tag{22-40}$$

we obtain the augmented state-variable model

$$\underbrace{\begin{pmatrix} \chi(k+1) \\ \nu(k+1) \end{pmatrix}}_{\mathbf{x}(k+1)} = \underbrace{\begin{pmatrix} a_1 & 0 \\ 0 & a_2 \end{pmatrix}}_{\mathbf{\Phi}} \underbrace{\begin{pmatrix} \chi(k) \\ \nu(k) \end{pmatrix}}_{\mathbf{x}(k)} + \underbrace{\begin{pmatrix} 1 & 0 \\ 0 & 1 \end{pmatrix}}_{\mathbf{\Gamma}} \underbrace{\begin{pmatrix} \omega(k) \\ n(k) \end{pmatrix}}_{\mathbf{w}(k)} \tag{22-41}$$

and

$$z(k+1) = \underbrace{\begin{pmatrix} h & 1 \end{pmatrix}}_{\mathbf{H}} \underbrace{\begin{pmatrix} \chi(k+1) \\ \nu(k+1) \end{pmatrix}}_{\mathbf{x}(k+1)} \tag{22-42}$$

Observe that a vector process noise now excites the augmented state equation and that there is no measurement noise in the measurement equation. This second observation can lead to serious numerical problems in our state estimators, because it means that we must set $\mathbf{R} = \mathbf{0}$ in those estimators, and, when we do this, covariance matrices become and remain singular. □

Let us examine what happens to $\mathbf{P}(k+1|k+1)$ when covariance matrix $\mathbf{R}$ is set equal to zero. From (17-14) and (17-12) (in which we set $\mathbf{R} = \mathbf{0}$), we find that

$$\mathbf{P}(k+1|k+1) = \mathbf{P}(k+1|k) - \mathbf{P}(k+1|k)\mathbf{H}'(k+1) \\ [\mathbf{H}(k+1)\mathbf{P}(k+1|k)\mathbf{H}'(k+1)]^{-1}\mathbf{H}(k+1)\mathbf{P}(k+1|k) \tag{22-43}$$

Multiplying both sides of (22-43) on the right by $\mathbf{H}'(k+1)$, we find that

$$\mathbf{P}(k+1|k+1)\mathbf{H}'(k+1) = \mathbf{0} \tag{22-44}$$

Because $\mathbf{H}'(k+1)$ is a nonzero matrix, (22-44) implies that $\mathbf{P}(k+1|k+1)$ must be a singular matrix. We leave it to the reader to show that once $\mathbf{P}(k+1|k+1)$ becomes singular it remains singular for all other values of $k$.

**EXAMPLE 22-4**

Consider the first-order system

$$x(k+1) = -\frac{x(k)}{8} + w(k) \tag{22-45}$$

$$z(k+1) = x(k+1) + \nu(k+1) \tag{22-46}$$

where $w(k)$ is white and Gaussian $[w(k) \sim N(w(k); 0, 1)]$ and $\nu(k)$ is the noise process that is summarized in Figure 22-1, in which noise $n(k)$ is also white and Gaussian $[n(k) \sim N(n(k); 0, \frac{1}{4})]$. Our objective is to obtain a filtered estimate of $x(k)$.

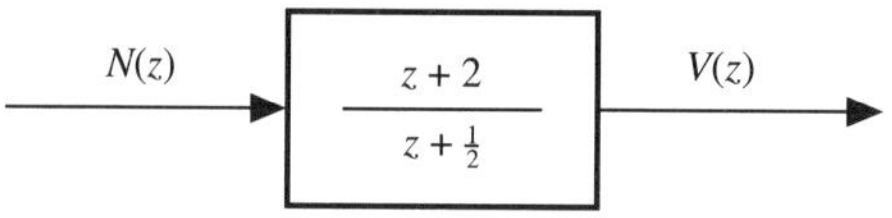


**Figure 22-1** Noise model.

To begin, we shall demonstrate that passing white noise $n(k)$ through the all-pass filter $(z+2)/(z+1/2)$ once again leads to another white noise sequence, $\nu(k)$. To show that $\nu(k)$ is white we compute its power spectrum, using the fact that

$$S_{VV}(\omega) = |H(\omega)|^2 S_{NN}(\omega) \tag{22-47}$$

in which

$$H(\omega) = \frac{z+2}{z+1/2}, \qquad \text{when } z = e^{j\omega} \tag{22-48}$$

so that

$$\begin{aligned} |H(\omega)|^2 &= H^*(\omega)H(\omega) = \frac{(e^{-j\omega}+2)(e^{j\omega}+2)}{4(2e^{-j\omega}+1)(2e^{j\omega}+1)} \\ &= \frac{4(5+2\cos\omega)}{(5+2\cos\omega)} = 4 \end{aligned} \tag{22-49}$$

Consequently,

$$S_{VV}(\omega) = 4S_{NN}(\omega) \tag{22-50}$$

This shows that the power spectrum of $v(k)$ is a constant; hence $v(k)$ is a white process.

Next, we must obtain a state-variable model for the Figure 22-1 system. From an input/output point of view, the additive noise system is given by

$$V(z) = \frac{z+2}{(z+1/2)}N(z) = N(z) + \frac{3}{2}\frac{1}{(z+1/2)}N(z) \tag{22-51}$$

Hence,

$$v(k) = n(k) + x_1(k) \tag{22-52}$$

where

$$x_1(k) = \mathcal{Z}^{-1}\left\{\frac{3}{2}\left(\frac{1}{z+1/2)}\right)N(z)\right\} \tag{22-53}$$

From this last equation, it follows that

$$x_1(k+1) = -\frac{x_1(k)}{2} + \frac{3}{2}n(k) \tag{22-54}$$

Equations (22-54) and (22-52) are the state-variable model for the Figure 22-1 system.

To obtain a filtered estimate for state $x(k)$, we must first combine the models in (22-45), (22-46), (22-54), and (22-52), by means of the state augmentation procedure. Letting $\mathbf{x}(k) \triangleq \text{col}\,[x(k), x_1(k)]$, we obtain the following augmented state-variable model:

$$\mathbf{x}(k+1) = \begin{bmatrix} -\frac{1}{8} & 0 \\ 0 & -\frac{1}{2} \end{bmatrix}\mathbf{x}(k) + \begin{bmatrix} w(k) \\ \frac{3}{2}n(k) \end{bmatrix} \tag{22-55}$$

$$z(k+1) = (1 \;\; 1)\mathbf{x}(k+1) + n(k+1) \tag{22-56}$$

This state-variable model has correlated process and measurement noises, because $n(k)$ appears in both noises; hence,

$$\mathbf{S}(k) = \mathbf{E}\left\{\begin{bmatrix} w(k) \\ \frac{3}{2}n(k) \end{bmatrix} n(k)\right\} = \begin{bmatrix} 0 \\ \frac{3}{8} \end{bmatrix} \tag{22-57}$$

Using Theorem 22-2, we can compute $\hat{\mathbf{x}}(k+1|k)$ and $\hat{\mathbf{x}}(k+1|k+1)$. To obtain the desired result, $\hat{x}(k+1|k+1)$, just pick out the first element of $\hat{\mathbf{x}}(k+1|k+1)$, that is $\hat{x}(k+1|k+1) = (1 \;\; 0)\hat{\mathbf{x}}(k+1|k+1)$. We could also use Corollary 22-2 directly to obtain $\hat{\mathbf{x}}(k+1|k+1)$.

The key quantities that are needed to implement Theorem 22-2 are $\mathbf{\Phi}_1$ and $\mathbf{Q}_1$, where

$$\mathbf{\Phi}_1 = \mathbf{\Phi} - \mathbf{\Gamma}\mathbf{S}\mathbf{R}^{-1}\mathbf{H} = \begin{bmatrix} -\frac{1}{8} & 0 \\ 0 & -\frac{1}{2} \end{bmatrix} - \begin{bmatrix} 0 \\ \frac{3}{8} \end{bmatrix} 4(1 \;\; 1) = \begin{bmatrix} -\frac{1}{8} & 0 \\ -\frac{3}{2} & -2 \end{bmatrix} \qquad (22\text{-}58)$$

and

$$\mathbf{Q}_1 = \mathbf{\Gamma}\mathbf{Q}\mathbf{\Gamma}' - \mathbf{\Gamma}\mathbf{S}\mathbf{R}^{-1}\mathbf{S}'\mathbf{\Gamma}' = \begin{bmatrix} 1 & 0 \\ 0 & \frac{9}{16} \end{bmatrix} - \begin{bmatrix} 0 \\ \frac{3}{8} \end{bmatrix} 4(0 \;\; \frac{3}{8}) = \begin{bmatrix} 1 & 0 \\ 0 & 0 \end{bmatrix} \quad \square \qquad (22\text{-}59)$$

## PERFECT MEASUREMENTS: REDUCED-ORDER ESTIMATORS

*What if you have perfect measurements?*

We have just seen that when $\mathbf{R} = \mathbf{0}$ (or, in fact, even if some, but not all, measurements are perfect) numerical problems can occur in the Kalman filter. One way to circumvent these problems is ad hoc, and that is to use small values for the elements of covariance matrix $\mathbf{R}$, even though measurements are thought to be perfect. Doing this has a stabilizing effect on the numerics of the Kalman filter.

A second way to circumvent these problems is to recognize that a set of "perfect" measurements reduces the number of states that have to be estimated. Suppose, for example, that there are $l$ perfect measurements and that state vector $\mathbf{x}(k)$ is $n \times 1$. Then, *we conjecture that we ought to be able to estimate* $\mathbf{x}(k)$ *by a Kalman filter whose dimension is no greater than* $n - l$. Such an estimator will be referred to as a *reduced-order estimator*. The payoff for using a reduced-order estimator is fewer computations and less storage.

To illustrate an approach to designing a reduced-order estimator, we limit our discussions in this section to the following time-invariant and stationary basic state-variable model in which $\mathbf{u}(k) \triangleq \mathbf{0}$ and *all* measurements are perfect,

$$\mathbf{x}(k + 1) = \mathbf{\Phi}\mathbf{x}(k) + \mathbf{\Gamma}\mathbf{w}(k) \qquad (22\text{-}60)$$

$$\mathbf{y}(k + 1) = \mathbf{H}\mathbf{x}(k + 1) \quad \textit{perfect measurement} \qquad (22\text{-}61)$$

In this model $\mathbf{y}$ is $l \times 1$. What makes the design of a reduced-order estimator challenging is the fact that the $l$ perfect measurements are linearly related to the $n$ states; i.e., $\mathbf{H}$ is rectangular.

To begin, we introduce a *reduced-order state vector*, $\mathbf{p}(k)$, whose dimension is $(n - l) \times 1$; $\mathbf{p}(k)$ is assumed to be a linear transformation of $\mathbf{x}(k)$; i.e.,

$$\mathbf{p}(k) \triangleq \mathbf{C}\mathbf{x}(k) \qquad (22\text{-}62)$$

Augmenting (22-62) to (22-61), we obtain

$$\begin{pmatrix} \mathbf{y}(k) \\ \mathbf{p}(k) \end{pmatrix} = \begin{pmatrix} \mathbf{H} \\ \hline \mathbf{C} \end{pmatrix} \mathbf{x}(k) \qquad (22\text{-}63)$$

Design matrix $\mathbf{C}$ is chosen so that $(\mathbf{H}/\mathbf{C})$ is invertible. Of course, many different choices of $\mathbf{C}$ are possible; thus, this first step of our reduced-order estimator design procedure is nonunique. Let

$$\begin{pmatrix} \mathbf{H} \\ \hline \mathbf{C} \end{pmatrix}^{-1} = (\mathbf{L}_1 | \mathbf{L}_2) \tag{22-64}$$

where $\mathbf{L}_1$ is $n \times l$ and $\mathbf{L}_2$ is $n \times (n - l)$; thus,

$$\mathbf{x}(k) = \mathbf{L}_1\mathbf{y}(k) + \mathbf{L}_2\mathbf{p}(k) \tag{22-65}$$

To obtain a filtered estimate of $\mathbf{x}(k)$, we operate on both sides of (22-65) with $\mathbf{E}\{\cdot|\mathbf{Y}(k)\}$, where

$$\mathbf{Y}(k) = \text{col}\,(\mathbf{y}(1), \mathbf{y}(2), \ldots, \mathbf{y}(k)) \tag{22-66}$$

Doing this, we find that

$$\hat{\mathbf{x}}(k|k) = \mathbf{L}_1\mathbf{y}(k) + \mathbf{L}_2\hat{\mathbf{p}}(k|k) \tag{22-67}$$

which is a *reduced-order estimator for* $\mathbf{x}(k)$. Of course, to evaluate $\hat{\mathbf{x}}(k|k)$, we must develop a reduced-order Kalman filter to estimate $\mathbf{p}(k)$. Knowing $\hat{\mathbf{p}}(k|k)$ and $\mathbf{y}(k)$, it is then a simple matter to compute $\hat{\mathbf{x}}(k|k)$, using (22-67).

To obtain $\hat{\mathbf{p}}(k|k)$, using our previously derived Kalman filter algorithm, we first must establish a state-variable model for $\mathbf{p}(k)$. A state equation for $\mathbf{p}$ is easily obtained, as follows:

$$\begin{aligned} \mathbf{p}(k+1) &= \mathbf{C}\mathbf{x}(k+1) = \mathbf{C}[\mathbf{\Phi}\mathbf{x}(k) + \mathbf{\Gamma}\mathbf{w}(k)] \\ &= \mathbf{C}\mathbf{\Phi}[\mathbf{L}_1\mathbf{y}(k) + \mathbf{L}_2\mathbf{p}(k)] + \mathbf{C}\mathbf{\Gamma}\mathbf{w}(k) \\ &= \mathbf{C}\mathbf{\Phi}\mathbf{L}_2\mathbf{p}(k) + \mathbf{C}\mathbf{\Phi}\mathbf{L}_1\mathbf{y}(k) + \mathbf{C}\mathbf{\Gamma}\mathbf{w}(k) \end{aligned} \tag{22-68}$$

Observe that this state equation is driven by white noise $\mathbf{w}(k)$ and the known *forcing function*, $\mathbf{y}(k)$.

A measurement equation is obtained from (22-61) as

$$\begin{aligned} \mathbf{y}(k+1) &= \mathbf{H}\mathbf{x}(k+1) = \mathbf{H}[\mathbf{\Phi}\mathbf{x}(k) + \mathbf{\Gamma}\mathbf{w}(k)] \\ &= \mathbf{H}\mathbf{\Phi}[\mathbf{L}_1\mathbf{y}(k) + \mathbf{L}_2\mathbf{p}(k)] + \mathbf{H}\mathbf{\Gamma}\mathbf{w}(k) \\ &= \mathbf{H}\mathbf{\Phi}\mathbf{L}_2\mathbf{p}(k) + \mathbf{H}\mathbf{\Phi}\mathbf{L}_1\mathbf{y}(k) + \mathbf{H}\mathbf{\Gamma}\mathbf{w}(k) \end{aligned} \tag{22-69}$$

At time $k + 1$, we know $\mathbf{y}(k)$; hence, we can reexpress (22-69) as

$$\mathbf{y}_1(k+1) = \mathbf{H}\mathbf{\Phi}\mathbf{L}_2\mathbf{p}(k) + \mathbf{H}\mathbf{\Gamma}\mathbf{w}(k) \tag{22-70}$$

where

$$\mathbf{y}_1(k+1) \triangleq \mathbf{y}(k+1) - \mathbf{H}\mathbf{\Phi}\mathbf{L}_1\mathbf{y}(k) \tag{22-71}$$

Before proceeding any further, we make some important observations about our state-variable model in (22-68) and (22-70). First, the new measurement $\mathbf{y}_1(k+1)$ represents a weighted difference between measurements $\mathbf{y}(k+1)$ and $\mathbf{y}(k)$. The technique for obtaining our reduced-order state-variable model is, therefore, sometimes referred to as a *measurement-differencing technique* (e.g., Bryson and Johansen, 1965). Because we have already used $\mathbf{y}(k)$ to reduce the dimension of $\mathbf{x}(k)$ from $n$ to $n - l$, we cannot again use $\mathbf{y}(k)$ alone as the measurements in our reduced-order state-variable model. As we have just seen, we must use both $\mathbf{y}(k)$ and $\mathbf{y}(k+1)$.

Second, measurement equation (22-70) appears to be a combination of signal and noise. Unless $\mathbf{H\Gamma} = \mathbf{0}$, the term $\mathbf{H\Gamma w}(k)$ will act as the measurement noise in our reduced-order state-variable model. Its covariance matrix is $\mathbf{H\Gamma Q\Gamma' H'}$. Unfortunately, it is possible for $\mathbf{H\Gamma}$ to equal the zero matrix. From linear system theory, we known that $\mathbf{H\Gamma}$ is the matrix of first Markov parameters for our original system in (22-60) and (22-61), and $\mathbf{H\Gamma}$ may equal zero. If this occurs, we must repeat all the above until we obtain a reduced-order state vector whose measurement equation is excited by white noise. We see, therefore, that depending on system dynamics, it is possible to obtain a reduced-order estimator of $\mathbf{x}(k)$ that uses a reduced-order Kalman filter of dimension less than $n - l$.

Third, the noises that appear in state equation (22-68) and measurement equation (22-70) are the same, $\mathbf{w}(k)$; hence, the reduced-order state-variable model involves the correlated noise case that we described before in this chapter in the section entitled Correlated Noises.

Finally, and most important, measurement equation (22-70) is nonstandard, in that it expresses $\mathbf{y}_1$ at $k+1$ in terms of $\mathbf{p}$ at $k$ rather than $\mathbf{p}$ at $k+1$. Recall that the measurement equation in our basic state-variable model is $\mathbf{z}(k+1) = \mathbf{Hx}(k+1) + \mathbf{v}(k+1)$. We cannot immediately apply our Kalman filter equations to (22-68) and (22-70) until we express (22-70) in the standard way.

To proceed, we let

$$\zeta(k) \triangleq \mathbf{y}_1(k+1) \tag{22-72}$$

so that

$$\zeta(k) = \mathbf{H\Phi L}_2\mathbf{p}(k) + \mathbf{H\Gamma w}(k) \tag{22-73}$$

Measurement equation (22-73) is now in the standard from; however, because $\zeta(k)$ equals a future value of $\mathbf{y}_1$, that is, $\mathbf{y}(k+1)$, we must be very careful in applying our estimator formulas to our reduced-order model (22-68) and (22-73).

To see this more clearly, we define the following two data sets:

$$\mathcal{D}_{\mathbf{y}_1}(k+1) = \{\mathbf{y}_1(1), \mathbf{y}_1(2), \ldots, \mathbf{y}_1(k+1), \ldots\} \tag{22-74}$$

and

$$\mathcal{D}_\zeta(k) = \{\zeta(0), \zeta(1), \ldots, \zeta(k), \ldots\} \tag{22-75}$$

Obviously,

$$\mathcal{D}_{\mathbf{y}_1}(k+1) = \mathcal{D}_\zeta(k) \tag{22-76}$$

Letting

$$\hat{\mathbf{p}}_{\mathbf{y}_1}(k+1|k+1) = \mathbf{E}\{\mathbf{p}(k+1)|\mathcal{D}_{\mathbf{y}_1}(k+1)\} \tag{22-77}$$

and

$$\hat{\mathbf{p}}_\zeta(k+1|k) = \mathbf{E}\{\mathbf{p}(k+1)|\mathcal{D}_\zeta(k)\} \tag{22-78}$$

we see that

$$\hat{\mathbf{p}}_{\mathbf{y}_1}(k+1|k+1) = \hat{\mathbf{p}}_\zeta(k+1|k) \tag{22-79}$$

Equation (22-79) tells us to obtain a recursive filter for our reduced-order model; that is, in terms of data set $\mathcal{D}_{\mathbf{y}_1}(k+1)$, we must first obtain a recursive predictor for that model, which is in terms of data set $\mathcal{D}_\zeta(k)$. Then, wherever $\zeta(k)$ appears in the recursive predictor, it can be replaced by $\mathbf{y}_1(k+1)$.

Using Corollary 22-1, applied to the reduced-order model in (22-68) and (22-73), we find that

$$\hat{\mathbf{p}}_\zeta(k+1|k) = \mathbf{C\Phi L}_2\hat{\mathbf{p}}_\zeta(k|k-1) + \mathbf{C\Phi L}_1\mathbf{y}(k) + \mathbf{L}(k)[\zeta(k) - \mathbf{H\Phi L}_2\hat{\mathbf{p}}_\zeta(k|k-1)] \tag{22-80}$$

Thus,

$$\hat{\mathbf{p}}_{\mathbf{y}_1}(k+1|k+1) = \mathbf{C\Phi L}_2\hat{\mathbf{p}}_{\mathbf{y}_1}(k|k) + \mathbf{C\Phi L}_1\mathbf{y}(k) + \mathbf{L}(k)[\mathbf{y}_1(k+1) - \mathbf{H\Phi L}_2\hat{\mathbf{p}}_{\mathbf{y}_1}(k|k)] \tag{22-81}$$

Equation (22-81) is our reduced-order Kalman filter. It provides filtered estimates of $\mathbf{p}(k+1)$ and is only of dimension $(n-l)\times 1$. Of course, when $\mathbf{L}(k)$ and $\mathbf{P}_{\mathbf{y}_1}(k+1|k+1)$ are computed using (22-24) and (22-25), respectively, we must make the following substitutions: $\mathbf{\Phi}(k+1,k) \rightarrow \mathbf{C\Phi L}_2$, $\mathbf{H}(k) \rightarrow \mathbf{H\Phi L}_2$, $\mathbf{\Gamma}(k+1,k) \rightarrow \mathbf{C\Gamma}$, $\mathbf{Q}(k) \rightarrow \mathbf{Q}$, $\mathbf{S}(k) \rightarrow \mathbf{Q\Gamma'H'}$, and $\mathbf{R}(k) \rightarrow \mathbf{H\Gamma Q\Gamma' H'}$.

## FINAL REMARK

To see the forest from the trees, we have considered each of our special cases separately. In actual practice, some or all of them may occur simultaneously. We have already seen one illustration of this in Example 22-4. As a guideline, you should handle the effects in your state-variable model in the order in which they were treated in this lesson. The exercises at the end of this lesson will permit you to gain experience with such cases.

## COMPUTATION

### Biases

Follow the procedure given in the text, which reduces a state-variable model that contains biases to our basic state-variable model. Then see the discussions in Lessons 17, 19, or 21 in order to do Kalman filtering or smoothing.

### Correlated Noises

There is no M-file to do Kalman filtering for a time-varying or nonstationary state-variable model in which process and measurement noises are correlated. It is possible to modify our Kalman filter M-file, given in Lesson 17, to this case; and we leave this to the reader.

Using the *Control System Toolbox*, you must use two M-files. One (**dlqew**) provides gain, covariance, and eigenvalue information about the steady-state KF; the other (**destim**) computes both the steady-state **predicted** and **filtered** values for $\mathbf{x}(k)$.

**dlqew**: Discrete linear quadratic estimator design. Computes the steady-state Kalman gain matrix, predictor, and filter steady-state covariance matrices and closed-loop eigenvalues of the predictor. Does this for the following model (here we use the notation used in the *Control System Toolbox* reference manual):

$$\mathbf{x}[n+1] = \mathbf{A}\mathbf{x}[n] + \mathbf{B}\mathbf{u}[n] + \mathbf{G}\mathbf{w}[n]$$
$$\mathbf{y}[n] = \mathbf{C}\mathbf{x}[n] + \mathbf{D}\mathbf{u}[n] + \mathbf{J}\mathbf{w}[n] + \mathbf{v}[n]$$

where $\mathbf{E}\{\mathbf{w}[n]\} = \mathbf{E}\{\mathbf{v}[n]\} = \mathbf{0}, \mathbf{E}\{\mathbf{w}[n]\mathbf{w}'[n]\} = \mathbf{Q}, \mathbf{E}\{\mathbf{v}[n]\mathbf{v}'[n]\} = \mathbf{R}$, and $\mathbf{E}\{\mathbf{w}[n]\mathbf{v}'[n]\} = \mathbf{0}$. The steady-state Kalman gain matrix, computed by **dlqew**, must be passed on to **destim**.

**destim**: Forms steady-state Kalman filter. Does this for the following model (here we use the notation used in the *Control System Toolbox* reference manual):

$$\mathbf{x}[n+1] = \mathbf{A}\mathbf{x}[n] + \mathbf{B}\mathbf{u}[n]$$
$$\mathbf{y}[n] = \mathbf{C}\mathbf{x}[n] + \mathbf{D}\mathbf{u}[n]$$

The state-variable model for the case of correlated noises can be put into the form of this model, by expressing it as

$$\mathbf{x}[n+1] = \mathbf{A}\mathbf{x}[n] + [\mathbf{B}_1|\mathbf{B}_2]\text{col}\,[\mathbf{u}[n], \mathbf{w}[n]]$$
$$\begin{bmatrix}\mathbf{y}[n]\\ \mathbf{z}[n]\end{bmatrix} = \begin{bmatrix}\mathbf{C}_1\\ \mathbf{C}_2\end{bmatrix}\mathbf{x}[n] + \begin{bmatrix}\mathbf{D}_{11} & \mathbf{D}_{12}\\ \mathbf{D}_{21} & \mathbf{D}_{22}\end{bmatrix}\begin{bmatrix}\mathbf{u}[n]\\ \mathbf{w}[n]\end{bmatrix}$$

where $\mathbf{y}[n]$ are the sensor outputs and $\mathbf{z}[n]$ are the remaining plant outputs (if there are none, then set $\mathbf{C}_2 = \mathbf{D}_{21} = \mathbf{D}_{22} = \mathbf{0}$). The partitioned matrices in this last system can then be equated to the plant matrices in the preceding basic state-variable model. Outputs steady-state predicted and filtered values for $\mathbf{x}(k)$, as well as the predicted value of the measurement vector $\mathbf{y}[n]$, all as functions of time.

### Colored Noises

State augmentation is the workhorse for handling colored noises. It can be automatically accomplished for continuous-time systems (but not for discrete-time systems) by using the following M-file from the *Control System Toolbox*:

**apend**: Combines dynamics of two (continuous-time) state-space systems

### Perfect Measurements

Here you must follow the text very closely. The state-variable model is given in Equations (22-68) and (22-73). It contains a known forcing function in the state equation and has correlated noises. See the subsection on Correlated Noises for how to proceed. Below Equation (22-79), it is stated that you need to obtain a recursive predictor for this model in order to compute the recursive filter for it.

## DERIVATION OF EQUATION (22-14)

The equation given in Theorem 22-2 for $\mathbf{P}(k+1|k)$ has a very pleasant form; i.e., it looks just like the usual prediction error covariance equation, which is given in (16-11), except that $\mathbf{\Phi}$ and $\mathbf{Q}$ have been replaced by $\mathbf{\Phi}_1$ and $\mathbf{Q}_1$, respectively. Of course, (22-14) can be obtained by a brute-force derivation that makes extensive use of (22-12) and (22-13). Here we provide a more interesting derivation of (22-14). To begin, we need the following:

**Theorem 22-3.** *The following identity is true:*

$$[\mathbf{H}(k)\mathbf{P}(k|k-1)\mathbf{H}'(k)+\mathbf{R}(k)]^{-1}\tilde{\mathbf{z}}(k|k-1) = \mathbf{R}^{-1}(k)[\mathbf{z}(k)-\mathbf{H}(k)\hat{\mathbf{x}}(k|k)] \qquad (22\text{-}82)$$

*Proof.* From (22-13), we know that

$$\hat{\mathbf{x}}(k|k) = \hat{\mathbf{x}}(k|k-1) + \mathbf{K}(k)\tilde{\mathbf{z}}(k|k-1); \qquad (22\text{-}83)$$

Hence, (22-82) can be written as

$$\mathbf{R}(k)[\mathbf{H}(k)\mathbf{P}(k|k-1)\mathbf{H}'(k)+\mathbf{R}(k)]^{-1}\tilde{\mathbf{z}}(k|k-1) = \mathbf{z}(k) - \mathbf{H}(k)\hat{\mathbf{x}}(k|k-1) - \mathbf{H}(k)\mathbf{K}(k)\tilde{\mathbf{z}}(k|k-1)$$

or

$$\{\mathbf{R}(k)[\mathbf{H}(k)\mathbf{P}(k|k-1)\mathbf{H}'(k)+\mathbf{R}(k)]^{-1} + \mathbf{H}(k)\mathbf{K}(k)\}\tilde{\mathbf{z}}(k|k-1) = \mathbf{z}(k) - \mathbf{H}(k)\hat{\mathbf{x}}(k|k-1) = \tilde{\mathbf{z}}(k|k-1)$$

or

$$\{\mathbf{R}(k)[\mathbf{H}(k)\mathbf{P}(k|k-1)\mathbf{H}'(k)+\mathbf{R}(k)]^{-1} + \mathbf{H}(k)\mathbf{P}(k|k-1)\mathbf{H}'(k)[\mathbf{H}(k)\mathbf{P}(k|k-1)\mathbf{H}'(k) + \mathbf{R}(k)]^{-1}\}\tilde{\mathbf{z}}(k|k-1) = \tilde{\mathbf{z}}(k|k-1)$$

Hence,

$$[\mathbf{H}(k)\mathbf{P}(k|k-1)\mathbf{H}'(k)+\mathbf{R}(k)][\mathbf{H}(k)\mathbf{P}(k|k-1)\mathbf{H}'(k)+\mathbf{R}(k)]^{-1}\tilde{\mathbf{z}}(k|k-1) = \tilde{\mathbf{z}}(k|k-1)$$

which reduces to the true statement that $\tilde{\mathbf{z}}(k|k-1) = \tilde{\mathbf{z}}(k|k-1)$. □

The importance of (22-82) is that it lets us reexpress the predictor, which is given in (22-12), as

$$\hat{\mathbf{x}}(k+1|k) = \mathbf{\Phi}_1(k+1,k)\hat{\mathbf{x}}(k|k) + \mathbf{\Psi}(k+1,k)\mathbf{u}(k) + \mathbf{\Gamma}(k+1,k)\mathbf{S}(k)\mathbf{R}^{-1}(k)\mathbf{z}(k) \qquad (22\text{-}84)$$

where

$$\mathbf{\Phi}_1(k+1,k) = \mathbf{\Phi}(k+1,k) - \mathbf{\Gamma}(k+1,k)\mathbf{S}(k)\mathbf{R}^{-1}(k)\mathbf{H}(k) \qquad (22\text{-}85)$$

Next, we show that there is a related state equation for which (22-84) is the natural predictor. The main ideas behind obtaining this related state equation are stated in Problem 22-2.

We begin by adding a convenient form of zero to state equation (15-17) in order to decorrelate the process noise in the modified basic state-variable model from the measurement noise $\mathbf{v}(k)$. We add $\mathbf{D}(k)[\mathbf{z}(k) - \mathbf{H}(k)\mathbf{x}(k) - \mathbf{v}(k)]$ to (15-17). The process noise, $\mathbf{w}_1(k)$, in the modified basic state-variable model, is equal to $\mathbf{\Gamma}(k+1,k)\mathbf{w}(k) - \mathbf{D}(k)\mathbf{v}(k)$. We choose the *decorrelation* matrix $\mathbf{D}(k)$ so that $\mathbf{E}\{\mathbf{w}_1(k)\mathbf{v}'(k)\} = \mathbf{0}$, from which we find that

$$\mathbf{D}(k) = \mathbf{\Gamma}(k+1,k)\mathbf{S}(k)\mathbf{R}^{-1}(k) \tag{22-86}$$

so that the modified basic state-variable model can be reexpressed as

$$\mathbf{x}(k+1) = \mathbf{\Phi}_1(k+1,k)\mathbf{x}(k) + \mathbf{\Psi}(k+1,k)\mathbf{u}(k) + \mathbf{D}(k)\mathbf{z}(k) + \mathbf{w}_1(k) \tag{22-87}$$

Obviously, $\hat{\mathbf{x}}(k+1|k)$, given in (22-84), can be written by inspection of (22-87).

Return to Lesson 16 to see how $\mathbf{P}(k+1|k)$ was obtained. It is straightforward, upon comparison of (22-87) and (15-17), to show that

$$\mathbf{P}(k+1|k) = \mathbf{\Phi}_1(k+1,k)\mathbf{P}(k|k)\mathbf{\Phi}_1'(k+1,k) + \mathbf{Q}_1(k) \tag{22-88}$$

Finally, we need a formula for $\mathbf{Q}_1(k)$, the covariance matrix of $\mathbf{w}_1(k)$; but

$$\mathbf{w}_1(k) = \mathbf{\Gamma}(k+1,k)\mathbf{w}(k) - \mathbf{D}(k)\mathbf{v}(k) \tag{22-89}$$

and a simple calculation reveals that

$$\mathbf{Q}_1(k) = \mathbf{\Gamma}(k+1,k)\mathbf{Q}(k)\mathbf{\Gamma}'(k+1,k) - \mathbf{\Gamma}(k+1,k)\mathbf{S}(k)\mathbf{R}^{-1}(k)\mathbf{S}'(k)\mathbf{\Gamma}'(k+1,k) \tag{22-90}$$

This completes the derivation of (22-14).

## SUMMARY QUESTIONS

1. Mathematicians would describe the technique we have adopted for handling biases as:
   **(a)** reductio ad absurdum
   **(b)** reducing the problem to one for which we already have a solution
   **(c)** induction
2. When colored noise is present in the state equation, the order of the final state-variable model:
   **(a)** is always larger than the original order
   **(b)** is sometimes larger than the original order
   **(c)** can be smaller than the original order
3. When $\mathbf{R}(k) = \mathbf{0}$:
   **(a)** $\mathbf{P}(k+1|k)$ becomes singular for all $k$
   **(b)** $\mathbf{P}(k+1|k+1)$ becomes singular for all $k$
   **(c)** $\mathbf{K}(k+1)$ cannot be computed
4. If $l$ perfect measurements are available, then we can estimate $\mathbf{x}(k)$ using a:
   **(a)** unique reduced-order estimator
   **(b)** nonunique KF
   **(c)** nonunique reduced-order estimator

5. Suppose a system is described by a second-order state-variable model, one that is excited by a white disturbance noise $\mathbf{w}(k)$. Noisy measurements are made of $\mathbf{y}(k+2)$. In this case:
   (a) biases are present
   (b) measurement noise is colored
   (c) measurement noise and disturbance noise will be correlated
6. A reduced-order Kalman filter is one that:
   (a) produces estimates of $\mathbf{x}(k)$
   (b) produces estimates of $\mathbf{p}(k)$
   (c) produces estimates of both $\mathbf{x}(k)$ and $\mathbf{p}(k)$
7. The term state augmentation means:
   (a) combine states in the original state-variable model
   (b) switch to a new set of states
   (c) append states to the original state vector
8. If $l$ perfect measurements are available, then a reduced-order estimator:
   (a) can be of dimension less than $n-l$
   (b) must be of dimension $n-l$
   (c) may be of dimension greater than $n-l$

## PROBLEMS

**22-1.** Derive the recursive predictor, given in (22-23), by expressing $\hat{\mathbf{x}}(k+1|k)$ as $\mathbf{E}\{\mathbf{x}(k+1)|\mathbf{Z}(k)\} = \mathbf{E}\{\mathbf{x}(k+1)|\mathbf{Z}(k-1), \tilde{\mathbf{z}}(k|k-1)\}$.

**22-2.** Here we derive the recursive filter, given in (22-26), by first adding a convenient form of zero to state equation (15-17), in order to decorrelate the process noise in this modified basic state-variable model from the measurement noise $\mathbf{v}(k)$. Add $\mathbf{D}(k)[\mathbf{z}(k) - \mathbf{H}(k)\mathbf{x}(k) - \mathbf{v}(k)]$ to (15-17). The process noise, $\mathbf{w}_1(k)$, in the modified basic state-variable model, is equal to $\mathbf{\Gamma}(k+1,k)\mathbf{w}(k) - \mathbf{D}(k)\mathbf{v}(k)$. Choose "decorrelation" matrix $\mathbf{D}(k)$ so that $\mathbf{E}\{\mathbf{w}_1(k)\mathbf{v}'(k)\} = \mathbf{0}$. Then complete the derivation of (22-26).

**22-3.** In solving Problem 22-2 we arrive at the following predictor equation:

$$\hat{\mathbf{x}}(k+1|k) = \mathbf{\Phi}_1(k+1,k)\hat{\mathbf{x}}(k|k) + \mathbf{\Psi}(k+1,k)\mathbf{u}(k) + \mathbf{D}(k)\mathbf{z}(k)$$

Beginning with this predictor equation and corrector equation (22-13), derive the recursive predictor given in (22-23).

**22-4.** Show that once $\mathbf{P}(k+1|k+1)$ becomes singular it remains singular for all other values of $k$.

**22-5.** Assume that $\mathbf{R} = \mathbf{0}$, $\mathbf{H\Gamma} = \mathbf{0}$, and $\mathbf{H\Phi\Gamma} \neq \mathbf{0}$. Obtain the reduced-order estimator and its associated reduced-order Kalman filter for this situation. Contrast this situation with the case given in the text, for which $\mathbf{H\Gamma} \neq \mathbf{0}$.

**22-6.** Develop a reduced-order estimator and its associated reduced-order Kalman filter for the case when $l$ measurements are perfect and $m-l$ measurements are noisy.

**22-7.** (Tom F. Brozenac, Spring 1992) (a) Develop the single-stage smoother in the correlated noise case. (b) Develop the double-stage smoother in the correlated noise case. (c) Develop formulas for $\hat{\mathbf{x}}(k|N)$ in the correlated noise case. (d) Refer to Example 21-1, but now $w(k)$ and $v(k)$ are correlated, and $\mathbf{E}\{w(i)v(j)\} = 0.5\delta(i,j)$. Compute the analogous quantities of Table 21-1 when $N = 4$.

**22-8.** Consider the first-order system $x(k+1) = \frac{3}{4}x(k) + w_1(k)$ and $z(k+1) = x(k+1) + \nu(k+1)$, where $\mathbf{E}\{w_1(k)\} = 3, \mathbf{E}\{\nu(k)\} = 0$, $w_1(k)$ and $\nu(k)$ are both white and Gaussian, $\mathbf{E}\{w_1^2(k)\} = 10, \mathbf{E}\{\nu^2(k)\} = 2$, and $w_1(k)$ and $\nu(k)$ are correlated; i.e., $\mathbf{E}\{w_1(k)\nu(k)\} = 1$.

**(a)** Obtain the steady-state recursive Kalman filter for this system.

**(b)** What is the steady-state filter error variance, and how does it compare with the steady-state predictor error variance?

**22-9.** Consider the first-order system $x(k+1) = \frac{1}{2}x(k) + w(k)$ and $z(k+1) = x(k+1) + \nu(k+1)$, where $w(k)$ is a first-order Markov process and $\nu(k)$ is Gaussian white noise with $\mathbf{E}\{\nu(k)\} = 4$, and $r = 1$.

**(a)** Let the model for $w(k)$ be $w(k+1) = \alpha w(k) + u(k)$, where $u(k)$ is a zero-mean white Gaussian noise sequence for which $\mathbf{E}\{u^2(k)\} = \sigma_u^2$. Additionally, $\mathbf{E}\{w(k)\} = 0$. What value must $\alpha$ have if $\mathbf{E}\{w^2(k)\} = W$ for all $k$?

**(b)** Suppose $W^2 = 2$ and $\sigma_u^2 = 1$. What are the Kalman filter equations for estimation of $x(k)$ and $w(k)$?

**22-10.** Obtain the equations from which we can find $\hat{x}_1(k+1|k+1)$, $\hat{x}_2(k+1|k+1)$, and $\hat{\nu}(k+1|k+1)$ for the following system:

$$\begin{aligned} x_1(k+1) &= -x_1(k) + x_2(k) \\ x_2(k+1) &= x_2(k) + w(k) \\ z(k+1) &= x_1(k+1) + \nu(k+1) \end{aligned}$$

where $\nu(k)$ is a colored noise process; i.e.,

$$\nu(k+1) = -\frac{1}{2}\nu(k) + n(k)$$

Assume that $w(k)$ and $n(k)$ are white processes and are mutually uncorrelated, and $\sigma_w^2(k) = 4$ and $\sigma_n^2(k) = 2$. Include a block diagram of the interconnected system and reduced-order KF.

**22-11.** Consider the system $\mathbf{x}(k+1) = \mathbf{\Phi}\mathbf{x}(k) + \boldsymbol{\gamma}\mu(k)$ and $z(k+1) = \mathbf{h}'\mathbf{x}(k+1) + \nu(k+1)$, where $\mu(k)$ is a colored noise sequence and $\nu(k)$ is zero-mean white noise. What are the formulas for computing $\hat{\mu}(k|k+1)$? Filter $\hat{\mu}(k|k+1)$ is a deconvolution filter.

**22-12.** Consider the scalar moving average (MA) time-series model

$$z(k) = r(k) + r(k-1)$$

where $r(k)$ is a unit variance, white Gaussian sequence. Show that the optimal one-step predictor for this model is [assume $P(0|0) = 1$]

$$\hat{z}(k+1|k) = \frac{k}{k+1}[z(k) - \hat{z}(k|k-1)]$$

(*Hint*: Express the MA model in state-space form.)

**22-13.** Consider the basic state-variable model for the stationary time-invariant case. Assume also that $\mathbf{w}(k)$ and $\mathbf{v}(k)$ are *correlated*; i.e., $\mathbf{E}\{\mathbf{w}(k)\mathbf{v}'(k)\} = \mathbf{S}$.

**(a)** Show, from first principles, that the single-stage smoother of $\mathbf{x}(k)$, i.e., $\hat{\mathbf{x}}(k|k+1)$, is given by

$$\hat{\mathbf{x}}(k|k+1) = \hat{\mathbf{x}}(k|k) + \mathbf{M}(k|k+1)\tilde{\mathbf{z}}(k+1|k)$$

where $\mathbf{M}(k|k+1)$ is an appropriate smoother gain matrix.

(b) Derive a closed-form solution for $\mathbf{M}(k|k+1)$ as a function of the correlation matrix $\mathbf{S}$ and other quantities of the basic state-variable model.

**22-14.** (Tony Hung-yao Wu, Spring 1992) In the section entitled Colored Noises we modeled colored noise processes by low-order Markov processes. We considered, by means of examples, the two cases when either the disturbance is colored, but the measurement noise is not, or the measurement noise is colored, but the disturbance is not. In this example, consider the case when both the disturbance and the measurement noises are colored, when the system is first-order. Obtain the augmented SVM for this situation, and comment on whether or not any difficulties occur when using this SVM as the starting point to estimate the system's states using a KF.

**22-15.** Given the state-variable model where now $\mathbf{w}(k) = \text{col}\,(\mathbf{w}_1(k), \mathbf{w}_2(k))$ in which $\mathbf{w}_1(k)$ is white with covariance matrix $\mathbf{Q}_1(k)$, but is nonzero mean, and $\mathbf{w}_2(k)$ is colored and is modeled as a vector first-order Markov sequence. Reformulate this not-so-basic state-variable model so that you can estimate $\mathbf{x}(k)$ by a Kalman filter designed for the basic SVM.

**22-16.** (Tony Hung-yao Wu, Spring 1992) Sometimes noise processes are more accurately modeled as second-order Markov sequences rather than as first-order sequences. Suppose both $w(k)$ and $v(k)$ are described by the following second-order sequences:

$$w(k+2) = b_1 w(k+1) + b_2 w(k) + n_w(k)$$

$$v(k+2) = c_1 v(k+1) + c_2 v(k) + n_v(k)$$

The system is the first-order system given in (22-31) and (22-32). Obtain the augmented SVM for this situation, and comment on whether or not any difficulties occur when using it as the starting point to estimate the system's states using a KF.

**22-17.** Consider the following first-order basic state-variable model:

$$x(k+1) = \frac{1}{\sqrt{2}}x(k) + \frac{1}{4}w(k)$$

$$z(k+1) = \frac{1}{3}x(k+1) + v(k+1)$$

where $\mathbf{E}\{x(0)\} = \sqrt{2}$, $\mathbf{E}\{w(k)\} = 1$, $\mathbf{E}\{v(k)\} = 2$, $\sigma_x^2 = 1$, $q = 4$, $r = \frac{1}{9}$, and $s = 3$.

(a) What is the steady-state value of the variance of $z(k+1)$?

(b) What is the structure of the Kalman filter in predictor–corrector format?

(c) It is claimed that the Kalman gain is the same as in part (c) of Problem 19-12, because Theorem 22-2 states "Kalman gain matrix, $\mathbf{K}(k+1)$, is given by (17-12) . . . ." Explain whether this claim is true or false.

**22-18.** Given the following second-order SVM:

$$x_1(k+1) = x_2(k)$$

$$x_2(k+1) = \frac{3}{4}x_1(k) - \frac{1}{8}x_2(k) + w(k)$$

$$z_1(k) = x_1(k)$$

$$z_2(k) = x_2(k) + v(k)$$

where $\mathbf{E}\{w(k)\} = 1/4$, $\mathbf{E}\{v(k)\} = 1/2$, $\sigma_w^2(k) = 1/16$, and $\sigma_v^2(k) = 1/64$. Obtain a first-order KF for $\hat{x}_2(k+1|k+1)$ expressed in predictor–corrector format. Include all equations necessary to implement this filter using numerical values from the given model.

# LESSON 23

# Linearization and Discretization of Nonlinear Systems


## SUMMARY

Many real-world systems are continuous-time in nature and quite a few are also nonlinear. The purpose of this lesson is to explain how to linearize and discretize a nonlinear differential equation model. We do this so that we will be able to apply our digital estimators to the resulting discrete-time system.

A nonlinear dynamical system is linearized about *nominal values* of its state vector and control input. For example, $\mathbf{x}^*(t)$ denotes the nominal value of $\mathbf{x}(t)$; it must provide a good approximation to the actual behavior of the system. The approximation is considered good if the difference between the nominal and actual solutions can be described by a set of linear differential equations, called *linear perturbation equations*. A linear perturbation state-variable model, which consists of a perturbation state equation and a perturbation measurement equation, is obtained in this lesson.

When a continuous-time dynamical system is excited by white noise, its discretization must be done with care, because white noise is not piecewise continuous. This means that we cannot assume a constant value for a continuous white noise process even in a very small time interval. We discretize such a system by first expressing its solution in closed form, then sampling the solution, and finally introducing a discrete-time disturbance that is statistically equivalent (through its first two moments) to an integrated white noise term that appears in the sampled solution. Tremendous simplifications of the discretized model occur if the plant matrices are approximately constant during a sampling interval.

When you complete this lesson, you will be able to linearize and discretize continuous-time nonlinear dynamical systems.

## INTRODUCTION

Many real-world systems are continuous-time in nature and quite a few are also nonlinear. For example, the state equations associated with the motion of a satellite of mass $m$ about a spherical planet of mass $M$, in a planet-centered coordinate system, are nonlinear, because the planet's force field obeys an inverse square law. Figure 23-1 depicts a situation where the measurement equation is nonlinear. The measurement is angle $\phi_i$ and is expressed in a rectangular coordinate system; i.e., $\phi_i = \tan^{-1}[y/(x - l_i)]$. Sometimes the state equation may be nonlinear and the measurement equation linear, or vice versa, or they may both be nonlinear. Occasionally, the coordinate system in which we choose to work causes the two former situations. For example, equations of motion in a polar coordinate system are nonlinear, whereas the measurement equations are linear. In a polar coordinate system, where $\phi$ is a state variable, the measurement equation for the situation depicted in Figure 23-1 is $z_i = \phi_i$, which is linear. In a rectangular coordinate system, on the other hand, equations of motion are linear, but the measurement equations are nonlinear.

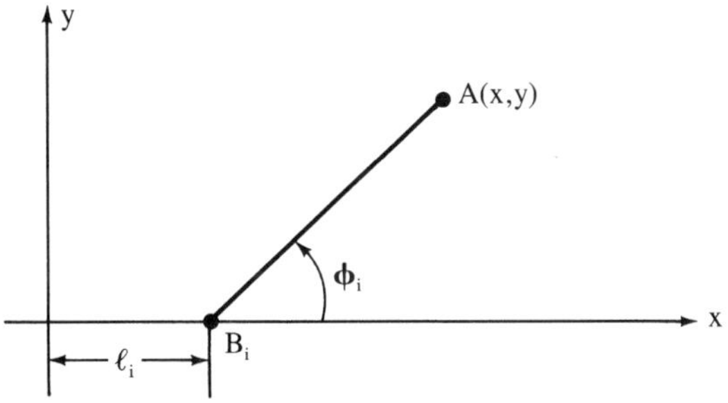


**Figure 23-1** Coordinate system for an angular measurement between two objects $A$ and $B$.

Finally, we may begin with a linear system that contains some unknown parameters. When these parameters are modeled as first-order Markov sequences, and these models are augmented to the original system, the augmented model is nonlinear, because the parameters that appeared in the original "linear" model are treated as states. We shall describe this situation in much more detail in Lesson 24.

The purpose of this lesson is to explain how to linearize and discretize a nonlinear differential equation model. We do this so that we will be able to apply our digital estimators to the resulting discrete-time system.

## A DYNAMICAL MODEL

The starting point for this lesson is the nonlinear state-variable model

$$\dot{\mathbf{x}}(t) = \mathbf{f}[\mathbf{x}(t), \mathbf{u}(t), t] + \mathbf{G}(t)\mathbf{w}(t) \tag{23-1}$$

$$\mathbf{z}(t) = \mathbf{h}[\mathbf{x}(t), \mathbf{u}(t), t] + \mathbf{v}(t) \tag{23-2}$$

We shall assume that measurements are only available at specific values of time, at $t = t_i$, $i = 1, 2, \ldots$; thus, our measurement equation will be treated as a discrete-time equation, whereas our state equation will be treated as a continuous-time equation. State vector $\mathbf{x}(t)$ is $n \times 1$; $\mathbf{u}(t)$ is an $l \times 1$ vector of known inputs; measurement vector $\mathbf{z}(t)$ is $m \times 1$; $\dot{\mathbf{x}}(t)$ is short for $d\mathbf{x}(t)/dt$; nonlinear functions $\mathbf{f}$ and $\mathbf{h}$ may depend both implicitly and explicitly on $t$, and we assume that both $\mathbf{f}$ and $\mathbf{h}$ are continuous and continuously differentiable with respect to all the elements of $\mathbf{x}$ and $\mathbf{u}$; $\mathbf{w}(t)$ is a continuous-time white noise process, i.e., $\mathbf{E}\{\mathbf{w}(t)\} = \mathbf{0}$, and

$$\mathbf{E}\{\mathbf{w}(t)\mathbf{w}'(\tau)\} = \mathbf{Q}(t)\delta(t - \tau) \tag{23-3}$$

$\mathbf{v}(t_i)$ is a discrete-time white noise sequence, i.e., $\mathbf{E}\{\mathbf{v}(t_i)\} = \mathbf{0}$ for $t = t_i$, $i = 1, 2, \ldots$, and

$$\mathbf{E}\{\mathbf{v}(t_i)\mathbf{v}'(t_j)\} = \mathbf{R}(t_i)\delta_{ij} \tag{23-4}$$

And $\mathbf{w}(t)$ and $\mathbf{v}(t_j)$ are mutually uncorrelated at all $t = t_i$; i.e.,

$$\mathbf{E}\{\mathbf{w}(t)\mathbf{v}'(t_i)\} = \mathbf{0}, \qquad \text{for } t = t_i \quad i = 1, 2, \ldots \tag{23-5}$$

Note that, whereas it is indeed correct to refer to $\mathbf{R}(t_i)$ as a covariance matrix, it is incorrect to refer to $\mathbf{Q}(t)$ as a covariance matrix. It is $\mathbf{Q}(t)\delta(t - \tau)$ that is the covariance matrix. In the special case when $\mathbf{w}(t)$ is stationary so that $\mathbf{Q}(t) = \mathbf{Q}$, then $\mathbf{Q}$ is a *spectral intensity matrix*, because the Fourier transform of $\mathbf{E}\{\mathbf{w}(t)\mathbf{w}'(\tau)\}$, which is the power spectrum of $\mathbf{w}(t)$, equals $\mathbf{Q}$. In general, matrix $\mathbf{Q}(t)$ can be referred to as an *intensity matrix*.

**EXAMPLE 23-1**

Here we expand upon the previously mentioned satellite–planet example. Our example is taken from Meditch (1969, pp. 60–61), who states, "Assuming that the planet's force field obeys an inverse square law, and that the only other forces present are the satellite's two thrust forces $u_r(t)$ and $u_\theta(t)$ (see Figure 23-2), and that the satellite's initial position and velocity vectors lie in the plane, we know from elementary particle mechanics that the satellite's motion is confined to the plane and is governed by the two equations

$$\ddot{r} = r\dot{\theta}^2 - \frac{\gamma}{r^2} + \frac{1}{m}u_r(t) \tag{23-6}$$

and

$$\ddot{\theta} = -\frac{2\dot{r}\dot{\theta}}{r} + \frac{1}{m}u_\theta(t) \tag{23-7}$$

where $\gamma = GM$ and $G$ is the universal gravitational constant.

"Defining $x_1 = r$, $x_2 = \dot{r}$, $x_3 = \theta$, $x_4 = \dot{\theta}$, $u_1 = u_r$, and $u_2 = u_\theta$, we have

$$\left.\begin{aligned} \dot{x}_1 &= x_2 \\ \dot{x}_2 &= x_1 x_4^2 - \frac{\gamma}{x_1^2} + \frac{1}{m}u_1(t) \\ \dot{x}_3 &= x_4 \\ \dot{x}_4 &= -\frac{2x_2x_4}{x_1} + \frac{1}{m}u_2(t) \end{aligned}\right\} \tag{23-8}$$

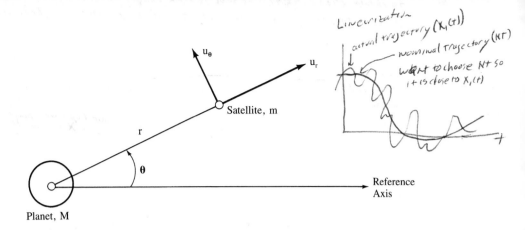

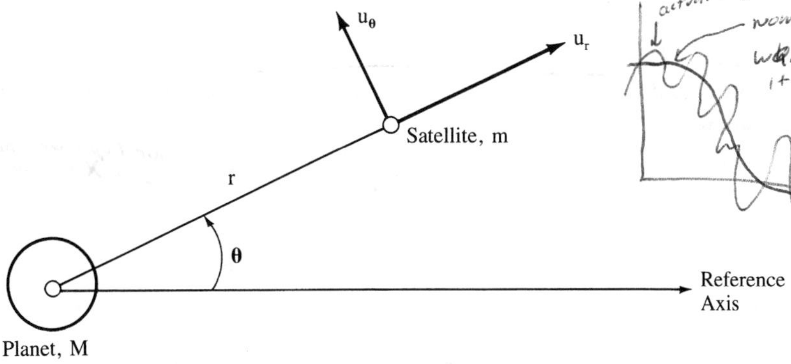


**Figure 23-2** Schematic for satellite–planet system (Copyright 1969, McGraw-Hill).

which is of the form in (23-1).... Assuming ... that the measurement made on the satellite during its motion is simply its distance from the surface of the planet, we have the scalar measurement equation

$$z(t) = r(t) - r_0 + \nu(t) = x_1(t) - r_0 + \nu(t) \tag{23-9}$$

where $r_0$ is the planet's radius."

Comparing (23-8) and (23-1), and (23-9) and (23-2), we conclude that

$$\mathbf{f}[\mathbf{x}(t), \mathbf{u}(t), t] = \text{col}\left[x_2, x_1x_4^2 - \frac{\gamma}{x_1^2} + \frac{1}{m}u_1, x_4, -\frac{2x_2x_4}{x_1} + \frac{1}{m}u_2\right] \tag{23-10}$$

and

$$\mathbf{h}[\mathbf{x}(t), \mathbf{u}(t), t] = x_1 - r_0 \tag{23-11}$$

Observe that, in this example, only the state equation is nonlinear. □

## LINEAR PERTURBATION EQUATIONS

In this section we shall linearize our nonlinear dynamical model in (23-1) and (23-2) about nominal values of $\mathbf{x}(t)$ and $\mathbf{u}(t)$, $\mathbf{x}^*(t)$ and $\mathbf{u}^*(t)$, respectively. If we are given a nominal input, $\mathbf{u}^*(t)$, then $\mathbf{x}^*(t)$ satisfies the nonlinear differential equation

$$\dot{\mathbf{x}}^*(t) = \mathbf{f}[\mathbf{x}^*(t), \mathbf{u}^*(t), t] \tag{23-12}$$

and associated with $\mathbf{x}^*(t)$ and $\mathbf{u}^*(t)$ is the following nominal measurement, $\mathbf{z}^*(t)$, where

$$\mathbf{z}^*(t) = \mathbf{h}[\mathbf{x}^*(t), \mathbf{u}^*(t)], \quad t = t_i, \quad i = 1, 2, \ldots \tag{23-13}$$

If $\mathbf{u}(t)$ is an input derived from a feedback control law, so that $\mathbf{u}(t) = \mathbf{u}[\mathbf{x}(t), t]$, then $\mathbf{u}(t)$ can differ from $\mathbf{u}^*(t)$, because $\mathbf{x}(t)$ will differ from $\mathbf{x}^*(t)$. On the other hand, if $\mathbf{u}(t)$ does not depend on $\mathbf{x}(t)$ then usually $\mathbf{u}(t)$ is the same as $\mathbf{u}^*(t)$, in which case $\delta\mathbf{u}(t) = \mathbf{0}$. Throughout this lesson, we shall assume that $\delta\mathbf{u}(t) \neq \mathbf{0}$ and that $\mathbf{x}^*(t)$ exists. We discuss two methods for choosing $\mathbf{x}^*(t)$ in Lesson 24. Obviously, one is just to solve (23-12) for $\mathbf{x}^*(t)$.

To obtain $x^*(t)$ (given $u^*(t)$) we solve 23-12 (maybe numerically) & get $z^*$ by straight substitution of $x^*$. This is (in theory) easy & straightforward.

Note that $\mathbf{x}^*(t)$ must provide a good approximation to the actual behavior of the system. The approximation is considered good if the difference between the nominal and actual solutions can be described by a system of linear differential equations, called *linear perturbation equations*. We derive these equations next. Let

we hope we choose $x^*(t)$ ∋ $\delta x(t)$ = small, small

$$\delta\mathbf{x}(t) = \mathbf{x}(t) - \mathbf{x}^*(t) \tag{23-14}$$

and

$$\delta\mathbf{u}(t) = \mathbf{u}(t) - \mathbf{u}^*(t) \tag{23-15}$$

then

$$\frac{d}{dt}\delta\mathbf{x}(t) = \delta\dot{\mathbf{x}}(t) = \dot{\mathbf{x}}(t) - \dot{\mathbf{x}}^*(t) = \mathbf{f}[\mathbf{x}(t), \mathbf{u}(t), t] + \mathbf{G}(t)\mathbf{w}(t) - \mathbf{f}[\mathbf{x}^*(t), \mathbf{u}^*(t), t] \tag{23-16}$$

**Fact 1.** *When* $\mathbf{f}[\mathbf{x}(t), \mathbf{u}(t), t]$ *is expanded in a Taylor series about* $\mathbf{x}^*(t)$ *and* $\mathbf{u}^*(t)$, *we obtain*

$$\mathbf{f}[\mathbf{x}(t), \mathbf{u}(t), t] = \mathbf{f}[\mathbf{x}^*(t), \mathbf{u}^*(t), t] + \mathbf{F}_{\mathbf{x}}[\mathbf{x}^*(t), \mathbf{u}^*(t), t]\delta\mathbf{x}(t) + \mathbf{F}_{\mathbf{u}}[\mathbf{x}^*(t), \mathbf{u}^*(t), t]\delta\mathbf{u}(t) + \text{higher-order terms} \tag{23-17}$$

*where* $\mathbf{F}_{\mathbf{x}}$ *and* $\mathbf{F}_{\mathbf{u}}$ *are* $n \times n$ *and* $n \times 1$ *Jacobian matrices; i.e.,*

$$\mathbf{F}_{\mathbf{x}}[\mathbf{x}^*(t), \mathbf{u}^*(t), t] = \begin{pmatrix} \partial f_1/\partial x_1^* & \cdots & \partial f_1/\partial x_n^* \\ \vdots & \ddots & \vdots \\ \partial f_n/\partial x_1^* & \cdots & \partial f_n/\partial x_n^* \end{pmatrix} \tag{23-18}$$

*and*

$$\mathbf{F}_{\mathbf{x}}[\mathbf{x}^*(t), \mathbf{u}^*(t), t] = \begin{pmatrix} \partial f_1/\partial u_1^* & \cdots & \partial f_1/\partial u_l^* \\ \vdots & \ddots & \vdots \\ \partial f_n/\partial u_1^* & \cdots & \partial f_n/\partial u_l^* \end{pmatrix} \tag{23-19}$$

*In these expressions* $\partial f_i/\partial x_j^*$ *and* $\partial f_i/\partial u_j^*$ *are short for*

$$\frac{\partial f_i}{\partial x_j^*} = \left.\frac{\partial f_i[\mathbf{x}(t), \mathbf{u}(t), t]}{\partial x_j(t)}\right|_{\mathbf{x}(t)=\mathbf{x}^*(t), \mathbf{u}(t)=\mathbf{u}^*(t)} \tag{23-20}$$

*and*

$$\frac{\partial f_i}{\partial u_j^*} = \left.\frac{\partial f_i[\mathbf{x}(t), \mathbf{u}(t), t]}{\partial u_j(t)}\right|_{\mathbf{x}(t)=\mathbf{x}^*(t), \mathbf{u}(t)=\mathbf{u}^*(t)} \tag{23-21}$$

*Proof.* The Taylor series expansion of the $i$th component of $\mathbf{f}[\mathbf{x}(t), \mathbf{u}(t), t]$ is

$$f_i[\mathbf{x}(t), \mathbf{u}(t), t] = f_i[\mathbf{x}^*(t), \mathbf{u}^*(t), t] + \frac{\partial f_i}{\partial x_1^*}[x_1(t) - x_1^*(t)] + \cdots$$

$$+ \frac{\partial f_i}{\partial x_n^*}[x_n(t) - x_n^*(t)] + \frac{\partial f_i}{\partial u_1^*}[u_1(t) - u_1^*(t)] + \cdots \qquad (23\text{-}22)$$

$$+ \frac{\partial f_i}{\partial u_l^*}[u_l(t) - u_l^*(t)] + \text{higher-order terms}$$

where $i = 1, 2, \ldots, n$. Collecting these $n$ equations together in vector-matrix format, we obtain (23-17), in which $\mathbf{F_x}$ and $\mathbf{F_u}$ are defined in (23-18) and (23-19), respectively. $\square$

Substituting (23-17) into (23-16) and neglecting the "higher-order terms", we obtain the following *perturbation state equation*:

$$\delta\dot{\mathbf{x}}(t) = \mathbf{F_x}[\mathbf{x}^*(t), \mathbf{u}^*(t), t]\delta\mathbf{x}(t)$$

$$+ \mathbf{F_u}[\mathbf{x}^*(t), \mathbf{u}^*(t), t]\delta\mathbf{u}(t) + \mathbf{G}(t)\mathbf{w}(t) \qquad (23\text{-}23)$$

*[handwritten margin note: which we can do only because $x^*(t) \sim x(t)$, if $x(t)$ is way off, this breaks down.]*

Observe that, even if our original nonlinear differential equation is not an explicit function of time {i.e., $\mathbf{f}[\mathbf{x}(t), \mathbf{u}(t), t] = \mathbf{f}[\mathbf{x}(t), \mathbf{u}(t)]$}, our perturbation state equation is always time-varying because Jacobian matrices $\mathbf{F_x}$ and $\mathbf{F_u}$ vary with time, because $\mathbf{x}^*$ and $\mathbf{u}^*$ vary with time.

Next, let

$$\delta\mathbf{z}(t) = \mathbf{z}(t) - \mathbf{z}^*(t) \qquad (23\text{-}24)$$

**Fact 2.** *When $\mathbf{h}[\mathbf{x}(t), \mathbf{u}(t), t]$ is expanded in a Taylor series about $\mathbf{x}^*(t)$ and $\mathbf{u}^*(t)$, we obtain*

$$\mathbf{h}[\mathbf{x}(t), \mathbf{u}(t), t] = \mathbf{h}[\mathbf{x}^*(t), \mathbf{u}^*(t), t] + \mathbf{H_x}[\mathbf{x}^*(t), \mathbf{u}^*(t), t]\delta\mathbf{x}(t)$$

$$+ \mathbf{H_u}[\mathbf{x}^*(t), \mathbf{u}^*(t), t]\delta\mathbf{u}(t) + \text{higher-order terms} \qquad (23\text{-}25)$$

*where $\mathbf{H_x}$ and $\mathbf{H_u}$ are $m \times n$ and $m \times l$ Jacobian matrices; i.e.,*

$$\mathbf{H_x}[\mathbf{x}^*(t), \mathbf{u}^*(t), t] = \begin{pmatrix} \partial h_1/\partial x_1^* & \cdots & \partial h_1/\partial x_n^* \\ \vdots & \ddots & \vdots \\ \partial h_m/\partial x_1^* & \cdots & \partial h_m/\partial x_n^* \end{pmatrix} \qquad (23\text{-}26)$$

and

$$\mathbf{H_u}[\mathbf{x}^*(t), \mathbf{u}^*(t), t] = \begin{pmatrix} \partial h_1/\partial u_1^* & \cdots & \partial h_1/\partial u_l^* \\ \vdots & \ddots & \vdots \\ \partial h_m/\partial u_1^* & \cdots & \partial h_m/\partial u_l^* \end{pmatrix} \qquad (23\text{-}27)$$

*In these expressions $\partial h_i/\partial x_j^*$ and $\partial h_i/\partial u_j^*$ are short for*

$$\frac{\partial h_i}{\partial x_j^*} = \frac{\partial h_i[\mathbf{x}(t), \mathbf{u}(t), t]}{\partial x_j(t)}\Bigg|_{\mathbf{x}(t)=\mathbf{x}^*(t), \mathbf{u}(t)=\mathbf{u}^*(t)} \qquad (23\text{-}28)$$

*and*

$$\frac{\partial h_i}{\partial u_j^*} = \left. \frac{\partial h_i[\mathbf{x}(t), \mathbf{u}(t), t]}{\partial u_j(t)} \right|_{\mathbf{x}(t)=\mathbf{x}^*(t),\mathbf{u}(t)=\mathbf{u}^*(t)} \quad \square \tag{23-29}$$

We leave the derivation of this fact to the reader, because it is analogous to the derivation of the Taylor series expansion of $\mathbf{f}[\mathbf{x}(t), \mathbf{u}(t), t]$.

Substituting (23-25) into (23-24) and neglecting the "higher-order terms," we obtain the following *perturbation measurement equation:*

$$\begin{aligned}\delta\mathbf{z}(t) &= \mathbf{H}_x[\mathbf{x}^*(t), \mathbf{u}^*(t), t]\delta\mathbf{x}(t) \\ &\quad + \mathbf{H}_\mathbf{u}[\mathbf{x}^*(t), \mathbf{u}^*(t), t]\delta\mathbf{u}(t) + \mathbf{v}(t), \qquad t = t_i, \quad i = 1, 2, \ldots\end{aligned} \tag{23-30}$$

Equations (23-23) and (23-30) constitute our linear perturbation equations, or our linear *perturbation state-variable model.*

An interesting article on the linearization of an equation very similar to (23-1), in which the derivatives needed in (23-18), (23-19), (23-26), and (23-27) are computed by numerical differentiation, is Taylor and Antoniotti (1993).

**EXAMPLE 23-2**

Returning to our satellite–planet Example 23-1, we find that

$$\mathbf{F}_\mathbf{x}[\mathbf{x}^*(t), \mathbf{u}^*(t), t] = \mathbf{F}_\mathbf{x}[\mathbf{x}^*(t)] = \begin{pmatrix} 0 & 1 & 0 & 0 \\ x_4^2(t) + \dfrac{2\gamma}{x_1^3(t)} & 0 & 0 & 2x_1(t)x_4(t) \\ 0 & 0 & 0 & 1 \\ \dfrac{2x_2(t)x_4(t)}{x_1^2(t)} & \dfrac{-2x_4(t)}{x_1(t)} & 0 & \dfrac{-2x_2(t)}{x_1(t)} \end{pmatrix}_*$$

$$\mathbf{F}_\mathbf{u}[\mathbf{x}^*(t), \mathbf{u}^*(t), t] = \mathbf{F}_\mathbf{u} = \begin{pmatrix} 0 & 0 \\ \dfrac{1}{m} & 0 \\ 0 & 0 \\ 0 & \dfrac{1}{m} \end{pmatrix}$$

$$\mathbf{H}_\mathbf{x}[\mathbf{x}^*(t), \mathbf{u}^*(t), t] = \mathbf{H}_\mathbf{x} = (1 \quad 0 \quad 0 \quad 0)$$

and

$$\mathbf{H}_\mathbf{u}[\mathbf{x}^*(t), \mathbf{u}^*(t), t] = \mathbf{0}$$

In the equation for $\mathbf{F}_\mathbf{x}[\mathbf{x}^*(t)]$, the notation $(\quad)_*$ means that all $x_i(t)$ within the matrix are nominal values, i.e., $x_i(t) = x_i^*(t)$.

Observe that the linearized satellite–planet system is time varying, because its linearized planet matrix, $\mathbf{F}_\mathbf{x}[\mathbf{x}^*(t)]$, depends on the nominal trajectory $\mathbf{x}^*(t)$. $\square$

# DISCRETIZATION OF A LINEAR TIME-VARYING STATE-VARIABLE MODEL

In this section we describe how we discretize the general linear, time-varying state-variable model

$$\dot{\mathbf{x}}(t) = \mathbf{F}(t)\mathbf{x}(t) + \mathbf{B}(t)\mathbf{u}(t) + \mathbf{G}(t)\mathbf{w}(t) \tag{23-31}$$

$$\mathbf{z}(t) = \mathbf{H}(t)\mathbf{x}(t) + \mathbf{v}(t), \qquad t = t_i \quad i = 1, 2, \ldots \tag{23-32}$$

The application of this section's results to the perturbation state-variable model is given in the following section.

In (23-31) and (23-32), $\mathbf{x}(t)$ is $n \times 1$, control input $\mathbf{u}(t)$ is $l \times 1$, process noise $\mathbf{w}(t)$ is $p \times 1$, and $\mathbf{z}(t)$ and $\mathbf{v}(t)$ are each $m \times 1$. Additionally, $\mathbf{w}(t)$ is a continuous-time white noise process, $\mathbf{v}(t_i)$ is a discrete-time white noise sequence, and $\mathbf{w}(t)$ and $\mathbf{v}(t_i)$ are mutually uncorrelated at all $t = t_i, i = 1, 2, \ldots$; i.e., $\mathbf{E}\{\mathbf{w}(t)\} = \mathbf{0}$ for all $t, \mathbf{E}\{\mathbf{v}(t_i)\} = \mathbf{0}$ for all $t_i$, and

$$\mathbf{E}\{\mathbf{w}(t)\mathbf{w}'(\tau)\} = \mathbf{Q}(t)\delta(t - \tau) \tag{23-33}$$

$$\mathbf{E}\{\mathbf{v}(t_i)\mathbf{v}'(t_j)\} = \mathbf{R}(t_i)\delta_{ij} \tag{23-34}$$

and

$$\mathbf{E}\{\mathbf{w}(t)\mathbf{v}'(t_i)\} = \mathbf{0}, \qquad \text{for } t = t_i, \quad i = 1, 2, \ldots \tag{23-35}$$

It is tempting to discretize (23-31) by setting $d\mathbf{x}(t) = [\mathbf{x}(t_{k+1}) - \mathbf{x}(t_k)]/\Delta t$, and setting $t = t_k$ on the right-hand side of (23-31). Whereas $\mathbf{u}(t_k)$ is a well-defined function at all time points, including $t_k$, $\mathbf{w}(t_k)$ is not, because $\mathbf{w}(t_k)$ is a continuous-time white noise process, which, in general is a mathematical fiction. See Lesson 26 for additional discussions on continuous-time white noise processes.

Consequently, our approach to discretizing state equation (23-31) begins with the solution of that equation.

**Theorem 23-1.** *The solution to state equation (23-31) can be expressed as*

$$\mathbf{x}(t) = \mathbf{\Phi}(t, t_0)\mathbf{x}(t_0) + \int_{t_0}^{t} \mathbf{\Phi}(t, \tau)[\mathbf{B}(\tau)\mathbf{u}(\tau) + \mathbf{G}(\tau)\mathbf{w}(\tau)]d\tau \tag{23-36}$$

*where state transition matrix $\mathbf{\Phi}(t, \tau)$ is the solution to the following matrix homogeneous differential equation:*

$$\left.\begin{array}{l} \dot{\mathbf{\Phi}}(t, \tau) = \mathbf{F}(t)\mathbf{\Phi}(t, \tau) \\ \mathbf{\Phi}(t, t) = \mathbf{I} \end{array}\right\} \quad \square \tag{23-37}$$

This result should be a familiar one to many readers of this book. See the Supplementary Material at the end of this lesson for its proof.

Next, we assume that $\mathbf{u}(t)$ is a piecewise constant function of time for $t \in [t_k, t_{k+1}]$ and set $t_0 = t_k$ and $t = t_{k+1}$ in (23-36), to obtain

$$\mathbf{x}(t_{k+1}) = \mathbf{\Phi}(t_{k+1}, t_k)\mathbf{x}(t_k) + \left[\int_{t_k}^{t_{k+1}} \mathbf{\Phi}(t_{k+1}, \tau)\mathbf{B}(\tau)d\tau\right]\mathbf{u}(t_k)$$

$$+ \int_{t_k}^{t_{k+1}} \mathbf{\Phi}(t_{k+1}, \tau)\mathbf{G}(\tau)\mathbf{w}(\tau)d\tau \qquad (23\text{-}38)$$

which can also be written as

$$\mathbf{x}(t_{k+1}) \equiv \mathbf{x}(k + 1) = \mathbf{\Phi}(k + 1, k)\mathbf{x}(k) + \mathbf{\Psi}(k + 1, k)\mathbf{u}(k) + \mathbf{w}_d(k) \qquad (23\text{-}39)$$

where

*remember*
*$T_{k+1} - T_k$ need*
*not be constant.*

$$\mathbf{\Phi}(k + 1, k) = \mathbf{\Phi}(t_{k+1}, t_k) \qquad (23\text{-}40)$$

$$\mathbf{\Psi}(k + 1, k) = \int_{t_k}^{t_{k+1}} \mathbf{\Phi}(t_{k+1}, \tau)\mathbf{B}(\tau)d\tau \qquad (23\text{-}41)$$

and $\mathbf{w}_d(k)$ is a discrete-time white Gaussian sequence that is *statistically equivalent through its first two moments* to

$$\int_{t_k}^{t_{k+1}} \mathbf{\Phi}(t_{k+1}, \tau)\mathbf{G}(\tau)\mathbf{w}(\tau)d\tau$$

Observe that the dependence of $\mathbf{\Phi}$ and $\mathbf{\Psi}$ on the two time points $t_k$ and $t_{k+1}$ rationalizes our use of double arguments for them, which is why we have used the double-argument notation for $\mathbf{\Phi}$ and $\mathbf{\Psi}$ in our basic state-variable model in Lesson 15. The mean and covariance matrices of $\mathbf{w}_d(k)$ are

$$\mathbf{E}\{\mathbf{w}_d(k)\} = \mathbf{E}\left\{\int_{t_k}^{t_{k+1}} \mathbf{\Phi}(t_{k+1}, \tau)\mathbf{G}(\tau)\mathbf{w}(\tau)d\tau\right\} = \mathbf{0} \qquad (23\text{-}42)$$

and

$$\mathbf{E}\{\mathbf{w}_d(k)\mathbf{w}_d'(k)\} \triangleq \mathbf{Q}_d(k + 1, k)$$

$$= \mathbf{E}\left\{\int_{t_k}^{t_{k+1}} \mathbf{\Phi}(t_{k+1}, \tau)\mathbf{G}(\tau)\mathbf{w}(\tau)d\tau\right.$$

$$\left.\int_{t_k}^{t_{k+1}} \mathbf{w}'(\xi)\mathbf{G}'(\xi)\mathbf{\Phi}'(t_{k+1}, \xi)d\xi\right\}$$

$$= \int_{t_k}^{t_{k+1}} \mathbf{\Phi}(t_{k+1}, \tau)\mathbf{G}(\tau)\mathbf{Q}(\tau)\mathbf{G}'(\tau)\mathbf{\Phi}'(t_{k+1}, \tau)d\tau \qquad (23\text{-}43)$$

respectively.

Observe, from the right-hand side of Equations (23-40), (23-41), and (23-43), that these quantities can be computed from knowledge about $\mathbf{F}(t), \mathbf{B}(t), \mathbf{G}(t)$, and $\mathbf{Q}(t)$. In general, we must compute $\mathbf{\Phi}(k + 1, k)$, $\mathbf{\Psi}(k + 1, k)$, and $\mathbf{Q}_d(k + 1, k)$ using numerical integration, and these matrices change from one time interval to the next because $\mathbf{F}(t), \mathbf{B}(t), \mathbf{G}(t)$, and $\mathbf{Q}(t)$ usually change from one time interval to the next.

Because our measurements have been assumed to be available only at sampled values of $t$ at $t = t_i, i = 1, 2, \ldots$, we can express (23-32) as

$$\mathbf{z}(k+1) = \mathbf{H}(k+1)\mathbf{x}(k+1) + \mathbf{v}(k+1) \tag{23-44}$$

Equations (23-39) and (23-44) constitute our discretized state-variable model.

**EXAMPLE 23-3**

Great simplifications of the calculations in (23-40), (23-41) and (23-43) occur if $\mathbf{F}(t), \mathbf{B}(t), \mathbf{G}(t)$, and $\mathbf{Q}(t)$ are approximately constant during the time interval $[t_k, t_{k+1}]$, i.e., if

$$\left.\begin{array}{llll} \mathbf{F}(t) \simeq \mathbf{F}_k, & \mathbf{B}(t) \simeq \mathbf{B}_k & \mathbf{G}(t) \simeq \mathbf{G}_k, & \text{and} \\ \mathbf{Q}(t) \simeq \mathbf{Q}_k, & \text{for } t \in [t_k, t_{k+1}] & & \end{array}\right\} \tag{23-45}$$

To begin, (23-37) is easily integrated to yield

$$\mathbf{\Phi}(t, \tau) = e^{\mathbf{F}_k(t-\tau)} \tag{23-46}$$

Hence,

$$\mathbf{\Phi}(k+1, k) = e^{\mathbf{F}_k T} = \mathbf{\Phi}(k) \tag{23-47}$$

where we have assumed that $t_{k+1} - t_k = T$. The matrix exponential is given by the infinite series

$$e^{\mathbf{F}_k T} = \mathbf{I} + \mathbf{F}_k T + \mathbf{F}_k^2 \frac{T^2}{2} + \mathbf{F}_k^3 \frac{T^3}{3!} + \cdots \tag{23-48}$$

and, for sufficiently small values of $T$,

$$e^{\mathbf{F}_k T} \simeq \mathbf{I} + \mathbf{F}_k T \tag{23-49}$$

We use this approximation for $e^{\mathbf{F}_k T}$ in deriving simpler expressions for $\mathbf{\Psi}(k+1, k)$ and $\mathbf{Q}_d(k+1, k)$. Comparable results can be obtained for higher-order truncations of $e^{\mathbf{F}_k T}$.

Substituting (23-46) into (23-41), we find that

$$\begin{aligned} \mathbf{\Psi}(k+1, k) &= \int_{t_k}^{t_{k+1}} \mathbf{\Phi}(t_{k+1}, \tau)\mathbf{B}_k d\tau = \int_{t_k}^{t_{k+1}} e^{\mathbf{F}_k(t_{k+1}-\tau)}\mathbf{B}_k d\tau \\ &\simeq \int_{t_k}^{t_{k+1}} [\mathbf{I} + \mathbf{F}_k(t_{k+1} - \tau)]\mathbf{B}_k d\tau \\ &\simeq \mathbf{B}_k T + \mathbf{F}_k \mathbf{B}_k t_{k+1} T - \mathbf{F}_k \mathbf{B}_k \int_{t_k}^{t_{k+1}} \tau d\tau \\ &\simeq \mathbf{B}_k T + \mathbf{F}_k \mathbf{B}_k \frac{T^2}{2} \simeq \mathbf{B}_k T = \mathbf{\Psi}(k) \end{aligned} \tag{23-50}$$

where we have truncated $\mathbf{\Psi}(k+1, k)$ to its first-order term in $T$. Proceeding in a similar manner for $\mathbf{Q}_d(k+1, k)$, it is straightforward to show that

$$\mathbf{Q}_d(k+1, k) \simeq \mathbf{G}_k \mathbf{Q}_k \mathbf{G}_k' T = \mathbf{Q}_d(k) \tag{23-51}$$

Note that (23-47), (23-49), (23-50), and (23-51), while much simpler than their original expressions, can still change in values from one time interval to another because of their dependence on $k$. Note, also, that under the piecewise constant condition the double arguments for $\mathbf{\Psi}$ and $\mathbf{Q}_d$ are not needed. □

## DISCRETIZED PERTURBATION STATE-VARIABLE MODEL

Applying the results of the preceding section to the perturbation state-variable model in (23-23) and (23-30), we obtain the following *discretized perturbation state-variable model*:

$$\delta\mathbf{x}(k+1) = \mathbf{\Phi}(k+1, k;^*)\delta\mathbf{x}(k) + \mathbf{\Psi}(k+1, k;^*)\delta\mathbf{u}(k) + \mathbf{w}_d(k) \tag{23-52}$$

$$\begin{aligned}\delta\mathbf{z}(k+1) = {} & \mathbf{H}_\mathbf{x}(k+1;^*)\delta\mathbf{x}(k+1) \\ & + \mathbf{H}_u(k+1;^*)\delta\mathbf{u}(k+1) + \mathbf{v}(k+1)\end{aligned} \tag{23-53}$$

The notation $\mathbf{\Phi}(k+1, k;^*)$, for example, denotes the fact that this matrix depends on $\mathbf{x}^*(t)$ and $\mathbf{u}^*(t)$. More specifically,

$$\mathbf{\Phi}(k+1, k;^*) = \mathbf{\Phi}(t_{k+1}, t_k;^*) \tag{23-54}$$

where

$$\begin{aligned}\dot{\mathbf{\Phi}}(t, \tau;^*) &= \mathbf{F}_\mathbf{x}[\mathbf{x}^*(t), \mathbf{u}^*(t), t]\mathbf{\Phi}(t, \tau;^*) \\ \mathbf{\Phi}(t, t;^*) &= \mathbf{I}\end{aligned} \tag{23-55}$$

Additionally,

$$\mathbf{\Psi}(k+1, k;^*) = \int_{t_k}^{t_{k+1}} \mathbf{\Phi}(t_{k+1}, \tau;^*)\mathbf{F}_\mathbf{u}[\mathbf{x}^*(\tau), \mathbf{u}^*(\tau), \tau]d\tau \tag{23-56}$$

and

$$\mathbf{Q}_d(k+1, k;^*) = \int_{t_k}^{t_{k+1}} \mathbf{\Phi}(t_{k+1}, \tau;^*)\mathbf{G}(\tau)\mathbf{Q}(\tau)\mathbf{G}'(\tau)\mathbf{\Phi}'(t_{k+1}, \tau;^*)d\tau \tag{23-57}$$

## COMPUTATION

To simulate the nonlinear dynamical system, given by (23-1) and (23-2), use Simulink®.

**Linearization** can be accomplished within SIMULINK.

There are no M-files available to **discretize** a time-varying differential equation model, such as the perturbation state equation in (23-31). When the matrices in the time-varying differential equation model are piecewise constant over the sampling interval, as described in Example 23-3, then discretization can be accomplished with the aid of the matrix exponential, an M-file for which can be found in MATLAB:

**expm**: Matrix exponential: $\mathbf{y} = \text{expm}(\mathbf{x})$

You will need to write an M-file to produce Equations (23-47), (23-41), and (23-43). You will then be able to implement the discrete-time state-equation in (23-39). Because we have assumed that measurements are given in sampled form, no discretization is needed in order to implement the discrete-time measurement equation in (23-44). If, however, your actual measurements are analog in nature, you must first digitize them (see Lesson 26 for a discussion on how to relate the statistics of continuous-time and discrete-time white noise).

## Supplementary Material

# PROOF OF THEOREM 23-1

Theorem 23-1 provides the solution to a time-varying state equation in terms of the state transition matrix $\mathbf{\Phi}(t, \tau)$. Our derivation of (23-36) for $\mathbf{x}(t)$ proceeds in three (classical) steps: (1) obtain the homogeneous solution to vector differential equation (23-31); (2) obtain the particular solution to (23-31); and (3) obtain the complete solution to (23-31).

## Homogeneous Solution to (23-31)

The homogeneous equation corresponding to (23-31) is

$$\frac{d\mathbf{x}_h(t)}{dt} = \mathbf{F}(t)\mathbf{x}_h(t), \qquad \text{for } t \geq t_0 \text{ and arbitrary } \mathbf{x}_0 \tag{23-58}$$

Let

$$\mathbf{x}_h(t) = \mathbf{X}(t)\mathbf{x}(t_0) \tag{23-59}$$

where $\mathbf{X}(t)$ is an arbitrary unknown $n \times n$ matrix. Then $d\mathbf{x}_h(t)/dt = [d\mathbf{X}(t)/dt]\mathbf{x}(t_0)$, so that $[d\mathbf{X}(t)/dt]\mathbf{x}(t_0) = \mathbf{F}(t)\mathbf{X}(t)\mathbf{x}(t_0)$, from which it follows that

$$\frac{d\mathbf{X}(t)}{dt} = \mathbf{F}(t)\mathbf{X}(t) \tag{23-60}$$

Letting $\mathbf{x}_h(t_0) = \mathbf{x}(t_0)$, we find, from (23-59), that $\mathbf{X}(t_0) = \mathbf{I}$, which is the initial condition for (23-60).

$\mathbf{X}(t)$ is called the *fundamental matrix* of the system in (23-31); it depends only on $\mathbf{F}(t)$.

## Particular Solution to (23-31)

We use the *Lagrange variation of parameter technique* as follows. Let

$$\mathbf{x}_p(t) = \mathbf{X}(t)\mathbf{y}(t) \tag{23-61}$$

where $\mathbf{y}(t)$ is an unknown $n \times 1$ vector. Differentiate $\mathbf{x}_p(t)$ and substitute the result, as well as (23-61), into (23-31) to see that

$$\frac{d\mathbf{X}(t)}{dt}\mathbf{y}(t) + \mathbf{X}(t)\frac{d\mathbf{y}(t)}{dt} = \mathbf{F}(t)\mathbf{X}(t)\mathbf{y}(t) + \mathbf{B}(t)\mathbf{u}(t) + \mathbf{G}(t)\mathbf{w}(t)$$

But $d\mathbf{X}(t)/dt = \mathbf{F}(t)\mathbf{X}(t)$, so this last equation reduces to

$$\frac{d\mathbf{y}(t)}{dt} = \mathbf{X}^{-1}(t)[\mathbf{B}(t)\mathbf{u}(t) + \mathbf{G}(t)\mathbf{w}(t)] \tag{23-62}$$

Meditch (1969, pp. 32-33) shows that $\mathbf{X}(t)$ is nonsingular so that $\mathbf{X}^{-1}(t)$ exists. Integrate (23-62) with respect to time to see that

$$\mathbf{y}(t) = \int_{t_0}^{t} \mathbf{X}^{-1}(\tau)[\mathbf{B}(\tau)\mathbf{u}(\tau) + \mathbf{G}(\tau)\mathbf{w}(\tau)]d\tau \tag{23-63}$$

so that

$$\mathbf{x}_p(t) = \mathbf{X}(t)\int_{t_0}^{t} \mathbf{X}^{-1}(\tau)[\mathbf{B}(\tau)\mathbf{u}(\tau) + \mathbf{G}(\tau)\mathbf{w}(\tau)]d\tau \tag{23-64}$$

### Complete Solution to (23-31)

The complete solution to (23-31) equals the sum of the homogeneous and particular solutions; i.e.,

$$\mathbf{x}(t) = \mathbf{x}_h(t) + \mathbf{x}_p(t) = \mathbf{X}(t)\mathbf{x}(t_0) + \mathbf{X}(t)\int_{t_0}^{t} \mathbf{X}^{-1}(\tau)[\mathbf{B}(\tau)\mathbf{u}(\tau) + \mathbf{G}(\tau)\mathbf{w}(\tau)]d\tau \tag{23-65}$$

Let

$$\mathbf{\Phi}(t,\tau) \triangleq \mathbf{X}(t)\mathbf{X}^{-1}(\tau) \tag{23-66}$$

Obviously, $\mathbf{\Phi}(t,t) = \mathbf{I}$, and $\mathbf{\Phi}(t,t_0) = \mathbf{X}(t)\mathbf{X}^{-1}(t_0) = \mathbf{X}(t)$. Putting these last three equations into (23-65), we find that

$$\mathbf{x}(t) = \mathbf{\Phi}(t,t_0)\mathbf{x}(t_0) + \int_{t_0}^{t} \mathbf{\Phi}(t,\tau)[\mathbf{B}(\tau)\mathbf{u}(\tau) + \mathbf{G}(\tau)\mathbf{w}(\tau)]d\tau \tag{23-67}$$

for $t \geq t_0$, which is the solution to (23-31) that is stated in the main body of this lesson as (23-36).

All that is left is to obtain the differential equation for generating $\mathbf{\Phi}(t,\tau)$. From (23-66), we find that

$$\begin{aligned}\frac{d\mathbf{\Phi}(t,\tau)}{dt} &= \frac{d\mathbf{X}(t)}{dt}\mathbf{X}^{-1}(\tau) = \mathbf{F}(t)\mathbf{X}(t)\mathbf{X}^{-1}(\tau)\\ &= \mathbf{F}(t)\mathbf{\Phi}(t,\tau)\end{aligned} \tag{23-68}$$

subject, of course, to $\mathbf{\Phi}(t,t) = \mathbf{I}$.

## SUMMARY QUESTIONS

1. Suppose our nonlinear state equation does not depend explicitly on $t$. Then the perturbation state equation will be:

(a) time invariant
(b) time varying
(c) stationary

2. Jacobian matrices $\mathbf{F_x}$ and $\mathbf{H_x}$:
(a) are constant matrices
(b) are square
(c) depend on $\mathbf{x}^*(t)$ and $\mathbf{u}^*(t)$

3. When we say that one random sequence is "statistically equivalent" to another random sequence, we mean that:
(a) their first two moments are equal
(b) all their moments are equal
(c) they are approximately Gaussian

4. If system matrices are approximately constant during time intervals $(t_k, t_{k+1})$, then:
(a) the system is time invariant
(b) the system is still time varying
(c) $\mathbf{\Phi}(t, \tau)$ equals a constant matrix

5. In the case of a time-invariant system, (23-36) becomes:
(a) $\mathbf{x}(t) = \mathbf{\Phi}(t - t_0)\mathbf{x}(t_0)$
(b) $\mathbf{x}(t) = \mathbf{\Phi}(t, t_0)\mathbf{x}(t_0) + \int_{t_0}^{t} \mathbf{\Phi}(t - \tau)[\mathbf{B}(\tau)\mathbf{u}(\tau) + \mathbf{G}(\tau)\mathbf{w}(\tau)]d\tau$
(c) $\mathbf{x}(t) = \mathbf{\Phi}(t - t_0)\mathbf{x}(t_0) + \int_{t_0}^{t} \mathbf{\Phi}(t - \tau)[\mathbf{B}(\tau)\mathbf{u}(\tau) + \mathbf{G}(\tau)\mathbf{w}(\tau)]d\tau$

6. Nominal state vector $\mathbf{x}^*(t)$ must:
(a) satisfy the original state equation
(b) be prespecified
(c) provide a good approximation to the actual behavior of the system

7. The perturbation state-variable model is:
(a) linear and time varying
(b) linear and time invariant
(c) nonlinear

## PROBLEMS

**23-1.** Derive the Taylor series expansion of $\mathbf{h}[\mathbf{x}(t), \mathbf{u}(t), t]$ given in (23-25).

**23-2.** Derive the formula for $\mathbf{Q}_d(k)$ given in (23-51).

**23-3.** Derive formulas for $\mathbf{\Psi}(k+1, k)$ and $\mathbf{Q}_d(k+1, k)$ that include first- and second-order effects of $T$, using the first three terms in the expansion of $e^{\mathbf{F}_k T}$.

**23-4.** Let a zero-mean stationary Gaussian random process $v(t)$ have the autocorrelation function $\phi_v(\tau)$ given by $\phi_v(\tau) = e^{-|\tau|} + e^{-2|\tau|}$.

(a) Show that this colored-noise process can be generated by passing white noise $\mu(t)$ through the linear system whose transfer function is

$$\sqrt{6}\,\frac{s + \sqrt{2}}{(s+1)(s+2)}$$

(b) Obtain a discrete-time state-variable model for this colored noise process (assume $T = 1$ msec).

**23-5.** This problem presents a model for estimation of the altitude, velocity, and constant ballistic coefficient of a vertically falling body (Athans et al., 1968). The measurement are made at discrete instants of time by a radar that measures range in the presence of discrete-time white Gaussian noise. The state equations for the falling body are

$$\begin{aligned}\dot{x}_1 &= -x_2 \\ \dot{x}_2 &= -e^{-\gamma x_1} x_2^2 x_3 \\ \dot{x}_3 &= 0\end{aligned}$$

where $\gamma = 5 \times 10^{-5}$, $x_1(t)$ is altitude, $x_2(t)$ is downward velocity, and $x_3$ is a constant ballistic parameter. Measured range is given by

$$z(k) = \sqrt{M^2 + [x_1(k) - H]^2} + \nu(k), \qquad k = 1, 2, \ldots,$$

where $M$ is horizontal distance and $H$ is radar altitude. Obtain the discretized perturbation state-variable model for this system.

**23-6.** Normalized equations of a stirred reactor are

$$\begin{aligned}\dot{x}_1 &= -(c_1 + c_4)x_1 + c_3(1 + x_4)^2 \exp \frac{K_1 x_1}{(1 + x_1)} + c_4 x_2 + c_1 u_1 \\ \dot{x}_2 &= -(c_5 + c_6)x_2 + c_5 x_1 + c_6 x_3 \\ \dot{x}_3 &= -(c_7 + c_8)x_3 + c_8 x_2 + c_7 u_2 \\ \dot{x}_4 &= -c_1 x_4 - c_2(1 + x_4)^2 \exp \frac{K_1 x_1}{(1 + x_1)} + c_1 u_3\end{aligned}$$

in which $u_1$, $u_2$, and $u_3$ control inputs. Measurements are

$$z_i(k) = x_i(k) + \nu_i(k), \qquad i = 1, 2, 3,$$

where $\nu_i(k)$ are zero-mean white Gaussian noises with variances $r_i$. Obtain the discretized perturbation state-variable model for this system.

**23-7.** Obtain the discretized perturbation-state equation for each of the following systems:

**(a)** Equation for the unsteady operation of a synchronous motor:

$$\ddot{x}(t) + C\dot{x}(t) + p \sin x(t) = L(t)$$

**(b)** Duffing's equation:

$$\ddot{x}(t) + C\dot{x}(t) + \alpha x(t) + \beta x^3(t) = F \cos \omega t$$

**(c)** Van der Pol's equation:

$$\ddot{x}(t) - \epsilon \dot{x}(t) \left[1 - \frac{1}{3}\dot{x}^2(t)\right] + x(t) = m(t)$$

**(d)** Hill's equation:

$$\ddot{x}(t) - a x(t) + b p(t) x(t) = m(t)$$

where $p(t)$ is a known periodic function.

**23-8.** (Keith M. Chugg, Fall 1991) (Computation Problem) Some nonlinear systems exhibit extreme sensitivity to initial conditions. When this is the case, there is no nominal path for all practical purposes.

**(a)** To illustrate this property, generate a computer program to iterate the following discrete-time nonlinear map (known as the *logistical map*, which is a chaotic system): $x(k+1) = 4x(k)[1 - x(k)]$ first with $x(0) = 0.213$ and then with $x(0) = 0.21300001$. Try at least 40 iterations.

**(b)** What does this property imply about the manner in which a linear perturbation model should be applied?

**23-9.** The following equations describe the dynamics for a *chaotic glycolytic oscillator* (Holden, 1986):

$$\frac{dx_1(t)}{dt} = -x_1(t)x_2^2(t) + 0.999 + 0.42\cos(1.75t)$$

$$\frac{dx_2(t)}{dt} = x_1(t)x_2^2(t) - x_2(t)$$

**(a)** Plot the phase plane trajectory for this system [i.e., $x_2(t)$ versus $x_1(t)$] from $t = 0$ to $t = 100$ with $x_1(0) = x_2(0) = 1.5$. Repeat this for $x_1(0) = x_2(0) = 1.4$.

**(b)** Obtain the discretized perturbation state-variable model for this system.

**23-10.** The ball and beam system, depicted in Figure P23-10, can be described by four states: $x_1 = r(t)$, $x_2 = dr(t)/dt$, $x_3 = \theta(t)$, and $x_4 = d\theta(t)/dt$. The beam is made to rotate in a vertical plane by applying a torque at the center of rotation and the ball is free to roll along the beam. Of course, the ball is required to remain in contact with the beam. This system can be represented by the following state-variable model:

$$\frac{dx_1(t)}{dt} = x_2(t)$$

$$\frac{dx_2(t)}{dt} = B[x_1(t)x_4^2(t) - G\sin x_3(t)]$$

$$\frac{dx_3(t)}{dt} = x_4(t)$$

$$\frac{dx_4(t)}{dt} = u(t)$$

$$y(t) = x_1(t)$$

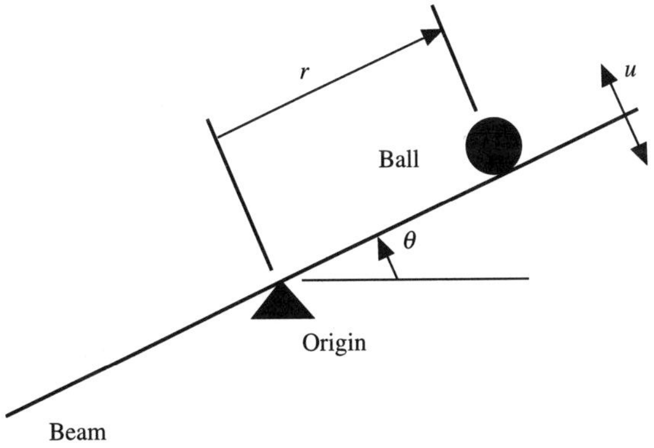


**Figure P23-10** Ball and beam.

where the control $u(t)$ is the acceleration of $\theta(t)$, and the parameters $B$ and $G$ are as defined in Hauser et al. (1992). Obtain the discretized perturbation state-variable model for this system.

**23-11.** (Gregg Isara and Loan Bui, Fall 1991) John Shuck is an engineer/entrepreneur who would like to build and operate a micro-brewery. John knows beer usually passes through several stages of fermentation, each stage in a different container before it is ready to be consumed. Since he is a busy man (being an engineer and a "brew-master"), he does not want to go through many tastings to find out if the batch is ready to be transferred. Because of his engineering background, he is able to model the rate of fermentation, $df/dt$, and the change in alcohol content, $dA/dt$, as

$$\frac{df(t)}{dt} = \left(\frac{A_y^2}{A_w}\right) f(t) \exp(-2f(t))$$

$$\frac{dA(t)}{dt} = A(t) + f^2(t)$$

John remembers, from estimation theory, that if he can obtain a discretized linear model he can apply various state prediction and filtering methods to estimate his fermentation state; but John can't remember how to do this. Help John obtain the discretized linear perturbation equations, assuming that the measurements are modeled as

$$z(t) = A(t) + \nu(t)$$

and that the states (or nominal values) are slowly varying over the time interval $[t_k, t_{k+1}]$.

**23-12.** (Roy T. Okida, Fall 1991) Determine an *exact* discretization of the following state equation for $T = 1$:

$$\frac{d\mathbf{x}(t)}{dt} = \mathbf{A}\mathbf{x}(t) + \mathbf{B}\mathbf{u}(t)$$

where

$$\mathbf{A} = \begin{bmatrix} 1 & 3 & -3 \\ 3 & 1 & -7 \\ 0 & 0 & -6 \end{bmatrix}$$

**23-13.** (Todd A. Parker and Mark A. Ranta, Fall 1991) The motion of a rocket is described by

$$\frac{dv}{dt} = -\frac{vw}{r} + \frac{(dm/dt)V_e \cos\phi}{m_0 + t\,dm/dt}$$

$$\frac{dw}{dt} = \frac{v^2}{r} - \frac{\mu}{r^2} + \frac{(dm/dt)V_e \sin\phi}{m_0 + t\,dm/dt}$$

where $v$ = tangential velocity, $w$ = radial velocity, $r$ = radius from the center of the earth, $m_0$ = initial mass, $dm/dt$ = rate of change in mass due to fuel burning ($dm/dt < 0$), $V_e$ = exhaust velocity, $\phi$ = thrust direction angle, and $\mu$ = constant. Define the state vector $\mathbf{x} = \text{col}(v, w, r, dm/dt)$. Let $dm/dt$ = constant and the input, $u$, be equal to $\phi$. Obtain the linearized perturbation equations for this system.

**23-14.** (Terry L. Bates, Fall 1991) The following equations can be used to describe a missile:

$$\frac{dx_1}{dt} = -\frac{x_2}{y} + yu_1(t)$$

$$\frac{dx_2}{dt} = -2x_1x_2^{-2}y + 3u_2(t)$$

$$z(t) = -2x_1y - x_2^2 + v(t)$$

where $y = e^{-T}$. Determine the perturbation state and measurement equations.

**23-15.** (Gregory Caso, Fall 1991) Consider the nonlinear baseband model for a type 2 second-order phase-lock loop (PLL) depicted in Figure P23-15. Note that $1/s$ denotes integration.

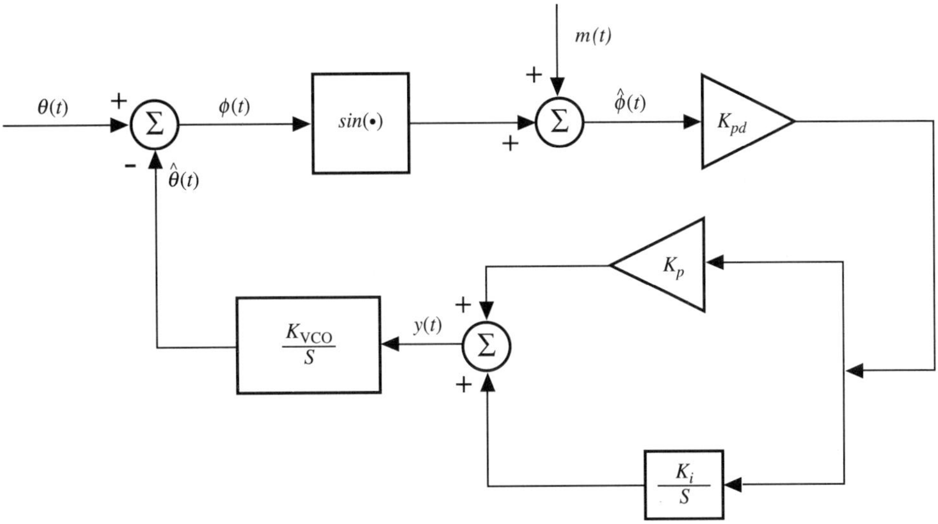


**Figure P23-15** Type 2 second-order phase-lock loop.

**(a)** Formulate the state-variable model for this system with states $x_1(t) = \phi(t)$, $x_2(t) = K_{\text{VCO}}y(t)$, and input $u(t) = d\theta(t)/dt$. Consider the system output to be $\hat{\phi}(t)$.

**(b)** Obtain the perturbation state-variable model for this system.

**23-16.** (John B. Benson, Spring 1992) Consider the circuit shown in Figure P23-16. Choose as the state variables the voltages across the capacitors, as shown. The ammeter in the circuit measures the current through the PN-junction diode; however, unavoidable disturbances cause slight errors to be made in the measurements. These errors may be modeled as additive white Gaussian uncertainty in the ammeter's reading.

**(a)** Find the nonlinear state-variable model for this circuit by writing Kirchhoff's current equations at the nodes between $R_1$ and the diode $D$, and $R_2$ and $D$.

**(b)** Find $\mathbf{f}[\mathbf{x}(t), \mathbf{u}(t), t]$ and $\mathbf{h}[\mathbf{x}(t), \mathbf{u}(t), t]$.

**23-17.** (Brad Verona, Spring 1992) Consider the simple pendulum depicted in Figure P23-17, where from Newton's laws we have

$$\frac{d^2\theta(t)}{dt^2} = \frac{g}{d}\sin\theta(t) + \frac{1}{md}u(t)$$

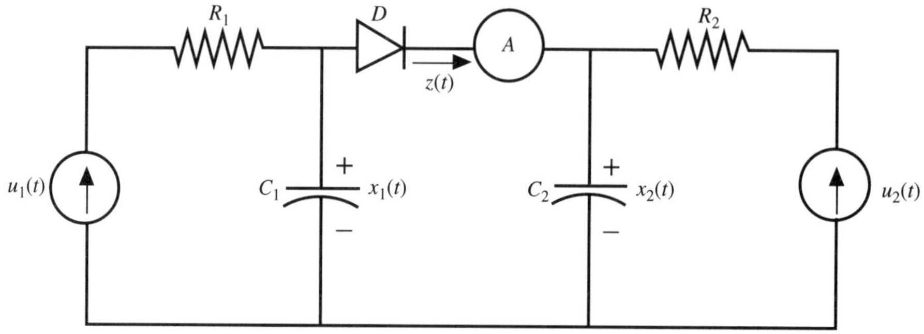


**Figure P23-16** Nonlinear electric network.

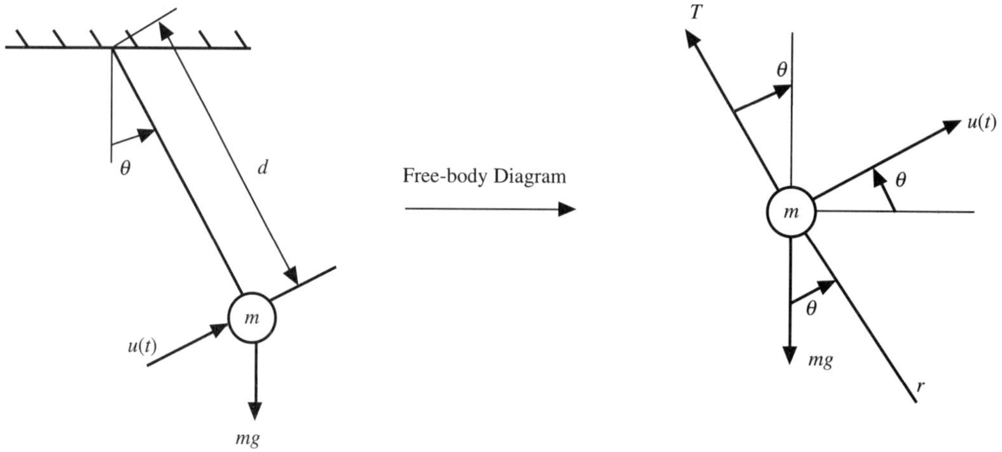


**Figure P23-17** Pendulum.

We are able to measure $\theta(t)$ subject to additive errors. Determine the linear perturbation state-variable model for this system.

**23-18.** The *inverted pendulum* depicted in Figure P23-18, is a landmark example that is used in many neural network studies. Its equations of motion are

$$\frac{dx_1(t)}{dt} = x_2(t)$$

$$\frac{dx_2(t)}{dt} = f(x_1, x_2) + g(x_1, x_2)u(t)$$

where

$$f(x_1, x_2) = \frac{g \sin x_1 - \dfrac{(mlx_2^2 \cos x_1 \sin x_1)}{(m_c + m)}}{l\left(\dfrac{4}{3} - \dfrac{m \cos^2 x_1}{(m_c + m)}\right)}$$

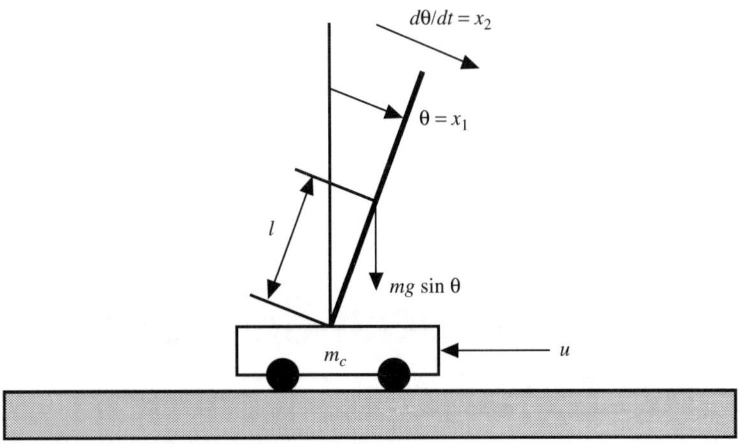


**Figure P23-18** Inverted pendulum (Wang, 1994).

and

$$g(x_1, x_2) = \frac{\dfrac{\cos x_1}{(m_c + m)}}{l\left(\dfrac{4}{3} - \dfrac{m\cos^2 x_1}{(m_c + m)}\right)}$$

in which $g = 9.8\,\text{m/s}^2$ is the acceleration due to gravity, $m_c$ is the mass of the cart, $m$ is the mass of the pole, $l$ is the half-length of the pole, and $u(t)$ is the applied force (control). We are able to measure $\theta(t)$ subject to additive noise. Determine the linear perturbation state-variable model for this system.

**23-19.** In the first-order system

$$\frac{dx(t)}{dt} = ax(t) + w(t)$$

$$z(t) = x(t) + \nu(t), \qquad t = t_i, \;\; i = 1, 2, \ldots, N$$

$a$ is an unknown parameter that is to be estimated. Processes $w(t)$ and $\nu(t)$ are, as usual, mutually uncorrelated, zero mean, and white, with intensities equal to 1 and 1/2, respectively. What is the perturbation SVM for this system? (*Note*: For additional discussions on continuous-time systems, see Lessons 24 and 26.)

LESSON 24

# Iterated Least Squares and Extended Kalman Filtering


## SUMMARY

This lesson is devoted primarily to the extended Kalman filter (EKF), which is a form of the Kalman filter "extended" to nonlinear dynamical systems of the form described in Lesson 23. We show that the EKF is related to the method of iterated least squares (ILS), the major difference being that the EKF is for dynamical systems, whereas ILS is not.

We explain ILS for the model $z(k) = f(\theta, k) + \nu(k), k = 1, 2, \ldots, N$, where the objective is to estimate $\theta$ from the measurements. A four-step procedure is given for ILS, from which we observe that in each complete cycle of this procedure *we use both the nonlinear and linearized models* and that ILS uses the estimate obtained from the linearized model to generate the nominal value of $\theta$ about which the nonlinear model is *relinearized.*

The notions of relinearizing about a filter output and using both the nonlinear and linearized models are also at the very heart of the EKF. The EKF is developed in predictor–corrector format. Its prediction equation is obtained by integrating the nominal differential equation associated with $\mathbf{x}^*(t)$, from $t_k$ to $t_{k+1}$. To do this, we learn that $\mathbf{x}^*(t_k) = \hat{\mathbf{x}}(k|k)$, whereas $\mathbf{x}^*(t_{k+1}) = \hat{\mathbf{x}}(k+1|k)$. The corrector equation is obtained from the Kalman filter associated with the discretized perturbation state-variable model derived in Lesson 23. The gain and covariance matrices that appear in the corrector equation depend on the nominal $\mathbf{x}^*(t)$ that results from prediction, $\hat{\mathbf{x}}(k+1|k)$.

An *iterated EKF* is an EKF in which the correction equation is iterated a fixed number of times. This improves the convergence properties of the EKF, because convergence is related to how close the nominal value of the state vector is to its actual value.

When you have completed this lesson, you will be able to (1) explain the method of iterated least squares (ILS); (2) derive and use the extended Kalman filter (EKF); (3) explain how the EKF is related to ILS; and (4) apply the EKF to estimate parameters in a state-variable model.

## INTRODUCTION

This lesson is primarily devoted to the extended Kalman filter (EKF), which is a form of the Kalman filter "extended" to nonlinear dynamical systems of the type described in Lesson 23. We shall show that the EKF is related to the method of iterated least squares (ILS), the major difference being that the EKF is for dynamical systems, whereas ILS is not.

## ITERATED LEAST SQUARES

*estimate const parameters, EKF will estimate Time varying params.*

We shall illustrate the method of ILS for the nonlinear model described in Example 2-5, i.e., for the model

*before LSF was linear*

$$z(k) = f(\theta, k) + v(k) \qquad (24\text{-}1)$$

*nonlinear function*

where $k = 1, 2, \ldots, N$.

Iterated least squares is basically a four-step procedure.

1. Linearize $f(\theta, k)$ about a nominal value of $\theta, \theta^*$. Doing this, we obtain the perturbation measurement equation

   $\theta^* = $ nominal value, (say $\theta_{ave}$ ?)

$$\delta z(k) = F_\theta(k; \theta^*)\delta\theta + v(k), \qquad k = 1, 2, \ldots, N \qquad (24\text{-}2)$$

   *linear wrt $\delta\theta$*

   *ok, or if you have a better estimate o x it. ($\theta = const$)*

   where

$$\delta z(k) = z(k) - z^*(k) = z(k) - f(\theta^*, k) \qquad (24\text{-}3)$$

$$\delta\theta = \theta - \theta^* \qquad (24\text{-}4)$$

   and

$$F_\theta(k; \theta^*) = \left.\frac{\partial f(\theta, k)}{\partial\theta}\right|_{\theta=\theta^*} \qquad (24\text{-}5)$$

2. Concatenate (24-2) and compute $\hat{\delta\theta}_{\text{WLS}}(N)$[or $\hat{\delta\theta}_{\text{LS}}(N)$] using our Lesson 3 formulas.

3. Solve the equation

   *known*          *known (from a. linearization)*

$$\hat{\delta\theta}_{\text{WLS}}(N) = \hat{\theta}_{\text{WLS}}(N) - \theta^* \qquad (24\text{-}6)$$

   for $\hat{\theta}_{\text{WLS}}(N)$, i.e.,

   *known*

$$\hat{\theta}_{\text{WLS}}(N) = \theta^* + \hat{\delta\theta}_{\text{WLS}}(N) \qquad (24\text{-}7)$$

4. Replace $\theta^*$ with $\hat{\theta}_{\text{WLS}}(N)$ and return to step 1. Iterate through these steps until convergence occurs. Let $\hat{\theta}^i_{\text{WLS}}(N)$ and $\hat{\theta}^{i+1}_{\text{WLS}}(N)$ denote estimates of $\theta$ obtained

at the $i$th and $(i + 1)$st iterations, respectively. Convergence of the ILS method occurs when

$$|\hat{\theta}^{i+1}_{\text{WLS}}(N) - \hat{\theta}^i_{\text{WLS}}(N)| < \epsilon \qquad (24\text{-}8)$$

where $\epsilon$ is a prespecified small positive number.

We observe, from this four-step procedure, that ILS uses the estimate obtained from the linearized model to generate the nominal value of $\theta$ about which the nonlinear model is *relinearized*. Additionally, in each complete cycle of this procedure, *we use both the nonlinear and linearized models*. The nonlinear model is used to compute $z^*(k)$ and subsequently $\delta z(k)$, using (24-3).

The notions of relinearizing about a filter output and using both the nonlinear and linearized models are also at the very heart of the EKF.

## EXTENDED KALMAN FILTER

*for nonlinear continuous (you must discretize this, already discreet system it already (cf p389)*

The nonlinear dynamical system of interest to us is the one described in Lesson 23. For convenience to the reader, we summarize aspects of that system next. The nonlinear state-variable model is

$$\dot{\mathbf{x}}(t) = \mathbf{f}[\mathbf{x}(t), \mathbf{u}(t), t] + \mathbf{G}(t)\mathbf{w}(t) \qquad (24\text{-}9)$$

$$\mathbf{z}(t) = \mathbf{h}[\mathbf{x}(t), \mathbf{u}(t), t] + \mathbf{v}(t), \qquad t = t_i, \quad i = 1, 2, \dots \qquad (24\text{-}10)$$

Given a nominal input $\mathbf{u}^*(t)$ and assuming that a nominal trajectory $\mathbf{x}^*(t)$ exists, $\mathbf{x}^*(t)$ and its associated nominal measurement satisfy the following nominal system model:

$$\dot{\mathbf{x}}^*(t) = \mathbf{f}[\mathbf{x}^*(t), \mathbf{u}^*(t), t] \quad \mathbf{x}^* = \int f(x^* u^* t) \qquad (24\text{-}11)$$

$$\mathbf{z}^*(t) = \mathbf{h}[\mathbf{x}^*(t), \mathbf{u}^*(t), t], \qquad t = t_i, \quad i = 1, 2, \dots \qquad (24\text{-}12)$$

Letting $\delta\mathbf{x}(t) = \mathbf{x}(t) - \mathbf{x}^*(t)$, $\delta\mathbf{u}(t) = \mathbf{u}(t) - \mathbf{u}^*(t)$, and $\delta\mathbf{z}(t) = \mathbf{z}(t) - \mathbf{z}^*(t)$, we also have the following *discretized perturbation state-variable model* that is associated with a linearized version of the original nonlinear state-variable model:

$$\delta\mathbf{x}(k + 1) = \mathbf{\Phi}(k + 1, k;^*)\delta\mathbf{x}(k) + \mathbf{\Psi}(k + 1, k;^*)\delta\mathbf{u}(k) + \mathbf{w}_d(k) \qquad (24\text{-}13)$$

$$\delta\mathbf{z}(k + 1) = \mathbf{H}_{\mathbf{x}}(k + 1;^*)\delta\mathbf{x}(k + 1)$$
$$+ \mathbf{H}_{\mathbf{u}}(k + 1;^*)\delta\mathbf{u}(k + 1) + \mathbf{v}(k + 1) \qquad (24\text{-}14)$$

In deriving (24-13) and (24-14), we made the important assumption that higher-order terms in the Taylor series expansions of $\mathbf{f}[\mathbf{x}(t), \mathbf{u}(t), t]$ and $\mathbf{h}[\mathbf{x}(t), \mathbf{u}(t), t]$ could be neglected. Of course, this is only correct as long as $\mathbf{x}(t)$ is "close" to $\mathbf{x}^*(t)$ and $\mathbf{u}(t)$ is "close" to $\mathbf{u}^*(t)$.

As discussed in Lesson 23, if $\mathbf{u}(t)$ is an input derived from a feedback control law, so that $\mathbf{u}(t) = \mathbf{u}[\mathbf{x}(t), t]$, then $\mathbf{u}(t)$ can differ from $\mathbf{u}^*(t)$, because $\mathbf{x}(t)$ will differ from $\mathbf{x}^*(t)$. On the other hand, if $\mathbf{u}(t)$ does not depend on $\mathbf{x}(t)$, then usually $\mathbf{u}(t)$ is the same as $\mathbf{u}^*(t)$, in which case $\delta\mathbf{u}(t) = \mathbf{0}$. We see, therefore, that

*Will Not Work (see p. 389)*

$\mathbf{x}^*(t)$ is the critical quantity in the calculation of our discretized perturbation state-variable model.

Suppose $\mathbf{x}^*(t)$ is given a priori; then we can compute predicted, filtered, or smoothed estimates of $\delta\mathbf{x}(k)$ by applying all our previously derived estimators to the discretized perturbation state-variable model in (24-13) and (24-14). We can precompute $\mathbf{x}^*(t)$ by solving the nominal differential equation (24-11). The Kalman filter associated with using a precomputed $\mathbf{x}^*(t)$ is known as *a relinearized KF.*

A relinearized KF usually gives poor results, because it relies on an open-loop strategy for choosing $\mathbf{x}^*(t)$. When $\mathbf{x}^*(t)$ is precomputed, there is no way of forcing $\mathbf{x}^*(t)$ to remain close to $\mathbf{x}(t)$, and this must be done or else the perturbation state-variable model is invalid. Divergence of the relinearized KF often occurs; hence, we do not recommend the relinearized KF.

The relinearized KF is based only on the discretized perturbation state-variable model. It does not use the nonlinear nature of the original system in an active manner. The extended Kalman filter relinearizes the nonlinear system about each new estimate as it becomes available; i.e., at $k = 0$, the system is linearized about $\hat{\mathbf{x}}(0|0)$. Once $\mathbf{z}(1)$ is processed by the EKF, so that $\hat{\mathbf{x}}(1|1)$ is obtained, the system is linearized about $\hat{\mathbf{x}}(1|1)$. By "linearize about $\hat{\mathbf{x}}(1|1)$," we mean $\hat{\mathbf{x}}(1|1)$ is used to calculate all the quantities needed to make the transition from $\hat{\mathbf{x}}(1|1)$ to $\hat{\mathbf{x}}(2|1)$ and subsequently $\hat{\mathbf{x}}(2|2)$. This phrase will become clear later. The purpose of relinearizing about the filter's output is to use a better reference trajectory for $\mathbf{x}^*(t)$. Doing this, $\delta\mathbf{x} = \mathbf{x} - \hat{\mathbf{x}}$ will be held as small as possible, so that our linearization assumptions are less likely to be violated than in the case of the relinearized KF.

The EKF is developed in predictor–corrector format (Jazwinski, 1970). Its prediction equation is obtained by integrating the nominal differential equation for $\mathbf{x}^*(t)$, from $t_k$ to $t_{k+1}$. To do this, we need to know how to choose $\mathbf{x}^*(t)$ for the entire interval of time $t \in [t_k, t_{k+1}]$. Thus far, we have only mentioned how $\mathbf{x}^*(t)$ is chosen at $t_k$, i.e., as $\hat{\mathbf{x}}(k|k)$.

**Theorem 24-1.** *As a consequence of relinearizing about* $\hat{\mathbf{x}}(k|k)(k = 0, 1, \ldots)$,

$$\delta\hat{\mathbf{x}}(t|t_k) = \mathbf{0}, \qquad \text{for } all\ t \in [t_k, t_{k+1}] \tag{24-15}$$

*This means that*

$$\mathbf{x}^*(t) = \hat{\mathbf{x}}(t|t_k) \qquad \text{for } all\ t \in [t_k, t_{k+1}] \tag{24-16}$$

Before proving this important result, we observe that it provides us with a choice of $\mathbf{x}^*(t)$ over the entire interval of time $t \in [t_k, t_{k+1}]$, and it states that at the left-hand side of this time interval $\mathbf{x}^*(t_k) = \hat{\mathbf{x}}(k|k)$, whereas at the right-hand side of this time interval $\mathbf{x}^*(t_{k+1}) = \hat{\mathbf{x}}(k+1|k)$. The transition from $\hat{\mathbf{x}}(k+1|k)$ to $\hat{\mathbf{x}}(k+1|k+1)$ will be made using the EKF's correction equation.

*Proof.* Let $t_l$ be an arbitrary value of $t$ lying in the interval between $t_k$ and $t_{k+1}$ (see Figure 24-1). For the purposes of this derivation, we can assume that $\delta\mathbf{u}(k) = \mathbf{0}$ [i.e., perturbation input $\delta\mathbf{u}(k)$ takes on no new values in the interval from $t_k$ to $t_{k+1}$; recall the piecewise-constant assumption made about $\mathbf{u}(t)$ in the derivation of (23-38)]; i.e.,

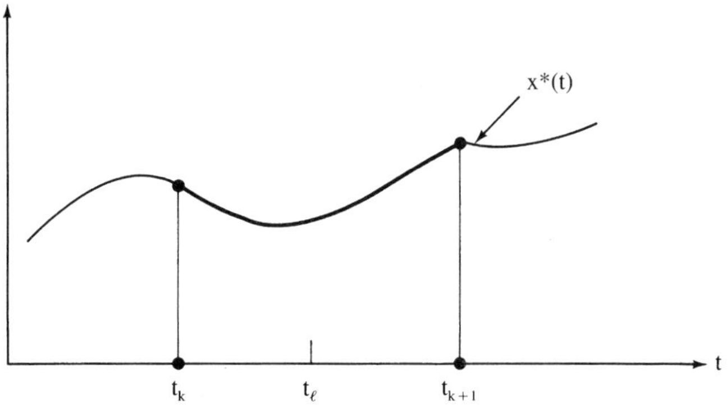

**Figure 24-1** Nominal state trajectory $x^*(t)$. over our trajectory

$$\delta \mathbf{x}(k+1) = \boldsymbol{\Phi}(k+1, k;^*) \delta \mathbf{x}(k) + \mathbf{w}_d(k) \tag{24-17}$$

Using our general state-predictor results given in (16-14), we see that (remember that $k$ is short for $t_k$, and that $t_{k+1} - t_k$ does not have to be a constant; this is true in all of our predictor, filter, and smoother formulas)

$$\delta \hat{\mathbf{x}}(t_l|t_k) = \boldsymbol{\Phi}(t_l, t_k;^*) \delta \hat{\mathbf{x}}(t_k|t_k)$$

produdule $l>k$ $\qquad = \boldsymbol{\Phi}(t_l, t_k;^*) [\hat{\mathbf{x}}(k|k) - \mathbf{x}^*(k)] \tag{24-18}$

In the EKF we set $\mathbf{x}^*(k) = \hat{\mathbf{x}}(k|k)$; thus, when this is done,

$$\delta \hat{\mathbf{x}}(t_l|t_k) = \mathbf{0} \tag{24-19}$$

and, because $t_l \in [t_k, t_{k+1}]$,

$$\delta \hat{\mathbf{x}}(t_l|t_k) = \mathbf{0}, \qquad \text{for } all \ t_l \in [t_k, t_{k+1}] \tag{24-20}$$

which is (24-15). Equation (24-16) follows from (24-20) and the fact that $\delta \hat{\mathbf{x}}(t_l|t_k) = \hat{\mathbf{x}}(t_l|t_k) - \mathbf{x}^*(t_l)$. $\square$ ∴ $x^*(t) = \hat{x}(k|k)$

We are now able to derive the EKF. As mentioned previously, the EKF must be obtained in predictor–corrector format. We begin the derivation by obtaining the predictor equation for $\hat{\mathbf{x}}(k+1|k)$.

Recall that $\mathbf{x}^*(t)$ is the solution of the nominal state equation (24-11). Using (24-16) in (24-11), we find that only valid between $T_k + T_{k+1}$

$$\frac{d}{dt}\hat{\mathbf{x}}(t|t_k) = \mathbf{f}[\hat{\mathbf{x}}(t|t_k), \mathbf{u}^*(t), t] \tag{24-21}$$

Integrating this equation from $t = t_k$ to $t = t_{k+1}$, we obtain

$\int_{t_k}^{t_{k+1}} \frac{d}{dt} \tilde{x}_{+/k} = \tilde{x}_{k+1|k} - \tilde{x}_{k|k}$

$$\hat{\mathbf{x}}(k+1|k) = \hat{\mathbf{x}}(k|k) + \int_{t_k}^{t_{k+1}} \mathbf{f}[\hat{\mathbf{x}}(t|t_k), \mathbf{u}^*(t), t]dt \tag{24-22}$$

which is the *EKF prediction equation*. Observe that the nonlinear nature of the system's state equation is used to determine $\hat{\mathbf{x}}(k+1|k)$. The integral in (24-22) is evaluated by means of numerical integration formulas that are initialized by $\mathbf{f}[\hat{\mathbf{x}}(t_k|t_k), \mathbf{u}^*(t_k), t_k]$.

The corrector equation for $\hat{\mathbf{x}}(k+1|k+1)$ is obtained from the Kalman filter associated with the discretized perturbation state-variable model in (24-13) and (24-14) and is

$$\delta\hat{\mathbf{x}}(k+1|k+1) = \delta\hat{\mathbf{x}}(k+1|k) + \mathbf{K}(k+1;^*)[\delta\mathbf{z}(k+1) - \mathbf{H}_{\mathbf{x}}(k+1;^*)\delta\hat{\mathbf{x}}(k+1|k) - \mathbf{H}_{\mathbf{u}}(k+1;^*)\delta\mathbf{u}(k+1)] \tag{24-23}$$

As a consequence of relinearizing about $\hat{\mathbf{x}}(k|k)$, we know that

$$\delta\hat{\mathbf{x}}(k+1|k) = \mathbf{0} \tag{24-24}$$

$$\begin{aligned}\delta\hat{\mathbf{x}}(k+1|k+1) &= \hat{\mathbf{x}}(k+1|k+1) - \mathbf{x}^*(k+1)\\ &= \hat{\mathbf{x}}(k+1|k+1) - \hat{\mathbf{x}}(k+1|k)\end{aligned} \tag{24-25}$$

and

$$\begin{aligned}\delta\mathbf{z}(k+1) &= \mathbf{z}(k+1) - \mathbf{z}^*(k+1)\\ &= \mathbf{z}(k+1) - \mathbf{h}[\mathbf{x}^*(k+1), \mathbf{u}^*(k+1), k+1]\\ &= \mathbf{z}(k+1) - \mathbf{h}[\hat{\mathbf{x}}(k+1|k), \mathbf{u}^*(k+1), k+1]\end{aligned} \tag{24-26}$$

Substituting these three equations into (24-23), we obtain

$$\hat{\mathbf{x}}(k+1|k+1) = \hat{\mathbf{x}}(k+1|k) + \mathbf{K}(k+1;^*)\{\mathbf{z}(k+1) - \mathbf{h}[\hat{\mathbf{x}}(k+1|k), \mathbf{u}^*(k+1), k+1] - \mathbf{H}_{\mathbf{u}}(k+1;^*)\delta\mathbf{u}(k+1)\} \tag{24-27}$$

which is the *EKF correction equation.* Observe that the nonlinear nature of the system's measurement equation is used to determine $\hat{\mathbf{x}}(k+1|k+1)$. One usually sees this equation for the case when $\delta\mathbf{u} = \mathbf{0}$, in which case the last term on the right-hand side of (24-27) is not present.

To compute $\hat{\mathbf{x}}(k+1|k+1)$, we must compute the EKF gain matrix $\mathbf{K}(k+1;^*)$. This matrix, as well as its associated $\mathbf{P}(k+1|k;^*)$ and $\mathbf{P}(k+1|k+1;^*)$ matrices, depend on the nominal $\mathbf{x}^*(t)$ that results from prediction, $\hat{\mathbf{x}}(k+1|k)$. Observe, from (24-16), that $\mathbf{x}^*(k+1) = \hat{\mathbf{x}}(k+1|k)$ and that the argument of $\mathbf{K}$ in the correction equation is $k+1$; hence, we are indeed justified to use $\hat{\mathbf{x}}(k+1|k)$ as the nominal value of $\mathbf{x}^*$ during the calculations of $\mathbf{K}(k+1;^*)$, $\mathbf{P}(k+1|k;^*)$, and $\mathbf{P}(k+1|k+1;^*)$. These three quantities are computed from

$$\mathbf{K}(k+1;^*) = \mathbf{P}(k+1|k;^*)\mathbf{H}'_{\mathbf{x}}(k+1;^*)[\mathbf{H}_{\mathbf{x}}(k+1;^*) \mathbf{P}(k+1|k;^*)\mathbf{H}'_{\mathbf{x}}(k+1;^*) + \mathbf{R}(k+1)]^{-1} \tag{24-28}$$

$$\mathbf{P}(k+1|k;^*) = \mathbf{\Phi}(k+1,k;^*)\mathbf{P}(k|k;^*)\mathbf{\Phi}'(k+1,k;^*) + \mathbf{Q}_d(k+1,k;^*) \tag{24-29}$$

$$\mathbf{P}(k+1|k+1;^*) = [\mathbf{I} - \mathbf{K}(k+1;^*)\mathbf{H}_{\mathbf{x}}(k+1;^*)]\mathbf{P}(k+1|k;^*) \tag{24-30}$$

Remember that in these three equations $^*$ denotes the use of $\hat{\mathbf{x}}(k+1|k)$.

The EKF is very widely used, especially in the aerospace industry; however, it does not provide an optimal estimate of $\mathbf{x}(k)$. The optimal estimate of $\mathbf{x}(k)$ is still $\mathbf{E}\{\mathbf{x}(k)|\mathbf{Z}(k)\}$, regardless of the linear or nonlinear nature of the system's model. The EKF is a first-order approximation of $\mathbf{E}\{\mathbf{x}(k)|\mathbf{Z}(k)\}$ that sometimes works quite well, but cannot be guaranteed always to work well. No convergence results are known for the EKF; hence, the EKF must be viewed as an ad hoc filter. Alternatives to the EKF, which are based on nonlinear filtering, are quite complicated and are rarely used.

A flow chart for implementation of the EKF is depicted in Figure 24-2.

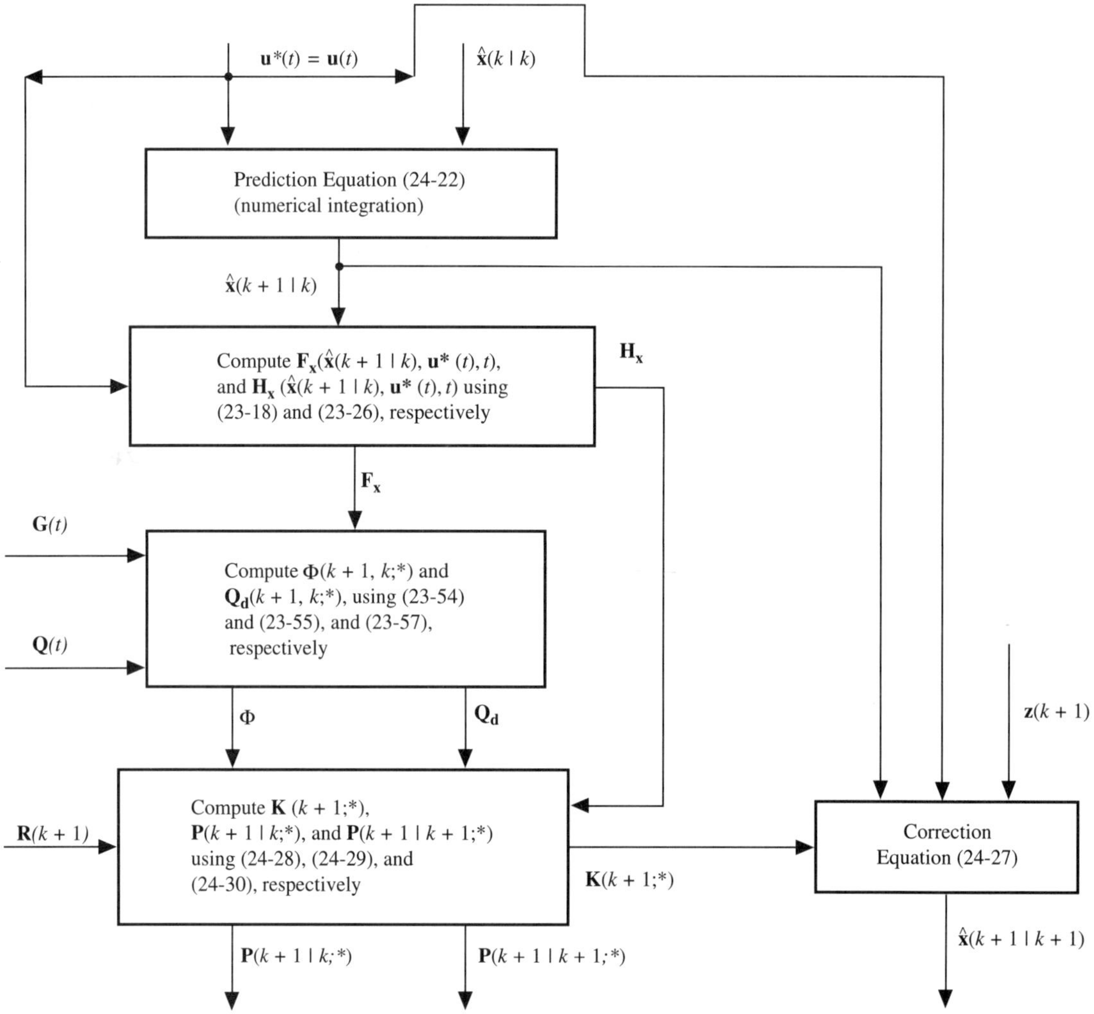


**Figure 24-2** EKF flow chart.

The EKF is designed to work well as long as $\delta\mathbf{x}(k)$ is "small". The *iterated EKF* (Jazwinski, 1970), depicted in Figure 24-3, is designed to keep $\delta\mathbf{x}(k)$ as small as possible. The iterated EKF differs from the EKF in that it iterates the correction equation $L$ times until $\|\hat{\mathbf{x}}_L(k+1|k+1) - \hat{\mathbf{x}}_{L-1}(k+1|k+1)\| \leq \epsilon$. Corrector 1 computes $\mathbf{K}(k+1;^*)$, $\mathbf{P}(k+1|k;^*)$, and $\mathbf{P}(k+1|k+1;^*)$ using $\mathbf{x}^* = \hat{\mathbf{x}}(k+1|k)$; corrector 2 computes these quantities using $\mathbf{x}^* = \hat{\mathbf{x}}_1(k+1|k+1)$; corrector 3 computes these quantities using $\mathbf{x}^* = \hat{\mathbf{x}}_2(k+1|k+1)$; etc.

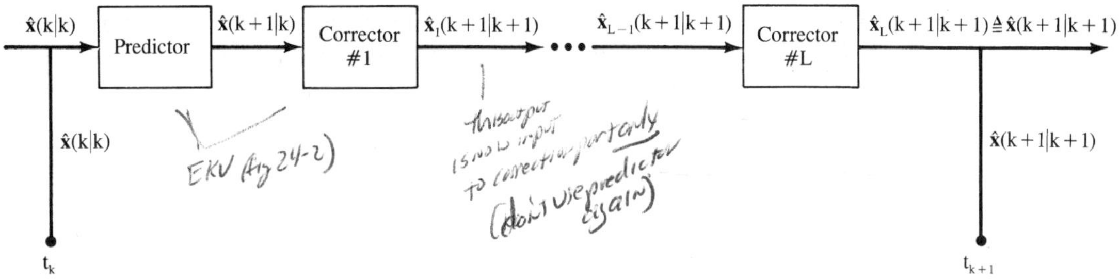


**Figure 24-3** Iterated EKF. All the calculations provide us with a refined estimate of $\mathbf{x}(k+1)$, $\hat{\mathbf{x}}(k+1|k+1)$, starting with $\hat{\mathbf{x}}(k|k)$.

Often, just adding one additional corrector (i.e., $L = 2$) leads to substantially better results for $\hat{\mathbf{x}}(k+1|k+1)$ than are obtained using the EKF.

## APPLICATION TO PARAMETER ESTIMATION

One of the earliest applications of the extended Kalman filter was to parameter estimation (Kopp and Orford, 1963). Consider the continuous-time linear system

$$\dot{\mathbf{x}}(t) = \mathbf{A}\mathbf{x}(t) + \mathbf{w}(t) \tag{24-31a}$$

$$\mathbf{z}(t) = \mathbf{H}\mathbf{x}(t) + \nu(t), \quad t = t_i, \quad i = 1, 2, \ldots \tag{24-31b}$$

Matrices $\mathbf{A}$ and $\mathbf{H}$ contain some unknown parameters, and our objective is to estimate these parameters from the measurements $\mathbf{z}(t_i)$ as they become available.

To begin, we assume differential equation models for the unknown parameters, i.e., either

$$\dot{a}_l(t) = 0, \qquad l = 1, 2, \ldots, l^* \tag{24-32a}$$

$$\dot{h}_j(t) = 0, \qquad j = 1, 2, \ldots, j^* \tag{24-32b}$$

or

$$\dot{a}_l(t) = c_l a_l(t) + n_l(t), \qquad l = 1, 2, \ldots, l^* \tag{24-33a}$$

$$\dot{h}_j(t) = d_j h_j(t) + \eta_j(t), \qquad j = 1, 2, \ldots, j^* \tag{24-33b}$$

In the latter models $n_l(t)$ and $\eta_j(t)$ are white noise processes, and we often choose $c_l = 0$ and $d_j = 0$. The noises $n_l(t)$ and $\eta_j(t)$ introduce uncertainty about the "constancy" of the $a_l$ and $h_j$ parameters.

Next, we augment the parameter differential equations to (24-31a) and (24-31b). The resulting system is nonlinear, because it contains products of states [e.g.,

$a_l(t)x_i(t)]$. The augmented system can be expressed as in (24-9) and (24-10), which means we have reduced the problem of parameter estimation in a linear system to state estimation in a nonlinear system.

Finally, we apply the EKF to the augmented state-variable model to obtain $\hat{a}_l(k|k)$ and $\hat{h}_j(k|k)$.

Ljung (1979) has studied the convergence properties of the EKF applied to parameter estimation and has shown that parameter estimates do not converge to their true values. He shows that another term must be added to the EKF corrector equation in order to guarantee convergence. For details, see his paper.

**EXAMPLE 24-1**

Consider the satellite and planet Example 23-1, in which the satellite's motion is governed by the two equations

$$\ddot{r} = r\dot{\theta}^2 - \frac{\gamma}{r^2} + \frac{l}{m}u_r(t) \tag{24-34}$$

and

$$\ddot{\theta} = -\frac{2\dot{r}\dot{\theta}}{r} + \frac{1}{m}u_\theta(t) \tag{24-35}$$

We shall assume that $m$ and $\gamma$ are unknown constants and shall model them as

$$\dot{m}(t) = 0 \tag{24-36}$$

$$\dot{\gamma}(t) = 0 \tag{24-37}$$

Defining $x_1 = r, x_2 = \dot{r}, x_3 = \theta, x_4 = \dot{\theta}, x_5 = m, x_6 = \gamma, u_1 = u_r$, and $u_2 = u_\theta$, we have

$$\left.\begin{aligned} \dot{x}_1 &= x_2 \\ \dot{x}_2 &= x_1x_4^2 - \frac{x_6}{x_1^2} + \frac{1}{x_5}u_1(t) \\ \dot{x}_3 &= x_4 \\ \dot{x}_4 &= -\frac{2x_2x_4}{x_1} + \frac{1}{x_5}u_2(t) \\ \dot{x}_5 &= 0 \\ \dot{x}_6 &= 0 \end{aligned}\right\} \tag{24-38}$$

Now,

$$f[\mathbf{x}(t), \mathbf{u}(t), t] = \text{col}\left[x_2, x_1x_4^2 - \frac{x_6}{x_1^2} + \frac{1}{x_5}u_1, x_4, -\frac{2x_2x_4}{x_1} + \frac{1}{x_5}u_2, 0, 0\right] \quad \square \tag{24-39}$$

We note, finally, that the modeling and augmentation approach to parameter estimation, just described, is not restricted to continuous-time linear systems. Additional situations are described in the exercises.

## COMPUTATION

There is no M-file available to do extended Kalman filtering. Using the flow chart that is given in Figure 24-2, it is straightforward to create such an M-file. It is best, in practice, to use an iterated EKF, with at least one iteration. Remember, though, that both the EKF and IEKF use a linearized and discretized system, which is time varying. See Lesson 17 for a Kalman filter for such a system.

## Supplementary Material

## EKF FOR A NONLINEAR DISCRETE-TIME SYSTEM

— so just linearize

Sometimes we begin with a description of a system directly in the discrete-time domain. An EKF can also be developed for such a system. Consider the nonlinear discrete-time system

$$\mathbf{x}(k+1) = \mathbf{f}[\mathbf{x}(k), k] + \mathbf{w}(k) \tag{24-40}$$

$$\mathbf{z}(k) = \mathbf{h}[\mathbf{x}(k), k] + \mathbf{v}(k), \quad k = 1, 2, \ldots \tag{24-41}$$

where $\mathbf{w}(k)$ and $\mathbf{v}(k)$ are Gaussian random sequences. Because this is already a discrete-time system, we only have to linearize it prior to obtaining the EKF formulas. We linearize the state equation about $\hat{\mathbf{x}}(k|k)$ because we will go from the filtered estimate, $\hat{\mathbf{x}}(k|k)$, to the predicted estimate $\hat{\mathbf{x}}(k+1|k)$. We linearize the measurement equation about $\hat{\mathbf{x}}(k+1|k)$ because we will go from the just-computed predicted value of $\mathbf{x}(k+1)$ to its filtered value, $\hat{\mathbf{x}}(k+1|k+1)$.

The linearized state and measurement equations are

$$\mathbf{x}(k+1) = \mathbf{f}[\hat{\mathbf{x}}(k|k), k] + \mathbf{F}_{\mathbf{x}}[\hat{\mathbf{x}}(k|k), k][\mathbf{x}(k) - \hat{\mathbf{x}}(k|k)] + \mathbf{w}(k) \tag{24-42}$$

$$\begin{aligned}\mathbf{z}(k+1) = \mathbf{h}[\hat{\mathbf{x}}(k+1|k), k+1] + \mathbf{H}_{\mathbf{x}}[\hat{\mathbf{x}}(k+1|k), k+1][\mathbf{x}(k+1)\\ - \hat{\mathbf{x}}(k+1|k)] + \mathbf{v}(k+1)\end{aligned} \tag{24-43}$$

The *predictor equation* is obtained by applying $\mathbf{E}\{\cdot|\mathbf{Z}(k)\}$, to the linearized state equation, i.e.,

$$\hat{\mathbf{x}}(k+1|k) = \mathbf{f}[\hat{\mathbf{x}}(k|k), k] \tag{24-44}$$

The *correction equation* is obtained by first reexpressing the linearized measurement equation as

$$\mathbf{z}_1(k+1) = \mathbf{H}_{\mathbf{x}}[\hat{\mathbf{x}}(k+1|k), k+1]\mathbf{x}(k+1) + \mathbf{v}(k+1) \tag{24-45}$$

where

$$\mathbf{z}_1(k+1) \triangleq \mathbf{z}(k+1) - \mathbf{h}[\hat{\mathbf{x}}(k+1|k), k+1] + \mathbf{H}_{\mathbf{x}}[\hat{\mathbf{x}}(k+1|k), k+1]\hat{\mathbf{x}}(k+1|k) \tag{24-46}$$

and then applying the regular KF correction equation to the linearized system, i.e.,

$$\begin{aligned}\hat{\mathbf{x}}(k+1|k+1) &= \hat{\mathbf{x}}(k+1|k) + \mathbf{K}(k+1;{}^*)\{\mathbf{z}_1(k+1)\\ &\quad - \mathbf{H}_{\mathbf{x}}[\hat{\mathbf{x}}(k+1|k), k+1]\hat{\mathbf{x}}(k+1|k)\}\\ &= \hat{\mathbf{x}}(k+1|k) + \mathbf{K}(k+1;{}^*)\{\mathbf{z}(k+1)\\ &\quad - \mathbf{h}[\hat{\mathbf{x}}(k+1|k), k+1]\}\end{aligned} \tag{24-47}$$

Equations (24-44) and (24-47) constitute the EKF equations for the original nonlinear discrete-time system. Note that $\mathbf{F}_{\mathbf{x}}[\hat{\mathbf{x}}(k|k), k]$ plays the role of $\mathbf{\Phi}$ and $\mathbf{H}_{\mathbf{x}}[\hat{\mathbf{x}}(k+1|k), k+1]$ plays the role of $\mathbf{H}$ in the matrix equations for $\mathbf{K}(k+1;^*)$, $\mathbf{P}(k+1|k;^*)$, and $\mathbf{P}(k+1|k+1;^*)$. Matrices $\mathbf{Q}$ and $\mathbf{R}$ do not change for this system because it is already a discrete-time system.

## SUMMARY QUESTIONS

**1.** The Kalman filter associated with precomputed $\mathbf{x}^*(t)$ is known as a:
  **(a)** nominal KF
  **(b)** extended KF
  **(c)** relinearized KF

**2.** The extended KF relinearizes the nonlinear system:
  **(a)** about each new estimate as it becomes available
  **(b)** only about filtered estimates
  **(c)** only about predicted estimates

**3.** The prediction equation of the EKF is obtained by:
  **(a)** applying the KF prediction equation to the perturbation state-variable model
  **(b)** integrating the nominal differential equation for $\mathbf{x}^*(t)$, from $t_k$ to $t_{k+1}$
  **(c)** setting $\delta\hat{\mathbf{x}}(t|t_k) = \mathbf{0}$ for all $t \in [t_k, t_{k+1}]$

**4.** The corrector equation of the EKF is obtained by:
  **(a)** integrating the measurement equation from $t_k$ to $t_{k+1}$
  **(b)** setting $\delta\hat{\mathbf{x}}(t|t_k) = \mathbf{0}$ for all $t \in [t_k, t_{k+1}]$
  **(c)** applying the KF corrector equation to the perturbation state-variable model

**5.** When some parameters that appear in the plant matrix of a linear system are unknown, and they are modeled by linear state equations:
  **(a)** the EKF converges to true values of the unknown parameters
  **(b)** the resulting augmented system is nonlinear
  **(c)** a KF can be used to estimate the parameters

**6.** In ILS, the nonlinear model is used to compute:
  **(a)** $\theta^*$
  **(b)** $\mathbf{z}^*(k)$
  **(c)** $\theta^*$ and $\mathbf{z}^*(k)$

**7.** In the iterated EKF, the:
  **(a)** corrector equation of the EKF is iterated
  **(b)** prediction equation of the EKF is iterated
  **(c)** prediction and correction equations of the EKF are iterated

## PROBLEMS

**24-1.** In the first-order system, $x(k+1) = ax(k) + w(k)$, and $z(k+1) = x(k+1) + \nu(k+1), k = 1, 2, \ldots, N$, $a$ is an unknown parameter that is to be estimated. Sequences $w(k)$ and $\nu(k)$ are, as usual, mutually uncorrelated and white, and $w(k) \sim N(w(k); 0, 1)$ and $\nu(k) \sim N(\nu(k); 0, \frac{1}{2})$. Explain, using equations and a flow chart, how parameter $a$ can be estimated using an EKF. (Hint: Use the EKF given in the Supplementary Material at the end of this lesson.)

**24-2.** Repeat the preceding problem where all conditions are the same except that now $w(k)$ and $v(k)$ are correlated, and $\mathbf{E}\{w(k)v(k)\} = \frac{1}{4}$.

**24-3.** The system of differential equations describing the motion of an aerospace vehicle about its pitch axis can be written as (Kopp and Orford, 1963)

$$\dot{x}_1(t) = x_2(t)$$

$$\dot{x}_2(t) = a_1(t)x_2(t) + a_2(t)x_1(t) + a_3(t)u(t)$$

where $x_1 = \dot{\theta}(t)$, which is the actual pitch rate. Sampled measurements are made of the pitch rate, i.e.,

$$z(t) = x_1(t) + \nu(t), \qquad t = t_i, \quad i = 1, 2, \ldots, N$$

Noise $\nu(t_i)$ is white and Gaussian, and $\sigma_\nu^2(t_i)$ is given. The control signal $u(t)$ is the sum of a desired control signal $u^*(t)$ and additive noise, i.e.,

$$u(t) = u^*(t) + \delta u(t)$$

The additive noise $\delta u(t)$ is a normally distributed random variable modulated by a function of the desired control signal, i.e.,

$$\delta u(t) = S[u^*(t)]w_0(t)$$

where $w_0(t)$ is zero-mean white noise with intensity $\sigma_{w0}^2$. Parameters $a_1(t)$, $a_2(t)$ and $a_3(t)$ may be unknown and are modeled as

$$\dot{a}_i(t) = \alpha_i(t)[a_i(t) - \bar{a}_i(t)] + w_i(t), \quad i = 1, 2, 3$$

In this model the parameters $\alpha_i(t)$ are assumed given, as are the a priori values of $a_i(t)$ and $\bar{a}_i(t)$, and $w_i(t)$ are zero-mean white noises with intensities $\sigma_{w_i}^2$.

**(a)** What are the EKF formulas for estimation of $x_1$, $x_2$, $a_1$, and $a_2$, assuming that $a_3$ is known?

**(b)** Repeat part (a) but now assume that $a_3$ is unknown.

**24-4.** Refer to Problem 23-7. Obtain the EKF for:

**(a)** Equation for the unsteady operation of a synchronous motor, in which $C$ and $p$ are unknown,

**(b)** Duffing's equation, in which $C$, $\alpha$, and $\beta$ are unknown,

**(c)** Van der Pol's equation, in which $\epsilon$ is unknown,

**(d)** Hill's equation, in which $a$ and $b$ are unknown.

**24-5.** (Richard W. Baylor, Spring 1992) Obtain the EKF for a series $RLC$ network in which noisy measurements are made of current, $i(t)$, and we wish to estimate $R$, $L$, and $C$ as well as the network's usual states. The equation for this system is

$$L\frac{d^2i(t)}{dt^2} + R\frac{di(t)}{dt} + \frac{1}{C}i(t) = \frac{dv(t)}{dt}$$

where $v(t)$ is voltage. Be sure to include the discretization of this continuous-time system.

**24-6.** (Patrick Lippert, Spring 1992) Given the RTT robotic arm with $r_3 = L_2$ (see also Problem 4-6, by Lippert), where $L_2$ is a constant length, we can use the theory of differential kinematics to relate the angular velocities of the robot arm to the Cartesian velocities of the tip of the robot arm. The equations of motion are

$$\frac{d\theta_1(t)}{dt} = \left(\frac{-1}{r_2 \cos 2\theta_1}\right)\left\{\frac{dx(t)}{dt} \times \cos\theta_1 + \frac{dy(t)}{dt}[(L_1 + L_2)\cos\theta_1 + r_2 \sin\theta_1]\right\}$$

$$\frac{dr_2(t)}{dt} = \left(\frac{-1}{r_2 \cos 2\theta_1}\right)\left\{\frac{dx(t)}{dt} \times \sin\theta_1 + \frac{dy(t)}{dt}[(L_1 + L_2)\sin\theta_1 + r_2 \cos\theta_1]\right\}$$

Assume that we have sampled measurements of the actuator signals $\theta_1(t)$ and $r_2(t)$. Define the state-variable model and the necessary quantities needed for computation by an EKF.

# LESSON 25

# Maximum-likelihood State and Parameter Estimation

## SUMMARY


The purpose of this lesson is to study the problem of obtaining maximum-likelihood estimates of a collection of parameters that appear in our basic state-variable model. A similar problem was studied in Lesson 11; there, state vector $\mathbf{x}(t)$ was deterministic, and the only source of uncertainty was the measurement noise. Here, state vector $\mathbf{x}(t)$ is random and measurement noise is present as well.

First, we develop a formula for the log-likelihood function for the basic state-variable model. It is in terms of the innovations process; hence, the Kalman filter acts as a constraint that is associated with the computation of the log-likelihood function for the basic state-variable model. We observe that, although we began with a parameter estimation problem, we wind up with a simultaneous state and parameter estimation problem. This is due to the uncertainties present in our state equation, which necessitate state estimation using a Kalman filter.

The only way to maximize the log-likelihood function is by means of mathematical programming. A brief description of the Levenberg–Marquardt algorithm for doing this is given. This algorithm requires the computation of the gradient of the log-likelihood function. The gradient vector is obtained from a bank of Kalman filter sensitivity systems, one such system for each unknown parameter. A steady-state approximation, which uses a clever transformation of variables, can be used to greatly reduce the computational effort that is associated with maximizing the log-likelihood function.

When you complete this lesson, you will be able to use the method of maximum likelihood to estimate parameters in our basic state-variable model.

## INTRODUCTION

In Lesson 11 we studied the problem of obtaining maximum-likelihood estimates of a collection of parameters, $\boldsymbol{\theta} = \text{col}$ (elements of $\boldsymbol{\Phi}$, $\boldsymbol{\Psi}$, $\mathbf{H}$, and $\mathbf{R}$), that appear in the state-variable model

$$\mathbf{x}(k+1) = \boldsymbol{\Phi}\mathbf{x}(k) + \boldsymbol{\Psi}\mathbf{u}(k) \tag{25-1}$$

$$\mathbf{z}(k+1) = \mathbf{H}\mathbf{x}(k+1) + \mathbf{v}(k+1), \qquad k = 0, 1, \ldots, N-1 \tag{25-2}$$

We determined the log-likelihood function to be

$$L(\boldsymbol{\theta}|\mathbf{Z}) = -\frac{1}{2}\sum_{i=1}^{N}[\mathbf{z}(i) - \mathbf{H}_{\theta}\mathbf{x}_{\theta}(i)]'\mathbf{R}_{\theta}^{-1}[\mathbf{z}(i) - \mathbf{H}_{\theta}\mathbf{x}_{\theta}(i)] - \frac{N}{2}\ln|\mathbf{R}_{\theta}| \tag{25-3}$$

where quantities that are subscripted $\boldsymbol{\theta}$ denote a dependence on $\boldsymbol{\theta}$. Finally, we pointed out that the state equation (25-1), written as

$$\mathbf{x}_{\theta}(k+1) = \boldsymbol{\Phi}_{\theta}\mathbf{x}_{\theta}(k) + \boldsymbol{\Psi}_{\theta}\mathbf{u}(k), \qquad \mathbf{x}_{\theta}(0) \text{ known} \tag{25-4}$$

acts as a constraint that is associated with the computation of the log-likelihood function. Parameter vector $\boldsymbol{\theta}$ must be determined by maximizing $L(\boldsymbol{\theta}|\mathbf{Z})$ subject to the constraint (25-4). This can only be done using mathematical programming techniques (i.e., an optimization algorithm such as steepest descent or Levenberg–Marquardt).

In this lesson we study the problem of obtaining maximum-likelihood estimates of a collection of parameters, also denoted $\boldsymbol{\theta}$, that appear in our basic state-variable model,

$$\mathbf{x}(k+1) = \boldsymbol{\Phi}\mathbf{x}(k) + \boldsymbol{\Gamma}\mathbf{w}(k) + \boldsymbol{\Psi}\mathbf{u}(k) \tag{25-5}$$

$$\mathbf{z}(k+1) = \mathbf{H}\mathbf{x}(k+1) + \mathbf{v}(k+1), \qquad k = 0, 1, \ldots, N-1 \tag{25-6}$$

Now, however,

$$\boldsymbol{\theta} = \text{col}\,(\text{elements of } \boldsymbol{\Phi}, \boldsymbol{\Gamma}, \boldsymbol{\Psi}, \mathbf{H}, \mathbf{Q}, \text{ and } \mathbf{R}) \tag{25-7}$$

and we assume that $\boldsymbol{\theta}$ is $d \times 1$. As in Lesson 11, *we shall assume that* $\boldsymbol{\theta}$ *is identifiable.*

Before we can determine $\hat{\boldsymbol{\theta}}_{ML}$, we must establish the log-likelihood function for our basic state-variable model.

## A LOG-LIKELIHOOD FUNCTION FOR THE BASIC STATE-VARIABLE MODEL

As always, we must compute $p(\mathbf{Z}|\boldsymbol{\theta}) = p(\mathbf{z}(1), \mathbf{z}(2), \ldots, \mathbf{z}(N)|\boldsymbol{\theta})$. This is difficult to do for the basic state-variable model, because

$$p(\mathbf{z}(1), \mathbf{z}(2), \ldots, \mathbf{z}(N)|\boldsymbol{\theta}) \neq p(\mathbf{z}(1)|\boldsymbol{\theta})p(\mathbf{z}(2)|\boldsymbol{\theta}) \cdots p(\mathbf{z}(N)|\boldsymbol{\theta}) \tag{25-8}$$

The measurements are all correlated due to the presence of either the process noise, $\mathbf{w}(k)$, or random initial conditions, or both. This represents the major difference between our basic state-variable model, (25-5) and (25-6), and the state-variable model studied earlier, in (25-1) and (25-2). Fortunately, the measurements and innovations are causally invertible, and the innovations are all uncorrelated, so it is still relatively easy to determine the log-likelihood function for the basic state-variable model.

**Theorem 25-1.** *The log-likelihood function for our basic state-variable model in (25-5) and (25-6) is*

$$L(\boldsymbol{\theta}|\mathbf{Z}) = -\frac{1}{2}\sum_{j=1}^{N}[\tilde{\mathbf{z}}'_{\theta}(j|j-1)\mathbf{P}^{-1}_{\tilde{\mathbf{z}}\tilde{\mathbf{z}}_{\theta}}(j|j-1)\tilde{\mathbf{z}}_{\theta}(j|j-1) + \ln|\mathbf{P}_{\tilde{\mathbf{z}}\tilde{\mathbf{z}}_{\theta}}(j|j-1)|] \tag{25-9}$$

*where* $\tilde{\mathbf{z}}_{\theta}(j|j-1)$ *is the innovations process, and* $\mathbf{P}_{\tilde{\mathbf{z}}\tilde{\mathbf{z}}_{\theta}}(j|j-1)$ *is the covariance of that process;*

$$\mathbf{P}_{\tilde{\mathbf{z}}\tilde{\mathbf{z}}_{\theta}}(j|j-1) = \mathbf{H}_{\theta}\mathbf{P}_{\theta}(j|j-1)\mathbf{H}'_{\theta} + \mathbf{R}_{\theta} \tag{25-10}$$

This theorem is also applicable to either time-varying or nonstationary systems or both. Within the structure of these more complicated systems there must still be a collection of unknown but constant parameters. It is these parameters that are estimated by maximizing $L(\boldsymbol{\theta}|\mathbf{Z})$.

*Proof* (Mendel, 1983, pp. 101-103). We must first obtain the joint density function $p(\mathbf{Z}|\boldsymbol{\theta}) = p(\mathbf{z}(1), \ldots, \mathbf{z}(N)|\boldsymbol{\theta})$. In Lesson 17 we saw that the innovations process $\tilde{\mathbf{z}}(i|i-1)$ and measurement $\mathbf{z}(i)$ are causally invertible; thus, the density function

$$\tilde{p}(\tilde{\mathbf{z}}(1|0), \tilde{\mathbf{z}}(2|1), \ldots, \tilde{\mathbf{z}}(N|N-1)|\boldsymbol{\theta})$$

contains the same data information as $p(\mathbf{z}(1), \ldots, \mathbf{z}(N)|\boldsymbol{\theta})$ does. Consequently, $L(\boldsymbol{\theta}|\mathbf{Z})$ can be replaced by $\tilde{L}(\boldsymbol{\theta}|\tilde{\mathbf{Z}})$, where

$$\tilde{\mathbf{Z}} = \text{col}\,(\tilde{\mathbf{z}}(1|0), \ldots, \tilde{\mathbf{z}}(N|N-1)) \tag{25-11}$$

and

$$\tilde{L}(\boldsymbol{\theta}|\tilde{\mathbf{Z}}) = \ln \tilde{p}(\tilde{\mathbf{z}}(1|0), \ldots, \tilde{\mathbf{z}}(N|N-1)|\boldsymbol{\theta}) \tag{25-12}$$

Now, however, we use the fact the innovations process is white noise to express $\tilde{L}(\boldsymbol{\theta}|\tilde{\mathbf{Z}})$ as

$$\tilde{L}(\boldsymbol{\theta}|\tilde{\mathbf{Z}}) = \ln \prod_{j=1}^{N} \tilde{p}_j(\tilde{\mathbf{z}}(j|j-1)|\boldsymbol{\theta}) \tag{25-13}$$

For our basic state-variable model, the innovations are identically distributed (and Gaussian), which means that $\tilde{p}_j(\tilde{\mathbf{z}}(j|j-1)|\boldsymbol{\theta}) = \tilde{p}(\tilde{\mathbf{z}}(j|j-1)|\boldsymbol{\theta})$ for $j = 1, \ldots, N$; hence,

$$\tilde{L}(\boldsymbol{\theta}|\tilde{\mathbf{Z}}) = \ln \prod_{j=1}^{N} \tilde{p}(\tilde{\mathbf{z}}(j|j-1)|\boldsymbol{\theta}) \tag{25-14}$$

From part (b) of Theorem 16-2, we know that

$$\tilde{p}(\tilde{\mathbf{z}}(j|j-1)|\boldsymbol{\theta}) = [(2\pi)^m |\mathbf{P}_{\tilde{\mathbf{z}}\tilde{\mathbf{z}}}(j|j-1)|]^{-1/2} \exp\left[-\frac{1}{2}\tilde{\mathbf{z}}'(j|j-1)\mathbf{P}_{\tilde{\mathbf{z}}\tilde{\mathbf{z}}}^{-1}(j|j-1)\tilde{\mathbf{z}}(j|j-1)\right] \quad (25\text{-}15)$$

Substitute (25-15) into (25-14) to show that

$$\tilde{L}(\boldsymbol{\theta}|\tilde{\mathbf{Z}}) = -\frac{1}{2}\sum_{j=1}^{N}[\tilde{\mathbf{z}}'(j|j-1)\mathbf{P}_{\tilde{\mathbf{z}}\tilde{\mathbf{z}}}^{-1}(j|j-1)\tilde{\mathbf{z}}(j|j-1) + \ln|\mathbf{P}_{\tilde{\mathbf{z}}\tilde{\mathbf{z}}}(j|j-1)|] \quad (25\text{-}16)$$

where by convention we have neglected the constant term $-\ln(2\pi)^{m/2}$ because it does not depend on $\boldsymbol{\theta}$.

Because $\tilde{p}(\#|\boldsymbol{\theta})$ and $p(\#|\boldsymbol{\theta})$ contain the same information about the data, $\tilde{L}(\boldsymbol{\theta}|\tilde{\mathbf{Z}})$ and $L(\boldsymbol{\theta}|\mathbf{Z})$ must also contain the same information about the data; hence, we can use $L(\boldsymbol{\theta}|\mathbf{Z})$ to denote the right-hand side of (25-16), as in (25-9). To indicate which quantities on the right-hand side of (25-9) may depend on $\boldsymbol{\theta}$, we have subscripted all such quantities with $\boldsymbol{\theta}$. □

- The innovations process $\tilde{\mathbf{z}}_{\boldsymbol{\theta}}(j|j-1)$ can be generated by a Kalman filter; hence, *the Kalman filter acts as a constraint that is associated with the computation of the log-likelihood function for the basic state-variable model.*

In the present situation, where the true values of $\boldsymbol{\theta}$, $\boldsymbol{\theta}_T$, are not known but are being estimated, the estimate of $\mathbf{x}(j)$ obtained from a Kalman filter will be suboptimal due to wrong values of $\boldsymbol{\theta}$ being used by that filter. In fact, we must use $\hat{\boldsymbol{\theta}}_{ML}$ in the implementation of the Kalman filter, because $\hat{\boldsymbol{\theta}}_{ML}$ will be the best information available about $\boldsymbol{\theta}_T$ at $t_j$. If $\hat{\boldsymbol{\theta}}_{ML} \to \boldsymbol{\theta}_T$ as $N \to \infty$, then $\tilde{\mathbf{z}}_{\hat{\boldsymbol{\theta}}_{ML}}(j|j-1) \to \tilde{\mathbf{z}}_{\boldsymbol{\theta}_T}(j|j-1)$ as $N \to \infty$, and the suboptimal Kalman filter will approach the optimal Kalman filter. This result is about the best that we can hope for in maximum-likelihood estimation of parameters in our basic state-variable model.

Note, also, that although we began with a parameter estimation problem we wound up with a simultaneous state and parameter estimation problem. This is due to the uncertainties present in our state equation, which necessitated state estimation using a Kalman filter.

## ON COMPUTING $\hat{\boldsymbol{\theta}}_{ML}$

How do we determine $\hat{\boldsymbol{\theta}}_{ML}$ for $L(\boldsymbol{\theta}|\mathbf{Z})$ given in (25-9) (subject to the constraint of the Kalman filter)? No simple closed-form solution is possible, because $\boldsymbol{\theta}$ enters into $L(\boldsymbol{\theta}|\mathbf{Z})$ in a complicated *nonlinear* manner. The only way presently known to obtain $\hat{\boldsymbol{\theta}}_{ML}$ is by means of mathematical programming.

The most effective optimization methods to determine $\hat{\boldsymbol{\theta}}_{ML}$ require the computation of the gradient of $L(\boldsymbol{\theta}|\mathbf{Z})$ as well as the Hessian matrix, or a pseudo-

Hessian matrix of $L(\theta|\mathbf{Z})$. The Levenberg–Marquardt algorithm (Bard, 1970; Marquardt, 1963), for example, has the form

$$\hat{\theta}_{ML}^{i+1} = \hat{\theta}_{ML}^{i} - (\mathbf{H}_i + \mathbf{D}_i)^{-1}\mathbf{g}_i, \qquad i = 0, 1, \ldots \tag{25-17}$$

where $\mathbf{g}_i$ denotes the gradient

$$\mathbf{g}_i = \left.\frac{\partial L(\theta|\mathbf{Z})}{\partial \theta}\right|_{\theta=\hat{\theta}_{ML}^{i}} \tag{25-18}$$

$\mathbf{H}_i$ denotes a pseudo-Hessian

$$\mathbf{H}_i = \left.\frac{\partial^2 L(\theta|\mathbf{Z})}{\partial \theta^2}\right|_{\theta=\hat{\theta}_{ML}^{i}} \tag{25-19}$$

and $\mathbf{D}_i$ is a diagonal matrix chosen to force $\mathbf{H}_i + \mathbf{D}_i$ to be positive definite so that $(\mathbf{H}_i + \mathbf{D}_i)^{-1}$ will always be computable.

Figure 25-1 depicts a flow chart for the Levenberg–Marquardt procedure. Initial values for $\hat{\theta}_{ML}^{0}$ must be specified as well as the order of the model, $n$ [for some discussions on how to do this, see Mendel (1990, p. 52), Mendel (1983, pp. 148–149) and Gupta and Mehra (1974)]. Matrix $\mathbf{D}_0$ is initialized to a diagonal matrix. The Levenberg–Marquardt algorithm in (25-17) computes $\hat{\theta}_{ML}^{1}$, after which $L(\hat{\theta}_{ML}^{1}|\mathbf{Z})$ is compared with $L(\hat{\theta}_{ML}^{0}|\mathbf{Z})$. If $L(\hat{\theta}_{ML}^{1}|\mathbf{Z}) < L(\hat{\theta}_{ML}^{0}|\mathbf{Z})$ we increase $\mathbf{D}_0$ and try again. If $L(\hat{\theta}_{ML}^{1}|\mathbf{Z}) \geq L(\hat{\theta}_{ML}^{0}|\mathbf{Z})$, then we accept this iteration and proceed to compute $\hat{\theta}_{ML}^{2}$ using (25-17). By accepting $\hat{\theta}_{ML}^{1}$ only if $L(\hat{\theta}_{ML}^{1}|\mathbf{Z}) \geq L(\hat{\theta}_{ML}^{0}|\mathbf{Z})$, we guarantee that each iteration will improve the likelihood of our estimates. This iterative procedure is terminated when the change in $L(\hat{\theta}_{ML}^{i}|\mathbf{Z})$ falls below some prespecified threshold.

Calculations of $\mathbf{g}_i$ and $\mathbf{H}_i$, which we describe next, are what is computationally intensive about this algorithm. The gradient of $L(\theta|\mathbf{Z})$ is (Problem 25-1)

$$\begin{aligned}\frac{\partial L(\theta|\mathbf{Z})}{\partial \theta} = -\sum\{&\tilde{\mathbf{z}}'(j|j-1)\mathbf{P}_{\tilde{\mathbf{z}}\tilde{\mathbf{z}}_\theta}^{-1}(j|j-1)\frac{\partial \tilde{\mathbf{z}}(j|j-1)}{\partial \theta} \\ &- \frac{1}{2}\tilde{\mathbf{z}}'(j|j-1)\mathbf{P}_{\tilde{\mathbf{z}}\tilde{\mathbf{z}}_\theta}^{-1}(j|j-1)\frac{\partial \mathbf{P}_{\tilde{\mathbf{z}}\tilde{\mathbf{z}}_\theta}(j|j-1)}{\partial \theta} \\ &\mathbf{P}_{\tilde{\mathbf{z}}\tilde{\mathbf{z}}_\theta}^{-1}(j|j-1)\tilde{\mathbf{z}}(j|j-1) \\ &+ \frac{1}{2}\operatorname{tr}(\mathbf{P}_{\tilde{\mathbf{z}}\tilde{\mathbf{z}}_\theta}^{-1}(j|j-1)\frac{\partial \mathbf{P}_{\tilde{\mathbf{z}}\tilde{\mathbf{z}}_\theta}(j|j-1)}{\partial \theta}\}\end{aligned} \tag{25-20}$$

Observe that $\partial L(\theta|\mathbf{Z})/\partial\theta$ requires the calculations of

$$\frac{\partial \tilde{\mathbf{z}}_\theta(j|j-1)}{\partial \theta} \quad \text{and} \quad \frac{\partial \mathbf{P}_{\tilde{\mathbf{z}}\tilde{\mathbf{z}}_\theta}(j|j-1)}{\partial \theta}$$

The innovations depend on $\hat{\mathbf{x}}_\theta(j|j-1)$; hence, to compute $\partial\tilde{\mathbf{z}}_\theta(j|j-1)/\partial\theta$, we must compute $\partial\hat{\mathbf{x}}_\theta(j|j-1)/\partial\theta$. A Kalman filter must be used to compute $\hat{\mathbf{x}}_\theta(j|j-1)$; but this filter requires the following sequence of calculations: $\mathbf{P}_\theta(k|k) \to \mathbf{P}_\theta(k+1|k) \to \mathbf{K}_\theta(k+1) \to \hat{\mathbf{x}}_\theta(k+1|k) \to \hat{\mathbf{x}}_\theta(k+1|k+1)$. Taking the partial derivative of the prediction equation with respect to $\theta_i$, we find that

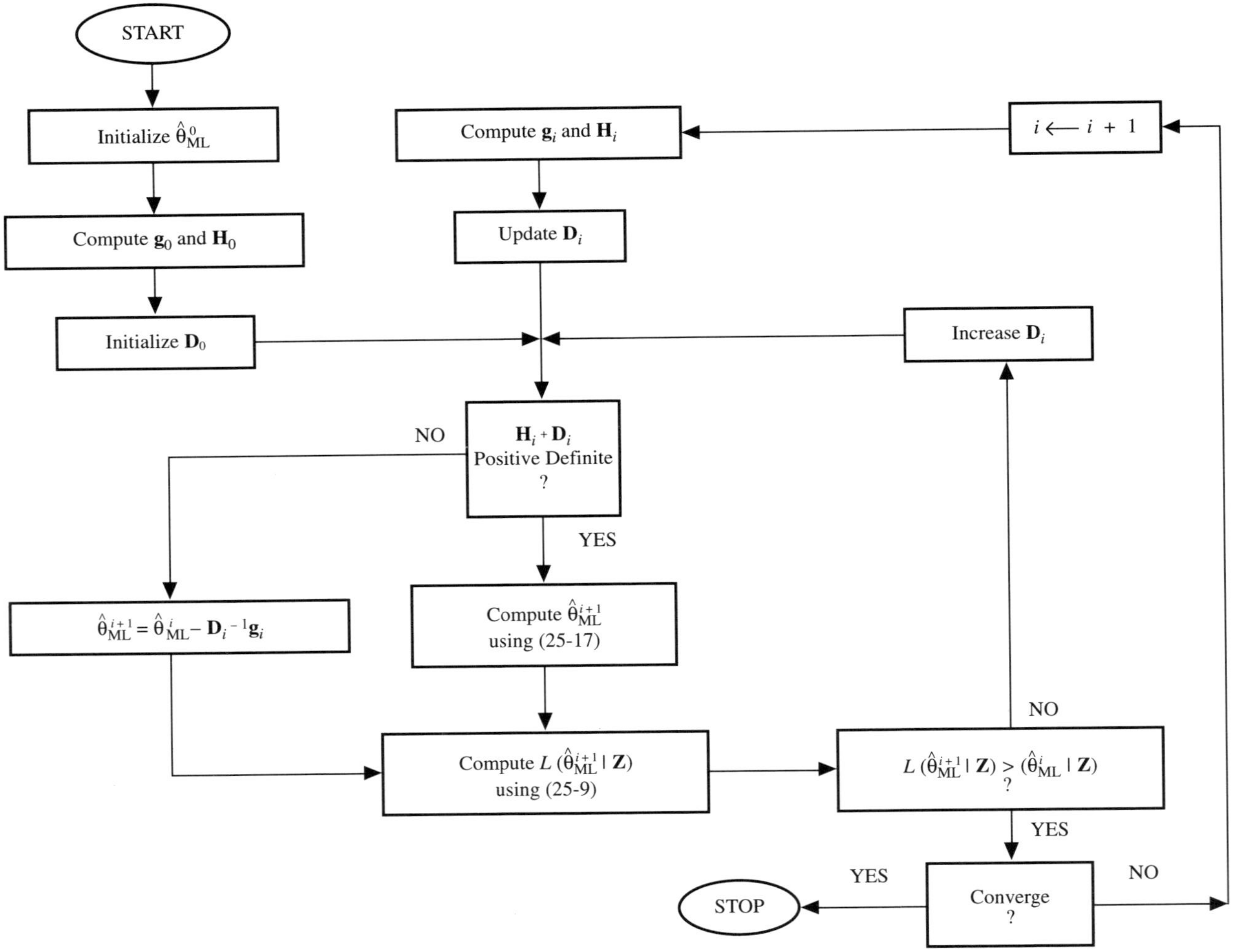


**Figure 25-1** Optimization strategy using the Levenberg–Marquardt algorithm. (Adopted from Figure 7.3-1 in J. M. Mendel (1983), *Optimal Seismic Deconvolution: An Estimation-based Approach*, Academic Press, New York).

$$\frac{\partial \hat{\mathbf{x}}_\theta(k+1|k)}{\partial \theta_i} = \mathbf{\Phi}_\theta \frac{\partial \hat{\mathbf{x}}_\theta(k|k)}{\partial \theta_i} + \frac{\partial \mathbf{\Phi}_\theta}{\partial \theta_i}\hat{\mathbf{x}}_\theta(k|k) + \frac{\partial \mathbf{\Psi}_\theta}{\partial \theta_i}\mathbf{u}(k), \qquad i = 1, 2, \ldots, d \tag{25-21}$$

We see that to compute $\partial \hat{\mathbf{x}}_\theta(k+1|k)/\partial \theta_i$, we must also compute $\partial \hat{\mathbf{x}}_\theta(k|k)/\partial \theta_i$. Taking the partial derivative of the correction equation with respect to $\theta_i$, we find that

$$\frac{\partial \hat{\mathbf{x}}_\theta(k+1|k+1)}{\partial \theta_i} = \frac{\partial \hat{\mathbf{x}}_\theta(k+1|k)}{\partial \theta_i} + \frac{\partial \mathbf{K}_\theta(k+1)}{\partial \theta_i}[\mathbf{z}(k+1) - \mathbf{H}_\theta \hat{\mathbf{x}}_\theta(k+1|k)] - \mathbf{K}_\theta(k+1)\left[\frac{\partial \mathbf{H}_\theta}{\partial \theta_i}\hat{\mathbf{x}}_\theta(k+1|k) + \mathbf{H}_\theta \frac{\partial \hat{\mathbf{x}}_\theta(k+1|k)}{\partial \theta_i}\right], \qquad i = 1, 2, \ldots, d \tag{25-22}$$

Observe that to compute $\partial \hat{\mathbf{x}}_\theta(k+1|k+1)/\partial \theta_i$, we must also compute $\partial \mathbf{K}_\theta(k+1)/\partial \theta_i$. We leave it to the reader (Problem 25-2) to show that the calculation of $\partial \mathbf{K}_\theta(k+1)/\partial \theta_i$ requires the calculation of $\partial \mathbf{P}_\theta(k+1|k)/\partial \theta_i$, which in turn requires the calculation of $\partial \mathbf{P}_\theta(k+1|k+1)/\partial \theta_i$.

The system of equations

$$\frac{\partial \hat{\mathbf{x}}_\theta(k+1|k)}{\partial \theta_i}, \frac{\partial \hat{\mathbf{x}}_\theta(k+1|k+1)}{\partial \theta_i},$$
$$\frac{\partial \mathbf{K}_\theta(k+1)}{\partial \theta_i}, \frac{\partial \mathbf{P}_\theta(k+1|k)}{\partial \theta_i}, \frac{\partial \mathbf{P}_\theta(k+1|k+1)}{\partial \theta_i}$$

is called a Kalman filter sensitivity system. It is a linear system of equations, just like the Kalman filter, which is not only driven by measurements $\mathbf{z}(k+1)$ [e.g., see (25-22)], but is also driven by the Kalman filter [e.g., see (25-21) and (25-22)]. We need a total of $d$ such sensitivity system, one for each of the $d$ unknown parameters in $\boldsymbol{\theta}$.

Each system of sensitivity equations requires about as much computation as a Kalman filter. Observe, however, that Kalman filter quantities are used by the sensitivity equations; hence, the Kalman filter must be run together with the $d$ sets of sensitivity equations. This procedure for recursively calculating the gradient $\partial L(\boldsymbol{\theta}|\mathbf{Z})/\partial \boldsymbol{\theta}$ therefore requires about as much computation as $d+1$ Kalman filters. The sensitivity systems are totally uncoupled and lend themselves quite naturally to parallel processing (see Figure 25-2).

The Hessian matrix of $L(\boldsymbol{\theta}|\mathbf{Z})$ is quite complicated, involving not only first derivatives of $\tilde{\mathbf{z}}_\theta(j|j-1)$ and $\mathbf{P}_{\tilde{\mathbf{z}}\tilde{\mathbf{z}}_\theta}(j|j-1)$, but also their second derivatives. The pseudo-Hessian matrix of $L(\boldsymbol{\theta}|\mathbf{Z})$ ignores all the second derivative terms; hence, it is relatively easy to compute because all the first derivative terms have already been computed in order to calculate the gradient of $L(\boldsymbol{\theta}|\mathbf{Z})$. Justification for neglecting the second derivative terms is given by Gupta and Mehra (1974), who show that as $\hat{\boldsymbol{\theta}}_{ML}$ approaches $\boldsymbol{\theta}_T$ the expected value of the dropped terms goes to zero.

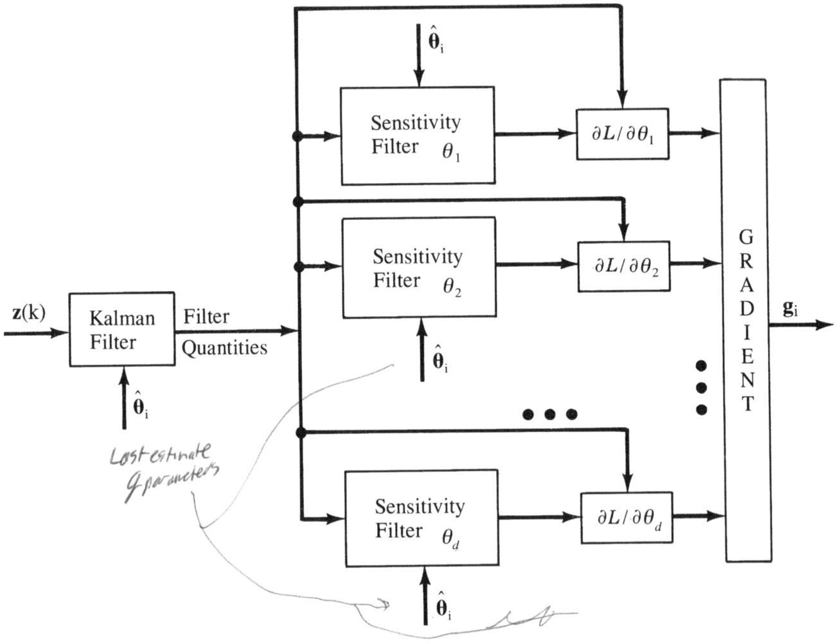


**Figure 25-2** Calculations needed to compute gradient vector $\mathbf{g}_i$. Note that $\theta_j$ denotes the $j$th component of $\boldsymbol{\theta}$ (Mendel, 1983, © 1983, Academic Press, Inc.).

The estimation literature is filled with many applications of maximum-likelihood state and parameter estimation. For example, Mendel (1983, 1990) applies it to seismic data processing, Mehra and Tyler (1973) apply it to aircraft parameter identification, and McLaughlin (1980) applies it to groundwater flow.

## A STEADY-STATE APPROXIMATION

Suppose our basic state-variable model is time invariant and stationary so that $\mathbf{P}_p = \lim_{j\to\infty} \mathbf{P}(j|j-1)$ exists. Let

$$\bar{\mathcal{N}} \triangleq \mathbf{H}\mathbf{P}_p\mathbf{H}' + \mathbf{R} \tag{25-23}$$

and

$$\bar{L}(\boldsymbol{\theta}|\mathbf{Z}) = -\frac{1}{2}\sum_{j=1}^{N} \tilde{\mathbf{z}}'(j|j-1)\bar{\mathcal{N}}^{-1}\tilde{\mathbf{z}}(j|j-1) - \frac{1}{2}N \ln|\bar{\mathcal{N}}| \tag{25-24}$$

Log-likelihood function $\bar{L}(\boldsymbol{\theta}|\mathbf{Z})$ is a steady-state approximation of $L(\boldsymbol{\theta}|\mathbf{Z})$. The steady-state Kalman filter used to compute $\tilde{\mathbf{z}}(j|j-1)$ and $\bar{\mathcal{N}}$ is

$$\hat{\mathbf{x}}(k+1|k) = \mathbf{\Phi}\hat{\mathbf{x}}(k|k) + \mathbf{\Psi}\mathbf{u}(k) \tag{25-25}$$

$$\hat{\mathbf{x}}(k+1|k+1) = \hat{\mathbf{x}}(k+1|k) + \bar{\mathbf{K}}[\mathbf{z}(k+1) - \mathbf{H}\hat{\mathbf{x}}(k+1|k)] \tag{25-26}$$

in which $\bar{\mathbf{K}}$ is the steady-state Kalman gain matrix.

Recall that

$$\theta = \text{col}\,(\text{elements of } \mathbf{\Phi}, \mathbf{\Gamma}, \mathbf{\Psi}, \mathbf{H}, \mathbf{Q}, \text{ and } \mathbf{R}) \tag{25-27}$$

We now make the following transformations of variables:

$$\left.\begin{array}{l} \mathbf{\Phi} \rightarrow \mathbf{\Phi} \\ \mathbf{\Psi} \rightarrow \mathbf{\Psi} \\ \mathbf{H} \rightarrow \mathbf{H} \\ (\mathbf{\Phi}, \mathbf{\Gamma}, \mathbf{\Psi}, \mathbf{H}, \mathbf{Q}, \mathbf{R}) \rightarrow \bar{\mathcal{N}} \\ (\mathbf{\Phi}, \mathbf{\Gamma}, \mathbf{\Psi}, \mathbf{H}, \mathbf{Q}, \mathbf{R}) \rightarrow \bar{\mathbf{K}} \end{array}\right\} \tag{25-28}$$

We ignore $\mathbf{\Gamma}$ (initially) for reasons that are explained following Equation (26-33). Let

$$\phi = \text{col}\,(\text{elements of } \mathbf{\Phi}, \mathbf{\Psi}, \mathbf{H}, \bar{\mathcal{N}}, \text{ and } \bar{\mathbf{K}}) \tag{25-29}$$

where $\phi$ is $p \times 1$ and view $\bar{L}$ as a function of $\phi$; i.e.,

$$\bar{L}(\phi|\mathbf{Z}) = -\frac{1}{2}\sum_{j=1}^{N} \tilde{\mathbf{z}}'_{\phi}(j|j-1)\bar{\mathcal{N}}_{\phi}^{-1}\tilde{\mathbf{z}}_{\phi}(j|j-1) - \frac{1}{2}N\ln|\bar{\mathcal{N}}_{\phi}| \tag{25-30}$$

Instead of finding $\hat{\theta}_{\text{ML}}$ that maximizes $\bar{L}(\theta|\mathbf{Z})$, subject to the constraints of a full-blown Kalman filter, we now propose to find $\hat{\phi}_{\text{ML}}$ that maximizes $\bar{L}(\phi|\mathbf{Z})$, subject to the constraints of the following filter:

$$\hat{\mathbf{x}}_{\phi}(k+1|k) = \mathbf{\Phi}_{\phi}\hat{\mathbf{x}}_{\phi}(k|k) + \mathbf{\Psi}_{\phi}\mathbf{u}(k) \tag{25-31}$$

$$\hat{\mathbf{x}}_{\phi}(k+1|k+1) = \hat{\mathbf{x}}_{\phi}(k+1|k) + \bar{\mathbf{K}}_{\phi}\tilde{\mathbf{z}}_{\phi}(k+1|k) \tag{25-32}$$

and

$$\tilde{\mathbf{z}}_{\phi}(k+1|k) = \mathbf{z}(k+1) - \mathbf{H}_{\phi}\hat{\mathbf{x}}_{\phi}(k+1|k) \tag{25-33}$$

Once we have computed $\hat{\phi}_{\text{ML}}$ we can compute $\hat{\theta}_{\text{ML}}$ by inverting the transformations in (25-28). Of course, when we do this, we are also using the invariance property of maximum-likelihood estimates.

Observe that $\bar{L}(\phi|\mathbf{Z})$ in (25-30) and the filter in (25-31)–(25-33) do not depend on $\mathbf{\Gamma}$; hence, we have not included $\mathbf{\Gamma}$ in any definition of $\phi$. We explain how to reconstruct $\mathbf{\Gamma}$ from $\hat{\phi}_{\text{ML}}$ following Equation (25-45).

Because maximum-likelihood estimates are asymptotically efficient (Lesson 11), once we have determined $\hat{\bar{\mathbf{K}}}_{\text{ML}}$, the filter in (25-31) and (25-32) will be the steady-state Kalman filter.

The major advantage of this steady-state approximation is that the filter sensitivity equations are greatly simplified. When $\bar{\mathbf{K}}$ and $\bar{\mathcal{N}}$ are treated as matrices of unknown parameters, we do not need the predicted and corrected error-covariance matrices to "compute" $\bar{\mathbf{K}}$ and $\bar{\mathcal{N}}$. The sensitivity equations for (25-33), (25-31), and (25-32) are

$$\frac{\partial \tilde{\mathbf{z}}_\phi(k+1|k)}{\partial \phi_i} = -\mathbf{H}_\phi \frac{\partial \hat{\mathbf{x}}_\phi(k+1|k)}{\partial \phi_i} - \frac{\partial \mathbf{H}_\phi}{\partial \phi_i}\hat{\mathbf{x}}_\phi(k+1|k) \tag{25-34}$$

$$\frac{\partial \hat{\mathbf{x}}_\phi(k+1|k)}{\partial \phi_i} = \mathbf{\Phi}_\phi \frac{\partial \hat{\mathbf{x}}_\phi(k|k)}{\partial \phi_i} + \frac{\partial \mathbf{\Phi}_\phi}{\partial \phi_i}\hat{\mathbf{x}}_\phi(k|k) + \frac{\partial \mathbf{\Psi}_\phi}{\partial \phi_i}\mathbf{u}(k) \tag{25-35}$$

and

$$\begin{aligned}\frac{\partial \hat{\mathbf{x}}_\phi(k+1|k+1)}{\partial \phi_i} = & \frac{\partial \hat{\mathbf{x}}_\phi(k+1|k)}{\partial \phi_i} + \frac{\partial \bar{\mathbf{K}}_\phi}{\partial \phi_i}[\mathbf{z}(k+1) - \mathbf{H}_\phi \hat{\mathbf{x}}_\phi(k|k+1)] \\ & - \bar{\mathbf{K}}_\phi \mathbf{H}_\phi \frac{\partial \hat{\mathbf{x}}_\phi(k+1|k)}{\partial \phi_i} \\ & - \bar{\mathbf{K}}_\phi \frac{\partial \mathbf{H}_\phi}{\partial \phi_i}\hat{\mathbf{x}}_\phi(k+1|k)\end{aligned} \tag{25-36}$$

where $i = 1, 2, \ldots, p$. Note that $\partial \bar{\mathbf{K}}_\phi / \partial \phi_i$ is zero for all $\phi_i$ not in $\bar{\mathbf{K}}_\phi$ and is a matrix filled with zeros and a single unity value for $\phi_i$ in $\bar{\mathbf{K}}_\phi$.

There are more elements in $\boldsymbol{\phi}$ than in $\boldsymbol{\theta}$, because $\bar{\mathcal{N}}$ and $\bar{\mathbf{K}}$ have more unknown elements in them than do $\mathbf{Q}$ and $\mathbf{R}$, i.e., $p > d$. Additionally, $\bar{\mathcal{N}}$ does not appear in the filter equations; it only appears in $\bar{L}(\boldsymbol{\phi}|\mathbf{Z})$. It is, therefore, possible to obtain a closed-form solution for covariance matrix $\hat{\bar{\mathcal{N}}}_{ML}$.

**Theorem 25-2.** *A closed-form solution for matrix* $\hat{\bar{\mathcal{N}}}_{ML}$ *is*

$$\hat{\bar{\mathcal{N}}}_{ML} = \frac{1}{N}\sum_{j=1}^{N} \tilde{\mathbf{z}}_\phi(j|j-1)\tilde{\mathbf{z}}'_\phi(j|j-1) \tag{25-37}$$

*Proof.* To determine $\hat{\bar{\mathcal{N}}}_{ML}$, we must set $\partial \bar{L}(\boldsymbol{\phi}|\mathbf{Z})/\partial \bar{\mathcal{N}}_\phi = \mathbf{0}$ and solve the resulting equation for $\hat{\bar{\mathcal{N}}}_{ML}$. This is most easily accomplished by applying gradient matrix formulas to (25-30), which are given in Schweppe (1974). Doing this, we obtain

$$\frac{\partial \bar{L}(\boldsymbol{\phi}|\mathbf{Z})}{\partial \bar{\mathcal{N}}} = \sum_{j=1}^{N}[\bar{\mathcal{N}}^{-1}\tilde{\mathbf{z}}_\phi(j|j-1)\tilde{\mathbf{z}}'_\phi(j|j-1)\bar{\mathcal{N}}^{-1}]' - N(\bar{\mathcal{N}}^{-1})' = \mathbf{0} \tag{25-38}$$

whose solution is $\hat{\bar{\mathcal{N}}}_{ML}$ in (25-37). □

Observe that $\hat{\bar{\mathcal{N}}}_{ML}$ is the sample steady-state covariance matrix of $\tilde{\mathbf{z}}_\phi$; i.e., as

$$\hat{\boldsymbol{\phi}}_{ML} \rightarrow \hat{\boldsymbol{\phi}}_T \quad \hat{\bar{\mathcal{N}}}_{ML} \rightarrow \lim_{j\to\infty} \text{cov}\,[\tilde{\mathbf{z}}_\phi(j|j-1)]$$

Suppose we are also interested in determining $\hat{\mathbf{Q}}_{ML}$ and $\hat{\mathbf{R}}_{ML}$. How do we obtain these quantities from $\hat{\boldsymbol{\phi}}_{ML}$?

As in Lesson 19, we let $\bar{\mathbf{K}}, \mathbf{P}_p$, and $\mathbf{P}_f$ denote the steady-state values of $\mathbf{K}(k+1), \mathbf{P}(k+1|k)$, and $\mathbf{P}(k|k)$, respectively, where

$$\bar{\mathbf{K}} = \mathbf{P}_p\mathbf{H}'(\mathbf{H}\mathbf{P}_p\mathbf{H}' + \mathbf{R})^{-1} = \mathbf{P}_p\mathbf{H}'\bar{\mathcal{N}}^{-1} \tag{25-39}$$

$$\mathbf{P}_p = \mathbf{\Phi}\mathbf{P}_f\mathbf{\Phi}' + \mathbf{\Gamma}\mathbf{Q}\mathbf{\Gamma}' \tag{25-40}$$

and

$$\mathbf{P}_f = (\mathbf{I} - \bar{\mathbf{K}}\mathbf{H})\mathbf{P}_p \tag{25-41}$$

Additionally, we know that

$$\bar{\mathcal{N}} = \mathbf{H}\mathbf{P}_p\mathbf{H}' + \mathbf{R} \tag{25-42}$$

By the invariance property of maximum-likelihood estimates, we know that

$$\hat{\bar{\mathcal{N}}}_{ML} = \widehat{(\mathbf{H}\mathbf{P}_p\mathbf{H}' + \mathbf{R})}_{ML} = \hat{\mathbf{H}}_{ML}\mathbf{P}_p\hat{\mathbf{H}}'_{ML} + \hat{\mathbf{R}}_{ML} \tag{25-43}$$

and

$$\hat{\bar{\mathbf{K}}}_{ML} = \widehat{(\mathbf{P}_p\mathbf{H}'\bar{\mathcal{N}}^{-1})}_{ML} = \mathbf{P}_p\hat{\mathbf{H}}'_{ML}(\hat{\bar{\mathcal{N}}}_{ML})^{-1} \tag{25-44}$$

Solving (25-44) for $\mathbf{P}_p\hat{\mathbf{H}}'_{ML}$ and substituting the resulting expression into (25-43), we obtain the following solution for $\hat{\mathbf{R}}_{ML}$:

$$\hat{\mathbf{R}}_{ML} = (\mathbf{I} - \hat{\mathbf{H}}_{ML}\hat{\bar{\mathbf{K}}}_{ML})\hat{\bar{\mathcal{N}}}_{ML} \tag{25-45}$$

No closed-form solution exists for $\hat{\mathbf{Q}}_{ML}$. Substituting $\hat{\mathbf{\Phi}}_{ML}, \hat{\mathbf{H}}_{ML}, \hat{\bar{\mathbf{K}}}_{ML}$, and $\hat{\bar{\mathcal{N}}}_{ML}$ into (25-39)–(25-41), and combining (25-40) and (25-41), we obtain

$$\mathbf{P}_p = \hat{\mathbf{\Phi}}_{ML}(\mathbf{I} - \hat{\bar{\mathbf{K}}}_{ML}\hat{\mathbf{H}}_{ML})\mathbf{P}_p\hat{\mathbf{\Phi}}'_{ML} + \widehat{(\mathbf{\Gamma}\mathbf{Q}\mathbf{\Gamma}')}_{ML} \tag{25-46}$$

and, from (25-44), we obtain

$$\mathbf{P}_p\hat{\mathbf{H}}'_{ML} = \hat{\bar{\mathbf{K}}}_{ML}\hat{\bar{\mathcal{N}}}_{ML} \tag{25-47}$$

These equations must be solved simultaneously for $\mathbf{P}_p$ and $\widehat{(\mathbf{\Gamma}\mathbf{Q}\mathbf{\Gamma}')}_{ML}$ using iterative numerical techniques. For details, see Mehra (1970a).

Note, finally, that the best for which we can hope by this approach is not $\hat{\mathbf{\Gamma}}_{ML}$ and $\hat{\mathbf{Q}}_{ML}$, but only $\widehat{(\mathbf{\Gamma}\mathbf{Q}\mathbf{\Gamma}')}_{ML}$. This is due to the fact that, when $\mathbf{\Gamma}$ and $\mathbf{Q}$ are both unknown, there will be an ambiguity in their determination; i.e., the term $\mathbf{\Gamma}\mathbf{w}(k)$ that appears in our basic state-variable model [for which $\mathbf{E}\{\mathbf{w}(k)\mathbf{w}'(k)\} = \mathbf{Q}$] cannot be distinguished from the term $\mathbf{w}_1(k)$, for which

$$\mathbf{E}\{\mathbf{w}_1(k)\mathbf{w}'_1(k)\} = \mathbf{Q}_1 = \mathbf{\Gamma}\mathbf{Q}\mathbf{\Gamma}' \tag{25-48}$$

This observation is also applicable to the original problem formulation wherein we obtained $\hat{\theta}_{ML}$ directly; i.e., when both $\mathbf{\Gamma}$ and $\mathbf{Q}$ are unknown, we should really choose

$$\theta = \text{col}\,(\text{elements of } \mathbf{\Phi}, \mathbf{\Psi}, \mathbf{H}, \mathbf{\Gamma}\mathbf{Q}\mathbf{\Gamma}', \text{and } \mathbf{R}) \tag{25-49}$$

In summary, when our basic state-variable model is time invariant and stationary, we can first obtain $\hat{\boldsymbol{\phi}}_{\text{ML}}$ by maximizing $\bar{L}(\boldsymbol{\phi}|\mathbf{Z})$ given in (25-30), subject to the constraints of the simple filter in (25-31), (25-32), and (25-33). A mathematical programming method must be used to obtain those elements of $\hat{\boldsymbol{\phi}}_{\text{ML}}$ associated with $\hat{\boldsymbol{\Phi}}_{\text{ML}}$, $\hat{\boldsymbol{\Psi}}_{\text{ML}}$, $\hat{\mathbf{H}}_{\text{ML}}$, and $\hat{\bar{\mathbf{K}}}_{\text{ML}}$. The closed-form solution, given in (25-37), is used for $\hat{\bar{\mathcal{N}}}_{\text{ML}}$. Finally, if we want to reconstruct $\hat{\mathbf{R}}_{\text{ML}}$ and $\widehat{(\boldsymbol{\Gamma}\mathbf{Q}\boldsymbol{\Gamma}')}_{\text{ML}}$, we use (25-45) for the former and must solve (25-46) and (25-47) for the latter.

**EXAMPLE 25-1** (Mehra, 1971)

The following fourth-order system, which represents the short period dynamics and the first bending mode of a missile, was simulated:

$$\begin{pmatrix} x_1(k+1) \\ x_2(k+1) \\ x_3(k+1) \\ x_4(k+1) \end{pmatrix} = \begin{pmatrix} 0 & 1 & 0 & 0 \\ 0 & 0 & 1 & 0 \\ 0 & 0 & 0 & 1 \\ -\alpha_1 & -\alpha_2 & -\alpha_3 & -\alpha_4 \end{pmatrix} \begin{pmatrix} x_1(k) \\ x_2(k) \\ x_3(k) \\ x_4(k) \end{pmatrix} + \begin{pmatrix} 0 \\ 1 \\ 0 \\ 1 \end{pmatrix} w(k) \tag{25-50}$$

$$z(k+1) = x_1(k+1) + \nu(k+1) \tag{25-51}$$

For this model, it was assumed that $\mathbf{x}(0) = \mathbf{0}, q = 1.0, r = 0.25, \alpha_1 = -0.656, \alpha_2 = 0.784, \alpha_3 = -0.18$, and $\alpha_4 = 1.0$.

Using measurements generated from the simulation, maximum-likelihood estimates were obtained for $\boldsymbol{\phi}$, where

$$\boldsymbol{\phi} = \text{col}\,(\alpha_1, \alpha_2, \alpha_3, \alpha_4, \bar{\mathcal{N}}, \bar{k}_1, \bar{k}_2, \bar{k}_3, \bar{k}_4) \tag{25-52}$$

In (25-52), $\bar{\mathcal{N}}$ is a scalar because, in this example, $z(k)$ is a scalar. Additionally, it was assumed that $\mathbf{x}(0)$ was known exactly. According to Mehra (1971, p. 30), "The starting values for the maximum likelihood scheme were obtained using a correlation technique given in Mehra (1970b). The results of successive iterations are shown in Table 25-1. The variances of the estimates obtained from the matrix of second partial derivatives (i.e., the Hessian matrix of $\bar{L}$) are also given. For comparison purposes, results obtained by using 1000 data points and 100 data points are given." □

## COMPUTATION

The *Optimization* and *Control System Toolboxes* are very useful to implement portions of maximum-likelihood state and parameter estimators; but neither toolbox has M-files that do everything that is needed for you to implement either a full-blown maximum-likelihood solution or the computationally simpler steady-state approximation.

To implement the full-blown maximum-likelihood solution, you will need to create M-files for Kalman filter sensitivity systems (see Figure 25-2), gradient, pseudo-Hessian, log-likelihood objective function in (25-9), and the Levenberg–Marquardt algorithm as described in the Figure 25-1 flow chart. Our Kalman filter M-file **kf** (see Appendix B) will also be needed. It is possible that the following M-file, which is in the *Optimization Toolbox* can be used to directly maximize the log-likelihood function:

**TABLE 25-1** PARAMETER ESTIMATES FOR MISSILE EXAMPLE

| Iteration | $L \times 10^{-3}$ | $\bar{N}$ | $\alpha_1$ | $\alpha_2$ | $\alpha_3$ | $\alpha_4$ | $\bar{k}_1$ | $\bar{k}_2$ | $\bar{k}_3$ | $\bar{k}_4$ |
|---|---|---|---|---|---|---|---|---|---|---|
| 0 | | | | | | | | | | |
| Results from correlation technique, Mehra (1970b). | | | | | | | | | | |
| 1 | −1.0706 | 2.3800 | −0.5965 | 0.8029 | −0.1360 | 0.8696 | 0.6830 | 0.2837 | 0.4191 | 0.8207 |
| ML estimates using 1000 points | | | | | | | | | | |
| 2 | −1.0660 | 2.3811 | −0.5938 | 0.8029 | −0.1338 | 0.8759 | 0.6803 | 0.2840 | 0.4200 | 0.8312 |
| 3 | −1.0085 | 2.4026 | −0.6054 | 0.7452 | −0.1494 | 0.9380 | 0.6304 | 0.2888 | 0.4392 | 1.0311 |
| 4 | −0.9798 | 2.4409 | −0.6036 | 0.8161 | −0.1405 | 0.8540 | 0.6801 | 0.3210 | 0.6108 | 1.1831 |
| 5 | −0.9785 | 2.4412 | −0.5999 | 0.8196 | −0.1370 | 0.8580 | 0.6803 | 0.3214 | 0.6107 | 1.1835 |
| 6 | −0.9771 | 2.4637 | −0.6014 | 0.8086 | −0.1503 | 0.8841 | 0.7068 | 0.3479 | 0.6059 | 1.2200 |
| 7 | −0.9769 | 2.4603 | −0.6023 | 0.8130 | −0.1470 | 0.8773 | 0.7045 | 0.3429 | 0.6106 | 1.2104 |
| 8 | −0.9744 | 2.5240 | −0.6313 | 0.8105 | −0.1631 | 0.9279 | 0.7990 | 0.3756 | 0.6484 | 1.2589 |
| 9 | −0.9743 | 2.5241 | −0.6306 | 0.8108 | −0.1622 | 0.9296 | 0.7989 | 0.3749 | 0.6480 | 1.2588 |
| 10 | −0.9734 | 2.5270 | −0.6374 | 0.7961 | −0.1630 | 0.9505 | 0.7974 | 0.3568 | 0.6378 | 1.2577 |
| 11 | −0.9728 | 2.5313 | −0.6482 | 0.7987 | −0.1620 | 0.9577 | 0.8103 | 0.3443 | 0.6403 | 1.2351 |
| 12 | −0.9720 | 2.5444 | −0.6602 | 0.7995 | −0.1783 | 0.9866 | 0.8487 | 0.3303 | 0.6083 | 1.2053 |
| 13 | −0.9714 | 2.5600 | −0.6634 | 0.7919 | −0.2036 | 1.0280 | 0.8924 | 0.3143 | 0.6014 | 1.2054 |
| 14 | −0.9711 | 2.5657 | −0.6624 | 0.7808 | −0.2148 | 1.0491 | 0.9073 | 0.3251 | 0.6122 | 1.2200 |
| ML estimates using 100 points | | | | | | | | | | |
| 30 | −0.9659 | 2.620 | −0.6094 | 0.7663 | −0.1987 | 1.0156 | 1.24 | 0.136 | 0.454 | 1.103 |
| Actual values | | | | | | | | | | |
| | −0.94 | 2.557 | −0.6560 | 0.7840 | −0.1800 | 1.0000 | 0.8937 | 0.2957 | 0.6239 | 1.2510 |
| Estimates of standard deviation using 1000 points | | | | | | | | | | |
| | | 0.0317 | 0.0277 | 0.0247 | 0.0275 | 0.0261 | 0.0302 | 0.0323 | 0.0302 | 0.029 |
| Estimates of standard deviation using 100 points | | | | | | | | | | |
| | | 0.149 | 0.104 | 0.131 | 0.084 | 0.184 | 0.303 | 0.092 | 0.082 | 0.09 |

Source: Mehra (1971, p. 30), © 1971, AIAA. Reprinted with permission.

**leastsq**: Solve nonlinear least squares optimization problems. Its default is the Levenberg–Marquardt algorithm; however, the implementation of this algorithm is very different from the description that we gave in Lesson 25. See the Algorithm section of the software reference manual for the *Optimization Toolbox* for its description.

To implement the steady-state approximation, you can use our M-file **sof** (see Appendix B) to implement (25-31) – (25-33), but you will need to create M-files for filter sensitivity systems in (25-34) – (25-36), gradient, pseudo-Hessian, log-likelihood objective function in (25-30), and a Levenberg–Marquardt algorithm that is similar to the one in Figure 25-1. It is also possible that the M-file **leastsq** can be used to directly maximize the log-likelihood function.

## SUMMARY QUESTIONS

**1.** Log-likelihood function $L(\boldsymbol{\theta}|\mathbf{Z})$ in (25-9) depends on $\boldsymbol{\theta}$:
**(a)** implicitly and explicitly

(b) implicitly
(c) explicitly

2. The log-likelihood function for the basic state-variable model is:
(a) unconstrained
(b) constrained by the state equation
(c) constrained by the Kalman filter

3. To determine $\hat{\theta}_{ML}$ for $L(\theta|\mathbf{Z})$ in (25-9), we must:
(a) solve a system of normal equations
(b) use mathematical programming
(c) use an EKF

4. The major advantage of the steady-state approximation is that:
(a) the log-likelihood function becomes unconstrained
(b) the filter sensitivity equations are greatly simplified
(c) all parameters can be estimated in closed form

5. For an $n$th-order system with $l$ unknown parameters, there will be ________ Kalman filter sensitivity systems:
(a) $l$
(b) $n$
(c) $(n+l)/2$

6. Matrices $\mathbf{\Gamma}$ and $\mathbf{Q}$, which appear in the basic state-variable model, can:
(a) always be uniquely identified
(b) only be identified to within a constant scale factor
(c) never be uniquely identified

7. The Levenberg–Marquardt algorithm, as presented here, uses:
(a) first- and second-derivative information
(b) only second-derivative information
(c) only first-derivative information

8. The Levenberg–Marquardt algorithm:
(a) converges to a global maximum of $L(\theta|\mathbf{Z})$
(b) will never converge
(c) converges to a local maximum of $L(\theta|\mathbf{Z})$

## PROBLEMS

**25-1.** Derive the formula for the derivative of the log-likelihood function that is given in (25-20). A useful reference is Kay (1993), pp. 73–74.

**25-2.** Obtain the sensitivity equations for $\partial\mathbf{K}_\theta(k+1)/\partial\theta_i$, $\partial\mathbf{P}_\theta(k+1|k)/\partial\theta_i$, and $\partial\mathbf{P}_\theta(k+1|k+1)/\partial\theta_i$. Explain why the sensitivity system for $\partial\hat{\mathbf{x}}_\theta(k+1|k)/\partial\theta_i$ and $\partial\hat{\mathbf{x}}_\theta(k+1|k+1)/\partial\theta_i$ is *linear*.

**25-3.** Compute a formula for $\mathbf{H}_i$. Then simplify $\mathbf{H}_i$ to a *pseudo-Hessian*.

**25-4.** In the first-order system $x(k+1) = ax(k)+w(k)$, and $z(k+1) = x(k+1)+\nu(k+1)$, $k = 1, 2, \ldots, N$, $a$ is an unknown parameter that is to be estimated. Sequences $w(k)$ and $\nu(k)$ are, as usual, mutually uncorrelated and white, and $w(k) \sim N(w(k); 0, 1)$ and $\nu(k) \sim N(\nu(k); 0, \frac{1}{2})$. Explain, using equations and a flow chart, how parameter $a$ can be estimated using a MLE.

**25-5.** Here we wish to modify the basic state-variable model to one in which $\mathbf{w}(k)$ and $\mathbf{v}(k)$ are correlated, i.e., $\mathbf{E}\{\mathbf{w}(k)\mathbf{v}'(k)\} = \mathbf{S}(k) \neq \mathbf{0}$.

**(a)** What is $\boldsymbol{\theta}$ for this system?

**(b)** What is the equation for the log-likelihood function $L(\boldsymbol{\theta}|\mathbf{Z})$ for this system?

**(c)** Provide the formula for the gradient of $L(\boldsymbol{\theta}|\mathbf{Z})$.

**(d)** Develop the *Kalman filter sensitivity system* for this system.

**(e)** Explain how the preceding results for the modified basic state-variable model are different from those for the basic state-variable model.

**25-6.** Repeat Problem 25-4 where all conditions are the same except that now $w(k)$ and $\nu(k)$ are correlated, and $\mathbf{E}\{w(k)\nu(k)\} = \frac{1}{4}$.

**25-7.** We are interested in estimating the parameters $a$ and $r$ in the following first-order system:

$$x(k+1) + ax(k) = w(k)$$

$$z(k) = x(k) + \nu(k), \qquad k = 1, 2, \ldots, N$$

Signals $w(k)$ and $\nu(k)$ are mutually uncorrelated, white, and Gaussian, and $\mathbf{E}\{w^2(k)\} = 1$ and $\mathbf{E}\{n^2(k)\} = r$.

**(a)** Let $\boldsymbol{\theta} = \text{col}\,(a, r)$. What is the equation for the log-likelihood function?

**(b)** Prepare a macro flow chart that depicts the sequence of calculations required to maximize $L(\boldsymbol{\theta}|\mathbf{Z})$. Assume an optimization algorithm is used that requires gradient information about $L(\boldsymbol{\theta}|\mathbf{Z})$.

**(c)** Write out the Kalman filter sensitivity equations for parameters $a$ and $r$.

**25-8.** (Gregory Caso, Fall 1991) For the first-order system

$$x(k+1) = ax(k) + w(k)$$

$$z(k+1) = x(k+1) + \nu(k+1)$$

where $a, q = \mathbf{E}\{w^2(k)\}$ and $\mathbf{E}\{\nu^2(k)\}$ are unknown:

**(a)** Write out the steady-state Kalman filter equations.

**(b)** Write out the steady-state Kalman filter sensitivity equations.

**(c)** Determine equations for $\hat{\mathbf{r}}_{\text{ML}}$ and $\hat{\mathbf{q}}_{\text{ML}}$ when the remaining parameters have been estimated.

**25-9.** (Todd A. Parker and Mark A. Ranta, Fall 1991) The "controllable canonical form" state-variable representation for the discrete-time autoregressive moving average (ARMA) was presented in Example 11-3. Find the log-likelihood function for estimating the parameters of the third-order ARMA system $H(z) = [\beta_1 z + \beta_2]/[z^2 + \alpha_1 z + \alpha_2]$.

**25-10.** Develop the sensitivity equations for the case considered in Lesson 11, i.e., for the case where the only uncertainty present in the state-variable model is measurement noise. Begin with $L(\boldsymbol{\theta}|\mathbf{Z})$ in (11-42).

**25-11.** Refer to Problem 23-7. Explain, using equations and a flow chart, how to obtain MLEs of the unknown parameters for:

**(a)** Equation for the unsteady operation of a synchronous motor, in which $C$ and $p$ are unknown.

**(b)** Duffing's equation, in which $C$, $\alpha$, and $\beta$ are unknown.

**(c)** Van der Pol's equation, in which $\epsilon$ is unknown.

**(d)** Hill's equation, in which $a$ and $b$ are unknown.

**25-12.** In this lesson, we formulated ML parameter estimation for a state-variable model. Formulate it for the single-input, single-output convolutional model that is described in Example 2-1, Equations (2-2) and (2-3). Assume that input $u(k)$ is zero mean and

Gaussian and that the unknown system is MA. You are to estimate the MA parameters as well as the variances of $u(k)$ and $n(k)$.

**25-13.** When a system is described by an ARMA model whose parameters are unknown and have to be estimated using a gradient-type optimization algorithm, derivatives of the system's impulse response, with respect to the unknown parameters, will be needed. Let the ARMA model be described by the following difference equation:

$$w(k+n) + a_1 w(k+n-1) + \cdots + a_{n-1} w(k+1) + a_n w(k)$$
$$= b_1 u(k+n-1) + b_2 u(k+n-2) + \cdots + b_{n-1} u(k+1) + b_n u(k)$$

The impulse response is obtained by replacing $u(k)$ by $\delta(k)$ and $w(k)$ by $h(k)$; i.e.,

$$h(k+n) + a_1 h(k+n-1) + \cdots + a_{n-1} h(k+1) + a_n h(k)$$
$$= b_1 \delta(k+n-1) + b_2 \delta(k+n-2) + \cdots + b_{n-1} \delta(k+1) + b_n \delta(k)$$

For notational simplicity, let $A(z) = z^n + a_1 z^{n-1} + \cdots + a_{n-1} z + a_n$ and $B(z) = b_1 z^{n-1} + b_2 z^{n-2} + \cdots + b_{n-1} z + b_n$.

**(a)** Show that $\partial h(k)/\partial a_1 (k = 1, 2, \ldots, N)$ can be obtained by solving the finite difference equation $A(z) s_{a1}(k) = -h(k+n-1)$, where *sensitivity coefficient* $s_{a1}(k)$ is defined as $s_{a1}(k) \triangleq \partial h(k)/\partial a_1$.

**(b)** Prove that $\partial h(k)/\partial a_j = \partial h(k-j+1) \partial a_1$ for all $j$ and $k$. This means that $\partial h(k)/\partial a_j$ is obtained from $\partial h(k)/\partial a_1$ simply by shifting the latter's arguments and properly storing the associated values. What is the significance of this result?

**(c)** Obtain comparable results for $\partial h(k)/\partial b_1$ and $\partial h(k)/\partial b_j$ ($j = 2, 3, \ldots, n$ and $k = 1, 2, \ldots, N$).

**(d)** Flow chart the entire procedure for obtaining $s_{aj}(k)$ and $s_{bj}(k)$ for $j = 1, 2, \ldots, n$ and $k = 1, 2, \ldots, N$.

**25-14.** In Example 14-2 we developed the essence of maximum-likelihood deconvolution. Review that example to see that we assumed that the channel was known. Now let us assume that the channel is unknown. Suppose that the channel is the ARMA model described in Problem 25-13 (the reference for this example is Mendel, 1990).

**(a)** Show that the results in Example 14-2 remain unchanged, except that objective function $M(\mathbf{q}|\mathbf{Z})$ in (14-35) also depends on the unknown ARMA parameters, $\mathbf{a} \triangleq \text{col}(a_1, a_2, \ldots, a_n)$, and $\mathbf{b} \triangleq \text{col}(b_1, b_2, \ldots, b_n)$; i.e., $M(\mathbf{q}|\mathbf{Z}) \to M(\mathbf{q}, \mathbf{a}, \mathbf{b}|\mathbf{Z})$.

**(b)** Explain why it is not possible to optimize $M(\mathbf{q}, \mathbf{a}, \mathbf{b}|\mathbf{Z})$ simultaneously with respect to $\mathbf{q}$, $\mathbf{a}$, and $\mathbf{b}$.

**(c)** Because of part (b), we usually maximize $M(\mathbf{q}, \mathbf{a}, \mathbf{b}|\mathbf{Z})$ using a block component method (BCM), in which we initially specify the ARMA parameters and then detect the events; then we optimize the ARMA parameters, after which we redetect the events; and so on until $M(\mathbf{q}, \mathbf{a}, \mathbf{b}|\mathbf{Z})$ does not change appreciably. Flow chart this BCM.

**(d)** Assume that a gradient algorithm is used to optimize $M(\mathbf{q}, \mathbf{a}, \mathbf{b}|\mathbf{Z})$ with respect to $\mathbf{a}$ and $\mathbf{b}$. Explain how you compute $\partial \mathbf{Z}'\boldsymbol{\Omega}^{-1}\mathbf{Z}/\partial a_j$ without computing the inverse of matrix $\boldsymbol{\Omega}$.

**(e)** Explain, using words, flow charts, and equations, how you compute $\partial M(\mathbf{q}, \mathbf{a}, \mathbf{b}|\mathbf{Z})/\partial \mathbf{a}$ and $\partial M(\mathbf{q}, \mathbf{a}, \mathbf{b}|\mathbf{Z})/\partial \mathbf{b}$.

**(f)** Explain why using the convolutional model to maximize $M(\mathbf{q}, \mathbf{a}, \mathbf{b}|\mathbf{Z})$ is a computational nightmare.

**(g)** Explain how you convert this problem into an equivalent state-variable model and why solving the resulting ML estimation problem is computationally efficient.

# LESSON 26

# Kalman–Bucy Filtering


## SUMMARY

The Kalman–Bucy filter (KBF) is the continuous-time counterpart to the Kalman filter. It is a continuous-time minimum-variance filter that provides state estimates for continuous-time dynamical systems that are described by linear, (possibly) time-varying, and (possibly) nonstationary ordinary differential equations.

The purpose of this lesson is to derive the KBF. We do this in two different ways: (1) use of a formal limiting procedure to obtain the KBF from the KF, and (2) assumption of a linear differential equation structure for the KBF, one that contains an unknown time-varying gain matrix that weights the difference between the measurement made at time $t$ and the estimate of that measurement, after which the gain matrix is chosen to minimize the mean-squared error.

The KBF plays an essential role in the solution of the linear-quadratic-Gaussian regulator problem in optimal control theory.

When you complete this lesson, you will be able to derive the continuous-time KBF from two points of view and understand its relationship to the discrete-time KF.


## INTRODUCTION

The Kalman–Bucy filter is the continuous-time counterpart to the Kalman filter. It is a continuous-time minimum-variance filter that provides state estimates for continuous-time dynamical systems that are described by linear, (possibly) time-varying, and (possibly) nonstationary ordinary differential equations.

The Kalman–Bucy filter (KBF) can be derived in a number of different ways, including the following three:

1. Use a formal limiting procedure to obtain the KBF from the KF (e.g., Meditch, 1969).
2. Begin by assuming the optimal estimator is a linear transformation of *all* measurements. Use a calculus of variations argument or the orthogonality principle to obtain the Wiener–Hopf integral equation. Embedded within this equation is the filter kernal. Take the derivative of the Wiener–Hopf equation to obtain a differential equation that is the KBF (Meditch, 1969).
3. Begin by assuming a linear differential equation structure for the KBF, one that contains an unknown time-varying gain matrix that weights the difference between the measurement made at time $t$ and the estimate of that measurement. Choose the gain matrix that minimizes the mean-squared error (Athans and Tse, 1967).

We shall describe the first approach in the main body of this lesson and the third approach in the Supplementary Material at the end of the lesson; but first we must define our continuous-time model and formally state the problem we wish to solve.

## SYSTEM DESCRIPTION

Our continuous-time system is described by the following state-variable model:

$$\dot{\mathbf{x}}(t) = \mathbf{F}(t)\mathbf{x}(t) + \mathbf{G}(t)\mathbf{w}(t) \tag{26-1}$$

$$\mathbf{z}(t) = \mathbf{H}(t)\mathbf{x}(t) + \mathbf{v}(t) \tag{26-2}$$

where $\mathbf{x}(t)$ is $n \times 1$, $\mathbf{w}(t)$ is $p \times 1$, $\mathbf{z}(t)$ is $m \times 1$, and $\mathbf{v}(t)$ is $m \times 1$. For simplicity, we have omitted a known forcing function term in state equation (26-1). Matrices $\mathbf{F}(t)$, $\mathbf{G}(t)$, and $\mathbf{H}(t)$ have dimensions that conform to the dimensions of the vector quantities in this state-variable model. Disturbance $\mathbf{w}(t)$ and measurement noise $\mathbf{v}(t)$ are zero-mean white noise processes, which are assumed to be uncorrelated; i.e., $\mathbf{E}\{\mathbf{w}(t)\} = \mathbf{0}$, $\mathbf{E}\{\mathbf{v}(t)\} = \mathbf{0}$,

$$\mathbf{E}\{\mathbf{w}(t)\mathbf{w}'(\tau)\} = \mathbf{Q}(t)\delta(t-\tau) \tag{26-3}$$

$$\mathbf{E}\{\mathbf{v}(t)\mathbf{v}'(\tau)\} = \mathbf{R}(t)\delta(t-\tau) \tag{26-4}$$

and

$$\mathbf{E}\{\mathbf{w}(t)\mathbf{v}'(\tau)\} = \mathbf{0} \tag{26-5}$$

Equations (26-3), (26-4), and (26-5) apply for $t \geq t_0$. Additionally, $\mathbf{R}(t)$ is continuous and positive definite, whereas $\mathbf{Q}(t)$ is continuous and positive semidefinite. Finally, we assume that the initial state vector $\mathbf{x}(t_0)$ may be random, and, if it is, it is uncorrelated with both $\mathbf{w}(t)$ and $\mathbf{v}(t)$. The statistics of a random $\mathbf{x}(t_0)$ are

$$\mathbf{E}\{\mathbf{x}(t_0)\} = \mathbf{m}_{\mathbf{x}}(t_0) \tag{26-6}$$

and

$$\text{cov}\{\mathbf{x}(t_0)\} = \mathbf{P_x}(t_0) \tag{26-7}$$

Measurements $\mathbf{z}(t)$ are assumed to be made for $t_0 \le t \le \tau$.

If $\mathbf{x}(t_0)$, $\mathbf{w}(t)$, and $\mathbf{v}(t)$ are jointly Gaussian for all $t \in [t_0, \tau]$, then the KBF will be the optimal estimator of state vector $\mathbf{x}(t)$. We will not make any distributional assumptions about $\mathbf{x}(t_0)$, $\mathbf{w}(t)$, and $\mathbf{v}(t)$ in this lesson, being content to establish the *linear optimal estimator* of $\mathbf{x}(t)$, which, of course, is the optimal estimator of $\mathbf{x}(t)$ when all sources of uncertainty are Gaussian.

## STATISTICS OF THE STATE VECTOR

Theorem 15-5 shows how the first- and second-order statistics of the state vector in our basic discrete-time state-variable model can be computed by means of vector difference equations. Theorem 15-6 then shows how the first- and second-order statistics of the measurement vector can be computed in terms of the first- and second-order statistics of the state vector, respectively. Let

$$\mathbf{m_x}(t) = \mathbf{E}\{\mathbf{x}(t)\} \tag{26-8}$$

and

$$\mathbf{P_x}(t) = \mathbf{E}\{[\mathbf{x}(t) - \mathbf{m_x}(t)][\mathbf{x}(t) - \mathbf{m_x}(t)]'\} \tag{26-9}$$

We now demonstrate that $\mathbf{m_x}(t)$ and $\mathbf{P_x}(t)$ can be computed by means of differential equations and that $\mathbf{m_z}(t)$ and $\mathbf{P_z}(t)$ can be computed from $\mathbf{m_x}(t)$ and $\mathbf{P_x}(t)$, respectively.

**Theorem 26-1.** *For the continuous-time basic state-variable model in (26-1) and (26-2),*

**a.** $\mathbf{m_x}(t)$ *can be computed from the vector differential equation* $\quad$ Diff EQ's replace difference eqn (continuous vs discrete)

$$\frac{d\mathbf{m_x}(t)}{dt} = \mathbf{F}(t)\mathbf{m_x}(t) \tag{26-10}$$

*where* $\mathbf{m_x}(t_0)$ *initializes (26-10).*

**b.** $\mathbf{P_x}(t)$ *can be computed from the matrix differential equation*

$$\frac{d\mathbf{P_x}(t)}{dt} = \mathbf{F}(t)\mathbf{P_x}(t) + \mathbf{P_x}(t)\mathbf{F}'(t) + \mathbf{G}(t)\mathbf{Q}(t)\mathbf{G}'(t) \tag{26-11}$$

*where* $\mathbf{P_x}(t_0)$ *initializes (26-11).*

**c.** $\mathbf{E}\{[\mathbf{x}(t_1) - \mathbf{m_x}(t_1)][\mathbf{x}(t_2) - \mathbf{m_x}(t_2)]'\} \triangleq \mathbf{P_x}(t_1, t_2)$ *can be computed from*

$$\mathbf{P_x}(t_1, t_2) = \begin{cases} \mathbf{\Phi}(t_1, t_2)\mathbf{P_x}(t_2), & \text{when } t_1 \ge t_2 \\ \mathbf{P_x}(t_1)\mathbf{\Phi}'(t_2, t_1), & \text{when } t_1 \le t_2 \end{cases} \tag{26-12}$$

**d.** $\mathbf{E}\{\mathbf{z}(t)\} = \mathbf{m_z}(t)$ *can be computed from*

$$\mathbf{m_z}(t) = \mathbf{H}(t)\mathbf{m_x}(t) \tag{26-13}$$

*where* $\mathbf{m_x}(t)$ *is computed from (26-10).*

**e.** $\mathbf{E}\{[\mathbf{z}(t) - \mathbf{m_z}(t)][\mathbf{z}(t) - \mathbf{m_z}(t)]'\} = \mathbf{P_z}(t)$ *can be computed from*

$$\mathbf{P_z}(t) = \mathbf{H}(t)\mathbf{P_x}(t)\mathbf{H}'(t) + \mathbf{R}(t) \tag{26-14}$$

*Proof.* Because the proof of this theorem is not central to the rest of this lesson, we provide it in the Supplementary Material at the end of the lesson. □

**EXAMPLE 26-1**

The *Dirac delta function* used in (26-3) and (26-4) is defined as

$$\delta(t) = \lim_{\varepsilon \to 0} \delta_\varepsilon(t) \tag{26-15}$$

where

$$\delta_\varepsilon(t) = \begin{cases} 0, & |t| > \varepsilon \\ \dfrac{1}{2\varepsilon}, & |t| < \varepsilon \end{cases} \tag{26-16}$$

In the Proof of Theorem 26-1, we need to evaluate the integral of $\mathbf{E}\{\mathbf{w}(t)\mathbf{w}'(\tau)\}$. Here we carry out that computation, as follows:

$$\int_{t_0}^{t} \mathbf{E}\{\mathbf{w}(t)\mathbf{w}'(\tau)\}d\tau = \int_{t_0}^{t} \mathbf{Q}(t)\delta(t-\tau)d\tau = \mathbf{Q}(t)\int_{t_0}^{t} \delta(t-\tau)d\tau = \frac{1}{2}\mathbf{Q}(t) \tag{26-17}$$

How did we arrive at this last result? We made use of (26-15) and (26-16), as depicted in Figure 26-1. Consequently,

$$\int_{t_0}^{t} \delta(t-\tau)d\tau = \lim_{\varepsilon \to 0} \int_{t_0}^{t} \delta_\varepsilon(t-\tau)d\tau = \int_{t-\varepsilon}^{t} \left(\frac{1}{2\varepsilon}\right)d\tau = \frac{1}{2} \;\square \tag{26-18}$$

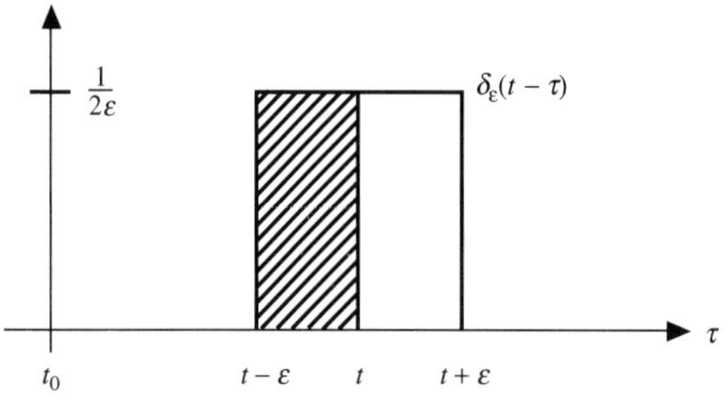


**Figure 26-1** Relationship of $\delta_\varepsilon(t-\tau)$ to $\tau = t$.

## NOTATION AND PROBLEM STATEMENT

Our notation for a continuous-time estimate of $\mathbf{x}(t)$ and its associated estimation error parallels our notation for the comparable discrete-time quantities; i.e., $\hat{\mathbf{x}}(t|t)$ denotes the optimal estimate of $\mathbf{x}(t)$ that uses all the measurements $\mathbf{z}(t)$, where $t \geq t_0$, and

$$\tilde{\mathbf{x}}(t|t) = \mathbf{x}(t) - \hat{\mathbf{x}}(t|t) \tag{26-19}$$

The mean-squared state estimation error is

$$J[\tilde{\mathbf{x}}(t|t)] = \mathbf{E}\{\tilde{\mathbf{x}}'(t|t)\tilde{\mathbf{x}}(t|t)\} \tag{26-20}$$

We shall determine $\hat{\mathbf{x}}(t|t)$ that minimizes $J[\tilde{\mathbf{x}}(t|t)]$, subject to the constraints of our state variable model and data set.

## THE KALMAN–BUCY FILTER

The solution to the problem stated in the preceding section is the Kalman–Bucy filter, the structure of which is summarized in the following:

**Theorem 26-2.** *The KBF is described by the vector differential equation*

$$\dot{\hat{\mathbf{x}}}(t|t) = \mathbf{F}(t)\hat{\mathbf{x}}(t|t) + \mathbf{K}(t)[\mathbf{z}(t) - \mathbf{H}(t)\hat{\mathbf{x}}(t|t)] \tag{26-21}$$

*where* $t \geq t_0$, $\hat{\mathbf{x}}(t_0|t_0) = \mathbf{m}_{\mathbf{x}}(t_0)$,

$$\mathbf{K}(t) = \mathbf{P}(t|t)\mathbf{H}'(t)\mathbf{R}^{-1}(t) \tag{26-22}$$

*and*

$$\dot{\mathbf{P}}(t|t) = \mathbf{F}(t)\mathbf{P}(t|t) + \mathbf{P}(t|t)\mathbf{F}'(t) - \mathbf{P}(t|t)\mathbf{H}'(t)\mathbf{R}^{-1}(t)\mathbf{H}(t)\mathbf{P}(t|t) + \mathbf{G}(t)\mathbf{Q}(t)\mathbf{G}'(t) \tag{26-23}$$

*Equation (26-23), which is a matrix Riccati differential equation, is initialized by* $\mathbf{P}(t_0|t_0) = \mathbf{P}_{\mathbf{x}}(t_0)$. □

Matrix $\mathbf{K}(t)$ is the *Kalman–Bucy gain matrix*, and $\mathbf{P}(t|t)$ is the state-estimation-error covariance matrix; i.e.,

$$\mathbf{P}(t|t) = \mathbf{E}\{\tilde{\mathbf{x}}(t|t)\tilde{\mathbf{x}}'(t|t)\} \tag{26-24}$$

Equation (26-21) can be rewritten as

$$\dot{\hat{\mathbf{x}}}(t|t) = [\mathbf{F}(t) - \mathbf{K}(t)\mathbf{H}(t)]\hat{\mathbf{x}}(t|t) + \mathbf{K}(t)\mathbf{z}(t) \tag{26-25}$$

which makes it very clear that the KBF is a time-varying filter that processes the measurements linearly to produce $\hat{\mathbf{x}}(t|t)$.

The solution to (26-25) is (see Theorem 23-1)

$$\hat{\mathbf{x}}(t|t) = \mathbf{\Phi}(t, t_0)\hat{\mathbf{x}}(t_0|t_0) + \int_{t_0}^{t} \mathbf{\Phi}(t, \tau)\mathbf{K}(\tau)\mathbf{z}(\tau)d\tau \tag{26-26}$$

where the state transition matrix $\mathbf{\Phi}(t, \tau)$ is the solution to the matrix differential equation

$$\left.\begin{aligned} \dot{\mathbf{\Phi}}(t, \tau) &= [\mathbf{F}(t) - \mathbf{K}(t)\mathbf{H}(t)]\mathbf{\Phi}(t, \tau) \\ \mathbf{\Phi}(t, t) &= \mathbf{I} \end{aligned}\right\} \tag{26-27}$$

If $\hat{\mathbf{x}}(t_0|t_0) = \mathbf{0}$, then

$$\hat{\mathbf{x}}(t|t) = \int_{t_0}^{t} \mathbf{\Phi}(t, \tau)\mathbf{K}(\tau)\mathbf{z}(\tau)d\tau = \int_{t_0}^{t} \mathbf{A}(t, \tau)\mathbf{z}(\tau)d\tau \tag{26-28}$$

where the filter kernel $\mathbf{A}(t, \tau)$ is

$$\mathbf{A}(t, \tau) = \mathbf{\Phi}(t, \tau)\mathbf{K}(\tau) \tag{26-29}$$

The second approach to deriving the KBF, mentioned in the introduction to this chapter, begins by assuming that $\hat{\mathbf{x}}(t|t)$ can be expressed as in (26-28), where $\mathbf{A}(t, \tau)$ is unknown. The mean-squared estimation error is minimized to obtain the following Wiener–Hopf integral equation:

$$\mathbf{E}\{\mathbf{x}(t)\mathbf{z}'(\sigma)\} - \int_{t_0}^{t} \mathbf{A}(t, \tau)\mathbf{E}\{\mathbf{z}(\tau)\mathbf{z}'(\sigma)\}d\tau = \mathbf{0} \tag{26-30}$$

where $t_0 \leq \sigma \leq t$. When this equation is converted into a differential equation, we obtain the KBF described in Theorem 26-2. For the details of this derivation, see Meditch, 1969, Chapter 8.

## DERIVATION OF KBF USING A FORMAL LIMITING PROCEDURE

Kalman filter Equation (17-11), expressed as

$$\hat{\mathbf{x}}(k+1|k+1) = \mathbf{\Phi}(k+1, k)\hat{\mathbf{x}}(k|k) + \mathbf{K}(k+1)[\mathbf{z}(k+1) - \mathbf{H}(k+1)\mathbf{\Phi}(k+1, k)\hat{\mathbf{x}}(k|k)] \tag{26-31}$$

can also be written as

$$\hat{\mathbf{x}}(t+\Delta t|t+\Delta t) = \mathbf{\Phi}(t+\Delta t, t)\hat{\mathbf{x}}(t|t) + \mathbf{K}(t+\Delta t)[\mathbf{z}(t+\Delta t) - \mathbf{H}(t+\Delta t)\mathbf{\Phi}(t+\Delta t, t)\hat{\mathbf{x}}(t|t)] \tag{26-32}$$

where we have let $t_k = t$ and $t_{k+1} = t + \Delta t$. In Example 23-3 we showed that, if $\mathbf{F}(t)$, $\mathbf{G}(t)$, and $\mathbf{Q}(t)$ are approximately constant during the time interval $[t_k, t_{k+1}]$, then [(23-47) and (23-49)]

$$\mathbf{\Phi}(t+\Delta t, t) \simeq \mathbf{I} + \mathbf{F}(t)\Delta t + O(\Delta t^2) \tag{26-33}$$

and [(23-51)]

$$\mathbf{Q}_d(t+\Delta t, t) \simeq \mathbf{G}(t)\mathbf{Q}(t)\mathbf{G}'(t)\Delta t + O(\Delta t^2) \tag{26-34}$$

Observe that $\mathbf{Q}_d(t+\Delta t, t)$ can also be written as

$$\mathbf{Q}_d(t+\Delta t, t) \simeq [\mathbf{G}(t)\Delta t]\left[\frac{\mathbf{Q}(t)}{\Delta t}\right][\mathbf{G}(t)\Delta t]' + O(\Delta t^2) \tag{26-35}$$

and, if we express $\mathbf{w}_d(k)$ as $\mathbf{\Gamma}(k+1, k)\mathbf{w}(k)$, so that $\mathbf{Q}_d(t+\Delta t, t) = \mathbf{\Gamma}(k+1, k)\mathbf{Q}(k)\mathbf{\Gamma}'(k+1, k)$, then

$$\mathbf{\Gamma}(t+\Delta t, t) \simeq \mathbf{G}(t)\Delta t + O(\Delta t^2) \tag{26-36}$$

and

$$\mathbf{Q}(k=t) \Rightarrow \frac{\mathbf{Q}(t)}{\Delta t} \tag{26-37}$$

Equation (26-37) means that we replace $\mathbf{Q}(k=t)$ in the KF by $\mathbf{Q}(t)/\Delta t$. Note that we have encountered a bit of a notational problem here, because we have used $\mathbf{w}(k)$ [and its associated covariance $\mathbf{Q}(k)$] to denote the disturbance in our discrete-time model, and $\mathbf{w}(t)$ [and its associated intensity $\mathbf{Q}$(t)] to denote the disturbance in our continuous-time model.

Without going into technical details, we shall also replace $\mathbf{R}(k+1)$ in the KF by $\mathbf{R}(t+\Delta t)/\Delta t$; i.e.,

$$\mathbf{R}(k+1=t+\Delta t) \Rightarrow \mathbf{R}(t+\Delta t)/\Delta t \tag{26-38}$$

See Meditch (1969, pp. 139–142) for an explanation. Substituting (26-33) into (26-32), and omitting all higher-order terms in $\Delta t$, we find that

$$\begin{aligned}\hat{\mathbf{x}}(t+\Delta t|t+\Delta t) = [\mathbf{I}+\mathbf{F}(t)\Delta t]\hat{\mathbf{x}}(t|t) + \mathbf{K}(t+\Delta t)\{\mathbf{z}(t+\Delta t)\\ -\mathbf{H}(t+\Delta t)[\mathbf{I}+\mathbf{F}(t)\Delta t]\hat{\mathbf{x}}(t|t)\}\end{aligned} \tag{26-39}$$

from which it follows that

$$\lim_{\Delta t\to 0}\frac{\hat{\mathbf{x}}(t+\Delta t|t+\Delta t)-\hat{\mathbf{x}}(t|t)}{\Delta t} = \mathbf{F}(t)\hat{\mathbf{x}}(t|t) + \lim_{\Delta t\to 0}\frac{\mathbf{K}(t+\Delta t)\{\mathbf{z}(t+\Delta t)-\mathbf{H}(t+\Delta t)[\mathbf{I}+\mathbf{F}(t)\Delta t]\hat{\mathbf{x}}(t|t)\}}{\Delta t}$$

or

$$\dot{\hat{\mathbf{x}}}(t|t) = \mathbf{F}(t)\hat{\mathbf{x}}(t|t) + \lim_{\Delta t\to 0}\frac{\mathbf{K}(t+\Delta t)\{\mathbf{z}(t+\Delta t)-\mathbf{H}(t+\Delta t)[\mathbf{I}+\mathbf{F}(t)\Delta t]\hat{\mathbf{x}}(t|t)\}}{\Delta t} \tag{26-40}$$

Under suitable regularity conditions, which we shall assume are satisfied here, we can replace the limit of a product of functions by the product of limits; i.e.;

$$\begin{aligned}&\lim_{\Delta t\to 0}\frac{\mathbf{K}(t+\Delta t)\{\mathbf{z}(t+\Delta t)-\mathbf{H}(t+\Delta t)[\mathbf{I}+\mathbf{F}(t)\Delta t]\hat{\mathbf{x}}(t|t)\}}{\Delta t}\\ &= \lim_{\Delta t\to 0}\frac{\mathbf{K}(t+\Delta t)}{\Delta t}\lim_{\Delta t\to 0}\{\mathbf{z}(t+\Delta t)-\mathbf{H}(t+\Delta t)[\mathbf{I}+\mathbf{F}(t)\Delta t]\hat{\mathbf{x}}(t|t)\}\end{aligned} \tag{26-41}$$

The second limit on the right-hand side of (26-41) is easy to evaluate; i.e.;

$$\lim_{\Delta t\to 0}\{\mathbf{z}(t+\Delta t)-\mathbf{H}(t+\Delta t)[\mathbf{I}+\mathbf{F}(t)\Delta t]\hat{\mathbf{x}}(t|t)\} = \mathbf{z}(t)-\mathbf{H}(t)\hat{\mathbf{x}}(t|t) \tag{26-42}$$

To evaluate the first limit on the right-hand side of (26-41), we first substitute $\mathbf{R}(t+\Delta t)/\Delta t$ for $\mathbf{R}(k+1=t+\Delta t)$ in (17-12), to obtain

$$\begin{aligned}\mathbf{K}(t+\Delta t) = &\ \mathbf{P}(t+\Delta t|t)\mathbf{H}'(t+\Delta t)\\ &[\mathbf{H}(t+\Delta t)\mathbf{P}(t+\Delta t|t)\mathbf{H}'(t+\Delta t)\Delta t+\mathbf{R}(t+\Delta t)]^{-1}\Delta t\end{aligned} \tag{26-43}$$

Then we substitute $\mathbf{Q}(t)/\Delta t$ for $\mathbf{Q}(k = t)$ in (17-13), to obtain

$$\mathbf{P}(t + \Delta t|t) = \mathbf{\Phi}(t + \Delta t, t)\mathbf{P}(t|t)\mathbf{\Phi}'(t + \Delta t, t) + \mathbf{\Gamma}(t + \Delta t, t)\frac{\mathbf{Q}(t)}{\Delta t}\mathbf{\Gamma}'(t + \Delta t, t) \tag{26-44}$$

Substitute (26-33) and (26-36) into (26-44) to see that the only term that does not depend on $\Delta t$ is $\mathbf{P}(t|t)$; hence,

$$\lim_{\Delta t \to 0} \mathbf{P}(t + \Delta t|t) = \mathbf{P}(t|t) \tag{26-45}$$

and, therefore,

$$\lim_{\Delta t \to 0} \frac{\mathbf{K}(t + \Delta t)}{\Delta t} = \mathbf{P}(t|t)\mathbf{H}'(t)\mathbf{R}^{-1}(t) \stackrel{\triangle}{=} \mathbf{K}(t) \tag{26-46}$$

Combining (26-40), (26-41), (26-42), and (26-46), we obtain the KBF in (26-21) and the KB gain matrix in (26-22).

To derive the matrix differential equation for $\mathbf{P}(t|t)$, we begin with (17-14), substitute (26-44) along with the expansions of $\mathbf{\Phi}(t + \Delta t, t)$ and $\mathbf{\Gamma}(t + \Delta t, t)$ into that equation, to show that

$$\begin{aligned} \mathbf{P}(t + \Delta t|t + \Delta t) &= \mathbf{P}(t + \Delta t|t) - \mathbf{K}(t + \Delta t)\mathbf{H}(t + \Delta t)\mathbf{P}(t + \Delta t|t) \\ &= \mathbf{P}(t|t) + [\mathbf{F}(t)\mathbf{P}(t|t) + \mathbf{P}(t|t)\mathbf{F}'(t) \\ &\quad + \mathbf{G}(t)\mathbf{Q}(t)\mathbf{G}'(t)]\Delta t \\ &\quad - \mathbf{K}(t + \Delta t)\mathbf{H}(t + \Delta t)\mathbf{P}(t + \Delta t|t) \end{aligned} \tag{26-47}$$

Consequently,

$$\lim_{\Delta t \to 0} \frac{\mathbf{P}(t + \Delta t|t + \Delta t) - \mathbf{P}(t|t)}{\Delta t} = \dot{\mathbf{P}}(t|t) = \mathbf{F}(t)\mathbf{P}(t|t) + \mathbf{P}(t|t)\mathbf{F}'(t) + \mathbf{G}(t)\mathbf{Q}(t)\mathbf{G}'(t) - \lim_{\Delta t \to 0} \frac{\mathbf{K}(t + \Delta t)\mathbf{H}(t + \Delta t)\mathbf{P}(t + \Delta t|t)}{\Delta t} \tag{26-48}$$

or finally, using (26-46),

$$\dot{\mathbf{P}}(t|t) = \mathbf{F}(t)\mathbf{P}(t|t) + \mathbf{P}(t|t)\mathbf{F}'(t) + \mathbf{G}(t)\mathbf{Q}(t)\mathbf{G}'(t) - \mathbf{P}(t|t)\mathbf{H}'(t)\mathbf{R}^{-1}(t)\mathbf{H}(t)\mathbf{P}(t|t) \tag{26-49}$$

This completes the derivation of the KBF using a formal limiting procedure. It is also possible to obtain continuous-time smoothers by means of this procedure (e.g., see Meditch, 1969, Chapter 7).

**EXAMPLE 26-2**

This is a continuation of Example 18-7. Our goal is to obtain the continuous-time counterpart to (18-35). Because a control is central to the development of this equation, we first introduce a control into state equation (26-1), as

$$\frac{d\mathbf{x}(t)}{dt} = \mathbf{F}(t)\mathbf{x}(t) + \mathbf{B}(t)\mathbf{u}(t) + \mathbf{G}(t)\mathbf{w}(t) \tag{26-50}$$

Beginning with (18-35), using (26-33), (26-36), and (23-50) expressed as

$$\mathbf{\Psi}(t + \Delta t, t) \approx \mathbf{B}(t)\Delta t + O(\Delta t^2) \tag{26-51}$$

and the fact that [see (18-34)]

$$\mathbf{\Phi}_c(t + \Delta t, t) = \mathbf{\Phi}(t + \Delta t, t) + \mathbf{\Psi}(t + \Delta t, t)\mathbf{C}(t) \approx \mathbf{I} + \mathbf{F}_c(t)\Delta t + O(\Delta t^2) \tag{26-52}$$

where

$$\mathbf{F}_c(t) = \mathbf{F}(t) + \mathbf{B}(t)\mathbf{C}(t) \tag{26-53}$$

it is straightforward, using the formal limiting procedure that was used to obtain the KBF from the KF (Problem 26-2), to show that $\mathbf{P_x}(t) \triangleq \mathbf{E}\{(\mathbf{x}(t)\mathbf{x}'(t)\}$ is described by the following matrix differential equation (Mendel, 1971a):

$$\frac{d\mathbf{P_x}(t)}{dt} = \mathbf{P_x}(t)\mathbf{F}_c'(t) + \mathbf{F}_c(t)\mathbf{P_x}(t) + \mathbf{G}(t)\mathbf{Q}(t)\mathbf{G}'(t) - \mathbf{P}(t|t)\mathbf{C}'(t)\mathbf{B}'(t) - \mathbf{B}(t)\mathbf{C}(t)\mathbf{P}(t|t) \tag{26-54}$$

Once again, as is evident from the last two terms in (26-54), we see that state estimation errors act as an additional plant disturbance. This example is continued in Example 26-4. □

## STEADY-STATE KBF

If our continuous-time system is time invariant and stationary, then, when certain system-theoretic conditions are satisfied (see, e.g., Kwakernaak and Sivan, 1972), $\dot{\mathbf{P}}(t|t) \to \mathbf{0}$ in which case $\mathbf{P}(t|t)$ has a steady-state value, denoted $\bar{\mathbf{P}}$. In this case, $\mathbf{K}(t) \to \bar{\mathbf{K}}$, where

$$\bar{\mathbf{K}} = \bar{\mathbf{P}}\mathbf{H}'\mathbf{R}^{-1} \tag{26-55}$$

$\bar{\mathbf{P}}$ is the solution of the algebraic Riccati equation

$$\mathbf{F}\bar{\mathbf{P}} + \bar{\mathbf{P}}\mathbf{F}' - \bar{\mathbf{P}}\mathbf{H}'\mathbf{R}^{-1}\mathbf{H}\bar{\mathbf{P}} + \mathbf{G}\mathbf{Q}\mathbf{G}' = \mathbf{0} \tag{26-56}$$

and the steady-state KBF is asymptotically stable; i.e., the eigenvalues of $\mathbf{F} - \bar{\mathbf{K}}\mathbf{H}$ all lie in the left-half of the complex $s$-plane.

**EXAMPLE 26-3**

Here we examine the steady-state KBF for the simplest second-order system, the double integrator,

$$\ddot{x}(t) = w(t) \tag{26-57}$$

and

$$z(t) = x(t) + \nu(t) \tag{26-58}$$

in which $w(t)$ and $\nu(t)$ are mutually uncorrelated white noise processes, with intensities $q$ and $r$, respectively. With $x_1(t) = x(t)$ and $x_2(t) = \dot{x}(t)$, this system is expressed in state-variable format as

$$\begin{pmatrix} \dot{x}_1 \\ \dot{x}_2 \end{pmatrix} = \begin{pmatrix} 0 & 1 \\ 0 & 0 \end{pmatrix} \begin{pmatrix} x_1 \\ x_2 \end{pmatrix} + \begin{pmatrix} 0 \\ 1 \end{pmatrix} w \tag{26-59}$$

$$z = (1 \;\; 0) \begin{pmatrix} x_1 \\ x_2 \end{pmatrix} + \nu \tag{26-60}$$

The algebraic Riccati equation for this system is

$$\begin{pmatrix} 0 & 1 \\ 0 & 0 \end{pmatrix}\begin{pmatrix} \bar{p}_{11} & \bar{p}_{12} \\ \bar{p}_{12} & \bar{p}_{22} \end{pmatrix} + \begin{pmatrix} \bar{p}_{11} & \bar{p}_{12} \\ \bar{p}_{12} & \bar{p}_{22} \end{pmatrix}\begin{pmatrix} 0 & 0 \\ 1 & 0 \end{pmatrix}$$

$$- \begin{pmatrix} \bar{p}_{11} & \bar{p}_{12} \\ \bar{p}_{12} & \bar{p}_{22} \end{pmatrix}\begin{pmatrix} 1 \\ 0 \end{pmatrix}\frac{1}{r}(1 \;\; 0)\begin{pmatrix} \bar{p}_{11} & \bar{p}_{12} \\ \bar{p}_{12} & \bar{p}_{22} \end{pmatrix} \tag{26-61}$$

$$+ \begin{pmatrix} 0 \\ 1 \end{pmatrix} q(0 \;\; 1) = \begin{pmatrix} 0 & 0 \\ 0 & 0 \end{pmatrix}$$

which leads to the following three algebraic equations:

$$2\bar{p}_{12} - \frac{1}{r}\bar{p}_{11}^2 = 0 \tag{26-62a}$$

$$\bar{p}_{22} - \frac{1}{r}\bar{p}_{11}\bar{p}_{12} = 0 \tag{26-62b}$$

and

$$-\frac{1}{r}\bar{p}_{12}^2 + q = 0 \tag{26-62c}$$

It is straightforward to show that the unique solution of these nonlinear algebraic equations, for which $\bar{\mathbf{P}} > 0$ is,

$$\bar{p}_{12} = (qr)^{1/2} \tag{26-63a}$$

$$\bar{p}_{11} = \sqrt{2}\, q^{1/4} r^{3/4} \tag{26-63b}$$

$$\bar{p}_{22} = \sqrt{2}\, q^{3/4} r^{1/4} \tag{26-63c}$$

The steady-state KB gain matrix is computed from (26-55) as

$$\bar{\mathbf{K}} = \bar{\mathbf{P}}\mathbf{H}'\frac{1}{r} = \frac{1}{r}\begin{pmatrix} \bar{p}_{11} \\ \bar{p}_{12} \end{pmatrix} = \begin{pmatrix} \sqrt{2}(q/r)^{1/4} \\ (q/r)^{1/2} \end{pmatrix} \tag{26-64}$$

Observe that, just as in the discrete-time case, the single-channel KBF depends only on the ratio $q/r$.

Although we only needed $\bar{p}_{11}$ and $\bar{p}_{12}$ to compute $\bar{\mathbf{K}}$, $\bar{p}_{22}$ is an important quantity, because

$$\bar{p}_{22} = \lim_{t\to\infty} \mathbf{E}\{[\dot{x}(t) - \hat{\dot{x}}(t|t)]^2\} \tag{26-65}$$

Additionally,

$$\bar{p}_{11} = \lim_{t\to\infty} \mathbf{E}\{[x(t) - \hat{x}(t|t)]^2\} \tag{26-66}$$

Using (26-63b) and (26-63c), we find that

$$\bar{p}_{22} = \left(\frac{q}{r}\right)^{1/2} \bar{p}_{11} \tag{26-67}$$

If $q/r > 1$ (i.e., $\overline{\text{SNR}}$ possibly greater than unity), we will always have larger errors in estimation of $\dot{x}(t)$ than estimation of $x(t)$. This is not too surprising because our measurement depends only on $x(t)$, and both $w(t)$ and $\nu(t)$ affect the calculation of $\hat{\dot{x}}(t|t)$.

The steady-state KBF is characterized by the eigenvalues of matrix $\mathbf{F} - \bar{\mathbf{K}}\mathbf{H}$, where

$$\mathbf{F} - \bar{\mathbf{K}}\mathbf{H} = \begin{pmatrix} -\sqrt{2}(q/r)^{1/4} & 1 \\ -(q/r)^{1/2} & 0 \end{pmatrix} \tag{26-68}$$

These eigenvalues are solutions of the equation

$$s^2 + \sqrt{2}\left(\frac{q}{r}\right)^{1/4} s + \left(\frac{q}{r}\right)^{1/2} = 0 \tag{26-69}$$

When this equation is expressed in the normalized form

$$s^2 + 2\zeta\omega_n s + \omega_n^2 = 0$$

we find that

$$\omega_n = \left(\frac{q}{r}\right)^{1/4} \tag{26-70}$$

and

$$\zeta = 0.707 \tag{26-71}$$

thus, the steady-state KBF for the simple double integrator system is damped at 0.707. The filter's poles lie on the 45° line depicted in Figure 26-2. They can be moved along this line by adjusting the ratio $q/r$; hence, once again, we may view $q/r$ as a filter tuning parameter. □

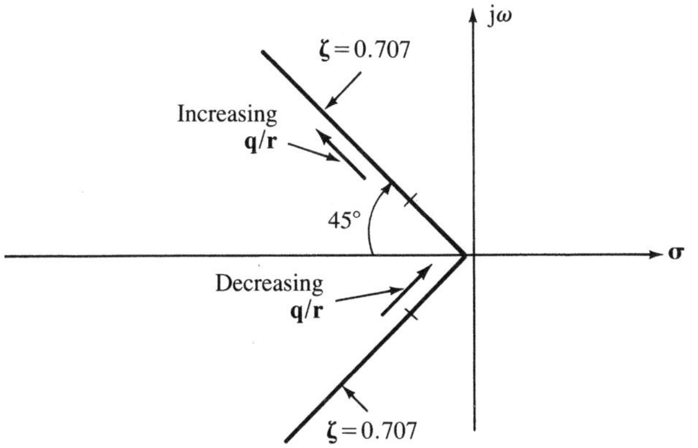


**Figure 26-2** Eigenvalues of steady-state KBF lie along ±45° lines. Increasing $q/r$ moves them farther away from the origin, whereas decreasing $q/r$ moves them closer to the origin.

## AN IMPORTANT APPLICATION FOR THE KBF

Consider the system

$$\begin{aligned} \dot{\mathbf{x}}(t) &= \mathbf{F}(t)\mathbf{x}(t) + \mathbf{B}(t)\mathbf{u}(t) + \mathbf{G}(t)\mathbf{w}(t) \\ \mathbf{x}(t_0) &= \mathbf{x}_0 \end{aligned} \tag{26-72}$$

for $t \geq t_0$, where $\mathbf{x}_0$ is a random initial condition vector with mean $\mathbf{m}_\mathbf{x}(t_0)$ and covariance matrix $\mathbf{P}_\mathbf{x}(t_0)$. Measurements are given by

$$\mathbf{z}(t) = \mathbf{H}(t)\mathbf{x}(t) + \mathbf{v}(t) \tag{26-73}$$

for $t \geq t_0$. The joint random process col $[\mathbf{w}(t), \mathbf{v}(t)]$ is a white noise process with intensity

$$\begin{pmatrix} \mathbf{Q}(t) & 0 \\ 0 & \mathbf{R}(t) \end{pmatrix}, \qquad t \geq t_0$$

The *controlled variable* can be expressed as

$$\boldsymbol{\xi}(t) = \mathbf{D}(t)\mathbf{x}(t), \qquad t \geq t_0 \tag{26-74}$$

The *stochastic linear optimal output feedback regulator problem* is the problem of finding the functional

$$\mathbf{u}(t) = \mathbf{f}[\mathbf{z}(\tau), \quad t_0 \leq \tau \leq t] \tag{26-75}$$

for $t_0 \leq t \leq t_1$ such that the objective function

$$J[\mathbf{u}] = \mathbf{E}\left\{\frac{1}{2}\mathbf{x}'(t_1)\mathbf{W}_1\mathbf{x}(t_1) + \frac{1}{2}\int_{t_0}^{t}[\boldsymbol{\xi}'(\tau)\mathbf{W}_3\boldsymbol{\xi}(\tau) + \mathbf{u}'(\tau)\mathbf{W}_2\mathbf{u}(\tau)]d\tau\right\} \tag{26-76}$$

is minimized. Here $\mathbf{W}_1, \mathbf{W}_2$, and $\mathbf{W}_3$ are symmetric weighting matrices, and $\mathbf{W}_1 \geq 0, \mathbf{W}_2 > 0$, and $\mathbf{W}_3 > 0$ for $t_0 \leq t \leq t_1$.

In the control theory literature, this problem is also known as the linear-quadratic-Gaussian regulator problem (i.e., the LQG problem; see Athans, 1971, for example). We state the structure of the solution to this problem, without proof, next.

The optimal control, $\mathbf{u}^*(t)$, that minimizes $J[\mathbf{u}]$ in (26-76) is

$$\mathbf{u}^*(t) = -\mathbf{F}^0(t)\hat{\mathbf{x}}(t|t) \tag{26-77}$$

where $\mathbf{F}^0(t)$ is an optimal gain matrix, computed as

$$\mathbf{F}^0(t) = \mathbf{W}_2^{-1}\mathbf{B}(t)\mathbf{P}_c(t) \tag{26-78}$$

where $\mathbf{P}_c(t)$ is the solution of the control Riccati equation

$$\begin{aligned} -\dot{\mathbf{P}}_c(t) = {} & \mathbf{F}'(t)\mathbf{P}_c(t) + \mathbf{P}_c(t)\mathbf{F}(t) - \mathbf{P}_c(t)\mathbf{B}(t)\mathbf{W}_2^{-1}\mathbf{B}'(t)\mathbf{P}_c(t) \\ & + \mathbf{D}'(t)\mathbf{W}_3\mathbf{D}(t) \end{aligned} \tag{26-79}$$

$\mathbf{P}_c(t_1)$ given

and $\hat{\mathbf{x}}(t|t)$ is the output of a KBF, properly modified to account for the control term in the state equation; i.e.,

$$\dot{\hat{\mathbf{x}}}(t|t) = \mathbf{F}(t)\hat{\mathbf{x}}(t|t) + \mathbf{B}(t)\mathbf{u}^*(t) + \mathbf{K}(t)[\mathbf{z}(t) - \mathbf{H}(t)\hat{\mathbf{x}}(t|t)] \tag{26-80}$$

We see that the KBF plays an essential role in the solution of the LQG problem.

Observe, from (26-78), that the control gain matrix $\mathbf{F}^0(t)$ will be a constant matrix if the original system is time invariant and if $\mathbf{P}_c(t)$ becomes a constant matrix. The former implies that all the matrices in the system's description are constant matrices. The latter implies that a steady-state solution of (26-79) must exist. For conditions when this occurs, see Anderson and Moore (1979).

**EXAMPLE 26-4**

Here we examine the effect of the KBF on the following first-order system (Mendel, 1971a):

$$\frac{dx(t)}{dt} = -ax(t) + u(t) + w(t) \tag{26-81}$$

and

$$z(t) = x(t) + \nu(t) \tag{26-82}$$

where $w(t)$ and $\nu(t)$ are mutually uncorrelated white noise processes, with intensities $q$ and $r$, respectively. A Kalman–Bucy filter is used to estimate $x(t)$, and the control $u(t)$ is designed to minimize

$$\mathbf{E}\left\{\frac{1}{2}\int_0^\infty [\rho x^2(\tau) + u^2(\tau)]d\tau\right\}$$

Details of the solution to this problem are given in Athans and Falb (1965); but the solution can be obtained directly from (26-74)–(26-80). The steady-state control gain, obtained by solving for the steady-state value of $p_c(t)$, $p_c(\infty)$, as described above, is

$$p_c(\infty) = -a + (a^2 + \rho)^{1/2} \tag{26-83}$$

so that

$$u^*(t) = [a - (a^2 + \rho)^{1/2}]\hat{x}(t|t) \tag{26-84}$$

Weighting parameter $\rho$ is chosen so that

$$\lim_{t\to\infty} \text{var}\,[x(t)] \stackrel{\triangle}{=} p_x = \epsilon \tag{26-85}$$

where $\epsilon$ is a design limit that is specified ahead of time.

Steady-state variance $p_{x_1}$, computed assuming that *perfect knowledge of the states is available* (i.e., $u^*(t) = [a - (a^2 + \rho)^{1/2}]x(t)$), is obtained from (26-11) by setting $d\mathbf{P_x}(t)/dt = \mathbf{0}$. Doing this, we find that

$$p_{x_1} = \frac{q}{2(a^2 + \rho)^{1/2}} \stackrel{\triangle}{=} \epsilon \tag{26-86}$$

Steady-state variance $p_{x_2}$, computed assuming that *a steady-state KBF is in the loop* [i.e., $u^*(t) = [a - (a^2 + \rho)^{1/2}]\hat{x}(t|t)$ and $\lim_{t\to\infty} p(t|t) = \pi$], is obtained by setting the right-hand side of (26-54) equal to zero. Doing this, we find

$$p_{x_2} = \epsilon + \left[1 - \frac{a}{(a^2 + \rho)^{1/2}}\right]\pi \tag{26-87}$$

Comparing (26-86) and (26-87), we see that, unless $q = 0$, in which case $\pi = 0$, $p_x$ with the optimal filter in the loop will always be larger than its value ($\epsilon$) determined by assuming that all states are known perfectly. We might be tempted to argue that the bracketed term in (26-87) can be made negative, which would then contradict this conclusion. For this to occur, $\rho < 0$; but this violates the optimal control design requirement that $\rho > 0$ [see the sentence just below (26-76); in this example, $w_3 = \rho$]. □

## COMPUTATION

This lesson is loaded with computational possibilities. The *Control System Toolbox* contains all the M-files listed below except **eig**, which is a MATLAB M-file.

Models come in different guises, such as transfer functions, zeros and poles, and state variable. Sometimes it is useful to be able to go from one type of model to another. The following Model Conversion M-files, which are applicable for continuous-time as well as discrete-time systems, will let you do this:

**ss2tf**: State-space to transfer-function conversion.
**ss2zp**: State-space to zero–pole conversion.
**tf2ss**: Transfer-function to state-space conversion.
**tf2zp**: Transfer-function to zero–pole conversion.
**zp2tf**: Zero–pole to transfer-function conversion.
**zp2ss**: Zero–pole to state-space conversion.

Once a model has been established, it can be used to generate time responses. The following "time response" M-files will let you do this:

**lsim**: Continuous-time simulation to arbitrary inputs.
**impulse**: Unit impulse response.
**initial**: Continuous-time initial condition response.
**step**: Unit step response.

Frequency responses can be obtained from:

**bode**: Bode frequency response plot. It computes the magnitude and phase response of continuous-time LTI systems.

Eigenvalues can be obtained from:

**ddamp**: Discrete damping factors and natural frequencies.
**eig**: Eigenvalues and eigenvectors.

There is no M-file that lets us implement the full-blown time-varying state-vector covariance equation (26-11). If, however, your system is time invariant and stationary so that all the matrices associated with the basic state-variable model are constant, then the steady-state covariance matrix of $\mathbf{x}(t)$ (i.e., the solution of $\mathbf{PF}' + \mathbf{FP} + \mathbf{GQG}' = \mathbf{0}$) can be computed using:

**covar**: (Steady-state) covariance response to white noise.

There is no M-file for a full-blown Kalman–Bucy filter.

The steady-state Kalman–Bucy filter can be computed using two M-files from the *Control System Toolbox*. One (**lqe**) provides gain, covariance, and eigenvalue information about the steady-state KBF; the other (**estim**) computes the steady-state filtered values for $\mathbf{x}(t)$.

**lqe**: Linear quadratic estimator design. Computes the steady-state Kalman–Bucy gain matrix, steady-state filter covariance matrix, and closed-loop eigenvalues of the filter. Does this for the basic state-variable model.

**estim**: Forms steady-state Kalman–Bucy filter. The steady-state Kalman–Bucy gain matrix, computed by **lqe**, must be passed on to **estim**. Outputs steady-state filtered values for $\mathbf{x}(t)$, as well for the measurement vector, all as functions of time.

M-file **estim** can also be used to provide suboptimal filtered state estimates, simply by providing it with a gain matrix other than the steady-state Kalman–Bucy gain matrix.

## Supplementary Material

## PROOF OF THEOREM 26-1

(a) Formally differentiate $\mathbf{m_x}(t) = \mathbf{E}\{\mathbf{x}(t)\}$, interchanging the order of the derivative and expectation operations, to obtain

$$\frac{d\mathbf{m_x}(t)}{dt} = \mathbf{E}\left\{\frac{d\mathbf{x}(t)}{dt}\right\} = \mathbf{E}\{\mathbf{F}(t)\mathbf{x}(t) + \mathbf{G}(t)\mathbf{w}(t)\} = \mathbf{F}(t)\mathbf{m_x}(t) + \mathbf{G}(t)\mathbf{E}\{\mathbf{w}(t)\}$$
$$= \mathbf{F}(t)\mathbf{m_x}(t) \tag{26-88}$$

where we have, of course, made use of the zero-mean nature of $\mathbf{w}(t)$.

(b) Let $\Xi(t) \triangleq \mathbf{x}(t) - \mathbf{m_x}(t)$, so that

$$\mathbf{P_x}(t) = \mathbf{E}\{\Xi(t)\Xi'(t)\} \tag{26-89}$$

Differentiating (26-89), we find

$$\frac{d\mathbf{P_x}(t)}{dt} = \mathbf{E}\left\{\frac{d\Xi(t)}{dt}\Xi'(t) + \Xi(t)\frac{d\Xi'(t)}{dt}\right. \tag{26-90}$$

We leave it to the reader to show that $\Xi(t)$ satisfies the following differential equation:

$$\frac{d\Xi(t)}{dt} = \mathbf{F}(t)\Xi(t) + \mathbf{G}(t)\mathbf{w}(t) \tag{26-91}$$

Hence,

$$\frac{d\mathbf{P_x}(t)}{dt} = \mathbf{E}\{[\mathbf{F}(t)\mathbf{\Xi}(t) + \mathbf{G}(t)\mathbf{w}(t)]\mathbf{\Xi}'(t) + \mathbf{\Xi}(t)[\mathbf{F}(t)\mathbf{\Xi}(t) + \mathbf{G}(t)\mathbf{w}(t)]'\}$$
$$= \mathbf{F}(t)\mathbf{P_x}(t) + \mathbf{P_x}(t)\mathbf{F}'(t) + \mathbf{G}(t)\mathbf{E}\{\mathbf{w}(t)\mathbf{\Xi}'(t)\}$$
$$+ \mathbf{E}\{\mathbf{\Xi}(t)\mathbf{w}'(t)\}\mathbf{G}'(t) \tag{26-92}$$

To proceed further, we need the solution to (26-91); i.e. (see Theorem 23-1),

$$\mathbf{\Xi}(t) = \mathbf{\Phi}(t, t_0)\mathbf{\Xi}(t_0) + \int_{t_0}^{t} \mathbf{\Phi}(t, \tau)\mathbf{G}(\tau)\mathbf{w}(\tau)d\tau \tag{26-93}$$

where $\mathbf{\Phi}(t, \tau)$ is the transition matrix for the system in (26-91). Let us now compute $\mathbf{E}\{\mathbf{\Xi}(t)\mathbf{w}'(t)\}$:

$$\mathbf{E}\{\mathbf{\Xi}(t)\mathbf{w}'(t)\} = \mathbf{\Phi}(t, t_0)\mathbf{E}\{\mathbf{\Xi}(t_0)\mathbf{w}'(t)\} + \int_{t_0}^{t} \mathbf{\Phi}(t, \tau)\mathbf{G}(\tau)\mathbf{E}\{\mathbf{w}(\tau)\mathbf{w}'(t)\}d\tau \tag{26-94}$$

The first term in this equation is zero because $\mathbf{\Xi}(t_0)$ and $\mathbf{w}(t)$ are independent and $\mathbf{w}(t)$ is zero mean; hence,

$$\mathbf{E}\{\mathbf{\Xi}(t)\mathbf{w}'(t)\} = \int_{t_0}^{t} \mathbf{\Phi}(t, \tau)\mathbf{G}(\tau)\mathbf{Q}(\tau)\delta(t-\tau)d\tau = \mathbf{\Phi}(t, t)\mathbf{G}(t)\mathbf{Q}(t)\int_{t_0}^{t} \delta(t-\tau)d\tau$$
$$= \frac{\mathbf{G}(t)\mathbf{Q}(t)}{2} \tag{26-95}$$

where we have made use of (26-18) and the fact that $\mathbf{\Phi}(t, t) = \mathbf{I}$. Obviously, $\mathbf{E}\{\mathbf{w}(t)\mathbf{\Xi}'(t)\} = [\mathbf{E}\{\mathbf{w}(t)\mathbf{\Xi}'(t)\}]' = \mathbf{Q}(t)\mathbf{G}'(t)/2$. Substituting this result, as well as (26-95) into (26-92), we obtain the matrix differential equation for $\mathbf{P_x}(t)$ in (26-11).

(c) The proof of (26-12) uses the following solution to (26-11), which is derived on page 102 of Kwakernaak and Sivan (1972):

$$\mathbf{P_x}(t) = \mathbf{\Phi}(t, t_0)\mathbf{P_x}(t_0)\mathbf{\Phi}'(t, t_0) + \int_{t_0}^{t} \mathbf{\Phi}(t, \tau)\mathbf{G}(\tau)\mathbf{Q}(\tau)\mathbf{G}'(\tau)\mathbf{\Phi}'(t, \tau)d\tau \tag{26-96}$$

We leave the details of the proof to the reader.

(d)/(e) We leave the details of the proofs of these two parts of Theorem 26-1 to the reader.

## DERIVATION OF THE KBF WHEN THE STRUCTURE OF THE FILTER IS PRESPECIFIED

In this derivation of the KBF, we begin by assuming that the filter has the following structure,

$$\dot{\hat{\mathbf{x}}}(t|t) = \mathbf{F}(t)\hat{\mathbf{x}}(t|t) + \mathbf{K}(t)[\mathbf{z}(t) - \mathbf{H}\hat{\mathbf{x}}(t|t)] \tag{26-97}$$

Our objective is to find the matrix function $\mathbf{K}(\tau)$, $t_0 \le t \le \tau$, that minimizes the following mean-squared error:

$$J[\mathbf{K}(\tau)] = \mathbf{E}\{\mathbf{e}'(\tau)\mathbf{e}(\tau)\}, \qquad \tau \ge t_0 \tag{26-98}$$

where

$$\mathbf{e}(\tau) = \mathbf{x}(\tau) - \hat{\mathbf{x}}(\tau|\tau) \tag{26-99}$$

This optimization problem is a fixed-time, free-end-point [i.e., $\tau$ is fixed but $\mathbf{e}(\tau)$ is not fixed] problem in the calculus of variations (e.g., Kwakernaak and Sivan, 1972; Athans and Falb, 1965; and Bryson and Ho, 1969).

It is straightforward to show that $\mathbf{E}\{\mathbf{e}(\tau)\} = \mathbf{0}$, so that $\mathbf{E}\{[\mathbf{e}(\tau) - \mathbf{E}\{\mathbf{e}(\tau)\}][\mathbf{e}(\tau) - \mathbf{E}\{\mathbf{e}(\tau)\}]'\} = \mathbf{E}\{\mathbf{e}(\tau)\mathbf{e}'(\tau)\}$. Letting

$$\mathbf{P}(t|t) = \mathbf{E}\{\mathbf{e}(t)\mathbf{e}'(t)\} \tag{26-100}$$

we know that $J[\mathbf{K}(\tau)]$ can be reexpressed as

$$J[\mathbf{K}(\tau)] = \operatorname{tr}\mathbf{P}(\tau|\tau) \tag{26-101}$$

We leave it to the reader (Problem 26-17) to derive the following state equation for $\mathbf{e}(t)$, and its associated covariance equation,

$$\dot{\mathbf{e}}(t) = [\mathbf{F}(t) - \mathbf{K}(t)\mathbf{H}(t)]\mathbf{e}(t) + \mathbf{G}(t)\mathbf{w}(t) - \mathbf{K}(t)\mathbf{v}(t) \tag{26-102}$$

and

$$\begin{aligned}\dot{\mathbf{P}}(t|t) = [\mathbf{F}(t) - \mathbf{K}(t)\mathbf{H}(t)]\mathbf{P}(t|t) + \mathbf{P}(t|t)[\mathbf{F}(t) - \mathbf{K}(t)\mathbf{H}(t)]' \\ + \mathbf{G}(t)\mathbf{Q}(t)\mathbf{G}'(t) + \mathbf{K}(t)\mathbf{R}(t)\mathbf{K}'(t)\end{aligned} \tag{26-103}$$

where $\mathbf{e}(t_0) = \mathbf{x}(t_0) - \mathbf{m}_\mathbf{x}(t_0)$ and $\mathbf{P}(t_0|t_0) = \mathbf{P}_\mathbf{x}(t_0)$.

Our optimization problem for determining $\mathbf{K}(t)$ is *given the matrix differential equation (26-103), satisfied by the error-covariance matrix* $\mathbf{P}(t|t)$, *a terminal time* $\tau$, *and the cost functional* $J[\mathbf{K}(\tau)]$ *in (26-101), determine the matrix* $\mathbf{K}(t)$, $t_0 \le t \le \tau$ *that minimizes* $J[\mathbf{K}(t)]$ (Athans and Tse, 1967).

The elements $p_{ij}(t|t)$ of $\mathbf{P}(t|t)$ may be viewed as the state variables of a dynamical system, and the elements $k_{ij}(t)$ of $\mathbf{K}(t)$ may be viewed as the control variables in an optimal control problem. The cost functional is then a terminal time penalty function on the state variables $p_{ij}(t|t)$. Euler–Lagrange equations associated with a free-end-point problem can be used to determine the optimal gain matrix $\mathbf{K}(t)$.

To do this, we define a set of costate variables $\sigma_{ij}(t)$ that correspond to the $p_{ij}(t|t), i, j = 1, 2, \ldots, n$. Let $\boldsymbol{\Sigma}(t)$ be an $n \times n$ costate matrix that is associated with $\mathbf{P}(t|t)$; i.e., $\boldsymbol{\Sigma}(t) = (\sigma_{ij}(t))_{ij}$. Next, we introduce the Hamiltonian function $\mathcal{H}(\mathbf{K}, \mathbf{P}, \boldsymbol{\Sigma})$, where for notational convenience we have omitted the dependence of $\mathbf{K}$, $\mathbf{P}$, and $\boldsymbol{\Sigma}$ on $t$, and

$$\mathcal{H}(\mathbf{K}, \mathbf{P}, \boldsymbol{\Sigma}) = \sum_{i=1}^{n}\sum_{j=1}^{n} \sigma_{ij}(t)\dot{p}_{ij}(t|t) = \operatorname{tr}[\dot{\mathbf{P}}(t|t)\boldsymbol{\Sigma}'(t)] \tag{26-104}$$

Substituting (26-103) into (26-104), we see that

$$
\begin{aligned}
\mathcal{H}(\mathbf{K},\mathbf{P},\mathbf{\Sigma}) &= \operatorname{tr}[\mathbf{F}(t)\mathbf{P}(t|t)\mathbf{\Sigma}'(t)] \\
&= -\operatorname{tr}[\mathbf{K}(t)\mathbf{H}(t)\mathbf{P}(t|t)\mathbf{\Sigma}'(t)] \\
&= +\operatorname{tr}[\mathbf{P}(t|t)\mathbf{F}'(t)\mathbf{\Sigma}'(t)] \\
&= -\operatorname{tr}[\mathbf{P}(t|t)\mathbf{H}'(t)\mathbf{K}'(t)\mathbf{\Sigma}'(t)] \\
&= +\operatorname{tr}[\mathbf{G}(t)\mathbf{Q}(t)\mathbf{G}'(t)\mathbf{\Sigma}'(t)] \\
&= +\operatorname{tr}[\mathbf{K}(t)\mathbf{R}(t)\mathbf{K}'(t)\mathbf{\Sigma}'(t)]
\end{aligned}
\tag{26-105}
$$

The Euler–Lagrange equations for our optimization problem are

$$\left.\frac{\partial\mathcal{H}(\mathbf{K},\mathbf{P},\mathbf{\Sigma})}{\partial\mathbf{K}}\right|_* = \mathbf{0} \tag{26-106}$$

$$\dot{\mathbf{\Sigma}}^*(t) = -\left.\frac{\partial\mathcal{H}(\mathbf{K},\mathbf{P},\mathbf{\Sigma})}{\partial\mathbf{P}}\right|_* \tag{26-107}$$

$$\dot{\mathbf{P}}^*(t|t) = \left.\frac{\partial\mathcal{H}(\mathbf{K},\mathbf{P},\mathbf{\Sigma})}{\partial\mathbf{\Sigma}}\right|_* \tag{26-108}$$

and

$$\mathbf{\Sigma}^*(\tau) = \frac{\partial}{\partial\mathbf{P}}\operatorname{tr}\mathbf{P}(\tau|\tau) \tag{26-109}$$

In these equations, starred quantities denote optimal quantities, and $|_*$ denotes the replacement of $\mathbf{K}, \mathbf{P}$, and $\mathbf{\Sigma}$ by $\mathbf{K}^*, \mathbf{P}^*$, and $\mathbf{\Sigma}^*$ *after* the appropriate derivative has been calculated. Note, also, that the derivatives of $\mathcal{H}(\mathbf{K}, \mathbf{P}, \mathbf{\Sigma})$ are derivatives of a scalar quantity with respect to a matrix (e.g., $\mathbf{K}$, $\mathbf{P}$, or $\mathbf{\Sigma}$). The calculus of gradient matrices (e.g., Schweppe, 1974; Athans and Schweppe, 1965) can be used to evaluate these derivatives.

Let $\mathbf{X}$ be an $r \times n$ matrix with elements $x_{ij}(i = 1, 2, \ldots, r$ and $j = 1, 2, \ldots, n)$. Let $f(\mathbf{X})$ be a scalar, real-valued function of the $x_{ij}$, i.e., $f(\mathbf{X}) = f(x_{11}, x_{12}, \ldots, x_{1n}, x_{21}, \ldots, x_{2n}, \cdots, x_{r1}, \ldots, x_{rn})$. The *gradient matrix* of $f(\mathbf{X})$ is the $r \times n$ matrix $\partial f(\mathbf{X})/\partial\mathbf{X}$ whose $ij$th element is given by $\partial f(\mathbf{X})/\partial x_{ij}$ for $i = 1, 2, \ldots, r$ and $j = 1, 2, \ldots, n$.

$f(\mathbf{X})$ is a *trace function* of matrix $\mathbf{X}$ if $f(\mathbf{X}) = \operatorname{tr}[\mathbf{F}(\mathbf{X})]$, where $\mathbf{F}(\mathbf{X})$ is a continuously differentiable mapping from the space of $r \times n$ matrices into the space of $n \times n$ matrices. $\mathbf{F}(\mathbf{X})$ is a square matrix, so its trace is well-defined. Some useful gradients of trace functions, which are needed to evaluate (26-106)–(26-109) are

$$\frac{\partial\operatorname{tr}[\mathbf{X}]}{\partial\mathbf{X}} = \mathbf{I} \tag{26-110}$$

$$\frac{\partial\operatorname{tr}[\mathbf{AXB}]}{\partial\mathbf{X}} = \mathbf{A}'\mathbf{B}' \tag{26-111}$$

$$\frac{\partial\operatorname{tr}[\mathbf{AX}'\mathbf{B}]}{\partial\mathbf{X}} = \mathbf{BA} \tag{26-112}$$

and

$$\frac{\partial \text{tr}\,[\mathbf{AXBX}']}{\partial \mathbf{X}} = \mathbf{A}'\mathbf{XB}' + \mathbf{AXB} \tag{26-113}$$

Applying (26-110)–(26-113) to (26-106)–(26-109), using (26-105), we obtain

$$-\mathbf{\Sigma}^*\mathbf{P}^{*\,\prime}\mathbf{H}' - \mathbf{\Sigma}^{*\,\prime}\mathbf{P}^*\mathbf{H}' + \mathbf{\Sigma}^*\mathbf{K}^*\mathbf{R} + \mathbf{\Sigma}^{*\,\prime}\mathbf{K}^*\mathbf{R} = \mathbf{0} \tag{26-114}$$

$$\dot{\mathbf{\Sigma}}^* = -\mathbf{\Sigma}^*(\mathbf{F} - \mathbf{K}^*\mathbf{H}) - (\mathbf{F} - \mathbf{K}^*\mathbf{H})'\mathbf{\Sigma}^* \tag{26-115}$$

$$\dot{\mathbf{P}}^* = (\mathbf{F} - \mathbf{K}^*\mathbf{H})\mathbf{P}^* + \mathbf{P}^*(\mathbf{F} - \mathbf{K}^*\mathbf{H})' + \mathbf{GQG}' + \mathbf{K}^*\mathbf{RK}^{*\,\prime} \tag{26-116}$$

and

$$\mathbf{\Sigma}^*(\tau) = \mathbf{I} \tag{26-117}$$

Our immediate objective is to obtain an expression for $\mathbf{K}^*(t)$.

**Fact.** *Matrix* $\mathbf{\Sigma}^*(\text{t})$ *is symmetric and positive definite.* □

We leave the proof of this fact as an exercise for the reader (Problem 26-18). Using this fact, and the fact that covariance matrix $\mathbf{P}^*(t)$ is symmetric, we are able to express (26-114) as

$$2\mathbf{\Sigma}^*(\mathbf{K}^*\mathbf{R} - \mathbf{P}^*\mathbf{H}') = \mathbf{0} \tag{26-118}$$

Because $\mathbf{\Sigma}^* > 0$, $(\mathbf{\Sigma}^*)^{-1}$ exists, so (26-118) has for its only solution

$$\mathbf{K}^*(t) = \mathbf{P}^*(t|t)\mathbf{H}'(t)\mathbf{R}^{-1}(t) \tag{26-119}$$

which is the Kalman–Bucy gain matrix stated in Theorem 26-2. To obtain the covariance equation associated with $\mathbf{K}^*(t)$, substitute (26-119) into (26-116). The result is (26-23).

This completes the derivation of the KBF when the structure of the filter is prespecified.

## SUMMARY QUESTIONS

1. In this lesson we derive the Kalman–Bucy filter in which two ways:
   **(a)** formal limiting procedure
   **(b)** calculus of variations
   **(c)** prespecify filter structure
2. The KBF is a:
   **(a)** linear time-invariant differential equation
   **(b)** linear time-varying differential equation
   **(c)** nonlinear differential equation
3. The matrix Riccati differential equation is:
   **(a)** asymmetric
   **(b)** linear
   **(c)** nonlinear

4. Matrices $\mathbf{Q}(t)$ and $\mathbf{R}(t)$ are called:
   **(a)** covariance matrices
   **(b)** intensity matrices
   **(c)** noise matrices
5. The first- and second-order statistics of the state vector are described by:
   **(a)** two vector differential equations
   **(b)** a vector differential equation and a matrix differential equation, respectively
   **(c)** two matrix differential equations
6. When a control system is designed in which the controller is a signal that is proportional to the output of a Kalman–Bucy filter:
   **(a)** state estimation errors act as an additional plant disturbance, but these disturbances do not have to be accounted for during the overall design
   **(b)** we can optimize the control gains by assuming that there are no state estimation errors
   **(c)** state estimation errors act as an additional plant disturbance, and these disturbances must be accounted for during the overall design

## PROBLEMS

**26-1.** Explain the replacement of covariance matrix $\mathbf{R}(k+1 = t + \Delta t)$ by $\mathbf{R}(t+\Delta t)/\Delta t$ in (26-38).

**26-2.** Derive Equation (26-47).

**26-3.** Derive a continuous-time useful fixed-interval smoother using a formal limiting procedure.

**26-4.** Our derivation of the KBF using a formal limiting procedure seems to depend heavily on (26-33) and (26-34). These results were derived in Example 23-3 and depended on truncating the matrix exponential at two terms [see (23-49)]. Show that we obtain the same KBF even if we include a third term in the truncation of the matrix exponential. Obviously, this generalizes to any truncation of the matrix exponential.

**26-5.** Repeat Example 26-4 for the following system: $d^2x(t)/dt^2 = u(t) + w(t), z(t) = x(t) + v(t)$, where $w(t)$ and $v(t)$ are mutually uncorrelated white noise processes, with intensities $q$ and $r$, respectively. Choose the control, $u(t)$, to minimize,

$$\mathbf{E}\left\{\frac{1}{2}\int_0^\infty [x^2(\tau) + 2bx(\tau)\frac{dx(\tau)}{d\tau} + \rho\left(\frac{dx(\tau)}{d\tau}\right)^2 + u^2(\tau)]d\tau\right\}$$

**26-6.** Repeat Example 26-4 for the following system: $d^2x(t)/dt^2 = -ax(t) + u(t) + w(t), z(t) = x(t) + v(t)$, where $w(t)$ and $v(t)$ are mutually uncorrelated white noise processes, with intensities $q$ and $r$, respectively. Choose the control $u(t)$ to minimize

$$\mathbf{E}\left\{\frac{1}{2}\int_0^\infty [\rho x^2(\tau) + u^2(\tau)]d\tau\right\}$$

**26-7.** (Bruce Rebhan, Spring 1992) You are surprised one morning to find the CEO of Major Defense Corporation standing in your doorway. "We have a problem," he says. "I'm supposed to give the final presentation on the new attack fighter at ten o'clock and I don't know how the inertial navigation system (INS) operates. You're the only one in the whole company who understands it." His sweaty palms and plaintive eyes remind you of a begging dog. The chairman gulps nervously. "Can you help me?"

You know that the following continuous-time system attempts to model the INS calculations of the attack fighter that you are helping to design:

$$\frac{dx}{dt} = f(t)x(t) + g(t)w(t)$$

$$z(t) = h(t)x(t) + v(t)$$

where $h(t) = 1/2, g(t) = 0, q(t) = 1, x(t_0) = 0$, and $p_x(t_0) = 0$. $x(t)$ is the state vector of the INS, while $z(t)$ is a measurement of ownship velocity. It is important to produce a continuous-time filter to model this system, and so you decide to employ the Kalman–Bucy filter.

Determine the closed-form KBF for the following cases, and give a macroscopic description of the system for each case:

**(a)** $r(t) = 1/4$ and $f(t) = 0$.

**(b)** $r(t) \gg 0$ and $f(t) = 1$.

**26-8.** (Kwan Tjia, Spring 1992) Rederive Theorem 26-2 when state equation (26-1) includes a known forcing function, $\mathbf{C}(t)\mathbf{u}(t)$.

**26-9.** (George Semeniuk, Spring 1992) (Computional Problem) Example 26-3 effectively models an inertial position measurement device, where the input is true acceleration and the process noise is the result of the acceleration sensing dynamics. Suppose the process noise, $w(t)$, is colored instead of white; i.e.,

$$\tau \frac{dw(t)}{dt} + w(t) = v(t)$$

where $\tau = 20$ and $v(t)$ is white noise with $r = 1$; additionally, $q = 1$.

A certain engineer attempts to derive the closed-form steady-state KBF variances for the augmented state-variable model (now $3 \times 1$). He quickly realizes that he cannot do this and hypothesizes the following simplified, but totally incorrect, solution.

His hypothesis is that at steady state the problem can be separated into two components. The first component is the colored noise filter and the second component is the two integrators. He claims that the steady-state variance of the colored noise filter drives the system of two integrators, so the new steady-state noise variance can be substituted into the already obtained closed-form solution and voilà!

**(a)** Show, by means of simulation, that this hypothesis is wrong.

**(b)** Explain, theoretically, the error in his thinking.

**26-10.** (Kiet D. Ngo, Spring, 1992) In linear quadratic Gaussian (LQG) optimal design problems, the states of the controlled plant are fed back to implement the optimal control law. Frequently, in practice, not all the states of the plant can be measured. In this situation, the KBF plays an essential role in the implementation of the LQG controller. Figure P26-10 illustrates the basic interconnection of plant and controller in the LQG problem. Consider the following first-order plant:

$$\frac{dx(t)}{dt} = -x(t) + u(t) + v(t), \qquad x(0) = 1$$

$$y(t) = x(t) + w(t)$$

where $v(t)$ and $w(t)$ are independent stationary Gaussian white noises with intensities $q$ and $r$, respectively.

**(a)** Design a KBF for this system. (*Hint:* $\int_a^b dx/x = \ln x|_a^b$.)

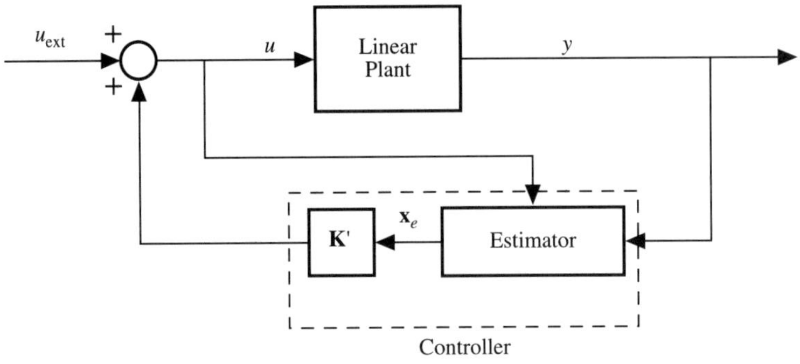


**Figure P26-10**

**(b)** Find the steady-state state-error variance $p(\infty)$.

**(c)** Express the steady-state gain $k(\infty)$ as a function of $q/r$.

**26-11.** (William A. Emanuelsen, Spring 1992) To use a Kalman–Bucy filter, we must find a solution of the matrix Riccati differential equation. Show how to obtain a solution using the substitution $\mathbf{P}(t|t) = \mathbf{X}(t|t)\mathbf{Z}^{-1}(t|t)$. Is this solution unique?

**26-12.** (Richard S. Lee, Jr., Spring 1992) Consider the following continuous-time state-variable model:

$$\frac{dx(t)}{dt} = -x(t) + w(t)$$
$$z(t) = x(t) + \nu(t)$$

where $\nu(t)$ and $w(t)$ are zero-mean, mutually uncorrelated, white noise processes, with $q = r = 1$. Given that $p(0|0) = 1$, show that

$$p(t|t) = \frac{\rho_1 - \rho_2\alpha \exp(-2\sqrt{2}t)}{1 - \alpha \exp(-2\sqrt{2}t)}$$

where $\rho_1 = -1 + \sqrt{2}$, $\rho_2 = -1 - \sqrt{2}$, and $\alpha = (1 - \rho_1)/(1 - \rho_2)$.

**26-13.** (Lance Kaplan, Spring 1992) Example 18-5 illustrates the "divergence phenomenon" of the Kalman filter when the process noise is nonexistent. We will now examine an example where the KBF suffers from the same phenomenon. Expressions similar to Eqs. (18-15)–(18-18) are set up in continuous time.

Suppose the true system is:

$$\frac{dx(t)}{dt} = b \tag{26-13.1}$$

$$z(t) = x(t) + \nu(t) \tag{26-13.2}$$

where $b$ is small, so small that we choose to ignore it in the KBF design. Consequently, the KBF is based on

$$\frac{dx_m(t)}{dt} = 0 \tag{26-13.3}$$

$$z(t) = x_m(t) + \nu(t) \tag{26-13.4}$$

Equations (26-13.1)–(26-13.4) are analogous to Eqs. (18-15)–(18-18).

**(a)** Show that $d\hat{x}_m(t|t)/dt = \{p(0)/[tp(0) + r]\}[z(t) - \hat{x}_m(t|t)]$.

**(b)** Let $\tilde{x}(t|t) = x(t) - \hat{x}_m(t|t)$ represent the KBF error. Show that

$$\tilde{x}(t|t) = \{1/[tp(0) + r]\}\{r\tilde{x}(0|0) - p(0)\int_0^t \nu(\tau)d\tau + [t^2/2 \cdot p(0) + tr]b\}$$

**(c)** What happens to the KBF error as $t$ gets large?

**(d)** Find an expression for $\tilde{x}_m(t|t) = x_m(t) - \hat{x}_m(t|t)$ in terms of $\tilde{x}(0|0)$ and $\nu(t)$.

**(e)** What happens to $\tilde{x}_m(t|t)$ as $t$ gets large?

**26-14.** (Iraj Manocheri, Spring 1992) Given the system block diagram in Figure P26-14, in which $\nu(t)$ is white and Gaussian with intensity $N_0/2$, and $u(t)$ is also white and Gaussian with unity intensity. Derive the KBF estimate of $y(t)$ and draw the filter's block diagram.

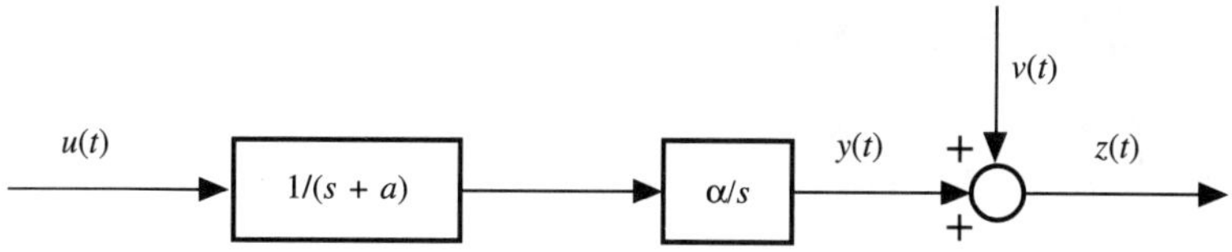


**Figure P26-14** System block diagram.

**26-15.** In this problem, you are asked to fill in some of the missing details in the proof of Theorem 26-1.

**(a)** Derive the differential equation for $\Xi(t)$ in (26-91).

**(b)** Derive (26-12) using (26-96).

**(c)** Derive (26-13) and (26-14).

**26-16.** Another way to obtain the results in parts (a) to (c) of Theorem 26-1 is by using a formal limiting procedure, as we did in the derivation of the KBF. Derive (26-10) and (26-11) using such a procedure.

**26-17.** Derive the state equation for error $\mathbf{e}(t)$, given in (26-102), and its associated covariance equation (26-103).

**26-18.** Prove that matrix $\mathbf{\Sigma}^*(t)$ is symmetric and positive definite.

# Sufficient Statistics and Statistical Estimation of Parameters

## SUMMARY

The concept of sufficient statistic is an elegant tool in estimation, hypothesis testing, and decision theories. Sufficient statistic is a compaction of the information contained in the original data so that estimation using the sufficient statistic is as good as using the entire original data. The dimension of the sufficient statistic is much less than the original data and is also independent of it.

A formal definition of a sufficient statistic is given in this lesson; however, it is not very easy to use. The *factorization theorem* gives a direct method for the identification of a sufficient statistic from the likelihood of observations. A general structure is given for the family of *exponential distributions.* Familiar distributions such as Gaussian, beta, binomial, Rayleigh, and gamma distributions are members of the exponential family. A nice feature of this family is that the sufficient statistic can be easily identified using the factorization theorem.

If the likelihood belongs to the exponential family, more elegant results can be obtained. For example, under some existence conditions for solutions to a set of algebraic equations, the maximum-likelihood estimator can be directly computed from the sufficient statistic without having to go through the differentiation of the likelihood function.

The concept of sufficient statistic is also useful for the computation of *uniformly minimum variance unbiased* (UMVU) estimates. The UMVU estimator is more general than BLUE, because UMVU estimates are not restricted to be linear. In general, UMVU estimates are difficult to compute; but, when a sufficient statistic exists, elegant methods for the computation of UMVU estimates are available. To accomplish this, in addition to being sufficient, the statistic must also be *complete*. The *Lehmann–Scheffe theorem* provides a method for computing the UMVU estimator for a scalar parameter;

this theorem is also extended to the vector parameter case. Both the Lehmann–Scheffe theorem and its extension use the notions of sufficient and complete statistics.

All the lesson's important points are illustrated by examples.

When you complete this lesson, you will be able to (1) describe the concept of a sufficient statistic; (2) compute the sufficient statistic using the factorization theorem; (3) compute the maximum likelihood estimate using the sufficient statistic; and (4) compute the UMVU estimator using sufficient and complete statistics.

## INTRODUCTION

In this lesson,* we discuss the usefulness of the notion of sufficient statistics in statistical estimation of parameters. Specifically, we discuss the role played by sufficient statistics and exponential families in maximum-likelihood and uniformly minimum-variance unbiased (UMVU) parameter estimation.

## CONCEPT OF SUFFICIENT STATISTICS

The notion of a sufficient statistic can be explained intuitively (Ferguson, 1967), as follows. We observe $\mathbf{Z}(N)$ ($\mathbf{Z}$ for short), where $\mathbf{Z} = \text{col}\,[\mathbf{z}(1), \mathbf{z}(2), \ldots, \mathbf{z}(N)]$, in which $\mathbf{z}(1), \ldots, \mathbf{z}(N)$ are independent and identically distributed random vectors, each having a density function $p(\mathbf{z}(i)|\boldsymbol{\theta})$, where $\boldsymbol{\theta}$ is unknown. Often the information in $\mathbf{Z}$ can be represented equivalently in a statistic, $T(\mathbf{Z})$, whose dimension is independent of $N$, such that $T(\mathbf{Z})$ contains all the information about $\boldsymbol{\theta}$ that is originally in $\mathbf{Z}$. Such a statistic is known as a *sufficient statistic*.

**EXAMPLE A-1**

Consider a sampled sequence of $N$ manufactured cars. For each car we record whether it is defective or not. The observed sample can be represented as $\mathbf{Z} = \text{col}\,[z(1), \ldots, z(N)]$, where $z(i) = 0$ if the $i$th car is not defective and $z(i) = 1$ if the $i$th car is defective. The total number of observed defective cars is

$$T(\mathbf{Z}) = \sum_{i=1}^{N} z(i)$$

This is a statistic that maps many different values of $z(1), \ldots, z(N)$ into the same value of $T(\mathbf{Z})$. It is intuitively clear that, if we are interested in estimating the proportion $\theta$ of defective cars, nothing is lost by simply recording and using $T(\mathbf{Z})$ in place of $z(1), \ldots, z(N)$. The particular sequence of ones and zeros is irrelevant. Thus, as far as estimating the proportion of defective cars, $T(\mathbf{Z})$ contains all the information contained in $\mathbf{Z}$. □

An advantage associated with the concept of a sufficient statistic is dimensionality reduction. In Example A-1, the dimensionality reduction is from $N$ to 1.

---

* This lesson was written by Dr. Rama Chellappa, Center for Automation Research, University of Maryland, College Park, Maryland.

**Definition A-1.** *A statistic* T(**Z**) *is sufficient for vector parameter* $\boldsymbol{\theta}$, *if and only if the distribution of* **Z**, *conditioned on* T(**Z**) = t, *does not involve* $\boldsymbol{\theta}$. □

**EXAMPLE A-2**

This example illustrates the application of Definition A-1 to identify a sufficient statistic for the model in Example A-1. Let $\theta$ be the probability that a car is defective. Then $z(1), z(2), \ldots, z(N)$ is a record of $N$ Bernoulli trials with probability $\theta$; thus, $\Pr(t) = \theta^t(1-\theta)^{N-t}, 0 < \theta < 1$, where $t = \sum_{i=1}^N z(i)$, and $z(i) = 1$ or 0. The conditional distribution of $z(1), \ldots, z(N)$, given $\sum_{i=1}^N z(i) = t$, is

$$\mathbf{P}[\mathbf{Z}|T = t] = \frac{\mathbf{P}[\mathbf{Z}, T = t]}{\mathbf{P}(T = t)} = \frac{\boldsymbol{\theta}^t(1-\theta)^{N-t}}{\binom{N}{t}\theta^t(1-\theta)^{N-t}} = \frac{1}{\binom{N}{t}}$$

which is independent of $\theta$; hence, $T(\mathbf{Z}) = \sum_{i=1}^N z(i)$ is sufficient. Any one-to-one function of $T(\mathbf{Z})$ is also sufficient. □

This example illustrates that deriving a sufficient statistic using Definition A-1 can be quite difficult. An equivalent definition of sufficiency, which is easy to apply, is given in the following:

**Theorem A-1** (Factorization Theorem). *A necessary and sufficient condition for* T(**Z**) *to be sufficient for* $\boldsymbol{\theta}$ *is that there exists a factorization*

$$p(\mathbf{Z}|\boldsymbol{\theta}) = g(T(\mathbf{Z}), \boldsymbol{\theta})h(\mathbf{Z}) \tag{A-1}$$

*where the first factor in (A-1) may depend on* $\boldsymbol{\theta}$, *but depends on* **Z** *only through* T(**Z**), *whereas the second factor is independent of* $\boldsymbol{\theta}$. □

The proof of this theorem is given in Ferguson (1967) for the continuous case and Duda and Hart (1973) for the discrete case.

**EXAMPLE A-3** (Continuation of Example A-2)

In Example A-2, the probability distribution of samples $z(1), \ldots, z(N)$ is

$$\mathbf{P}[\mathbf{Z}|\theta] = \theta^t(1-\theta)^{N-t} \tag{A-2}$$

where the total number of defective cars is $t = \sum_{i=1}^N z(i)$ and $z(i)$ is either 0 or 1. Equation (A-2) can be written equivalently as

$$\mathbf{P}[\mathbf{Z}|\theta] = \exp\left[t \ln \frac{\theta}{1-\theta} + N \ln(1-\theta)\right] \tag{A-3}$$

Comparing (A-3) with (A-1), we conclude that

$$h(\mathbf{Z}) = 1$$

$$g(T(\mathbf{Z}), \theta) = \exp\left[t \ln \frac{\theta}{1-\theta} + N \ln(1-\theta)\right]$$

and

$$T(\mathbf{Z}) = t = \sum_{i=1}^{N} z(i)$$

Using the Factorization Theorem, it was easy to determine $T(\mathbf{Z})$. □

**EXAMPLE A-4**

Let $\mathbf{Z} = \text{col}\,[z(1), \ldots, z(N)]$ be a random sample drawn from a univariate Gaussian distribution, with unknown mean $\mu$ and known variance $\sigma^2 > 0$. Then

$$p(\mathbf{Z}) = \exp\left[\frac{\mu}{\sigma^2}\sum_{i=1}^{N} z(i) - \frac{N\mu^2}{2\sigma^2}\right] h(\mathbf{Z})$$

where

$$h(\mathbf{Z}) = \exp\left[-\frac{N}{2}\ln 2\pi\sigma^2 - \frac{1}{2\sigma^2}\sum_{i=1}^{N} z^2(i)\right]$$

Based on the factorization theorem, we identify $T(\mathbf{Z}) = \sum_{i=1}^{N} z(i)$ as a sufficient statistic for $\mu$. □

Because the concept of sufficient statistics involves reduction of data, it is worthwhile to know how far such a reduction can be done for a given problem. The dimension of the smallest set of statistics that is still sufficient for the parameters is called a *minimal sufficient statistic*. See Barankin and Katz (1959) for techniques useful in identifying a minimal sufficient statistic.

## EXPONENTIAL FAMILIES OF DISTRIBUTIONS

It is of interest to study families of distributions, $p(z(i)|\boldsymbol{\theta})$ for which, irrespective of the sample size $N$, there exists a sufficient statistic of fixed dimension. The exponential families of distributions have this property. For example, the family of normal distributions $N(\mu, \sigma^2)$, with $\sigma^2$ known and $\mu$ unknown, is an exponential family that, as we have seen in Example A-4, has a one-dimensional sufficient statistic for $\mu$ that is equal to $\sum_{i=1}^{N} z(i)$. As Bickel and Doksum (1977) state:

**Definition A-2** (Bickel and Doksum, 1977). *If there exist real-valued functions* $\mathbf{a}(\boldsymbol{\theta})$ *and* $\mathrm{b}(\boldsymbol{\theta})$ *on parameter space* $\boldsymbol{\Theta}$*, and real-valued vector functions* $\mathbf{T}(\mathbf{z})$[$\mathbf{z}$ *is short for* $\mathbf{z}(\mathrm{i})$] *and* $\mathrm{h}(\mathbf{z})$ on $\mathbf{R}^{\mathrm{N}}$, *such that the density function* $\mathrm{p}(\mathbf{z}|\boldsymbol{\theta})$ *can be written as*

$$p(\mathbf{z}|\boldsymbol{\theta}) = \exp[\mathbf{a}'(\boldsymbol{\theta})\mathbf{T}(\mathbf{z}) + b(\boldsymbol{\theta}) + h(\mathbf{z})] \tag{A-4}$$

*then* $\mathrm{p}(\mathbf{z}|\boldsymbol{\theta}), \boldsymbol{\theta} \in \boldsymbol{\Theta}$, *is said to be a one-parameter exponential family of distributions.* □

The Gaussian, binomial, beta, Rayleigh, and gamma distributions are examples of such one-parameter exponential families.

In a one-parameter exponential family, $\mathbf{T}(\mathbf{z})$ is sufficient for $\boldsymbol{\theta}$. The family of distributions obtained by sampling from one-parameter exponential families is also a one-parameter exponential family. For example, suppose that $\mathbf{z}(1), \ldots, \mathbf{z}(N)$ are independent and identically distributed with common density $p(\mathbf{z}|\theta)$; then

$$p(\mathbf{Z}|\theta) = \exp\left[\mathbf{a}'(\theta)\sum_{i=1}^{N}\mathbf{T}(\mathbf{z}(i)) + Nb(\theta) + \sum_{i=1}^{N}h(\mathbf{z}(i))\right] \tag{A-5}$$

By Theorem A-1, the sufficient statistic $\mathbf{T}(\mathbf{Z})$ for this situation is

$$\mathbf{T}(\mathbf{Z}) = \sum_{i=1}^{N}\mathbf{T}(\mathbf{z}(i))$$

**EXAMPLE A-5**

Let $\mathbf{z}(1), \ldots, \mathbf{z}(N)$ be a random sample from a multivariate Gaussian distribution (see Lesson 12) with unknown $d \times 1$ mean vector $\boldsymbol{\mu}$ and known covariance matrix $\mathbf{P}_\mu$. Then [$\mathbf{z}$ is short for $\mathbf{z}(i)$]

$$p(\mathbf{z}|\boldsymbol{\mu}) = \exp[\mathbf{a}'(\boldsymbol{\mu})\mathbf{T}(\mathbf{z}) + b(\boldsymbol{\mu}) + h(\mathbf{z})]$$

where

$$\begin{aligned}\mathbf{a}'(\boldsymbol{\mu}) &= \boldsymbol{\mu}'\mathbf{P}_\mu^{-1}\\ b(\boldsymbol{\mu}) &= -\frac{1}{2}\boldsymbol{\mu}'\mathbf{P}_\mu^{-1}\boldsymbol{\mu}\\ \mathbf{T}(\mathbf{z}) &= \mathbf{z}\end{aligned}$$

and

$$h(\mathbf{z}) = \exp\left[-\frac{1}{2}\mathbf{z}'\mathbf{P}_\mu^{-1}\mathbf{z} - \frac{d}{2}\ln 2\pi - \frac{1}{2}\ln\det\mathbf{P}_\mu\right]$$

Additionally [(see (A-5)],

$$p(\mathbf{Z}|\boldsymbol{\mu}) = \exp[\boldsymbol{\mu}'\mathbf{P}_\mu^{-1}\mathbf{T}(\mathbf{Z}) + Nb(\boldsymbol{\mu}) + h(\mathbf{Z})]$$

where

$$\mathbf{T}(\mathbf{Z}) = \sum_{i=1}^{N}\mathbf{z}(i)$$

and

$$h(\mathbf{Z}) = \exp\left[-\frac{1}{2}\sum_{i=1}^{N}\mathbf{z}'(i)\mathbf{P}_\mu^{-1}\mathbf{z}(i) - \frac{Nd}{2}\ln 2\pi - \frac{N}{2}\ln\det\mathbf{P}_\mu\right] \quad \square$$

The notion of a one-parameter exponential family of distributions as stated in Bickel and Doksum (1977) can easily be extended to $m$ parameters and vector observations in a straightforward manner.

**Definition A-3.** *If there exist real matrices* $\mathbf{A}_1, \ldots, \mathbf{A}_m$, *a real function* b *of* $\theta$, *where* $\theta \in \Theta$, *real matrices* $\mathbf{T}_i(\mathbf{z})$ *and a real function* $h(\mathbf{z})$, *such that the density function* $p(\mathbf{z}|\theta)$ *can be written as*

$$p(\mathbf{z}|\boldsymbol{\theta}) = \exp\left\{\text{tr}\left[\sum_{i=1}^{m} \mathbf{A}_i(\boldsymbol{\theta})\mathbf{T}_i(\mathbf{z})\right]\right\} \exp\{b(\boldsymbol{\theta}) + h(\mathbf{z})\} \tag{A-6}$$

*then* $p(\mathbf{z}|\boldsymbol{\theta})$, $\boldsymbol{\theta} \in \boldsymbol{\Theta}$, *is said to be an* m*-parameter exponential family of distributions.* □

**EXAMPLE A-6**

The family of $d$-variate normal distributions $N(\boldsymbol{\mu}, \mathbf{P}_\mu)$, where both $\boldsymbol{\mu}$ and $\mathbf{P}_\mu$ are unknown, is an example of a two-parameter exponential family in which $\boldsymbol{\theta}$ contains $\boldsymbol{\mu}$ and the elements of $\mathbf{P}_\mu$. In this case

$$\begin{aligned}
\mathbf{A}_1(\boldsymbol{\theta}) &= \mathbf{a}_1(\boldsymbol{\theta}) = \mathbf{P}_\mu^{-1}\boldsymbol{\mu} \\
\mathbf{T}_1(\mathbf{z}) &= \mathbf{z}' \\
\mathbf{A}_2(\boldsymbol{\theta}) &= -\frac{1}{2}\mathbf{P}_\mu^{-1} \\
\mathbf{T}_2(\mathbf{z}) &= \mathbf{z}\mathbf{z}' \\
b(\boldsymbol{\theta}) &= -\frac{1}{2}\boldsymbol{\mu}'\mathbf{P}_\mu^{-1}\boldsymbol{\mu} - \frac{d}{2}\ln 2\pi - \frac{1}{2}\ln\det\mathbf{P}_\mu
\end{aligned}$$

and

$$h(\mathbf{z}) = 0 \quad \square$$

As is true for a one-parameter exponential family of distributions, if $\mathbf{z}(1), \ldots, \mathbf{z}(N)$ are drawn randomly from an $m$-parameter exponential family, then $p[\mathbf{z}(1), \ldots, \mathbf{z}(N)|\boldsymbol{\theta}]$ form an $m$-parameter exponential family with sufficient statistics $\mathbf{T}_1(\mathbf{Z}), \ldots, \mathbf{T}_m(\mathbf{Z})$, where $\mathbf{T}_i(\mathbf{Z}) = \sum_{j=1}^{N} \mathbf{T}_i[\mathbf{z}(j)]$.

## EXPONENTIAL FAMILIES AND MAXIMUM-LIKELIHOOD ESTIMATION

Let us consider a vector of unknown parameters $\boldsymbol{\theta}$ that describes a collection of $N$ independent and identically distributed observations $\mathbf{Z} = \text{col}\,[\mathbf{z}(1), \ldots, \mathbf{z}(N)]$. The maximum-likelihood estimate (MLE) of $\boldsymbol{\theta}$ is obtained by maximizing the likelihood of $\boldsymbol{\theta}$ given the observations $\mathbf{Z}$. Likelihood is defined in Lesson 11 to be proportional to the value of the probability density of the observations, given the parameters; i.e.,

$$l(\boldsymbol{\theta}|\mathbf{Z}) \propto p(\mathbf{Z}|\boldsymbol{\theta})$$

As discussed in Lesson 11, a sufficient condition for $l(\boldsymbol{\theta}|\mathbf{Z})$ to be maximized is

$$\mathbf{J}_o(\hat{\boldsymbol{\theta}}_{ML}|\mathbf{Z}) < 0 \tag{A-7}$$

where

$$\mathbf{J}_o(\hat{\boldsymbol{\theta}}_{ML}|\mathbf{Z}) = \left(\frac{\partial^2 L(\boldsymbol{\theta}|\mathbf{Z})}{\partial\theta_i\partial\theta_j}\right)_{\boldsymbol{\theta}=\hat{\boldsymbol{\theta}}_{ML}}, \qquad i, j = 1, 2, \ldots, n$$

and $L(\theta|\mathbf{Z}) = \ln l(\theta|\mathbf{Z})$. Maximum-likelihood estimates of $\theta$ are obtained by solving the system of $n$ equations

$$\frac{\partial L(\theta|\mathbf{Z})}{\partial \theta_i} = 0, \qquad i = 1, 2, \ldots, n \tag{A-8}$$

for $\hat{\theta}_{ML}$ and checking whether the solution to (A-8) satisfies (A-7).

When this technique is applied to members of exponential families, $\hat{\theta}_{ML}$ can be obtained by solving a set of algebraic equations. The following theorem paraphrased from Bickel and Doksum (1977) formalizes this technique for vector observations.

**Theorem A-2** (Bickel and Doksum, 1977). *Let* $p(\mathbf{z}|\theta) = \exp[\mathbf{a}'(\theta)\mathbf{T}(\mathbf{z}) + b(\theta) + h(\mathbf{z})]$ *and let* $\mathcal{A}$ *denote the interior of the range of* $\mathbf{a}(\theta)$. *If the equation*

$$\mathbf{E}_{\theta}\{\mathbf{T}(\mathbf{z})\} = \mathbf{T}(\mathbf{z}) \tag{A-9}$$

*has a solution* $\hat{\theta}(\mathbf{z})$ *for which* $\mathbf{a}[\hat{\theta}(\mathbf{z})] \in \mathcal{A}$, *then* $\hat{\theta}(\mathbf{z})$ *is the unique MLE of* $\theta$. □

The proof of this theorem can be found in Bickel and Doksum (1977).

**EXAMPLE A-7** (Continuation of Example A-5)

In this case

$$\mathbf{T}(\mathbf{Z}) = \sum_{i=1}^{N} \mathbf{z}(i)$$

and

$$\mathbf{E}_{\mu}\{\mathbf{T}(\mathbf{Z})\} = N\mu$$

Hence (A-9) becomes

$$\sum_{i=1}^{N} \mathbf{z}(i) = N\mu$$

whose solution, $\hat{\mu}$, is

$$\hat{\mu} = \frac{1}{N}\sum_{i=1}^{N} \mathbf{z}(i)$$

which is the well-known MLE of $\mu$. □

Theorem A-2 can be extended to the $m$-parameter exponential family case by using Definition A-3. We illustrate the applicability of this extension in the following example.

**EXAMPLE A-8** (see, also, Example A-6)

Let $\mathbf{Z} = \text{col}\,[\mathbf{z}(1), \ldots, \mathbf{z}(N)]$ be randomly drawn from $p(\mathbf{z}|\theta) = N(\mu, \mathbf{P}_{\mu})$, where both $\mu$ and $\mathbf{P}_{\mu}$ are unknown, so that $\theta$ contains $\mu$ and the elements of $\mathbf{P}_{\mu}$. Vector $\mu$ is $d \times 1$ and matrix $\mathbf{P}_{\mu}$ is $d \times d$, symmetric and positive definite. We express $p(\mathbf{Z}|\theta)$ as

$$p(\mathbf{Z}|\boldsymbol{\theta}) = (2\pi)^{-Nd/2}(\det \mathbf{P}_\mu)^{-N/2} \exp\left[-\frac{1}{2}\sum_{i=1}^{N}(\mathbf{z}(i)-\boldsymbol{\mu})'\mathbf{P}_\mu^{-1}(\mathbf{z}(i)-\boldsymbol{\mu})\right]$$

$$= (2\pi)^{-Nd/2}(\det \mathbf{P}_\mu)^{-N/2} \exp\left\{-\frac{1}{2}\mathrm{tr}\left[\mathbf{P}_\mu^{-1}\left(\sum_{i=1}^{N}(\mathbf{z}(i)-\boldsymbol{\mu})(\mathbf{z}(i)-\boldsymbol{\mu})'\right)\right]\right\}$$

$$= (2\pi)^{-Nd/2}(\det \mathbf{P}_\mu)^{-N/2} \exp\left\{-\frac{1}{2}\mathrm{tr}\left[\mathbf{P}_\mu^{-1}\left(\sum_{i=1}^{N}\mathbf{z}(i)\mathbf{z}'(i) - 2\boldsymbol{\mu}\sum_{i=1}^{N}\mathbf{z}'(i) + N\boldsymbol{\mu}\boldsymbol{\mu}'\right)\right]\right\}$$

Using Theorem A-1 or Definition A-3, it can be seen that $\sum_{i=1}^{N}\mathbf{z}'(i)$ and $\sum_{i=1}^{N}\mathbf{z}(i)\mathbf{z}'(i)$ are sufficient for $(\boldsymbol{\mu}, \mathbf{P}_\mu)$. Letting

$$\mathbf{T}_1(\mathbf{Z}) = \sum_{i=1}^{N}\mathbf{z}'(i)$$

and

$$\mathbf{T}_2(\mathbf{Z}) = \sum_{i=1}^{N}\mathbf{z}(i)\mathbf{z}'(i)$$

we find that

$$\mathbf{E}_\theta\{\mathbf{T}_1(\mathbf{Z})\} = N\boldsymbol{\mu}'$$

and

$$\mathbf{E}_\theta\{\mathbf{T}_2(\mathbf{Z})\} = N(\mathbf{P}_\mu + \boldsymbol{\mu}\boldsymbol{\mu}')$$

Applying (A-9) to both $\mathbf{T}_1(\mathbf{Z})$ and $\mathbf{T}_2(\mathbf{Z})$, we obtain

$$N\boldsymbol{\mu} = \sum_{i=1}^{N}\mathbf{z}(i)$$

and

$$N(\mathbf{P}_\mu + \boldsymbol{\mu}\boldsymbol{\mu}') = \sum_{i=1}^{N}\mathbf{z}(i)\mathbf{z}'(i)$$

whose solutions, $\hat{\boldsymbol{\mu}}$ and $\hat{\mathbf{P}}_\mu$, are

$$\hat{\boldsymbol{\mu}} = \frac{1}{N}\sum_{i=1}^{N}\mathbf{z}(i)$$

and

$$\hat{\mathbf{P}}_\mu = \frac{1}{N}\sum_{i=1}^{N}[\mathbf{z}(i)-\hat{\boldsymbol{\mu}}][\mathbf{z}(i)-\hat{\boldsymbol{\mu}}]'$$

which are the MLEs of $\boldsymbol{\mu}$ and $\mathbf{P}_\mu$. □

**EXAMPLE A-9 (Generic Linear Model)**

Consider the linear model

$$\mathbf{Z}(k) = \mathbf{H}(k)\boldsymbol{\theta} + \mathbf{V}(k)$$

in which $\boldsymbol{\theta}$ is an $n \times 1$ vector of deterministic parameters, $\mathbf{H}(k)$ is deterministic, and $\mathbf{V}(k)$ is a zero-mean white noise sequence, with known covariance matrix $\mathbf{R}(k)$. From (11-25), we can express $p(\mathbf{Z}(k)|\boldsymbol{\theta})$ as

$$p(\mathbf{Z}(k)|\boldsymbol{\theta}) = \exp[\mathbf{a}'(\boldsymbol{\theta})\mathbf{T}(\mathbf{Z}(k)) + b(\boldsymbol{\theta}) + h(\mathbf{Z}(k))]$$

where

$$\begin{aligned}
\mathbf{a}'(\boldsymbol{\theta}) &= \boldsymbol{\theta}' \\
\mathbf{T}(\mathbf{Z}(k)) &= \mathbf{H}'(k)\mathbf{R}^{-1}(k)\mathbf{Z}(k) \\
b(\boldsymbol{\theta}) &= -\frac{N}{2}\ln 2\pi - \frac{1}{2}\ln\det\mathbf{R}(k) - \frac{1}{2}\boldsymbol{\theta}'\mathbf{H}'(k)\mathbf{R}^{-1}(k)\mathbf{H}(k)\boldsymbol{\theta}
\end{aligned}$$

and

$$h(\mathbf{Z}(k)) = -\frac{1}{2}\mathbf{Z}'(k)\mathbf{R}^{-1}(k)\mathbf{Z}(k)$$

Observe that

$$\mathbf{E}_{\boldsymbol{\theta}}\{\mathbf{H}(k)\mathbf{R}^{-1}(k)\mathbf{Z}(k)\} = \mathbf{H}'(k)\mathbf{R}^{-1}(k)\mathbf{H}(k)\boldsymbol{\theta}$$

Hence, applying (A-9), we obtain

$$\mathbf{H}(k)\mathbf{R}^{-1}(k)\mathbf{H}(k)\hat{\boldsymbol{\theta}} = \mathbf{H}'(k)\mathbf{R}^{-1}(k)\mathbf{Z}(k)$$

whose solution, $\hat{\boldsymbol{\theta}}(k)$, is

$$\hat{\boldsymbol{\theta}}(k) = [\mathbf{H}'(k)\mathbf{R}^{-1}(k)\mathbf{H}(k)]^{-1}\mathbf{H}'(k)\mathbf{R}^{-1}(k)\mathbf{Z}(k)$$

which is the well-known expression for the MLE of $\boldsymbol{\theta}$ (see Theorem 11-3). The case when $\mathbf{R}(k) = \sigma^2\mathbf{I}$, where $\sigma^2$ is unknown, can be handled in a manner very similar to that in Example A-8. □

## SUFFICIENT STATISTICS AND UNIFORMLY MINIMUM-VARIANCE UNBIASED ESTIMATION

In this section we discuss how sufficient statistics can be used to obtain uniformly minimum-variance unbiased (UMVU) estimates. Recall, from Lesson 6, that an estimate $\hat{\theta}$ of parameter $\theta$ is said to be unbiased if

$$\mathbf{E}\{\hat{\theta}\} = \theta \tag{A-10}$$

Among such unbiased estimates, we can often find one estimate, denoted $\theta^*$, that improves all other estimates in the sense that

$$\text{var}\,(\theta^*) \leq \text{var}\,(\hat{\theta}) \tag{A-11}$$

When (A-11) is true for all (admissible) values of $\theta$, $\theta^*$ is known as the UMVU estimate of $\theta$. The UMVU estimator is obtained by choosing the estimator that has the minimum variance among the class of unbiased estimators. If the estimator is constrained further to be a *linear* function of the observations, then it becomes the BLUE, which was discussed in Lesson 9.

Suppose we have an estimate, $\hat{\theta}(\mathbf{Z})$, of parameter $\theta$ that is based on observations $\mathbf{Z} = \text{col}\,[z(1), \ldots, z(N)]$. Assume further that $p(\mathbf{Z}|\theta)$ has a finite-dimensional sufficient statistic, $T(\mathbf{Z})$, for $\theta$. Using $T(\mathbf{Z})$, we can construct an estimate $\theta^*(\mathbf{Z})$ that is at least as good as or even better than $\hat{\theta}$ by the celebrated Rao–Blackwell theorem (Bickel and Doksum, 1977). We do this by computing the conditional expectation of $\hat{\theta}(\mathbf{Z})$, i.e.,

$$\theta^*(\mathbf{Z}) = \mathbf{E}\{\hat{\theta}(\mathbf{Z})|T(\mathbf{Z})\} \tag{A-12}$$

Estimate $\theta^*(\mathbf{Z})$ is "better than $\hat{\theta}$" in the sense that $\mathbf{E}\{[\theta^*(\mathbf{Z}) - \theta]^2\} < \mathbf{E}\{[\hat{\theta}(\mathbf{Z}) - \theta]^2\}$. Because $T(\mathbf{Z})$ is sufficient, the conditional expectation $\mathbf{E}\{\hat{\theta}(\mathbf{Z})|T(\mathbf{Z})\}$ will not depend on $\theta$; hence, $\theta^*(\mathbf{Z})$ is a function of $\mathbf{Z}$ only. Application of this conditioning technique can only improve an estimate such as $\hat{\theta}(\mathbf{Z})$; it does not guarantee that $\theta^*(\mathbf{Z})$ will be the UMVU estimate. To obtain the UMVU estimate using this conditioning technique, we need the additional concept of *completeness*.

**Definition A-4** (Lehmann, 1959,1980; Bickel and Doksum, 1977). *A sufficient statistic* $\mathrm{T}(\mathbf{Z})$ *is said to be complete, if the only real-valued function,* g, *defined on the range of* $\mathrm{T}(\mathbf{Z})$, *that satisfies* $\mathbf{E}_\theta\{\mathrm{g}(\mathrm{T})\} = 0$ *for all* $\theta$ *is the function* $\mathrm{g}(\mathrm{T}) = 0$. □

Completeness is a property of the family of distributions of $T(\mathbf{Z})$ generated as $\theta$ varies over its range. The concept of a complete sufficient statistic, as stated by Lehmann (1980), can be viewed as an extension of the notion of sufficient statistics in reducing the amount of useful information required for the estimation of $\theta$. Although a sufficient statistic achieves data reduction, it may contain some additional information not required for the estimation of $\theta$. For instance, it may be that $\mathbf{E}_\theta[g(T(\mathbf{Z}))]$ is a constant independent of $\theta$ for some nonconstant function $g$. If so, we would like to have $\mathbf{E}_\theta[g(T(\mathbf{Z}))] = c$ (constant independent of $\theta$) imply that $g(T(\mathbf{Z})) = c$. By subtracting $c$ from $\mathbf{E}_\theta[g(T(\mathbf{Z}))]$, we arrive at Definition A-4. Proving completeness using Definition A-4 can be cumbersome. In the special case when $p(z(k)|\theta)$ is a one-parameter exponential family, i.e., when

$$p(z(k)|\theta) = \exp\,[a(\theta)T(z(k)) + b(\theta) + h(z(k))] \tag{A-13}$$

the completeness of $T(z(k))$ can be verified by checking if the range of $a(\theta)$ has an open interval (Lehmann, 1959).

**EXAMPLE A-10**

Let $\mathbf{Z} = \text{col}\,[z(1), \ldots, z(N)]$ be a random sample drawn from a univariate Gaussian distribution whose mean $\mu$ is unknown and whose variance $\sigma^2 > 0$ is known. From Example A-5, we know that the distribution of $\mathbf{Z}$ forms a one-parameter exponential family, with $T(\mathbf{Z}) = \sum_{i=1}^N z(i)$ and $a(\mu) = \mu/\sigma^2$. Because $a(\mu)$ ranges over an open interval as $\mu$ varies from $-\infty$ to $+\infty$, $T(\mathbf{Z}) = \sum_{i=1}^N z(i)$ is complete and sufficient.

The same conclusion can be obtained using Definition A-4 as follows. We must show that the Gaussian family of probability distributions (with $\mu$ unknown and $\sigma^2$ fixed) is complete. Note that the sufficient statistic $T(\mathbf{Z}) = \sum_{i=1}^N z(i)$ (see Example A-5) is Gaussian with mean $N\mu$ and variance $N^2\sigma^2$. Suppose $g$ is a function such that $\mathbf{E}_\mu\{g(T)\} = 0$ for all $-\infty < \mu < \infty$; then

$$\int_{-\infty}^{\infty} \frac{g(T)}{N\sigma\sqrt{2\pi}} \exp\left[-\frac{1}{2\sigma^2 N^2}(T - N\mu)^2\right] dT$$

$$= \int_{-\infty}^{\infty} \sqrt{2\pi}\, g(\nu\sigma N + N\mu) \exp\left(-\frac{\nu^2}{2}\right) d\nu = 0 \qquad \text{(A-14)}$$

implies $g(\cdot) = 0$ for all values of the argument of $g$. □

Other interesting examples that prove completeness for families of distributions are found in Lehmann (1959).

Once a complete and sufficient statistic $T(\mathbf{Z})$ is known for a given parameter estimation problem, the Lehmann–Scheffe theorem, given next, can be used to obtain a unique UMVU estimate. This theorem is paraphrased from Bickel and Doksum (1977).

**Theorem A-3** [Lehmann–Scheffe Theorem (e.g., Bickel and Doksum, 1977)]. *If a complete and sufficient statistic,* $T(\mathbf{Z})$, *exists for* $\theta$, *and* $\hat{\theta}$ *is an unbiased estimator of* $\theta$, *then* $\theta^*(\mathbf{Z}) = \mathbf{E}\{\hat{\theta}|T(\mathbf{Z})\}$ *is an UMVU estimator of* $\theta$. *If variance* $[\theta^*(\mathbf{Z})] < \infty$ *for all* $\theta$, *then* $\theta^*(\mathbf{Z})$ *is the unique UMVU estimate of* $\theta$. □

A proof of this theorem can be found in Bickel and Doksum (1977).

This theorem can be applied in two ways to determine an UMVU estimator [Bickel and Doksum (1977), and Lehmann (1959)].

**Method 1.** Find a statistic of the form $h(T(\mathbf{Z})$ such that

$$\mathbf{E}\{h(T(\mathbf{Z}))\} = \theta \qquad \text{(A-15)}$$

where $T(\mathbf{Z})$ is a complete and sufficient statistic for $\theta$. Then, $h(T(\mathbf{Z}))$ is an UMVU estimator of $\theta$. This follows from the fact that

$$\mathbf{E}\{h(T(\mathbf{Z}))|T(\mathbf{Z})\} = h(T(\mathbf{Z}))$$

**Method 2.** Find an unbiased estimator, $\hat{\theta}$, of $\theta$; then $\mathbf{E}\{\hat{\theta}|T(\mathbf{Z})\}$ is an UMVU estimator of $\theta$ for a complete and sufficient statistic $T(\mathbf{Z})$.

**EXAMPLE A-11** (Continuation of Example A-10)

We know that $T(\mathbf{Z}) = \sum_{i=1}^{N} z(i)$ is a complete and sufficient statistic for $\mu$. Furthermore, $1/N \sum_{i=1}^{N} z(i)$ is an unbiased estimator of $\mu$; hence, we obtain the well-known result from Method 1 that the sample mean, $1/N \sum_{i=1}^{N} z(i)$, is an UMVU estimate of $\mu$. Because this estimator is linear, it is also the BLUE of $\mu$. □

**EXAMPLE A-12 (Generic Linear Model)**

As in Example A-9, consider the linear model

$$\mathbf{Z}(k) = \mathbf{H}(k)\boldsymbol{\theta} + \mathbf{V}(k) \qquad \text{(A-16)}$$

where $\boldsymbol{\theta}$ is a deterministic but unknown $n \times 1$ vector of parameters, $\mathbf{H}(k)$ is deterministic, and $\mathbf{E}\{\mathbf{V}(k)\} = \mathbf{0}$. Additionally, assume that $\mathbf{V}(k)$ is Gaussian with known covariance matrix

$\mathbf{R}(k)$. Then the statistic $\mathbf{T}(\mathbf{Z}(k)) = \mathbf{H}'(k)\mathbf{R}^{-1}\mathbf{Z}(k)$ is sufficient (see Example A-9). That it is also complete can be seen by using Theorem A-4. To obtain UMVU estimate $\boldsymbol{\theta}$, we need to identify a function $h[\mathbf{T}(\mathbf{Z}(k))]$ such that $\mathbf{E}\{h[\mathbf{T}(\mathbf{Z}(k))]\} = \boldsymbol{\theta}$. The structure of $h[\mathbf{T}(\mathbf{Z}(k))]$ is obtained by observing that

$$\mathbf{E}\{\mathbf{T}(\mathbf{Z}(k))\} = \mathbf{E}\{\mathbf{H}'(k)\mathbf{R}^{-1}(k)\mathbf{Z}(k)\} = \mathbf{H}'(k)\mathbf{R}^{-1}(k)\mathbf{H}(k)\boldsymbol{\theta}$$

Hence

$$[\mathbf{H}'(k)\mathbf{R}^{-1}(k)\mathbf{H}(k)]^{-1}\mathbf{E}\{\mathbf{T}(\mathbf{Z}(k))\} = \boldsymbol{\theta}$$

Consequently, the UMVU estimator of $\boldsymbol{\theta}$ is

$$[\mathbf{H}'(k)\mathbf{R}^{-1}(k)\mathbf{H}(k)]^{-1}\mathbf{H}'(k)\mathbf{R}^{-1}(k)\mathbf{Z}(k)$$

which agrees with Equation (9-22). □

We now generalize the discussions given above to the case of an $m$-parameter exponential family and scalar observations. This theorem is paraphrased from Bickel and Doksum (1977).

**Theorem A-4** [Bickel and Doksum (1977) and Lehmann (1959)]. *Let* $p(z|\theta)$ *be an* m*-parameter exponential family given by*

$$p(z|\theta) = \exp\left[\sum_{i=1}^{m} a_i(\theta)T_i(z) + b(\theta) + h(z)\right]$$

*where* $a_1, \ldots, a_m$ *and* b *are real-valued functions of* $\theta$, *and* $T_1 \ldots, T_m$ *and* h *are real-valued functions of* z. *Suppose that the range of* $\mathbf{a} = col\,[a_1(\theta), \ldots, a_m(\theta)]$ *has an open* m*-rectangle* [*if* $(x_1, y_1), \ldots, (x_m, y_m)$ *are* m *open intervals, the set* $\{(s_1, \ldots, s_m) : x_i < s_i < y_i, 1 \le i \le m\}$ *is called the open* m*-rectangle*]; *then* $\mathbf{T}(z) = col\,[T_1(z), \ldots, T_m(z)]$ *is complete as well as sufficient.* □

**EXAMPLE A-13** (This example is taken from Bickel and Doksum, 1977, pp. 123–124)

As in Example A-6, let $\mathbf{Z} = \text{col}\,[z(1), \ldots, z(N)]$ be a sample from a $N(\mu, \sigma^2)$ population where both $\mu$ and $\sigma^2$ are unknown. As a special case of Example A-6, we observe that the distribution of $\mathbf{Z}$ forms a two-parameter exponential family where $\boldsymbol{\theta} = \text{col}\,(\mu, \sigma^2)$. Because $\text{col}\,[a_1(\boldsymbol{\theta}), a_2(\boldsymbol{\theta})] = \text{col}\,(\mu/\sigma^2, -\frac{1}{2}\sigma^2)$ ranges over the lower half-plane, as $\boldsymbol{\theta}$ ranges over $\text{col}\,[(-\infty, \infty), (0, \infty)]$, the conditions of Theorem A-4 are satisfied. As a result, $\mathbf{T}(\mathbf{Z}) = \text{col}\,[\sum_{i=1}^{N} z(i), \sum_{i=1}^{N} z^2(i)]$ is complete and sufficient. □

Theorem A-3 also generalizes in a straightforward manner to:

**Theorem A-5.** *If a complete and sufficient statistic* $\mathbf{T}(\mathbf{Z}) = col\,(T_1(\mathbf{Z}), \ldots, T_m(\mathbf{Z}))$ *exists for* $\boldsymbol{\theta}$, *and* $\hat{\boldsymbol{\theta}}$ *is an unbiased estimator of* $\boldsymbol{\theta}$, *then* $\boldsymbol{\theta}^*(\mathbf{Z}) = \mathbf{E}\{\hat{\boldsymbol{\theta}}|\mathbf{T}(\mathbf{Z})\}$ *is an UMVU estimator of* $\boldsymbol{\theta}$. *If the elements of the covariance matrix of* $\boldsymbol{\theta}^*(\mathbf{Z})$ *are* $< \infty$ *for all* $\boldsymbol{\theta}$, *then* $\boldsymbol{\theta}^*(\mathbf{Z})$ *is the unique UMVU estimate of* $\boldsymbol{\theta}$. □

The proof of this theorem is a straightforward extension of the proof of Theorem A-3, which can be found in Bickel and Doksum (1977).

**EXAMPLE A-14** (Continuation of Example A-13)

In Example A-13 we saw that $\text{col}\,[T_1(\mathbf{Z}), T_2(\mathbf{Z})] = \text{col}\,[\sum_{i=1}^{N} z(i), \sum_{i=1}^{N} z^2(i)]$ is sufficient and complete for both $\mu$ and $\sigma^2$. Furthermore, since

$$\bar{z} = \frac{1}{N}\sum_{i=1}^{N} z(i)$$

and

$$\bar{\sigma}^2 = \frac{1}{N-1}\sum_{i=1}^{N}[z(i) - \bar{z}]^2$$

are unbiased estimators of $\mu$ and $\sigma^2$, respectively, we use the extension of Method 1 to the vector parameter case to conclude that $\bar{z}$ and $\bar{\sigma}^2$ are UMVU estimators of $\mu$ and $\sigma^2$. □

It is not always possible to identify a function $h(T(\mathbf{Z}))$ that is an unbiased estimator of $\boldsymbol{\theta}$. Examples that use the conditioning Method 2 to obtain UMVU estimators are found, for example, in Bickel and Doksum (1977) and Lehman (1980).

## SUMMARY QUESTIONS

1. Check off two methods that help to compute a sufficient statistic:
   **(a)** identify a statistic such that the given observations conditioned on the statistic are independent of the parameters that characterize the probability distribution of the observations
   **(b)** factorization theorem
   **(c)** maximum of the observed data
2. The dimension of a sufficient statistic:
   **(a)** increases linearly with the number of observations
   **(b)** decreases linearly with the number of observations
   **(c)** is independent of the number of observations
3. When a sufficient statistic exists:
   **(a)** the maximum likelihood estimate and sufficient statistic are identical
   **(b)** the maximum likelihood estimate is a function of the sufficient statistic
   **(c)** the maximum likelihood estimate is independent of the sufficient statistic
4. In general, the UMVU estimator is:
   **(a)** more general than BLUE
   **(b)** the same as BLUE
   **(c)** the same as the maximum likelihood estimator
5. By conditioning on the sufficient statistic, an unbiased estimate:
   **(a)** can be made biased
   **(b)** can be made to be consistent
   **(c)** can be improved
6. To compute the UMVU estimator, in addition to sufficiency, we need the statistic to be:
   **(a)** complete
   **(b)** necessary
   **(c)** unique
7. A one-to-one function of a sufficient statistic is:

**(a)** necessary
**(b)** sufficient
**(c)** complete

**8.** Sufficient statistics are useful in Bayesian estimation for:
**(a)** deriving conjugate priors
**(b)** identifying the loss function
**(c)** reducing the bias and variance of estimates

**9.** Check off things that are true for members of the exponential family:
**(a)** they admit sufficient statistics for parameters that characterize the distributions
**(b)** UMVU estimates of parameters are generally obtainable
**(c)** they obey the factorization theorem

## PROBLEMS

**A-1.** Suppose $z(1), \ldots, z(N)$ are independent random variables, each uniform on $[0, \theta]$, where $\theta > 0$ is unknown. Find a sufficient statistic for $\theta$.

**A-2.** Suppose we have two independent observations from the Cauchy distribution

$$p(z) = \frac{1}{\pi}\frac{1}{1+(z-\theta)^2}, \qquad -\infty < z < \infty$$

Show that no sufficient statistic exists for $\theta$.

**A-3.** Let $z(1), z(2), \ldots, z(N)$ be generated by the first-order autoregressive process

$$z(i) = \theta z(i-1) + \sqrt{\beta} w(i)$$

where $\{w(i), i = 1, \ldots, N\}$ is an independent and identically distributed Gaussian noise sequence with zero mean and unit variance. Find a sufficient statistic for $\theta$ and $\beta$.

**A-4.** Suppose that $T(\mathbf{Z})$ is sufficent for $\boldsymbol{\theta}$ and that $\hat{\boldsymbol{\theta}}(\mathbf{Z})$ is a maximum-likelihood estimate of $\boldsymbol{\theta}$. Show that $\hat{\boldsymbol{\theta}}(\mathbf{Z})$ depends on $\mathbf{Z}$ only through $T(\mathbf{Z})$.

**A-5.** Using Theorem A-2, derive the maximum-likelihood estimator of $\theta$ when observations $z(1), \ldots, z(N)$ denote a sample from

$$p(z(i)|\theta) = \theta e^{-\theta z(i)}, \qquad z(i) \geq 0, \theta > 0$$

**A-6.** Show that the family of Bernoulli distributions, with unknown probability of success $p(0 \leq p \leq 1)$, is complete.

**A-7.** Show that the family of uniform distributions on $(0, \theta)$, where $\theta > 0$ is unknown, is complete.

**A-8.** Let $z(1), \ldots, z(N)$ be independent and identically distributed samples, where $p(z(i)|\theta)$ is a Bernoulli distribution with unknown probability of success $p(0 \leq p \leq 1)$. Find a complete sufficient statistic, $T$; the UMVU estimate $\phi(T)$ of $p$; and the variance of $\phi(T)$.

**A-9.** (Taken from Bickel and Doksum, 1977). Let $z(1), z(2), \ldots, z(N)$ be an independent and identically distributed sample from $N(\mu, 1)$. Find the UMVU estimator of $p_\mu[z(1) \geq 0]$.

**A-10.** (Taken from Bickel and Doksum, 1977). Suppose that $T_1$ and $T_2$ are two UMVU estimates of $\theta$ with finite variances. Show that $T_1 = T_2$.

**A-11.** In Example A-12 prove that $T(\mathbf{Z}(k))$ is complete.

## LESSON B

# Introduction to Higher-order Statistics


## SUMMARY

When signals are non-Gaussian, higher than second-order statistics can be very useful. These higher-order statistics are called *cumulants* and are related to higher-order moments. We prefer to use cumulants rather than higher-order moments because cumulants have some very desirable properties, which moments do not have, that let us treat cumulants as operators.

Higher-order statistics are defined in this lesson. In general, they are multidimensional functions; e.g., third-order statistics are functions of two variables and fourth-order statistics are functions of three variables. Cumulants, which are analogous to correlation functions, work directly with signals in the time domain. Polyspectra, which are analogous to the power spectrum, are multidimensional Fourier transforms of cumulants. Fortunately, many important problems can be solved using one or more one-dimensional slices of a higher-order statistic. Additionally, cumulants and polyspectra of stationary random sequences are rich in symmetries, so they only have to be determined over a rather small region of their domains of support.

Six important properties are given for cumulants. Their use lets us apply cumulants to new situations, just as we apply the familiar expectation operator to new situations. One of the most important situations is that of a measurement that equals signal plus noise, where the noise is Gaussian (colored or white). Using the properties of cumulants, we show that the cumulant of this measurement equals the cumulant of just the signal; hence, cumulants boost signal-to-noise ratio. The same is not true for the correlation of this measurement because it is adversely affected by the correlation of the additive noise.

When you complete this lesson, you will be able to (1) define and determine formulas for cumulants; (2) define and determine formulas for polyspectra; (3)

understand the relationships between cumulants and moments and why we prefer to work with cumulants; (4) understand the symmetries of cumulants and polyspectra; (5) understand the notion of a one-dimensional slice cumulant; (6) understand and use six important properties of cumulants; and (7) understand why cumulants tend to boost signal-to-noise ratio.

## INTRODUCTION

Gaussian signals are completely characterized by first- and second-order statistics, because a Gaussian probability density function is completely described by the mean and variance of the signal. We also know that linear transformations of Gaussian signals result in new Gaussian signals. These facts are the basis for many of the techniques described in this book. When signals are non-Gaussian, which is frequently the case in real-world applications, things are not so simple. Even if we knew the signal's original non-Gaussian probability density function, a linear transformation of the non-Gaussian signal would change the entire nature of the resulting probability density function.

Ideally, we would like to work with the signal's probability density function. To do so would require knowledge of all the moments of the probability density function, and, in general, this means that we would need to compute an infinite number of moments, something that is impossible to do. What are the alternatives? We could choose to work with just the first- and second-order moments and then stay within the framework of optimal linear estimators, an approach that has been described in Lesson 13; or we could choose to estimate the entire probability density function, an approach that has been briefly described in the Supplementary Material to Lesson 13; or we could choose to work not only with first- and second-order moments, but with some higher-order moments as well. Doing the latter is based on the premise that, when dealing with non-Gaussian signals, we ought to be able to do a better job of estimation (or other types of signal processing) using more than just first- and second-order statistics. This lesson, as well as the next, focuses on the latter approach.

Before we define higher-order moments and their close relations, cumulants, and study important properties of these statistics, let us pause to explain that these lessons are *not* about the so-called "method of moments." Around 1900, the statistician Karl Pearson (1936, and references therein) developed the method of moments, which is conceptually simple, but which has very little theoretical justification; i.e., it may give acceptable results sometimes, but frequently it gives very unacceptable results. The method of moments is based on the fact that sample moments (e.g., the sample mean, sample variance, etc.) are consistent estimators of population moments (e.g., Bury, 1975); hence, for large samples, it is reasonable to expect that the sample moments provide good approximations to the true (population) moments of the probability density function. If we knew the underlying probability density function, which may depend on some unknown parameters, and could express the population moments in terms of these parameters, then we could equate the sample moments to the population moments. If the density function depends on $n$ parameters (e.g., a Gamma density function depends on two parameters), then, if we do this for the first $n$ nonzero moments, we will obtain a system of $n$ equations in $n$ unknowns, which must be solved for the $n$ unknown parameters. Frequently, in signal-processing applications, the probability density function is not known ahead of time; hence, it is impossible to follow this procedure. Additionally, it is not uncommon for the $n$ equations to be nonlinear, in which case their solution is not unique and is difficult to obtain. See Sorenson (1980) and Gallant (1987) for very extensive discussions on the method of moments.

## DEFINITIONS OF HIGHER-ORDER STATISTICS

Higher-order statistics can be defined for both stochastic and deterministic signals. Because of our great interest in this book on the former types of signals, we only define higher-order statistics for them. See Nikias and Petropulu (1993) for definitions of higher-order statistics of deterministic signals. Some of the material in this section is taken from Nikias and Mendel (1993) and Mendel (1991).

If $\{x(k)\}$, $k = 0, \pm 1, \pm 2, \ldots$ is a real *stationary* discrete-time random sequence and its moments up to order $n$ exist, then (Papoulis, 1991) the $n$th-order moment function of $\{x(k)\}$, is given by

$$m_{n,x}(\tau_1, \tau_2, \ldots, \tau_{n-1}) = \mathbf{E}\{x(k)x(k + \tau_1) \cdots x(k + \tau_{n-1})\} \qquad \text{(B-1)}$$

Observe that, because $\{x(k)\}$ is a stationary sequence, the $n$th-order moment depends only on the time differences $\tau_1, \tau_2, \ldots, \tau_{n-1}$, where $\tau_i = 0, \pm 1, \pm 2, \ldots$, for all $i$. The second-order moment function $m_{2,x}(\tau_1)$ is the autocorrelation function, whereas $m_{3,x}(\tau_1, \tau_2)$ and $m_{4,x}(\tau_1, \tau_2, \tau_3)$ are the third- and fourth-order moments, respectively.

Note that higher-order moments of the sum of two (or more) statistically independent random sequences do not necessarily equal the sum of the higher-order moments of the two (or more) sequences. This makes it difficult to work with higher-order moments in applications, such as signal processing, where random sequences are combined additively. Cumulants, which are related to moments and are defined next, have the very desirable property that cumulants of the sum of two (or more) statistically independent random sequences equal the sum of the cumulants of the two (or more) sequences.

**Definition B-1.** *Let* $\mathbf{v} = \text{col}(v_1, v_2, \ldots, v_k)$ *and* $\mathbf{x} = \text{col}(x_1, x_2, \ldots, x_k)$, *where* $(x_1, x_2, \ldots, x_k)$ *denotes a collection of random variables. The kth-order cumulant of these random variables is defined (Priestley, 1981) as the coefficient of* $(v_1 v_2 \cdots v_k) \times j^k$ *in the Taylor series expansion (provided it exists) of the cumulant-generating function*

$$K(\mathbf{v}) = \ln \mathbf{E}\{\exp(j\mathbf{v}'\mathbf{x})\} \qquad \qquad \text{(B-2)}$$

For zero-mean *real* random variables, the second-, third- and fourth-order cumulants are given by

$$\text{cum}(x_1, x_2) = \mathbf{E}\{x_1 x_2\} \qquad = R_{x_1 x_2} \qquad \text{(B-3a)}$$

$$\text{cum}(x_1, x_2, x_3) = \mathbf{E}\{x_1 x_2 x_3\} \qquad = R_{x_1, x_2 x_3} \qquad \text{(B-3b)}$$

(why re use cumulants)

used to define moments (by taylor series)

$\Phi(v) = E\{e^{jv'x}\}$ used to define moments (by taylor series)

$1^{st}$ characteristic function

$K(v) = \ln \Phi(v)$ used to define cumulants (by taylor series)

$2^{nd}$ characteristic function

$$\text{cum}\,(x_1, x_2, x_3, x_4) = \mathbf{E}\{x_1x_2x_3x_4\} - \mathbf{E}\{x_1x_2\}\mathbf{E}\{x_3x_4\} \qquad \neq R_{x_1x_2x_3x_4}$$

$$- \mathbf{E}\{x_1x_3\}\mathbf{E}\{x_2x_4\} - \mathbf{E}\{x_1x_4\}\mathbf{E}\{x_2x_3\} \qquad (B\text{-}3c)$$

In the case of nonzero-mean real random variables, we replace $x_i$ by $x_i - \mathbf{E}\{x_i\}$ in these formulas. The case of **complex** signals is treated in the references given above and in Problem B-8.

## EXAMPLE B-1

*remember the R.V. has zero mean.*

Here we derive (B-3a) using (B-2). We leave the comparable derivations of (B-3b) and (B-3c) as a problem for the reader (Problem B-1). According to Definition B-1,

$$\text{cum}\,(x_1, x_2) = -\left.\frac{\partial^2 K(\mathbf{v})}{\partial v_1 \partial v_2}\right|_{v_1=v_2=0} = -\left.\frac{\partial^2 \ln \Phi(v_1, v_2)}{\partial v_1 \partial v_2}\right|_{v_1=v_2=0} \qquad (B\text{-}4)$$

where

$$\Phi(v_1, v_2) = \mathbf{E}\{\exp[j(v_1x_1 + v_2x_2)]\} \qquad (B\text{-}5)$$

Consequently,

$$\text{cum}\,(x_1, x_2) = -\frac{\partial}{\partial v_1}\left[\frac{1}{\Phi(v_1, v_2)} \times \frac{\partial \Phi(v_1, v_2)}{\partial v_2}\right]\bigg|_{v_1=v_2=0}$$

$$= \left[-\frac{1}{\Phi^2(v_1, v_2)} \times \frac{\partial \Phi(v_1, v_2)}{\partial v_1} \times \frac{\partial \Phi(v_1, v_2)}{\partial v_2}\right.$$

$$\left.+\frac{1}{\Phi(v_1, v_2)} \times \frac{\partial^2 \Phi(v_1, v_2)}{\partial v_1 \partial v_2}\right]\bigg|_{v_1=v_2=0} \qquad (B\text{-}6)$$

where

$$\frac{\partial \Phi(v_1, v_2)}{\partial v_i} = \mathbf{E}\{jx_i \exp[j(v_1x_1 + v_2x_2)]\}, \qquad i = 1, 2 \qquad (B\text{-}7)$$

Setting $v_i = 0$ in (B-7), we see that

$$\left.\frac{\partial \Phi(v_1, v_2)}{\partial v_i}\right|_{v_1=v_2=0} = j\mathbf{E}\{x_i\} = 0 \qquad (B\text{-}8)$$

because the random variables, $x_i$ are zero mean. Note, also, that

$$\Phi|_{v_1=v_2=0} = 1 \qquad (B\text{-}9)$$

Substituting (B-8) and (B-9) into (B-6), we find

$$\text{cum}\,(x_1, x_2) = \left.\frac{\partial^2 \Phi(v_1, v_2)}{\partial v_1 \partial v_2}\right|_{v_1=v_2=0} = \mathbf{E}\{x_1x_2\} \qquad (B\text{-}10)$$

where this last result follows (B-7). □ *probability thm note: representation factor Σφ.*

The $k$th-order cumulant is therefore defined in terms of its joint moments of orders up to $k$. See the Supplementary Material at the end of this lesson for explicit relationships between cumulants and moments.

**Definition B-2.** *Let* $\{x(t)\}$ *be a zero-mean* k*th-order* **stationary** *random process. The* k*th-order cumulant of this process, denoted* $C_{k,x}(\tau_1, \tau_2, \ldots, \tau_{k-1})$ *is defined as the joint* k*th-order cumulant of the random variables* $x(t), x(t+\tau_1), \ldots, x(t+\tau_{k-1})$; *i.e.,*

$$C_{k,x}(\tau_1, \tau_2, \ldots, \tau_{k-1}) = \text{cum}\,(x(t), x(t+\tau_1), \ldots, x(t+\tau_{k-1})). \ \square \qquad \text{(B-11)}$$

Because of stationarity, the $k$th-order cumulant is only a function of the $k-1$ lags $\tau_1, \tau_2, \ldots, \tau_{k-1}$. If $\{x(t)\}$ is **nonstationary**, then the $k$th-order cumulant depends explicitly on $t$ as well as on $\tau_1, \tau_2, \ldots, \tau_{k-1}$, and the notation $C_{k,x}(t; \tau_1, \tau_2, \ldots, \tau_{k-1})$ can be used.

The second-, third-, and fourth-order cumulants of zero-mean $x(t)$, which follow from (B-3) and (B-11), are

$$C_{2,x}(\tau) = \mathbf{E}\{x(t)x(t+\tau)\} \qquad \text{(B-12a)}$$

$$C_{3,x}(\tau_1, \tau_2) = \mathbf{E}\{x(t)x(t+\tau_1)x(t+\tau_2)\} \qquad \text{(B-12b)}$$

$$\begin{aligned} C_{4,x}(\tau_1, \tau_2, \tau_3) = {} & \mathbf{E}\{x(t)x(t+\tau_1)x(t+\tau_2)x(t+\tau_3)\} \\ & - C_{2,x}(\tau_1)C_{2,x}(\tau_2 - \tau_3) - C_{2,x}(\tau_2)C_{2,x}(\tau_3 - \tau_1) \\ & - C_{2,x}(\tau_3)C_{2,x}(\tau_1 - \tau_2) \end{aligned} \qquad \text{(B-12c)}$$

Of course, the second-order cumulant $C_{2,x}(\tau)$ is just the autocorrelation of $x(t)$. We shall use the more familiar notation for autocorrelation, $r_x(\tau)$, interchangeably with $C_{2,x}(\tau)$.

The $\tau_1 - \tau_2 - \cdots - \tau_{k-1}$ space constitutes the **domain of support** for $C_{k,x}(\tau_1, \tau_2, \ldots, \tau_{k-1})$; i.e., it is where the $(k-1)$-dimensional function $C_{k,x}(\tau_1, \tau_2, \ldots, \tau_{k-1})$ must be evaluated. The domain of support for the autocorrelation function is the one-dimensional real axis. The domain of support for the third-order cumulant function is the two-dimensional $\tau_1 - \tau_2$ plane, and the domain of support for the fourth-order cumulant function is the three-dimensional $\tau_1 - \tau_2 - \tau_3$ volume. Note that a plot of autocorrelation is two-dimensional, a plot of a third-order cumulant is three-dimensional, and a plot of a fourth-order cumulant would be four-dimensional. Naturally, we cannot display the fourth-order cumulant by usual means. Color or sound could be used for the fourth dimension.

### EXAMPLE B-2

Here we compute the second- through fourth-order cumulants of the *Bernoulli–Gaussian sequence*, which we have used in earlier lessons in connection with minimum-variance and maximum-likelihood deconvolution and which is described in Example 13-1. That sequence is

$$\mu(k) = r(k)q(k) \qquad \text{(B-13)}$$

where $r(k)$ is a Gaussian white random sequence, $r(k) = N(r(k); 0, \sigma_r^2)$, and $q(k)$ is a Bernoulli sequence for which $Pr[q(k) = 1] = \lambda$. Additionally, $r(k)$ and $q(k)$ are statistically independent. Obviously, $\mathbf{E}\{\mu(k)\} = 0$. Due to the whiteness of both $r(k)$ and $q(k)$, as well as their statistical independence, the only higher-order statistics that *may* be nonzero are those for which all lags are zero. Consequently,

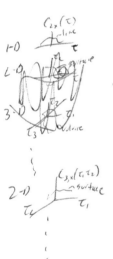

$$C_{2,\mu}(0) = \mathbf{E}\{\mu^2(k)\} = \mathbf{E}\{r^2(k)\}\mathbf{E}\{q^2(k)\} = \sigma_r^2 \mathbf{E}\{q(k)\} = \sigma_r^2 \lambda \qquad \text{(B-14)}$$

$$C_{3,\mu}(0, 0) = \mathbf{E}\{\mu^3(k)\} = \mathbf{E}\{r^3(k)\}\mathbf{E}\{q^3(k)\} = 0 \qquad \text{(B-15)}$$

because $\mathbf{E}\{r^3(k)\} = 0$, and $= E(r^4)E(q^4) = 3\sigma_r^4\lambda$ *since* $E\{q^4\} = E\{q\} = \lambda$

$$C_{4,\mu}(0, 0, 0) = \mathbf{E}\{\mu^4(k)\} - 3[\mathbf{E}\{\mu^2(k)\}]^2 = 3\sigma_r^4\lambda - 3(\sigma_r^2\lambda)^2 = 3\lambda(1-\lambda)\sigma_r^4 \qquad \text{(B-16)}$$

The Bernoulli–Gaussian sequence is one whose third-order cumulant is zero, so we must consequently use fourth-order cumulants when working with it. It is also an example of a *higher-order white noise sequence*, i.e., a white noise sequence whose second- and higher-order cumulants are multidimensional impulse functions. Its correlation function is nonzero only at the origin of the line of support, with amplitude given by (B-14). Its third-order cumulant is zero everywhere in its plane of support (we could also say that its only possible nonzero value, which occurs at the origin of its two-dimensional domain of support, is zero). Its fourth-order cumulant is nonzero only at the origin of its three-dimensional volume of support, and its value at the origin is given by (B-16). □

Equations (B-12) can be used in two different ways: given a data-generating model, they can be used to determine exact theoretical formulas for higher-order statistics of signals in that model; or, given data, they can be used as the basis for estimating higher-order statistics, usually by replacing expectations by sample averages. We shall discuss both of these uses in Lesson C.

For a zero-mean stationary random process and for $k = 3, 4$, the $k$th-order cumulant of $\{x(t)\}$ can also be defined as (Problem B-2)

$$C_{k,x}(\tau_1, \tau_2, \ldots, \tau_{k-1}) = M_{k,x}[x(\tau_1), \cdots, x(\tau_{k-1})]$$

$$-M_{k,g}[g(\tau_1), \cdots, g(\tau_{k-1})] \qquad \text{(B-17)}$$

where $M_{k,x}[x(\tau_1), \cdots, x(\tau_{k-1})]$ is the $k$th-order moment function of $x(t)$ and $M_{k,g}$ $[g(\tau_1), \cdots, g(\tau_{k-1})]$ is the $k$th-order moment function of an equivalent Gaussian process, $g(t)$, that has the same mean value and autocorrelation function as $x(t)$. *Cumulants, therefore, not only display the amount of higher-order correlation, but also provide a measure of the distance of the random process from Gaussianity.* Clearly, if $x(t)$ is Gaussian, then the cumulants are all zero; this is not only true for $k = 3$ and 4, but for all $k$.

A *one-dimensional slice* of the $k$th-order cumulant is obtained by freezing $k - 2$ of its $k - 1$ indexes. Many types of one-dimensional slices are possible, including radial, vertical, horizontal, diagonal, and offset diagonal. A *diagonal slice* is obtained by setting $\tau_i = \tau, i = 1, 2, \ldots, k - 1$. Some examples of domains of support for such slices are depicted in Figure B-1. All these one-dimensional slices are very useful in applications of cumulants in signal processing.

Many symmetries exist in the arguments of $C_{k,x}(\tau_1, \tau_2, \ldots, \tau_{k-1})$, which make the calculations of cumulants manageable. In practical applications, we only work with third- or fourth-order cumulants, because the variances of their estimation errors increase with the order of the cumulant (Lesson C and Nikias and Petropulu, 1993); hence, we only give symmetry conditions for third- and fourth-order cumulants here.

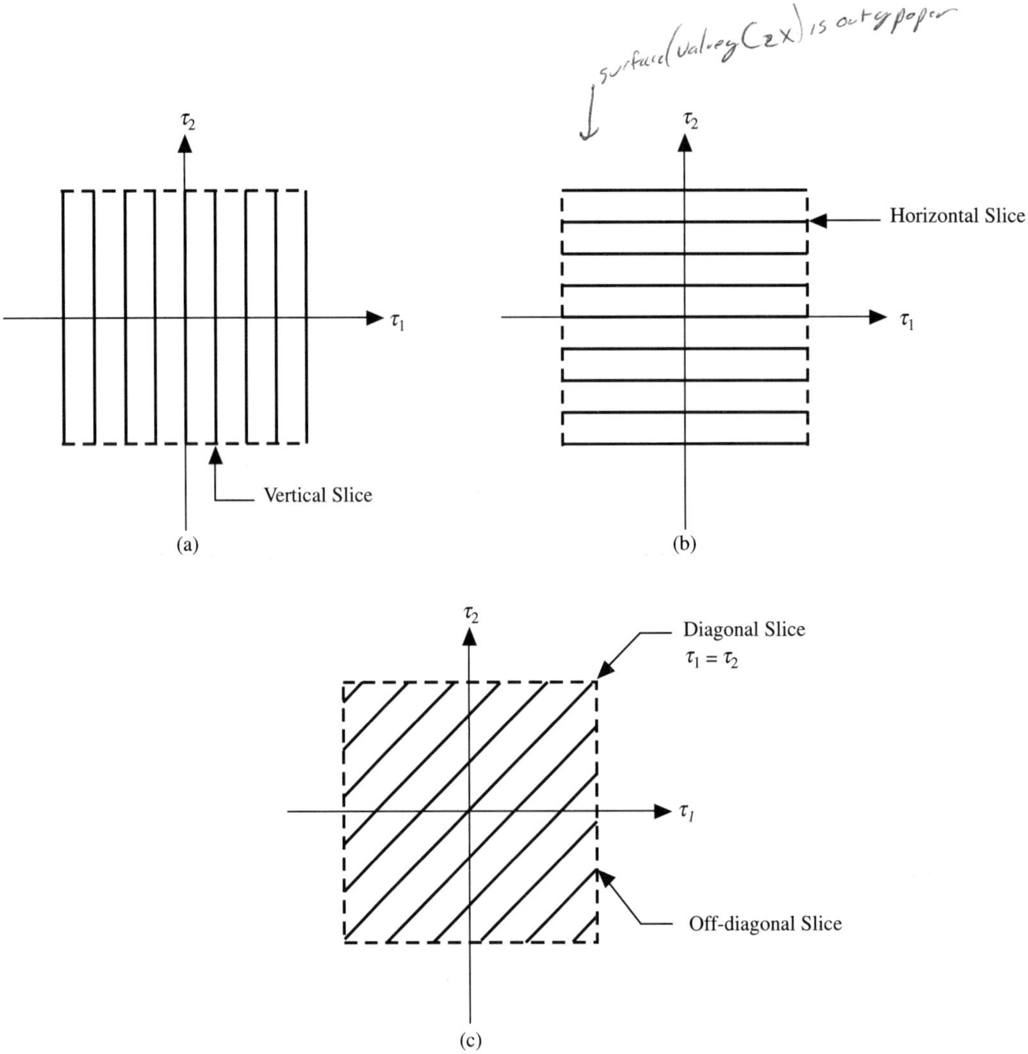


**Figure B-1** Domains of support for one-dimensional cumulant slices of third-order cumulants: (a) vertical slices, (b) horizontal slices, and (c) diagonal and off-diagonal slices.

**Theorem B-1.** *(a) Third-order cumulants are symmetric in the following five ways:*

$$\begin{aligned} C_{3,x}(\tau_1, \tau_2) &= C_{3,x}(\tau_2, \tau_1) = C_{3,x}(-\tau_2, \tau_1 - \tau_2) = C_{3,x}(-\tau_1, \tau_2 - \tau_1) \\ &= C_{3,x}(\tau_2 - \tau_1, -\tau_1) = C_{3,x}(\tau_1 - \tau_2, -\tau_2) \end{aligned} \quad \text{(B-18)}$$

*(b) Fourth-order cumulants are symmetric in the following 23 ways:*

$$\begin{aligned} C_{4,x}(\tau_1, \tau_2, \tau_3) &= C_{4,x}(-\tau_1, \tau_2 - \tau_1, \tau_3 - \tau_1) = C_{4,x}(-\tau_2, \tau_1 - \tau_2, \tau_3 - \tau_2) \\ &= C_{4,x}(-\tau_3, \tau_1 - \tau_3, \tau_2 - \tau_3) \end{aligned} \quad \text{(B-19)}$$

$$\begin{aligned} C_{4,x}(\tau_1, \tau_2, \tau_3) &= C_{4,x}(\tau_1, \tau_3, \tau_2) = C_{4,x}(\tau_2, \tau_1, \tau_3) \\ &= C_{4,x}(\tau_2, \tau_3, \tau_1) = C_{4,x}(\tau_3, \tau_1, \tau_2) = C_{4,x}(\tau_3, \tau_2, \tau_1) \end{aligned} \quad \text{(B-20)}$$

*The remaining 15 equalities are obtained by permuting the arguments in the three equalities on the right-hand side of (B-19) five times as in (B-20).* □

The proof of this theorem works directly with the definitions of the third- and fourth- order cumulants, in (B-12b) and (B-12c), and is left to the reader (Problems B-3 and B-4). Note that the 23 equalities for the fourth-order cumulant are obtained from the 3 equalities in (B-19) plus the $4 \times 5 = 20$ equalities obtained by applying the 5 equalities in (B-20) to the 4 cumulants in (B-19). See Pflug et al. (1992) for very comprehensive discussions about these symmetry regions.

Using the five equations in (B-18), we can divide the $\tau_1 - \tau_2$ plane into six regions, as depicted in Figure B-2. Knowing the cumulants in any one of these regions, we can calculate the cumulants in the other five regions using these equations. The principal region of support is the first-quadrant 45° sector, $0 < \tau_2 \leq \tau_1$. See Pflug et al. (1992, Fig. 9) for a comparable diagram for the 24 regions of symmetry of the fourth-order cumulant. The symmetry relationships for cumulants do not hold in the nonstationary case.

If a random sequence is symmetrically distributed, then its third-order cumulant equals zero; hence, for such a process we must use fourth-order cumulants. For example, Laplace, uniform, Gaussian, and Bernoulli–Gaussian distributions are symmetric, whereas exponential, Rayleigh, and *K*-distributions are nonsymmetric (Problems B-6 and B-7).

### EXAMPLE B-3

What do cumulants look like? Figure B-3 depicts the impulse response for a tenth-order system, as well as the autocorrelation function, third-order diagonal-slice cumulant, and fourth-order diagonal slice cumulant for that system. Exactly how these results were obtained will be discussed in Lesson C. Whereas autocorrelation is symmetric about the origin [i.e., $r_y(-\tau) = r_y(\tau)$], third- and fourth-order diagonal-slice cumulants are not symmetric about their origins. Observe that positive-lag values of the diagonal-slice cumulants have a higher frequency content than do negative lag values and that positive-lag frequency content is greater for fourth-order cumulants than it is for third-order cumulants. The latter is a distinguishing feature of these cumulants, which always differentiates them from each other.

Fourth-order cumulants for this tenth-order system are depicted in Figure B-4. These plots were obtained by first fixing $\tau_3$ and then varying $\tau_1$ and $\tau_2$ from 1 to 9.

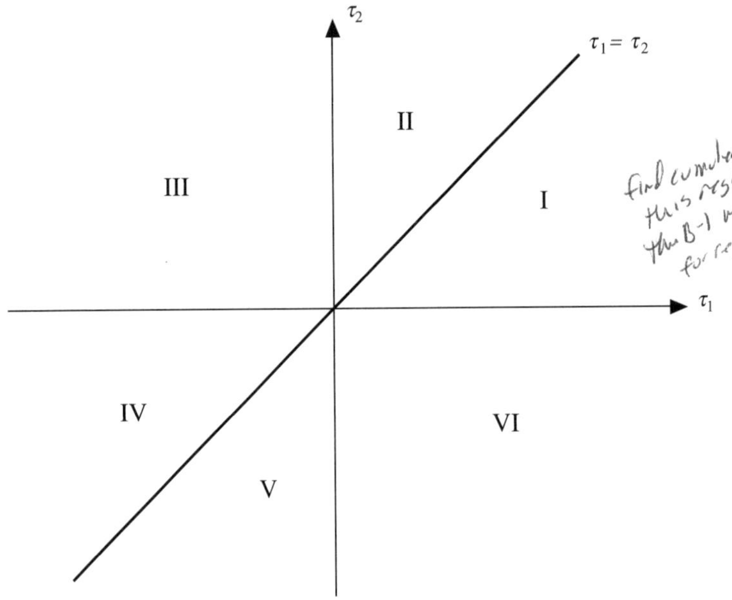


**Figure B-2** Symmetry regions for third-order cumulants.

As a second example, Figure B-5 depicts $C_{3,y}(\tau_1, \tau_2)$ for the ARMA (2, 1) system $H(z) = (z-2)/(z^2 - 1.5z + 0.8)$. The contour plot clearly demonstrates the nonsymmetrical nature of the complete third-order cumulant about the origin of the domain of support. The three-dimensional plot demonstrates that the largest value of the third-order cumulant occurs at the origin of its domain of support and that it can have negative as well as positive values. □

**Definition B-3.** *Assuming that* $C_{k,x}(\tau_1, \tau_2, \ldots, \tau_{k-1})$ *is absolutely summable, the* kth-*order* polyspectrum *is defined as the* (k − 1) *-dimensional discrete-time Fourier transform of the* kth-*order cumulant; i.e.,*

$$S_{k,x}(\omega_1, \omega_2, \ldots, \omega_{k-1}) = \sum_{\tau_1=-\infty}^{\infty} \cdots \sum_{\tau_{k-1}=-\infty}^{\infty} C_{k,x}(\tau_1, \tau_2, \ldots, \tau_{k-1}) \times \exp\left[-j \sum_{i=1}^{k-1} \omega_i \tau_i\right] \tag{B-21}$$

*where* $|\omega_i| \leq \pi$ *for* $i = 1, 2, \ldots, k-1$, *and* $|\omega_1 + \omega_2 + \cdots + \omega_{k-1}| \leq \pi$. □

The $\omega_1 - \omega_2 - \cdots - \omega_{k-1}$ space is the *domain of support* for $S_{k,x}(\omega_1, \omega_2, \ldots, \omega_{k-1})$. $S_{3,x}(\omega_1, \omega_2)$ is known as the *bispectrum* [in many papers the notation $B_x(\omega_1, \omega_2)$ is used to denote the bispectrum], whereas $S_{4,x}(\omega_1, \omega_2, \omega_3)$ is known as the *trispectrum*.

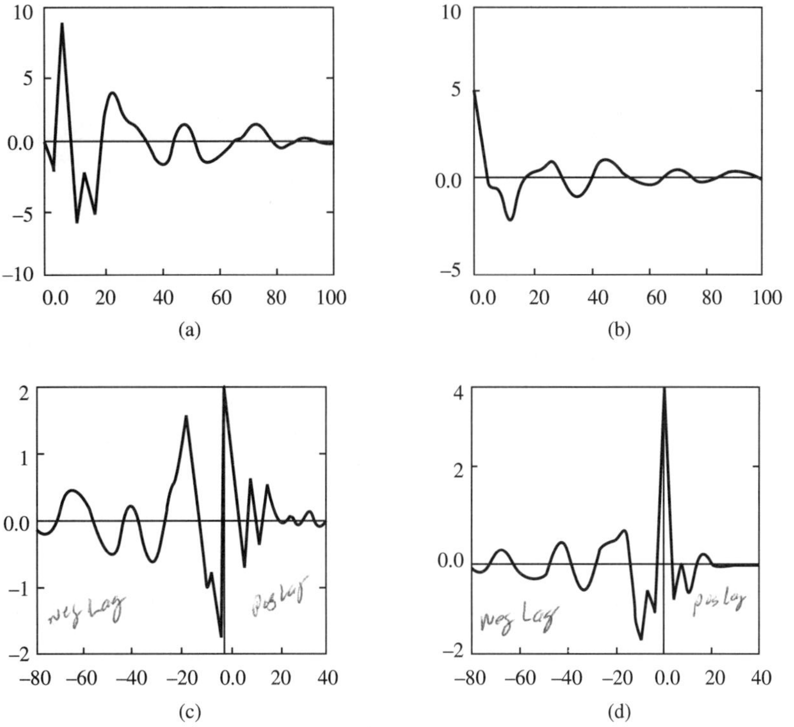


**Figure B-3** (a) Impulse response for a SISO system that is excited by non-Gaussian white noise and its associated (b) output autocorrelation, (c) output diagonal-slice third-order cumulant, and (d) output diagonal-slice fourth-order cumulant.

**Theorem B-2.** *(a) The bispectrum is symmetric in the following 11 ways:*

$$\begin{aligned} S_{3,x}(\omega_1, \omega_2) &= S_{3,x}(\omega_2, \omega_1) = S_{3,x}(\omega_1, -\omega_1 - \omega_2) = S_{3,x}(\omega_2, -\omega_1 - \omega_2) \\ &= S_{3,x}(-\omega_1 - \omega_2, \omega_1) = S_{3,x}(-\omega_1 - \omega_2, \omega_2) = S^*_{3,x}(-\omega_1, -\omega_2) \\ &= S^*_{3,x}(-\omega_2, -\omega_1) = S^*_{3,x}(-\omega_1, \omega_1 + \omega_2) = S^*_{3,x}(-\omega_2, \omega_1 + \omega_2) \\ &= S^*_{3,x}(\omega_1 + \omega_2, -\omega_1) = S^*_{3,x}(\omega_1 + \omega_2, -\omega_2) \end{aligned} \quad \text{(B-22)}$$

*where* $|\omega_1| \le \pi$, $|\omega_2| \le \pi$, *and* $|\omega_1 + \omega_2| \le \pi$. *(b) The trispectrum is symmetric in 95 ways, all of which can be found in Pflug et al. (1992), where* $|\omega_1| \le \pi$, $|\omega_2| \le \pi$, $|\omega_3| \le \pi$, *and* $|\omega_1 + \omega_2 + \omega_3| \le \pi$. □

The proof of this theorem works directly with the definitions of bispectrum and trispectrum, which are obtained from (B-21). The reader is asked to prove (B-22) in Problem B-5. Note that if $x(t)$ is complex then $S_{3,x}(\omega_1, \omega_2)$ is symmetric in only 5 ways [the five unconjugated equations in (B-22)]. When $x(t)$ is *real*, then $S_{3,x}(\omega_1, \omega_2) = S^*_{3,x}(-\omega_1, -\omega_2)$ so that symmetry occurs in 11 ways.

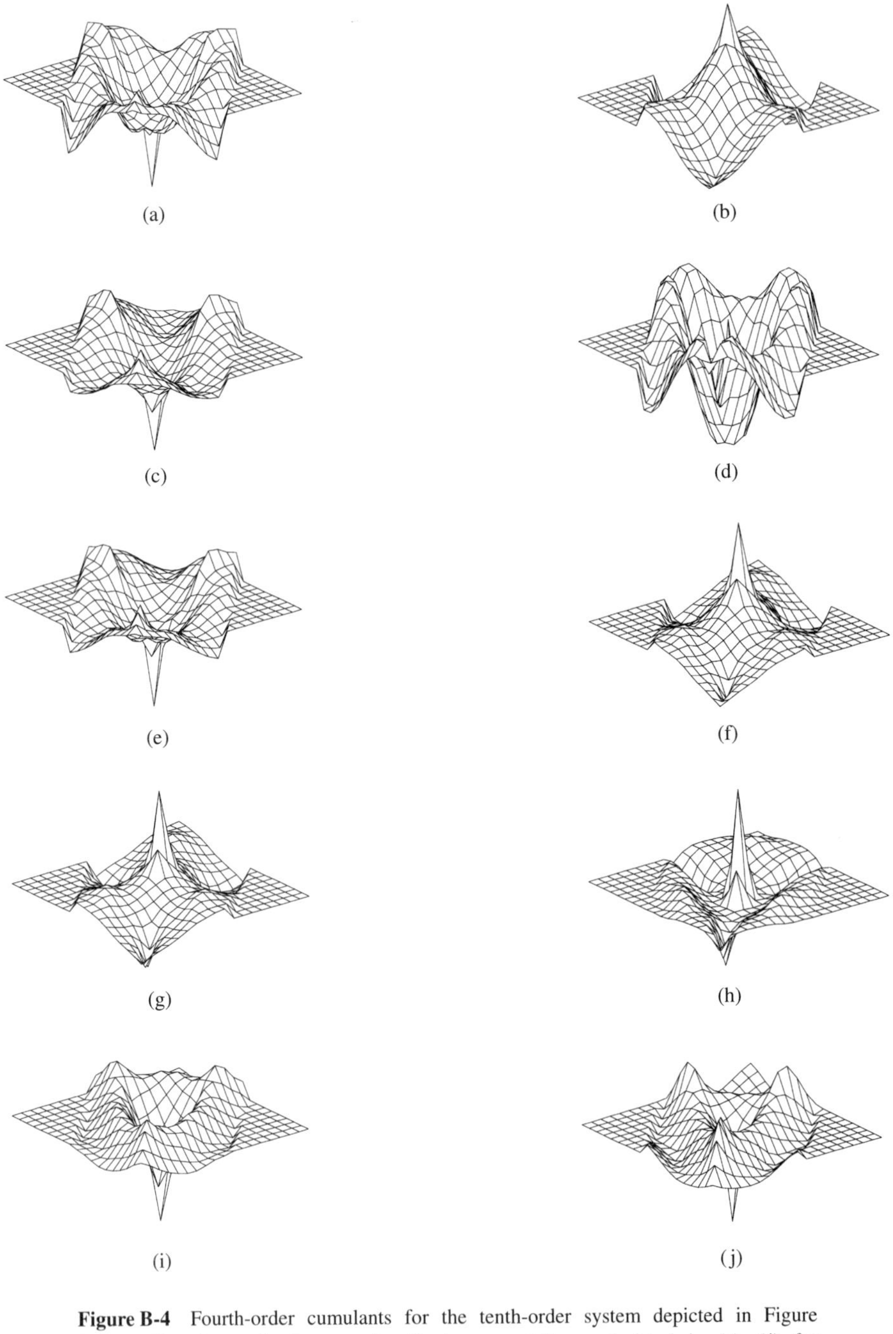


**Figure B-4** Fourth-order cumulants for the tenth-order system depicted in Figure B-3a. Cumulants $C_{4,y}(\tau_1, \tau_2, \tau_3)$, $-9 \leq \tau_1, \tau_2 \leq 9$ are depicted in (a)–(j) for $\tau_3 = 0, 1, \ldots, 9$. The orientations of the axes are as follows: origin is at the center of the grid; vertical axis is $C_{4,y}(\tau_1, \tau_2, \tau_3)$; $\tau_1$ axis points to the right; and $\tau_2$ axis points to the left (Swami and Mendel, 1990a, © 1990, IEEE).

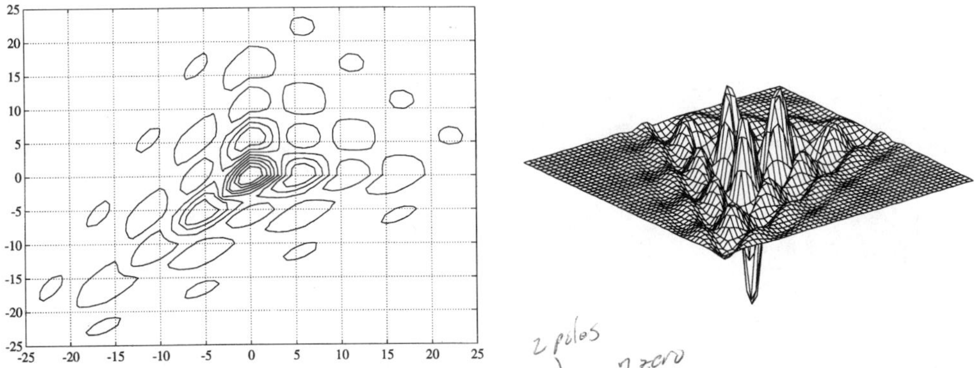


**Figure B-5** Third-order output cumulant for an ARMA(2, 1) SISO system that is excited by non-Gaussian white noise.

Using the 11 equations in (B-22), we can divide the $\omega_1 - \omega_2$ plane into 12 regions, as depicted in Figure B-6. Note that these regions are bounded because of the three inequalities on $\omega_1$ and $\omega_2$, which are stated just after (B-22). Knowing the bispectrum in any one of these regions, we can calculate the bispectrum in the other 11 regions using (B-22).

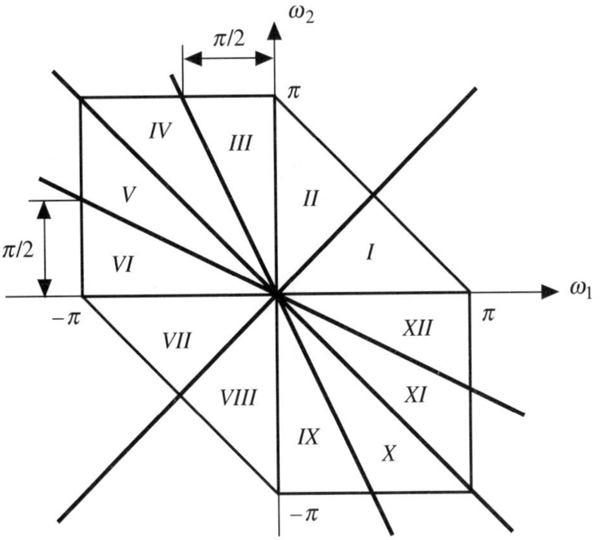


**Figure B-6** Symmetry regions for the bispectrum.

### EXAMPLE B-4

Figure B-7 depicts two different signals that have identical autocorrelation but different third-order statistics. Consequently, these two signals have identical power spectra and different bispectra. Figure (a-1) depicts the zeros of a minimum-phase system; i.e., all the zeros lie inside of the unit circle. Figure (b-1) depicts the zeros of a *spectrally equivalent* nonminimum-phase system; i.e., some of the zeros from the (a-1) constellation of zeros have

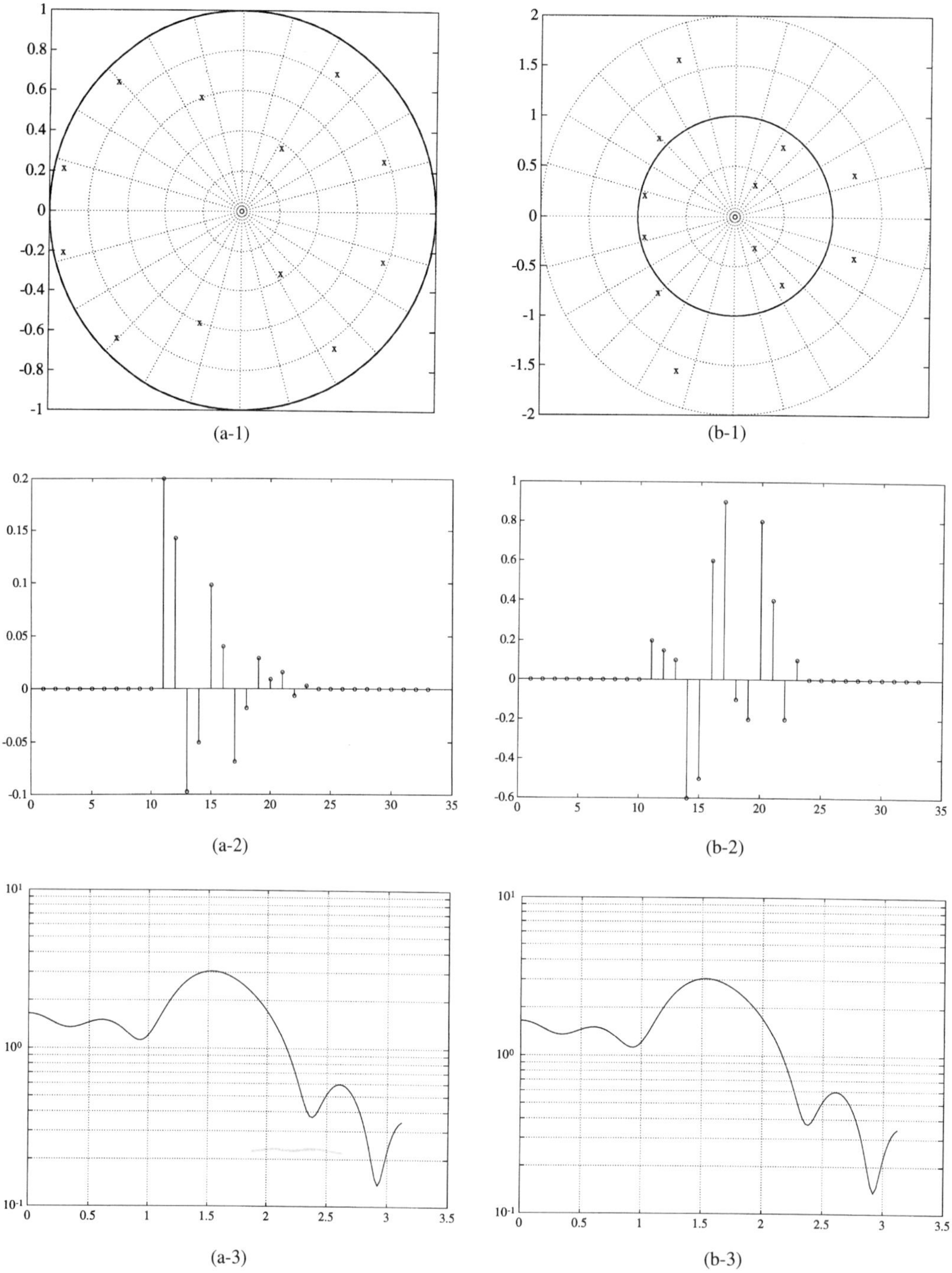


**Figure B-7** Comparisons of two systems that have the same power spectra but different bispectra. One system is minimum phase and the other is a spectrally equivalent nonminimum-phase system. See the text of Example B-4 for a description of the many parts of this figure.

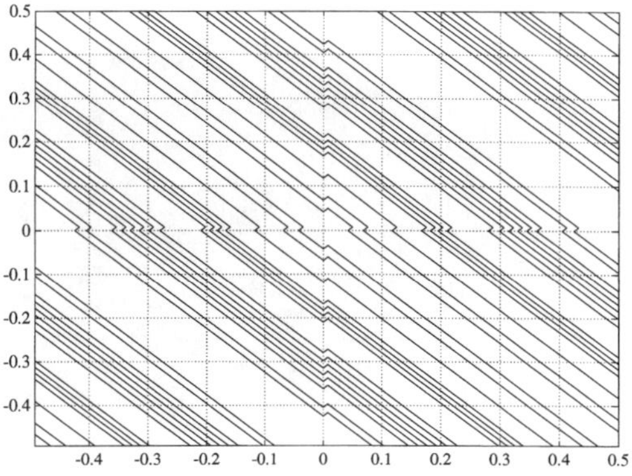


(a-4)

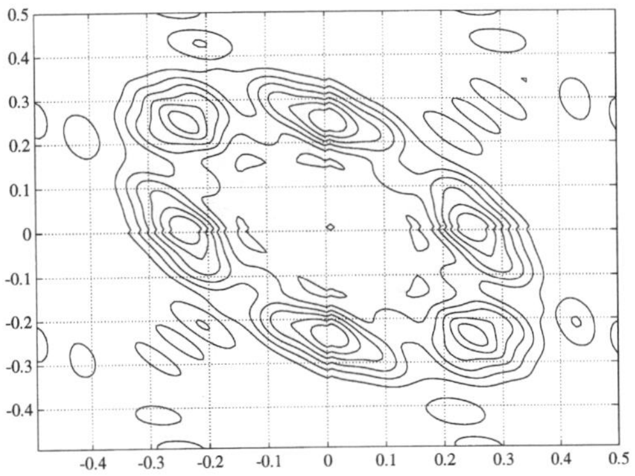


(b-4)

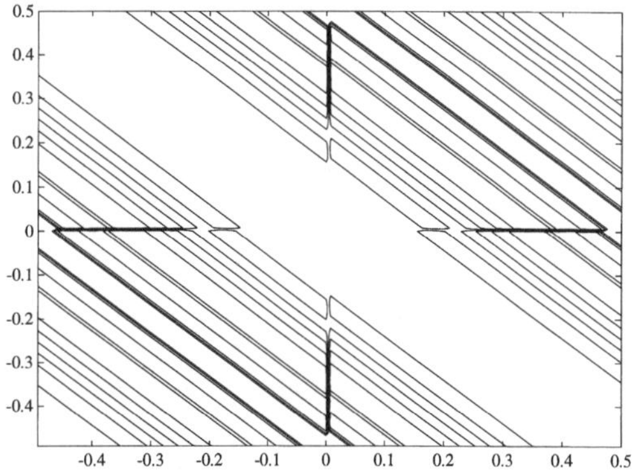


(a-5)

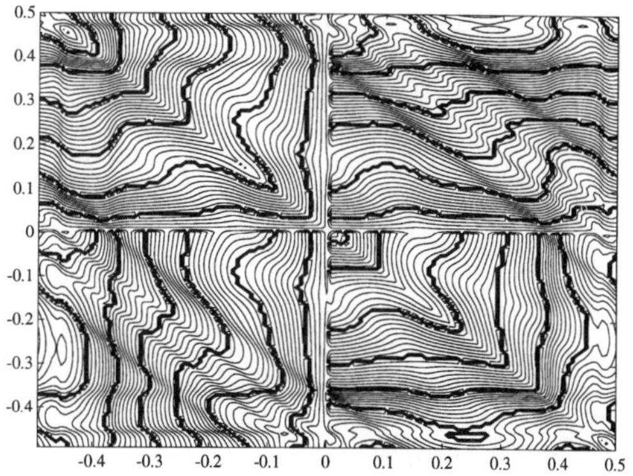


(b-5)

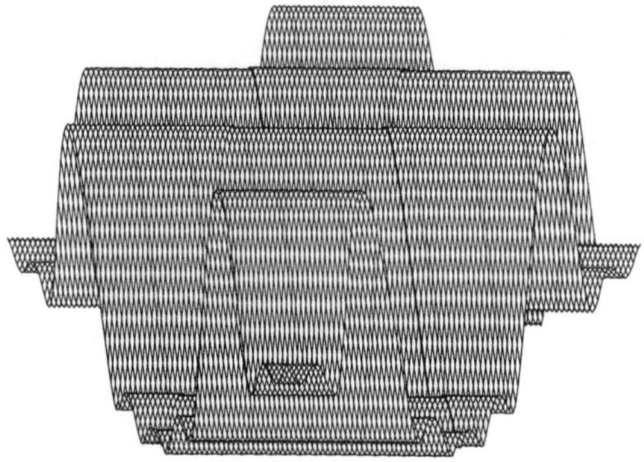

(a-6)

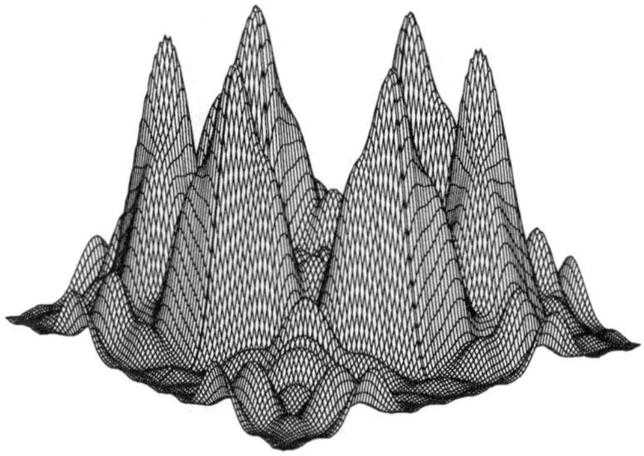

(b-6)

**Figure B-7** *(continued).*

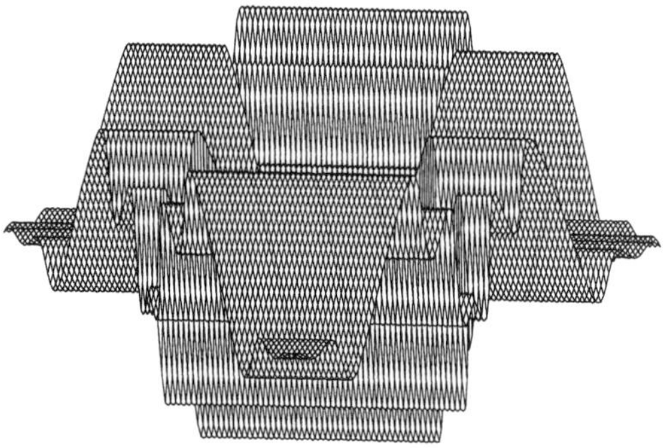

(a-7)

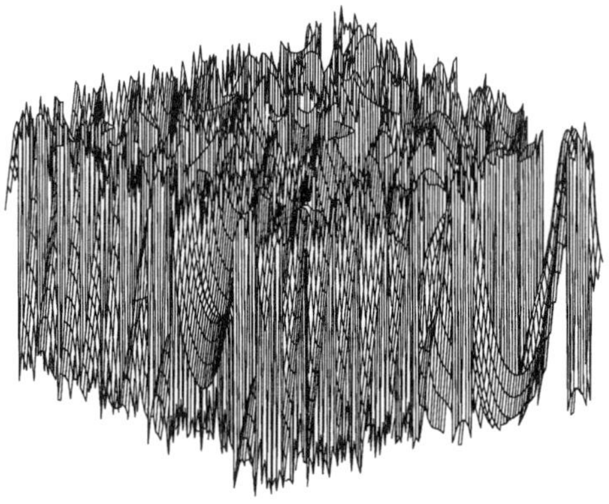

(b-7)

**Figure B-7** *(continued).*

been reflected to their reciprocal locations outside of the unit circle. The term spectrally equivalent denotes the fact that the spectrum, which is proportional to $|H(z)H(z^{-1})|$, is unchanged when $z$ is replaced by $z^{-1}$; hence, when some or all of a minimum-phase $H(z)$'s zeros are reflected (i.e., $z \rightarrow z^{-1}$) outside the unit circle (minimum phase $\rightarrow$ nonminimum phase), $|H(z)H(z^{-1})|$ remains unchanged.

Figures (a-2) and (b-2) depict the impulse responses for the minimum- and nonminimum-phase MA systems. Observe that the IR of the minimum-phase system is "front-loaded"; i.e., its maximum value occurs at its front end, whereas the maximum value of the IR of the nonminimum-phase system occurs between its initial and final values. This is a distinguishing feature of minimum- and nonminimum-phase systems.

Figures (a-3) and (b-3) depict the power spectra for the two systems, which of course are identical. We see, therefore, that it is impossible to discriminate between these two systems on the basis of their power spectra, because power spectrum is "phase blind."

The bispectrum is a complex function of two frequencies, $\omega_1$ and $\omega_2$. A contour of the magnitude of the bispectrum is depicted in (a-4) and (b-4) for the two systems. Examination of these two figures reveals significant differences. A contour of the phase of the bispectrum is depicted in (a-5) and (b-5) for the two systems. Three-dimensional plots of the magnitude and phase of the bispectra for the two systems are given in (a-6) and (a-7) and (b-6) and (b-7), respectively. It is very clear that the phase plots are also markedly different for the two systems. This suggests that it should be possible to discriminate between the two systems on the basis of their bispectra (or their third-order cumulants). We shall demonstrate the truth of this in Lesson C. □

## PROPERTIES OF CUMULANTS

One of the most powerful features of expectation is that it can be treated as a linear operator; i.e., $\mathbf{E}\{ax(k) + by(k)\} = a\mathbf{E}\{x(k)\} + b\mathbf{E}\{y(k)\}$. Once we have proved this fact, using the probabilistic definition of expectation, we do not have to reprove it every time we are faced with a new situation to which we apply it. Cumulants can also be treated as an operator; hence, they can be applied in a straightforward manner to many new situations. The next theorem provides the bases for doing this.

**Theorem B-3.** *Cumulants enjoy the following properties:*
**[CP1]** *If* $\lambda_i$, i = 1, ..., k, *are constants, and* $x_i$, i = 1, ..., k, *are random variables, then*

$$\text{Cum}(\lambda_1 x_1, \ldots, \lambda_k x_k) = \left(\prod_{i=1}^{k} \lambda_i\right) \text{Cum}(x_1, \ldots, x_k) \tag{B-23}$$

**[CP2]** *Cumulants are symmetric in their arguments; i.e.,*

$$\text{Cum}(x_1, \ldots, x_k) = \text{Cum}(x_{i_1}, \ldots, x_{i_k}) \tag{B-24}$$

*where* $(i_1, \ldots, i_k)$ *is a permutation of* (1, ..., k).
**[CP3]** *Cumulants are additive in their arguments; i.e.,*

$$\text{Cum}(x_0 + y_0, z_1, \ldots, z_k) = \text{Cum}(x_0, z_1, \ldots, z_k) + \text{Cum}(y_0, z_1, \ldots, z_k) \tag{B-25}$$

*This means that cumulants of sums equal sums of cumulants (hence, the name "cumulant").*
**[CP4]** *If* $\alpha$ *is a constant, then*

$$\text{Cum}(\alpha + z_1, z_2, \ldots, z_k) = \text{Cum}(z_1, \ldots, z_k) \tag{B-26}$$

**[CP5]** *If the random variables* $\{x_i\}$ *are independent of the random variables* $\{y_i\}$, i = 1, 2, ..., k, *then*

$$\text{Cum}(x_1 + y_1, \ldots, x_k + y_k) = \text{Cum}(x_1, \ldots, x_k) + \text{Cum}(y_1, \ldots, y_k) \tag{B-27}$$

**[CP6]** *If a subset of the* k *random variables* $\{x_i\}$ *is independent of the rest, then*

$$\text{Cum}(x_1, \ldots, x_k) = 0 \tag{B-28}$$

*Proof.* A complete proof of this theorem is given in the Supplementary Material at the end of this lesson. □

**EXAMPLE B-5** (Mendel, 1991)

Suppose $z(n) = y(n) + v(n)$, where $y(n)$ and $v(n)$ are independent; then, from [CP5],

$$C_{k,z}(\tau_1, \tau_2, \ldots, \tau_{k-1}) = C_{k,y}(\tau_1, \tau_2, \ldots, \tau_{k-1}) + C_{k,v}(\tau_1, \tau_2, \ldots, \tau_{k-1}) \tag{B-29}$$

If $v(n)$ is Gaussian (colored or white) and $k \geq 3$, then

$$C_{k,z}(\tau_1, \tau_2, \ldots, \tau_{k-1}) = C_{k,y}(\tau_1, \tau_2, \ldots, \tau_{k-1}) \tag{B-30}$$

whereas

$$C_{2,z}(\tau) = C_{2,y}(\tau) + C_{2,n}(\tau) \tag{B-31}$$

This makes higher-order statistics more robust to additive measurement noise than correlation, even if that noise is colored. In essence, cumulants can draw non-Gaussian signals out of Gaussian noise, thereby boosting their signal-to-noise ratios.

It is important to understand that (B-30) is a theoretical result. When cumulants are estimated from data using sample averages, the variances of these estimates are affected by the statistics of the additive noise $\nu(n)$. See Lesson C for some discussions on estimating cumulants. □

Additional applications of Theorem B-3 are given in Lesson C.

### Supplementary Material

## RELATIONSHIPS BETWEEN CUMULANTS AND MOMENTS (Mendel, 1991)

Let $\mathbf{x}$ denote a collection of random variables, i.e., $\mathbf{x} = \text{col}\,(x_1, x_2, \ldots, x_k)$, and $I_{\mathbf{x}} = \{1, 2, \ldots, k\}$, denote the set of indexes of the components of $\mathbf{x}$. If $I \subseteq I_{\mathbf{x}}$, then $\mathbf{x}_I$ is the vector consisting of those components of $\mathbf{x}$ whose indexes belong to $I$. We denote the simple moment and cumulant of the subvector $\mathbf{x}_I$ of the vector $\mathbf{x}$ as $m_{\mathbf{x}}(I)$ [i.e., $m_{\mathbf{x}}(I)$ is the expectation of the product of the elements in $\mathbf{x}_I$] and $C_{\mathbf{x}}(I)$. The *partition* of the set $I$ is the unordered collection of nonintersecting nonempty sets $I_p$ such that $\bigcup_p I_p = I$. For example, the set of partitions corresponding to $k = 3$ is $\{(1,2,3)\}, \{(1),(2,3)\}, \{(2),(1,3)\}, \{(3),(1,2)\}, \{(1),(2),(3)\}$.

The moment-to-cumulant (i.e., M–C) formula is [Leonov and Shiryaev (1959)]

$$C_{\mathbf{x}}(I) = \sum_{\bigcup_{p=1}^{q} I_p = I} (-1)^{q-1}(q-1)! \prod_{p=1}^{q} m_{\mathbf{x}}(I_p) \tag{B-32}$$

where $\bigcup_{p=1}^{q} I_p = I$ denotes summation over all partitions of set $I$. In the preceding example, $q = 1$ for $\{(1,2,3)\}$, $q = 2$ for $\{(1),(2,3)\}, \{(2),(1,3)\}$, and $\{(3),(1,2)\}$, and $q = 3$ for $\{(1),(2),(3)\}$. The cumulant-to-moment (i.e., C–M) equation is [Leonov and Shiryaev (1959)]

$$m_{\mathbf{x}}(I) = \sum_{\bigcup_{p=1}^{q} I_p = I} \prod_{p=1}^{q} C_{\mathbf{x}}(I_p) \tag{B-33}$$

An example that illustrates the use of (B-32) and (B-33) for $I = \{1, 2, 3, 4\}$ is given in Table B-1. In its bottom row, $\sum$ means add all preceding rows to obtain either cum $\{(x_1, x_2, x_3, x_4)\}$, using (B-32), or $\mathbf{E}\{(x_1 x_2 x_3 x_4)\}$ using (B-33).

## PROOF OF THEOREM B-3 (Mendel, 1991)

**Property [CP1]**: Let $\mathbf{y} = \text{col}\,(\lambda_1 x_1, \ldots, \lambda_k x_k)$, and $\mathbf{x} = \text{col}\,(x_1, \ldots, x_k)$. Note that (see previous section) $I_{\mathbf{x}} = I_{\mathbf{y}}$. From (B-32), we see that

$$C_{\mathbf{y}}(I_{\mathbf{y}}) = \sum_{\bigcup_{p=1}^{q} I_p = I_{\mathbf{y}}} (-1)^{q-1}(q-1)! \prod_{p=1}^{q} m_{\mathbf{y}}(I_p) \tag{B-34}$$

**TABLE B-1** CALCULATIONS OF FOURTH-ORDER CUMULANTS IN TERMS OF MOMENTS AND VICE VERSA (MENDEL, 1991)

| $I_1$ | $I_2$ | $I_3$ | $I_4$ | $q$ | M−C equation $(-1)^{q-1}(q-1)!\prod_{p=1}^{q} m_{\mathbf{x}}(I_p)$ | C−M equation $\prod_{p=1}^{q} C_{\mathbf{x}}(I_p)$ |
|---|---|---|---|---|---|---|
| 1 | 2 | 3 | 4 | 4 | $-6\mathbf{E}\{x_1\}\mathbf{E}\{x_2\}\mathbf{E}\{x_3\}\mathbf{E}\{x_4\}$ | $C(x_1)C(x_2)C(x_3)C(x_4)$ |
| 1, 2 | 3 | 4 | | 3 | $2\mathbf{E}\{x_1x_2\}\mathbf{E}\{x_3\}\mathbf{E}\{x_4\}$ | $C(x_1, x_2)C(x_3)C(x_4)$ |
| 1, 3 | 2 | 4 | | 3 | $2\mathbf{E}\{x_1x_3\}\mathbf{E}\{x_2\}\mathbf{E}\{x_4\}$ | $C(x_1, x_3)C(x_2)C(x_4)$ |
| 1, 4 | 2 | 3 | | 3 | $2\mathbf{E}\{x_1x_4\}\mathbf{E}\{x_2\}\mathbf{E}\{x_3\}$ | $C(x_1, x_4)C(x_2)C(x_3)$ |
| 2, 3 | 1 | 4 | | 3 | $2\mathbf{E}\{x_2x_3\}\mathbf{E}\{x_1\}\mathbf{E}\{x_4\}$ | $C(x_2, x_3)C(x_1)C(x_4)$ |
| 2, 4 | 1 | 3 | | 3 | $2\mathbf{E}\{x_2x_4\}\mathbf{E}\{x_1\}\mathbf{E}\{x_3\}$ | $C(x_2, x_4)C(x_1)C(x_3)$ |
| 3, 4 | 1 | 2 | | 3 | $2\mathbf{E}\{x_3x_4\}\mathbf{E}\{x_1\}\mathbf{E}\{x_2\}$ | $C(x_3, x_4)C(x_1)C(x_2)$ |
| 1, 2 | 3, 4 | | | 2 | $-\mathbf{E}\{x_1x_2\}\mathbf{E}\{x_3x_4\}$ | $C(x_1, x_2)C(x_3, x_4)$ |
| 1, 3 | 2, 4 | | | 2 | $-\mathbf{E}\{x_1x_3\}\mathbf{E}\{x_2, x_4\}$ | $C(x_1, x_3)C(x_2, x_4)$ |
| 1, 4 | 2, 3 | | | 2 | $-\mathbf{E}\{x_1x_4\}\mathbf{E}\{x_2x_3\}$ | $C(x_1, x_4)C(x_2, x_3)$ |
| 1, 2, 3 | 4 | | | 2 | $-\mathbf{E}\{x_1x_2x_3\}\mathbf{E}\{x_4\}$ | $C(x_1, x_2, x_3)C(x_4)$ |
| 1, 2, 4 | 3 | | | 2 | $-\mathbf{E}\{x_1x_2x_4\}\mathbf{E}\{x_3\}$ | $C(x_1, x_2, x_4)C(x_3)$ |
| 1, 3, 4 | 2 | | | 2 | $-\mathbf{E}\{x_1x_3x_4\}\mathbf{E}\{x_2\}$ | $C(x_1, x_3, x_4)C(x_2)$ |
| 2, 3, 4 | 1 | | | 2 | $-\mathbf{E}\{x_2x_3x_4\}\mathbf{E}\{x_1\}$ | $C(x_2, x_3, x_4)C(x_1)$ |
| 1, 2, 3, 4 | | | | 1 | $\mathbf{E}\{x_1x_2x_3x_4\}$ | $C(x_1, x_2, x_3, x_4)$ |
| | | | | $\sum$ | cum $\{x_1, x_2, x_3, x_4\}$ | $\mathbf{E}\{x_1x_2x_3x_4\}$ |

where [for example, apply the results in Table B-1 to $\mathbf{y} = \text{col}\,(\lambda_1 x_1, \lambda_2 x_2, \lambda_3 x_3, \lambda_4 x_4)$]

$$\prod_{p=1}^{q} m_{\mathbf{y}}(I_p) = \left(\prod_{p=1}^{k} \lambda_p\right)\left(\prod_{p=1}^{q} m_{\mathbf{x}}(I_p)\right) \tag{B-35}$$

Consequently,

$$C_{\mathbf{y}}(I_{\mathbf{y}}) = \left(\prod_{p=1}^{k} \lambda_p\right) C_{\mathbf{x}}(I_{\mathbf{x}}) \tag{B-36}$$

which is (B-23).

**Property [CP2]**: Referring to (B-32), since the partition of the set $I_{\mathbf{x}}$ is an unordered collection of nonintersecting nonempty sets of $I_p$ such that $\bigcup_{p=1}^{q} I_p = I_{\mathbf{x}}$, the order in the cumulant's argument is irrelevant to the value of the cumulant. As a result, cumulants are symmetric in their arguments.

**Property [CP3]**: Let $\mathbf{x} = \text{col}\,(u_1+v_1, x_2, \ldots, x_k)$, where $\mathbf{u} = \text{col}\,(u_1, x_2, \ldots, x_k)$ and $\mathbf{v} = \text{col}\,(v_1, x_2, \ldots, x_k)$. Observe that [because $m_{\mathbf{x}}(I_i)$ is the expectation of the product of the elements in $I_i$, and $u_1 + v_1$ appears only raised to the unity power]

$$\prod_{i=1}^{q} m_{\mathbf{x}}(I_i) = \prod_{i=1}^{q} m_{\mathbf{u}}(I_i) + \prod_{i=1}^{q} m_{\mathbf{v}}(I_i) \tag{B-37}$$

Substitute (B-37) into (B-32) to obtain the result in (B-25).

**Property [CP4]**: Let $\mathbf{y} = \text{col}\,(\alpha + z_1, z_2, \ldots, z_k)$; then, from (B-2), we see that

$$\begin{aligned} K(\mathbf{v}) &= \ln \mathbf{E}\{\exp(j\mathbf{v}'\mathbf{y})\} = \ln \mathbf{E}\{\exp(j[\nu_1(\alpha + z_1) + \nu_2 z_2 + \cdots + \nu_k z_k])\} \\ &= \ln \mathbf{E}\{\exp(j\nu_1\alpha)\} + \ln \mathbf{E}\{\exp(j\nu_1 z_1 + \cdots + j\nu_k z_k)\} \end{aligned} \tag{B-38}$$

According to the paragraph that precedes (B-2), we know that

$$\text{cum}\,(\alpha + z_1, z_2, \ldots, z_k) = \frac{1}{k!}\left[\frac{\partial^k}{\partial \nu_1 \partial \nu_2 \ldots \partial \nu_k}\right] K(\mathbf{v})\Big|_{\mathbf{v}=\mathbf{0}} \tag{B-39}$$

but, from (B-38), we see that

$$\begin{aligned} \frac{1}{k!}\left[\frac{\partial^k}{\partial \nu_1 \partial \nu_2 \ldots \partial \nu_k}\right] K(\mathbf{v})\Big|_{\mathbf{v}=\mathbf{0}} &= \frac{1}{k!}\left[\frac{\partial^k}{\partial \nu_1 \partial \nu_2 \ldots \partial \nu_k}\right] [\ln \mathbf{E}\{\exp(j\nu_1\alpha)\} \\ &\quad + \ln \mathbf{E}\{\exp(j\nu_1 z_1 + \cdots + j\nu_k z_k)\}]|_{\mathbf{v}=\mathbf{0}} \\ &= \frac{1}{k!}\left[\frac{\partial^k}{\partial \nu_1 \partial \nu_2 \ldots \partial \nu_k}\right] \\ &\quad \times \ln \mathbf{E}\{\exp(j\nu_1 z_1 + \cdots + j\nu_k z_k)\}]|_{\mathbf{v}=\mathbf{0}} \\ &= \text{cum}\,(z_1, z_2, \ldots, z_k) \end{aligned}$$

which is (B-26).

**Property [CP5]**: Let $\mathbf{z} = \text{col}\,(x_1 + y_1, \ldots, x_k + y_k) = \mathbf{x} + \mathbf{y}$, where $\mathbf{x} = \text{col}\,(x_1, \ldots, x_k)$ and $\mathbf{y} = \text{col}\,(y_1, \ldots, y_k)$. Using the independence of the $\{x_i\}$ and $\{y_i\}$, it follows that

$$\begin{aligned} K_{\mathbf{z}}(\mathbf{v}) &= \ln \mathbf{E}\{\exp[j\nu_1(x_1 + y_1) + \cdots + j\nu_k(x_k + y_k)]\} \\ &= \ln \mathbf{E}\{\exp(j\nu_1 x_1 + \cdots + j\nu_k x_k)\} + \ln \mathbf{E}\{\exp(j\nu_1 y_1 + \cdots + j\nu_k y_k)\} \\ &= K_{\mathbf{x}}(\mathbf{v}) + K_{\mathbf{y}}(\mathbf{v}) \end{aligned} \tag{B-40}$$

from which the result in (B-27) follows directly.

**Property [CP6]**: Assume that $(x_1, \ldots, x_i)$ is independent of $(x_{i+1}, \ldots, x_k)$; hence,

$$K(\mathbf{v}) = \ln \mathbf{E}\{\exp(j\nu_1 x_1 + \cdots + j\nu_i x_i)\} + \ln \mathbf{E}\{\exp[j\nu_{i+1} x_{i+1} + \cdots + j\nu_k x_k)\} \tag{B-41}$$

Now

$$\begin{aligned} \frac{1}{k!}\left[\frac{\partial^k}{\partial \nu_1 \partial \nu_2 \ldots \partial \nu_k}\right] K(\mathbf{v})\Big|_{\mathbf{v}=\mathbf{0}} &= \frac{1}{k!}\left[\frac{\partial^k}{\partial \nu_1 \partial \nu_2 \ldots \partial \nu_k}\right] \\ &\quad [\ln \mathbf{E}\{\exp(j\nu_1 x_1 + \cdots + j\nu_i x_i)\} \\ &\quad + \ln \mathbf{E}\{\exp(j\nu_{i+1} x_{i+1} + \cdots + j\nu_k x_k)\}]|_{\mathbf{v}=\mathbf{0}} = 0 \end{aligned}$$

which is (B-28).

## SUMMARY QUESTIONS

1. Cumulants and moments are:
   (a) always the same
   (b) never the same
   (c) sometimes the same
2. The domain of support for $C_{k,x}(\tau_1, \tau_2, \ldots, \tau_{k-1})$ is the:
   (a) $\tau_2 - \tau_1, \tau_3 - \tau_1, \ldots, \tau_{k-1} - \tau_1$ space
   (b) $t, \tau_1, \tau_2, \ldots, \tau_{k-1}$ space
   (c) $\tau_1, \tau_2, \ldots, \tau_{k-1}$ space
3. Cumulants not only display the amount of higher-order correlation, but also provide a measure of:
   (a) non-Gaussianity
   (b) the distance of the random process from Gaussianity
   (c) causality
4. Symmetries, which exist in the arguments of $C_{k,x}(\tau_1, \tau_2, \ldots, \tau_{k-1})$:
   (a) are only useful in theoretical analyses
   (b) are only valid for $k = 3$
   (c) make the calculations of cumulants manageable
5. The bispectrum is the discrete-time Fourier transform of the:
   (a) fourth-order cumulant
   (b) third-order cumulant
   (c) correlation function
6. In general, cumulants are preferrable to higher-order moments, because:
   (a) cumulants of two (or more) statistically independent random sequences equal the sum of the cumulants of the two (or more) sequences, whereas the same is not necessarily true for higher-order moments
   (b) cumulants are easier to compute than higher-order moments
   (c) odd-order cumulants of symmetrically distributed sequences equal zero
7. Cumulants are said to "boost signal-to-noise ratio" because:
   (a) cum(signal + non-Gaussian noise) = cum(signal) + cum(non-Gaussian noise)
   (b) cum(signal + Gaussian noise) = cum(signal)
   (c) cum(signal + Gaussian noise) = cum(signal) × cum(Gaussian noise)
8. Cumulants and moments can:
   (a) never be determined from one another
   (b) sometimes be determined from one another
   (c) always be determined from one another
9. Cumulant properties:
   (a) let us treat the cumulant as an operator
   (b) always have to be reproved
   (c) are interesting, but are of no practical value

## PROBLEMS

**B-1.** Derive Equations (B-3b) and (B-3c) using Equation (B-2).

**B-2.** Beginning with Equation (B-17), derive the third- and fourth-order cumulant formulas that are given in Equations (B-12b) and (B-12c).

**B-3.** Derive the symmetry conditions given in Equation (B-18) for the third-order cumulant $C_{3,x}(\tau_1, \tau_2)$. Then explain how Figure B-2 is obtained from these conditions.

**B-4.** Derive the symmetry conditions given in Equations (B-19) and (B-20) for the fourth-order cumulant $C_{4,x}(\tau_1, \tau_2, \tau_3)$.

**B-5.** Derive the bispectrum symmetry conditions that are given in Equation (B-22). Then explain how Figure B-5 is obtained from these conditions.

**B-6.** Consider a random variable $x$ that is exponentially distributed; i.e., its probability density function is $p(x) = \lambda e^{-\lambda x} u_{-1}(x)$. Let $m_{i,x}$ and $C_{i,x}$ denote the $i$th-order moment and cumulant, respectively, for $x$. Show that (Nikias and Petropulu, 1993): $m_{1,x} = 1/\lambda, m_{2,x} = 2/\lambda^2, m_{3,x} = 6/\lambda^3, m_{4,x} = 24/\lambda^4$, and $C_{1,x} = 1/\lambda, C_{2,x} = 1/\lambda^2, C_{3,x} = 2/\lambda^3, C_{4,x} = 6/\lambda^4$. For comparable results about Rayleigh or $K$-distributed random variables, see page 11 of Nikias and Petropulu (1993).

**B-7.** Consider a random variable $x$ that is Laplace distributed; i.e., its probability density function is $p(x) = 0.5e^{-|x|}$. Let $m_{i,x}$ and $C_{i,x}$ denote the $i$th-order moment and cumulant, respectively, for $x$. Show that (Nikias and Petropulu, 1993): $m_{1,x} = 0, m_{2,x} = 2, m_{3,x} = 0, m_{4,x} = 24$, and $C_{1,x} = 0, C_{2,x} = 2, C_{3,x} = 0, C_{4,x} = 12$. For comparable results about Gaussian or uniformly distributed random variables, see page 10 of Nikias and Petropulu (1993). Note that for these three distributions $x$ is "symmetrically distributed" so that all odd-order moments or cumulants equal zero.

**B-8.** Let $a = \exp(j\phi)$, where $\phi$ is uniformly distributed over $[-\pi, \pi]$.

**(a)** Show that all third-order cumulants of complex harmonic $a$ are always zero.

**(b)** Show that, of the three different ways (different in the sense of which of the variables should be conjugated) to define a fourth-order cumulant of a complex harmonic, only one always yields a nonzero value; i.e., $\text{cum}\,(a, a, a, a) = 0$, $\text{cum}\,(a^*, a, a, a) = 0$, but $\text{cum}\,(a^*, a^*, a, a) = -1$. These results suggest that the fourth-order cumulant of the complex random sequence $\{y(n)\}$, should be defined as

$$C_{4,y}(\tau_1, \tau_2, \tau_3) = \text{cum}\,(y^*(n), y^*(n + \tau_1), y(n + \tau_2), y(n + \tau_3))$$

**(c)** Explain why it doesn't matter which two of the four arguments in this definition are conjugated.

**B-9.** Let $S = \exp(j\phi)$, where $\phi$ is uniformly distributed over $[-\pi, \pi]$, and let $a_l (l = 0, 1, 2, 3)$ be constants. Prove that (Swami, 1988)

$$\text{cum}(a_0 S + a_0^* S^*, a_1 S + a_1^* S^*, a_2 S + a_2^* S^*, a_3 S + a_3^* S^*)$$
$$= -2\,\text{Re}\,(a_0 a_1 a_2^* a_3^* + a_0 a_1^* a_2^* a_3 + a_0 a_1^* a_2 a_3^*)$$

**B-10.** A model that can be used to describe a wide range of problems is (Prasad et al., 1988)

$$y(n) = \sum_{i=1}^{p} a_i(n) S_n(\omega_i)$$

where the $S_n(\cdot)$'s are signal waveshapes, $\omega_i$'s are constants, and the $a_i(n)$'s are zero mean and mutually independent, with fourth-order cumulant $C_{4,a_i}(\tau_1, \tau_2, \tau_3)$. Show that

$$\text{cum}\,(y^*(n), y^*(n + \tau_1), y(n + \tau_2), y(n + \tau_3)) =$$
$$\sum_{i=1}^{p} S_n^*(\omega_i) S_{n+\tau_1}^*(\omega_i) S_{n+\tau_2}(\omega_i) S_{n+\tau_3}(\omega_i) \times C_{4,a_i}(\tau_1, \tau_2, \tau_3)$$

**B-11.** In the harmonic retrieval problem for complex signals, the model is

$$y(n) = \sum_{i=1}^{p} \alpha_i \exp\{j(\omega_i n + \phi_i)\}$$

where the $\phi_i$'s are independent identically distributed random variables uniformly distributed over $[-\pi, \pi]$, $\omega_i \neq \omega_j$ for $i \neq j$, and the $\alpha_i$'s and $\omega_i$'s are constants (i.e., not random). Given measurements of $y(n)$, the objective is to determine the number of harmonics, $p$, their frequencies, $\omega_i$, and their amplitudes, $\alpha_i$.

**(a)** Derive the following formula for the autocorrelation of $y(n)$:

$$C_{2,y}(\tau) = \sum_{k=1}^{p} \alpha_k^2 \exp(j\omega_k \tau)$$

**(b)** Show that $C_{3,y}(\tau_1, \tau_2) = 0$; hence, we must use fourth-order cumulants in this application.

**(c)** Derive the following formula for $C_{4,y}(\tau_1, \tau_2, \tau_3)$ (Swami and Mendel, 1991):

$$C_{4,y}(\tau_1, \tau_2, \tau_3) = -\sum_{k=1}^{p} \alpha_k^4 \exp\{j\omega_k(-\tau_1 + \tau_2 + \tau_3)\}$$

[*Hint*: Use the result of Problem B-10, with $S_n(\omega_i) = \exp(jn\omega_i)$ and $a_i(n) = a_i = \alpha_i \exp(j\phi_i)$.]

**(d)** Now assume that only the noise-corrupted measurements $z(n) = y(n) + v(n)$ are available, where $v(n)$ is Gaussian colored noise. Determine formulas for $C_{2,z}(\tau)$ and $C_{4,z}(\tau_1, \tau_2, \tau_3)$. Explain why any method for determining the unknown parameters that is based on fourth-order cumulants will still work when we have access only to $z(n)$, whereas a method based on second-order statistics will not work when we have access only to $z(n)$.

Observe that when $\tau_1 = \tau_2 = \tau_3$ the equation for $C_{4,y}(\tau, \tau, \tau)$ looks very similar to the equation for $C_{2,y}(\tau)$. This suggests that a method for determining $p, \omega_i$ and $\alpha_i$ that is based on second-order statistics can also be applied to $C_{4,y}(\tau, \tau, \tau)$. See Swami and Mendel (1991) for details on how to do this.

**B-12.** In the harmonic retrieval problem for real signals, the model is

$$y(n) = \sum_{i=1}^{p} \alpha_i \cos(\omega_i n + \phi_i)$$

where the $\phi_i$'s are independent identically distributed random variables uniformly distributed over $[-\pi, \pi]$, the frequencies are distinct, and the $\alpha_i$'s and $\omega_i$'s are constants (i.e., not random). Given measurements of $y(n)$, the objective is to determine the number of harmonics, $p$, their frequencies, $\omega_i$, and their amplitudes, $\alpha_i$.

**(a)** Derive the following formula for the autocorrelation of $y(n)$:

$$C_{2,y}(\tau) = \frac{1}{2} \sum_{k=1}^{p} \alpha_k^2 \cos(\omega_k \tau)$$

**(b)** Show that $C_{3,y}(\tau_1, \tau_2) = 0$; hence, we must use fourth-order cumulants in this application.

**(c)** Derive the following formula for $C_{4,y}(\tau_1, \tau_2, \tau_3)$ (Swami and Mendel, 1991):

$$C_{4,y}(\tau_1, \tau_2, \tau_3) = -\frac{1}{8}\sum_{k=1}^{p} \alpha_k^4[\cos\omega_k(\tau_1 - \tau_2 - \tau_3) + \cos\omega_k(\tau_2 - \tau_3 - \tau_1) + \cos\omega_k(\tau_3 - \tau_1 - \tau_2)]$$

[*Hint*: Express each cosine function as a sum of two complex exponentials, and use the results of Problems B-10 and B-9.]

**(d)** Now assume that only the noise-corrupted measurements $z(n) = y(n) + \nu(n)$ are available, where $\nu(n)$ is Gaussian colored noise. Determine formulas for $C_{2,z}(\tau)$ and $C_{4,z}(\tau_1, \tau_2, \tau_3)$. Explain why any method for determining the unknown parameters that is based on fourth-order cumulants will still work when we have access only to $z(n)$, whereas a method based on second-order statistics will not work when we have access only to $z(n)$.

Observe that when $\tau_1 = \tau_2 = \tau_3$ the equation for $C_{4,y}(\tau, \tau, \tau)$ looks very similar to the equation for $C_{2,y}(\tau)$. This suggests that a method for determining $p, \omega_i$, and $\alpha_i$ that is based on second-order statistics can also be applied to $C_{4,y}(\tau, \tau, \tau)$. See Swami and Mendel (1991) for details on how to do this.

**B-13.** We are given two measurements, $x(n) = S(n) + w_1(n)$ and $y(n) = AS(n-D) + w_2(n)$, where $S(n)$ is an unknown random signal and $w_1(n)$ and $w_2(n)$ are Gaussian noise sources. The objective in this *time delay estimation* problem is to estimate the unknown time delay, $D$ [Nikias and Pan (1988)].

**(a)** Assume that $w_1(n)$ and $w_2(n)$ are uncorrelated, and show that $r_{xy}(\tau) = \mathbf{E}\{x(n)y(n+\tau)\} = r_{ss}(\tau - D)$. This suggests that we can determine $D$ by locating the time at which the cross-correlation function $r_{xy}(\tau)$ is a maximum.

**(b)** Now assume that $w_1(n)$ and $w_2(n)$ are correlated and Gaussian, and show that $r_{xy}(\tau) = r_{ss}(\tau - D) + r_{12}(\tau)$, where $r_{12}(\tau)$ is the cross-correlation function between $w_1(n)$ and $w_2(n)$. Because this cross-correlation is unknown, the method stated in part (a) for determining $D$ won't work in the present case.

**(c)** Again assume that $w_1(n)$ and $w_2(n)$ are correlated and Gaussian, and show that the third-order cross-cumulant $C_{xyx}(\tau, \rho) \triangleq \mathbf{E}\{x(n)y(n+\tau)x(n+\rho)\} = r_{ss}(\tau - D, \rho)$, and $C_{3,x}(\tau, \rho) \triangleq \mathbf{E}\{x(n)x(n+\tau)x(n+\rho)\} = r_{ss}(\tau, \rho)$. Then show that $S_{xyx}(\omega_1, \omega_2) = S_{3,s}(\omega_1, \omega_2)\exp(-j\omega_1 D)$ and $S_{xxx}(\omega_1, \omega_2) = S_{3,s}(\omega_1, \omega_2)$. Suggest a method for determining time-delay $D$ from these results. Observe that the approach in this part of the problem has been able to handle the case of correlated Gaussian noises, whereas the second-order statistics-based approach of part (b) was not able to do it.

# LESSON C

# Estimation and Applications of Higher-order Statistics

from Real data + apply them to problems

## SUMMARY

Higher-order statistics can either be estimated from data or computed from models. In this lesson we learn how to do both. Cumulants are estimated from data by sample-averaging techniques. Polyspectra are estimated from data either by taking multidimensional discrete-time Fourier transforms of estimated cumulants or by combining the multidimensional discrete Fourier transform of the data using windowing techniques. Because cumulant and polyspectra estimates are random, their statistical properties are important. These properties are described.

Just as it is important to be able to compute the second-order statistics of the output of a single-input, single-output dynamical system, it is important to be able to compute the higher-order statistics of the output for such a system. This can be accomplished using the Bartlett–Brillinger–Rosenblatt formulas. One formula lets us compute the $k$th-order cumulant of the system's output from knowledge about the system's impulse response. Another formula lets us compute the $k$th-order polyspectrum of the system's output from knowledge about the Fourier transform of the system's impulse response.

Not only are the Bartlett–Brillinger–Rosenblatt formulas useful for computing cumulants and polyspectra for given systems, but they are also the starting points for many important related results, including a closed-form formula (the $C(q, k)$ formula) for determining the impulse response of an MA($q$) system; a closed-form formula (the $q$-slice formula) for determining the impulse response of an ARMA($p, q$) system, given that we know the AR coefficients of the system; cumulant-based normal equations for determining the AR coefficients of either an AR($p$) system or an ARMA($p, q$) system; a nonlinear system of equations (the GM equations) that relates second-order statistics to higher-order statistics, which can be used to determine the coefficients of an MA($q$) system; and equations that relate a system and its inverse, which lead to new optimal deconvolution filters that are based on higher-order statistics.

When you complete this lesson, you will be able to (1) estimate cumulants from data; (2) estimate polyspectra from data; (3) understand the statistical properties of cumulant and polyspectral estimators; (4) derive the Bartlett–Brillinger–Rosenblatt formulas; (4) apply the Bartlett–Brillinger–Rosenblatt formulas to compute cumulants and polyspectra for single-input and single-output dynamical systems; and (5) understand many applications of the Bartlett–Brillinger–Rosenblatt formulas to system identification and deconvolution problems.

## ESTIMATION OF CUMULANTS (Nikias and Mendel, 1993)

In many practical situations we are given data and want to estimate cumulants from the data. Later in this lesson we describe applications where this must be done in order to extract useful information about non-Gaussian signals from the data. Cumulants involve expectations and cannot be computed in an exact manner from real data; *they must be estimated*, in much the same way that correlations are estimated from real data. Cumulants are estimated by replacing expectations by sample averages; i.e.,

$$C_{3,x}(\tau_1, \tau_2) \simeq \hat{C}_{3,x}(\tau_1, \tau_2) = \frac{1}{N_R} \sum_{t \in R} x(t)x(t+\tau_1)x(t+\tau_2) \qquad \text{(C-1)}$$

where $N_R$ is the number of samples in region $R$. A similar but more complicated equation can be obtained for $\hat{C}_{4,x}(\tau_1, \tau_2, \tau_3)$ by beginning with (B-12c) (Problem C-1). It will not only involve a sum of a product of four terms [analogous to (C-1)], but it will also involve products of sample correlations, where

$$C_{2,x}(\tau) \simeq \hat{C}_{2,x}(\tau) = \frac{1}{N_R} \sum_{t \in R} x(t)x(t+\tau) \qquad \text{(C-2)}$$

A practical algorithm for estimating $C_{3,x}(\tau_1, \tau_2)$ is (Nikias and Mendel, 1993): Let $\{x(1), x(2), \ldots, x(N)\}$ be the given data set; then (1) segment the data into $K$ records of $M$ samples each, i.e., $N = KM$; (2) subtract the average value of each record; (3) assuming that $\{x^{(i)}(k), k = 0, 1, \ldots, M-1\}$, is the zero-mean data set per segment $i = 1, 2, \ldots, K$, obtain an estimate of the third-moment sequence

$$C_{3,x}^{(i)}(\tau_1, \tau_2) = \frac{1}{M} \sum_{j=s_1}^{s_2} x^{(i)}(j)x^{(i)}(j+\tau_1)x^{(i)}(j+\tau_2) \qquad \text{(C-3)}$$

where $i = 1, 2, \ldots, K$; $s_1 = \max(0, -\tau_1, -\tau_2)$; and, $s_2 = \min(M-1, M-1-\tau_1, M-1-\tau_2)$ (Problem C-2); (4) average $C_{3,x}^{(i)}(\tau_1, \tau_2)$ over all segments to obtain the final estimate

$$\hat{C}_{3,x}(\tau_1, \tau_2) = \frac{1}{K} \sum_{i=1}^{K} C_{3,x}^{(i)}(\tau_1, \tau_2) \qquad \text{(C-4)}$$

Of course, we should use the many symmetry properties of third (fourth)-order cumulants, which are given in Lesson B, to reduce the number of computations.

It is well known (e.g., Ljung and Soderstrom, 1983, and Priestley, 1981) that exponential stability of the underlying channel model guarantees the convergence in probability of the sampled autocorrelation function to the true autocorrelation function. A correlation estimator that uses several independent realizations is very reliable, since doing this reduces the variance of the estimator. It is preferrable to use a biased unsegmented correlation estimator to an unbiased unsegmented correlation estimator, because the biased estimator leads to a positive-definite covariance matrix, whereas the unbiased estimator does not (Priestley, 1981). When we only have a single realization, as is often the case in signal-processing applications, the biased segmented estimator gives poorer correlation estimates than both the unbiased segmented estimator and the biased unsegmented estimator. Some of these issues are explored in Problems C-3 and C-4 at the end of this lesson.

The convergence of sampled third-order cumulants to true third-order cumulants is studied in Rosenblatt and Van Ness (1965). Essentially, if the underlying channel model is exponentally stable, input random process $v(n)$ is stationary, and its first six (eight) cumulants are absolutely summable, then the sampled third-order (fourth-order) cumulants converge in probability to the true third-order (fourth-order) cumulants. Additionally, sampled third-order and fourth-order cumulants are asymptotically Gaussian (Giannakis and Swami, 1990; Lii and Rosenblatt, 1982, 1990; and Van Ness, 1966). Consequently, we are able to begin with an equation like

$$\text{estimate of cumulant} = \text{cumulant} + \text{estimation error} \tag{C-5}$$

in which estimation error is asymptotically Gaussian and to extend traditional classification or detection procedures to this formulation.

Formulas for estimating the covariances of higher-order moment estimates can be found in Friedlander and Porat (1990) and Porat and Friedlander (1989). Although they are quite complicated, they may be of value in certain methods where estimates of such quantities are needed.

The reason it is important to know that sample estimates of cumulants converge in probability to their true values (i.e., are consistent) is that functions of these estimates are used in many of the techniques that have been developed to solve a wide range of signal-processing problems. From Lesson 7, we know that arbitrary functions of consistent estimates are also consistent; hence, we are assured of convergence in probability (or sometimes with probability 1) when using these techniques.

From Lesson 11, we know that sampled estimates of Gaussian processes are also optimal in a maximum-likelihood sense; hence, they inherit all the properties of such estimates. Unfortunately, sampled estimates of non-Gaussian processes are not necessarily optimal in any sense; hence, it may be true that estimates other than the conventional sampled estimates provide "better" results.

## ESTIMATION OF BISPECTRUM (Nikias and Mendel, 1993)

There are two popular approaches for estimating the bispectrum, the *indirect method* and the *direct method*, both of which can be viewed as approximations of the definition of the bispectrum. Although these approximations are straightforward, sometimes the required computations may be expensive despite the use of fast Fourier transform (FFT) algorithms. We will briefly describe both conventional methods. Their extensions to trispectrum estimation is described in Nikias and Petropulu (1993).

### Indirect Method

Let $\{x(1), x(2), \ldots, x(N)\}$, be the given data set; then (1) estimate cumulants using segmentation and averaging, as described in the preceding section, to obtain $\hat{C}_{3,x}(\tau_1, \tau_2)$ in (C-4); and (2) generate the bispectrum estimate

$$\hat{S}_{3,x}(\omega_1, \omega_2) = \sum_{\tau_1=-L}^{L} \sum_{\tau_2=-L}^{L} \hat{C}_{3,x}(\tau_1, \tau_2) W(\tau_1, \tau_2) \exp\{-j(\omega_1\tau_1 + \omega_2\tau_2)\} \qquad \text{(C-6)}$$

where $L < M - 1$ and $W(\tau_1, \tau_2)$ is a two-dimensional window function. For discussions on how to choose the window function, see Nikias and Petropulu (1993). □

This method is "indirect" because we first estimate cumulants, after which we take their two-dimensional discrete-time Fourier transforms (DTFT). Of course, the computations of the bispectrum estimate may be substantially reduced if the symmetry properties of third-order cumulants are taken into account for the calculations of $\hat{C}_{3,x}(\tau_1, \tau_2)$ and if the symmetry properties of the bispectrum, as discussed in Lesson B, are also taken into account.

### Direct Method

Let $\{x(1), x(2), \ldots, x(N)\}$ be the given data set. Assume that $f_s$ is the sampling frequency and $\Delta_0 = f_s/N_0$ is the required spacing between frequency samples in the bispectrum domain along horizontal or vertical directions; thus, $N_0$ is the total number of frequency samples. Then (1) segment the data into $K$ segments of $M$ samples each, i.e., $N = KM$, and subtract the average value of each segment. If necessary, add zeros at the end of each segment to obtain a convenient (even) length $M$ for the FFT; (2) assuming that $\{x^{(i)}(k), k = 0, 1, \ldots, M-1\}$, is the data set per segment $i = 1, 2, \ldots, K$, generate the discrete Fourier transform (DFT) coefficients (Brigham, 1988) $Y^{(i)}(\lambda), i = 1, 2, \ldots, K$, where

$$Y^{(i)}(\lambda) = \frac{1}{M} \sum_{k=0}^{M-1} x^{(i)}(k) \exp\left(\frac{-j2\pi k\lambda}{M}\right) \qquad \text{(C-7)}$$

and $\lambda = 0, 1, \ldots, M/2$; (3) form the bispectrum estimate [motivated by (C-17) for $k = 3$] based on the DFT coefficients

$$\hat{b}_i(\lambda_1, \lambda_2) = \frac{1}{\Delta_0^2} \sum_{k_1=-L_1}^{L_1} \sum_{k_2=-L_1}^{L_1} Y^{(i)}(\lambda_1 + k_1) Y^{(i)}(\lambda_2 + k_2) Y^{(i)^*}(\lambda_1 + \lambda_2 + k_1 + k_2) \quad \text{(C-8)}$$

over the triangular region $0 \leq \lambda_2 \leq \lambda_1, \lambda_1 + \lambda_2 = f_s/2$. In general, $M = M_1 \times N_0$, where $M_1$ is a positive integer (assumed odd); i.e., $M_1 = 2L_1 + 1$. Since $M$ is even and $M_1$ is odd, we compromise on the value of $N_0$ (closest integer). In the special case where no averaging is performed in the bispectrum domain, $M_1 = 1$ and $L_1 = 0$, in which case

$$\hat{b}_i(\lambda_1, \lambda_2) = \frac{1}{\Delta_0^2} Y^{(i)}(\lambda_1) Y^{(i)}(\lambda_2) Y^{(i)^*}(\lambda_1 + \lambda_2) \quad \text{(C-9)}$$

(4) the bispectrum estimate of the given data is the average over the $K$ segments; i.e.,

$$\hat{S}_{3,x}(\omega_1, \omega_2) = \frac{1}{K} \sum_{i=1}^{K} \hat{b}_i(\omega_1, \omega_2) \quad \text{(C-10)}$$

where

$$\omega_1 = \lambda_1 \times (2\pi \Delta_0) \quad \text{and} \quad \omega_2 = \lambda_2 \times (2\pi \Delta_0) \; \square \quad \text{(C-11)}$$

This method is called "direct" because all calculations are in the frequency domain.

The statistical properties of the direct and indirect conventional methods for bispectrum estimation have been studied by Rosenblatt and Van Ness (1965), Van Ness (1966), Brillinger and Rosenblatt (1967a), and Rao and Gabr (1984). Their implications in signal processing have been studied by Nikias and Petropulu (1993). The key results are that both methods are asymptotically unbiased, consistent, and asymptotically complex Gaussian. Formulas for their asymptotic variances can be found in Nikias and Petropulu (1993, p. 143). From these formulas, we can conclude that the variance of these conventional estimators can be reduced by (1) increasing the number of records ($K$), (2) reducing the size of support of the window in the cumulant domain ($L$) or increasing the size of the frequency smoothing window ($M_1$), and (3) increasing the record size ($M$). Increasing the number of records ($K$) is demanding on computer time and may introduce potential nonstationarities. Frequency-domain averaging over large rectangles of size $M_1$ or use of cumulant-domain windows with small values of $L_0$ reduces frequency resolution and may increase bias. In the case of "short length" data, $K$ could be increased by using overlapping records (Nikias and Raghuveer, 1987).

Both conventional methods have the advantage of ease of implementation (FFT algorithms can be used) and provide good estimates for sufficiently long data; however, because of the "uncertainty principle" of the Fourier transform, the ability of the conventional methods to resolve harmonic components in the bispectral domain is limited.

# APPLICATIONS OF HIGHER-ORDER STATISTICS TO LINEAR SYSTEMS

A familiar starting point (some of the material in this section is taken from Mendel, 1991) for many problems in signal-processing and system theory is the single-input, single-output (SISO) linear and time-invariant (LTI) model depicted in Figure C-1, in which $v(k)$ is white Gaussian noise with finite variance $\sigma_v^2$, $H(z)[h(k)]$ is causal and stable, $n(k)$ is also white Gaussian noise with variance $\sigma_n^2$, and $v(k)$ and $n(k)$ are statistically independent. Letting $r(\cdot)$ and $S(\cdot)$ denote correlation and power spectrum, respectively, then it is well known (Problem C-5) that (e.g., Papoulis, 1991)

*classical 2nd order results* →

$$r_z(k) = r_y(k) + r_n(k) = \sigma_v^2 \sum_{i=0}^{\infty} h(i)h(i+k) + \sigma_n^2 \delta(k) \qquad (C\text{-}12)$$

*autocorrelation*

$$S_z(\omega) = \sigma_v^2 |H(\omega)|^2 + \sigma_n^2 = \int r_z(k) e^{-j\omega k} dk \qquad (C\text{-}13)$$

and

$$r_{vz}(k) \triangleq E\{v(n)z(n+k)\} = \sigma_v^2 h(k) \qquad \Rightarrow h(k) = \frac{1}{\sigma_v^2} r_{vz}(k) \qquad (C\text{-}14)$$

*note: the additive noise (n(k)) effect the 2nd order statistics, (rds) of our.*

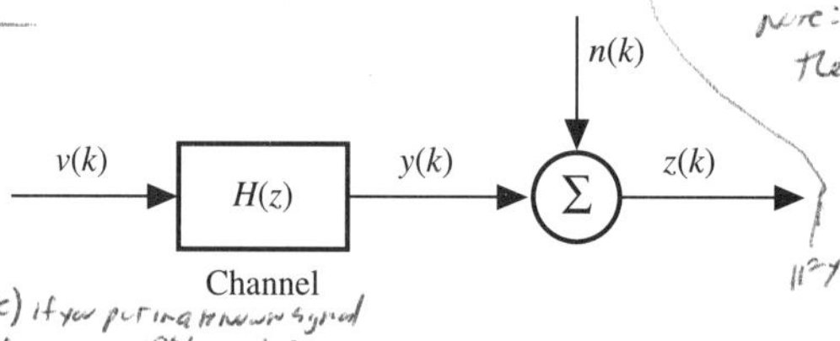

**Figure C-1**   Single-channel system.

*estimate*
① *so you can find $h(k)$ if you're using known signal + measure (estimate) $r_{vz}$*

From (C-13) we see that all phase information has been lost in the spectrum (or in the autocorrelation); hence, we say that *correlation or spectra are phase blind.*

Bartlett (1955) and Brillinger and Rosenblatt (1967b) established a major generalization to Equations (C-12) and (C-13). In their case the system in Figure C-1 is assumed to be causal and exponentially stable, and $\{v(k)\}$ is assumed to be independent, identically distributed (i.i.d.), and *non-Gaussian*; i.e.,

$$C_{k,v}(\tau_1, \tau_2, \ldots, \tau_{k-1}) = \begin{cases} \gamma_{k,v}, & \text{if } \tau_1 = \tau_2 = \cdots = \tau_{k-1} = 0 \\ 0, & \text{otherwise} \end{cases} \qquad (C\text{-}15)$$

where $\gamma_{k,v}$ denotes the $k$th-order cumulant of $v(i)$. Additive noise $n(k)$ is assumed to be Gaussian, but need not be white.

**Theorem C-1.**   *For the single-channel system depicted in Figure C-1, if $\{v(i)\}$ is independent, identically distributed (i.i.d.), and non-Gaussian, as in (C-15), and n(k) is Gaussian (colored or white), then the kth-order cumulant and polyspectra of measured signal z(k) are*

$$C_{k,z}(\tau_1, \tau_2, \ldots, \tau_{k-1}) = \gamma_{k,v} \sum_{n=0}^{\infty} h(n)h(n+\tau_1) \cdots h(n+\tau_{k-1}) \qquad (C\text{-}16)$$

*$k^{th}$ order φ lag cumulant of v(i).*

*and*

*N(k) not here finite cumulants are φ for k≥3*

$$S_{k,z}(\omega_1, \omega_2, \ldots, \omega_{k-1}) = \gamma_{k,v} H(\omega_1)H(\omega_2)\cdots H(\omega_{k-1})H(-\sum_{i=1}^{k-1}\omega_i) \qquad \text{(C-17)}$$

*where* k $\geq$ 3.

This theorem generalizes our statistical characterization and understanding of the system in Figure C-1, from second- to higher-order statistics. It is even valid for $k = 2$ by replacing $z(k)$ by $y(k)$. In that case, (C-16) and (C-17) reduce to $C_{2,y}(\tau_1) = \gamma_{2,v} \sum h(n)h(n+\tau_1)$ and $S_{2,y}(\omega_1) = \gamma_{2,v} H(\omega_1)H(-\omega_1) = \sigma_v^2 |H(\omega_1)|^2$. Note that (C-12) and (C-13) are the correct second-order statistics for noisy measurement $z(k)$. Equations (C-16) and (C-17) have been the starting points for many nonparametric and parametric higher-order statistics techniques that have been developed during the past 10 years [e.g., Nikias and Raghuveer (1987) and Mendel (1988)].

*Proof.* (Mendel, 1991) Because $n(k)$ is assumed to be Gaussian, the $k$th-order cumulant of $z(k)$ equals the $k$th-order cumulant of $y(k)$, where

$$y(k) = \sum_{i=-\infty}^{\infty} v(i)h(k-i) \qquad \text{(C-18)}$$

Note that the following derivation is expedited by working with the more general form in (C-18), where $i$ ranges from $-\infty$ to $\infty$, instead of the form associated with a causal IR for which $i$ ranges from 0 to $k$. Changes of variables then do not change the ranges of the summation. Consequently,

$$\begin{aligned}
C_{k,z}(\tau_1, \ldots, \tau_{k-1}) &= C_{k,y}(\tau_1, \ldots, \tau_{k-1}) = \text{cum}\,(y(l), y(l+\tau_1), \ldots, y(l+\tau_{k-1})) \\
&= \text{cum}\,[\sum_{i_0} v(i_0)h(l-i_0), \sum_{i_1} v(i_1)h(l-i_1+\tau_1), \ldots, \\
&\quad \sum_{i_{k-1}} v(i_{k-1})h(l-i_{k-1}+\tau_{k-1})] \\
&= \sum_{i_0}\sum_{i_1}\cdots\sum_{i_{k-1}} \text{cum}\,[v(i_0)h(l-i_0), v(i_1)h(l-i_1+\tau_1), \ldots, \\
&\quad v(i_{k-1})h(l-i_{k-1}+\tau_{k-1})] \\
&= \sum_{i_0}\sum_{i_1}\cdots\sum_{i_{k-1}} h(l-i_0)h(l-i_1+\tau_1)\cdots h(l-i_{k-1}+\tau_{k-1}) \\
&\quad \times \text{cum}\,[v(i_0), v(i_1), \ldots, v(i_{k-1})]
\end{aligned} \qquad \text{(C-19)}$$

To arrive at the third line of this derivation, we have used cumulant property [CP3]; and to arrive at the fourth line, we have used cumulant property [CP1] (see Lesson B).

In the case of a *white noise input*, (C-19) simplifies considerably to

$$\begin{aligned}
C_{k,z}(\tau_1, \ldots, \tau_{k-1}) &= \gamma_{k,v} \sum_{i_0} h(l-i_0)h(l-i_0+\tau_1)\cdots h(l-i_0+\tau_{k-1}) \\
&= \gamma_{k,v} \sum_{n=0}^{\infty} h(n)h(n+\tau_1)\cdots h(n+\tau_{k-1})
\end{aligned}$$

which is (C-16). To arrive at the first line, we have used (C-15), noting that

$$\text{cum}\,[\nu(i_0), \nu(i_1), \ldots, \nu(i_{k-1})] = \text{cum}\,[\nu(i_0), \nu(i_0 + i_1 - i_0), \ldots, \nu(i_0 + i_{k-1} - i_0)]$$

$$= C_{k,\nu}(i_1 - i_0, \ldots, i_{k-1} - i_0)$$

$$= \gamma_{k,\nu}, \quad \text{only if } i_1 = i_0, \ldots, i_{k-1} = i_0 \qquad \text{(C-20)}$$

*But $C_{k,\nu}(0,0,0\ldots) = \delta_{k\nu}$*

And to arrive at the second line of (C-19), we have made a simple substitution of variables and invoked the stationarity of $y(n)$ and the causality of $h(n)$. The former tells us that $C_{k,y}(\tau_1, \ldots, \tau_{k-1}) = C_{k,z}(\tau_1, \ldots, \tau_{k-1})$ will not depend on time $n$; the latter tells us that $h(n) = 0$ for $n < 0$.

The polyspectrum in (C-17) is easily obtained by taking the $(k-1)$-dimensional discrete-time Fourier transform of (C-16):

$$S_{k,z}(\omega_1, \ldots, \omega_{k-1}) = \sum_{\tau_1=-\infty}^{\infty} \cdots \sum_{\tau_{k-1}=-\infty}^{\infty} C_{k,z}(\tau_1, \ldots, \tau_{k-1})$$

$$\exp[-j(\omega_1 \tau_1 + \ldots + \omega_{k-1} \tau_{k-1})] \qquad \text{(C-21)}$$

Substitute (C-16) into (C-21) to obtain (C-17) (Problem C-6). □

From this proof, we observe that (C-16) and (C-17) are also valid for noncausal systems, in which case $N$ ranges from $-\infty$ to $\infty$ in (C-16).

The generalization of (C-16) to the case of a *colored* non-Gaussian input $\nu(i)$ is [Bartlett (1955) only considers the $k = 2, 3, 4$ cases; Brillinger and Rosenblatt (1967b) provide results for all $k$]

$$C_{k,z}(\tau_1, \tau_2, \ldots, \tau_{k-1}) = \sum_{m_1} \sum_{m_2} \cdots \sum_{m_{k-1}} C_{k,\nu}(\tau_1 - m_1, \tau_2 - m_2, \ldots,$$

$$\tau_{k-1} - m_{k-1}) \times C_{k,h}(m_1, m_2, \ldots, m_{k-1}) \qquad \text{(C-22)}$$

where

$$C_{k,h}(m_1, m_2, \ldots, m_{k-1}) = \sum_{k_1} h(k_1) h(k_1 + m_1) \cdots h(k_1 + m_{k-1}) \qquad \text{(C-23)}$$

A derivation of this result is requested in Problem C-7. Observe that the right-hand side of (C-22) is a *multidimensional convolution* of $C_{k,\nu}(\tau_1, \tau_2, \ldots, \tau_{k-1})$ with the deterministic correlation function of $h(n)$, which we have denoted $C_{k,h}(m_1, m_2, \ldots, m_{k-1})$. Consequently, the generalization of (C-17) to the case of a *colored* non-Gaussian input $\nu(i)$ is

$$S'_{k,z}(\omega_1, \omega_2, \ldots, \omega_{k-1}) = S_{k,\nu}(\omega_1, \omega_2, \ldots, \omega_{k-1}) H(\omega_1) H(\omega_2) \cdots H(\omega_{k-1})$$

$$\times H\left(-\sum_{i=1}^{k-1} \omega_i\right) \qquad \text{(C-24)}$$

*Colored noise results!*

One very important use for (C-16) is to compute cumulants for models of systems. The procedure for doing this is (1) determine the model's IR, $h(k), k = 0, 1, 2, \ldots, K$; (2) fix the $\tau_j$ values at integers, and evaluate (C-16); and (3) repeat step 2 for all $\tau_j$ values in the domain of support of interest (be sure to

$\frac{1}{3}$ of the column

$\tau_2$

for all those pts compute using the $C-1$
then use symmetry.

$\tau_1$

use the symmetry properties of cumulants to reduce computations). This is how the results in Figures B-3(c) and (d), (B-4), and (B-5) were obtained.

Similarily, a very important use of (C-17) is to compute polyspectra for models. The procedure for doing this is (1) determine the model's IR, $h(k)$, $k = 0, 1, 2, \ldots, K$; (2) compute the DTFT of $h(k)$, $H(\omega)$; (3) fix the $\omega_j$ values and evaluate (C-17); and (4) repeat step 3 for all $\omega_j$ values in the domain of support of interest (be sure to use the symmetry properties of polyspectra to reduce computations). This is how the results in Figures B-7 were obtained. Of course, another way to compute the polyspectra is to first compute the cumulants, as just described, and then compute their multidimensional DTFT's.

### EXAMPLE C-1  MA($q$) Model (Mendel, 1991)

Suppose that $h(k)$ is the impulse response of a causal *moving average* (MA) system. Such a system has a finite IR and is described by the following model:

$$y(k) = \sum_{i=0}^{q} b(i)v(k - i) \qquad h(k) = \sum b(i)\delta(k+i) = b(n) \text{ here} \qquad \text{(C-25)}$$

The MA parameters are $b(0), b(1), \ldots, b(q)$, where $q$ is the order of the MA model and $b(0)$ is usually assumed equal to unity [the scaling is absorbed into the statistics of $v(k)$]. It is well known that for this model $h(i) = b(i), i = 0, 1, \ldots, q$; hence, when $\{v(l)\}$ is i.i.d., we find from (C-16) that

$$C_{3,y}(\tau_1, \tau_2) = \gamma_{3,v} \sum_{l=0}^{q} b(l)b(l + \tau_1)b(l + \tau_2) \qquad \text{(C-26)}$$

An interesting question is, "For which values of $\tau_1$ and $\tau_2$ is $C_{3,y}(\tau_1, \tau_2)$ nonzero?" The answer to this question is depicted in Figure C-2. The domain of support for $C_{3,y}(\tau_1 = m, \tau_2 = n)$ is the six-sided region. This is due to the FIR nature of the MA system. The triangular region in the first quadrant is the *principal region*. In the stationary case, we only have to determine third-order cumulant values in this region, **R**, where

$$\mathbf{R} = \{m, n : 0 \le n \le m \le q\} \qquad \text{(C-27)}$$

Observe, from Figure C-2, that the third-order cumulant equals zero for either one or both of its arguments equal to $q + 1$. This suggests that it should be possible to determine the order of the MA model, $q$, by testing, in a statistical sense, the smallness of a third-order cumulant such as $C_{3,y}(q+1, 0)$. See Giannakis and Mendel (1990) for how to do this. $\square$

### EXAMPLE C-2  $C(q, k)$ Formula

Equation (C-14) is a "classical" way for computing the impulse response of a SISO system; however, it requires that we have access to both the system's input and output in order to be able to compute the cross-correlation $r_{vz}(k)$. In many applications, we have access only to a noise-corrupted version of the system's output (e.g., blind equalization, reflection seismology). A natural question to ask is, "Can the system's IR be determined just from output measurements?" Using output cumulants, the answer to this question is yes. Giannakis (1987b) was the first to show that the IR of a $q$th-order MA system can be calculated just from the system's output cumulants.

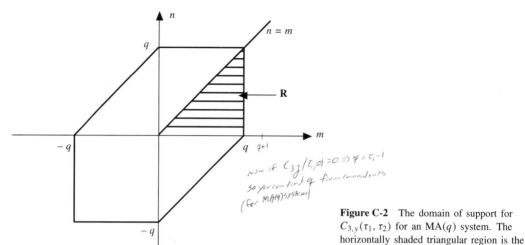

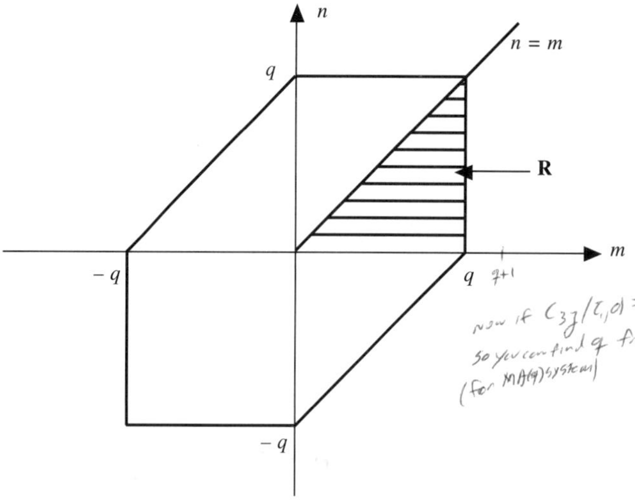


**Figure C-2** The domain of support for $C_{3,y}(\tau_1, \tau_2)$ for an MA($q$) system. The horizontally shaded triangular region is the principal region.

We begin with (C-16) for $k = 3$; i.e.,

$$C_{3,z}(\tau_1, \tau_2) = \gamma_{3,v} \sum_{n=0}^{\infty} h(n)h(n+\tau_1)h(n+\tau_2) \qquad \text{(C-28)}$$

in which $h(0) = 1$, for normalization purposes. Set $\tau_1 = q$ and $\tau_2 = k$ in (C-28), and use the fact that for an MA($q$) system $h(j) = 0\,\forall j > q$, to see that

$$C_{3,z}(q, k) = \gamma_{3,v} h(q)h(k) \qquad \text{(C-29)}$$

Next, set $\tau_1 = q$ and $\tau_2 = 0$ in (C-28) to see that

$$C_{3,z}(q, 0) = \gamma_{3,v} h(q) \qquad \text{(C-30)}$$

Dividing (C-29) by (C-30), we obtain the "$C(q, k)$ formula,"

$$h(k) = \frac{C_{3,y}(q, k)}{C_{3,y}(q, 0)}, \qquad k = 0, 1, \ldots, q \qquad \text{(C-31)}$$

This is a rather remarkable theoretical result in that it lets us determine a system's IR just from system output measurements. Observe that it only uses values of third-order cumulants along the right-hand side of the principal region of support (see Figure C-2).

The $C(q, k)$ formula can be generalized to arbitrary-order cumulants (Problem C-8), and, as a byproduct, we can also obtain formulas to compute the cumulant of the white input noise (Problem C-9). □

**Corollary C-1.** *Suppose that* h(k) *is associated with the following ARMA* (p, q) *model*

$$\sum_{k=0}^{p} a(k)y(n-k) = \sum_{k=0}^{q} b(k)v(n-k) \qquad \text{(C-32)}$$

*Then*

$$\sum_{k=0}^{p} a(k)C_{m,y}(\tau - k, t, 0, \ldots, 0) = \gamma_{m,v} \sum_{j=0}^{q} h^{(m-2)}(j - \tau)h(j - \tau + t)b(j) \quad \text{(C-33)}$$

This rather strange looking result (Swami and Mendel, 1990b) has some very interesting applications, as we shall demonstrate later.

*Proof.* To begin, we define the scalar function $f_m(t; \tau)$ as

$$f_m(t; \tau) = \sum_{k=0}^{p} a(k)C_{m,y}(\tau - k, t, 0, \ldots, 0) \quad \text{(C-34)}$$

If we know both the AR coefficients and the output cumulants, then we can compute $f_m(t; \tau)$ for different values of $t$ and $\tau$. Next, substitute (C-16) into the right-hand side of (C-34) to see that

$$f_m(t; \tau) = \gamma_{m,v} \sum_{i=0}^{\infty} h^{(m-2)}(i)h(i + t) \sum_{k=0}^{p} a(k)h(i + \tau - k) \quad \text{(C-35)}$$

Impulse response $h(n)$ of the ARMA $(p, q)$ model in (C-32) satisfies the recursion

$$\sum_{k=0}^{p} a(k)h(n - k) = \sum_{k=0}^{q} b(k)\delta(n - k) = b(n) \quad \text{(C-36)}$$

Hence, we recognize that the sum on the far right of (C-35) equals $b(i + \tau)$, so (C-35) becomes

$$f_m(t; \tau) = \gamma_{m,v} \sum_{i=0}^{\infty} h^{(m-2)}(i)h(i + t)b(i + \tau) \quad \text{(C-37)}$$

Finally, to obtain the right-hand side of (C-33) from (C-37); let $i + \tau = j$; truncate the upper limit in the summation from $\infty$ to $q$, because $b(j) = 0 \; \forall j > q$; and extend the lower limit in the summation from $j = \tau$ to $j = 0$, because $h(l)$ is causal [so that $h(-\tau) = h(-\tau + 1) = \cdots = h(-1) = 0$]. □

**EXAMPLE C-3** ***q*-Slice Formula**

From (C-37) and the fact that $b(j) = 0$ for $\forall j > q$, it is straightforward to show that

$$f_m(0; q) = \gamma_{m,v}b(q) \quad \text{(C-38)}$$

and

$$f_m(t; q) = \gamma_{m,v}b(q)h(t) \quad \text{(C-39)}$$

Consequently,

$$\begin{aligned} h(t) &= \frac{f_m(t; q)}{f_m(0; q)} \\ &= \frac{\sum_{k=0}^{p} a(k)C_{m,y}(q - k, t, 0, \ldots, 0)}{\sum_{k=0}^{p} a(k)C_{m,y}(q - k, 0, 0, \ldots, 0)} \end{aligned} \quad \text{(C-40)}$$

where $t = 0, 1, \ldots$. This is a closed-form formula for the IR of an ARMA$(p, q)$ system (Swami and Mendel, 1990b). When $p = 0$, then (C-40) reduces to the $C(q, k)$ formula. To use (C-40), the AR parameters must be known, and we must estimate the cumulants in its numerator and denominator.

It is easy to show that (C-36) can be used to compute the MA coefficients $b(1), b(2), \ldots, b(q)$ if we are given all the AR coefficients and IR values $h(0), h(1), \ldots, h(q)$ (Problem C-10). Because of normalization, $h(0) = 1$; hence, we can use (C-40) to compute $h(1), \ldots, h(q)$. For each value of $t$, the numerator cumulants must be estimated along a horizontal slice, ranging from $\tau_1 = q - p$ (which will be negative, because $p > q$) to $\tau_1 = q$. The denominator of (C-40) only needs to be computed one time, and it uses cumulants along a horizontal slice along the $\tau_1$ axis, where again $\tau_1$ ranges from $\tau_1 = q - p$ to $\tau_1 = q$. Because the numerator slice changes with each calculation of $h(t)$, and we need to compute $q$ values of $h(t)$, (C-40) is known as the $q$-slice algorithm (Swami and Mendel, 1990b). The $q$-slice algorithm can be used as part of a procedure to estimate the coefficients in an ARMA$(p, q)$ system (Problem C-21). □

### EXAMPLE C-4 Cumulant-based Normal Equations

Many ways exist for determining the AR coefficients of either the ARMA$(p, q)$ model in (C-32) or the following AR$(p)$ model:

$$\sum_{k=0}^{p} a(k)y(n-k) = v(n) \tag{C-41}$$

where $a(0) = 1$. One of the most popular ways is to use the following *correlation-based normal equations* (Box and Jenkins, 1970) (Problem C-11):

$$\sum_{k=0}^{p} a(k)r_y(k-m) = 0, \quad \text{for } m > 0 \tag{C-42}$$

If AR order $p$ is known, then (C-42) is collected for $p$ values of $m$, from which the AR coefficients $a(1), a(2), \ldots, a(p)$ can be determined. Note that (C-42) is also true for an ARMA$(p, q)$ model when $m > q$. The correlation-based normal equations do not lead to very good results when only a noise-corrupted version of $y(k)$ is available, because additive Gaussian measurement noise seriously degrades the estimation of $r_y(k-m)$; hence, we are motivated to seek alternatives that are able to suppress additive Gaussian noise. Cumulants are such an alternative.

Starting with (C-33), it follows that

$$\sum_{k=0}^{p} a(k)C_{m,y}(\tau - k, k_0, 0, \ldots, 0) = \gamma_{m,v} \sum_{j=0}^{q} h^{(m-2)}(j-\tau)h(j-\tau+k_0)b(j) \tag{C-43}$$

From the causality of $h(l)$, it follows that the right-hand side of (C-43) is zero for $\tau > q$; hence,

$$\sum_{k=0}^{p} a(k)C_{m,y}(\tau - k, k_0, 0, \ldots, 0) = 0, \quad \text{for } \tau > q \tag{C-44}$$

Of course, in the AR case $q = 0$, so in this case the right-hand side of (C-44) is zero for $\tau > 0$.

Equation (C-44) is a *cumulant-based normal equation* and should be compared with (C-42). Parzen (1967) credits Akaike (1966) for the "idea of extending" correlation-based normal equations to cumulants. See, also, Giannakis (1987a), Giannakis and Mendel (1989), Swami (1988), Swami and Mendel (1992), and Tugnait (1986a, b).

For an AR($p$) system, suppose that we concatenate (C-44) for $\tau = 1, 2, \ldots, p + M$ (or $\tau = q + 1, q + 2, \ldots, q + p + M$ in the ARMA case), where $M \geq 0$ (when $M > 0$, we will have an overdetermined system of equations, which is helpful for reducing the effects of noise and cumulant estimation errors) and $k_0$ is arbitrary. Then we obtain

$$\mathbf{C}(k_0)\mathbf{a} = \mathbf{0} \tag{C-45}$$

where the cumulant entries in Toeplitz matrix $\mathbf{C}(k_0)$ are easily deduced, and $\mathbf{a} = \text{col}\,(1, a(1), \ldots, a(p-1), a(p))$. If $\mathbf{C}(k_0)$ has rank $p$, then the corresponding one-dimensional slice (parameterized by $k_0$) of the $m$th-order cumulant is a full rank slice and the $p$ AR coefficients can be solved for from (C-45). If $\mathbf{C}(k_0)$ does not have rank $p$, then some of the AR coefficients cannot be solved for from (C-45), and those that can be solved for do not equal their true values. In essence, some poles of the AR model, and subsequently some of the AR coefficients, are invisible to (C-45).

Interestingly enough, $\mathbf{C}(k_0)$ does not necessarily have rank $p$ for an arbitrary value of $k_0$. This issue has been studied in great detail by Swami and Mendel (1992), Giannakis (1989, 1990), and Tugnait (1989). For example, Swami and Mendel (1992) show that (1) *every one-dimensional cumulant slice need not be a full rank slice, and* (2) *a full rank cumulant slice may not exist.* Swami and Mendel, as well as Giannakis, have shown that the *AR coefficients of an ARMA model can be determined when at least* $p + 1$ *slices of the* $m$*th-order cumulant are used.* Furthermore, these cannot be arbitrary slices. Equation (C-44) must be concatenated for $\tau = 1, 2, \ldots, p + M$ and (at least) $k_0 = -p, \ldots, 0$ [or $\tau = q + 1, q + 2, \ldots, q + p + M$ and (at least) $k_0 = q - p, \ldots, q$ in the ARMA case], where $M \geq 0$. The resulting linear system of equations is overdetermined and should be solved using a combination of SVD and TLS. □

**Corollary C-2.** *Let* $C_{k,y}(\tau)$ *denote the diagonal slice of the* $k$*th-order cumulant, e.g.,* $C_{3,y}(\tau) = C_{3,y}(\tau_1 = \tau, \tau_2 = \tau)$ *and* $S_{k,y}(\omega)$ *is its discrete-time Fourier transform (DTFT). Additionally, let*

$$H_{k-1}(\omega) = H(\omega) \circledast H(\omega) \circledast \cdots \circledast H(\omega), \quad (k-2 \text{ complex convolutions}) \tag{C-46}$$

*Then*

$$H_{k-1}(\omega) S_y(\omega) = \frac{\sigma_v^2}{\gamma_{k,v}} H(\omega) S_{k,y}(\omega) \tag{C-47}$$

This is a very interesting equation that relates the usual spectrum of a system's output to a special polyspectrum, $S_{k,y}(\omega)$. It was developed by Giannakis (1987a) and Giannakis and Mendel (1989). Some applications of (C-47) are given later.

*Proof.* We derive (C-47) for the case when $k = 4$, leaving the general derivation as an exercise (Problem C-12). From (C-16), we find that

$$C_{4,y}(\tau) = \gamma_{4,v} \sum_{n=0}^{\infty} h(n) h^3(n + \tau) \tag{C-48}$$

Hence,

$$S_{4,y}(\omega) = \gamma_{4,v} \sum_{n=0}^{\infty} \sum_{\tau=-\infty}^{\infty} h(n) h^3(n + \tau) e^{-j\omega\tau} \tag{C-49}$$

which can be expressed as

$$S_{4,y}(\omega) = \gamma_{4,v} \sum_{n=0}^{\infty} h(n)e^{j\omega n} \sum_{\tau=-\infty}^{\infty} h^3(n+\tau)e^{-j\omega(\tau+n)} \tag{C-50}$$

Using DTFT properties, it is straightforward to show that (C-50) can be written as

$$S_{4,y}(\omega) = \gamma_{4,v} H(-\omega)[H(\omega) * H(\omega) * H(\omega)] = \gamma_{4,v} H(-\omega)H_3(\omega) \tag{C-51}$$

Recall that the spectrum of the system's output is

$$S_y(\omega) = \sigma_v^2 |H(\omega)|^2 = \sigma_v^2 H(\omega)H(-\omega) \tag{C-52}$$

Multiply both sides of (C-51) by $H(\omega)$, and substitute (C-52) into the resulting equation to see that

$$H(\omega)S_{4,y}(\omega) = \frac{\gamma_{4,v}}{\sigma_v^2} H_3(\omega)S_y(\omega) \tag{C-53}$$

which can be reexpressed as (C-47). □

**EXAMPLE C-5 GM Equations for MA($q$) Systems**

For the MA($q$) system in (C-25), $h(n) = b(n)(n = 1, 2, \ldots, q)$, $H_2(\omega)$ is the DTFT of $b^2(n)$, and $H_3(\omega)$ is the DTFT of $b^3(n)$, so (C-47) can be expressed in the time domain as

$$\sum_{k=0}^{q} b^2(k)r_y(m-k) = \frac{\sigma_v^2}{\gamma_{3,v}} \sum_{k=0}^{q} b(k)C_{3,y}(m-k, m-k) \tag{C-54}$$

and

$$\sum_{k=0}^{q} b^3(k)r_y(m-k) = \frac{\sigma_v^2}{\gamma_{4,v}} \sum_{k=0}^{q} b(k)C_{4,y}(m-k, m-k, m-k) \tag{C-55}$$

where $-q \le m \le 2q$ (Problem C-13). These formulas, especially (C-54), have been used to estimate the MA coefficients $b(1), b(2), \ldots, b(q)$ using least-squares or adaptive algorithms (Giannakis and Mendel, 1989; Friedlander and Porat, 1989). Friedlander and Porat refer to these formulas as the "GM equations." Because they are special cases of (C-47), we prefer to call (C-47) the GM equation.

See Problems C-13 and C-14 for additional aspects of using the GM equations to estimate MA coefficients and an even more general version of the GM equations. □

**EXAMPLE C-6 Deconvolution**

Our system of interest is depicted in Figure C-3. It is single input and single output, causal, linear, and time invariant, in which $\nu(k)$ is stationary, white, and non-Gaussian with finite variance $\sigma_v^2$, nonzero third- or fourth-order cumulant, $\gamma_{3,v}$ or $\gamma_{4,v}$, (communication signals are usually symmetrically distributed and, therefore, have zero third-order cumulants); $n(k)$ is Gaussian noise (colored or white); $\nu(k)$ and $n(k)$ are statistically independent; and $H(z)$ is an unknown asymptotically stable system. Our objective is to design a deconvolution filter in order to obtain an estimate of input $\nu(k)$. The deconvolution filter is given by $\theta(z)$; it may be noncausal.

Using the fact that $S_y(\omega) = \sigma_v^2 H(\omega)H(-\omega)$, GM equation (C-47) can be expressed as

$$H_{k-1}(\omega)H(-\omega) = \frac{1}{\gamma_{k,v}} S_{k,y}(\omega), \qquad k = 2, 3, \ldots \tag{C-56}$$

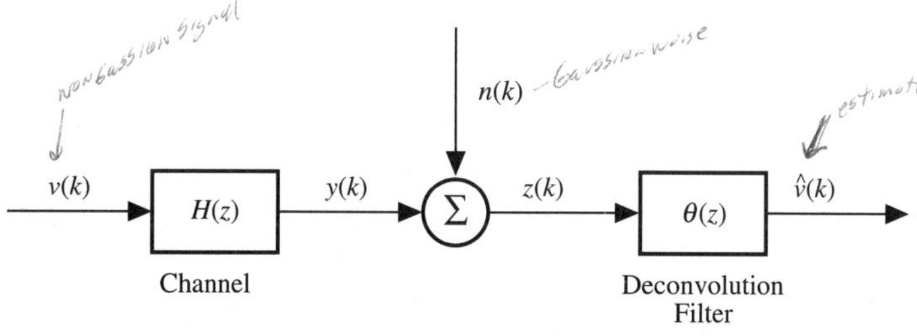


**Figure C-3** Single-channel system and deconvolution filter.

Hence,

$$H_{k-1}(\omega) = \frac{1}{\gamma_{k,v}} \frac{S_{k,y}(\omega)}{H(-\omega)} = \frac{1}{\gamma_{k,v}} S_{k,y}(\omega)\theta(-\omega), \qquad k = 2, 3, \ldots \tag{C-57}$$

where we have used the fact that, ideally, $\theta(\omega) = 1/H(\omega)$. Taking the inverse DTFT of (C-57), using the formula for $H_{k-1}(\omega)$ in (C-46), we obtain the following interesting relation between $H(\omega)$ and its inverse filter $\theta(\omega)$ (Dogan and Mendel, 1994):

$$\frac{1}{\gamma_{k,v}} \sum_{j=r_1}^{r_2} \theta(j) C_{k,y}(i+j) = h^{k-1}(i), \qquad k = 2, 3, \ldots \tag{C-58}$$

where $r_1$ and $r_2$ are properly selected orders of the noncausal and causal parts of $\theta(z)$, respectively.

When we have access only to noisy measurements, $z(k) = y(k) + n(k)$, where $n(k)$ is Gaussian and is independent of $v(k)$, then (C-58) becomes

$$\frac{1}{\sigma_v^2} \sum_{j=r_1}^{r_2} \theta(j)[r_z(i+j) - r_n(i+j)] = h(i) \tag{C-59a}$$

and

$$\frac{1}{\gamma_{k,v}} \sum_{j=r_1}^{r_2} \theta(j) C_{k,z}(i+j) = h^{k-1}(i), \qquad k = 3, 4, \ldots \tag{C-59b}$$

Equation (C-59a) is obtained by using the fact that $r_z(i+j) = r_y(i+j) + r_n(i+j)$. Equation (C-59b) is obtained by using the fact that $C_{k,n}(i+j) = 0$, for all $k \geq 3$, because $n(k)$ is Gaussian.

Equations (C-58) and (C-59) are applicable to any SISO system, including MA, AR, and ARMA systems. When noisy measurements are available, we must be very careful about using (C-59a), because usually $r_n(i+j)$ is unknown. If (C-59a) is used simply by ignoring the $r_n(i+j)$ term, errors will be incurred. Such errors do not occur in (C-59b) because the higher-order cumulants have (theoretically) suppressed the additive Gaussian noise.

In the case of an MA($q$) system, (C-58) can be expressed as

$$\frac{1}{\sigma_v^2} \sum_{j=r_1}^{r_2} \theta(j) r_y(i+j) = \begin{cases} b(i), & \text{if } i \in [0, q] \\ 0, & \text{otherwise} \end{cases} \tag{C-60}$$

$$\frac{1}{\gamma_{3,v}} \sum_{j=r_1}^{r_2} \theta(j) C_{3,y}(i+j) = \begin{cases} b^2(i), & \text{if } i \in [0, q] \\ 0, & \text{otherwise} \end{cases} \tag{C-61}$$

and

$$\frac{1}{\gamma_{4,v}} \sum_{j=r_1}^{r_2} \theta(j) C_{4,y}(i+j) = \begin{cases} b^3(i), & \text{if } i \in [0, q] \\ 0, & \text{otherwise} \end{cases} \tag{C-62}$$

These equations were first derived by Zheng et al. (1993), but in a very different way than our derivation (Problem C-15). They can be used to solve for the deconvolution filter's coefficients; by choosing $i > q$, (C-60)–(C-62) become linear in these coefficients. The resulting equations can be solved using least squares. For many numerical examples, see Zheng et al. (1993); their examples demonstrate that in the noisy measurement case it is better to use (C-62) [or (C-61) when third-order cumulants are nonzero] than (C-60), for the reasons described above in connection with (C-59a) and (C-59b). □

Finally, we are able to connect the *IIR digital Wiener deconvolution filter*, described in the Supplementary Material at the end of Lesson 21 (see the section entitled Relationship between Steady-state MVD Filter and an Infinite Impulse Response Digital Wiener Deconvolution Filter) and the cumulant-based deconvolution filter of Example C-6.

**Theorem C-2** (Dogan and Mendel, 1994). *Under the condition of* perfect deconvolution, *the deconvolution coefficients determined from (C-59) are optimal in the sense that they are the coefficients that minimize the* MSE $\mathbf{E}\{[\hat{\mathrm{v}}(\mathrm{k}) - \mathrm{v}(\mathrm{k})]^2\}$, *when*

$$\hat{v}(k) = \sum_{i=-\infty}^{\infty} \theta(i) y(k-i) \tag{C-63}$$

*and only noisy measurements are available; i.e.,*

$$z(k) = y(k) + n(k) \tag{C-64}$$

*regardless of the color of the Gaussian noise,* n(k).

*Proof.* The condition of *perfect deconvolution* is

$$H(\omega)\theta(\omega) = 1 \quad (\text{or } \beta \neq 0) \tag{C-65}$$

Consider the problem of minimizing $\mathbf{E}\{[\hat{v}(k) - v(k)]^2\}$, when $\hat{v}(k)$ is given by (C-63). This problem was studied in the Supplementary Material at the end of Lesson 21 (see the section entitled Relationship between Steady-state MVD Filter and an Infinite Impulse Response Digital Wiener Deconvolution Filter). Using (21-93), in which $z(k)$ is replaced by $y(k)$, we see that $\theta(\omega)$ must be chosen so that

$$\theta(\omega) S_y(\omega) = \sigma_v^2 H^*(\omega) \tag{C-66}$$

Because only $z(k)$ [and not $y(k)$] is available, we cannot estimate $S_y(\omega)$; the best we can do is to estimate $S_z(\omega)$, where

$$S_z(\omega) = S_y(\omega) + S_n(\omega) \tag{C-67}$$

Of course, $S_n(\omega)$ is unknown to us.

Substitute (C-67) into (C-66) to see that

$$\theta(\omega)[S_z(\omega) - S_n(\omega)] = \sigma_v^2 H^*(\omega) \tag{C-68}$$

We now use the GM equation, solving it for $S_n(\omega)$. Beginning with (C-47), using (C-67) for $S_y(\omega)$, and the fact that $S_{k,y}(\omega) = S_{k,z}(\omega)$ for all $k \geq 3$ [due to the Gaussian nature of $n(k)$], (C-47) can be expressed as

$$H_{k-1}(\omega)[S_z(\omega) - S_n(\omega)] = \frac{\sigma_\nu^2}{\gamma_{k,\nu}} H(\omega)S_{k,z}(\omega) \tag{C-69}$$

Hence,

$$S_n(\omega) = S_z(\omega) - \frac{\sigma_\nu^2}{\gamma_{k,\nu}} \frac{H(\omega)S_{k,z}(\omega)}{H_{k-1}(\omega)} \tag{C-70}$$

Substituting (C-70) into (C-68), we find that

$$\frac{\theta(\omega)H(\omega)S_{k,z}(\omega)}{\gamma_{k,\nu}} = H_{k-1}(\omega)H^*(\omega) \tag{C-71}$$

Applying (C-65) to (C-71), we see that $S_{k,z}(\omega)/\gamma_{k,\nu} = H_{k-1}(\omega)H^*(\omega)$, or

$$\frac{S_{k,z}(\omega)}{\gamma_{k,\nu}H^*(\omega)} = H_{k-1}(\omega) \tag{C-72}$$

From (C-65), we find that $1/H(-\omega) = \theta(-\omega)$. For real signals, $H^*(\omega) = H(-\omega)$; hence, (C-72) becomes

$$\frac{\theta(-\omega)S_{k,z}(\omega)}{\gamma_{k,\nu}} = H_{k-1}(\omega) \tag{C-73}$$

which is exactly the same as (C-57), in which $S_{k,y}(\omega)$ is replaced by $S_{k,z}(\omega)$. From (C-73), we find that

$$\frac{1}{\gamma_{k,\nu}} \sum_j \theta(j)C_{k,z}(i+j) = h^{k-1}(i), \qquad k = 3, 4, \ldots \tag{C-74}$$

which is (C-59b).

We have now shown that the IIR WF design equation (C-66) leads to the cumulant-based design equation (C-59b); hence, the cumulant-based deconvolution design is optimal in the mean-squared sense given in the statement of this theorem. □

So far all our results are based on (C-16). What about starting with (C-17)? Two approaches that have done this are worth mentioning. Giannakis and Swami (1987, 1990) have developed a "double $C(q, k)$ algorithm" for estimating the AR and MA coefficients in an ARMA$(p, q)$ model. Their algorithm is derived by starting with (C-17) and inserting the fact that, for an ARMA model, $H(z) = B(z)/A(z)$. By means of some very clever analysis, they then show that both the AR and MA parameters can be determined from a $C(q, k)$ type of equation. See Problem C-16 for some of the details.

The *complex cepstrum* is widely known in digital signal-processing circles (e.g., Oppenheim and Schafer, 1989, Chapter 12). We start with the transfer function $H(z)$, take its logarithm, $\hat{H}(z) = \ln H(z)$, and then take the inverse $z$-transform of $\hat{H}(z)$ to obtain the complex cepstrum $\hat{h}(k)$. Associated with $\hat{h}(k)$ are cepstral coefficients. If they can be computed, then it is possible to reconstruct $h(k)$ to within a scale factor. Starting with (C-17), it is possible to derive a linear system of equations for the cepstral coefficients (Nikias and Pan, 1987; Pan and Nikias, 1988). This method lets us determine the magnitude and phase of a system's transfer functon using cepstral techniques without the need for phase unwrapping. See Problems C-17 and C-18 for some of the details.

All the higher-order statistical results in Lessons B and C have been for scalar random variables and processes. Extensions of many of these results to vector random variables and processes can be found in Swami and Mendel (1990a) and Mendel (1991). Among their results is a generalization of the Bartlett–Brillinger–Rosenblatt formula (C-16), a generalization in which multiplication [e.g., $h(n)h(n+\tau_1)$] is replaced by the Kronecker product.

Some additional applications of cumulants to system identification can be found in the problems at the end of this lesson.

## COMPUTATION

All the M-files that are described next are in the *Hi-Spec*™ *Toolbox* (Hi-Spec is a trademark of United Signals and Systems, Inc.). To estimate cumulants or the bispectrum from real data, use:

**cumest**: Computes sample estimates of cumulants using the overlapped segment method.

**bispecd**: Bispectrum estimation using the direct (FFT-based) method.

**bispeci**: Bispectrum estimation using the indirect method.

To compute cumulants for a given single-channel model, use:

**cumtrue**: Computes theoretical (i.e., true) cumulants of a linear process. This is done using the Bartlett–Brillinger–Rosenblatt formula (C-16). If you also want to compute the theoretical bispectrum of a linear process, first compute the cumulants of that process and then compute the two-dimensional DTFT of the cumulants; or evaluate (C-17) directly.

Determining parameters in AR, MA, or ARMA models can be accomplished using the following Hi-Spec M-files:

**arrcest**: Estimates AR parameters using the normal equations based on autocorrelations and/or cumulants. See Example C-4.

**maest**: Estimates MA parameters using the modified GM method. See Example C-5 and Problems C-13 and C-14.

**armaqs**: Estimates ARMA parameters using the $q$-slice method. See Example C-3 and Problem C-21.

**armarts**: Estimates ARMA parameters using the residual time series method. See Problem C-20.

The impulse response of a linear process can be estimated using the bicepstrum method, as described in Problem C-18. This can be accomplished using:

**biceps:** Nonparametric system identification or signal reconstruction using the bicepstrum method. Computation of the complex cepstrum, by third-order cumulants, without phase unwrapping.

## SUMMARY QUESTIONS

**1.** Cumulants are estimated from data by:
**(a)** numerical integration, as required by the formal definition of expectation
**(b)** passing the data through a low-pass filter
**(c)** replacing expectation by sample average

**2.** The reason it is important to know that sample estimates of cumulants converge in probability to their true values is:
**(a)** we use functions of cumulants in solutions to many signal-processing problems and can therefore use the *consistency carry-over* property for these functions
**(b)** we use functions of cumulants in solutions to many signal-processing problems and can therefore conclude that these functions converge in mean square
**(c)** it lets us work with short data lengths when we estimate cumulants

**3.** Which bispectrum estimation method does not first estimate cumulants?
**(a)** segmentation and averaging
**(b)** direct method
**(c)** indirect method

**4.** The Bartlett–Brillinger–Rosenblatt formulas are important because they:
**(a)** provide a new way to compute second-order statistics
**(b)** relate cumulants and polyspectra
**(c)** generalize our statistical characterization and understanding of the Figure C-1 system from second- to higher-order statistics

**5.** The Bartlett–Brillinger–Rosenblatt formulas:
**(a)** only need knowledge of a system's impulse response
**(b)** need knowledge of a system's impulse response and the cumulant of the system's input
**(c)** need knowledge of a system's impulse response and the correlation function of the system's input

**6.** The $C(q, k)$ formula lets us determine a channel's impulse response from:
**(a)** input and output measurements
**(b)** only output measurements
**(c)** only input measurements

**7.** The output of an AR($p$) system is measured in the presence of additive Gaussian noise. Which equations would you use to estimate the $p$ AR coefficients?
**(a)** correlation-based normal equations

(b) Wiener–Hopf equations
(c) cumulant-based normal equations

**8.** The GM equation relates:
(a) power spectrum to the DTFT of a diagonal-slice cumulant
(b) power spectrum to the DTFT of the full-blown cumulant
(c) bispectrum to trispectrum

**9.** A diagonal-slice cumulant relates:
(a) the impulse response of a channel to powers of an inverse filter's coefficients
(b) inverse filter's coefficients to powers of the channel's impulse response
(c) signal to noise

## PROBLEMS

**C-1.** Explain in detail how to estimate fourth-order cumulants.

**C-2.** Here we examine different aspects of $C_{3,x}^{(i)}(\tau_1, \tau_2)$ given in Equation (C-3).

**(a)** Verify and explain the meanings of the ranges for $s_1$ and $s_2$ given below Equation (C-3).

**(b)** Observe that $M$ in the factor $1/M$ is not equal to the total number of elements in the summation; hence, (C-4) will be biased. Change this factor to $1/M(\tau_1, \tau_2)$, where $M(\tau_1, \tau_2) = s_2 - s_1 + 1$. Prove that $\hat{C}_{3,x}(\tau_1, \tau_2)$ is then an unbiased estimator of $C_{3,x}(\tau_1, \tau_2)$.

**(c)** Prove that $M(\tau_1, \tau_2) = M - \max(0, \tau_1, \tau_2) + \min(0, \tau_1, \tau_2)$. [*Hint:* $\max(-a, -b) = -\min(a, b)$.]

**C-3.** In this problem we consider *biased* correlation estimators. Problem C-4 considers unbiased correlation estimators. Assume that our data are obtained either from $M$ independent realizations, each with a finite number of $K$ samples, or from $MK$ samples from one realization.

**(a)** Let $r_1^b(k)$ denote the following biased *segmented* correlation estimator:

$$r_1^b(k) = \frac{1}{M} \sum_{i=1}^{M} \tilde{r}_i^b(k)$$

where $\tilde{r}_i^b(k)$ denotes the estimated correlation function in the $i$th segment; i.e.,

$$\tilde{r}_i^b(k) = \frac{1}{K} \sum_{n=1}^{K-|k|} x_i(n) x_i(n+k), \qquad i = 1, 2, \ldots, M$$

Show that $\mathbf{E}\{r_1^b(k)\} = (K - |k|)/K \times r(k)$, where $r(k) = \mathbf{E}\{x_i(n)x_i(n+k)\}$, and explain why this estimator is neither unbiased nor asymptotically unbiased.

**(b)** Let $r_2^b(k)$ denote an estimator that uses all $MK$ data without segmentation; i.e.,

$$r_2^b(k) = \frac{1}{MK} \sum_{n=1}^{MK-|k|} x(n) x(n+k)$$

Show that $\mathbf{E}\{r_2^b(k)\} = (MK - |k|)/MK \times r(k)$ and explain why this estimator is biased but is asymptotically unbiased.

**(c)** Compare the biases for $r_1^b(k)$ and $r_2^b(k)$.

**(d)** Determine formulas for cov $[r_1^b(k), r_1^b(k+\nu)]$, and cov $[r_2^b(k), r_2^b(k+\nu)]$. Note that this part of the problem will take a very significant effort.

**C-4.** In this problem we consider *unbiased* correlation estimators. Problem C-3 considers biased correlation estimators. Assume that our data are obtained either from $M$ independent realizations, each with a finite number of $K$ samples, or from $MK$ samples from one realization.

**(a)** Let $r_1^u(k)$ be the unbiased *segmented* correlation estimator

$$r_1^u(k) = \frac{1}{M}\sum_{i=1}^{M} \tilde{r}_i^u(k)$$

where $\tilde{r}_i^u(k)$ denotes the estimated correlation function in the $i$th segment; i.e.,

$$\tilde{r}_i^u(k) = \frac{1}{K-|k|}\sum_{n=1}^{K-|k|} x_i(n)x_i(n+k), \qquad i = 1, 2, \ldots, M$$

Show that $\mathbf{E}\{r_1^b(k)\} = r(k)$, where $r(k) = \mathbf{E}\{x_i(n)x_i(n+k)\}$.

**(b)** Let $r_2^u(k)$ denote an estimator that uses all $MK$ data without segmentation; i.e.,

$$r_2^u(k) = \frac{1}{MK-|k|}\sum_{n=1}^{MK-|k|} x(n)x(n+k)$$

Show that $\mathbf{E}\{r_2^b(k)\} = r(k)$.

**(c)** If you have also completed Problem C-3, then compare $r_1^b(k)$ and $r_1^u(k)$ and $r_2^b(k)$ and $r_2^u(k)$.

**(d)** Determine formulas for cov $[r_1^u(k), r_1^u(k+\nu)]$ and cov $[r_2^u(k), r_2^u(k+\nu)]$. Note that this part of the problem will take a very significant effort.

**C-5.** Derive the second-order statistics given in Equations (C-12)–(C-14).

**C-6.** Complete the details of the derivation of (C-17), starting with (C-16) and (C-21).

**C-7.** Derive the Bartlett–Brillinger–Rosenblatt equation (C-22), which is valid for the case of a *colored* non-Gaussian system input, $\nu(k)$.

**C-8.** Show that the $C(q,k)$ formula for fourth-order cumulants is

$$h(k) = \frac{C_{4,y}(q,0,k)}{C_{4,y}(q,0,0)}, \qquad k = 0, 1, \ldots, q$$

What is the structure of the $C(q,k)$ formula for $k$th-order cumulants?

**C-9.** Obtain the following formulas for $\gamma_{3,\nu}$ and $\gamma_{4,\nu}$, which are by-products of the $C(q,k)$ formula:

$$\gamma_{3,\nu} = \frac{C_{3,y}(0,0)}{\sum_{k=0}^{q}[C_{3,y}(q,k)/C_{3,y}(q,0)]^3}$$

and

$$\gamma_{4,\nu} = \frac{C_{4,y}(0,0,0)}{\sum_{k=0}^{q}[C_{4,y}(q,0,k)/C_{4,y}(q,0,0)]^4}$$

**C-10.** Show that (C-36) can be used to compute the MA coefficients $b(1), b(2), \ldots, b(q)$ if we are given all the AR coefficients and only the IR values $h(0), h(1), \ldots, h(q)$.

**C-11.** This problem explores *correlation-based normal equations.*

**(a)** Derive the correlation-based normal equations for the case of noise-free measurements, as stated in (C-42). Do this for both an AR($p$) model and an ARMA($p,q$) model. Be sure your derivation includes an explanation of the condition $m > 0$ for the case of an AR($p$) model. What is the comparable condition for the ARMA($p,q$) model?

**(b)** Repeat part (a), but for noisy measurements. For an ARMA($p,q$) model and the case of additive white noise, show that

$$\sum_{k=0}^{p} a(k) r_z(k-m) = \begin{cases} \sigma_v^2 a(m), & \text{for } 0 < m \leq p \\ 0, & \text{for } m > p \end{cases}$$

**(c)** Repeat part (b) but for the case of additive colored noise.

**(d)** Using the results from parts (b) and (c), explain why measurement noise vastly complicates the utility of the correlation-based normal equations.

[*Hints:* Operate on both sides of (C-41) with $\mathbf{E}\{\cdot y(n-m)\}$ or $\mathbf{E}\{\cdot z(n-m)\}$; use the fact that $y(n-m)$ or $z(n-m)$ depends at most on $\nu(n-m)$.]

**C-12.** Derive the GM equation (C-47) for arbitrary values of $k$.

**C-13.** One method for determining MA coefficients is to use the GM equations of Example C-5, treating both $b(i)$ and $b^2(i)$ [or $b^3(i)$] as *independent parameters*, concatenating (C-54) or (C-55), and then solving for all the unknown $b(i)$ and $b^2(i)$ [or $b^3(i)$] coefficients using least squares. Here we focus on (C-54); comparable results can be stated for (C-55).

**(a)** Explain why, at most, $-q \leq m \leq 2q$.

**(b)** Once $b(i)$ and $b^2(i)$ have been estimated using least squares, explain two strategies for obtaining $\hat{b}(i), i = 1, 2, \ldots, q$. Also, explain why this general approach to estimating MA parameters may not give credible results.

**(c)** Suppose only *white* noisy measurements are available; then GM equation (C-54) becomes

$$\sum_{k=0}^{q} b^2(k)[r_z(m-k) - r_n(m-k)] = \frac{\sigma_v^2}{\gamma_{3,v}} \sum_{k=0}^{q} b(k) C_{3,y}(m-k, m-k)$$

Because we don't know the variance of the additive white measurement noise, we must remove all the equations to which it contributes. Show that by doing this we are left with an underdetermined system of equations (Tugnait, 1990).

This problem is continued in Problem C-14.

**C-14.** Tugnait (1990) and Friedlander and Porat (1990) have generalized the GM equation from a diagonal slice to another one-dimensional slice result. We illustrate their results for third-order cumulants.

**(a)** Set $\tau_1 = \tau$ and $\tau_2 = \tau + l$ in (C-16) and obtain the following *generalized GM equation* ($z = e^{j\omega T}$):

$$[H(z) * z^l H(z)] S_y(z) = \frac{\sigma_v^2}{\gamma_{3,v}} H(z) S_{3,y}(z)$$

Observe that when $l = 0$ this equation reduces to (C-47) (for third-order cumulants).

**(b)** Show that for an MA($q$) system the generalized GM equation can be expressed as

$$\sum_{k=0}^{q} b(k)b(k+l)r_y(m-k) = \frac{\sigma_\nu^2}{\gamma_{3,\nu}} \sum_{k=0}^{q} b(k)C_{3,y}(m-k, m-k+l)$$

Observe that when $l = 0$ this equation reduces to (C-54).

**(c)** Set $l = q$ in the formula derived in part (b) to obtain an equation that is almost linear in the unknown MA coefficients. Show that there are $2q+1$ such equations.

**(d)** How many equations survive when only white noisy measurements are available?

**(e)** Explain how the remaining equations from part (d) can be combined with those from part (c) of Problem C-13 to provide an overdetermined system of equations that can then be simultaneously solved for $b(k)$ and $b^2(k)$, $k = 1, 2, \ldots, q$.

**C-15.** In this problem you will derive (C-60) and (C-61) using a very different approach than the GM equation. The approach (Zheng et al., 1993) is to obtain two formulas for $\Phi_{\nu y}(i) = \mathbf{E}\{\nu(k)y(k+i)\}$ (or $\Phi_{\nu yy}(i) = \mathbf{E}\{\nu(k)y^2(k+i)\}$) and to then equate the two formulas. The first formula is one in which $y(k)$ is replaced by its convolutional model, $y(k) = \nu(k) * h(k)$, whereas the second formula is one in which $\nu(k)$ is replaced by $\hat{\nu}(k)$, where $\hat{\nu}(k) = \theta(k) * y(k)$.

**(a)** Derive Equation (C-60).

**(b)** Derive Equation (C-61).

Note that it is also possible to derive (C-62) by computing $\Phi_{\nu yyy}(i) = \mathbf{E}\{\nu(k)y^3(k+i)\}$, in two ways.

**C-16.** In this problem we examine the double $C(q, k)$ algorithm, which was developed by Giannakis and Swami (1987, 1990) for estimating the parameters of the ARMA$(p, q)$ system $H(z) = B(z)/A(z)$, where

$$B(z) = 1 + \sum_{i=1}^{q} b(i)z^{-i} \quad \text{and} \quad A(z) = 1 + \sum_{i=1}^{p} a(i)z^{-i}$$

The double $C(q, k)$ algorithm is a very interesting application of (C-17).

**(a)** Let $D(z_1, z_2) = A(z_1)A(z_2)A(z_1^{-1}z_2^{-1})$ and show that

$$\sum_{i=0}^{p} \sum_{j=0}^{p} d(i, j)C_{3,y}(m-i, n-j) = 0 \qquad \text{(C-16.1)}$$

for $(m, n)$ outside the Figure C-2 six-sided region, where

$$d(i, j) = \sum_{k=0}^{p} a(k)a(k+i)a(k+j)$$

**(b)** Starting with (C-16.1) when $m = n$, obtain a linear system of equations for $d(i, j)/d(p, 0)$ [be sure to use the symmetry properties of $d(i, j)$]. Show that the number of variables $d(i, j)/d(p, 0)$ is $p(p+3)/2$. Explain how you would solve for these variables.

**(c)** Explain how the $C(q, k)$ algorithm can then be used to extract the AR coefficients.

**(d)** Show that we can also compute another two-dimensional array of numbers, $\gamma_{3,\nu}b(m, n)$, where

$$\gamma_{3,\nu}b(m, n) = \sum_{i=0}^{p} \sum_{j=0}^{p} d(i, j)C_{3,y}(m-i, n-j)$$

and the $b(m, n)$ are related to the MA coefficients of the ARMA model as

$$b(m, n) = \sum_{i=0}^{q} b(i)b(i+m)b(i+n)$$

**(e)** Explain how the MA coefficients can then be computed using the $C(q, k)$ algorithm.

The practical application of this double $C(q, k)$ algorithm requires a modified formulation so that steps (c) and (e) can be performed using least squares. See Giannakis and Swami for details on how to do this. Note, finally, that this algorithm is applicable to noncausal as well as to causal channels.

**C-17.** Different representations of ARMA$(p, q)$ processes are possible. Here we focus on a cascade representation, where $H(z) = B(z)/A(z) = cz^{-r}I(z^{-1})O(z)$, $c$ is a constant, $r$ is an integer,

$$I(z^{-1}) = \frac{\prod_{i=1}^{L_1}(1 - a_i z^{-1})}{\prod_{j=1}^{L_3}(1 - c_j z^{-1})}$$

is the minimum phase component of $H(z)$, with poles $\{c_i\}$ and zeros $\{a_i\}$ inside the unit circle, i.e., $|c_i| < 1$ and $|a_i| < 1$ for all $i$, and

$$O(z) = \prod_{i=1}^{L_2}(1 - b_i z)$$

is the maximum phase component of $H(z)$, with zeros outside the unit circle at $1/|b_i|$, where $|b_i| < 1$ for all $i$. Clearly, $h(k) = ci(k) * o(k) * \delta(k - r)$. In this problem we show how to estimate $o(k)$ and $i(k)$, after which $h(k)$ can be reconstructed to within a scale factor and time delay.

Oppenheim and Schafer [1989, Eqs. (12.77) and (12.78)] show $i(k)$ and $o(k)$ can be computed recursively from knowledge of so-called A and B cepstral coefficients, where

$$A^{(i)} = \sum_{j=1}^{L_1} a_j^i - \sum_{j=1}^{L_3} c_j^i \quad \text{and} \quad B^{(i)} = \sum_{j=1}^{L_2} b_j^i$$

In this problem we are interested in how to estimate the cepstral coefficients using higher-order statistics (Nikias and Pan, 1987; Pan and Nikias, 1988).

**(a)** Beginning with (C-17) for the bispectrum, expressed in the $z_1 - z_2$ domain as $S_{3,y}(z_1, z_2) = \gamma_{3,v}H(z_1)H(z_2)H(z_1^{-1}z_2^{-1})$, take the logarithm of $S_{3,y}(z_1, z_2)$, $\hat{S}_{3,y}(z_1, z_2) = \ln S_{3,y}(z_1, z_2)$, and then take the inverse two-dimensional transform of $\hat{S}_{3,y}(z_1, z_2)$ to obtain the *complex bicepstrum* $b_y(m, n)$. To do this, you will need the following power series expansions:

$$\ln(1 - \alpha z^{-1}) = -\sum_{n=1}^{\infty}\left(\frac{\alpha^n}{n}\right)z^{-n}, \quad |z| > |\alpha|$$

and

$$\ln(1 - \beta z) = -\sum_{n=1}^{\infty}\left(\frac{\beta^n}{n}\right)z^{n}, \quad |z| < |\beta^{-1}|$$

Show that $b_y(m, n)$ has nonzero values only at $m = n = 0$, integer values along the $m$ and $n$ axes and at the intersection of these values along the 45° line $m = n$; i.e.,

$$
\begin{aligned}
b_y(m,n) = \{&\ln|c\gamma_{3,v}| \text{ when } m = n = 0;\\
&-1/n \times A^{(n)} \text{ when } m = 0 \text{ and } n > 0;\\
&-1/m \times A^{(m)} \text{ when } n = 0 \text{ and } m > 0;\\
&1/m \times B^{(-m)} \text{ when } n = 0 \text{ and } m < 0;\\
&1/n \times B^{(-n)} \text{ when } m = 0 \text{ and } n < 0;\\
&-1/n \times B^{(n)} \text{ when } m = n > 0;\\
&1/n \times A^{(-n)} \text{ when } m = n < 0;\\
&0, \text{ otherwise}\}
\end{aligned}
$$

Plot the locations of these complex bicepstral values.

**(b)** Beginning with (C-17) for the bispectrum, expressed in the $z_1 - z_2$ domain as $S_{3,y}(z_1, z_2) = \gamma_{3,v}H(z_1)H(z_2)H(z_1^{-1}z_2^{-1})$, show that

$$\sum_{k=-\infty}^{\infty}\sum_{l=-\infty}^{\infty} kb_y(k,l)C_{3,y}(m-k, n-l) = mC_{3,y}(m,n) \tag{C-17.1}$$

This is a very interesting equation that links the third-order cumulant and the bicepstral values. [*Hint:* Take the two-dimensional inverse $z$-transform of $\partial\hat{S}_{3,y}(z_1, z_2)/\partial z_1 = [1/S_{3,y}(z_1, z_2)] \times \partial S_{3,y}(z_1, z_2)/\partial z_1$.]

**(c)** Using the results from parts (a) and (b), show that

$$
\begin{aligned}
\sum_{k=1}^{\infty}\{&A^{(k)}[C_{3,y}(m-k, n) - C_{3,y}(m+k, n+k)]\\
&+B^{(k)}[C_{3,y}(m-k, n-k) - C_{3,y}(m+k, n)]\} = -mC_{3,y}(m,n)
\end{aligned} \tag{C-17.2}
$$

**(d)** Explain why $A^{(k)}$ and $B^{(k)}$ decay exponentially so that $A^{(k)} \to 0$ for $k > p_1$ and $B^{(k)} \to 0$ for $k > q_1$.

**(e)** Truncate (C-17.2) and explain how the $A$ and $B$ cepstral coefficients can be computed using least squares or TLS.

**C-18.** Here we shall determine second- and third-order statistics for the following three systems (Nikias and Raghuveer, 1987): $H_1(z) = (1 - az^{-1})(1 - bz^{-1})$, $H_2(z) = (1 - az)(1 - bz)$, and $H_3(z) = (1 - az)(1 - bz^{-1})$. Note that $H_1(z)$ is minimum phase, $H_2(z)$ is maximum phase, and $H_3(z)$ is mixed phase.

**(a)** Determine the three autocorrelations $r(0)$, $r(1)$, and $r(2)$ for the three systems. Explain why you obtained the same results for the three systems.

**(b)** Determine the following third-order cumulants for the three systems: $C(0,0)$, $C(1,0)$, $C(1,1)$, $C(2,0)$, $C(2,1)$, and $C(2,2)$.

**C-19.** The *residual time-series method* for estimating ARMA coefficients is (Giannakis, 1987a, and Giannakis and Mendel, 1989) a three-step procedure: (1) estimate the AR coefficients; (2) compute the residual time series; and (3) estimate the MA parameters from the residual time series. The AR parameters can be estimated using cumulant-based normal equations, as described in Example C-4. The "residual time series," denoted $\tilde{y}(n)$, equals $y(n) - \hat{y}(n)$, where

$$\hat{y}(n) = -\sum_{k=1}^{p}\hat{a}(k)y(n-k)$$

**(a)** Letting $\tilde{a}(k) = a(k) - \hat{a}(k)$, show that the residual satisfies the following equation:

$$\tilde{y}(n) = \sum_{k=0}^{q} b(k)v(n-k) - \sum_{k=1}^{p} \tilde{a}(k)y(n-k)$$

**(b)** Explain why $\tilde{a}(k)y(n-k)$ is a highly nonlinear function of the measurements.

**(c)** At this point in the development of the residual time-series method, it is assumed that $\tilde{a}(k) \approx 0$, so

$$\tilde{y}(n) \approx \sum_{k=0}^{q} b(k)v(n-k)$$

Explain how to determine the MA coefficients.

**(d)** Explain the sources of error in this method.

**C-20.** The $q$-slice formula in Example C-3 is the basis for the following *q-slice algorithm* for determining the coefficients in an ARMA model (Swami, 1988, and Swami and Mendel, 1990b): (1) estimate the AR coefficients; (2) determine the first $q$ IR coefficients using $q$ one-dimensional cumulant slices; and (3) determine the MA coefficients using (C-36). The AR parameters can be estimated using cumulant-based normal equations, as described in Example C-4.

**(a)** Show that (C-40) can be reexpressed as

$$\sum_{k=1}^{p} a(k)C_{m,y}(q-k, t, 0, \ldots, 0) - f_m(0; q)h(t) = -C_{m,y}(q, t, 0, \ldots, 0)$$

**(b)** Using the cumulant-based normal equations from Example C-4 and the result in part (a), explain how the AR parameters and scaled values of the IR can be *simultaneously* estimated.

**(c)** Explain how we can obtain an estimate of the unscaled IR response values.

**(d)** Discuss the strengths and weaknesses of this method.

# LESSON D

# Introduction to State-variable Models and Methods

## SUMMARY

State-variable models are widely used in controls, communications, and signal processing. This lesson provides a brief introduction to such models. To begin, it explains the notions of state, state variables, and state space. This culminates in what is meant by a *state-variable model*, a set of equations that describes the unique relations among a system's input, output, and state.

Many state-variable representations can result in the same input/output system behavior. We obtain a number of important and useful state-variable representations for single-input, single-output, discrete, linear time-invariant systems, including moving average, autoregressive, and autoregressive moving average.

What can we do with a state-variable model? We can compute a system's output, given its inputs and initial conditions. A closed-form formula is obtained for doing this. We can also determine the system's transfer function in terms of the matrices that appear in the state-variable model.

Because a linear time-invariant system can be described in many equivalent ways, including a state-variable model or a transfer function model, it is important to relate the properties of these two representations. We do this at the end of this lesson.

When you complete this lesson, you will be able to (1) express a linear time-invariant system in state-variable format; (2) understand the equivalence of difference equation and state-variable representations; (3) compute the solution to a state-variable model; (4) evaluate the transfer function of a state-variable model; and (5) understand the connections between a state-variable model and a transfer function model.

## NOTIONS OF STATE, STATE VARIABLES, AND STATE SPACE

The *state* of a dynamic system (the material in this lesson is taken from Mendel, 1983, pp. 26–41 and 14–16) at time $t = t_0$ is the amount of information at $t_0$ that, together with the inputs defined for all values of $t \geq t_0$, determines uniquely the behavior of the system for all $t \geq t_0$. For systems of interest to us, the state can be represented by a column vector $\mathbf{x}$ called the *state vector*, whose dimension is $n \times 1$.

When the system input is real valued and $\mathbf{x}$ is finite dimensional, our *state space*, defined as an $n$-dimensional space in which $x_1(t), x_2(t), \ldots, x_n(t)$ are coordinates, is a finite-dimensional real vector space. The state at time $t$ of our system will be defined by $n$ equations and can be represented as a point in $n$-dimensional state space.

**EXAMPLE D-1**

The following second-order differential equation models many real devices:

$$m\ddot{x}(t) + c\dot{x}(t) + kx(t) = m\omega(t) \tag{D-1}$$

The solution to this differential equation requires two initial conditions in addition to knowledge of input $\omega(t)$ for $t \geq t_0$. Any two initial conditions will do, although $x(t_0)$ and $\dot{x}(t_0)$ are most often specified. We can choose $\mathbf{x}(t) = \text{col}\,[x_1(t), x_2(t)] = \text{col}\,[x(t), \dot{x}(t)]$ as our $2 \times 1$ state vector. The state space is two dimensional and is easily visualized as in Figure D-1. Observe how $x(t_i)$ is represented as a point in the two-dimensional state space. When we connect all these points together, we obtain the trajectory of the system in state space. □

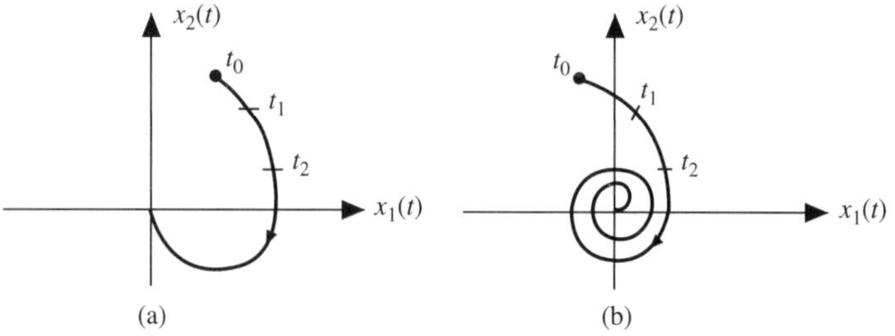


**Figure D-1** Trajectories in two-dimensional state space for (a) overdamped and (b) underdamped systems. (After J. M. Mendel et al. (1981), *Geophysics*, **46**, 1399.)

Visualization is easy in two- and three-dimensional state spaces. In higher-dimensional spaces, it is no longer possible to depict the trajectory of the system, but it is possible to abstract the notions of a point and trajectory in such spaces.

By a *state-variable model* we mean the set of equations that describes the unique relations among the input, output, and state. It is composed of a *state equation* and an *output equation*. A continuous-time state-variable model is

$$\dot{\mathbf{x}}(t) = \mathbf{A}\mathbf{x}(t) + \mathbf{B}\mathbf{u}(t) \tag{D-2}$$

$$\mathbf{y}(t) = \mathbf{C}\mathbf{x}(t) + \mathbf{D}\mathbf{u}(t) \tag{D-3}$$

State equation (D-2) governs the behavior of the state vector $\mathbf{x}(t)$ ($\mathbf{x}$ for short); it is a vector first-order differential equation. Output equation (D-3) relates the outputs to the state vector and inputs. In (D-2) and (D-3), $\mathbf{x}(t)$ is an $n \times 1$ state vector, $\mathbf{u}(t)$ is an $r \times 1$ control input vector, and $\mathbf{y}(t)$ is an $m \times 1$ observation vector; $\mathbf{A}$ is an $n \times n$ state transition matrix, $\mathbf{B}$ and $\mathbf{D}$ are $n \times r$ and $m \times r$ input distribution matrices, respectively (they distribute the elements of $\mathbf{u}$ into the proper state and output equations), and $\mathbf{C}$ is an $m \times n$ observation matrix.

It is possible to have two distinctly different types of inputs acting on a system: controlled and uncontrolled inputs. A controlled input is either known ahead of time (e.g., a sinusoidal function with known amplitude, frequency, and phase) or can be measured (e.g., the output of a random number generator). An uncontrolled input cannot be measured (e.g., a gust of wind acting on an airplane). Usually, an uncontrolled input is referred to as a *disturbance input.* Such inputs are included in our basic state-variable model. Here we assume that our state-variable model includes only controlled inputs.

**EXAMPLE D-2**

Here we demonstrate how to express the second-order model stated in Example D-1 in state-variable format. To begin, we normalize (D-1) by making the coefficient of the second derivative unity:

$$\ddot{x}(t) + 2\zeta\omega_n\dot{x}(t) + \omega_n^2 x(t) = u(t) \tag{D-4}$$

where $\omega_n = \sqrt{k/m}$ and $\zeta = (c/2m)\sqrt{k/m}$. Choosing $x_1(t) = x(t)$ and $x_2(t) = \dot{x}(t)$ as our state variables, we observe that

$$\dot{x}_1 = \dot{x} = x_2$$

and

$$\dot{x}_2 = \ddot{x} = -2\zeta\omega_n\dot{x}(t) - \omega_n^2 x(t) + u(t) = -2\zeta\omega_n x_2 - \omega_n^2 x_1 + u(t)$$

These two state equations can be expressed [as in (D-2)] as

$$\begin{pmatrix} \dot{x}_1(t) \\ \dot{x}_2(t) \end{pmatrix} = \begin{pmatrix} 0 & 1 \\ -\omega_n^2 & -2\zeta\omega_n \end{pmatrix} \begin{pmatrix} x_1(t) \\ x_2(t) \end{pmatrix} + \begin{pmatrix} 0 \\ 1 \end{pmatrix} u(t)$$

Here $\mathbf{x}(t) = \text{col}\,[x_1(t), x_2(t)]$,

$$\mathbf{A} = \begin{pmatrix} 0 & 1 \\ -\omega_n^2 & -2\zeta\omega_n \end{pmatrix}, \qquad \mathbf{B} = \begin{pmatrix} 0 \\ 1 \end{pmatrix}$$

and $\mathbf{u}(t)$ is a scalar input $u(t)$.

To complete the description of the second-order differential equation as a state-variable model, we need an output equation. If displacement is recorded, then $y_1(t) = x(t)$, which can be expressed in terms of state vector $\mathbf{x}(t)$ as

$$y_1(t) = (1 \quad 0)\mathbf{x}(t)$$

On the other hand, if velocity is recorded, then $y_2(t) = \dot{x}(t)$, or $y_2(t) = (0 \quad 1)\mathbf{x}(t)$.

Observe that there is no direct transmission of input $u(t)$ into these measurements; hence, $\mathbf{D} = \mathbf{0}$ for both $y_1(t)$ and $y_2(t)$. The same is not true if acceleration is recorded, for in that case

$$y_3(t) = \ddot{x}(t) = (-\omega_n^2 \quad -2\zeta\omega_n)\mathbf{x}(t) + u(t)$$

Finally, if two or more signals can be measured simultaneously, then the measurements can be collected (as in (D-3)) as a vector of measurements. For example, if $x(t)$ and $\ddot{x}(t)$ can both be recorded, then

$$\mathbf{y}(t) = \begin{pmatrix} y_1(t) \\ y_2(t) \end{pmatrix} = \begin{pmatrix} 1 & 0 \\ -\omega_n^2 & -2\zeta\omega_n \end{pmatrix} \begin{pmatrix} x_1(t) \\ x_2(t) \end{pmatrix} + \begin{pmatrix} 0 \\ 1 \end{pmatrix} u(t) \quad \square$$

A state-variable model can also be defined for discrete-time systems using a similar notion of state vector. Indeed, discrete-time state-variable models are now more popular than continuous-time models because of their applicability to digital computer systems and digital filtering. In many applications it is therefore necessary to convert a continuous-time physical system into an equivalent discrete-time model. Among many discretization methods, one direct approach is to transform a differential equation into a compatible difference equation. See, also, Lesson 23.

**EXAMPLE D-3**

We discretize the second-order differential equation in Example D-1 by letting

$$\dot{x}(t) \approx \frac{x(t+\delta) - x(t)}{\delta}$$

and

$$\ddot{x}(t) \approx \frac{\dot{x}(t+\delta) - \dot{x}(t)}{\delta} = \frac{x(t+2\delta) - 2x(t+\delta) + x(t)}{\delta^2}$$

Making the appropriate substitutions, we obtain

$$mx(t+2\delta) + c_1 x(t+\delta) + k_1 x(t) = m_1 u(t)$$

where $c_1 = c\delta - 2m$, $k_1 = k\delta^2 - c\delta + m$, and $m_1 = m\delta^2$. A commonly used discretization of time is $t = n\delta$; it is also common to use a normalized time unit, achieved by setting $\delta = 1$. The preceding equation can then be further simplified to

$$mx(n+2) + c_1 x(n+1) + k_1 x(n) = m_1 \omega(n)$$

a second-order difference equation. For other ways to approximate $\dot{x}(t)$ and $\ddot{x}(t)$, see Taylor and Antoniotti (1993). $\square$

In the next section we shall demonstrate how a state-variable model for a difference equation is easily derived. Here we note that a discrete-time state-variable model is again composed of a state equation and an output equation:

$$\mathbf{x}(k+1) = \mathbf{\Phi}\mathbf{x}(k) + \mathbf{\Psi}\mathbf{u}(k) \tag{D-5}$$

$$\mathbf{y}(k) = \mathbf{H}\mathbf{x}(k) + \mathbf{D}\mathbf{u}(k) \tag{D-6}$$

State equation (D-5) is a vector first-order difference equation. Equations (D-5) and (D-6) are usually defined for $k = 0, 1, \ldots$. When $\mathbf{y}(0)$ is not physically available or is undefined, it is customary to express the output equation as

$$\mathbf{y}(k+1) = \mathbf{H}\mathbf{x}(k+1) + \mathbf{D}\mathbf{u}(k+1) \tag{D-6$'$}$$

Equations (D-5) and (D-6$'$) then constitute our state-variable model for $k = 0, 1, \ldots$. Quantities $\mathbf{x}(k)$, $\mathbf{u}(k)$, $\mathbf{y}(k)$, $\mathbf{\Phi}$, $\mathbf{\Psi}$, $\mathbf{D}$, and $\mathbf{H}$ have the same definitions as given above for the continuous-time state-variable model.

So as to see the forest from the trees, all our remaining discussions are limited to single-input, single-output linear time-invariant systems. The more general case is treated in the main body of Lesson 15, in the section The Basic State-variable Model.

## CONSTRUCTING STATE-VARIABLE REPRESENTATIONS

Many state-variable representations can result in the same input/output system behavior. In this section we obtain a number of important and useful state-variable representations for single-input, single-output (SISO) discrete, linear time-invariant (LTI) systems. Derivations are by means of examples; then general results are stated.

How we obtain a state-variable model depends on our starting point. There are four possibilities: we may be given (1) a collection of differential or difference equations, obtained as in Example D-1 from physical principles, (2) a transfer function, (3) an impulse response function, or (4) impulse response data. We shall discuss case 2 in detail. Case 1 has been illustrated in Example D-1. Case 3 can be reduced to case 4 simply by sampling the impulse response function. Case 4 is related to the field of approximation and is beyond the scope of this book.

Our objective is to find state-variable representations for LTI SISO discrete-time systems. We wish to represent such systems by a pair of equations of the form

$$\mathbf{x}(k+1) = \mathbf{\Phi x}(k) + \boldsymbol{\psi} u(k) \tag{D-7}$$

$$y(k) = \mathbf{h}'\mathbf{x}(k) + du(k) \tag{D-8}$$

We must learn not only how to choose the state vector $\mathbf{x}$, but also how to specify the elements of the matrices that appear in the state-variable model in terms of the parameters that appear in the transfer function models.

### All-zero (Moving Average) Models

**EXAMPLE D-4**

The MA model

$$y(k) = \beta_3 u(k-3) + \beta_2 u(k-2) + \beta_1 u(k-1) \tag{D-9}$$

is easily converted to a third-order state-variable model by choosing state variables $x_1, x_2$, and $x_3$ as follows: $x_1(k) = u(k-1), x_2(k) = u(k-2)$, and $x_3(k) = u(k-3)$. Equation (D-9), which is treated as an output equation [i.e., Equation (D-8)], is reexpressed in terms of $x_1, x_2$, and $x_3$ as

$$y(k) = \beta_3 x_3(k) + \beta_2 x_2(k) + \beta_1 x_1(k) \tag{D-10}$$

To obtain the associated state equation, we observe that

$$\begin{aligned} x_1(k+1) &= u(k) \\ x_2(k+1) &= u(k-1) = x_1(k) \\ x_3(k+1) &= u(k-2) = x_2(k) \end{aligned} \tag{D-11}$$

These three equations are grouped in vector-matrix form to give

$$\begin{pmatrix} x_1(k+1) \\ x_2(k+1) \\ x_3(k+1) \end{pmatrix} = \begin{pmatrix} 0 & 0 & 0 \\ 1 & 0 & 0 \\ 0 & 1 & 0 \end{pmatrix} \begin{pmatrix} x_1(k) \\ x_2(k) \\ x_3(k) \end{pmatrix} + \begin{pmatrix} 1 \\ 0 \\ 0 \end{pmatrix} u(k) \tag{D-12}$$

Additionally, $y(k)$ can be written as

$$y(k) = (\beta_1, \beta_2, \beta_3) \begin{pmatrix} x_1(k) \\ x_2(k) \\ x_3(k) \end{pmatrix} \tag{D-13}$$

Equations (D-12) and (D-13) constitute the state-variable representation of the MA model in (D-9).

If $y(k)$ also contains the term $\beta_0 u(k)$, we proceed exactly as before, obtaining state equation (D-12). In this case, observation equation (D-13) is modified to

$$y(k) = (\beta_1, \beta_2, \beta_3) \begin{pmatrix} x_1(k) \\ x_2(k) \\ x_3(k) \end{pmatrix} + \beta_0 u(k) \tag{D-14}$$

The term $\beta_0 u(k)$ acts as a direct throughput of input $u(k)$ into the observation. □

Using this example, we observe that *a state-variable representation for the discrete-time MA model*

$$y(k) = \beta_0 u(k) + \beta_1 u(k-1) + \cdots + \beta_n u(k-n) \tag{D-15}$$

is

$$\begin{pmatrix} x_1(k+1) \\ x_2(k+1) \\ x_3(k+1) \\ \vdots \\ x_n(k+1) \end{pmatrix} = \underbrace{\begin{pmatrix} 0 & 0 & 0 & \cdots & 0 & 0 \\ 1 & 0 & 0 & \cdots & 0 & 0 \\ 0 & 1 & 0 & \cdots & 0 & 0 \\ \vdots & \vdots & \vdots & \ddots & \vdots & \vdots \\ 0 & 0 & 0 & \cdots & 1 & 0 \end{pmatrix}}_{\mathbf{\Phi}} \begin{pmatrix} x_1(k) \\ x_2(k) \\ x_3(k) \\ \vdots \\ x_n(k) \end{pmatrix} + \underbrace{\begin{pmatrix} 1 \\ 0 \\ 0 \\ \vdots \\ 0 \end{pmatrix}}_{\boldsymbol{\psi}} u(k) \tag{D-16}$$

$$y(k) = \underbrace{(\beta_1, \beta_2, \cdots, \beta_n)}_{\mathbf{h}'} \mathbf{x}(k) + \underbrace{\beta_0}_{d} u(k) \tag{D-17}$$

*Quantities* $\mathbf{\Phi}$, $\boldsymbol{\psi}$, $\mathbf{h}'$, *and* $d$ *are indicated in (D-16) and (D-17).*

## All-pole (Autoregressive) Models

### EXAMPLE D-5

The discrete-time AR model

$$y(k+3) + \alpha_1 y(k+2) + \alpha_2 y(k+1) + \alpha_3 y(k) = \beta u(k) \tag{D-18}$$

associated with the transfer function

$$H(z) = \frac{\beta}{z^3 + \alpha_1 z^2 + \alpha_2 z + \alpha_3} \tag{D-19}$$

is easily converted to a third-order state-variable model by choosing state variables $x_1, x_2$, and $x_3$ as follows: $x_1(k) = y(k), x_2(k) = y(k+1)$, and $x_3(k) = y(k+2)$. Then

$$\begin{aligned}
x_1(k+1) &= y(k+1) = x_2(k) \\
x_2(k+1) &= y(k+2) = x_3(k) \\
x_3(k+1) &= y(k+3) = -\alpha_1 y(k+2) - \alpha_2 y(k+1) - \alpha_3 y(k) + \beta u(k) \\
&= -\alpha_1 x_3(k) - \alpha_2 x_2(k) - \alpha_3 x_1(k) + \beta u(k)
\end{aligned}$$

or

$$\begin{pmatrix} x_1(k+1) \\ x_2(k+1) \\ x_3(k+1) \end{pmatrix} = \begin{pmatrix} 0 & 1 & 0 \\ 0 & 0 & 1 \\ -\alpha_3 & -\alpha_2 & -\alpha_1 \end{pmatrix} \begin{pmatrix} x_1(k) \\ x_2(k) \\ x_3(k) \end{pmatrix} + \begin{pmatrix} 0 \\ 0 \\ \beta \end{pmatrix} u(k) \tag{D-20}$$

Additionally, because $y(k) = x_1(k)$

$$y(k) = (1, 0, 0) \begin{pmatrix} x_1(k) \\ x_2(k) \\ x_3(k) \end{pmatrix} \tag{D-21}$$

Equations (D-20) and (D-21) constitute the state-variable representation for the AR model (D-18). □

More generally, *for the discrete-time AR transfer function*

$$H(z) = \frac{\beta}{z^n + \alpha_1 z^{n-1} + \cdots + \alpha_{n-1} z + \alpha_n} \tag{D-22}$$

*which implies the AR difference equation*

$$y(k+n) + \alpha_1 y(k+n-1) + \cdots + \alpha_n y(k) = \beta u(k), \tag{D-23}$$

$$\begin{aligned}
\mathbf{x}(k) &= \text{col}\,[x_1(k), x_2(k), \ldots, x_n(k)] \\
&= \text{col}\,[y(k), y(k+1), \ldots, y(k+n-1)]
\end{aligned} \tag{D-24}$$

*qualifies as a state vector. The AR system is described in state variable form as*

$$\begin{pmatrix} x_1(k+1) \\ x_2(k+1) \\ \vdots \\ x_{n-1}(k+1) \\ x_n(k+1) \end{pmatrix} = \underbrace{\begin{pmatrix} 0 & 1 & 0 & \cdots & 0 \\ 0 & 0 & 1 & \cdots & 0 \\ \vdots & \vdots & \vdots & \ddots & \vdots \\ 0 & 0 & 0 & \cdots & 1 \\ -\alpha_n & -\alpha_{n-1} & -\alpha_{n-2} & \cdots & -\alpha_1 \end{pmatrix}}_{\mathbf{\Phi}}$$

$$\times \begin{pmatrix} x_1(k) \\ x_2(k) \\ \vdots \\ x_{n-1}(k) \\ x_n(k) \end{pmatrix} + \underbrace{\begin{pmatrix} 0 \\ 0 \\ \vdots \\ 0 \\ \beta \end{pmatrix}}_{\boldsymbol{\psi}} u(k) \tag{D-25}$$

$$y(k) = \underbrace{(1 \quad 0 \quad \cdots \quad 0 \quad 0)}_{\mathbf{h}'} \mathbf{x}(k) \tag{D-26}$$

### Pole–Zero (Autoregressive Moving Average) Models

The literature on linear systems contains many state-variable models for pole–zero models [Chen, 1984, and Kailath, 1980, for example]. In this section we describe a few of them.

*The controllable canonical form state-variable representation for the discrete-time ARMA model*

$$H(z) = \frac{\beta_1 z^{n-1} + \beta_2 z^{n-2} + \cdots + \beta_{n-1} z + \beta_n}{z^n + \alpha_1 z^{n-1} + \cdots + \alpha_{n-1} z + \alpha_n} \tag{D-27}$$

*which implies the ARMA difference equation*

$$y(k+n) + \alpha_1 y(k+n-1) + \cdots + \alpha_n y(k) = \beta_1 u(k+n-1) + \cdots + \beta_n u(k) \tag{D-28}$$

*is*

$$\begin{pmatrix} x_1(k+1) \\ x_2(k+1) \\ \vdots \\ x_{n-1}(k+1) \\ x_n(k+1) \end{pmatrix} = \underbrace{\begin{pmatrix} 0 & 1 & 0 & \cdots & 0 \\ 0 & 0 & 1 & \cdots & 0 \\ \vdots & \vdots & \vdots & \ddots & \vdots \\ 0 & 0 & 0 & \cdots & 1 \\ -\alpha_n & -\alpha_{n-1} & -\alpha_{n-2} & \cdots & -\alpha_1 \end{pmatrix}}_{\mathbf{\Phi}}$$

$$\times \begin{pmatrix} x_1(k) \\ x_2(k) \\ \vdots \\ x_{n-1}(k) \\ x_n(k) \end{pmatrix} + \underbrace{\begin{pmatrix} 0 \\ 0 \\ \vdots \\ 0 \\ 1 \end{pmatrix}}_{\boldsymbol{\psi}} \tag{D-29}$$

*and*

$$y(k) = \underbrace{(\beta_n, \beta_{n-1}, \ldots, \beta_1)}_{\mathbf{h}'} \mathbf{x}(k) \tag{D-30}$$

A very nice feature of this representation is that the $\alpha$ and $\beta$ coefficients from the difference equation appear unaltered in (D-29) and (D-30). The same cannot be said about all state-variable representations.

**EXAMPLE D-6**

Rather than prove this result in general, we illustrate the proof for the case when $n = 3$, for which the third-order ARMA model is

$$y(k+3) + \alpha_1 y(k+2) + \alpha_2 y(k+1) + \alpha_3 y(k) = \beta_1 u(k+2) + \beta_2 u(k+1) + \beta_3 u(k) \tag{D-31}$$

The associated transfer function is

$$H(z) = \frac{\beta_1 z^2 + \beta_2 z + \beta_3}{z^3 + \alpha_1 z^2 + \alpha_2 z + \alpha_3} = \frac{Y(z)}{U(z)} \tag{D-32}$$

Introduce intermediate variable $y_1(k)$ whose $z$ transform is $Y_1(z)$:

$$H(z) = \frac{\beta_1 z^2 + \beta_2 z + \beta_3}{z^3 + \alpha_1 z^2 + \alpha_2 z + \alpha_3} \frac{Y_1(z)}{Y_1(z)} = \frac{Y(z)}{U(z)} \frac{Y_1(z)}{Y_1(z)} = \frac{Y(z)}{U(z)} \tag{D-33}$$

Doing this is equivalent to treating the ARMA transfer function $H(z)$ as a cascade of two transfer functions; i.e., $H(z) = Y(z)/U(z) = H_1(z)H_2(z)$, where $H_1(z) = Y_1(z)/U(z) = 1/(z^3 + \alpha_1 z^2 + \alpha_2 z + \alpha_3)$ and $H_2(z) = Y(z)/Y_1(z) = \beta_1 z^2 + \beta_2 z + \beta_3$. Equate the numerator and denominator terms of (D-33) to see that

$$y(k) = \beta_1 y_1(k+2) + \beta_2 y_1(k+1) + \beta_3 y_1(k) \tag{D-34}$$

and

$$y_1(k+3) + \alpha_1 y_1(k+2) + \alpha_2 y_1(k+1) + \alpha_3 y_1(k) = u(k) \tag{D-35}$$

Equation (D-35) is an autoregressive model, and (D-34) relates intermediate variable $y_1(k)$ to $y(k)$.

Using results from Example (D-5), we choose $x_1(k) = y_1(k), x_2(k) = y_1(k+1)$, and $x_3(k) = y_1(k+2)$; hence, (D-34) and (D-35) can be expressed as

$$\begin{pmatrix} x_1(k+1) \\ x_2(k+1) \\ x_3(k+1) \end{pmatrix} = \begin{pmatrix} 0 & 1 & 0 \\ 0 & 0 & 1 \\ -\alpha_3 & -\alpha_2 & -\alpha_1 \end{pmatrix} \begin{pmatrix} x_1(k) \\ x_2(k) \\ x_3(k) \end{pmatrix} + \begin{pmatrix} 0 \\ 0 \\ 1 \end{pmatrix} u(k) \tag{D-36}$$

and

$$y(k) = (\beta_3, \beta_2, \beta_1)\mathbf{x}(k) \tag{D-37}$$

where $\mathbf{x}(k) = \text{col}\,(x_1(k), x_2(k), x_3(k))$. The results in (D-36) and (D-37) are precisely those obtained by setting $n = 3$ in (D-29) and (D-30). □

A completely different state-variable model is obtained when we begin by expanding the transfer function of the system into a partial fraction expansion.

### EXAMPLE D-7

Here we express $H(z)$, given in (D-32), as a sum of three terms:

$$H(z) = \frac{\beta_1 z^2 + \beta_2 z + \beta_3}{z^3 + \alpha_1 z^2 + \alpha_2 z + \alpha_3} = \frac{e_1}{z - \lambda_1} + \frac{e_2}{z - \lambda_2} + \frac{e_3}{z - \lambda_3} \tag{D-38}$$

where $\lambda_1, \lambda_2$, and $\lambda_3$, the poles of $H(z)$, are the roots of the characteristic equation $z^3 + \alpha_1 z^2 + \alpha_2 z + \alpha_3 = 0$. For the purposes of this example, we assume that $\lambda_1, \lambda_2$, and $\lambda_3$ are real and unequal. Equation (D-38) constitutes the partial fraction expansion of $H(z)$.

We now view $y(k)$ as the sum of the outputs from three systems that are in parallel (see Figure D-2). Consider the input/output equation for $y_1(k)$,

$$y_1(k+1) - \lambda_1 y_1(k) = u(k) \tag{D-39}$$

Let $x_1(k) = y_1(k)$ so that (D-39) can be written in terms of state variable $x_1(k)$ as

$$x_1(k+1) - \lambda_1 x_1(k) = u(k) \tag{D-40}$$

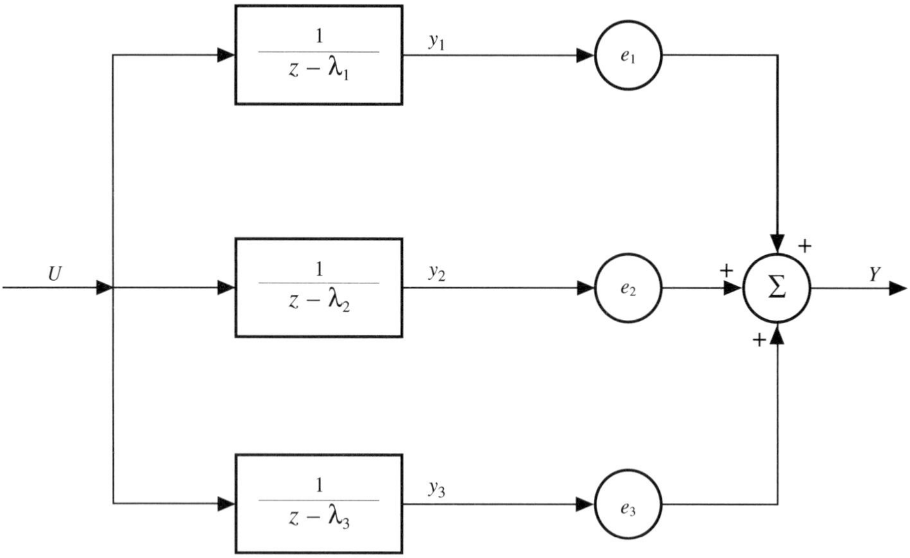


**Figure D-2** Parallel connection of three first-order systems.

Proceeding similarly for $y_2(k)$ and $y_3(k)$, letting $x_2(k) = y_2(k)$ and $x_3(k) = y_3(k)$, we find that

$$x_2(k+1) - \lambda_2 x_2(k) = u(k) \tag{D-41}$$

and

$$x_3(k+1) - \lambda_3 x_3(k) = u(k) \tag{D-42}$$

Combining (D-40) – (D-42), we obtain our final state-variable model:

$$\begin{pmatrix} x_1(k+1) \\ x_2(k+1) \\ x_3(k+1) \end{pmatrix} = \underbrace{\begin{pmatrix} \lambda_1 & 0 & 0 \\ 0 & \lambda_2 & 0 \\ 0 & 0 & \lambda_3 \end{pmatrix}}_{\mathbf{\Phi}} \begin{pmatrix} x_1(k) \\ x_2(k) \\ x_3(k) \end{pmatrix} + \underbrace{\begin{pmatrix} 1 \\ 1 \\ 1 \end{pmatrix}}_{\boldsymbol{\psi}} u(k) \tag{D-43}$$

where

$$y(k) = \underbrace{(e_1, e_2, e_3)}_{\mathbf{h}'} \mathbf{x}(k) \tag{D-44}$$

In this state-variable model, matrix $\mathbf{\Phi}$ is diagonal, which is useful in some computations. If $\lambda_1$ is complex, then $\lambda_2 = \lambda_1^*$, and $\mathbf{\Phi}$ and $\mathbf{h}$ will contain complex numbers. We explain how to handle this situation in Example D-8. Finally, if two or more of the poles are equal, the structure of $\mathbf{\Phi}$ will be different, because the structure of the partial fraction expansion is not that given in (D-38). □

The approach to developing a state-variable model illustrated by this example leads to a representation known as the *Jordan canonical form*. We do not state its structure for an arbitrary transfer function. It can be obtained for any AR or ARMA model by the following procedure: (1) calculate the poles of the system; (2) perform a partial fraction expansion; (3) for real (single or repeated) poles, treat each term in the partial fraction expansion as a single-input, single-output (SISO) system and obtain its state-variable representation (as in Example D-7); (4) for complex poles, treat the conjugate pole pairs together as a second-order system and obtain a state-variable representation for it that involves only real numbers; and (5) combine the representations obtained in steps 3 and 4 in the output equation.

**EXAMPLE D-8**

Here we give a very useful state-variable model for a subsystem composed of two complex conjugate poles:

$$H_1(z) = \frac{\beta}{(z-\alpha)^2 + \beta^2} \tag{D-45}$$

This transfer function is associated with the second-order difference equation

$$y_1(k+2) - 2\alpha y_1(k+1) + (\alpha^2 + \beta^2) y_1(k) = \beta u(k) \tag{D-46}$$

which can be expressed in state-variable form as

$$\begin{pmatrix} x_1(k+1) \\ x_2(k+1) \end{pmatrix} = \begin{pmatrix} \alpha & \beta \\ -\beta & \alpha \end{pmatrix} \begin{pmatrix} x_1(k) \\ x_2(k) \end{pmatrix} + \begin{pmatrix} 0 \\ 1 \end{pmatrix} u(k) \tag{D-47}$$

where

$$y_1(k) = (1 \quad 0)\mathbf{x}(k) \tag{D-48}$$

We leave the verification of this representation to the reader. □

Many other equivalent state-variable models are possible for ARMA systems. The equivalence of different state-variable models is demonstrated later under Miscellaneous Properties.

## SOLUTIONS OF STATE EQUATIONS FOR TIME-INVARIANT SYSTEMS

Now that we know how to construct state-variable models, what can we do with them? In this section we show how to obtain $y(k)$ given $u(k)$ and initial conditions $\mathbf{x}(0)$. We also show how easy it is to obtain an expression for the transfer function $Y(z)/U(z)$ from a state-variable model.

State equation (D-7) is recursive and, as we demonstrate next, easily solved for $\mathbf{x}(k)$ as an explicit function of $\mathbf{x}(0), u(0), u(1), \ldots,$ and $u(k-1)$. Once we know $\mathbf{x}(k)$, it is easy to solve for $y(k)$, because $y(k) = \mathbf{h}'\mathbf{x}(k)$ (we assume $d = 0$ here).

To solve (D-7) for $\mathbf{x}(k)$, we proceed as follows:

$$
\begin{aligned}
\mathbf{x}(1) &= \mathbf{\Phi x}(0) + \boldsymbol{\psi} u(0) \\
\mathbf{x}(2) &= \mathbf{\Phi x}(1) + \boldsymbol{\psi} u(1) = \mathbf{\Phi}[\mathbf{\Phi x}(0) + \boldsymbol{\psi} u(0)] + \boldsymbol{\psi} u(1) \\
&= \mathbf{\Phi}^2\mathbf{x}(0) + \mathbf{\Phi}\boldsymbol{\psi} u(0) + \boldsymbol{\psi} u(1) \\
\mathbf{x}(3) &= \mathbf{\Phi x}(2) + \boldsymbol{\psi} u(2) = \mathbf{\Phi}[\mathbf{\Phi}^2\mathbf{x}(0) + \mathbf{\Phi}\boldsymbol{\psi} u(0) + \boldsymbol{\psi} u(1)] + \boldsymbol{\psi} u(2) \\
&= \mathbf{\Phi}^3\mathbf{x}(0) + \mathbf{\Phi}^2\boldsymbol{\psi} u(0) + \mathbf{\Phi}\boldsymbol{\psi} u(1) + \boldsymbol{\psi} u(2)
\end{aligned}
$$

etc.

By this iterative procedure we see that a pattern has emerged for expressing $\mathbf{x}(k)$ as a function of $\mathbf{x}(0), u(0), u(1), \ldots, u(k-1)$; i.e.,

$$
\mathbf{x}(k) = \mathbf{\Phi}^k\mathbf{x}(0) + \sum_{i=1}^{k} \mathbf{\Phi}^{k-i}\boldsymbol{\psi} u(i-1) \tag{D-49}
$$

where $k = 1, 2, \ldots$. An inductive proof of (D-49) is found in Mendel (1973, pp. 21–23).

Assuming that $y(k) = \mathbf{h}'\mathbf{x}(k)$, we see that

$$
y(k) = \mathbf{h}'\mathbf{\Phi}^k\mathbf{x}(0) + \sum_{i=1}^{k} \mathbf{h}'\mathbf{\Phi}^{k-i}\boldsymbol{\psi} u(i-1) \tag{D-50}
$$

The first term on the right-hand side of (D-50) is owing to the initial conditions; it is the transient (homogeneous) response. The second term is owing to the forcing function $u$; it is the forced response and is a convolution summation.

Equation (D-49) provides us with directions for going from an initial state $\mathbf{x}(0)$ directly to state $\mathbf{x}(k)$. For linear systems we can also reach $\mathbf{x}(k)$ by first going from $\mathbf{x}(0)$ to state $\mathbf{x}(j)$ and then going from $\mathbf{x}(j)$ to $\mathbf{x}(k)$. In fact, $\mathbf{x}(k)$ can be expressed in terms of $\mathbf{x}(j)$ as

$$
\mathbf{x}(k) = \mathbf{\Phi}^{k-j}\mathbf{x}(j) + \sum_{i=j+1}^{k} \mathbf{\Phi}^{k-i}\boldsymbol{\psi} u(i-1) \tag{D-51}
$$

where $k \geq j + 1$.

Matrix $\mathbf{\Phi}^{k-i}$ in (D-49)–(D-51) is called a *state transition matrix* for the homogeneous system $\mathbf{x}(k+1) = \mathbf{\Phi x}(k)$. We denote this matrix $\mathbf{\Phi}(k, i)$. It has the following properties:

**a.** *Identity property:* $\quad \mathbf{\Phi}(k, k) = \mathbf{I} \quad$ for all $k = 0, 1, \ldots$. (D-52)

**b.** *Semigroup property:* $\quad \mathbf{\Phi}(k, j)\mathbf{\Phi}(j, i) = \mathbf{\Phi}(k, i) \quad$ for all $k \geq j \geq i$. (D-53)

**c.** *Inverse property:* $\quad \mathbf{\Phi}^{-1}(k, i) = \mathbf{\Phi}(i, k)$ (D-54)

Proofs of these properties use the definition of $\mathbf{\Phi}(k, i)$ and are left to the reader. The semigroup property was already used when we showed that $\mathbf{x}(k)$ could be reached from $\mathbf{x}(0)$ by going from $\mathbf{x}(0)$ to $\mathbf{x}(j)$ and then from $\mathbf{x}(j)$ to $\mathbf{x}(k)$.

Next we direct our attention to computing the transfer function $Y(z)/U(z)$.

**EXAMPLE D-9**

Consider a second-order ARMA model written in controllable canonical form as

$$\begin{pmatrix} x_1(k+1) \\ x_2(k+1) \end{pmatrix} = \begin{pmatrix} 0 & 1 \\ -\alpha_2 & -\alpha_1 \end{pmatrix} \begin{pmatrix} x_1(k) \\ x_2(k) \end{pmatrix} + \begin{pmatrix} 0 \\ 1 \end{pmatrix} u(k) \tag{D-55}$$

$$y(k) = (\beta_2\ \beta_1)\mathbf{x}(k) \tag{D-56}$$

To compute $Y(z)/U(z)$, we must express $X_1(z)$ and $X_2(z)$ as functions of $U(z)$, after which it is easy to express $Y(z)$ as a function of $U(z)$, using (D-56).

Take the $z$-transform of the two state equations (D-55), to show that

$$zX_1(z) = X_2(z)$$

$$zX_2(z) = -\alpha_2 X_1(z) - \alpha_1 X_2(z) + U(z)$$

Treat these equations as a system of two equations in the two unknowns $X_1(z)$ and $X_2(z)$, and rewrite them as

$$zX_1(z) - X_2(z) = 0$$

$$zX_2(z) + \alpha_2 X_1(z) + \alpha_1 X_2(z) = U(z)$$

or

$$\left[\begin{pmatrix} z & 0 \\ 0 & z \end{pmatrix} - \begin{pmatrix} 0 & 1 \\ -\alpha_2 & -\alpha_1 \end{pmatrix}\right] \begin{pmatrix} X_1(z) \\ X_2(z) \end{pmatrix} = \begin{pmatrix} 0 \\ 1 \end{pmatrix} U(z) \tag{D-57}$$

Recognize that (D-57) can also be written, symbolically, as

$$(z\mathbf{I} - \mathbf{\Phi})\mathbf{X}(z) = \boldsymbol{\psi} U(z) \tag{D-58}$$

in which $\mathbf{I}$ is the $2 \times 2$ identity matrix and $\mathbf{X}(z)$ is the $z$-transform of $\mathbf{x}(k)$ [i.e., $X_i(z) = \mathcal{Z}\{x_1(k)\}, i = 1, 2$].

This example demonstrates that we can take the $z$-transform of state equation (D-7) *without* having to do it equation by equation. □

Proceeding formally to determine $Y(z)/U(z)$, we first take the $z$-transform of state equation (D-7)

$$\mathcal{Z}\{\mathbf{x}(k+1) = \mathbf{\Phi}\mathbf{x}(k) + \boldsymbol{\psi} u(k)\}$$

$$z\mathbf{X}(z) = \mathbf{\Phi}\mathbf{X}(z) + \boldsymbol{\psi} U(z) \tag{D-59}$$

$$\mathbf{X}(z) = (\mathbf{I} - z^{-1}\mathbf{\Phi})^{-1} z^{-1} \boldsymbol{\psi} U(z)$$

Then we take the $z$-transform of the output equation $y(k) = \mathbf{h}'\mathbf{x}(k)$.

$$\mathcal{Z}\{y(k) = \mathbf{h}'\mathbf{x}(k)\}$$

or

$$Y(z) = \mathbf{h}'\mathbf{X}(z) \tag{D-60}$$

Finally, we combine (D-59) and (D-60) to give us the desired transfer function.

$$\frac{Y(z)}{U(z)} = \mathbf{h}'(\mathbf{I} - z^{-1}\mathbf{\Phi})^{-1} z^{-1} \boldsymbol{\psi} \triangleq H(z) \tag{D-61}$$

We denote this transfer function as $H(z)$ to remind us of the well-known fact (Chen, 1984) that the transfer function is the $z$-transform of the unit spike (impulse) response of the system $h(k)$. Observe that we have just determined the transfer function for a LTI SISO discrete-time system in terms of the parameters $\mathbf{\Phi}$, $\boldsymbol{\psi}$, and $\mathbf{h}$ of a state-variable representation of that system.

Matrix $(\mathbf{I} - z^{-1}\mathbf{\Phi})^{-1}$, which appears in $H(z)$, has the power-series expansion

$$(\mathbf{I} - z^{-1}\mathbf{\Phi})^{-1} = \mathbf{I} + z^{-1}\mathbf{\Phi} + z^{-2}\mathbf{\Phi}^2 + \cdots + z^{-j}\mathbf{\Phi}^j + \cdots \tag{D-62}$$

Hence,

$$H(z) = \mathbf{h}'\boldsymbol{\psi}z^{-1} + \mathbf{h}'\mathbf{\Phi}\boldsymbol{\psi}z^{-2} + \mathbf{h}'\mathbf{\Phi}^2\boldsymbol{\psi}z^{-3} + \cdots + \mathbf{h}'\mathbf{\Phi}^j\boldsymbol{\psi}z^{-(j+1)} + \cdots \tag{D-63}$$

Denoting the coefficients of $H(z)$ as $h(j)$, we see that

$$h(j) = \mathbf{h}'\mathbf{\Phi}^{j-1}\boldsymbol{\psi}, \qquad j = 1, 2, \ldots \tag{D-64}$$

These coefficients are commonly known as *Markov parameters.*

Equation (D-64) is important and can be used in two ways: (1) Given $(\mathbf{\Phi}, \boldsymbol{\psi}, \mathbf{h})$, compute the sampled impulse response $\{h(1), h(2), \ldots\}$, and (2) given $\{h(1), h(2), \ldots\}$, compute $(\mathbf{\Phi}, \boldsymbol{\psi}, \mathbf{h})$.

## MISCELLANEOUS PROPERTIES

In this section we present some important facts about stability, nonuniqueness, and flexibility of state-variable models.

Recall that the asymptotic stability of a LTI SISO discrete-time system is guaranteed if all the poles of $H(z)$ lie inside the unit circle in the complex $z$ domain. For $H(z) = N(z)/D(z)$, we know that the poles of $H(z)$ are the roots of $D(z)$. We also know that

$$H(z) = \mathbf{h}'(\mathbf{I} - z^{-1}\mathbf{\Phi})^{-1}z^{-1}\boldsymbol{\psi} = \mathbf{h}'(z\mathbf{I} - \mathbf{\Phi})^{-1}\boldsymbol{\psi} = \frac{\mathbf{h}'\mathbf{Q}(z)\boldsymbol{\psi}}{\det(z\mathbf{I} - \mathbf{\Phi})} \tag{D-65}$$

where $\mathbf{Q}(z)$ is an $n \times n$ matrix whose elements are simple polynomials in $z$ whose degree is at most $n-1$, and $\det(z\mathbf{I} - \mathbf{\Phi})$, the characteristic polynomial of $\mathbf{\Phi}$, is a scalar $n$th-degree polynomial. *The poles of $H(z)$ are therefore the roots of the characteristic polynomial of $\mathbf{\Phi}$.*

Recall from matrix theory [Franklin, 1968, for example] that $\boldsymbol{\xi} \neq \mathbf{0}$ is called an eigenvector of matrix $\mathbf{\Phi}$ if there is a scalar $\lambda$ such that

$$\mathbf{\Phi}\boldsymbol{\xi} = \lambda\boldsymbol{\xi} \tag{D-66}$$

In this equation $\lambda$ is called an eigenvalue of $\mathbf{\Phi}$ corresponding to $\boldsymbol{\xi}$. Equation (D-66) can be written as

$$(\lambda\mathbf{I} - \mathbf{\Phi})\boldsymbol{\xi} = \mathbf{0} \tag{D-67}$$

Because $\boldsymbol{\xi} \neq \mathbf{0}$, by definition of an eigenvector, matrix $(\lambda\mathbf{I} - \mathbf{\Phi})$ must be singular; hence,

$$\det(\lambda\mathbf{I} - \mathbf{\Phi}) = 0 \tag{D-68}$$

We see, therefore, that the eigenvalues of $\mathbf{\Phi}$ are the roots of $\det(\lambda\mathbf{I} - \mathbf{\Phi})$, the characteristic polynomial of $\mathbf{\Phi}$. By means of this line of reasoning, we have demonstrated the truth of:

**Property 1.** The poles of $H(z)$ are the eigenvalues of matrix $\mathbf{\Phi}$. □

The asymptotic stability of a LTI SISO discrete-time system is therefore guaranteed if all the eigenvalues of $\mathbf{\Phi}$ lie inside the unit circle.

Next we direct our attention to the nonuniqueness and flexibility of state-variable representations. We know that many state-variable representations are possible for the same LTI SISO system, for example, the controllable and Jordan canonical forms for ARMA systems. One form may be better for numerical calculations, whereas a different form may be better for parameter estimation, and so on. Regardless of the form used, we must be sure that from an input/output point of view there are no theoretical differences (numerical differences may occur owing to word length and computer roundoff).

Suppose we make a change (transformation) of state variables, as follows:

$$\mathbf{x}(k) = \mathbf{P}\mathbf{z}(k) \quad \text{or} \quad \mathbf{z}(k) = \mathbf{P}^{-1}\mathbf{x}(k) \tag{D-69}$$

where $\mathbf{P}$ is a constant nonsingular $n \times n$ matrix. Then

$$\mathbf{x}(k+1) = \mathbf{\Phi}\mathbf{x}(k) + \boldsymbol{\psi}u(k) \tag{D-70}$$

$$y(k) = \mathbf{h}'\mathbf{x}(k) \tag{D-71}$$

can be expressed in terms of $\mathbf{z}(k)$ as

$$\mathbf{z}(k+1) = \mathbf{P}^{-1}\mathbf{\Phi}\mathbf{P}\mathbf{z}(k) + \mathbf{P}^{-1}\boldsymbol{\psi}u(k) \tag{D-72}$$

$$y(k) = \mathbf{h}'\mathbf{P}\mathbf{z}(k) \tag{D-73}$$

We shall show that the $\mathbf{x}$ system in (D-70) and (D-71) and the $\mathbf{z}$ system in (D-72) and (D-73) are equivalent from an input/output point of view. First we show that $\mathbf{P}^{-1}\mathbf{\Phi}\mathbf{P}$ has the same eigenvalues as $\mathbf{\Phi}$. Then we show that $H(z)$ is the same for both systems.

**Property 2.** The eigenvalues of $\mathbf{P}^{-1}\mathbf{\Phi}\mathbf{P}$ are those of $\mathbf{\Phi}$.

*Proof.* Let $\mathbf{M} \triangleq \mathbf{P}^{-1}\mathbf{\Phi}\mathbf{P}$; $\mathbf{\Phi}$ and $\mathbf{M}$ are said to be *similar* matrices. Let $p_{\mathbf{\Phi}}(z) = \det(z\mathbf{I} - \mathbf{\Phi})$ and $p_{\mathbf{M}}(z) = \det(z\mathbf{I} - \mathbf{M})$. Consequently,

$$\begin{aligned} p_{\mathbf{M}}(z) &= \det(z\mathbf{I} - \mathbf{M}) = \det(z\mathbf{I} - \mathbf{P}^{-1}\mathbf{\Phi}\mathbf{P}) = \det(z\mathbf{P}^{-1}\mathbf{P} - \mathbf{P}^{-1}\mathbf{\Phi}\mathbf{P}) \\ &= \det(\mathbf{P}^{-1}(z\mathbf{I} - \mathbf{\Phi})\mathbf{P}) = \det\mathbf{P}^{-1}\det(z\mathbf{I} - \mathbf{\Phi})\det\mathbf{P} \\ &= (1/\det\mathbf{P})\det(z\mathbf{I} - \mathbf{\Phi})\det\mathbf{P} = p_{\mathbf{\Phi}}(z) \end{aligned}$$

where we have used the following well-known facts about determinants: (1) the determinant of a product of matrices equals the product of the determinants of the matrices that comprise the product matrix, and (2) the determinant of a matrix inverse equals 1 divided by the determinant of the matrix. □

**Property 3.** Let $H_{\mathbf{x}}(z)$ and $H_{\mathbf{z}}(z)$ denote $Y(z)/U(z)$ for the **x** system in (D-70) and (D-71) and the **z** system in (D-72) and (D-73), respectively. Then

$$H_{\mathbf{z}}(z) = H_{\mathbf{x}}(z) \tag{D-74}$$

*Proof.* From (D-61),

$$H_{\mathbf{x}}(z) = \mathbf{h}'(z\mathbf{I} - \mathbf{\Phi})^{-1}\boldsymbol{\psi} \tag{D-75}$$

Repeating the analysis that led to $H_{\mathbf{x}}(z)$, but for (D-72) and (D-73), we find that

$$H_{\mathbf{z}}(z) = \mathbf{h}'\mathbf{P}(z\mathbf{I} - \mathbf{P}^{-1}\mathbf{\Phi}\mathbf{P})^{-1}\mathbf{P}^{-1}\boldsymbol{\psi} \tag{D-76}$$

Consequently,

$$\begin{aligned} H_{\mathbf{z}}(z) &= \mathbf{h}'\mathbf{P}(\mathbf{P}^{-1}(z\mathbf{I} - \mathbf{\Phi})\mathbf{P})^{-1}\mathbf{P}^{-1}\boldsymbol{\psi} = \mathbf{h}'\mathbf{P}\mathbf{P}^{-1}(z\mathbf{I} - \mathbf{\Phi})^{-1}\mathbf{P}\mathbf{P}^{-1}\boldsymbol{\psi} \\ &= \mathbf{h}'(z\mathbf{I} - \mathbf{\Phi})^{-1}\boldsymbol{\psi} = H_{\mathbf{x}}(z) \end{aligned}$$

To arrive at this result, we have used the fact, from matrix theory, that $(\mathbf{GHL})^{-1} = \mathbf{L}^{-1}\mathbf{H}^{-1}\mathbf{G}^{-1}$. □

## COMPUTATION

See the discussion on computation in Lesson 15.

## SUMMARY QUESTIONS

1. A state-variable model:
   **(a)** must be linear and time invariant
   **(b)** occurs in the frequency domain
   **(c)** is composed of a state equation and an output equation
2. Matrix $\mathbf{\Phi}(t, \tau)$ satisfies a linear:
   **(a)** time-varying matrix differential equation
   **(b)** time-invariant matrix differential equation
   **(c)** time-varying matrix difference equation
3. An $n$th-order difference equation can be expressed as a vector first-order state equation with:
   **(a)** $n + m$ states, where $m$ equals the number of measurements
   **(b)** exactly $n$ states
   **(c)** less than $n$ states
4. In a state-variable model, the choice of state variables is:
   **(a)** unique
   **(b)** done by solving an optimization problem
   **(c)** not unique
5. The Jordan canonical form has a plant matrix that is:
   **(a)** block diagonal
   **(b)** always diagonal
   **(c)** full

6. The *semigroup property* of state transition matrix $\mathbf{\Phi}$ means that we can:
   **(a)** stop at $\mathbf{x}(j)$ and return to starting point $\mathbf{x}(i)$
   **(b)** stop at $\mathbf{x}(j)$ and continue to $\mathbf{x}(i)$
   **(c)** return to starting point $\mathbf{x}(k)$ after having reached $\mathbf{x}(i)$
7. The $z$-transform of a state equation:
   **(a)** must be taken equation by equation
   **(b)** does not exist
   **(c)** does not have to be taken equation by equation
8. The poles of the system's transfer function are the:
   **(a)** eigenvalues of matrix $\mathbf{\Phi}^{-1}$
   **(b)** eigenvalues of matrix $\mathbf{\Phi}$
   **(c)** roots of $\det(z\mathbf{I} - \psi\psi')$
9. For a linear transformation of state variables, i.e., $\mathbf{z}(k) = \mathbf{P}^{-1}(k)\mathbf{x}(k)$, which of the following are true?
   **(a)** the $\mathbf{z}$ system's eigenvalues are scaled versions of the $\mathbf{x}$ system's eigenvalues
   **(b)** the eigenvalues of $\mathbf{P}^{-1}\mathbf{\Phi}\mathbf{P}$ are those of $\mathbf{\Phi}$
   **(c)** if the $\mathbf{x}$ system is minimum phase, then the $\mathbf{z}$ system is nonminimum phase
   **(d)** the transfer function remains unchanged
   **(e)** if the $\mathbf{x}$ system is minimum phase, then the $\mathbf{z}$ system is minimum phase

## PROBLEMS

**D-1.** Verify the state-variable model given in Example D-8 when a subsystem has a pair of complex conjugate poles.

**D-2.** Prove the truth of the identity property, semigroup property, and inverse property, given in (D-52), (D-53) and (D-54), respectively, for the state transition matrix. Provide a *physical explanation* for each of these important properties.

**D-3.** Determine a state-variable model for the system depicted in Figure PD-3.

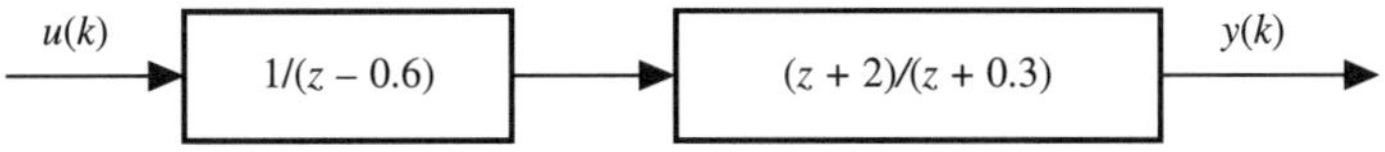


**Figure PD-3.**

**D-4.** Determine a state-variable model for the system depicted in Figure PD-4.

**Figure PD-4.**

**D-5.** Determine a state-variable model for the system depicted in Figure PD-5.

**D-6.** Determine a state-variable model for the system depicted in Figure PD-6.

**D-7.** Determine a state-variable model for the system depicted in Figure PD-7.

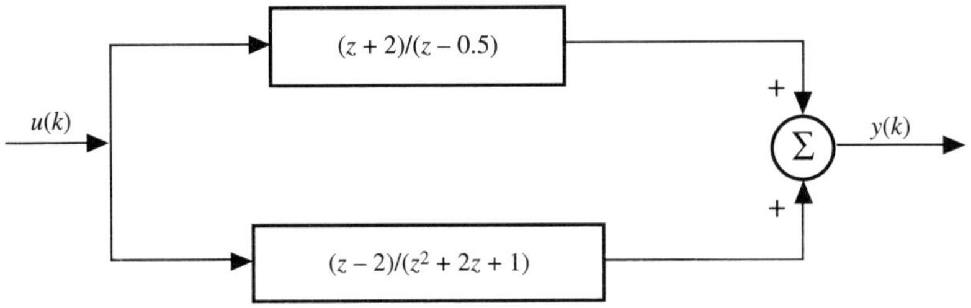


**Figure PD-5.**

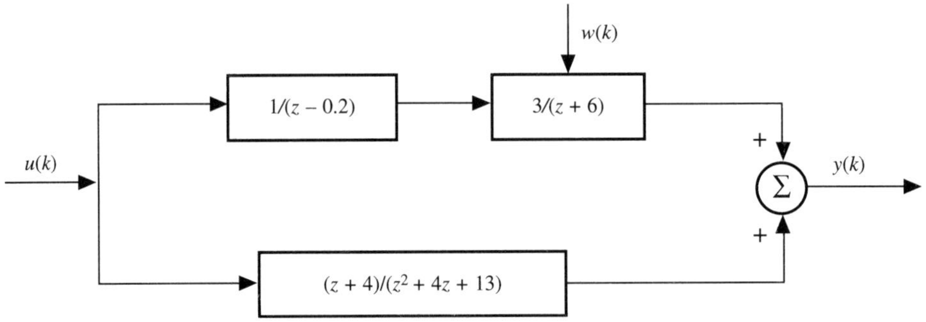


**Figure PD-6.**

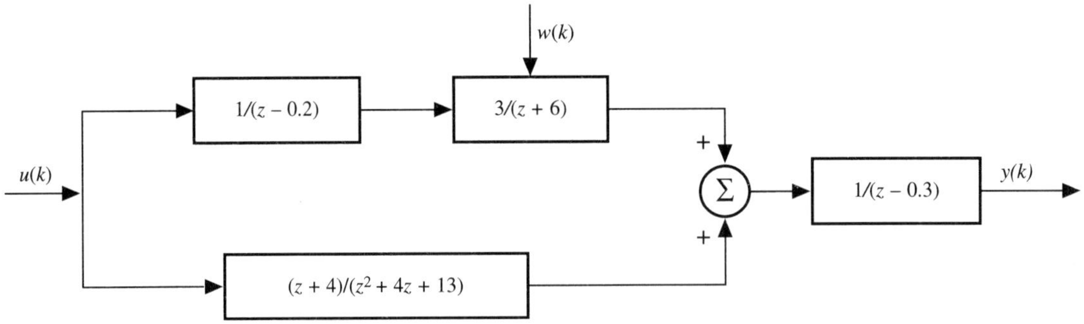


**Figure PD-7.**

**D-8.** Depicted in Figure PD-8 is the interconnection of recording equipment, coloring filter, and band-pass filter to a channel model. Noise $\nu(k)$ could be additive measurement noise or could be used to model digitization effects associated with a digital band-pass filter. Assume that the four subsystems are described by the following state vectors: $S_1$: $\mathbf{x}_1(n_1 \times 1)$; $S_2$: $\mathbf{x}_2(n_2 \times 1)$; $S_3$: $\mathbf{x}_3(n_3 \times 1)$; and $S_4$: $\mathbf{x}_4(n_4 \times 1)$. Obtain a state-variable model for this system that is of dimension $n_1 + n_2 + n_3 + n_4$.

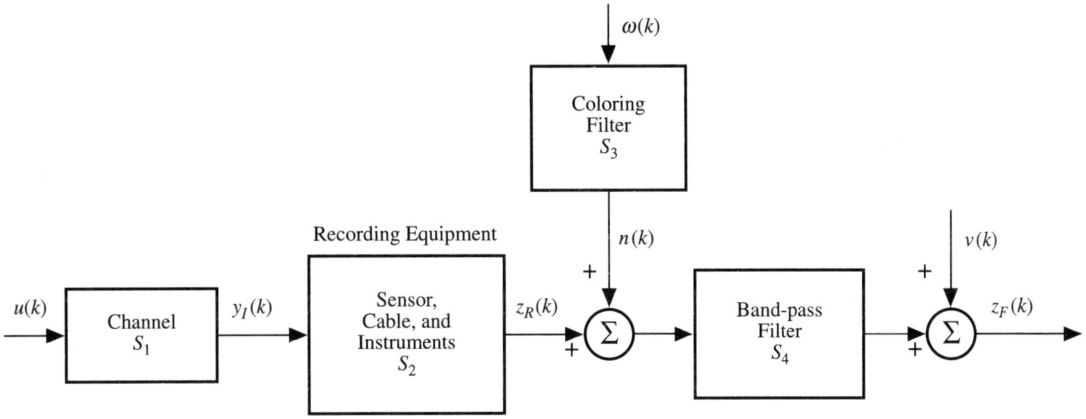


**Figure PD-8.**

**D-9.** Determine the eigenvalues of the system in:
**(a)** Problem D-3
**(b)** Problem D-4
**(c)** Problem D-5
**(d)** Problem D-6
**(e)** Problem D-7

**D-10.** (Computational Problem) For the system in Figure PD-4, compute $y(k)$ in two different ways for 25 time units, one of which must be by solution of the system's state equation. Sketch the results.

**D-11.** Consider the system depicted in Figure PD-11. Note that a pole–zero cancellation occurs as a result of the cascade of the two subsystems. Explain whether or not this is a stable system.

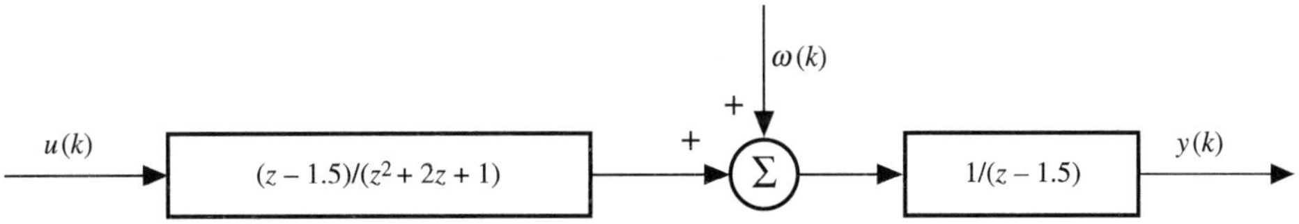


**Figure PD-11.**

# APPENDIX A

# Glossary of Major Results

| | |
|---|---|
| Equations (3-10) and (3-11) | Batch formulas for $\hat{\theta}_{WLS}(k)$ and $\hat{\theta}_{LS}(k)$. |
| Theorem 4-1 | Singular value decomposition of a rectangular matrix. |
| Theorem 4-2 | How to compute $\hat{\theta}_{LS}(k)$ by singular value decomposition. |
| Theorem 4-3 | How to compute $\hat{\theta}_{LS}(k)$ by singular value decomposition in the overdetermined case. |
| Theorem 4-4 | Pseudoinverse of a full-rank matrix. |
| Theorem 4-5 | Computation of the pseudoinverse by singular value decomposition. |
| Theorem 5-1 | Information form of recursive LSE. |
| Lemma 5-1 | Matrix inversion lemma. |
| Theorem 5-2 | Covariance form of recursive LSE. |
| Theorem 6-1 | Sufficient conditions for a linear batch estimator to be unbiased. |
| Theorem 6-2 | Sufficient condition for a linear recursive estimator to be unbiased. |
| Theorem 6-3 | Cramer–Rao inequality for a scalar parameter. |
| Corollary 6-1 | Achieving the Cramer–Rao lower bound. |
| Theorem 6-4 | Cramer–Rao inequality for a vector of parameters. |

| | |
|---|---|
| Corollary 6-2 | Lower bound for error-variance of $i$th parameter. |
| Theorem 7-1 | Mean-squared convergence implies convergence in probability. |
| Theorem 7-2 | Conditions under which $\hat{\theta}(k)$ is a consistent estimator of $\theta$. |
| Theorem 8-1 | Sufficient conditions for $\hat{\boldsymbol{\theta}}_{WLS}(k)$ to be an unbiased estimator of $\boldsymbol{\theta}$. |
| Theorem 8-2 | A formula for cov $[\tilde{\boldsymbol{\theta}}_{WLS}(k)]$. |
| Corollary 8-1 | A formula for cov $[\tilde{\boldsymbol{\theta}}_{LS}(k)]$ under special conditions on the measurement noise, and when $\mathbf{H}(k)$ is deterministic. |
| Theorem 8-3 | Efficiency of the least-squares estimator within the class of linear estimators. |
| Theorem 8-4 | An unbiased estimator of $\sigma_\nu^2$. |
| Theorem 8-5 | Sufficient conditions for $\hat{\boldsymbol{\theta}}_{LS}(k)$ to be a consistent estimator of $\boldsymbol{\theta}$. |
| Theorem 8-6 | Consistency of the instrumental variable least-squares estimator. |
| Theorem 8-7 | Sufficient conditions for $\hat{\sigma}_\nu^2(k)$ to be a consistent estimator of $\sigma_\nu^2$. |
| Equation (9-24) | Batch formula for $\hat{\boldsymbol{\theta}}_{BLU}(k)$. |
| Theorem 9-1 | The relationship between $\hat{\boldsymbol{\theta}}_{BLU}(k)$ and $\hat{\boldsymbol{\theta}}_{WLS}(k)$. |
| Corollary 9-1 | When all the results obtained in Lessons 3, 4, and 5 for $\hat{\boldsymbol{\theta}}_{WLS}(k)$ can be applied to $\hat{\boldsymbol{\theta}}_{BLU}(k)$. |
| Theorem 9-2 | When $\hat{\boldsymbol{\theta}}_{BLU}(k)$ equals $\hat{\boldsymbol{\theta}}_{LS}(k)$ (Gauss–Markov theorem). |
| Theorem 9-3 | A formula for cov $[\tilde{\boldsymbol{\theta}}_{BLU}(k)]$. |
| Corollary 9-2 | The equivalence between $\mathbf{P}(k)$ and cov $[\tilde{\boldsymbol{\theta}}_{BLU}(k)]$. |
| Theorem 9-4 | Most efficient estimator property of $\hat{\boldsymbol{\theta}}_{BLU}(k)$, within the class of linear estimators. |
| Corollary 9-3 | When $\hat{\boldsymbol{\theta}}_{LS}(k)$ is a most efficient estimator of $\boldsymbol{\theta}$. |
| Theorem 9-5 | Invariance of $\hat{\boldsymbol{\theta}}_{BLU}(k)$ to scale changes. |
| Theorem 9-6 | Information form of recursive BLUE. |
| Theorem 9-7 | Covariance form of recursive BLUE. |
| Definition 10-1 | Likelihood defined. |
| Theorem 10-1 | Likelihood ratio of combined data from statistically independent sets of data. |

| | |
|---|---|
| Theorem 11-1 | Large-sample properties of maximum-likelihood estimates. |
| Theorem 11-2 | Invariance property of MLEs. |
| Theorem 11-3 | Condition under which $\hat{\theta}_{ML}(k) = \hat{\theta}_{BLU}(k)$, and resulting estimator properties. |
| Corollary 11-1 | Conditions under which $\hat{\theta}_{ML}(k) = \hat{\theta}_{BLU}(k) = \hat{\theta}_{LS}(k)$, and resulting estimator properties. |
| Theorem 12-1 | A formula for $p(\mathbf{x}\|\mathbf{y})$ when $\mathbf{x}$ and $\mathbf{y}$ are jointly Gaussian. |
| Theorem 12-2 | Properties of $\mathbf{E}\{\mathbf{x}\|\mathbf{y}\}$ when $\mathbf{x}$ and $\mathbf{y}$ are jointly Gaussian. |
| Theorem 12-3 | Expansion formula for $\mathbf{E}\{\mathbf{x}\|\mathbf{y}, \mathbf{z}\}$ when $\mathbf{x}, \mathbf{y}$, and $\mathbf{z}$ are jointly Gaussian and $\mathbf{y}$ and $\mathbf{z}$ are statistically independent. |
| Theorem 12-4 | Expansion formula for $\mathbf{E}\{\mathbf{x}\|\mathbf{y}, \mathbf{z}\}$ when $\mathbf{x}, \mathbf{y}$, and $\mathbf{z}$ are jointly Gaussian and $\mathbf{y}$ and $\mathbf{z}$ are not necessarily statistically independent. |
| Theorem 13-1 | A formula for $\hat{\theta}_{MS}(k)$ (the fundamental theorem of estimation theory). |
| Corollary 13-1 | A formula for $\hat{\theta}_{MS}(k)$ when $\theta$ and $\mathbf{Z}(k)$ are jointly Gaussian. |
| Corollary 13-2 | A linear mean-squared estimator of $\theta$ in the non-Gaussian case. |
| Corollary 13-3 | Orthogonality principle. |
| Theorem 13-2 | Mean-squared estimator for the generic linear and Gaussian model. |
| Theorem 13-3 | Conditions under which $\hat{\theta}_{MS}(k) = \hat{\theta}_{BLU}(k)$. |
| Theorem 13-4 | Conditions under which $\hat{\theta}_{MS}(k) = \hat{\theta}^a_{BLU}(k)$. |
| Theorem 14-1 | When $\hat{\theta}_{MAP}(k) = \hat{\theta}_{MS}(k)$. |
| Theorem 14-2 | Conditions under which $\hat{\theta}_{MAP}(k) = \hat{\theta}^a_{BLU}(k)$. |
| Theorem 14-3 | Equality of probability of detection and unconditional maximum-likelihood detection rules. |
| Theorem 15-1 | Expansion of a joint probability density function for a first-order Markov process. |
| Theorem 15-2 | Calculation of conditional expectation for a first-order Markov process. |
| Theorem 15-3 | Interpretation of Gaussian white noise as a special first-order Markov process. |

| | |
|---|---|
| Equations (15-17) and (15-18) | The basic state-variable model. |
| Theorem 15-4 | Conditions under which $\mathbf{x}(k)$ is a Gauss–Markov sequence. |
| Theorem 15-5 | Recursive equations for computing $\mathbf{m}_\mathbf{x}(k)$ and $\mathbf{P}_\mathbf{x}(k)$. |
| Theorem 15-6 | Formulas for computing $\mathbf{m}_\mathbf{z}(k)$ and $\mathbf{P}_\mathbf{z}(k)$. |
| Equations (16-4) and (16-11) | Single-stage predictor formulas for $\hat{\mathbf{x}}(k\|k-1)$ and $\mathbf{P}(k\|k-1)$. |
| Theorem 16-1 | Formula for and properties of general state predictor $\hat{\mathbf{x}}(k\|j), k > j$. |
| Theorem 16-2 | Representations and properties of the innovations process. |
| Theorem 17-1 | Kalman filter formulas and properties of resulting estimates and estimation error. |
| Theorem 19-1 | Steady-state Kalman filter. |
| Theorem 19-2 | Equivalence of steady-state Kalman filter and infinite-length digital Wiener filter. |
| Supplement to Lesson 19 | Levinson algorithm. |
| Theorem 20-1 | Single-state smoother formula for $\hat{\mathbf{x}}(k\|k+1)$. |
| Corollary 20-1 | Relationship between single-stage smoothing gain matrix and Kalman gain matrix. |
| Corollary 20-2 | Another way to express $\hat{\mathbf{x}}(k\|k+1)$. |
| Theorem 20-2 | Double-stage smoother formula for $\hat{\mathbf{x}}(k\|k+2)$. |
| Corollary 20-3 | Relationship between double-stage smoothing gain matrix and Kalman gain matrix. |
| Corollary 20-4 | Two other ways to express $\hat{\mathbf{x}}(k\|k+2)$. |
| Theorem 21-1 | Formulas for a useful fixed-interval smoother of $\mathbf{x}(k)$, $\hat{\mathbf{x}}(k\|N)$, and its error-covariance matrix, $\mathbf{P}(k\|N)$. |
| Theorem 21-2 | Formulas for a most useful two-pass fixed-interval smoother of $\mathbf{x}(k)$, $\hat{\mathbf{x}}(k\|N)$, and its associated error-covariance matrix, $\mathbf{P}(k\|N)$. |
| Theorem 21-3 | Formulas for a most useful fixed-point smoothed estimator of $\mathbf{x}(k)$, $\hat{\mathbf{x}}(k\|k+l)$, where $l = 1, 2, \ldots$, and its associated error-covariance matrix, $\mathbf{P}(k\|k+l)$. |
| Theorem 21-4 | Conditions under which a single-channel state-variable model is equivalent to a convolutional sum model. |
| Theorem 21-5 | Recursive minimum-variance deconvolution formulas. |

| | |
|---|---|
| Theorem 21-6 | Steady-state MVD filter and zero phase nature of estimated signal component of system input. |
| Theorem 21-7 | Equivalence between steady-state MVD filter and Berkhout's infinite impulse response digital Wiener deconvolution filter. |
| Theorem 21-8 | Maximum-likelihood deconvolution results. |
| Theorem 22-1 | How to handle biases that may be present in a state-variable model. |
| Theorem 22-2 | Predictor–corrector Kalman filter for the correlated noise case. |
| Corollary 22-1 | Recursive predictor formulas for the correlated noise case. |
| Corollary 22-2 | Recursive filter formulas for the correlated noise case. |
| Theorem 22-3 | An identity involving the innovations. |
| Equations (23-1) and (23-2) | Nonlinear state-variable model. |
| Equations (23-23) and (23-30) | Perturbation state-variable model. |
| Theorem 23-1 | Solution to a time-varying continuous-time state equation. |
| Equations (23-39) and (23-44) | Discretized state-variable model. |
| Theorem 24-1 | A consequence of relinearizing about $\hat{\mathbf{x}}(k\|k)$. |
| Equations (24-22) and (24-27) | Extended Kalman filter prediction and correction equations for a nonlinear continuous-time system. |
| Equations (24-44) and (24-47) | Extended Kalman filter prediction and correction equations for a nonlinear discrete-time system. |
| Theorem 25-1 | Formula for the log-likelihood function of the basic state-variable model. |
| Theorem 25-2 | Closed-form formula for the maximum-likelihood estimate of the steady-state value of the innovation's covariance matrix. |
| Theorem 26-1 | Differential equations for first- and second-order statistics of the continuous-time basic state-variable model. |
| Theorem 26-2 | Kalman–Bucy filter equations. |
| Definition A-1 | Sufficient statistic defined. |
| Theorem A-1 | Factorization theorem. |
| Theorem A-2 | A method for computing the unique maximum-likelihood estimator of $\boldsymbol{\theta}$ that is associated with exponential families of distributions. |

| | |
|---|---|
| Theorem A-3 | Lehmann–Scheffe theorem. Provides a uniformly minimum-variance unbiased estimator of $\theta$. |
| Theorem A-4 | Method for determining whether or not $T(\mathbf{z})$ is complete as well as sufficient when $p(\mathbf{z}\|\boldsymbol{\theta})$ is an $m$-parameter exponential family. |
| Theorem A-5 | Provides a uniformly minimum-variance unbiased estimator of vector $\boldsymbol{\theta}$. |
| Definition B-1 | The $k$th-order cumulant for $k$ random variables defined. |
| Theorem B-1 | Symmetry conditions for third- and fourth-order cumulants. |
| Definition B-3 | The $k$th-order polyspectrum defined. |
| Theorem B-2 | Symmetry conditions for the bispectrum and trispectrum. |
| Theorem B-3 | Six cumulant properties. |
| Theorem C-1 | The $k$th-order statistics of a linear time-invariant system (Bartlett–Brillinger–Rosenblatt theorem). |
| Corollary C-1 | A cumulant formula for an ARMA$(p, q)$ model. |
| Corollary C-2 | Relationship between the usual spectrum of a system's output and a special polyspectrum (GM equation). |
| Theorem C-2 | Mean-squared optimality of a cumulant-based linear deconvolution filter. |

# APPENDIX B

# Estimation Algorithm M-files

## INTRODUCTION

This appendix contains the listings of six estimation algorithm M-files that were written by Mitsuru Nakamura. These M-files do not appear in any MathWorks toolboxes or in MATLAB. To see the forest from the trees, we first provide a brief description of these M-files:

**rwlse**: a recursive weighted least-squares algorithm. When the weighting matrix is set equal to the identity matrix, we obtain a recursive least-squares algorithm; and when the weighting matrix is set equal to the inverse of the covariance matrix of the measurement noise vector, we obtain a recursive BLUE.

**kf**: a recursive Kalman filter that can be used to obtain mean-squared *filtered* estimates of the states of our basic state-variable model. This model can be time varying or nonstationary.

**kp**: a recursive Kalman predictor that can be used to obtain mean-squared *predicted* estimates of the states of our basic state-variable model. This model can be time varying or nonstationary.

**sof**: a recursive *suboptimal filter* in which the gain matrix must be prespecified. It implements all the Kalman filter equations except (17-12) and uses (17-9) to compute $\mathbf{P}(k+1|k+1)$.

**sop**: a recursive *suboptimal predictor* in which the gain matrix must be prespecified. It implements all the Kalman filter equations except (17-12) and uses (17-9) to compute $\mathbf{P}(k+1|k+1)$.

**fis**: obtains the mean-squared fixed-interval smoothed estimate of the state vector of our basic state-variable model. This model can be time varying or nonstationary.

The listings of these M-files are given in the following sections. We have not included a diskette with this book, because the actual listing of each M-file is quite short. Most of what is given in the following sections is the descriptions of the M-files, including important notes and footnotes that will aid the end user.

## RECURSIVE WEIGHTED LEAST-SQUARES ESTIMATOR

```
function  [THETA,P] = rwlse(Z,H,THETAi,Pi,W)
%
%   Recursive Weighted Least-Squares Estimator
%
%    << Purpose >>
%         Perform weighted or nonweighted least-squares estimation
%         using the covariance form of recursive estimator.
%         See Lesson 5 for algorithm.
%
%    << Synopsis >>
%         [THETA,P] = rwlse(Z,H,THETAi,Pi)
%         [THETA,P] = rwlse(Z,H,THETAi,Pi,W)
%
%    << Input Parameters >>
%      Z      : measurement vector    : N x m
%               See note-2.
%      H      : observation matrix    : Nm x n
%      THETAi : initial estimate for theta   : n x 1
%      Pi     : initial value for P (only diagonal elements) : n x 1
%      W      : weighting matrix (must be diagonal) for rwlse
%              (diagonal elements in vector form)   : N x m
%               W(i,j) stands for the weight of jth component of
%               measurement vector at ith time step.
%               See also note-1 and note-2.
%
%    << Output Parameters >>
%      THETA  : estimated parameters : n x N
%      P      : covariance matrix (only diagonal elements) : n x N
%
%    << Internal Parameters (not required by the input) >>
%      N      : the number of available measurements (time steps)
%      n      : the number of the parameters in theta
%      m      : the number of measurements (vector measurement case)
%
%    << Notes >>
%    note-1 : If user specifies matrix-W in the input, this routine selects
%             weighted estimation automatically; otherwise, nonweighted
%             estimation is performed.
%
%    note-2 : For the vector measurement case, a special arrangement for
```

```
%           vector-Z and matrix-W is required.  Note that the arrangement
%           is different from the one that is explained in Table 5-1.
%
n = length(THETAi);
[N,m] = size(Z);

THETA = zeros(n,N);  % See footnote
P = zeros(n,N);      % See footnote

if nargin == 4    % -- RLSE --
 Pt = diag(Pi);
 K = Pt*H(1:m,:)'/(H(1:m,:)*Pt*H(1:m,:)'+eye(m));
 THETA(:,1) = THETAi+K*(Z(1,:)'-H(1:m,:)*THETAi);
 Pt = (eye(n)-K*H(1:m,:))*Pt;
 P(:,1) = diag(Pt);
 for L = 2:N,
  Ht = H((L-1)*m+1:L*m,:);
  K = Pt*Ht'/(Ht*Pt*Ht'+eye(m));                              % (5-31)
  THETA(:,L) = THETA(:,L-1)+K*(Z(L,:)'-Ht*THETA(:,L-1));      % (5-30)
  Pt = (eye(n)-K*Ht)*Pt;                                      % (5-32)
  P(:,L) = diag(Pt);
 end

elseif nargin == 5   % -- RWLSE --
 Pt = diag(Pi);
 K = Pt*H(1:m,:)'/(H(1:m,:)*Pt*H(1:m,:)'+diag(ones(1,m)./W(1,:)));
 THETA(:,1) = THETAi+K*(Z(1,:)'-H(1:m,:)*THETAi);
 Pt = (eye(n)-K*H(1:m,:))*Pt;
 P(:,1) = diag(Pt);
 for L = 2:N,
  Ht = H((L-1)*m+1:L*m,:);
  K = Pt*Ht'/(Ht*Pt*Ht'+diag(ones(1,m)./W(1,:)));             % (5-31)
  THETA(:,L) = THETA(:,L-1)+K*(Z(L,:)'-Ht*THETA(:,L-1));      % (5-30)
  Pt = (eye(n)-K*Ht)*Pt;                                      % (5-32)
  P(:,L) = diag(Pt);
 end

end
%
%    << Footnote : Computation performance >>
%         For large N, this line improves the computation performance
%         significantly.  See MATLAB manual for details.
```

## KALMAN FILTER

```
function  [Xf,Pf,Kg] = kf(F,G,D,H,U,Z,Q,R,Xi,Pi)
%
%   Kalman Filter
%
%    << Purpose >>
%         Perform Kalman filtering for the following basic
%         state-variable model:
%
```

```
%        X(k+1) = F(k+1,k)X(k)+G(k+1,k)W(k)+D(k+1,k)U(k)
%        Z(k+1) = H(k+1)X(k+1)+V(k+1)
%
%        See Lesson 17 for the Kalman filter algorithm.
%
%    << Synopsis >>
%        [Xf,Pf] = kf(F,G,D,H,U,Z,Q,R,Xi,Pi)
%        [Xf,Pf,Kg] = kf(F,G,D,H,U,Z,Q,R,Xi,Pi)
%
%    << Input Parameters >>
%      F  : transition matrix  : n x nN
%           (equivalent to phi, used in the text)
%      G  : disturbance distribution matrix : n x pN
%           (equivalent to gamma, used in the text)
%      D  : known input distribution matrix : n x lN
%           (equivalent to psi, used in the text)
%           kth F i.e., F(:,(n*(k-1)+1):n*k) stands for F(k+1,k);
%           similar for G and D
%      H  : observation matrix              : m x nN
%
%      U  : known system input vector       : l x N
%      Z  : measurement vector              : m x N
%
%      Q  : covariance matrix of disturbance
%           (only diagonal elements)        : p x N
%      R  : covariance matrix of measurement noise
%           (only diagonal elements)        : m x N
%
%      Xi : initial value for filtered X (Xf(1|1);See note-1) : n x 1
%      Pi : initial value for Pf (Pf(1|1);See note-1)         : n x n
%
%    << Output Parameters >>
%      Xf : filtered estimates of X         : n x N
%      Pf : filtered error covariance matrix : n x nN
%      Kg : Kalman gain                     : n x mN
%           See note-2.
%
%    << Internal Parameters (not required by the input) >>
%      n  : dimension of state vector
%      m  : dimension of measurement vector
%      p  : dimension of unknown disturbance vector
%      l  : dimension of known system input vector
%      N  : the number of time steps
%
%    << Notes >>
%    note-1 : Since MATLAB cannot treat zero as a matrix or vector index,
%             the user must consider k = 1 in this M-file as k = 0 in the
%             text's notation.
%
%    note-2 : If the user specifies the gain matrix in the output, this
%             routine stores the Kalman gain that is calculated at each time
%             step and the user obtains the gain matrix as an output.
%             Otherwise, this routine does not store the gain matrix.
%
n = min(size(F));
```

```
[m,N] = size(Z);
[l,N] = size(U);
[p,N] = size(Q);

if nargout == 3      % -- keeping Kg as output --
 Xf = zeros(n,N);     % See footnote-a
 Pf = zeros(n,n*N);   % See footnote-a
 Kg = zeros(n,m*N);   % See footnote-a

 Xf(:,1) = Xi;
 Pf(:,1:n) = Pi;

 for L = 1:N-1,
  Ft = F(:,(n*(L-1)+1):(n*L));
  Gt = G(:,(p*(L-1)+1):(p*L));
  Dt = D(:,(l*(L-1)+1):(l*L));
  Ht = H(:,(n*L+1):(n*(L+1)));
  Qt = diag(Q(:,L));
  Rt = diag(R(:,L+1));

  Xp = Ft*Xf(:,L)+Dt*U(:,L);                          % (16-4)
  Pp = Ft*Pf(:,(n*(L-1)+1):(n*L))*Ft'+Gt*Qt*Gt';      % (17-13)
  HP = Ht*Pp;
  Kt = HP'/(HP*Ht'+Rt);                               % (17-12)
  Kg(:,(m*L+1):(m*(L+1))) = Kt;
  Xf(:,L+1) = (eye(n)-Kt*Ht)*Xp+Kt*Z(:,L+1);          % (17-11)
  tmp = Pp-Kt*HP;
  Pt = tmp-tmp*Ht'*Kt'+Kt*Rt*Kt';                     % (17-9)
                                                      % See footnote-b
  tmp = (triu(Pt)+triu(Pt'))/2.0;                         % See footnote-c
  Pf(:,(n*L+1):(n*(L+1))) = -diag(diag(Pt))+tmp+tmp'; % See footnotes-c and d
 end

elseif nargout == 2    %  -- not keeping Kg as output --
 Xf = zeros(n,N);     % See footnote-a
 Pf = zeros(n,n*N);   % See footnote-a

 Xf(:,1) = Xi;
 Pf(:,1:n) = Pi;

 for L = 1:N-1,
  Ft = F(:,(n*(L-1)+1):(n*L));
  Gt = G(:,(p*(L-1)+1):(p*L));
  Dt = D(:,(l*(L-1)+1):(l*L));
  Ht = H(:,(n*L+1):(n*(L+1)));
  Qt = diag(Q(:,L));
  Rt = diag(R(:,L+1));

  Xp = Ft*Xf(:,L)+Dt*U(:,L);                          % (16-4)
  Pp = Ft*Pf(:,(n*(L-1)+1):(n*L))*Ft'+Gt*Qt*Gt';      % (17-13)
  HP = Ht*Pp;
  Kg = HP'/(HP*Ht'+Rt);                               % (17-12)
  Xf(:,L+1) = (eye(n)-Kg*Ht)*Xp+Kg*Z(:,L+1);          % (17-11)
  tmp = Pp-Kg*HP;
  Pt = tmp-tmp*Ht'*Kg'+Kg*Rt*Kg';                     % (17-9)
```

```
                                                       % See footnote-b
  tmp = (triu(Pt)+triu(Pt'))/2.0;                      % See footnote-c
  Pf(:,(n*L+1):(n*(L+1))) = -diag(diag(Pt))+tmp+tmp'; % See footnote-c
 end
end
%
%    << Footnotes : Computation performance >>
%   note-a : For large N, this line improves the computation performance
%            significantly.  See MATLAB manual for details.
%
%   note-b : The so-called stabilized algorithm is implemented to calculate
%            matrix-Pf.  See Observation 6, in Lesson 17's section
%            "Observation about the Kalman Filter," and also Problem 17-17.
%
%   note-c : These lines are implemented to ensure the symmetry of
%            matrix-Pf.
%
%   note-d : The matrix Pf requires large storage space for large N.
%            If the user does not need Pf as the output, the user
%            can modify this routine so that Pf is used recursively
%            in the routine and is not stored at every time step.
%            Similarly, if the user needs only the diagonal elements
%            of Pf, the user can modify this routine to save the
%            storage space.
```

## KALMAN PREDICTOR

```
function  [Xp,Pp,Kg] = kp(F,G,D,H,U,Z,Q,R,Xi,Pi)
%
%   Kalman Predictor
%
%    << Purpose >>
%         Perform Kalman prediction for the following basic
%         state-variable model:
%
%         X(k+1) = F(k+1,k)X(k)+G(k+1,k)W(k)+D(k+1,k)U(k)
%         Z(k+1) = H(k+1)X(k+1)+V(k+1)
%
%         See Lesson 17 for the Kalman filter algorithm.  It includes
%         prediction
%
%    << Synopsis >>
%         [Xp,Pp] = kp(F,G,D,H,U,Z,Q,R,Xi,Pi)
%         [Xp,Pp,Kg] = kp(F,G,D,H,U,Z,Q,R,Xi,Pi)
%
%    << Input Parameters >>
%      F  : transition matrix  : n x nN
%           (equivalent to phi, used in the text)
%      G  : disturbance distribution matrix : n x pN
%           (equivalent to gamma, used in the text)
%      D  : known input distribution matrix : n x lN
%           (equivalent to psi, used in the text)
%           kth F i.e., F(:,(n*(k-1)+1):n*k) stands for F(k + 1, k);
```

```
%          similar for G and D
%     H  : observation matrix                : m x nN
%
%     U  : known system input vector         : l x N
%     Z  : measurement vector                : m x N
%
%     Q  : covariance matrix of disturbance
%          (only diagonal elements)          : p x N
%     R  : covariance matrix of measurement noise
%          (only diagonal elements)          : m x N
%
%     Xi : initial value for filtered X (Xf(1|1);See note-1) : n x 1
%     Pi : initial value for Pf (Pf(1|1);See note-1)         : n x n
%
%   << Output Parameters >>
%     Xp : one step prediction of X             : n x N
%     Pp : prediction error covariance matrix : n x nN
%     Kg : Kalman gain                          : n x mN
%          See note-2.
%
%   << Internal Parameters (not required by the input) >>
%     n  : dimension of state vector
%     m  : dimension of measurement vector
%     p  : dimension of unknown disturbance vector
%     l  : dimension of known system input vector
%     N  : the number of time steps
%
%   << Notes >>
%   note-1 : Since MATLAB cannot treat zero as a matrix or vector index,
%            the user must consider k = 1 in this M-file as k=0 in the
%            text's notation.
%
%   note-2 : If the user specifies the gain matrix in the output, this
%            routine stores the Kalman gain that is calculated at each time
%            step and the user obtains the gain matrix as an output.
%            Otherwise, this routine does not store the gain matrix.
%
n = min(size(F));
[m,N] = size(Z);
[l,N] = size(U);
[p,N] = size(Q);

if nargout == 3     % -- keeping Kg as output --
 Xp = zeros(n,N);     % See footnote-a
 Pp = zeros(n,n*N);  % See footnote-a
 Kg = zeros(n,m*N);  % See footnote-a

 Xf = Xi;
 Pf = Pi;

 for L=1:N-1,
  Ft = F(:,(n*(L-1)+1):(n*L));
  Gt = G(:,(p*(L-1)+1):(p*L));
  Dt = D(:,(l*(L-1)+1):(l*L));
  Ht = H(:,(n*L+1):(n*(L+1)));
```

```
  Qt = diag(Q(:,L));
  Rt = diag(R(:,L+1));

  Xp(:,L) = Ft*Xf+Dt*U(:,L);                          % (16-4)
  Pp(:,(n*(L-1)+1):(n*L)) = Ft*Pf*Ft'+Gt*Qt*Gt';   % (17-13)
  HP = Ht*Pp(:,(n*(L-1)+1):(n*L));
  Kt = HP'/(HP*Ht'+Rt);                               % (17-12)
  Kg(:,(m*L+1):(m*(L+1))) = Kt;
  Xf = (eye(n)-Kt*Ht)*Xp(:,L)+Kt*Z(:,L+1);           % (17-11)
  tmp = Pp(:,(n*(L-1)+1):(n*L))-Kt*HP;
  Pt = tmp-tmp*Ht'*Kt'+Kt*Rt*Kt';                     % (17-9)
                                                      % See footnote-b
  tmp = (triu(Pt)+triu(Pt'))/2.0;                     % See footnote-c
  Pf = -diag(diag(Pt))+tmp+tmp';                      % See footnote-c
 end
 Ft = F(:,(n*(N-1)+1):(n*N));
 Gt = G(:,(p*(N-1)+1):(p*N));
 Dt = D(:,(l*(N-1)+1):(l*N));
 Qt = diag(Q(:,N));
 Xp(:,N) = Ft*Xf+Dt*U(:,N);
 Pp(:,(n*(N-1)+1):(n*N)) = Ft*Pf*Ft'+Gt*Qt*Gt';    % See footnote-d

elseif nargout == 2   % -- not keeping Kg as output --
 Xp = zeros(n,N);     % See footnote-a
 Pp = zeros(n,n*N);   % See footnote-a

 Xf = Xi;
 Pf = Pi;

 for L = 1:N-1,
  Ft = F(:,(n*(L-1)+1):(n*L));
  Gt = G(:,(p*(L-1)+1):(p*L));
  Dt = D(:,(l*(L-1)+1):(l*L));
  Ht = H(:,(n*L+1):(n*(L+1)));
  Qt = diag(Q(:,L));
  Rt = diag(R(:,L+1));

  Xp(:,L) = Ft*Xf+Dt*U(:,L);                          % (16-4)
  Pp(:,(n*(L-1)+1):(n*L)) = Ft*Pf*Ft'+Gt*Qt*Gt';   % (17-13)
  HP = Ht*Pp(:,(n*(L-1)+1):(n*L));
  Kg = HP'/(HP*Ht'+Rt);                               % (17-12)
  Xf = (eye(n)-Kg*Ht)*Xp(:,L)+Kg*Z(:,L+1);           % (17-11)
  tmp = Pp(:,(n*(L-1)+1):(n*L))-Kg*HP;
  Pt = tmp-tmp*Ht'*Kg'+Kg*Rt*Kg';                     % (17-9)
                                                      % See footnote-b
  tmp = (triu(Pt)+triu(Pt'))/2.0;                     % See footnote-c
  Pf = -diag(diag(Pt))+tmp+tmp';                      % See footnote-c
 end
 Ft = F(:,(n*(N-1)+1):(n*N));
 Gt = G(:,(p*(N-1)+1):(p*N));
 Dt = D(:,(l*(N-1)+1):(l*N));
 Qt = diag(Q(:,N));
 Xp(:,N) = Ft*Xf+Dt*U(:,N);
 Pp(:,(n*(N-1)+1):(n*N)) = Ft*Pf*Ft'+Gt*Qt*Gt';
end
```

```
%
%    << Footnotes : Computation performance >>
%   note-a : For large N, this line improves the computation performance
%            significantly.  See MATLAB manual for details.
%
%   note-b : The so-called stabilized algorithm is implemented to calculate
%            matrix-Pf.  See Observation 6, in Lesson 17's section
%            "Observation about the Kalman Filter," and also Problem 17-17.
%
%   note-c : These lines are implemented to ensure the symmetry
%            of matrix-Pf.
%
%   note-d : The matrix-Pp requires large storage space for large N.
%            If the user does not need Pp as the output, the user
%            can modify this routine so that Pp is used recursively
%            in the routine and is not stored at every time step.
%            Similarly, if the user needs only the diagonal elements
%            of Pp, the user can modify this routine to save the
%            storage space.
```

## SUBOPTIMAL FILTER

```
function [Xf,Pf] = sof(F,G,D,H,U,Z,Q,R,Xi,Pi,Kg)
%
%   Suboptimal Filter
%
%    << Purpose >>
%         Perform suboptimal filtering for the following basic
%         state-variable model:
%
%         X(k+1) = F(k+1,k)X(k)+G(k+1,k)W(k)+D(k+1,k)U(k)
%         Z(k+1) = H(k+1)X(k+1)+V(k+1)
%
%         The only difference between sof and kf is that the gain matrix
%         must be provided ahead of time in sof, whereas it is calculated
%         in kf.
%
%    << Synopsis >>
%         [Xf,Pf] = sof(F,G,D,H,U,Z,Q,R,Xi,Pi,Kg)
%
%    << Input Parameters >>
%      F  : transition matrix   : n x nN
%           (equivalent to phi, used in the text)
%      G  : disturbance distribution matrix : n x pN
%           (equivalent to gamma, used in the text)
%      D  : known input distribution matrix : n x lN
%           (equivalent to psi, used in the text)
%           kth F i.e., F(:,(n*(k-1)+1):n*k) stands for F(k+1,k);
%           similar for G and D
%      H  : observation matrix              : m x nN
%
%      U  : known system input vector       : l x N
%      Z  : measurement vector              : m x N
%
```

```
%      Q  : covariance matrix of disturbance
%           (only diagonal elements)        : p x N
%      R  : covariance matrix of measurement noise
%           (only diagonal elements)        : m x N
%
%      Xi : initial value for filtered X (Xf(1|1);See note) : n x 1
%      Pi : initial value for Pf (Pf(1|1);See note)        : n x n
%      Kg : gain matrix                                : n x mN
%
%    << Output Parameters >>
%      Xf : filtered estimates of X         : n x N
%      Pf : filtered error covariance matrix : n x nN
%
%    << Internal Parameters (not required by the input) >>
%      n  : dimension of state vector
%      m  : dimension of measurement vector
%      p  : dimension of unknown disturbance vector
%      l  : dimension of known system input vector
%      N  : the number of time steps
%
%    << Note >>
%           Since MATLAB cannot treat zero as a matrix or vector index,
%           the user must consider k = 1 in this M-file as k = 0 in the
%           text's notation.
%
n = min(size(F));
[m,N] = size(Z);
[l,N] = size(U);
[p,N] = size(Q);

Xf = zeros(n,N);    % See footnote-a
Pf = zeros(n,n*N);  % See footnote-a

Xf(:,1) = Xi;
Pf(:,1:n) = Pi;

for L = 1:N-1,
 Ft = F(:,(n*(L-1)+1):(n*L));
 Gt = G(:,(p*(L-1)+1):(p*L));
 Dt = D(:,(l*(L-1)+1):(l*L));
 Ht = H(:,(n*L+1):(n*(L+1)));
 Qt = diag(Q(:,L));
 Rt = diag(R(:,L+1));

 Xp = Ft*Xf(:,L)+Dt*U(:,L);                          % (16-4)
 Pp = Ft*Pf(:,(n*(L-1)+1):(n*L))*Ft'+Gt*Qt*Gt';  % (17-13)
 Kt = Kg(:,(m*L+1):(m*(L+1)));
 Xf(:,L+1) = (eye(n)-Kt*Ht)*Xp+Kt*Z(:,L+1);      % (17-11)
 tmp = Pp-Kt*Ht*Pp;
 Pt = tmp-tmp*Ht'*Kt'+Kt*Rt*Kt';                  % (17-9)
                                                  % See footnote-b
 tmp = (triu(Pt)+triu(Pt'))/2.0;                      % See footnote-c
 Pf(:,(n*L+1):(n*(L+1))) = -diag(diag(Pt))+tmp+tmp'; % See footnotes-c and d
end
%
```

```
%   << Footnotes : Computation performance >>
%  note-a : For large N, this line improves the computation performance
%           significantly.  See MATLAB manual for details.
%
%  note-b : The so-called stabilized algorithm is implemented to calculate
%           matrix-Pf.  See Observation 6, in Lesson 17's section
%           "Observation about the Kalman Filter," and also Problem 17-17.
%
%  note-c : These lines are implemented to ensure the symmetry of
%           matrix-Pf.
%
%  note-d : The matrix-Pf requires large storage space for large N.
%           If the user does not need Pf as the output, the user
%           can modify this routine so that Pf is used recursively
%           in the routine and is not stored at every time step.
%           Similarly, if the user needs only the diagonal elements
%           of Pf, the user can modify this routine to save the
%           storage space.
```

## SUBOPTIMAL PREDICTOR

```
function [Xp,Pp] = sop(F,G,D,H,U,Z,Q,R,Xi,Pi,Kg)
%
%  Suboptimal Predictor
%
%   << Purpose >>
%       Perform suboptimal prediction for the following basic
%       state-variable model:
%
%       X(k+1) = F(k+1,k)X(k)+G(k+1,k)W(k)+D(k+1,k)U(k)
%       Z(k+1) = H(k+1)X(k+1)+V(k+1)
%
%       The only difference between sop and kp is that the gain matrix
%       must be provided ahead of time in sop, whereas it is calculated
%       in kp.
%
%   << Synopsis >>
%       [Xp,Pp] = sop(F,G,D,H,U,Z,Q,R,Xi,Pi,Kg)
%
%   << Input Parameters >>
%     F  : transition matrix   : n x nN
%          (equivalent to phi, used in the text)
%     G  : disturbance distribution matrix : n x pN
%          (equivalent to gamma, used in the text)
%     D  : known input distribution matrix : n x lN
%          (equivalent to psi, used in the text)
%          kth F i.e., F(:,(n*(k-1)+1):n*k) stands for F(k+1,k);
%          similar for G and D
%     H  : observation matrix              : m x nN
%
%     U  : known system input vector       : l x N
%     Z  : measurement vector              : m x N
%
%     Q  : covariance matrix of disturbance
```

```
%           (only diagonal elements)         : p x N
%      R  : covariance matrix of measurement noise
%           (only diagonal elements)         : m x N
%
%      Xi : initial value for filtered X (Xf(1|1); see note) : n x 1
%      Pi : initial value for Pf (Pf(1|1); see note)         : n x n
%      Kg : gain matrix                                : n x mN
%
%    << Output Parameters >>
%      Xp : one step prediction of X           : n x N
%      Pp : prediction error covariance matrix : n x nN
%
%    << Internal Parameters (not required by the input) >>
%      n  : dimension of state vector
%      m  : dimension of measurement vector
%      p  : dimension of unknown disturbance vector
%      l  : dimension of known system input vector
%      N  : the number of time steps
%
%    << Note >>
%           Since MATLAB cannot treat zero as a matrix or vector index,
%           the user must consider k = 1 in this M-file as k = 0 in the
%           text's notation.
%
n = min(size(F));
[m,N] = size(Z);
[l,N] = size(U);
[p,N] = size(Q);

Xp = zeros(n,N);    % See footnote-a
Pp = zeros(n,n*N);  % See footnote-a

Xf = Xi;
Pf = Pi;

for L=1:N-1,
 Ft = F(:,(n*(L-1)+1):(n*L));
 Gt = G(:,(p*(L-1)+1):(p*L));
 Dt = D(:,(l*(L-1)+1):(l*L));
 Ht = H(:,(n*L+1):(n*(L+1)));
 Qt = diag(Q(:,L));
 Rt = diag(R(:,L+1));

 Xp(:,L) = Ft*Xf+Dt*U(:,L);                       % (16-4)
 Pp(:,(n*(L-1)+1):(n*L)) = Ft*Pf*Ft'+Gt*Qt*Gt';   % (17-13)
 Kt = Kg(:,(m*L+1):(m*(L+1)));
 Xf = (eye(n)-Kt*Ht)*Xp(:,L)+Kt*Z(:,L+1);         % (17-11)
 tmp = Pp(:,(n*(L-1)+1):(n*L))-Kt*Ht*Pp(:,(n*(L-1)+1):(n*L));
 Pt = tmp-tmp*Ht'*Kt'+Kt*Rt*Kt';                  % (17-9)
                                                  % See footnote-b
 tmp = (triu(Pt)+triu(Pt'))/2.0;                  % See footnote-c
 Pf = -diag(diag(Pt))+tmp+tmp';                   % See footnote-c
end
Ft = F(:,(n*(N-1)+1):(n*N));
Gt = G(:,(p*(N-1)+1):(p*N));
```

```
Dt = D(:,(l*(N-1)+1):(l*N));
Qt = diag(Q(:,N));
Xp(:,N) = Ft*Xf+Dt*U(:,N);
Pp(:,(n*(N-1)+1):(n*N)) = Ft*Pf*Ft'+Gt*Qt*Gt';
end
%
%    << Footnotes : Computation performance >>
%   note-a : For large N, this line improves the computation performance
%            significantly.  See MATLAB manual for details.
%
%   note-b : The so-called stabilized algorithm is implemented to calculate
%            matrix-Pf.  See Observation 6, in Lesson 17's section
%            "Observation about the Kalman Filter," and also Problem 17-17.
%
%   note-c : These lines are implemented to ensure the symmetry of
%            matrix-Pf.
%
%   note-d : The matrix-Pp requires large storage space for large N.
%            If the user does not need Pp as the output, the user
%            can modify this routine so that Pp is used recursively
%            in the routine and is not stored at every time step.
%            Similarly, if the user needs only the diagonal elements
%            of Pp, the user can modify this routine to save the
%            storage space.
```

## FIXED-INTERVAL SMOOTHER

```
function  [Xs,Ps] = fis(F,G,D,H,U,Z,Q,R,Xi,Pi,Kg)
%
%   Fixed-Interval Smoother
%
%    << Purpose >>
%         Compute fixed-interval smoothed estimates using the prediction
%         obtained either by the optimal predictor (kp) or by the suboptimal
%         predictor (sop), for the following basic state-variable model:
%
%         X(k+1)  = F(k+1,k)X(k)+G(k+1,k)W(k)+D(k+1,k)U(k)
%         Z(k+1)  = H(k+1)X(k+1)+V(k+1)
%
%    << Synopsis >>
%         [Xs,Ps] = fis(F,G,D,H,U,Z,Q,R,Xi,Pi)
%         [Xs,Ps] = fis(F,G,D,H,U,Z,Q,R,Xi,Pi,Kg)
%
%    << Input Parameters >>
%      F   : transition matrix   : n x nN
%            (equivalent to phi, used in the text)
%      G   : disturbance distribution matrix : n x pN
%            (equivalent to gamma, used in the text)
%      D   : known input distribution matrix : n x lN
%            (equivalent to psi, used in the text)
%            kth F i.e., F(:,(n*(k-1)+1):n*k) stands for F(k+1,k);
%            similar for G and D
%      H   : observation matrix              : m x nN
%
```

```
%      U  : known system input vector        : l x N
%      Z  : measurement vector               : m x N
%
%      Q  : covariance matrix of disturbance
%           (only diagonal elements)         : p x N
%      R  : covariance matrix of measurement noise
%           (only diagonal elements)         : m x N
%
%      Xi : initial value for filtered X (Xf(1|1);See note-1) : n x 1
%      Pi : initial value for Pf (Pf(1|1);See note-1)         : n x n
%
%      Kg : gain matrix                           : n x mN
%           See notes-2 and 3.
%
%    << Output Parameters >>
%      Xs  : smoothed estimates of x            : n x N
%      Ps  : Smoothing error-covariance matrix  : n x nN
%
%    << Internal Parameters (not required by the input) >>
%      n  : dimension of state vector
%      m  : dimension of measurement vector
%      p  : dimension of unknown disturbance vector
%      l  : dimension of known system input vector
%      N  : the number of time steps
%
%    << Notes >>
%    note-1 : Since MATLAB cannot treat zero as a matrix or vector index,
%             the user must consider k = 1 in this M-file as k = 0 in the
%             text's notation.
%
%    note-2 : If the user specifies the gain matrix in the input, this
%             routine uses the gain in smoothing process.  Otherwise,
%             this routine calculates the Kalman gain using function-kp,
%             which is previously described.
%
%    note-3 : Since this routine calls either function-kp or function-sop
%             internally, the user also has to code at least one of these
%             routines.

if nargin == 10      % -- calling function-kp to get Kalman gain --
 [Xp,Pp,Kg] = kp(F,G,D,H,U,Z,Q,R,Xi,Pi);
elseif nargin == 11  % -- calling function-sop --
 [Xp,Pp]    = sop(F,G,D,H,U,Z,Q,R,Xi,Pi,Kg);
end

n = min(size(F));
[m,N] = size(Z);

%  -- Innovations process --
Zi = zeros(m,N);     % See footnote-a
Pz = zeros(m,m*N);  % See footnote-a

for L = 1:N-1
  Ht = H(:,(n*L+1):(n*(L+1)));
  Rt = diag(R(:,L+1));
```

```
  Zi(:,L) = Z(:,L+1)-Ht*Xp(:,L);                                  % (16-31)
  Pz(:,(m*(L-1)+1):(m*L)) = Ht*Pp(:,(n*(L-1)+1):(n*L))*Ht'+Rt;   % (16-33)
end

%  -- Backward-running filter --
Xs = zeros(n,N);    % See footnote-a
Ps = zeros(n,n*N);  % See footnote-a
r = zeros(n,1);
S = zeros(n,n);

for L = N:-1:2
  Ft = F(:,(n*(L-1)+1):(n*L));
  Ht = H(:,(n*(L-1)+1):(n*L));
  Pt = Pp(:,(n*(L-2)+1):(n*(L-1)));
  tmp = Ht'/Pz(:,(m*(L-2)+1):(m*(L-1)));
  Fp = Ft*(eye(n)-Kg(:,(m*(L-1)+1):(m*L))*Ht);   % (21-33)

  r = Fp'*r+tmp*Zi(:,L-1);                       % (21-25)
  Xs(:,L) = Xp(:,L-1)+Pt*r;                      % (21-24)

  S = Fp'*S*Fp+tmp*Ht;                           % (21-27)
  Ps(:,(n*(L-1)+1):(n*L)) = Pt-Pt*S*Pt;          % (21-26); see footnote-b
end
%
%    << Footnotes : Computation performance >>
%   note-a : For large N, this line improves the computation performance
%            significantly.  See MATLAB manual for details.
%
%   note-b : The matrix-Ps requires large storage space for large N.
%            If the user does not need Ps as the output, the user
%            can modify this routine so that Ps is used recursively
%            in the routine and is not stored at every time step.
%            Similarly, if the user needs only the diagonal elements
%            of Ps, the user can modify this routine to save the
%            storage space.
%
%    << Special note for implementation of MVD  >>
%        If the user intends to implement MVD using this routine,
%        the user must modify several lines of this routine so that
%        r and S can be stored at every time step.  The user must be
%        careful about indexes.
%
%        Following are the modified lines for MVD implementation.
%
%  function  [Xs,Ps,r,S] = fis(F,G,D,H,U,Z,Q,R,Xi,Pi,Kg)
%  .....
%  r = zeros(n,N+1);
%  S = zeros(n,n*(N+1));
%  .....
%   r(:,L) = Fp'*r(:,L+1)+tmp*Zi(:,L-1);                         % (21-25)
%   Xs(:,L) = Xp(:,L-1)+Pt*r(:,L);                               % (21-24)
%
%   S(:,(n*(L-1)+1):(n*L)) = Fp'*S(:,(n*L+1):(n*(L+1)))*Fp+tmp*Ht; % (21-27)
%   Ps(:,(n*(L-1)+1):(n*L)) = Pt-Pt*S(:,(n*(L-1)+1):(n*L))*Pt;     % (21-26)
%  .....
```

# APPENDIX C

# Answers to Summary Questions

### Lesson 1

**1.** c; **2.** a; **3.** c; **4.** b

### Lesson 2

**1.** b; **2.** a, b, c, e, g; **3.** b; **4.** b; **5.** a; **6.** c; **7.** c; **8.** b; **9.** a; **10.** b; **11.** b; **12.** a; **13.** c

### Lesson 3

**1.** c; **2.** b; **3.** c; **4.** b; **5.** a; **6.** a; **7.** b; **8.** b; **9.** b, c, f, g

### Lesson 4

**1.** a; **2.** a, d, f; **3.** c; **4.** a; **5.** b; **6.** a, c, d; **7.** c

### Lesson 5

**1.** b; **2.** b; **3.** a; **4.** b; **5.** c; **6.** a; **7.** c; **8.** c; **9.** c

### Lesson 6

**1.** b; **2.** a; **3.** c; **4.** b; **5.** c; **6.** a; **7.** b; **8.** b; **9.** c

### Lesson 7

**1.** a; **2.** c; **3.** b; **4.** a; **5.** b; **6.** c; **7.** b; **8.** a

### Lesson 8

**1.** c; **2.** b; **3.** a; **4.** b; **5.** c; **6.** c; **7.** c; **8.** c; **9.** b

### Lesson 9

**1.** c; **2.** b, c; **3.** b; **4.** b; **5.** a; **6.** b; **7.** a, c; **8.** b; **9.** c; **10.** a

### Lesson 10

**1.** c; **2.** c; **3.** b; **4.** a; **5.** a; **6.** b

### Lesson 11

**1.** a, c; **2.** b; **3.** a; **4.** c; **5.** c; **6.** b; **7.** c; **8.** b; **9.** c

### Lesson 12

**1.** c; **2.** a, b; **3.** c; **4.** c; **5.** b; **6.** c; **7.** a

### Lesson 13

**1.** b; **2.** b; **3.** a, c, d, e, f; **4.** b; **5.** a; **6.** a, c; **7.** c; **8.** a; **9.** a; **10.** a; **11.** b; **12.** c

### Lesson 14

**1.** b; **2.** c; **3.** c; **4.** a; **5.** b; **6.** c; **7.** a, c, e, f; **8.** c; **9.** a, d; **10.** b

### Lesson 15

**1.** b; **2.** b; **3.** a; **4.** c; **5.** a; **6.** b; **7.** a; **8.** a, c, e; **9.** b; **10.** c; **11.** c; **12.** a, c

### Lesson 16

**1.** c; **2.** b; **3.** c; **4.** a; **5.** b; **6.** a; **7.** b; **8.** a, c, e; **9.** a, c

### Lesson 17

**1.** b; **2.** a; **3.** b; **4.** c; **5.** b; **6.** b; **7.** b; **8.** c; **9.** c; **10.** b; **11.** a; **12.** a; **13.** c; **14.** b

## Lesson 18

**1.** c; **2.** b; **3.** a; **4.** b; **5.** a; **6.** a, c, e; **7.** b; **8.** c; **9.** b

## Lesson 19

**1.** b; **2.** c, e; **3.** c; **4.** b; **5.** b; **6.** a; **7.** b; **8.** a

## Lesson 20

**1.** a; **2.** b; **3.** b; **4.** a; **5.** a; **6.** c; **7.** a; **8.** c; **9.** b

## Lesson 21

**1.** b; **2.** b; **3.** c; **4.** a; **5.** c; **6.** b; **7.** c; **8.** a; **9.** a; **10.** b; **11.** a

## Lesson 22

**1.** b; **2.** a; **3.** b; **4.** c; **5.** c; **6.** b; **7.** c; **8.** a

## Lesson 23

**1.** b; **2.** c; **3.** a; **4.** b; **5.** c; **6.** c; **7.** a

## Lesson 24

**1.** c; **2.** a; **3.** b; **4.** c; **5.** b; **6.** b; **7.** a

## Lesson 25

**1.** a; **2.** c; **3.** b; **4.** b; **5.** a; **6.** c; **7.** c; **8.** c

## Lesson 26

**1.** a, c; **2.** b; **3.** c; **4.** b; **5.** b; **6.** c

## Lesson A

**1.** a, b; **2.** c; **3.** b; **4.** a; **5.** c; **6.** a; **7.** b; **8.** a; **9.** a, c

## Lesson B

**1.** c; **2.** c; **3.** b; **4.** c; **5.** b; **6.** a; **7.** b; **8.** c; **9.** a

## Lesson C

**1.** c; **2.** a; **3.** b; **4.** c; **5.** b; **6.** b; **7.** c; **8.** a; **9.** b

## Lesson D

**1.** c; **2.** a; **3.** b; **4.** c; **5.** a; **6.** b; **7.** c; **8.** b; **9.** b, d, e

# References


ABATZOGLOU, T. J., J. M. MENDEL, and G. A. HARADA. 1991. "The constrained total least squares technique and its application to harmonic superresolution." *IEEE Trans. on Signal Processing*, Vol. 39, pp. 1070–1087.

AITKEN, A. C. 1935. "On least squares and linear combinations of observations." *Proc. of the Royal Society*, Edinburgh, Vol. 55, pp. 42–48.

AKAIKE, H. 1966. "Note on higher order spectra." *Annals Inst. Statistical Math.*, Tokyo, Vol. 18, pp. 123–126.

ANDERSON, B. D. O., and J. B. MOORE. 1979. *Optimal Filtering*. Englewood Cliffs, NJ: Prentice Hall.

AOKI, M. 1967. *Optimization of Stochastic Systems—Topics in Discrete-time Systems*. New York: Academic Press.

ARNOLD, S. F. 1990. *Mathematical Statistics*. Englewood Cliffs, NJ: Prentice Hall.

ÅSTRÖM, K. J. 1968. "Lectures on the identification problem—the least squares method." Rept. No. 6806, Lund Institute of Technology, Division of Automatic Control.

ATHANS, M. 1971. "The role and use of the stochastic linear-quadratic-Gaussian problem in control system design." *IEEE Trans. on Automatic Control*, Vol. AC-16, pp. 529–552.

ATHANS, M., and P. L. FALB. 1965, *Optimal Control: An Introduction to the Theory and Its Applications*. New York: McGraw-Hill.

ATHANS, M., and F. SCHWEPPE. 1965. "Gradient matrices and matrix calculations." MIT Lincoln Labs., Lexington, MA, Tech. Note 1965-53.

ATHANS, M., and E. TSE. 1967. "A direct derivation of the optimal linear filter using the maximum principle." *IEEE Trans. on Automatic Control*, Vol. AC-12, pp. 690–698.

ATHANS, M., R. P. WISHNER, and A. BERTOLINI. 1968. "Suboptimal state estimation for continuous-time nonlinear systems from discrete noisy measurements." *IEEE Trans. on Automatic Control*, Vol. AC-13, pp. 504–514.

AUTONNE, L. 1902. "Sur les groupes lineaires, reels et orthogonaux. *Bull. Soc. Math. France*, Vol. 30, pp. 121–133.

BARANKIN, E. W. 1961. "Application to exponential families of the solution to the minimal dimensionality problem for sufficient statistics," *Bull. Inst. Internat. Stat.*, Vol. 38, pp. 141–150.

BARANKIN, E. W., and M. KATZ, JR. 1959. "Sufficient statistics of minimal dimension." *Sankhya*, Vol. 21, pp. 217–246.

BARD, Y. 1970. "Comparison of gradient methods for the solution of nonlinear parameter estimation problems." *SIAM J. Numerical Analysis*, Vol. 7, pp. 157–186.

BARTLETT, M. S. 1955. *An Introduction to Stochastic Processes*. New York: Cambridge University Press.

BELL, E. T. 1937. "Gauss, the prince of mathematicians." In E. T. BELL, *Men of Mathematics*, Vol. 1. London: Penguin.

BERG, R. F. 1983. "Estimation and prediction for maneuvering target trajectories." *IEEE Trans. on Auto. Control*, Vol. AC-28, pp. 294–304.

BERKHOUT, A. G. 1977. "Least-squares inverse filtering and wavelet deconvolution." *Geophysics*, Vol. 42, pp. 1369–1383.

BICKEL, P. J., and K. A. DOKSUM. 1977. *Mathematical Statistics: Basic Ideas and Selected Topics*. San Francisco: Holden-Day.

BIERMAN, G. J. 1973a. "A comparison of discrete linear filtering algorithms." *IEEE Trans. on Aerospace and Electronic Systems*, Vol. AES-9, pp. 28–37.

BIERMAN, G. J. 1973b. "Fixed-interval smoothing with discrete measurements." *Int. J. Control*, Vol. 18, pp. 65–75.

BIERMAN, G. J. 1977. *Factorization Methods for Discrete Sequential Estimation*. New York: Academic Press.

BJERHAMMAR, A. 1951a. "Application of calculus of matrices to method of least squares; with special reference to geodetic calculations." *Trans. Roy. Inst. Tech. Stockholm*, No. 39, pp. 1–86.

BJERHAMMAR, A. 1951b. "Rectangular reciprocal matrices with special reference to geodetic calculations." *Bull. Geodesique*, pp. 188–220.

BLAIR, W. D., and D. KAZAKOS. 1993. "Tracking maneuvering targets with multiple intermittant sensors." *Proc. 27th Annual Asilomar Conf. on Signals, Systems, and Computers*, Pacific Grove, CA.

BOX, G. E. P., and G. M. JENKINS. 1970. *Time Series Analysis, Forecasting and Control*. San Francisco: Holden-Day.

BOX, J. F., 1978. *R. A. Fisher: The Life of a Scientist*. New York: Wiley.

BRAMMER, R. F., R. P. PASS, and J. V. WHITE. 1983. "Bathymetric and oceanographic applications of Kalman filtering techniques." *IEEE Trans. on Auto. Control*, Vol. AC-28, pp. 363–371.

BRIGHAM, E. O. 1988. *The Fast Fourier Transform and Its Applications*. Englewood Cliffs, NJ: Prentice Hall.

BRILLINGER, D. R., and M. ROSENBLATT. 1967a. "Asymptotic theory of estimates of $k$th order spectra." In *Spectral Analysis of Time Series*, B. Harris, ed. New York: Wiley, pp. 153–188.

BRILLINGER, D. R., and M. ROSENBLATT. 1967b. "Computation and interpretation of $k$th-order spectra." In *Spectral Analysis of Time Series*, B. Harris, ed. New York: Wiley, pp. 189–232.

BRYSON, A. E., JR., and M. FRAZIER. 1963. "Smoothing for linear and nonlinear dynamic systems." TDR 63-119, pp. 353–364, Aero. Sys. Div., Wright-Patterson Air Force Base, Ohio.

BRYSON, A. E., JR., and Y.C. HO. 1969. *Applied Optimal Control*. Waltham, MA: Blaisdell.

BRYSON, A. E., JR., and D. E. JOHANSEN. 1965. "Linear filtering for time-varying systems using measurements containing colored noise." *IEEE Trans. on Automatic Control*, Vol. AC-10, pp. 4–10.

BURY, K. V. 1975. *Statistical Methods in Applied Science*. New York: Wiley.

CAMPBELL, J. K., S. P. SYNNOTT, and G. J. BIERMAN. 1983. "Voyager orbit determination at Jupiter." *IEEE Trans. on Auto. Control*, Vol. AC-28, pp. 254–255.

CACOULLOS, T. 1966, "Estimation of a multivariate density." *Annals Inst. Statist. Math.* (Tokyo), Vol. 18, pp. 179–189.

CHEN, C. T. 1984. *Introduction to Linear System Theory*. New York: Holt.

CHI, C. Y. 1983. "Single-channel and multichannel deconvolution." Ph.D. dissertation, University of Southern California, Los Angeles, CA.

CHI, C. Y., and J. M. MENDEL. 1984. "Performance of minimum-variance deconvolution filter." *IEEE Trans. on Acoustics, Speech and Signal Processing*, Vol. ASSP-32, pp. 1145–1153.

COURANT, R. 1969. "Gauss and the present situation of the exact sciences." In T. L. Saaty and F. J. Weyl (eds.), *The Spirit and Use of Mathematical Sciences*. New York: McGraw-Hill, pp. 141–155.

CRAMER, H. 1946. *Mathematical Methods of Statistics*. Princeton, NJ: Princeton University Press.

DEUTSCH, R., 1965. *Estimation Theory*. Englewood Cliffs, NJ: Prentice Hall,

DOGAN, M. C., and J. M. MENDEL. 1993. "On blind optimum-beamforming for gated signals," submitted for publication.

DOGAN, M. C., and J. M. MENDEL. 1994. "Blind deconvolution (equalization): some new results," submitted for publication.

DONGARRA, J. J., J. R. BUNCH, C. B. MOLER, and G. W. STEWART. 1979. *LINPACK User's Guide*. Philadelphia: SIAM.

DOOB, J. L. 1966. "Wiener's work in probability theory." *Bull. Amer. Math. Soc.*, Vol. 72, No. 1, Part II, pp. 69–72.

DUDA, R. D., and P. E. HART. 1973. *Pattern Classification and Scene Analysis*. New York: Wiley-Interscience.

DURBIN, J. 1954. "Errors in variables." *Rev. Int. Statist. Inst.*, 22, pp. 23–32.

ECKART, C., and G. YOUNG. 1939. "A principal axis transformation for non-Hermitian matrices." *Bull. Am. Math. Soc.*, Vol. 45, pp. 118–121.

EDWARDS, A. W. F. 1972. *Likelihood*. London: Cambridge University Press.

EYKHOFF, P. 1974. *System Identification: Parameter and State Estimation*. New York: Wiley.

FAURRE, P. L. 1976. "Stochastic Realization Algorithms," in *System Identification: Advances and Case Studies* (eds., R. K. Mehra and D. G. Lainiotis), pp. 1–25. New York: Academic Press.

FERGUSON, T. S. 1967. *Mathematical Statistics: A Decision Theoretic Approach*. New York: Academic Press.

FOMBY, T. B., R. C. HILL, and S. R. JOHNSON. 1984. *Advanced Econometric Methods*. New York: Springer-Verlag.

FRANKLIN, J. N. 1968. *Matrix Theory*. Englewood Cliffs, NJ: Prentice Hall.

FRASER, D. 1967. "Discussion of optimal fixed-point continuous linear smoothing (by J. S. Meditch)." *Proc. 1967 Joint Automatic Control Conf.*, p. 249, University of Pennsylvania, Philadelphia.

FRIEDLAND, B. 1986. *Control System Design: An Introduction to State-space Methods*. New York: McGraw-Hill.

FRIEDLANDER, B., and B. PORAT. 1989. "Adaptive IIR algorithms based on high-order statistics." *IEEE Trans. on Acoustics, Speech and Signal Processing*, Vol. 37, pp. 485–495.

FRIEDLANDER, B., and B. PORAT. 1990. "Asymptotically optimal estimation of MA and ARMA parameters of non-Gaussian processes from high-order moments." *IEEE Trans. on Automatic Control*, Vol. 35, pp. 27–35.

GALLANT, A. R. 1987. *Nonlinear Statistical Models*. New York: Wiley.

GESING, W. S., and D. B. REID. 1983. "An integrated multisensor aircraft track recovery system for remote sensing." *IEEE Trans. on Auto. Control*, Vol. AC-28, pp. 356–363.

GIANNAKIS, G. B. 1987a. *Signal Processing Using Higher-order Statistics*, Ph. D. dissertation, Dep. Elec. Eng., University of Southern California, Los Angeles, CA.

GIANNAKIS, G. B. 1987b. "Cumulants: a powerful tool in signal processing." *Proc. IEEE*, Vol. 75, pp. 1333–1334.

GIANNAKIS, G. B. 1989. "Wavelet parameter and phase estimation using cumulant slices." *IEEE Trans. on Geoscience and Remote Sensing*, Vol. 27, pp. 452–455.

GIANNAKIS, G. B. 1990. "On the identifiability on non-Gaussian ARMA models using cumulants." *IEEE Trans. on Automatic Control*, Vol. 35, pp. 18–26.

GIANNAKIS, G. B., and J. M. MENDEL. 1989. "Identification of non-minimum phase systems using higher-order statistics." *IEEE Trans. on Acoustics, Speech and Signal Processing*, Vol. 37, pp. 360–377.

GIANNAKIS, G. B., and J. M. MENDEL. 1990. "Cumulant-based order determination of non-Gaussian ARMA models." *IEEE Trans. on Acoustics, Speech and Signal Processing*, Vol. 38, pp. 1411–1423.

GIANNAKIS, G. B., and A. SWAMI. 1987. "New results on state-space and input–output identification of non-Gaussian processes using cumulants." In Proc. SPIE-87, *Advanced Algorithms and Architectures for Signal Processing II*, San Diego, CA, pp. 199–204.

GIANNAKIS, G. B., and A. SWAMI. 1990. "On estimating noncausal nonminimum phase ARMA models of non-Gaussian processes." *IEEE Trans. Acoustics, Speech and Signal Processing*, Vol. 38, pp. 478–495.

GOLDBERGER, A. S. 1964. *Econometric Theory*. New York: John Wiley.

GOLUB, G. H., and C. F. VAN LOAN. 1980. "An analysis of the total least squares problem." *SIAM Journal of Numerical Analysis*, Vol. 17, pp. 883–893.

GOLUB, G. H., and C. F. VAN LOAN. 1989. *Matrix Computations*, 2nd ed. Baltimore, MD: Johns Hopkins University Press.

GRAYBILL, F. A. 1961. *An Introduction to Linear Statistical Models*, Vol. 1. New York: McGraw-Hill.

GREWAL, M. S., and A. P. ANDREWS. 1993. *Kalman Filtering: Theory and Practice*. Englewood Cliffs, NJ: Prentice Hall.

GREVILLE, T. N. E. 1959. "The pseudoinverse of a rectangular or singular matrix and its application to the solution of systems of linear equations." *SIAM Review*, Vol. 1, pp. 38–43.

GREVILLE, T. N. E. 1960. "Some applications of the pseudoinverse of a matrix." *SIAM Review*, Vol. 2, pp. 15–22.

GRIMMETT, G. R., and D. R. STIRZAKER. 1982. *Probability and Random Processes*. Oxford, Great Britain: Oxford Science Publications.

GUPTA, N. K., and R. K. MEHRA. 1974. "Computational aspects of maximum likelihood estimation and reduction of sensitivity function calculations." *IEEE Trans. on Automatic Control*, Vol. AC-19, pp. 774–783.

GURA, I. A., and A. B. BIERMAN. 1971. "On computational efficiency of linear filtering algorithms." *Automatica*, Vol. 7, pp. 299–314.

HAMMING, R. W. 1983. *Digital Filters*, 2nd ed. Englewood Cliffs, NJ: Prentice-Hall.

HAUSER, J., S. SASTRY, and P. KOKOTOVIC. 1992. "Nonlinear control via approximate input–output linearization: the ball and beam example." *IEEE Trans. on Automatic Control*, Vol. AC-37, pp. 392–398.

HAYKIN, S. 1991. *Adaptive Filter Theory*, 2nd ed. Englewood Cliffs, NJ: Prentice Hall.

HO, Y. C. 1963. "On the stochastic approximation method and optimal filtering." *J. Math. Anal. and Appl.*, Vol. 6, pp. 152–154.

HOLDEN, A. V. 1986. *Chaos*. Princeton, NJ: Princeton University Press.

HSIA, T. C. 1977. *System Identification*. Lexington MA: D. C. Heath.

JAZWINSKI, A. H. 1970. *Stochastic Processes and Filtering Theory*. New York: Academic Press.

KAILATH, T. 1968. "An innovations approach to least-squares estimation—Part 1: Linear filtering in additive white noise." *IEEE Trans. on Automatic Control*, Vol. AC-13, pp. 646–655.

KAILATH, T. 1974. "A view of three decades of linear filtering theory." *IEEE Trans. on Information Theory*, Vol. IT-20, pp. 146–181.

KAILATH, T. K. 1980. *Linear Systems*. Englewood Cliffs, NJ: Prentice Hall.

KALMAN, R. E. 1960. "A new approach to linear filtering and prediction problems." *Trans. ASME J. Basic Eng. Series D*, Vol. 82, pp. 35–46.

KALMAN, R. 1963. "Mathematical description of linear dynamic systems." *SIAM J. on Control*, ser. A, Vol. 1, No. 2, pp. 152–192.

KALMAN, R. E., and R. BUCY. 1961. "New results in linear filtering and prediction theory." *Trans. ASME, J. Basic Eng. Series D*, Vol. 83, pp. 95–108.

KASHYAP, R. L., and A. R. RAO. 1976. *Dynamic Stochastic Models from Empirical Data*. New York: Academic Press.

KAY, S. M. 1993. *Fundamentals of Statistical Signal Processing: Estimation Theory*. Englewood Cliffs, NJ: Prentice Hall.

KELLY, C. N., and B. D. O. ANDERSON. 1971. "On the stability of fixed-lag smoothing algorithms." *J. Franklin Inst.*, Vol. 291, pp. 271–281.

KENDALL, M. G., and A. STUART. 1961. *The Advanced Theory of Statistics*. Vol. 2. London: Griffin.

KLEMA, V. C., and A. J. LAUB. 1980. "The singular-value decomposition: its computation and some applications." *IEEE Trans. Autom. Control*, Vol. AC-25, pp. 164–176.

KMENTA, J. 1971. *Elements of Econometrics*. New York: Macmillan.

KOPP, R. E., and R. J. ORFORD. 1963. "Linear regression applied to system identification for adaptive control systems." *AIAA J.*, Vol. 1, p. 2300.

KORMYLO, J. 1979. *Maximum-likelihood Seismic Deconvolution*, Ph. D. dissertation, Department of Electrical Engineering, University of Southern California, Los Angeles.

KORMYLO, J., and J. M. MENDEL. 1982. "Maximum-likelihood detection and estimation of Bernoulli–Gaussian processes. *IEEE Trans. on Info. Theory*, Vol. IT-28, pp. 482–488.

KUNG, S. Y. 1978. "A new identification and model reduction algorithm via singular value decomposition." Paper presented at the 12th Annual Asilomar Conference on Circuits, Systems, and Computers, Pacific Grove, CA.

KWAKERNAAK, H., and R. SIVAN. 1972. *Linear Optimal Control Systems*. New York: Wiley-Interscience.

LARSON, H. J., and B. O. SHUBERT. 1979. *Probabilistic Models in Engineering Sciences*, Vol. 1. New York: Wiley.

LAUB, A. J. 1979. "A Schur method for solving algebraic Riccati equations." *IEEE Trans. on Automatic Control*, Vol. AC-24, pp. 913–921.

LEHMANN, E. L. 1959. *Testing Statistical Hypotheses*. New York: Wiley.

LEHMANN, E. L. 1980. *Theory of Point Estimation*. New York: Wiley.

LEIBUNDGUT, B. G., A. RAULT, and F. GENDREAU. 1983. "Application of Kalman filtering to demographic models." *IEEE Trans. on Auto. Control*, Vol. AC-28, pp. 427–434.

LEONOV, V. P., and A. N. SHIRYAEV. 1959. "On a method of calculation of semi-invariants." *Theory Prob. Appl.*, Vol. 4, pp. 319–328.

LEVINSON, N. 1966. "Wiener's life." *Bull Amer. Math. Soc.*, Vol. 72, No. 1, Part II, pp. 1–32.

LII, K. S., and M. ROSENBLATT. 1982. "Deconvolution and estimation of transfer function phase and coefficients for non-Gaussian linear processes." *Ann. Statistics*, Vol. 10, pp. 1195–1208.

LII, K. S., and M. ROSENBLATT. 1990. "Asymptotic normality of cumulant spectral estimates." *J. Theoretical Prob.*, Vol. 3, pp. 367–385.

LJUNG, L. 1976. "Consistency of the Least-squares Identification Method," *IEEE Trans. on Automatic Control*, Vol. AC-21, pp. 779–781.

LJUNG, L. 1979. "Asymptotic behavior of the extended Kalman filter as a parameter estimator for linear systems," *IEEE Trans. on Automatic Control*, Vol. AC-24, pp. 36–50.

LJUNG, L. 1987. *System Identification: Theory for the User*. Englewood Cliffs, NJ: Prentice Hall.

LJUNG, L., and T. SODERSTROM. 1983. *Theory and Practice of Recursive Identification*. Cambridge, MA: MIT Press.

Lumelsky, V. J. 1983. "Estimation and prediction of unmeasureable variables in the steel mill soaking pit control system." *IEEE Trans. on Auto. Control*, Vol. AC-28, pp. 388–400.

MacDuffee, C. C. 1933. *The Theory of Matrices*. New York: Springer.

Marquardt, D. W. 1963. "An algorithm for least-squares estimation of nonlinear parameters." *J. Soc. Indust. Appl. Math.*, Vol. 11, pp. 431–441.

McLaughlin, D. B. 1980. "Distributed systems—notes." *Proc. 1980 Pre-JACC Tutorial Workshop on Maximum-Likelihood Identification*, San Francisco, CA.

Mealy, G. L., and W. Tang. 1983. "Application of multiple model estimation to a recursive terrain height correlation system." *IEEE Trans. on Auto. Control*, Vol. AC-28, pp. 323–331.

Meditch, J. S. 1969. *Stochastic Optimal Linear Estimation and Control*. New York: McGraw-Hill.

Mehra, R. K. 1970a. "An algorithm to solve matrix equations $\mathbf{PH}^{T} = \mathbf{G}$ and $\mathbf{P} = \mathbf{\Phi P \Phi}^{T} + \mathbf{\Gamma\Gamma}^{T}$." *IEEE Trans. on Automatic Control*, Vol. AC-15, p. 600.

Mehra, R. K. 1970b. "On-line identification of linear dynamic systems with applications to Kalman filtering." *Proc. Joint Automatic Control Conference*, Atlanta, GA, pp. 373–382.

Mehra, R. K. 1971. "Identification of stochastic linear dynamic systems using Kalman filter representation." *AIAA J.*, Vol. 9, pp. 28–31.

Mehra, R. K., and J. S. Tyler. 1973. "Case studies in aircraft parameter identification." *Proc. 3rd IFAC Symposium on Identification and System Parameter Estimation*, North Holland, Amsterdam.

Melsa, J. L., and D. L. Cohn 1978. *Decision and Estimation Theory*. New York: McGraw-Hill.

Mendel, J. M. 1971a. "On the need for and use of a measure of state estimation errors in the design of quadratic-optimal control gains." *IEEE Trans. on Automatic Control*, Vol. AC-16, pp. 500–503.

Mendel, J. M. 1971b. "Computational requirements for a discrete Kalman filter." *IEEE Trans. on Automatic Control*, Vol. AC-16, pp. 748–758.

Mendel, J. M. 1973. *Discrete Techniques of Parameter Estimation: The Equation Error Formulation*. New York: Marcel Dekker.

Mendel, J. M. 1975. "Multi-stage least squares parameter estimators." *IEEE Trans. on Automatic Control*, Vol. AC-20, pp. 775–782.

Mendel, J. M. 1977a. "White noise estimators for seismic data processing in oil exploration." *IEEE Trans. on Automatic Control*, Vol. AC-22, pp. 694–706.

Mendel, J. M. 1977b. "A quantitative evaluation of Ott and Meder's prediction error filter." *Geophysical Prospecting*, Vol. 25, pp. 692–698.

Mendel, J. M. 1981. "Minimum-variance deconvolution." *IEEE Trans. on Geoscience and Remote Sensing*, Vol. GE-19, pp. 161–171.

Mendel, J. M. 1983. *Optimal Seismic Deconvolution: An Estimation Based Approach*. New York: Academic Press.

Mendel, J. M. 1988. "Use of higher-order statistics in signal processing and system theory: an update." in *Proc. SPIE Conf. on Advanced Algorithms and Architectures for Signal Processing III*. San Diego, CA, pp. 126–144.

Mendel, J. M. 1990. *Maximum-likelihood Deconvolution*. New York: Springer.

MENDEL, J. M. 1991. "Tutorial on higher-order statistics (spectra) in signal processing and system theory: theoretical results and some applications." *Proc. IEEE*, Vol. 79, pp. 278–305.

MENDEL, J. M., and D. L. GIESEKING. 1971. "Bibliography on the linear-quadratic-gaussian problem." *IEEE Trans. on Automatic Control*, Vol. AC-16, pp. 847–869.

MENDEL, J. M., and J. KORMYLO. 1978. "Single channel white noise estimators for deconvolution." *Geophysics*, Vol. 43, pp. 102–124.

MENDEL, J. M., J. KORMYLO, F. AMINZADEH, J. S. LEE, and F. HABIBI-ASHRAFI. 1981. "A novel approach to seismic signal processing and modeling." *Geophysics*, Vol. 46, pp. 1398–1414.

MOORE, E. H. 1920. *Bull. Amer. Math. Soc.*, Vol. 26, pp. 394–395.

MOORE, E. H. 1935. "General analysis I." *Mem. Am. Phil. Soc.*, Vol. 1, especially pp. 197–209.

MORRISON, N. 1969. *Introduction to Sequential Smoothing and Prediction*. New York: McGraw-Hill.

NAHI, N. E. 1969. *Estimation Theory and Applications*. New York: Wiley.

NIKIAS, C. L., and J. M. MENDEL. 1993. "Signal processing with higher-order spectra." *IEEE Signal Processing Magazine*, Vol. 10, pp. 10–37.

NIKIAS, C. L., and R. PAN. 1987. "Non-minimum phase system identification via cepstrum modeling of high-order moments." In *Proc. ICASSP-87*, Dallas, TX, pp. 980–983.

NIKIAS, C. L., and R. PAN. 1988. "Time delay estimation in unknown Gaussian spatially correlated noise." *IEEE Trans. on Acoustics, Speech and Signal Processing*, Vol. 7, pp. 291–325.

NIKIAS C. L., and A. PETROPULU. 1993. *Higher-Order Spectral Analysis: A Nonlinear Signal Processing Framework*. Englewood Cliffs, NJ: Prentice Hall.

NIKIAS, C. L., and M. R. RAGHUVEER. 1987. "Bispectrum estimation: a digital signal processing framework." *Proc. IEEE*, Vol. 75, pp. 869–891.

OPPENHEIM, A. V., and R. W. SHAFER. 1989. *Discrete-Time Signal Processing*. Englewood Cliffs, NJ: Prentice Hall.

OTT, N., and H. G. MEDER. 1972. "The Kalman filter as a prediction error filter." *Geophysical Prospecting*, Vol. 20, pp. 549–560.

PAEZ, M. D., and T. H. GLISON. 1972. "Minimum mean squared-error quantization in speech." *IEEE Trans. Commun.*, Vol. COM-20, pp. 225–230.

PAN, R., and C. L. NIKIAS. 1988. "The complex cepstrum of higher-order moments." *IEEE Trans. on Acoustics, Speech and Signal Processing*, Vol. 36, pp. 186–205.

PAPOULIS, A. 1991. *Probability, Random Variables, and Stochastic Processes*, 3rd ed. New York: McGraw-Hill.

PARZEN, E. 1962. "On estimation of a probability density function and mode." *Ann. Math. Statist.*, Vol. 33, pp. 1065–1076.

PARZEN, E. 1967. "Time series analysis for models of signal plus white noise." In B. Harris, ed., *Spectral Analysis of Time Series*, pp. 233–257. New York: Wiley.

PEARSON, K. 1936. "Method of moments and method of maximum likelihood." *Biometrika*, Vol. 28, pp. 34–59.

PELED, A., and B. LIU. 1976. *Digital Signal Processing: Theory, Design, and Implementation*, New York: Wiley.

PENROSE, R. 1955. "A generalized inverse for matrices." *Proc. Cambridge Philos. Soc.*, Vol. 51, pp. 406–413.

PFLUG, A. L., G. E. IOUP, J. W. IOUP, and R. L. FIELD. 1992. "Properties of higher-order correlation and spectra for band-limited, deterministic transients." *J. Acoustics Soc. Am.*, Vol. 91(2), pp. 975–988.

PORAT, B., and B. FRIEDLANDER. 1989. "Performance analysis of parameter estimation algorithms based on high-order moments." *Int. J. Adaptive Control and Signal Processing*, Vol. 3, pp. 191–229.

POWELL, M. J. D. 1987. "Radial basis functions for multivariable interpolation: a review." In *Algorithms for Approximation*, J. C. Mason and M. G. Cox, eds. New York. Oxford University Press, pp. 143–167.

PRASAD, S., R. T. WILLIAMS, A. K. MAHALANABIS, and L. H. SIBUL. 1988. "A transform-based covariance differencing approach for some classes of parameter estimation problems." *IEEE Trans. Acoustics, Speech and Signal Processing*, Vol. 36, pp. 631–641.

PRIESTLEY, M. B. 1981. *Spectral Analysis and Time Series*. New York: Academic Press.

RAO, T. S., and M. M. GABR. 1984. *An Introduction to Bispectral Analysis and Bilinear Time Series Models*. Lecture Notes in Statistics, 24. New York: Springer.

RAUCH, H. E., F. TUNG, and C. T. STRIEBEL. 1965. "Maximum-likelihood estimates of linear dynamical systems." *AIAA J.*, Vol. 3, pp. 1445–1450.

ROHATGI, V. K. 1976. *An Introduction to Probability Theory and Mathematical Statistics*. New York: Wiley.

ROOT, W. L. 1966. "Contributions of Norbert Wiener to communication theory." *Bull. Amer. Math. Soc.*, Vol. 72, No. 1, Part II, pp. 126–134.

ROSENBLATT, M., and J. W. VAN NESS. 1965. "Estimation of the bispectrum." *Ann. Math. Stat.*, Vol. 36, pp. 420–436.

RUCKEBUSCH, G. 1983. "A Kalman filtering approach to natural gamma ray spectroscopy in well logging." *IEEE Trans. on Auto. Control*, Vol. AC-28, pp. 372–380.

SAELID, S., N. A. JENSSEN, and J. G. BALCHEN. 1983. "Design and analysis of a dynamic positioning system based on Kalman filtering and optimal control." *IEEE Trans. on Auto. Control*, Vol. AC-28, pp. 331–338.

SAGE, A. P., and J. L. MELSA. 1971. *Estimation Theory with Applications to Communications and Control*. New York: McGraw-Hill.

SCHETZEN, M. 1974. "A theory of non-linear system identification." *Int. J. of Control*, Vol. 20, pp. 577–592.

SCHWEPPE, F. C. 1965. "Evaluation of likelihood functions for gaussian signals." *IEEE Trans. on Information Theory*, Vol. IT-11, pp. 61–70.

SCHWEPPE, F. C. 1974. *Uncertain Dynamic Systems*. Englewood Cliffs, NJ: Prentice Hall.

SERFLING, R. J. 1980. *Approximation Theorems of Mathematical Statistics*. New York: Wiley.

SHERMAN, S. 1958. "Non-mean-square error criteria." *IRE Trans. Inform. Theory*, Vol. IT-4, p. 125.

SIDAR, M. M., and B. F. DOOLIN. 1983. "On the feasibility of real-time prediction of aircraft carrier motion at sea." *IEEE Trans. on Auto. Control*, Vol. AC-28, pp. 350–355.

SORENSON, H. W. 1970. "Least-squares estimation: from Gauss to Kalman." *IEEE Spectrum*, Vol. 7, pp. 63–68.

SORENSON, H. W. 1980. *Parameter Estimation: Principles and Problems*. New York: Marcel Dekker.

SORENSON, H. W. (ed.). 1985. *Kalman Filtering: Theory and Applications*. New York: IEEE Press.

SORENSON, H. W., and D. L. ALSPACH. 1971. "Recursive Bayesian estimation using Gaussian sums." *Automatica*, Vol. 7, pp. 465–479.

SORENSON, H. W., and J. E. SACKS. 1971. "Recursive fading memory filtering." *Information Science*, Vol. 3, pp. 101–119.

SPECHT, D. F. 1967. "Generation of polynomial discriminant functions for pattern recognition." *IEEE Trans. Electron. Comput.*, Vol. EC-16, pp. 308–319.

SPECHT, D. F. 1991. "A general regression network." *IEEE Trans. on Neural Networks*, Vol. 2, pp. 568–576.

STARK, H., and J. W. WOODS. 1986. *Probability, Random Processes, and Estimation Theory for Engineers*. Englewood Cliffs, NJ: Prentice Hall.

STEFANI, R. T. 1967. "Design and simulation of a high performance, digital, adaptive, normal acceleration control system using modern parameter estimation techniques." Rept. No. DAC-60637, Douglas Aircraft Co., Santa Monica, CA.

STEPNER, D. E., and R. K. MEHRA. 1973. "Maximum likelihood identification and optimal input design for identifying aircraft stability and control derivatives." Ch. IV, NASA-CR-2200.

STEWART, G. W. 1973. *Introduction to Matrix Computations*. New York: Academic Press.

STOICA, P., and R. L. MOSES. 1990. "On biased estimators and the unbiased Cramer–Rao lower bound." *Signal Processing*, Vol. 21, pp. 349–350.

STRANG, G. 1988. *Linear Algebra and its Applications*, 3rd. ed. New York: Academic Press.

SWAMI, A. 1988. *System Identification Using Cumulants*, Ph. D. dissertation, USC SIPI Report 140, Dep. Elec. Eng.-Syst., University of Southern California, Los Angeles.

SWAMI, A., and J. M. MENDEL. 1990a. "Time and lag recursive computation of cumulants from a state space model." *IEEE Trans. on Automatic Control*, Vol. 35, pp. 4–17.

SWAMI, A., and J. M. MENDEL. 1990b. "ARMA parameter estimation using only output cumulants." *IEEE Trans. on Acoustics, Speech and Signal Processing*, Vol. 38, pp. 1257–1265.

SWAMI, A., and J. M. MENDEL. 1991. "Cumulant-based approach to the harmonic retrieval and related problems." *IEEE Trans. on Signal Processing*, Vol. 39, pp. 1099–1109.

SWAMI, A., and J. M. MENDEL. 1992. "Identifiability of the parameters of an ARMA process using cumulants." *IEEE Trans. on Automatic Control*, Vol. 37, pp. 268–273.

TAYLOR, J. H., and A. J. ANTONIOTTI. 1993. "Linearization algorithm for computer-aided control engineering." *IEEE Control Systems Magazine*, Vol. 13, pp. 58–64.

THERRIEN, C. W. 1992. *Discrete Random Signals and Statistical Signal Processing*. Englewood Cliffs, NJ: Prentice Hall.

TREITEL, S. 1970. "Principles of digital multichannel filtering." *Geophysics*, Vol. 35, pp. 785–811.

TREITEL, S., and E. A. ROBINSON. 1966. "The design of high-resolution digital filters." *IEEE Trans. on Geoscience and Electronics*, Vol. GE-4, pp. 25–38.

TUCKER, H. G. 1962. *An Introduction to Probability and Mathematical Statistics*. New York: Academic Press.

TUCKER, H. G. 1967. *A Graduate Course in Probability*. New York: Academic Press.

TUGNAIT, J. 1986a. "Identification of nonminimum phase linear stochastic systems." *Automatica*, Vol. 22, pp. 454–464.

TUGNAIT, J. 1986b. "Order reduction of SISO nonminimum phase stochastic systems." *IEEE Trans. on Automatic Control*, Vol. 31, pp. 623–632.

TUGNAIT, J. 1989. "Recovering the poles from third-order cumulants of system output." *IEEE Trans on Automatic Control*, Vol. 34, pp. 1085–1089.

TUGNAIT, J. 1990. "Approaches to FIR system identification with noisy data using higher-order statistics." *IEEE Trans. on Acoustics, Speech and Signal Processing*, Vol. 38, pp. 1307–1317.

TYLEE, J. L. 1983. "On-line failure detection in nuclear power plant instrumentation." *IEEE Trans. on Auto. Control*, Vol. AC-28, pp. 406–415.

VAN HUFFEL, S. and J. VANDEWALLE. 1991. *The Total Least Squares Problem, Computational Aspects and Analysis*, SIAM, Philadelphia.

VAN NESS, J. W. 1966. "Asymptotic normality of bispectral estimates." *Ann. Math. Statistics*, Vol. 37, pp. 1257–1272.

VAN TREES, H. L. 1968. *Detection, Estimation and Modulation Theory*, Vol. 1. New York: Wiley.

VERHAEGEN, M., and P. VAN DOOREN. 1986. "Numerical aspects of different Kalman filter implementations." *IEEE Trans. Auto. Control*, Vol. AC-31, pp. 907–917.

VOLTERRA, T. 1959. *Theory of Functionals and of Integrals and Integro-Differential Equations*. New York: Dover.

WANG, L.-X. 1994. *Adaptive Fuzzy Systems and Control: Design and Stability Analysis*. Englewood Cliffs, NJ: Prentice Hall.

WANG, L.-X., and J. M. MENDEL. 1992. "Fuzzy basis functions, universal approximation, and orthogonal least-squares learning." *IEEE Trans. on Neural Networks*, Vol. 3, pp. 807–814.

YOUNG, P. C. 1984. *Recursive Estimation and Time-series Analysis*. New York: Springer.

ZACKS, S. 1971. *The Theory of Statistical Inference*. New York: Wiley.

ZHENG, F.-C., S. MCLAUGHLIN, and B. MULGREW. 1993. "Cumulant-based deconvolution and identification: several new families of linear equations." *Signal Processing*, Vol. 30, pp. 199–219.

# Index

## A



## B

## F



## G

## M

## N



## O



## P

## T



## U



## V



## W